THE ENCYCLOPEDIA OF Cultivated Palms

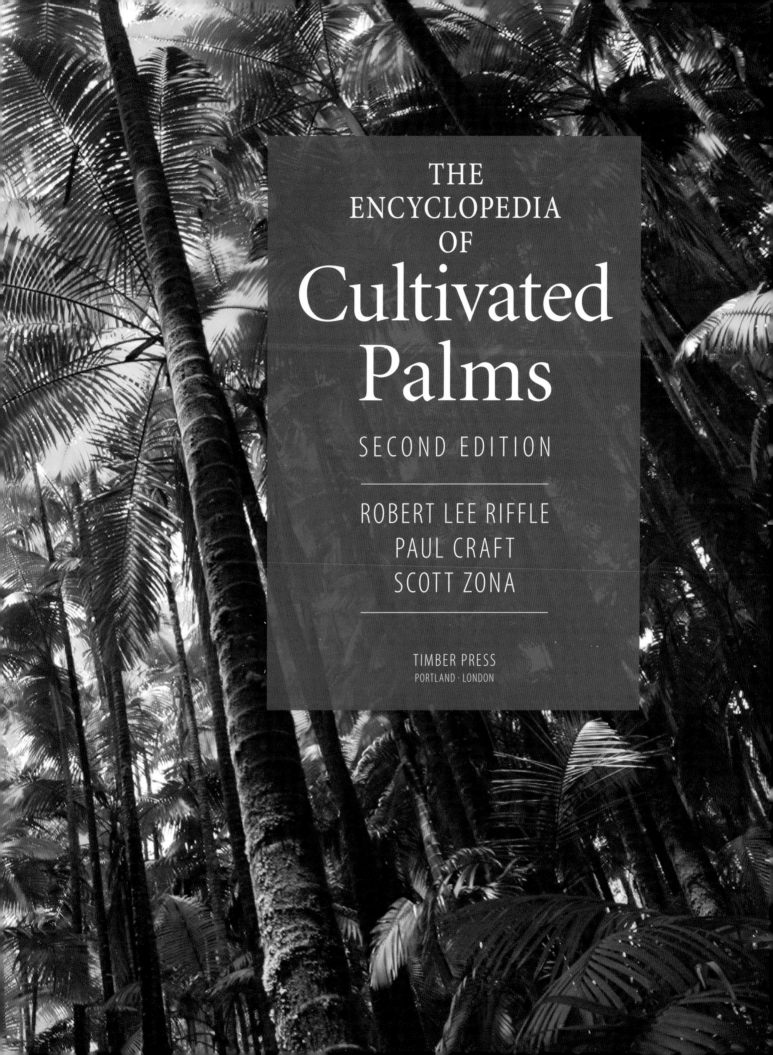

THE ENCYCLOPEDIA OF Cultivated Palms

SECOND EDITION

ROBERT LEE RIFFLE

PAUL CRAFT

SCOTT ZONA

TIMBER PRESS
PORTLAND · LONDON

To the two Ds in my life,
Diana Gabaldon and Diane Laird.
—RLR

To my wife, Patty,
who has stood by me throughout my palm lunacy.
—PC

To Tsyr Han:
I am the lucky one.
—SZ

Page 1: *Copernicia fallaensis*. Pages 2–3: *Archontophoenix alexandrae*. Naturalized in Hawaii.
Page 5: *Licuala mattanensis* 'Mapu'. Page 15: *Daemonorops curranii*.

Copyright © 2012 by Robert Lee Riffle, Paul Craft, and Scott Zona. All rights reserved.

Published in 2012 by Timber Press, Inc.

The Haseltine Building
133 S.W. Second Avenue, Suite 450
Portland, Oregon 97204
timberpress.com

2 The Quadrant
135 Salusbury Road
London NW6 6RJ
timberpress.co.uk

Printed in China
Designed by Susan Applegate

Library of Congress Cataloging-in-Publication Data

Riffle, Robert Lee.
An encyclopedia of cultivated palms/Robert Lee Riffle, Paul Craft, Scott Zona.—2nd ed.
p. cm.
Includes bibliographical references and index.
ISBN 978-1-60469-205-1
1. Palms—Encyclopedias. I. Craft, Paul, 1950– II. Zona, Scott. III. Title.
SB317.P3R54 2012
634.9'74—dc23   2011029273
A catalog record for this book is also available from the British Library.

CONTENTS

- 6 Preface
- 7 Introduction
- 15 Gallery of Palms
- 257 Palm Descriptions A to Z
- 495 Landscape Lists
- 501 Internet Sources for Information on Palms
- 502 Botanical Gardens and Public Collections with Significant Palm Collections
- 506 Glossary
- 509 Bibliography
- 512 Acknowledgments
- 513 Photography Credits
- 514 Index of Synonyms and Common Names

PREFACE

This book is intended primarily for gardeners and horticulturists. Although it includes the latest taxonomic thinking about the family and the delineation of species, it is not a systematic treatment of the palms. Rather, it is scientific only to the extent that it needs to be for an adequate understanding of the plant descriptions. Some technical jargon is unavoidable and the glossary is, we believe, mercifully short.

The Authors' Biases

No nonfiction work is free of its author's biases, even if the biases are evident only in the way the material is organized, and we think we should declare ours up front. First, we are not fond of common names and, although we have included them for reference, they are probably the least dependable part of a palm's description. Doubtless many vernacular names are missing as we endeavored to include only English names.

Second, given the limited amount of space in this work and the inevitability of having to sometimes choose what to include and what to eschew, we have opted for the species of larger landscape proportions to the exclusion of smaller species, where such a choice existed.

Third, we are not fond of mutations and sports and believe that most variegated examples of palm species are less than esthetically thrilling. Furthermore, the last-mentioned prejudice also extends to man-made hybrids. It is our conviction that palm lovers should first be concerned with knowing, saving, and maintaining what is found in nature, which in some cases teeters on the edge of extinction. Palm horticulture, unlike for example rose or orchid cultivation, is just beginning to advance to the state where cultivars and hybrid "improvements" can be recognized.

Hardiness Zones

Although they are inadequate, we have resorted to using the United States Department of Agriculture zones for general indications of a given palm's cold hardiness. These zones are based on average minimum winter temperatures. The inadequacies of these artificial zones have mostly to do with factors that cannot be included in simplified terms, factors such as microclimates, the amount of summer heat, the frequency and duration of cold outbreaks for a given zone, and the accompanying conditions of such outbreaks, such as the amount of precipitation. We have tried to overcome these limitations of the zones by amplifying the relevant and specific text. The zone numbers and their corresponding temperatures are as shown on right.

Pronunciation Guides

Guides to the pronunciation of scientific names are provided. The transcriptions are based on how English speakers in the United States pronounce the names. European, South African, and Australian readers may balk at some of the formulae. Uppercase letters indicate the primary stress syllable of a word. The ′ mark indicates the syllable that receives the secondary stress. In all cases, only one syllable is to be pronounced between each hyphen. Vowels and diphthongs are to be rendered as shown on right.

Alternative pronunciations illustrate the dictum that there is no one and only correct way to pronounce scientific names.

Taxonomy

The scientific names used in this book are those accepted by Kew's *World Checklist of Selected Plant Families*, which is available on-line at http://apps.kew.org/wcsp/home.do. This freely accessible database is widely, although not universally, regarded as the definitive working list of accepted palm names and reflects the latest taxonomic conclusions of taxonomists working with palms. Common synonyms of the genera and species are listed in the text.

That Which Is Lacking

We have not included separate chapters on the structural biology, economic importance, natural history (pollination, seed dispersal, plant-animal interactions), cultivation, diseases and pests, indoor cultivation, and the art of propagation. For further reading, we recommend *Evolution and Ecology of Palms* (Henderson 2002), *Genera Palmarum* (2d ed., Dransfield et al. 2008), *Ornamental Palm Horticulture* (Broschat and Meerow 2000), and *Palms Throughout the World* (Jones 1995).

Hardiness Zones

TEMP °F	ZONE	TEMP °C
−15 to −20	5a	−26 to −29
−10 to −15	5b	−23 to −26
−5 to −10	6a	−21 to −23
0 to −5	6b	−18 to −21
5 to 0	7a	−15 to −18
10 to 5	7b	−12 to −15
15 to 10	8a	−10 to −12
20 to 15	8b	−7 to −10
25 to 20	9a	−4 to −7
30 to 25	9b	−1 to −4
35 to 30	10a	2 to −1
40 to 35	10b	4 to 2
40 and above	11	4.5 and above

Pronunciation

a	is short as in *cat* or even shorter when at the end of a word
e	is short as in *yes*
i	is short as in *pin*
o	is long as in *go*
u	is short as in *up*
g	is hard as in *go*
ee	is long as in *see*
oo	as in *boost*, not as in *book*
ow	as in *cow*, not as in *show*
th	as in *thing*, not as in *the*
y	as a vowel, as in *fly*, not as in *puppy*

INTRODUCTION

> "Palms are a feature of the tropics and are so associated in the minds of all folk. . . . All are beautiful and possessed of characters by which the veriest tyro recognizes a palm immediately, never mistaking it for some other plant."
>
> —Ernest Henry "Chinese" Wilson

Palms are among those garden plants that are instantly recognizable. Even the casual gardener recognizes a palm when he or she first encounters one, although establishing the identity of the species and knowing where it fits in the classification hierarchy of the plant kingdom may require deeper contemplation. A palm is, first of all, a flowering plant, or angiosperm, as opposed to a cone-bearing plant (for example, cycads or pines) or a spore-bearing plant (ferns). Plants that produce flowers include species as tiny as *Wolffia*, an almost microscopic, aquatic floating plant related to the duckweeds (*Lemna*), and as large as *Eucalyptus regnans* of Tasmania at 350 feet (110 m) tall.

Furthermore, a palm is a monocot (short for monocotyledon), which is a specialized lineage of the flowering plants. Monocots have one seedling leaf (cotyledon), mostly non-woody tissues, parallel venation in their leaves, and flower parts in threes or multiples of three. Palms are among the few monocot families with woody tissues. Within the monocots, palms all belong to the family Palmae or Arecaceae (both names are correct). The palm family is related, albeit not closely, to many other familiar monocot families including grasses, bromeliads, gingers, cannas, and heliconias. Some other monocot families are the lilies, onions, aroids (for example, philodendrons, dieffenbachias, anthuriums), orchids, and pandanus, but they are only distantly related to palms.

What is not a palm? The answer is cycads, tree ferns, yuccas, cordylines, pandanus, and similar-looking large woody plants, all of which outwardly resemble palms. Many have common names that include the word *palm*, such as sago palm (*Cycas revoluta*) or traveler's palm (*Ravenea madagascariensis*). They are not palms and are banished from further discussion in this work.

Palm Structure

Palms may be groundcovers, shrubs, trees, or climbers (vines), growing in swamps, grasslands, forests, and even deserts. The only growth form not represented in the family is that of the true epiphyte. Most gardeners are surprised to learn of climbing palms, but such plants exist in the tropics. A few palm species naturally produce aerially branching trunks. Many examples of abnormally branched palms exist, such as ordinarily single-trunked palms forking or branching to bear multiple crowns, and these are sometimes prized by collectors for their rarity and uniqueness. Abnormal branching is usually caused by an injury to the terminal bud.

Trunks, Stems, and Stilt Roots

Some palm species have a solitary trunk, others clustering or clumping stems. Ultimate stem height ranges from 12 inches (30 cm) to at least 150 feet (46 m), and stem diameter ranges from less than ½ inch to 6 feet (13 mm to 1.8 m). The trunks are graced with the scars of former leaves in the crown; these scars may or may not be prominent and are circular in general form. Otherwise the stems may be more or less smooth or variously adorned with the bottom parts of old and dead leaves, as well as spines or other protrusions. The trunks may be straight as an arrow, leaning (usually necessarily the case in densely clumping species), creeping along the ground, or even subterranean.

As monocots, palms do not increase their trunk size by growing new wood. Rather, if the trunk enlarges, it does so by expanding the existing tissues. This characteristic has one important implication for gardeners: injuries to palm trunks are permanent and not repaired by the plants. Also, since few palms with trunks naturally branch, killing or removing the growing point means the death of that stem and, in the case of solitary-trunked species, the death of the plant.

Similar to other tropical rainforest plants, some palms develop aboveground stilt roots to support the main trunk. These structures are usually associated with species which begin their lives as undergrowth subjects and grow quickly upwards to attain the sunlight they need to mature. Most palms have relatively shallow root systems compared to those of other trees. This fact is important to remember when growing palms: they are mainly surface feeders and need regular applications of moisture.

Leaves

Palm leaves are produced from a terminal growing point. Each stem has only one growing point, and growers must take care not to damage it. Unlike some other garden plants, palm stems cannot be pruned to encourage branching. Damage to palms from cold weather can sometimes lead to fungal infection of the growing point, which can be lethal to the plant, even if the cold temperature was not.

A palm leaf consists of three parts: the sheath (the part that clasps the stem), the petiole (leafstalk), and the blade (lamina). After the leaf emerges, the sheath may

INTRODUCTION

Roystonea regia has a single trunk and a single growing point from which all the leaves are produced. The green crownshaft formed by the sheathing leaf bases is a prominent feature of these palms.

remain more or less tubular or may split. In some species, the sheath disintegrates into shaggy fibers or spines. Beyond the sheath is the petiole, which may be long or so short as to not be apparent, and beyond it is the blade, the most obvious part of the palm leaf. The blade may be entire (unsegmented), feather-shaped (pinnate), or fan-shaped (palmate). In pinnate palms, the central stalk that bears the segments is called the *rachis*.

Many fan palms leaves are actually costapalmate, meaning that they are essentially palmate but have a riblike extension of the petiole into the blade. *Sabal palmetto* is a good example of a costapalmate leaf. A final leaf form is that of the genus *Caryota*. In this genus, and only in this genus, leaves are bipinnate. Each leaf has a central rachis, which instead of bearing segments bears secondary rachises, and these in turn bear the segments.

Segments making up the feather of pinnate leaves may be completely free of one another or may be united or partly united, or sometimes united in pairs or small groups. The individual segments of this type of leaf are referred to as the leaflets. They may grow in one plane from the rachis to create a flat or nearly flat leaf, or they may grow from the rachis at an angle that creates a V-shaped leaf in cross-section, or they may diverge at various angles, creating a plumose or fox-tail leaf. In the palmate leaf, the lamina may be entire or divided into segments of various widths and lengths. These segments may, as in the pinnate leaf, be united in groups, but they are always basally united to one another. Segments in palmate leaves are never entirely free from one another, as they can be in pinnate leaves.

Leaf length (including sheath, petiole, and lamina) ranges from less than 1 foot (30 cm) to more than 80 feet (24 m), the largest leaf in the plant kingdom. Like the stem, the leaf of a palm may bear spines or other protrusions. Palmate- and costapalmate-leaved species may have a protrusion termed a *hastula*. This flap of tissue occurs at the junction between the petiole and the blade, usually on the upper surface of the leaf, but sometimes on both sides of the leaf.

The sheaths of palm leaves may form a crownshaft at the top of the trunk. This cylindrical, pillarlike organ may be tiny or grand, green or beautifully colored. It may also bear hairs, threads, or spines, exactly like the rest of the leaf.

Flowers and Fruits

Palm flowers are small, most being significantly less than 1 inch (2.5 cm) in diameter; however, they are invariably formed in large clusters or inflorescences, and one palm genus (*Corypha*) boasts the largest inflorescences in the world. The inflorescences may be borne among the leaves or below the leaves at nodes on the trunks, or may grow directly from the top of the stem (terminal inflorescence) as they do in *Corypha*. A terminal inflorescence results in the slow death of the stem from which it grows. The inflorescence may be variously branched, and the ultimate, flower-bearing branches are called *rachillae* (singular "*rachilla*").

Some palms have bisexual flowers (having both male and female parts), and these palms are called *hermaphroditic*. Many palms have unisexual flowers (male or female); when these are borne on the same plant the palm is said to be *monoecious*, when these are borne on separate plants, the palm is said to be *dioecious*.

Fruits arise from female flowers or the female parts of bisexual flowers. Fruits cannot form from male flowers. For this reason male plants may be preferred over female plants in certain landscape applications where falling fruits may create a maintenance problem.

Palm fruits range from small to gigantic. They may be smooth or covered in scales or spines, and their surfaces may be soft or hard. Their flesh may be wet and juicy or dry and fibrous. In most palm fruits, the innermost layer, called the *endocarp* (stone), is hard and bony; it can be very thick and can impede germination of the seed within it. Many palms have germination pores (for example, the three pores on a coconut) that allow the embryo to germinate from within the confines of the endocarp. Horticulturists often use endocarp and seed interchangeably, and in this book, we shall continue the convenient practice. Fruits normally contain only one seed in the center of the flesh, but there may be as many as 10 seeds in one fruit. The seed of one palm species, *Lodoicea maldivica*, is the largest seed in the plant kingdom.

Palm Seed Germination

The seeds of all palm species benefit from bottom heat, which speeds germination. During the day, the soil temperature should be kept at 80°F (27°C) or higher, cooling off to the upper 70s (around 25°C) at night. Most palm seeds are viable for only a short time and cannot be stored well. Unless otherwise stated in the plant descriptions, seeds should be kept slightly moist until they are ready to sow and should be planted soon after they are harvested. In almost all cases, seeds that are allowed to dry out completely during storage or transport will germinate poorly.

Some genera are remote germinating, meaning the seed sends a germination stalk down into the soil. When the stalk reaches a certain depth, roots are sent down farther and the first leaf is sent up. A deep container should be used when germinating seeds of these species.

Almost all seeds are planted with the endocarp (if present) intact. Some growers have success in speeding the germination process by removing the hard, bony endocarps of some species (such as *Jubaea chilensis* and other Cocoseae) to circumvent the *physical dormancy*. Despite even these efforts, most palms are comparatively slow to germinate, because the seed is dispersed or sown while the embryo is immature. This is *physiological dormancy*. Embryo maturation and growth occur only after the seed is sown in a moist, warm medium, and the seed will germinate only when the embryo is large enough. Chemical or mechanical seed treatments cannot speed the growth of the embryo and so have little effect on physiological dormancy.

Palm Habitats

Palms are indigenous to every continent except Antarctica. The northernmost naturally occurring species (*Chamaerops humilis*) is in southern Europe and the southernmost naturally occurring species (*Rhopalostylis sapida*) is found in New Zealand.

Palm habitats include all types except cold montane regions and polar regions. Palms grow in mangrove coastal environments, estuaries, and fresh water swamps, the oases of deserts, tropical and subtropical coastal plains and grasslands, deciduous tropical forests, rain forest (both lowland and montane and both tropical and warm temperate), and even in the drier regions of mountains. Several species begin their lives as submerged aquatic plants for at least part of the year.

The greatest diversity of palms occurs in tropical forests, especially humid and moist forests. Here many smaller palm species are undergrowth subjects, and the larger species are at home in clearings and along riverbanks.

Relations in the Palm Family

Present taxonomy puts the number of genera (singular "genus") in the palm family at 184, and the number of palm species (singular and plural) at about 2500 (see pages 10–11). Of course, these numbers may change with new discoveries or new techniques of investigation. In trying to represent better the relations between the species, modern taxonomists have divided the family into subfamilies, tribes, subtribes, and genera, all of which are natural groups with progressively closer affinities. Since most of the relations involve the structure not only of the leaves and trunks but also more importantly the flowering parts of the palms, the affinities are not always that obvious to nonbotanists. For example, the subfamily Coryphoideae includes only palmate-shaped leaves, with one salient exception, *Phoenix* (date palms).

While an understanding of the tribes into which palms are divided may seem superfluous for the gardener, it does have a practical application for anyone who wants to grow palms from seed. The genera of a given tribe all germinate similarly. Thus, for example, species in the tribe Corypheae push the axis of the seedling out of the seed and into the soil, and when it reaches a certain depth, the axis becomes the starting point for both the roots (which continue downward from that point) and the first leaf (which sprouts up from this depth). This fact informs the gardener to start such seed in a deep pot.

Cyrtostachys renda has pinnate leaves with bright red leaf sheaths, petioles, and rachises.

Until the advent of airplane travel and nearly instant communication, the palm family was among the most poorly known large families of flowering plants. Taxonomy is based on specimen collection, and palms boast the biggest leaves and inflorescences among the flowering plants, making the specimens very difficult to collect and preserve. Add to this the fact that plant classification was, until the twentieth century, mostly a product of the Western world and the fact that most of the species in the family originate in remote (from Europe and North America) areas of the globe, and it is not hard to understand why most of the advancement in palm taxonomy has occurred recently. The advent of gene sequencing technology now allows botanists to classify palms on the basis of shared segments of DNA. The new classification used in this edition reflects the relationships revealed by gene sequencing.

Palms in the Landscape

Palms are the most underused design elements in nearly every garden, even most of those in the tropics or subtropics.

ARECACEAE, the palm family

Subfamily Arecoideae

Tribe Areceae
  Bentinckia
  Clinostigma
  Cyrtostachys
  Dictyosperma
  Dransfieldia
  Heterospathe
  Hydriastele
  Iguanura
  Loxococcus
  Rhopaloblaste
  Subtribe Archontophoenicinae
    Actinokentia
    Actinorhytis
    Archontophoenix
    Chambeyronia
    Kentiopsis
  Subtribe Arecinae
    Areca
    Nenga
    Pinanga
  Subtribe Basseliniinae
    Basselinia
    Burretiokentia
    Cyphophoenix
    Cyphosperma
    Lepidorrhachis
    Physokentia
  Subtribe Carpoxylinae
    Carpoxylon
    Neoveitchia
    Satakentia
  Subtribe Clinospermatinae
    Clinosperma
    Cyphokentia
  Subtribe Dypsidinae
    Dypsis
    Lemurophoenix
    Marojejya
    Masoala
  Subtribe Laccospadicinae
    Calyptrocalyx
    Howea
    Laccospadix
    Linospadix
  Subtribe Oncospermatinae
    Acanthophoenix
    Deckenia
    Oncosperma
    Tectiphiala
  Subtribe Ptychospermatinae
    Adonidia
    Balaka
    Brassiophoenix
    Carpentaria
    Drymophloeus
    Normanbya
    Ponapea
    Ptychococcus
    Ptychosperma
    Solfia
    Veitchia
    Wodyetia
  Subtribe Rhopalostylidinae
    Hedyscepe
    Rhopalostylis
  Subtribe Verschaffeltiinae
    Nephrosperma
    Phoenicophorium
    Roscheria
    Verschaffeltia

Tribe Chamaedoreeae
  Chamaedorea
  Gaussia
  Hyophorbe
  Synechanthus
  Wendlandiella

Tribe Cocoseae
  Subtribe Attaleinae
    Allagoptera
    Attalea
    Beccariophoenix
    Butia
    Cocos
    Jubaea
    Jubaeopsis
    Lytocaryum
    Parajubaea
    Syagrus
    Voanioala
  Subtribe Bactridinae
    Acrocomia
    Aiphanes
    Astrocaryum
    Bactris
    Desmoncus
  Subtribe Elaeidinae
    Barcella
    Elaeis

Tribe Euterpeae
  Euterpe
  Hyospathe
  Neonicholsonia
  Oenocarpus
  Prestoea

Tribe Geonomateae
  Asterogyne
  Calyptrogyne
  Calyptronoma
  Geonoma
  Pholidostachys
  Welfia

Tribe Iriarteeae
  Dictyocaryum
  Iriartea
  Iriartella
  Socratea
  Wettinia

Tribe Leopoldinieae
  Leopoldinia

Tribe Manicarieae
  Manicaria

Tribe Oranieae
  Orania

Tribe Pelagodoxeae
  Pelagodoxa
  Sommieria

Tribe Podococceae
  Podococcus

Tribe Reinhardtieae
  Reinhardtia

Tribe Roystoneeae
  Roystonea

Tribe Sclerospermeae
  Sclerosperma

Subfamily Calamoideae

Tribe Calameae
  Subtribe Calaminae
    Calamus
    Ceratolobus
    Daemonorops
    Pogonotium
    Retispatha
  Subtribe Korthalsiinae
    Korthalsia
  Subtribe Metroxylinae
    Metroxylon
  Subtribe Pigafettinae
    Pigafetta
  Subtribe Plectocomiinae
    Myrialepis
    Plectocomia
    Plectocomiopsis
  Subtribe Salaccinae
    Eleiodoxa
    Salacca

Tribe Eugeissonae
  Eugeissona

Tribe Lepidocaryeae
  Subtribe Ancistrophyllinae
    Eremospatha
    Laccosperma
    Oncocalamus
  Subtribe Mauritiinae
    Lepidocaryum
    Mauritia
    Mauritiella
  Subtribe Raphiinae
    Raphia

Subfamily Ceroxyloideae

Tribe Ceroxyleae
   Ceroxylon
   Juania
   Oraniopsis
   Ravenea

Tribe Cyclospatheae
   Pseudophoenix

Tribe Phytelepheae
   Ammandra
   Aphandra
   Phytelephas

Subfamily Coryphoideae

Tribe Borasseae
   Subtribe Hyphaeninae
      Bismarckia
      Hyphaene
      Medemia
      Satranala
   Subtribe Lataniinae
      Borassodendron
      Borassus
      Latania
      Lodoicea

Tribe Caryoteae
   Arenga
   Caryota
   Wallichia

Tribe Chuniophoeniceae
   Chuniophoenix
   Kerriodoxa
   Nannorrhops
   Tahina

Tribe Corypheae
   Corypha

Tribe Cryosophileae
   Chelyocarpus
   Coccothrinax
   Cryosophila
   Hemithrinax
   Itaya
   Leucothrinax
   Schippia
   Thrinax
   Trithrinax
   Zombia

Tribe Phoeniceae
   Phoenix

Tribe Sabaleae
   Sabal

Tribe Trachycarpeae
   Acoelorrhaphe
   Brahea
   Colpothrinax
   Copernicia
   Pritchardia
   Serenoa
   Washingtonia
   Subtribe Livistoninae
      Johannesteijsmannia
      Lanonia
      Licuala
      Livistona
      Pholidocarpus
      Saribus
   Subtribe Rhapidinae
      Chamaerops
      Guihaia
      Maxburretia
      Rhapidophyllum
      Rhapis
      Trachycarpus

Subfamily Nypoideae

Nypa

INTRODUCTION

A composition of a pinnate-leaf palm (*Phoenix canariensis*) and a palmate-leaf palm (*Chamaerops humilis*) exhibits variety in form, height, and texture and is esthetically pleasing.

Fortunate indeed are gardeners who live in regions where at least a few palms grow, as their unique variety of forms cannot begin to be simulated by anything other than massive ferns, yuccas, and cordylines, which themselves are tropical or subtropical.

The biggest reason these princes of the plant world are eschewed even where they can be grown is probably lack of space: most palm species are relatively large. This problem is compounded by the gardener's desire for color. The desire is, of course, an important one, but color is greatly overused at the expense of variety of form. The smaller the garden, the more the use of color alone tends to become overwhelming and even tiring, somewhat analogous to eating a diet of only cake and ice cream, or listening to only one type of music. *Variety* is the operative word and is what palms excel at with their ineluctably different forms. In addition, palms are often colorful, especially the tropical species. Their crownshafts, inflorescences, and leaf colors are sometimes extraordinary. Palms lend to the landscape a more controlled and subtle color palette than that of most "flowering plants."

A few points should be considered when incorporating palms into the landscape. First, palms don't look good planted in straight lines, but what type of plant does? This arrangement is unnatural in the sense that it doesn't occur in nature. The larger palms are magnificent when lining streets or avenues, but the dictum still applies: a curving street, path, or driveway is infinitely more esthetically pleasing than a straight one where variety is the missing element.

Second, palms generally look their best when planted in small groups or groves rather than as a single tree surrounded by space. Again, the reason is that palms do not occur that way in nature. Furthermore the discrete groups look best when the number of individuals therein is three, five, or more and the individuals are of varying heights; if each palm is the same height, the crowns visually "fight." Variety is the missing element in groups of same-height palms.

Third, a landscape whose horizon is

basically at one level is incredibly less interesting and beautiful than one of varying levels. Nothing fixes the imbalance better than using palms as canopy-scapes, where the crowns of trees float above the general level of the surrounding vegetation. Such palms substitute remarkably well for a lack of mountains. Again, variety of form.

Fourth, the wall of vegetation that constitutes the horizon of the garden is so much less appealing if it is of one form or of one texture. No plants are better suited to fix this problem than palms, whether large or small, fan leaved or feather leaved. Palms are the *sine-qua-non* elements to create the needed form and texture. Again, variety of form.

A few palm species are so large and impressive that they can be advantageously planted alone as specimen plants surrounded by space and still look good. They would look even better if planted in groups, but the size or other limitations of a given landscape or garden can often make this difficult or impossible.

Finally, no palm species has an uninteresting or ugly silhouette, so palms can be planted in front of walls or other structures, especially if the structure has a contrasting color. In this situation even small palms can look wonderful planted singly.

Pruning

While it is true that palms are underutilized in nearly every landscaping situation, it is also the case that, once planted, they tend to be overpruned. This is partly because they are unlike other garden trees and shrubs with which most gardeners are more familiar. Most people agonize all too much about whether or not to prune a palm tree, thinking, naturally enough, that a palm is like a hydrangea bush, an oak tree, or a morning-glory vine that needs diligent guidance from youth to old age to make a presentable landscape specimen. This is not the case.

A palm is, except for some special concerns of esthetics, human safety, or the health of the plant itself, a low-maintenance landscape subject. It occasionally requires some pruning, but the type and frequency of pruning requirements are unlike those of other landscape subjects. Furthermore, all pruning tools must be disinfected to avoid spreading diseases from one palm to another when living stems and leaves are cut.

In California, a deadly and incurable fungal disease (*Fusarium* wilt) is transmitted by pruning tools, as the fungus organism resides in the sap of the tree. In all cases, but especially in California, pruning tools should be disinfected between pruning one palm tree and another. Chain saws should not be used for pruning palms, because they cannot be properly cleaned and disinfected. Pruners, loppers, and hand saws should be thoroughly cleaned after each use by brushing or wiping off all sawdust. Then the blades should be soaked for 10 minutes in diluted pine oil (1 part pine oil to 3 parts water) or 5 minutes in diluted bleach (1 part bleach to 1 part water). Alternatively, the blades can be flamed for at least 10 seconds with a butane or propane torch.

Esthetic Concerns

Several palm species, like *Washingtonia filifera* and *W. robusta*, have a characteristic and mostly picturesque shag, sometimes called a *petticoat*, consisting of the adherent dead leaves whose leaf bases often refuse to fall from the trunk. These masses of dead and dry leaves can be a habitat for all manner of wildlife, both desirable (owls, moths) and undesirable (roaches, rats).

In the palms' native desert and arid grassland regions, natural fires commonly burn off the old leaves. In cultivation, the palms seldom retain enough dead leaves in moist and humid climes to form the petticoat but, in more arid regions, they usually do. Some palm owners love the shag, while others detest the "ugly haystack." The easiest way to remove the shag is, of course, one leaf at a time, as the individual leaves die and become pendent.

Then there is the manicurist mentality of those who want palms like *Washingtonia robusta* to have clean and smooth trunks from the bottom all the way to the leaf crown. This look is popular, especially in fantasy parks and Las Vegas. Removing all adherent leaf bases (or boots) from these palms creates an undeniably handsome look because of the elegantly thin and smooth, tall trunks, but the procedure is both time consuming and labor intensive, especially for tall specimens. Moreover, the procedure can create ugly wounds on the stem, which can be the entry points for diseases.

The leaf bases of some palms fall off at various times. In other palms, the blades fall off but the leaf bases themselves, and sometimes most of the petiole, adhere to the trunk for some time. A prime example is *Sabal palmetto* in the southeastern United States. The dead leaf bases may remain for most of the life of this slow-growing palm, turning woody and much lighter colored than the actual "bark" of the palm, and giving a picturesque wicker-weave look to the trunks. Often these woody boots are trimmed with a saw to conform one to the other in appearance and to create an even more elegant and manicured look.

Some popular palms like *Phoenix canariensis* produce immense, rounded canopies of long pinnate leaves. We personally consider this aspect of the species its most desirable trait, but others feel the palm looks better when the pendent leaves are removed, so that only the leaves that do not fall below the horizontal plane remain on the trunk. Removing the pendent leaves is no more difficult than removing the petticoats of *Washingtonia* palms and, since *P. canariensis* grows more slowly than either *Washingtonia* species, the task need not be done as often.

Most pruners leave the boots of the just-cut crop of leaves, which practice results in a picturesque, large rounded knobby cluster (informally called the "bulb" or "pineapple") of leaf bases directly beneath the leaf crown; by the time of the next leaf pruning, the bulb of boots will have mostly fallen away from the trunk naturally. In some parts of southern Europe, each boot is sculpted and shaped to add a decorative pattern to the bulb below the crown. Other pruners remove the bulb also, which results in an even more picturesque tree with a more tropical and more airy aspect.

Many palm owners, or the landscapers they hire, go far beyond these pruning practices: they remove enough leaves to create leaf crowns that look like feather dusters, leaving only the most erect living leaves therein. This truly abominable practice completely ruins the wonderful appearance of these species. It is also inimical to the overall health of the tree, which must struggle to photosynthesize enough food to maintain itself. Alas, this misguided practice is not limited to *Phoenix* species. It is all too common to see *Washingtonia*, *Syagrus romanzoffiana,* and many other species cropped in this manner, especially by plant maintenance contractors who, of course, make more money by recommending the monthly butchering of these beauties.

Several clustering species have unique esthetic and maintenance characteristics. A prime example is *Phoenix reclinata*, which forms many suckers or subsidiary trunks throughout its life. A single specimen may have as many as two dozen trunks. The trunks are not straight but lean gracefully outward from their points of origin. Few natural phenomena are as beautiful as a large "tuft" of this palm, but the clumps are made even more graceful and dramatic if a few trunks are thinned out as the mass develops, leaving trunks of differing heights. This allows the individual beauty of each trunk and the exquisitely graceful tableau of trunks and crowns to be seen and appreciated to their fullest.

The trunks should be thinned when they are young and short, using a large pair of lopping shears (with the blades disinfected) to cut off the growing point; when the trunk decays sufficiently, it can be easily pulled apart. Alternately, a hand saw (properly disinfected) can be used to carefully cut at the base of the developing stem.

For palms planted around swimming pools and walkways, fallen fruits can be a maintenance problem. Fruit clusters can be trimmed from palms before the fruit ripen (or even earlier, before the flowers open), but in many cases, doing so eliminates one of the most decorative aspects of the palm. *Carpentaria acuminata* with clusters of vivid red fruits is a wondrous sight and one that would be a shame to miss.

Human Safety

For most palms the rapidity with which the dead leaf parts fall from the trunks is a matter of environmental conditions. The factors which influence the fall are wind, abrasion, fire, rain, and humidity, the latter two of which precipitate the breakdown of the dead leaf; and then there is vandalism. In many parts of the southwestern United States, juvenile (and sometimes not so juvenile) miscreants are wont to create "fireworks" by setting ablaze the shags of *Washingtonia* and other petticoat-forming species. The conflagration may be induced not by vandalism alone, but also by the careless tossing of live cigarettes and cigars by passersby as well as lightning strikes and even (in rare instances) damaged live electrical wires. Pruning dead leaves removes the fire hazard.

Of course, it is better not to plant any tree where it can become a nuisance or hazard, crowd out other plants, or obstruct a desirable view. In practice, however, it often happens because the person who planted the palm did not know the ultimate dimensions of the particular species or just wanted its juvenile look for a certain period of time.

In many instances the desirable, ultimate, and overall landscape use of the palm warrants temporary pruning of its leaves, so that, in time, it becomes a positive component of the landscape. In such instances it should be remembered that, as with any other type of tree, the palm needs a certain amount of functioning leaves for proper photosynthesis and health. If too many leaves are constantly removed over a given period of time, the palm's trunk is subject to exhibiting varying trunk calipers because of the varying amounts of sugar it could produce at a given point in time; and the sight of a palm trunk with constrictions and bulges in what should be a columnar or naturally tapering stem can ruin an otherwise beautiful landscape subject.

Few palms produce large enough leaves or grow tall enough for their falling leaves to be much of a danger, but some, like the coconut (*Cocos nucifera*) and large date palms (for example, *Phoenix canariensis*), attain great heights and have large, heavy leaves with massive petioles and rachises. In addition, the coconut produces clusters of large, heavy fruit whose natural abscission (upon completely ripening) from the fruit stalk and consequent rapid descent can (in very rare instances) be deadly. Removing the leaves and fruits from tall specimens is labor intensive, especially if there are many trees to deal with. In Florida and the Caribbean, the removal of coconuts is recommended at the start of the hurricane season so that the nuts do not become dangerous projectiles in high winds.

A few tropical palm species have natural rings of large spines at regular intervals around their trunks. Among these are the star-nut palms (*Astrocaryum* spp.) and gru-gru palms (*Acrocomia* spp.). These spines are easily removed with shears or small branch cutters. Much more common in the United States are those palm species whose leaves have vicious spines for a part of their length. Prime examples are all date palms (*Phoenix* spp.) whose lower (basal) leaflets are actually spines of varying lengths and viciousness. When young, the older, lower, and more spreading leaves of these palms are often removed if they are near human pathways. Workers should always wear protective gear when pruning these palms, as the spines can cause severe and lasting injuries.

Health of the Palm

In only two instances is pruning involved in the good health of a palm: transplantation and disease cure or prevention. Leaf pruning (with disinfected shears or hand saw), especially of bare-root or balled-and-burlapped plants, is usually a necessity when transplanting. Removing at least half the leaf crown is almost always recommended in these cases to reduce moisture loss through transpiration because of the inevitable root loss. Indeed, with all *Sabal* species the recommendation is to remove all leaves except the "spear" or newest unfurled leaf, as the roots of this palm die under such circumstances and the palm will have to grow an entirely new set from the trunk.

Localized infections of fungal diseases can sometimes be controlled simply by removing the infected leaf. The excised leaf should be removed from the garden or burned so that it cannot be a source of infection for other palms.

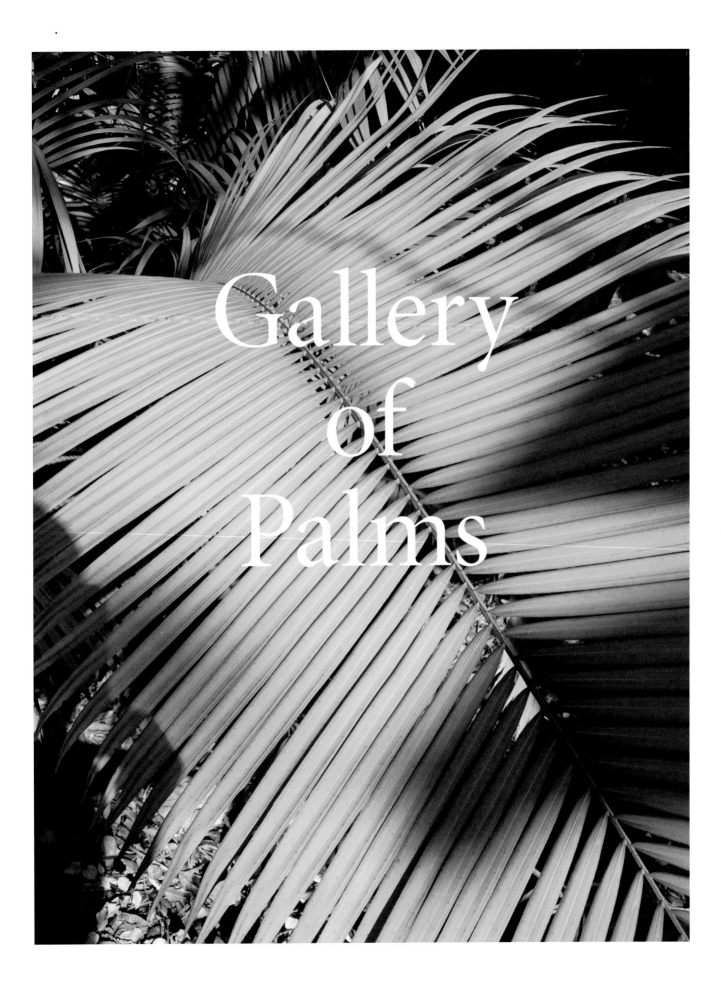

Gallery of Palms

Plate 1. *Acanthophoenix crinita*. Jeff Marcus garden, Hawaii.

Plate 2. *Acanthophoenix rubra*. Jeff Marcus garden, Hawaii.

Plate 3. *Acoelorrhaphe wrightii*. Jill Menzel garden, Brazil.

Plate 4. *Acrocomia aculeata*. Montgomery Botanical Center, Florida.

Plate 5. *Acrocomia crispa*. Juvenile, showing extreme spininess. Montgomery Botanical Center, Florida.

Plate 6. *Acrocomia crispa*. Young adult. In habitat, Cuba.

Plate 7. *Acrocomia crispa*. Fairchild Tropical Botanic Garden, Florida.

Plate 8. *Actinokentia divaricata*. Ho'omaluhia Garden, Hawaii.

Plate 9. *Actinokentia divaricata*. New leaf color. Jerry Andersen garden, Hawaii.

Plate 10. *Actinorhytis calapparia*. Montgomery Botanical Center, Florida.

Plate 11. *Adonidia merrillii*. Mounts Botanical Garden, Florida.

Plate 12. *Adonidia merrillii*. Yellow crownshaft cultivar. Nong Nooch Tropical Botanical Garden.

Plate 13. *Aiphanes horrida*. Pana'ewa Zoo, Hawaii.

Plate 14. *Aiphanes minima*. Jardín Botánico Nacional, Cuba.

Plate 15. *Aiphanes ulei*. Fairchild Tropical Botanic Garden, Florida.

Plate 16. *Allagoptera arenaria*. Private garden, Hawaii.

Plate 17. *Allagoptera brevicalyx*. Private garden, Florida.

Plate 18. *Allagoptera campestris*. Private garden, Florida.

Plate 19. *Allagoptera caudescens*. Jill Menzel garden, Brazil.

Plate 20. *Allagoptera caudescens*. Rio de Janeiro Botanical Garden, Brazil.

Plate 21. *Allagoptera leucocalyx*. Fairchild Tropical Botanic Garden, Florida.

Plate 22. *Aphandra natalia*. In habitat, Ecuador.

Plate 23. *Archontophoenix alexandrae*. Jerry Andersen garden, Hawaii.

Plate 24. *Archontophoenix cunninghamiana*. Balboa Park, California.

Plate 25. *Archontophoenix cunninghamiana*. Inflorescences. California.

Plate 26. *Archontophoenix maxima*. In habitat, Australia.

Plate 27. *Archontophoenix myolensis*. Fairchild Tropical Botanic Garden, Florida.

Plate 28. *Archontophoenix purpurea*. Waimea Valley Audubon Center, Hawaii.

Plate 29. *Archontophoenix purpurea*. Crownshaft. Herminio nursery, Brazil.

Plate 30. *Archontophoenix tuckeri*. Private garden, Hawaii.

Plate 31. *Areca caliso*. Jerry Andersen garden, Hawaii.

Plate 32. *Areca catechu*. Ho'omaluhia Botanical Garden, Hawaii.

Plate 33. *Areca catechu*. Yellow crownshaft cultivar. Jill Menzel garden, Brazil.

Plate 34. *Areca guppyana*. Cairns Botanical Gardens, Australia.

Plate 35. *Areca ipot*.

Plate 36. *Areca macrocalyx*. Red crownshaft form. Jeff Marcus garden, Hawaii.

Plate 37. *Areca macrocalyx*. Red crownshaft form. Jerry Andersen garden, Hawaii.

Plate 38. *Areca macrocarpa*. Paul Humann garden, Florida.

Plate 39. *Areca minuta*. Private garden, Australia.

Plate 40. *Areca montana*. Pana'ewa Zoo, Hawaii.

Plate 41. *Areca multifida*. Jerry Andersen garden, Hawaii.

Plate 42. *Areca ridleyana*. In habitat, Malaysia.

Plate 43. *Areca triandra*. Jill Menzel garden, Brazil.

Plate 44. *Areca tunku*. In habitat, Malaysia.

Plate 45. *Areca tunku*. Crownshaft and colorful inflorescence. In habitat, Malaysia.

Plate 46. *Areca vestiaria*. Rio de Janeiro Botanical Garden, Brazil.

Plate 47. *Areca vestiaria*. Infructescence. Private nursery, Brazil.

Plate 48. *Areca vestiaria*. Private garden, Australia.

Plate 49. *Areca vestiaria*. Maroon form. Private garden, Hawaii.

Plate 50. *Areca vidaliana*. Herminio nursery, Brazil.

Plate 51. *Arenga australasica*. Queensland, Australia.

Plate 52. *Arenga brevipes*. Private garden, California.

Plate 53. *Arenga brevipes*. Private garden, California.

Plate 54. *Arenga caudata*. Bob and Marita Bobick garden, Florida.

Plate 55. *Arenga caudata*. Fine-leaf form. Nong Nooch Tropical Botanical Garden, Thailand.

Plate 56. *Arenga engleri*. Montgomery Botanical Center, Florida.

Plate 57. *Arenga engleri*. Inflorescence.

Plate 58. *Arenga hastata*. In habitat, Sarawak.

Plate 59. *Arenga hookeriana*. Fairchild Tropical Botanic Garden, Florida.

Plate 60. *Arenga hookeriana*. Private garden, New Caledonia.

Plate 61. *Arenga microcarpa*. Montgomery Botanical Center, Florida.

Plate 62. *Arenga obtusifolia*. Fairchild Tropical Botanic Garden, Florida.

Plate 63. *Arenga pinnata*. Private garden, Australia.

Plate 64. *Arenga porphyrocarpa*.

Plate 65. *Arenga tremula*. Flamingo Gardens, Florida.

Plate 66. *Arenga undulatifolia*. Fairchild Tropical Botanic Garden, Florida.

Plate 67. *Arenga westerhoutii*. In habitat, Malaysia.

Plate 68. *Asterogyne martiana*. Jill Menzel garden, Brazil.

Plate 69. *Asterogyne martiana*. Infructescence. In habitat, Belize.

Plate 70. *Asterogyne spicata*. In habitat, Venezuela.

Plate 71. *Astrocaryum aculeatissimum*. In habitat, Brazil.

Plate 72. *Astrocaryum alatum*. Fairchild Tropical Botanic Garden, Florida.

Plate 73. *Astrocaryum mexicanum*. Fairchild Tropical Botanic Garden, Florida.

Plate 74. *Astrocaryum murumuru*. Rio de Janeiro Botanical Garden, Brazil.

Plate 75. *Astrocaryum standleyanum*. Ho'omaluhia Botanical Garden, Hawaii.

Plate 76. *Attalea allenii*. Ann Norton Sculpture Gardens, Florida.

Plate 77. *Attalea amygdalina*. Townsville Palmetum, Australia.

Plate 78. *Attalea butyracea*. In habitat, Costa Rica.

Plate 79. *Attalea cohune*. Private nursery, Belize.

Plate 80. *Attalea crassispatha*. Montgomery Botanical Center, Florida.

Plate 81. *Attalea guacuyule*. Montgomery Botanical Center, Florida.

Plate 82. *Attalea humilis*. Montgomery Botanical Center, Florida.

Plate 83. *Attalea maripa*. Cienfuegos Botanical Garden, Cuba.

Plate 84. *Attalea phalerata*. Fairchild Tropical Botanic Garden, Florida.

Plate 85. *Attalea speciosa*. Fairchild Tropical Botanic Garden, Florida.

Plate 86. *Bactris brongniartii*. Venezuelan form. Montgomery Botanical Center, Florida.

Plate 87. *Bactris brongniartii*. Inflorescence. Montgomery Botanical Center, Florida.

Plate 88. *Bactris concinna*. Rio de Janeiro Botanical Garden, Brazil.

Plate 89. *Bactris gasipaes*. Fairchild Tropical Botanic Garden, Florida.

Plate 90. *Bactris gasipaes*. Heart-of-palm, ready for market.

Plate 91. *Bactris grayumi*. Jeff Marcus garden, Hawaii.

Plate 92. *Bactris hondurensis*. Jerry Andersen garden, Hawaii.

Plate 93. *Bactris major*. Montgomery Botanical Center, Florida.

Plate 94. *Bactris militaris*. Jeff Marcus garden, Hawaii.

Plate 95. *Bactris plumeriana*. Jardín Botánico Nacional Dr. Rafael M. Moscoso, Dominican Republic.

Plate 96. *Balaka diffusa*. Jeff Marcus garden, Hawaii.

Plate 97. *Balaka longirostris*. Ho'omaluhia Botanical Garden, Hawaii.

Plate 98. *Balaka microcarpa*. Jeff Marcus garden, Hawaii.

Plate 99. *Balaka seemannii*. Jerry Andersen garden, Hawaii.

Plate 100. *Basselinia deplanchei*. In habitat, New Caledonia.

Plate 101. *Basselinia favieri*. In habitat, New Caledonia.

Plate 102. *Basselinia glabrata*. Ho'omaluhia Botanical Garden, Hawaii.

Plate 103. *Basselinia gracilis*. In habitat, New Caledonia.

Plate 104. *Basselinia gracilis*. Crownshaft. Private garden, Hawaii.

Plate 105. *Basselinia humboldtiana*. Jeff Marcus garden, Hawaii.

Plate 106. *Basselinia pancheri*. In habitat, New Caledonia.

Plate 107. *Basselinia pancheri*. In habitat, New Caledonia.

Plate 108. *Basselinia tomentosa*. Jeff Marcus garden, Hawaii.

Plate 109. *Basselinia velutina*. Jeff Marcus garden, Hawaii.

Plate 110. *Basselinia vestita*. Jeff Marcus garden, Hawaii.

Plate 111. *Beccariophoenix alfredii*. A natural stand in Madagascar.

Plate 112. *Beccariophoenix alfredii*. Infructescences borne below the crown and showing the short peduncle. In habitat, Madagascar.

Plate 113. *Beccariophoenix madagascariensis*. Ho'omaluhia Botanical Garden, Hawaii.

Plate 114. *Beccariophoenix madagascariensis*. Young plant. Private garden, Hawaii.

Plate 115. *Bentinckia condapanna*. Private garden, Hawaii.

Plate 116. *Bentinckia nicobarica*. Juveniles. Waimea Valley Audubon Center, Hawaii.

Plate 117. *Bentinckia nicobarica*. Fairchild Tropical Botanic Garden, Florida.

Plate 118. *Bismarckia nobilis*. Fairchild Tropical Botanic Garden, Florida.

Plate 119. *Bismarckia nobilis*. Leaf bases and fruits. Mounts Botanical Garden, Florida.

Plate 120. *Borassodendron borneense*. In habitat, Sarawak.

Plate 121. *Borassodendron machadonis*. Fairchild Tropical Botanic Garden, Florida.

Plate 122. *Borassodendron machadonis*. Male flowers. Fairchild Tropical Botanic Garden, Florida.

Plate 123. *Borassus aethiopum*. Flamingo Gardens, Florida.

Plate 124. *Borassus aethiopum*. Black leaf bases. Montgomery Botanical Center, Florida.

Plate 125. *Borassus flabellifer*. Townsville Palmetum, Australia.

Plate 126. *Borassus madagascariensis*. Private garden, Florida Keys.

Plate 127. *Brahea aculeata*. Huntington Botanical Gardens, California.

Plate 128. *Brahea armata*. Balboa Park, California.

Plate 129. *Brahea brandegeei*. California.

Plate 130. *Brahea calcarea*. Huntington Botanical Gardens, California.

Plate 131. *Brahea decumbens*. Huntington Botanical Gardens, California.

Plate 132. *Brahea dulcis*. Fairchild Tropical Botanic Garden, Florida.

Plate 133. *Brahea edulis*. California.

Plate 134. *Brassiophoenix drymophloeoides*. Arden Dearden nursery, Australia.

Plate 135. *Burretiokentia dumasii*. Jeff Marcus garden, Hawaii.

Plate 136. *Burretiokentia grandiflora*. Jeff Marcus garden, Hawaii.

Plate 137. *Burretiokentia hapala*. Ho'omaluhia Botanical Garden, Hawaii.

Plate 138. *Burretiokentia koghiensis*. Jeff Marcus garden, Hawaii.

Plate 139. *Burretiokentia vieillardii*. Ho'omaluhia Botanical Garden, Hawaii.

Plate 140. *Burretiokentia vieillardii*. Private garden, Hawaii.

Plate 141. *Butia archeri*. In habitat, Brazil.

Plate 142. *Butia campicola*. Private garden, Brazil.

Plate 143. *Butia capitata*. Private garden, Florida Keys.

Plate 144. *Butia capitata*. With Rolf Kyburz. Australia.

Plate 145. *Butia eriospatha*. Jill Menzel garden, Brazil.

Plate 146. *Butia microspadix*. In habitat, Brazil.

Plate 148. *Butia odorata*. Fruits. In habitat, Uruguay.

Plate 147. *Butia odorata*. In habitat, Uruguay.

Plate 149. ×*Butiagrus nabonnandii*. Oakland Palmetum, California.

Plate 150. ×*Jubautia*. The hybrid between *Jubaea* and *Butia*. Florida.

Plate 151. *Butia paraguayensis*. In habitat, Paraguay.

Plate 152. *Butia paraguayensis*. Huntington Botanical Gardens, California.

Plate 153. *Butia purpurascens*. In habitat, Brazil.

Plate 154. *Butia purpurascens*. Herminio nursery, Brazil.

Plate 155. *Butia yatay*. Huntington Botanical Gardens, California.

Plate 156. *Calamus australis*. In habitat, Australia.

Plate 157. *Calamus caryotoides*. Royal Botanic Gardens, Sydney, Australia.

Plate 158. *Calamus ciliaris*. Jerry Andersen garden, Hawaii.

Plate 159. *Calamus erectus*. Jeff Marcus garden, Hawaii.

Plate 160. *Calamus javensis*. Forest Research Institute of Malaysia.

Plate 161. *Calamus moti*. In habitat, Australia.

Plate 162. *Calamus viminalis*. Private garden, Florida.

Plate 163. *Calyptrocalyx albertisianus*. Jill Menzel garden, Brazil.

Plate 164. *Calyptrocalyx arfakiensis*. Flecker Botanic Gardens, Australia.

Plate 165. *Calyptrocalyx awa*. Private garden, Australia.

Plate 166. *Calyptrocalyx doxanthus*. Private garden, Bogor, Indonesia.

Plate 167. *Calyptrocalyx forbesii*. Rio de Janeiro Botanical Garden, Brazil.

Plate 168. *Calyptrocalyx hollrungii*. Jerry Andersen garden, Hawaii.

Plate 169. *Calyptrocalyx hollrungii*. New leaf. Herminio Nursery, Brazil.

Plate 170. *Calyptrocalyx micholitzii*. New leaf. Jeff Marcus garden, Hawaii.

Plate 171. *Calyptrocalyx micholitzii*. Jeff Marcus garden, Hawaii.

Plate 172. *Calyptrocalyx pachystachys*. Private garden, Australia.

Plate 173. *Calyptrocalyx pachystachys*. Colorful new leaf. Private garden, Australia.

Plate 174. *Calyptrocalyx pauciflorus*. Australia.

Plate 175. *Calyptrocalyx polyphyllus*. Private garden, Australia.

Plate 176. *Calyptrocalyx spicatus*. In habitat, Ambon, Indonesia.

Plate 177. *Calyptrogyne ghiesbreghtiana*. Marco Herrero Nursery, Costa Rica.

Plate 178. *Calyptrogyne ghiesbreghtiana*. Flowers. Fairchild Tropical Botanic Garden, Florida.

Plate 179. *Calyptronoma occidentalis*. Fairchild Tropical Botanic Garden, Florida.

Plate 180. *Calyptronoma occidentalis*. Leaf bases and inflorescences. Fairchild Tropical Botanic Garden, Florida.

Plate 181. *Calyptronoma plumeriana*. In habitat, Cuba.

Plate 182. *Calyptronoma rivalis*. Fairchild Tropical Botanic Garden, Florida.

Plate 183. *Carpentaria acuminata*. Montgomery Botanical Center, Florida.

Plate 184. *Carpoxylon macrospermum*. Private garden, Hawaii.

Plate 185. *Caryota maxima*. In bloom, California.

Plate 186. *Caryota mitis*. Fairchild Tropical Botanic Garden, Florida.

Plate 187. *Caryota monostachya*. Fairchild Tropical Botanic Garden, Florida.

Plate 188. *Caryota no*. Flecker Botanic Gardens, Australia.

Plate 189. *Caryota obtusa*. Jeff Brusseau garden, California.

Plate 190. *Caryota ophiopellis*. Jeff Marcus garden, Hawaii.

Plate 191. *Caryota rumphiana*. Montgomery Botanical Center, Florida.

Plate 192. *Caryota urens*. Australia.

Plate 193. *Caryota zebrina*. Herminio nursery, Brazil.

Plate 194. *Caryota zebrina*. Petioles. Herminio nursery, Brazil.

Plate 195. *Ceratolobus glaucescens*. Lyon Arboretum, Hawaii.

Plate 196. *Ceroxylon alpinum*. In habitat, Venezuela.

Plate 197. *Ceroxylon amazonicum*. In habitat, Ecuador.

Plate 199. *Ceroxylon parvifrons*. Darold Petty garden, California.

Plate 198. *Ceroxylon ceriferum*. In habitat, Venezuela.

Plate 200. *Ceroxylon parvum*. In habitat, Ecuador.

Plate 201. *Ceroxylon quindiuense*. Jardín Botánico José Celestino Mutis, Colombia.

Plate 202. *Ceroxylon ventricosum*. In habitat, Ecuador.

Plate 203. *Ceroxylon vogelianum*. (foreground) and *C. quindiuense* (background). Jardín Botánico José Celestino Mutis, Colombia.

Plate 204. *Chamaedorea adscendens*. Fairchild Tropical Botanic Garden, Florida.

Plate 205. *Chamaedorea amabilis*. In habitat, Costa Rica.

Plate 206. *Chamaedorea brachypoda*. Dale Holton garden, Florida.

Plate 207. *Chamaedorea cataractarum*. Private Nursery, Belize.

Plate 208. *Chamaedorea correae*. Jerry Andersen garden, Hawaii.

Plate 209. *Chamaedorea costaricana*. San Francisco Botanic Garden, California.

Plate 210. *Chamaedorea deckeriana*. Private garden, Florida.

Plate 211. *Chamaedorea deneversiana*. Jerry Andersen garden, Hawaii.

Plate 212. *Chamaedorea elatior*. Espaliered against a wall, Orto Botanico "Giardino dei Semplici," Italy.

Plate 213. *Chamaedorea elatior*. Juvenile showing modified leaflets. Anne Norton Sculpture Garden, Florida.

Plate 214. *Chamaedorea elegans*. Sítio Roberto Burle Marx, Brazil.

Plate 215. *Chamaedorea ernesti-augusti*. Sítio Roberto Burle Marx, Brazil.

Plate 216. *Chamaedorea fragrans*. Herminio nursery Brazil.

Plate 217. *Chamaedorea frondosa*. Jeff Marcus garden, Hawaii.

Plate 218. *Chamaedorea geonomiformis*. Jeff Marcus garden, Hawaii.

Plate 219. *Chamaedorea glaucifolia*. Private garden, Florida.

Plate 220. *Chamaedorea graminifolia*. In habitat, Belize.

Plate 222. *Chamaedorea metallica*. Private garden, Florida.

Plate 221. *Chamaedorea hooperiana*. Dale Holton garden, Florida.

Plate 223. *Chamaedorea metallica*. Private Nursery, Belize.

Plate 224. *Chamaedorea microspadix*. University of Florida, Ft. Lauderdale, Florida.

Plate 225. *Chamaedorea microspadix*. Dick Douglas garden, California.

Plate 226. *Chamaedorea neurochlamys*. Belize.

Plate 227. *Chamaedorea oblongata*. Tikal, Guatemala.

Plate 228. *Chamaedorea oreophila*. Jerry Andersen garden, Hawaii.

Plate 229. *Chamaedorea pinnatifrons*. In habitat, Costa Rica.

Plate 230. *Chamaedorea plumosa*. Darold Petty garden, California.

Plate 231. *Chamaedorea pumila*. In habitat, Costa Rica.

Plate 232. *Chamaedorea radicalis*. Jardín Botánico de Caracas, Venezuela.

Plate 233. *Chamaedorea radicalis*. Aerial-trunked form. John DeMott nursery, Florida.

Plate 234. *Chamaedorea sartorii*. In habitat, Mexico.

Plate 235. *Chamaedorea seifrizii*. Broad-leaflet form. Fairchild Tropical Botanic Garden, Florida.

Plate 236. *Chamaedorea seifrizii*. Private garden, Florida.

Plate 237. *Chamaedorea stolonifera*. Bob & Marita Bobick garden, Florida.

Plate 238. *Chamaedorea tenerrima*. Jerry Andersen garden, Hawaii.

Plate 239. *Chamaedorea tepejilote*. Broward College, Florida.

Plate 240. *Chamaedorea tepejilote*. Clustering form. Tikal, Guatemala.

Plate 242. *Chamaedorea warscewiczii*. Private garden, Florida.

Plate 241. *Chamaedorea tuerckheimii*. Private garden, Australia.

Plate 243. *Chamaedorea woodsoniana*. Mexico.

Plate 244. *Chamaerops humilis* var. *argentea*. Jeff Marcus garden, Hawaii.

Plate 245. *Chamaerops humilis* var. *humilis*. Balboa Park, California.

Plate 246. *Chamaerops humilis* var. *humilis*. Single-trunked form. Florida International University, Florida.

Plate 247. *Chamaerops humilis* 'Vulcano'. Private garden, United Kingdom.

Plate 248. *Chambeyronia lepidota*. Jeff Marcus garden, Hawaii.

Plate 249. *Chambeyronia macrocarpa*. Jerry Andersen garden, Hawaii.

Plate 250. *Chambeyronia macrocarpa*. Yellow crownshaft ('Hookeri') form, Jeff Marcus garden, Hawaii.

Plate 251. *Chambeyronia macrocarpa*. "Watermelon" crownshaft form.

Plate 252. *Chelyocarpus chuco*. Jeff Marcus garden, Hawaii.

Plate 253. *Chelyocarpus ulei*. Jeff Marcus garden, Hawaii.

Plate 254. *Chuniophoenix hainanensis*. Private garden, Hawaii.

Plate 255. *Chuniophoenix humilis*. Jeff Marcus garden, Hawaii.

Plate 256. *Chuniophoenix nana*. Dale Holton garden, Florida.

Plate 257. *Clinosperma bracteale*. In habitat, New Caledonia.

Plate 258. *Clinosperma lanuginosa*. Jeff Marcus garden, Hawaii.

Plate 259. *Clinostigma exorrhizum*. Jeff Marcus garden, Hawaii.

Plate 260. *Clinostigma harlandii*. Jerry Andersen garden, Hawaii.

Plate 262. *Clinostigma samoense*. Pana'ewa Rainforest Zoo & Gardens, Hawaii.

Plate 261. *Clinostigma ponapense*. Three plants growing at Pana'ewa Rainforest Zoo & Gardens, Hawaii.

Plate 263. *Clinostigma savoryanum*. Fairchild Tropical Botanic Garden, Florida.

Plate 264. *Coccothrinax alexandri*. Baracoa, Cuba.

Plate 265. *Coccothrinax argentata*. Fairchild Tropical Botanic Garden, Florida.

Plate 266. *Coccothrinax argentea*. Jardín Botánico Nacional Dr. Rafael M. Moscoso, Dominican Republic.

Plate 267. *Coccothrinax* "azul." In habitat, Cuba.

Plate 268. *Coccothrinax barbadensis*. Ho'omaluhia Botanical Garden, Hawaii.

Plate 269. *Coccothrinax borhidiana*. Montgomery Botanical Center, Florida.

Plate 270. *Coccothrinax boschiana*. University of Florida, Homestead, Florida.

Plate 271. *Coccothrinax boschiana*. In habitat, Dominican Republic.

Plate 272. *Coccothrinax clarensis*. In habitat, Cuba.

Plate 273. *Coccothrinax crinita* subsp. *crinita*. Private garden, Florida Keys.

Plate 274. *Coccothrinax crinita* subsp. *brevicrinis*. Jardín Botánico Nacional, Cuba.

Plate 275. *Coccothrinax ekmanii*. In habitat, Cuba.

Plate 276. *Coccothrinax gracilis*. In habitat, Dominican Republic.

Plate 277. *Coccothrinax gundlachii*. In habitat, Cuba.

Plate 278. *Coccothrinax hiorami*. In habitat, Cuba.

Plate 279. *Coccothrinax litoralis*. In habitat, Cuba.

Plate 280. *Coccothrinax miraguama* subsp. *arenicola*. Private garden, Florida.

Plate 281. *Coccothrinax miraguama* subsp. *havanensis*. Private garden, Florida.

Plate 282. *Coccothrinax moaensis*. In habitat, Cuba.

Plate 283. *Coccothrinax montana*. Dale Holton garden, Florida.

Plate 284. *Coccothrinax pseudorigida*. In habitat, Cuba.

Plate 285. *Coccothrinax salvatoris*. In habitat, Cuba.

Plate 286. *Coccothrinax scoparia*. Montgomery Botanical Center, Florida.

Plate 287. *Coccothrinax spissa*. In habitat, Dominican Republic.

Plate 288. *Cocos nucifera*. Rio de Janeiro, Brazil.

Plate 289. *Cocos nucifera* 'Golden Malayan Dwarf'. Paul Humann garden, Florida.

Plate 290. *Colpothrinax cookii*. Ho'omaluhia Botanical Garden, Hawaii.

Plate 291. *Colpothrinax cookii*. Detail of leaf base fibers. Ho'omaluhia Botanical Garden, Hawaii.

Plate 292. *Colpothrinax wrightii*. In habitat, Cuba.

Plate 293. *Copernicia alba*. Private garden, Florida Keys.

Plate 294. *Copernicia baileyana*. Fairchild Tropical Botanic Garden, Florida.

Plate 295. *Copernicia berteroana*. Montgomery Botanical Center, Florida.

Plate 296. *Copernicia brittonorum*. In habitat, Cuba.

Plate 297. *Copernicia cowellii*. In habitat, Cuba.

Plate 298. *Copernicia cowellii*. Leaf. In habitat, Cuba.

Plate 299. *Copernicia curtissii*. In habitat, Cuba.

Plate 300. *Copernicia ekmanii*. Fairchild Tropical Botanic Garden, Florida.

Plate 301. *Copernicia fallaensis*. Fairchild Tropical Botanic Garden, Florida.

Plate 302. *Copernicia gigas*. In habitat, Cuba.

Plate 303. *Copernicia glabrescens*. Jardín Botánico Nacional, Cuba.

Plate 304. *Copernicia hospita*. Jardín Botánico Nacional, Cuba.

Plate 305. *Copernicia macroglossa*. Fairchild Tropical Botanic Garden, Florida.

Plate 306. *Copernicia prunifera*. Fairchild Tropical Botanic Garden, Florida.

Plate 307. *Copernicia rigida*. Juvenile. Cuba.

Plate 308. *Copernicia rigida*. Jardín Botánico Nacional, Cuba.

Plate 309. *Copernicia tectorum*. Fairchild Tropical Botanic Garden, Florida.

Plate 310. *Copernicia ×textilis*. In habitat, Cuba.

Plate 311. *Corypha umbraculifera*. Young plant. Ho'omaluhia Botanical Garden, Hawaii.

Plate 312. *Corypha umbraculifera*. In bloom, Panama.

Plate 313. *Corypha umbraculifera*. Fruiting palm on left, Rio de Janeiro, Brazil.

Plate 314. *Corypha utan*. Leafages showing spiral pattern. Australia.

Plate 315. *Corypha utan*. Private garden, Florida Keys.

Plate 316. *Cryosophila guagara*. Fairchild Tropical Botanic Garden, Florida.

Plate 317. *Cryosophila nana*. In habitat, Guerrero, Mexico.

Plate 318. *Cryosophila stauracantha*. Pana'ewa Rainforest Zoo & Gardens, Hawaii.

Plate 319. *Cryosophila stauracantha*. Roots pines and stilt roots. Jill Menzel garden, Brazil.

Plate 320. *Cryosophila warscewiczii*. Wilson Botanic Garden, Costa Rica.

Plate 321. *Cryosophila williamsii*. Fairchild Tropical Botanic Garden, Florida.

Plate 322. *Cyphokentia cerifera*. In habitat, New Caledonia.

Plate 323. *Cyphokentia macrostachya*. Jeff Marcus garden, Hawaii.

Plate 324. *Cyphophoenix alba*. In habitat, New Caledonia.

Plate 325. *Cyphophoenix alba*. Crownshaft. Jeff Marcus garden, Hawaii.

Plate 326. *Cyphophoenix elegans*. Ho'omaluhia Botanical Garden, Hawaii.

Plate 327. *Cyphophoenix elegans*. Crownshaft. Jerry Andersen garden, Hawaii.

Plate 328. *Cyphophoenix fulcita*. In habitat, New Caledonia.

Plate 329. *Cyphophoenix fulcita*. Crownshaft. In habitat, New Caledonia.

Plate 330. *Cyphophoenix nucele*. Jerry Andersen garden, Hawaii.

Plate 331. *Cyphosperma balansae*. Ho'omaluhia Botanical Garden, Hawaii.

Plate 332. *Cyphosperma balansae*. Leafages. Jerry Andersen garden, Hawaii.

Plate 333. *Cyphosperma tanga*. In habitat, Fiji.

Plate 334. *Cyphosperma trichospadix*. Jeff Marcus garden, Hawaii.

Plate 335. *Cyrtostachys elegans*. Pana'ewa Rainforest Zoo & Gardens, Hawaii.

Plate 336. *Cyrtostachys renda*. Private garden, Australia.

Plate 337. *Cyrtostachys renda*. Dark maroon form, Jeff Marcus garden, Hawaii.

Plate 338. *Daemonorops curranii*. Fairchild Tropical Botanic Garden, Florida.

Plate 339. *Daemonorops jenkinsiana*. In habitat, Bengal, India.

Plate 340. *Daemonorops melanochaetes*. Australia.

Plate 341. *Deckenia nobilis*. Ho'omaluhia Botanical Garden, Hawaii.

Plate 342. *Deckenia nobilis*. Crownshaft and inflorescences. Ho'omaluhia Botanical Garden, Hawaii.

Plate 343. *Deckenia nobilis*. Spines on petioles and leaf sheaths of juvenile. Redland Nursery, Florida.

101

Plate 344. *Desmoncus orthacanthos*. Montgomery Botanical Center, Florida.

Plate 345. *Desmoncus orthacanthos*. In habitat, Belize.

Plate 346. *Desmoncus polyacanthos*. National Tropical Botanical Garden, Hawaii.

Plate 347. *Dictyocaryum lamarckianum*. In habitat, Ecuador.

Plate 348. *Dictyosperma album*. Private garden, Florida.

Plate 349. *Dictyosperma album* var. *conjugatum*. Private garden, Florida Keys.

Plate 350. *Dransfieldia micrantha*. Fairchild Tropical Botanic Garden, Florida.

Plate 351. *Drymophloeus hentyi*. Jill Menzel garden, Brazil.

Plate 352. *Drymophloeus hentyi*. Detail of leaf. Jill Menzel garden, Brazil.

Plate 353. *Drymophloeus litigiosus*. Private garden, Hawaii.

Plate 354. *Drymophloeus oliviformis*. Fairchild Tropical Botanic Garden, Florida.

Plate 355. *Drymophloeus pachycladus*. Fairchild Tropical Botanic Garden, Florida.

Plate 356. *Drymophloeus subdistichus*. Rio de Janeiro Botanical Garden, Brazil.

Plate 357. *Dypsis albofarinosa*. Private garden, Hawaii.

Plate 358. *Dypsis ambositrae*. In habitat, Madagascar.

Plate 359. *Dypsis arenarum*. Jeff Marcus garden, Hawaii.

Plate 360. *Dypsis arenarum*. Detail of crownshaft. Jeff Marcus garden, Hawaii.

Plate 361. *Dypsis baronii*. Private garden, Hawaii.

Plate 362. *Dypsis beentjei*. Jeff Marcus garden, Hawaii.

Plate 363. *Dypsis bejofo*. Jeff Marcus garden, Hawaii.

Plate 364. *Dypsis bosseri*. Jeff Marcus garden, Hawaii.

Plate 365. *Dypsis cabadae*. Florida.

Plate 366. *Dypsis carlsmithii*. Private garden, Hawaii.

Plate 367. *Dypsis catatiana*. In habitat, Madagascar.

Plate 368. *Dypsis coriacea*. Hawaii.

Plate 369. *Dypsis crinita*. Private garden, Hawaii.

Plate 370. *Dypsis crinita*. Red new leaf. Fairchild Tropical Botanic Garden, Florida.

Plate 371. *Dypsis decaryi*. Private garden, Florida Keys.

Plate 372. *Dypsis decaryi*. In habitat, Madagascar.

Plate 374. *Dypsis dransfieldii*. Jeff Marcus garden, Hawaii.

Plate 373. *Dypsis decipiens*. Ho'omaluhia Botanical Garden, Hawaii.

Plate 375. *Dypsis faneva*. Jeff Marcus garden, Hawaii.

Plate 376. *Dypsis fibrosa*. Ho'omaluhia Botanical Garden, Hawaii.

Plate 377. *Dypsis fibrosa*. Detail of stem. Jeff Marcus garden, Hawaii.

Plate 378. *Dypsis heteromorpha*. Jeff Marcus garden, Hawaii.

Plate 379. *Dypsis heteromorpha*. Detail of leaf bases. Jeff Marcus garden, Hawaii.

Plate 380. *Dypsis hildebrandtii*. Jeff Marcus garden, Hawaii.

Plate 381. *Dypsis hildebrandtii*. Juveniles. In habitat, Madagascar.

Plate 382. *Dypsis hovomantsina*. Jeff Marcus garden, Hawaii.

Plate 383. *Dypsis lanceolata*. Montgomery Botanical Center, Florida.

Plate 384. *Dypsis lantzeana*. Jeff Marcus garden, Hawaii.

Plate 386. *Dypsis lastelliana*. Detail of crownshaft. Ho'omaluhia Botanical Garden, Hawaii.

Plate 385. *Dypsis lastelliana*. Ho'omaluhia Botanical Garden, Hawaii.

Plate 387. *Dypsis leptocheilos*. Flamingo Gardens, Florida.

Plate 388. *Dypsis leptocheilos*. Crownshaft.

Plate 389. *Dypsis louvelii*. Jeff Marcus garden, Hawaii.

Plate 390. *Dypsis lutescens*. Brazil.

Plate 391. *Dypsis madagascariensis*. Private garden, Hawaii.

Plate 392. *Dypsis madagascariensis*. Detail of crownshafts. Herminio Nursery, Brazil.

Plate 393. *Dypsis madagascariensis* var. *lucubensis*. Jerry Andersen garden, Hawaii.

Plate 394. *Dypsis malcomberi*. Jeff Marcus garden, Hawaii.

Plate 395. *Dypsis malcomberi*. In habitat, Madagascar.

Plate 396. *Dypsis mananjarensis*. In habitat, Madagascar.

Plate 397. *Dypsis mananjarensis*. Detail of scales on petioles. Jeff Marcus garden, Hawaii.

Plate 398. *Dypsis marojejyi*. Jeff Marcus garden, Hawaii.

Plate 399. *Dypsis nauseosa*. In habitat, Madagascar.

Plate 400. *Dypsis nodifera*. Fairchild Tropical Botanic Garden, Florida.

Plate 401. *Dypsis onilahensis*. In habitat, Madagascar.

Plate 402. *Dypsis pachyramea*. In habitat, Madagascar.

Plate 403. *Dypsis paludosa*. Jerry Andersen garden, Hawaii.

Plate 404. *Dypsis pembana*. Solitary form. Ho'omaluhia Botanical Garden, Hawaii.

Plate 405. *Dypsis pembana*. Clustering form. Jerry Andersen garden, Hawaii.

Plate 406. *Dypsis pilulifera*. Private garden, Hawaii.

Plate 407. *Dypsis pilulifera*. Crownshaft. Private garden, Hawaii.

Plate 408. *Dypsis pinnatifrons*. Marco Herrero nursery, Costa Rica.

Plate 409. *Dypsis plumosa*. Jerry Andersen garden, Hawaii.

Plate 410. *Dypsis prestoniana*. Jeff Marcus garden, Hawaii.

Plate 411. *Dypsis procera*. Private garden, Hawaii.

Plate 412. *Dypsis psammophila*. Jeff Marcus garden, Hawaii.

Plate 413. *Dypsis remotiflora*. Jeff Marcus garden, Hawaii.

Plate 414. *Dypsis rivularis*. Private garden, Hawaii.

Plate 415. *Dypsis rivularis*. Crownshaft. Jill Menzel garden, Brazil.

Plate 416. *Dypsis robusta*. Jeff Marcus garden, Hawaii.

Plate 417. *Dypsis saintelucei*. Jeff Marcus garden, Hawaii.

Plate 418. *Dypsis sanctaemariae*. Undivided leaf form. In habitat, Madagascar.

Plate 419. *Dypsis scottiana*. Jeff Marcus garden, Hawaii.

Plate 420. *Dypsis simianensis*. Jeff Marcus garden, Hawaii.

Plate 421. *Dypsis* sp. 'Mayotte'. Private garden, Florida.

Plate 422. *Dypsis* sp. 'Pink Crownshaft'. Private garden, Hawaii.

Plate 423. *Dypsis* sp. 'Pink Crownshaft'. Young inflorescences. Hawaii.

Plate 424. *Dypsis tokoravina*. Jeff Marcus garden, Hawaii.

Plate 425. *Dypsis tokoravina*. Crownshaft. Jeff Marcus garden, Hawaii.

Plate 426. *Dypsis tsaravoasira*. Private garden, Hawaii.

Plate 427. *Dypsis utilis*. Ivoloina Zoo, Madagascar.

Plate 428. *Elaeis guineensis*. Flamengo Park, Rio de Janeiro, Brazil.

Plate 429. *Elaeis guineensis*. Fruits. Jill Menzel garden, Brazil.

Plate 431. *Eugeissona insignis*. Private garden, Australia.

Plate 430. *Elaeis oleifera*. Waimea Valley Audubon Center, Hawaii.

Plate 432. *Eugeissona tristis*. Pana'ewa Rainforest Zoo & Gardens, Hawaii.

Plate 433. *Euterpe catinga*. Private garden, Hawaii.

Plate 434. *Euterpe edulis*. In habitat, Brazil.

Plate 435. *Euterpe edulis*. Orange crownshaft form. Jill Menzel garden, Brazil.

Plate 436. *Euterpe oleracea*. Pana'ewa Rainforest Zoo & Gardens, Hawaii.

Plate 437. *Euterpe oleracea*. Inflorescence. Flamingo Gardens, Florida.

Plate 438. *Euterpe precatoria* var. *longivaginata*. Jerry Andersen garden, Hawaii.

Plate 439. *Euterpe precatoria* var. *precatoria*. Herminio nursery, Brazil.

Plate 440. *Euterpe precatoria* var. *precatoria*. Detail of crownshaft. Private garden, Hawaii.

Plate 441. *Gaussia attenuata*. Fairchild Tropical Botanic Garden, Florida.

Plate 442. *Gaussia gomez-pompae*. Dale Holton garden, Florida.

Plate 443. *Gaussia maya*. Private garden, Florida.

Plate 444. *Gaussia maya*. Private garden, Florida.

Plate 445. *Gaussia princeps*. In habitat, Cuba.

Plate 446. *Gaussia princeps*. Fairchild Tropical Botanic Garden, Florida.

Plate 447. *Gaussia spirituana*. In habitat, Cuba.

Plate 448. *Geonoma congesta*. In habitat, Costa Rica.

Plate 449. *Geonoma densa*. Australia.

Plate 450. *Geonoma deversa*. In habitat, Costa Rica.

Plate 451. *Geonoma epetiolata*. Jeff Marcus garden, Hawaii.

Plate 452. *Geonoma ferruginea*. In habitat, Costa Rica.

Plate 453. *Geonoma interrupta*. Ho'omaluhia Botanical Garden, Hawaii.

Plate 454. *Geonoma longivaginata*. Australia.

Plate 455. *Geonoma schottiana*. In habitat, Brazil.

Plate 456. *Geonoma undata*. Darold Petty garden, California.

Plate 457. *Guihaia argyrata*. Fairchild Tropical Botanic Garden, Florida.

Plate 458. *Guihaia grossifibrosa*. Fairchild Tropical Botanic Garden, Florida.

Plate 459. *Hedyscepe canterburyana*. New Zealand.

Plate 460. *Hemithrinax ekmaniana*. Montgomery Botanical Center, Florida.

Plate 461. *Hemithrinax ekmaniana*. In habitat, Cuba.

Plate 462. *Hemithrinax rivularis*. In habitat, Cuba.

Plate 463. *Heterospathe cagayanensis*. Jeff Marcus garden, Hawaii.

Plate 464. *Heterospathe delicatula*. Australia.

Plate 465. *Heterospathe elata*. Paul Humann garden, Florida.

Plate 466. *Heterospathe elmeri*. Fairchild Tropical Botanic Garden, Florida.

Plate 467. *Heterospathe glauca*. Jeff Marcus garden, Hawaii.

Plate 468. *Heterospathe longipes*. Jeff Marcus garden, Hawaii.

Plate 469. *Heterospathe minor*. Private garden, Hawaii.

Plate 470. *Heterospathe negrosensis*. Fairchild Tropical Botanic Garden, Florida.

Plate 471. *Heterospathe philippinensis*. Jeff Marcus garden, Hawaii.

Plate 472. *Heterospathe scitula*. Makiling Botanic Gardens, Laguna, Philippines.

Plate 473. *Heterospathe woodfordiana*. Australia.

Plate 474. *Heterospathe woodfordiana*. Colorful new leaf.

Plate 475. *Howea belmoreana*. Balboa Park, California.

Plate 476. *Howea forsteriana*. Balboa Park, California.

Plate 477. *Hydriastele affinis*. Jeff Marcus garden, Hawaii.

Plate 478. *Hydriastele beguinii*. Jerry Andersen garden, Hawaii.

Plate 479. *Hydriastele costata*. Ho'omaluhia Botanical Garden, Hawaii.

Plate 480. *Hydriastele cylindrocarpa*. Ho'omaluhia Botanical Garden, Hawaii.

Plate 481. *Hydriastele dransfieldii*. Jeff Marcus garden, Hawaii.

Plate 482. *Hydriastele kasesa*. Flecker Botanic Gardens, Australia.

Plate 484. *Hydriastele longispatha*. Jeff Marcus garden, Hawaii.

Plate 483. *Hydriastele ledermanniana*. Private garden, Hawaii.

Plate 485. *Hydriastele macrospadix*. Private garden, Hawaii.

Plate 486. *Hydriastele microcarpa*. Jerry Andersen garden, Hawaii.

Plate 487. *Hydriastele microspadix*. Ho'omaluhia Botanical Garden, Hawaii.

Plate 488. *Hydriastele palauensis*. In habitat, Palau.

Plate 489. *Hydriastele pinangoides*. Australia.

Plate 490. *Hydriastele pinangoides*. New leaf. Private nursery, Brazil.

Plate 491. *Hydriastele rheophytica*. Jeff Marcus garden, Hawaii.

Plate 492. *Hydriastele rostrata*. Fairchild Tropical Botanic Garden, Florida.

Plate 494. *Hydriastele vitiensis*. In habitat, Fiji.

Plate 493. *Hydriastele valida*. Jeff Marcus garden, Hawaii.

Plate 495. *Hydriastele wendlandiana*. Jerry Andersen garden, Hawaii.

Plate 496. *Hyophorbe indica*. Montgomery Botanical Center, Florida.

Plate 497. *Hyophorbe indica*. Infructescences.

Plate 498. *Hyophorbe lagenicaulis*. Waimea Valley Audubon Center, Hawaii.

Plate 499. *Hyophorbe lagenicaulis*. Juvenile. Mike Harris garden, Florida.

Plate 500. *Hyophorbe verschaffeltii*. Jill Menzel garden, Brazil.

Plate 501. *Hyospathe elegans*. Jeff Marcus garden, Hawaii.

Plate 502. *Hyphaene compressa*. Montgomery Botanical Center, Florida.

Plate 503. *Hyphaene coriacea*. Montgomery Botanical Center, Florida.

Plate 504. *Hyphaene dichotoma*. Fairchild Tropical Botanic Garden, Florida.

Plate 505. *Hyphaene petersiana*. Private garden, Florida Keys.

Plate 506. *Hyphaene thebaica*. Foster Botanical Garden, Hawaii.

Plate 507. *Hyphaene thebaica*. Exceptionally large specimen. Cienfuegos Botanical Garden, Cuba.

Plate 508. *Hyphaene thebaica*. Fruits. Townsville Palmetum, Australia.

Plate 509. *Iguanura bicornis*. Jeff Marcus garden, Hawaii.

Plate 510. *Iguanura elegans*. Private garden, Australia.

Plate 511. *Iguanura geonomiformis*. Forest Research Institute Malaysia.

Plate 512. *Iguanura palmuncula*. In habitat, Sarawak.

Plate 513. *Iguanura polymorpha*. Jeff Marcus garden, Hawaii.

Plate 514. *Iguanura polymorpha*. Forest Research Institute Malaysia.

Plate 515. *Iguanura wallichiana* var. *major*. Jerry Andersen garden, Hawaii.

Plate 516. *Iguanura wallichiana* var. *wallichiana*. Jeff Marcus garden, Hawaii.

Plate 517. *Iriartea deltoidea*. Note stilt roots. Costa Rica.

Plate 518. *Iriartea deltoidea*. Private garden, Hawaii.

Plate 519. *Iriartea deltoidea*. Leaf. In habitat, Costa Rica.

Plate 520. *Iriartella setigera*. In habitat, Brazil.

Plate 521. *Itaya amicorum*. Jeff Marcus garden, Hawaii.

Plate 522. *Itaya amicorum*. Detail of leaf. Jeff Marcus garden, Hawaii.

Plate 523. *Johannesteijsmannia altifrons*. Jerry Andersen garden, Hawaii.

Plate 524. *Johannesteijsmannia lanceolata*. Kepong Botanic Gardens, Malaysia.

Plate 525. *Johannesteijsmannia magnifica*. Jerry Andersen garden, Hawaii.

Plate 526. *Johannesteijsmannia magnifica*. Detail of leaf. Jerry Andersen garden, Hawaii.

Plate 527. *Johannesteijsmannia perakensis*. In habitat, Malaysia.

Plate 528. *Juania australis*. Lakeside Palmetum of Oakland, California.

Plate 529. *Jubaea chilensis*. California.

Plate 530. *Jubaea chilensis*. Tresco Abbey Gardens, Isles of Scilly, United Kingdom.

Plate 531. *Jubaeopsis caffra*. California.

Plate 532. *Kentiopsis magnifica*. Ho'omaluhia Botanical Garden, Hawaii.

Plate 534. *Kentiopsis piersoniorum*. Jeff Marcus garden, Hawaii.

Plate 533. *Kentiopsis oliviformis*. Private garden, Florida.

Plate 535. *Kentiopsis pyriformis*. Juvenile. New Caledonia.

Plate 537. *Kerriodoxa elegans*. Showing underside, petiole, and female flowers.

Plate 536. *Kerriodoxa elegans*. Australia.

Plate 538. *Korthalsia* sp. Detail of leaf. Huntington Botanical Gardens, California.

Plate 539. *Korthalsia scortechinii*. In habitat, Malaysia.

Plate 540. *Laccospadix australasicus*. With fruits. In habitat, Australia.

Plate 541. *Laccospadix australasicus*. Jerry Andersen garden, Hawaii.

Plate 542. *Lanonia dasyantha*. Jerry Andersen garden, Hawaii.

Plate 543. *Latania loddigesii*. Private garden, Florida.

Plate 544. *Latania lontaroides*. Juvenile showing red petioles. Florida.

Plate 545. *Latania lontaroides*. Fairchild Tropical Botanic Garden, Florida.

Plate 546. *Latania verschaffeltii*. Franco D'Ascanio garden, Florida Keys.

Plate 547. *Lemurophoenix halleuxii*. Young plant. Jeff Marcus garden, Hawaii.

Plate 548. *Leopoldinia piassaba*. Rio de Janeiro Botanical Garden, Brazil.

Plate 549. *Lepidorrhachis mooreana*. California.

Plate 550. *Leucothrinax morrisii*. Private garden, Florida Keys.

Plate 551. *Licuala beccariana*.

Plate 552. *Licuala bintulensis*. In habitat, Sarawak.

Plate 553. *Licuala cordata*. Australia.

Plate 554. *Licuala cordata*. Segmented leaf form. Australia.

Plate 555. *Licuala distans*. Jeff Marcus garden, Hawaii.

Plate 556. *Licuala glabra* var. *glabra*. Jeff Marcus garden, Hawaii.

Plate 557. *Licuala glabra* var. *selangorensis*. Forest Research Institute Malaysia.

Plate 558. *Licuala grandis*. Jill Menzel garden, Brazil.

Plate 559. *Licuala grandis*. Fruits. Hawaii Tropical Garden, Hawaii.

Plate 560. *Licuala lauterbachii*. Fairchild Tropical Botanic Garden, Florida.

Plate 561. *Licuala longipes*. In habitat, Malaysia.

Plate 562. *Licuala malajana*. Jeff Marcus garden, Hawaii.

Plate 563. *Licuala mattanensis*. Herminio nursery, Brazil.

Plate 564. *Licuala mattanensis* 'Mapu'. Herminio nursery, Brazil.

Plate 565. *Licuala orbicularis*. Private garden, Hawaii.

Plate 566. *Licuala paludosa*. National Tropical Botanical Garden, Hawaii.

Plate 567. *Licuala peekelii*. Dale Holton garden, Florida.

Plate 568. *Licuala peltata* var. *peltata*. Jeff Marcus garden, Hawaii.

Plate 569. *Licuala peltata* var. *sumawongii*. Jill Menzel garden, Brazil.

Plate 570. *Licuala petiolulata*. Jeff Marcus garden, Hawaii.

Plate 571. *Licuala platydactyla*. Paul Humann garden, Hawaii.

Plate 572. *Licuala poonsakii*. Private garden, Hawaii.

Plate 573. *Licuala ramsayi*. Private garden, Hawaii.

Plate 574. *Licuala ramsayi*. Private garden, Costa Rica.

Plate 575. *Licuala sallehana* var. *sallehana*. Forest Research Institute Malaysia.

Plate 576. *Licuala sarawakensis*. In habitat, Sarawak.

Plate 577. *Licuala scortechinii*. In habitat, Malaysia.

Plate 578. *Licuala spinosa*. Hilo, Hawaii.

Plate 579. *Licuala spinosa*. Detail of leaves. Private garden, Florida Keys.

Plate 580. *Licuala triphylla*. Sítio Roberto Burle Marx.

Plate 581. *Linospadix microcaryus*. In habitat, Australia.

Plate 582. *Linospadix minor*. Australia.

Plate 583. *Linospadix monostachyos*. Private garden, Hawaii.

Plate 584. *Linospadix palmerianus*. In habitat, Australia.

Plate 585. *Livistona australis*. Huntington Botanical Gardens, California.

Plate 586. *Livistona benthamii*. Darwin, Australia.

Plate 587. *Livistona carinensis*. Fairchild Tropical Botanic Garden, Florida.

Plate 588. *Livistona chinensis*. Jardín Botánico Nacional Dr. Rafael M. Moscoso, Dominican Republic.

Plate 589. *Livistona concinna*. Montgomery Botanical Center, Florida.

Plate 591. *Livistona drudei*. Townsville Palmetum, Australia.

Plate 590. *Livistona decora*. Huntington Botanical Gardens, California.

Plate 592. *Livistona endauensis*. In habitat, Malaysia.

Plate 593. *Livistona fulva*. Flamingo Gardens, Florida.

Plate 594. *Livistona humilis*. George Brown Darwin Botanic Gardens, Australia.

Plate 595. *Livistona inermis*. George Brown Darwin Botanic Gardens, Australia.

Plate 596. *Livistona jenkinsiana*. Private garden, Hawaii.

Plate 598. *Livistona mariae*. Fairchild Tropical Botanic Garden, Florida.

Plate 597. *Livistona lanuginosa*. Montgomery Botanical Center, Florida.

Plate 599. *Livistona muelleri*. Private garden, Hawaii.

Plate 600. *Livistona nitida*. Townsville Palmetum, Australia.

Plate 601. *Livistona rigida*. Darwin, Australia.

Plate 602. *Livistona saribus*. Fairchild Tropical Botanic Garden, Florida.

Plate 603. *Livistona saribus*. Juvenile in shade.

Plate 604. *Livistona victoriae*. In habitat, Western Australia.

Plate 605. *Lodoicea maldivica*. Juvenile. Ho'omaluhia Botanical Garden, Hawaii.

Plate 606. *Lodoicea maldivica*. Sri Lanka.

Plate 607. *Lodoicea maldivica*. Fruits. Sri Lanka.

Plate 608. *Lodoicea maldivica*. Endocarp, varnished.

Plate 609. *Loxococcus rupicola*. Jeff Marcus garden, Hawaii.

Plate 610. *Lytocaryum hoehnei*. Jeff Marcus garden, Hawaii.

Plate 611. *Lytocaryum insigne*. In habitat, Brazil.

Plate 612. *Lytocaryum weddellianum*. Jerry Andersen garden, Hawaii.

Plate 613. *Manicaria saccifera*. Jeff Marcus garden, Hawaii.

Plate 614. *Manicaria saccifera*. Red new leaf. Jeff Marcus garden, Hawaii.

Plate 615. *Marojejya darianii*. Pana'ewa Rainforest Zoo & Gardens, Hawaii.

Plate 616. *Marojejya darianii*. Juvenile. Jerry Andersen garden, Hawaii.

Plate 617. *Marojejya insignis*. Juvenile. Jeff Marcus looking on. Jeff Marcus garden, Hawaii.

Plate 618. *Masoala madagascariensis*. Juvenile. Jeff Marcus garden, Hawaii.

Plate 620. *Mauritia flexuosa*. Fruits. Rio de Janeiro Botanical Garden, Brazil.

Plate 619. *Mauritia flexuosa*. Private garden, Australia.

Plate 621. *Mauritiella aculeata*. Rio de Janeiro Botanical Garden, Brazil.

Plate 622. *Mauritiella armata*. Pana'ewa Rainforest Zoo & Gardens, Hawaii.

Plate 623. *Mauritiella armata*. Detail of stem. Pana'ewa Rainforest Zoo & Gardens, Hawaii.

Plate 624. *Maxburretia furtadoana*. Jeff Marcus garden, Hawaii.

Plate 625. *Maxburretia furtadoana*. Detail of stem. Jeff Marcus garden, Hawaii.

Plate 626. *Medemia argun*. In habitat, Nubian Desert, Sudan.

Plate 627. *Medemia argun*. Montgomery Botanical Center, Florida.

Plate 628. *Metroxylon amicarum*. Private garden, Hawaii.

Plate 629. *Metroxylon sagu*. Australia.

Plate 630. *Metroxylon salomonense*. Australia.

Plate 631. *Metroxylon salomonense*. In fruit. Australia.

Plate 632. *Metroxylon vitiense*. Australia.

Plate 633. *Metroxylon vitiense*. Leaf sheaths. Fairchild Tropical Botanic Garden, Florida.

Plate 634. *Metroxylon warburgii*. In habitat, Samoa.

Plate 635. *Myrialepis paradoxa*. The Retreat, Nassau, Bahamas.

Plate 636. *Nannorrhops ritchieana*. Fairchild Tropical Botanic Garden, Florida.

Plate 637. *Nannorrhops ritchieana*. Fairchild Tropical Botanic Garden, Florida.

Plate 638. *Nenga pumila*. Herminio nursery, Brazil.

Plate 639. *Neonicholsonia watsonii*. Fairchild Tropical Botanic Garden, Florida.

Plate 640. *Neoveitchia storckii*. Pana'ewa Rainforest Zoo & Gardens, Hawaii.

Plate 641. *Neoveitchia storckii*. Detail of crownshaft. Pana'ewa Rainforest Zoo & Gardens, Hawaii.

Plate 643. *Normanbya normanbyi*. Australia.

Plate 642. *Nephrosperma vanhoutteanum*. Ho'omaluhia Botanical Garden, Hawaii.

Plate 644. *Nypa fruticans*. Montgomery Botanical Center, Florida.

Plate 645. *Nypa fruticans*. Fruits. Montgomery Botanical Center, Florida.

Plate 646. *Oenocarpus bacaba*. In habitat, Brazil.

Plate 647. *Oenocarpus bataua*. In habitat, Ecuador.

Plate 648. *Oenocarpus distichus*. Herminio nursery, Brazil.

Plate 649. *Oenocarpus mapora*. Costa Rica.

Plate 650. *Oncosperma gracilipes*. Makiling Botanic Gardens, Philippines.

Plate 651. *Oncosperma tigillarium*. Fairchild Tropical Botanic Garden, Florida.

Plate 652. *Oncosperma tigillarium*. Detail of stem. Fairchild Tropical Botanic Garden, Florida.

Plate 653. *Orania disticha*.

Plate 654. *Orania lauterbachiana*. Townsville Palmetum, Australia.

Plate 655. *Orania palindan*. Cienfuegos Botanical Garden, Cuba.

Plate 656. *Orania ravaka*. Jeff Marcus garden, Hawaii.

Plate 657. *Orania sylvicola*.

Plate 658. *Orania trispatha*. In habitat, Madagascar.

Plate 659. *Oraniopsis appendiculata*. In habitat, Australia.

Plate 660. *Oraniopsis appendiculata*. Juvenile. Oakland Palmetum, California.

Plate 661. *Parajubaea cocoides*. Ecuador.

Plate 662. *Parajubaea sunkha*. In habitat, Bolivia.

Plate 663. *Parajubaea torallyi*. Bolivia.

Plate 664. *Pelagodoxa henryana*. Foster Botanical Garden, Hawaii.

Plate 665. *Pelagodoxa henryana*. Juvenile. Hawaii Tropical Garden, Hawaii.

Plate 666. *Pelagodoxa henryana*. Fruit. Ho'omaluhia Botanical Garden, Hawaii.

Plate 667. *Phoenicophorium borsigianum*. Australia.

Plate 668. *Phoenicophorium borsigianum*. Juvenile leaf. Jill Menzel garden, Brazil.

Plate 669. *Phoenix acaulis*. Huntington Botanical Gardens, California.

Plate 670. *Phoenix canariensis*. Fairchild Tropical Botanic Garden, Florida.

Plate 671. *Phoenix dactylifera*. Private garden, Florida Keys.

Plate 672. *Phoenix dactylifera*. Fruit. Moody Gardens, Texas.

Plate 673. *Phoenix loureiroi* var. *loureiroi*. Florida.

Plate 674. *Phoenix paludosa*. Private garden, Florida Keys.

Plate 675. *Phoenix pusilla*. Fairchild Tropical Botanic Garden, Florida.

Plate 676. *Phoenix reclinata*. Fairchild Tropical Botanic Garden, Florida.

Plate 677. *Phoenix roebelenii*. Private garden, Hawaii.

Plate 678. *Phoenix roebelenii*. Characteristic leaf scars on stem. Fairchild Tropical Botanic Garden, Florida.

Plate 679. *Phoenix rupicola*. California.

Plate 680. *Phoenix sylvestris*. Private garden, Florida Keys.

Plate 681. *Phoenix theophrasti*. Suckers have been removed. Huntington Botanical Gardens, California.

Plate 683. *Pholidocarpus macrocarpus*. Detail of petioles. Jeff Marcus garden, Hawaii.

Plate 682. *Pholidocarpus macrocarpus*. Juvenile. Jeff Marcus garden, Hawaii.

Plate 684. *Pholidocarpus majadan*. In habitat, Sarawak.

Plate 685. *Pholidostachys dactyloides*. In habitat, Ecuador.

Plate 686. *Pholidostachys kalbreyeri*. Lyon Arboretum, Hawaii.

Plate 687. *Pholidostachys pulchra*. In habitat, Costa Rica.

Plate 688. *Pholidostachys synanthera*. In habitat, Ecuador.

Plate 689. *Physokentia insolita*. Private garden, Hawaii.

Plate 691. *Phytelephas aequatorialis*. Private garden, Hawaii.

Plate 690. *Physokentia petiolata*. Jerry Andersen garden, Hawaii.

Plate 692. *Phytelephas macrocarpa*. Rio de Janeiro Botanical Garden, Brazil.

Plate 693. *Phytelephas macrocarpa*. Fruits. Rio de Janeiro Botanical Garden, Brazil.

Plate 694. *Phytelephas seemannii*. Australia.

Plate 695. *Phytelephas tenuicaulis*. Flecker Botanic Gardens, Australia.

Plate 696. *Pigafetta elata*. Townsville Palmetum, Australia.

Plate 697. *Pigafetta elata*. Leaf bases covered in brown spines. Jardín Botánico Nacional, Cuba.

Plate 698. *Pigafetta filaris*. In habitat, West Papua Province, Indonesia.

Plate 699. *Pinanga adangensis*. Waimea Valley Audubon Center, Hawaii.

Plate 700. *Pinanga aristata*. Private garden, Australia.

Plate 701. *Pinanga batanensis*. Private garden, Hawaii.

Plate 702. *Pinanga bicolana*. Makiling Botanic Gardens, Philippines.

Plate 703. *Pinanga caesia*. Jeff Marcus garden, Hawaii.

Plate 704. *Pinanga caesia*. Flowers and fruits. Costa Rica.

Plate 705. *Pinanga chaiana*. Flecker Botanic Gardens, Australia.

Plate 706. *Pinanga copelandii*. Makiling Botanic Gardens, Philippines.

Plate 707. *Pinanga coronata*. Pana'ewa Rainforest Zoo & Gardens, Hawaii.

Plate 708. *Pinanga coronata*. Flowers. Cuba.

Plate 709. *Pinanga crassipes*. Jeff Marcus garden, Hawaii.

Plate 710. *Pinanga crassipes*. Leaf mottling. Jerry Andersen garden, Hawaii.

Plate 711. *Pinanga curranii*. Jeff Marcus garden, Hawaii.

Plate 712. *Pinanga densiflora*. Jerry Andersen garden, Hawaii.

Plate 713. *Pinanga dicksonii*. Private garden, Florida.

Plate 714. *Pinanga disticha*. In habitat, Malaysia.

Plate 715. *Pinanga disticha*. Jeff Marcus garden, Hawaii.

Plate 716. *Pinanga geonomiformis*. Jeff Marcus garden, Hawaii.

Plate 717. *Pinanga gracilis*. Private garden, Hawaii.

Plate 718. *Pinanga heterophylla*. Jeff Marcus garden, Hawaii.

Plate 719. *Pinanga heterophylla*. Crownshaft. Jeff Marcus garden, Hawaii.

Plate 720. *Pinanga insignis*. Ho'omaluhia Botanical Garden, Hawaii.

Plate 721. *Pinanga javana*. Jeff Marcus garden, Hawaii.

Plate 722. *Pinanga javana*. Crownshaft and fruits.

Plate 723. *Pinanga limosa*. In habitat, Malaysia.

Plate 724. *Pinanga maculata*. Australia.

Plate 725. *Pinanga maculata*. Detail of leaf. Herminio nursery, Brazil.

Plate 726. *Pinanga malaiana*. Jeff Marcus garden, Hawaii.

Plate 727. *Pinanga modesta*. Pana'ewa Rainforest Zoo & Gardens, Hawaii.

Plate 728. *Pinanga negrosensis*. Fairchild Tropical Botanic Garden, Florida.

Plate 729. *Pinanga paradoxa*. Jerry Andersen garden, Hawaii.

Plate 730. *Pinanga patula*. Rio de Janeiro Botanical Garden, Brazil.

Plate 731. *Pinanga patula*. Rio de Janeiro Botanical Garden, Brazil.

Plate 732. *Pinanga philippinensis*. Jeff Marcus garden, Hawaii.

Plate 733. *Pinanga rumphiana*. Jeff Marcus garden, Hawaii.

Plate 734. *Pinanga salicifolia*. Jeff Marcus garden, Hawaii.

Plate 735. *Pinanga sclerophylla*. Roth garden, Australia.

Plate 736. *Pinanga scortechinii*. Townsville Palmetum, Australia.

Plate 737. *Pinanga simplicifrons*. Jeff Marcus garden, Hawaii.

Plate 738. *Pinanga sobolifera*. Jeff Marcus garden, Hawaii.

Plate 739. *Pinanga speciosa*. Private garden, Costa Rica.

Plate 740. *Pinanga urosperma*. Private garden, Hawaii.

Plate 741. *Pinanga veitchii*. Arden Dearden nursery, Australia.

Plate 742. *Pinanga watanaiana*. Jeff Marcus garden, Hawaii.

Plate 743. *Plectocomia elongata*. Jeff Marcus garden, Hawaii.

Plate 744. *Plectocomia himalayana*. In habitat, India.

Plate 745. *Plectocomia himalayana*. San Francisco Botanical Garden, California.

Plate 746. *Plectocomiopsis corneri*. In habitat, Malaysia.

Plate 747. *Ponapea hosinoi*. In habitat, Micronesia.

Plate 748. *Ponapea ledermanniana*. Fairchild Tropical Botanic Garden, Florida.

Plate 749. *Prestoea acuminata* var. *montana*. in habitat, Puerto Rico.

Plate 750. *Prestoea acuminata* var. *montana*. Pana'ewa Rainforest Zoo & Gardens, Hawaii.

Plate 751. *Prestoea decurrens*. In habitat, Costa Rica.

Plate 752. *Pritchardia arecina*. Jeff Marcus garden, Hawaii.

Plate 753. *Pritchardia beccariana*. Waimea Valley Audubon Center, Hawaii.

Plate 754. *Pritchardia glabrata*. Ho'omaluhia Botanical Garden, Hawaii.

Plate 755. *Pritchardia hardyi*. National Tropical Botanic Garden, Hawaii.

Plate 756. *Pritchardia hillebrandii*. Private garden, Florida Keys.

Plate 757. *Pritchardia hillebrandii*. Blue form. Ho'omaluhia Botanical Garden, Hawaii.

Plate 758. *Pritchardia kaalae*. Waimea Valley Audubon Center, Hawaii.

Plate 759. *Pritchardia lanigera*. Private garden, Hawaii.

Plate 760. *Pritchardia maideniana*. Waimea Valley Audubon Center, Hawaii.

Plate 761. *Pritchardia martii*. Ho'omaluhia Botanical Garden, Hawaii.

Plate 762. *Pritchardia minor*. Waimea Valley Audubon Center, Hawaii.

Plate 763. *Pritchardia munroi*. Waimea Valley Audubon Center, Hawaii.

Plate 764. *Pritchardia pacifica*. Hawaii Tropical Garden, Hawaii.

Plate 765. *Pritchardia pacifica*. Flamengo Park, Rio de Janeiro, Brazil.

Plate 766. *Pritchardia remota*. Ann Norton Sculpture Gardens, Florida.

Plate 767. *Pritchardia schattaueri*. Jeff Marcus garden, Hawaii.

Plate 768. *Pritchardia thurstonii*. Private garden, Florida Keys.

Plate 769. *Pritchardia thurstonii*. Inflorescence.

Plate 770. *Pritchardia viscosa*. Jeff Marcus garden, Hawaii.

Plate 772. *Pseudophoenix ekmanii*. In habitat, Dominican Republic.

Plate 771. *Pritchardia waialealeana*. Waimea Valley Audubon Center, Hawaii.

Plate 773. *Pseudophoenix lediniana*. Fairchild Tropical Botanic Garden, Florida.

Plate 774. *Pseudophoenix sargentii*. Franco D'Ascanio garden, Florida Keys.

Plate 776. *Pseudophoenix vinifera*. In habitat, Dominican Republic.

Plate 775. *Pseudophoenix sargentii*. Navassa Island form. Private garden, Florida Keys.

Plate 777. *Pseudophoenix vinifera*. Fairchild Tropical Botanic Garden, Florida.

Plate 778. *Ptychococcus lepidotus*. Australia.

Plate 779. *Ptychococcus paradoxus*. In habitat, Papua New Guinea.

Plate 780. *Ptychosperma burretianum*. Jerry Andersen garden, Hawaii.

Plate 781. *Ptychosperma caryotoides*. Ho'omaluhia Botanical Garden, Hawaii.

Plate 782. *Ptychosperma cuneatum*. Private garden, Florida.

Plate 783. *Ptychosperma elegans*. Paul Humann garden, Florida.

Plate 784. *Ptychosperma furcatum*. Fairchild Tropical Botanic Garden, Florida.

Plate 785. *Ptychosperma keiense*. Fairchild Tropical Botanic Garden, Florida.

Plate 786. *Ptychosperma lauterbachii*. Fairchild Tropical Botanic Garden, Florida.

Plate 787. *Ptychosperma lineare*. Private garden, Florida.

Plate 788. *Ptychosperma macarthurii*. Pana'ewa Rainforest Zoo & Gardens, Hawaii.

Plate 789. *Ptychosperma microcarpum*. Private garden, Florida.

Plate 790. *Ptychosperma salomonense*. Fairchild Tropical Botanic Garden, Florida.

Plate 792. *Ptychosperma sanderianum*. Cultivar with undivided leaves. Nong Nooch Tropical Botanical Garden.

Plate 791. *Ptychosperma sanderianum*. Ho'omaluhia Botanical Garden, Hawaii.

Plate 793. *Ptychosperma schefferi*. Fairchild Tropical Botanic Garden, Florida.

Plate 794. *Ptychosperma waitianum*. Campus, University of Miami, Florida.

Plate 795. *Ptychosperma waitianum*. Red new leaf. Fairchild Tropical Botanic Garden, Florida.

Plate 796. *Raphia australis*. Australia.

Plate 797. *Raphia farinifera*. Australia.

Plate 798. *Raphia farinifera*. Fruits. Private garden, Hawaii.

Plate 799. *Raphia hookeri*. Venezuela.

Plate 800. *Raphia hookeri*. Stem, showing fibers. Private garden, Hawaii.

Plate 801. *Raphia sudanica*. Fairchild Tropical Botanic Garden, Florida.

Plate 802. *Ravenea albicans*. Jeff Marcus garden, Hawaii.

Plate 803. *Ravenea dransfieldii*. Jeff Marcus garden, Hawaii.

Plate 804. *Ravenea glauca*. Jerry Andersen garden, Hawaii.

Plate 805. *Ravenea hildebrandtii*. Fairchild Tropical Botanic Garden, Florida.

Plate 806. *Ravenea hildebrandtii*. With fruit. Jerry Andersen garden, Hawaii.

Plate 807. *Ravenea julietiae*. In habitat, Madagascar.

Plate 808. *Ravenea lakatra*. Jeff Marcus garden, Hawaii.

Plate 809. *Ravenea madagascariensis*. In habitat, Madagascar.

Plate 810. *Ravenea moorei*. Jerry Andersen garden, Hawaii.

Plate 811. *Ravenea musicalis*. In habitat, Madagascar.

Plate 812. *Ravenea musicalis*. Jeff Marcus garden, Hawaii.

Plate 813. *Ravenea rivularis*. Fairchild Tropical Botanic Garden, Florida.

Plate 814. *Ravenea rivularis*. Inflorescence. Florida International University, Florida.

Plate 815. *Ravenea robustior*. In habitat, Madagascar.

Plate 816. *Ravenea sambiranensis*. In habitat, Madagascar.

Plate 817. *Ravenea xerophila*. Juvenile. Montgomery Botanical Center, Florida.

Plate 818. *Reinhardtia gracilis*. In habitat, Belize.

Plate 819. *Reinhardtia latisecta*. Private garden, Hawaii.

Plate 820. *Reinhardtia paiewonskiana*. Juvenile. In habitat, Dominican Republic.

Plate 821. *Retispatha dumetosa*. Lyon Arboretum, Hawaii.

Plate 822. *Rhapidophyllum hystrix*. Fairchild Tropical Botanic Garden, Florida.

Plate 823. *Rhapis excelsa*. Waimea Valley Audubon Center, Hawaii.

Plate 824. *Rhapis excelsa*. Variegated cultivar. Jerry Andersen garden, Hawaii.

Plate 825. *Rhapis humilis*. Bob & Marita Bobick garden, Florida.

Plate 826. *Rhapis loasensis*. Private garden, Florida.

Plate 827. *Rhapis multifida*. Bob & Marita Bobick garden, Florida.

Plate 828. *Rhapis subtilis*. Australia.

Plate 829. *Rhopaloblaste augusta*. Australia.

Plate 830. *Rhopaloblaste ceramica*. Herminio nursery, Brazil.

Plate 831. *Rhopaloblaste elegans*. Ho'omaluhia Botanical Garden, Hawaii.

Plate 832. *Rhopaloblaste singaporensis*. Pana'ewa Rainforest Zoo & Gardens, Hawaii.

Plate 833. *Rhopalostylis baueri*. Balboa Park, California.

Plate 834. *Rhopalostylis sapida*. Balboa Park, California.

Plate 835. *Roscheria melanochaetes*. Private garden, Hawaii.

Plate 836. *Roystonea altissima*. Castleton Botanical Garden, Jamaica.

Plate 837. *Roystonea borinquena*. Dominican Republic.

Plate 838. *Roystonea maisiana*. In habitat, Cuba.

Plate 840. *Roystonea oleracea*. National Tropical Botanic Garden, Hawaii.

Plate 839. *Roystonea oleracea*. Rio de Janeiro Botanical Garden, Brazil.

Plate 841. *Roystonea princeps*. Fairchild Tropical Botanic Garden, Florida.

Plate 842. *Roystonea regia*. Fairchild Tropical Botanic Garden, Florida.

Plate 843. *Roystonea violacea*. In habitat, Cuba.

Plate 844. *Sabal bermudana*. Fairchild Tropical Botanic Garden, Florida.

Plate 845. *Sabal causiarum*. For scale, note people to the right of the trunk. Rio de Janeiro Botanical Garden, Brazil.

Plate 847. *Sabal etonia*. Fairchild Tropical Botanic Garden, Florida.

Plate 846. *Sabal domingensis*. Jardín Botánico Nacional Dr. Rafael M. Moscoso, Dominican Republic.

Plate 848. *Sabal maritima*. In habitat, Cuba.

Plate 849. *Sabal mauritiiformis*. Private garden, Florida Keys.

Plate 850. *Sabal mauritiiformis*. Underside of leaf. Private garden, Florida.

Plate 851. *Sabal mexicana*. Florida.

Plate 852. *Sabal minor*. University of Florida campus, Gainesville, Florida.

Plate 853. *Sabal palmetto*. Myakka River State Park, Florida.

Plate 854. *Sabal pumos*. Fairchild Tropical Botanic Garden, Florida.

Plate 855. *Sabal rosei*. In habitat, Mexico.

Plate 856. *Sabal uresana*. Montgomery Botanical Center, Florida.

Plate 857. *Sabal yapa*. Fairchild Tropical Botanic Garden, Florida.

Plate 858. *Sabal yapa*. Underside of leaf. Waimea Valley Audubon Center, Hawaii.

Plate 859. *Salacca magnifica*. Jerry Andersen garden, Hawaii.

Plate 860. *Salacca wallichiana*. Geoffrey Fowler garden, Australia.

Plate 861. *Salacca zalacca*. Larry Dieterich garden, Florida.

Plate 862. *Salacca zalacca*. Fruits in market, Bogor, Indonesia.

Plate 863. *Saribus jeanneneyi*. New Caledonia.

Plate 864. *Saribus merrillii*. Fairchild Tropical Botanic Garden, Florida.

Plate 873. *Sclerosperma mannii*. In habitat, Democratic Republic of the Congo.

Plate 874. *Serenoa repens*. Private garden, Florida.

Plate 875. *Serenoa repens*. Private garden, Florida.

Plate 876. *Serenoa repens*. Detail of branching, upright trunk. In habitat, Florida.

Plate 877. *Socratea exorrhiza*. Wilson Botanic Garden, Costa Rica.

Plate 879. *Sommieria leucophylla*. Fruits. Jerry Andersen garden, Hawaii.

Plate 878. *Sommieria leucophylla*. Jerry Andersen garden, Hawaii.

Plate 880. *Syagrus amara*. Montgomery Botanical Center, Florida.

Plate 881. *Syagrus botryophora*. Montgomery Botanical Center, Florida.

Plate 883. *Syagrus cardenasii*. Private garden, Bolivia.

Plate 882. *Syagrus botryophora*. Mature form. Larry Dieterich garden, Florida.

Plate 884. *Syagrus cearensis*. Private garden, Florida Keys.

Plate 885. *Syagrus cocoides*. Sítio Roberto Burle Marx, Brazil.

Plate 886. *Syagrus comosa*. In habitat, Brazil.

Plate 887. *Syagrus coronata*. Private garden, Florida.

Plate 888. *Syagrus coronata*. Detail of trunk. Fairchild Tropical Botanic Garden, Florida.

Plate 889. *Syagrus* ×*costae*. Fairchild Tropical Botanic Garden, Florida.

Plate 891. *Syagrus glaucescens*. In habitat, Brazil.

Plate 890. *Syagrus flexuosa*. Sítio Roberto Burle Marx, Brazil.

Plate 892. *Syagrus harleyi*. In habitat, Brazil.

Plate 893. *Syagrus macrocarpa*. Sítio Roberto Burle Marx, Brazil.

Plate 894. *Syagrus oleracea*. Herminio nursery, Brazil.

Plate 895. *Syagrus orinocensis*. Jardín Botánico de Caracas, Venezuela.

Plate 896. *Syagrus orinocensis*. Multiple trunk form, formerly known as *S. stenopetala*. Montgomery Botanical Center, Florida.

Plate 898. *Syagrus pseudococos*. In habitat, Brazil.

Plate 897. *Syagrus picrophylla*. Montgomery Botanical Center, Florida.

Plate 899. *Syagrus romanzoffiana*. Paul Humann garden, Florida.

Plate 900. *Syagrus ruschiana*. In habitat, Brazil.

Plate 901. *Syagrus sancona*. Flamingo Gardens, Florida.

Plate 902. *Syagrus schizophylla*. Flamengo Park, Rio de Janeiro, Brazil.

Plate 903. *Syagrus vagans*. Dale Holton garden, Florida.

Plate 904. *Syagrus vermicularis*. Montgomery Botanical Center, Florida.

Plate 905. *Synechanthus fibrosus*. Australia.

Plate 906. *Synechanthus warscewiczianus*. Private garden, Hawaii.

Plate 907. *Tahina spectabilis*. In habitat, Madagascar. Soejatmi Dransfield provides scale.

Plate 908. *Tahina spectabilis*. Juvenile. Dale Holton garden, Florida.

Plate 909. *Thrinax excelsa*. Fairchild Tropical Botanic Garden, Florida.

Plate 911. *Thrinax radiata*. Private garden, Florida Keys.

Plate 910. *Thrinax parviflora*. Curly leaf form. Fairchild Tropical Botanic Garden, Florida.

Plate 912. *Trachycarpus fortunei*. Wakehurst Place, England.

Plate 913. *Trachycarpus fortunei* 'Wagnerianus'. California.

Plate 914. *Trachycarpus latisectus*. In habitat, India.

Plate 915. *Trachycarpus martianus*. India.

Plate 916. *Trachycarpus nanus*. In habitat, China.

Plate 917. *Trachycarpus oreophilus*. In habitat, Thailand.

Plate 918. *Trachycarpus princeps*. In habitat, China.

Plate 919. *Trachycarpus princeps*. Showing underside of leaf. In habitat, China.

Plate 920. *Trithrinax brasiliensis*. Jill Menzel garden, Brazil.

Plate 921. *Trithrinax campestris*. Dick Douglas garden, California.

Plate 922. *Trithrinax campestris*. Detail of leaf segment tips. Lakeside Palmetum of Oakland, California.

Plate 923. *Trithrinax schizophylla*. Montgomery Botanical Center, Florida.

Plate 924. *Veitchia arecina*. Fairchild Tropical Botanic Garden, Florida.

Plate 925. *Veitchia filifera*. Private garden, Hawaii.

Plate 926. *Veitchia joannis*. Pana'ewa Rainforest Zoo & Gardens, Hawaii.

Plate 927. *Veitchia metiti*. Fairchild Tropical Botanic Garden, Florida.

Plate 928. *Veitchia spiralis*. Montgomery Botanical Center, Florida.

Plate 929. *Veitchia vitiensis*. Fairchild Tropical Botanic Garden, Florida.

Plate 930. *Veitchia vitiensis*. Crownshaft. Jeff Marcus garden, Hawaii.

Plate 931. *Veitchia winin*. Fairchild Tropical Botanic Garden, Florida.

Plate 933. *Verschaffeltia splendida*. Jill Menzel garden, Brazil.

Plate 932. *Verschaffeltia splendida*. Ho'omaluhia Botanical Garden, Hawaii.

Plate 934. *Voanioala gerardii*. Juvenile. Private garden, Hawaii.

Plate 935. *Wallichia disticha*. University of Miami campus, Florida.

Plate 936. *Wallichia oblongifolia*. Private garden, Florida.

Plate 937. *Washingtonia filifera*. Huntington Botanical Gardens, California.

Plate 938. *Washingtonia robusta*. Fairchild Tropical Botanic Garden, Florida.

Plate 939. *Welfia regia*. In habitat, Costa Rica.

Plate 940. *Welfia regia*. Red new leaf. Jerry Andersen garden, Hawaii.

Plate 941. *Wendlandiella gracilis* var. *polyclada*. Fairchild Tropical Botanic Garden, Florida.

Plate 942. *Wettinia augusta*. Jeff Marcus garden, Hawaii.

Plate 943. *Wettinia hirsuta*. Jeff Marcus garden, Hawaii.

Plate 944. *Wettinia maynensis*. In habitat, Ecuador.

Plate 945. *Wettinia praemorsa*. Juvenile. Private garden, Hawaii.

Plate 946. *Wettinia quinaria*. Pana'ewa Rainforest Zoo & Gardens, Hawaii.

Plate 947. *Wodyetia bifurcata*. Townsville Palmetum, Australia.

Plate 948. *Wodyetia bifurcata*. Flecker Botanic Gardens, Australia.

Plate 949. *Zombia antillarum*. Montgomery Botanical Center, Florida.

Plate 950. *Zombia antillarum*. Detail of stem covered with spiny leaf sheaths. Fairchild Tropical Botanic Garden, Florida.

PALM DESCRIPTIONS A to Z

Acanthophoenix
a-kanth'-o-FEE-nix

Acanthophoenix comprises 3 solitary-trunked, pinnate-leaved, monoecious palms. The only common name is barbel palm. The genus name translates from the Greek as "spine" and "date palm," but is used in the general sense of "palm," as in many palm names that are compounded with *Phoenix*. *Acanthophoenix* seeds germinate within 60 days. For further seed germination details, see *Areca*.

Acanthophoenix crinita
a-kanth'-o-FEE-nix · kri-NEET-a
PLATE 1

Acanthophoenix crinita is restricted to the island of Réunion in montane rain forest and in wet *Pandanus* thickets above 2700 feet (820 m). The species was recently resurrected from synonymy under *A. rubra* when observation plants in habitat revealed a number of distinctive features. The epithet means "hairy," probably in reference to the bristly spines on the leaf sheath.

The stem is dark brown and rough, rising to about 35 feet (11 m) and only 4–8 inches (10–20 cm) in diameter; it is marked with prominent distantly spaced leaf scars. The crownshaft is 24–31 inches (60–79 cm) tall and covered with stiff, brown hair said to resemble the fur of tenrecs, the hedgehoglike mammals of Madagascar and Africa. The short leaves are about 6 feet (1.8 m) long and straight, with evenly spaced, dark green leaflets hanging gracefully from either side of the rachis. The petiole often bears a few bristly spines on its edge. The inflorescences are borne below the crownshaft and have short spines on the main stalk. The flowers are ivory white, sometime suffused with pink or orange. Fruits are about ¼ inch (6 mm) long, curved, and black when mature.

Overall this palm is stockier than the more commonly cultivated *A. rubra*. It needs cool temperatures, moist but well-drained soil, and tolerates occasional frost. Even under optimum conditions it is slow growing. It has potential as an ornamental palm in New Zealand, Hawaii, or even the Bay Area of California, and may also thrive in a cool temperate greenhouse, as long as it receives abundant moisture and protection from hard freezes.

Acanthophoenix rubra
a-kanth'-o-FEE-nix · ROOB-ra
PLATE 2

Acanthophoenix rubra is found on the islands of Mauritius and Réunion in the Indian Ocean, where it is nearly extinct. It grows in lowland rain forest from sea level to 2640 feet (800 m) in the wetter (eastern) parts of the islands, on sloping ground. The epithet means "red."

The straight, columnar, light-colored trunk can attain a height of 60 or more feet (18 m) and often is spiny, especially when young, on the newer growth. Older parts of the stem are light gray to almost white and conspicuously ringed except for the oldest part. The large rounded crown of coconut-like leaves is 20 feet (6 m) wide and 2–3 feet (60–90 cm) tall, bulging at the base, and covered with greenish gray to reddish brown tomentum. The leaves are 6–10 feet (1.8–3 m) long, usually 8 feet (2.4 m). The many dark green leaflets are 2 feet (60 cm) long or more, with pointed tips, and are borne in a flat plane off the rachis, which itself is often twisted at its midsection to make the leaflets vertical in orientation from that point on. The leaflets of young plants often have a red midvein with small prickles, while those of older plants have a yellow midvein and often no spines; leaves of any age have lighter-colored, waxy undersides. The short petioles are less than 1 foot (30 cm) long and clothed in light brown spines in younger palms but unarmed in mature palms. The inflorescences grow from beneath the crownshaft and often carry black spines, especially near the base of the main stalk. Numerous light pink, red, or purplish male and female flowers are borne on the same tree in the same inflorescence. The globose fruits are less than ½ inch (13 mm) in diameter and are always black when mature.

This palm needs at least a half day of sun to thrive, although seedlings require protection from midday sun, especially in hot climates. It also needs regular moisture and a neutral or slightly acidic, well-drained soil. It is a reasonably fast grower, especially when young, if given proper soil and moisture, but is not hardy to cold; it is impossible outside of zone 11 and even marginal in 10b. It does well but grows slowly in the warmest parts of southern coastal California (Geoff Stein, pers. comm.).

While not closely related to *Archontophoenix*, *Dictyosperma*, or *Veitchia* species, *Acanthophoenix* has many visual similarities to these genera, with a leaf that is as beautiful as that of the coconut. It is large and imposing enough to be planted singly, especially as a canopy-scape, but is even more glorious in groups of 3 or more individuals of varying heights.

Acoelorrhaphe
a-see'-lo-RAY-fee

Acoelorrhaphe is a monotypic genus of clustering, palmate-leaved, hermaphroditic palm in North and Central America. Common names include Everglades palm and paurotis palm. The genus name is derived from Greek words meaning "without," "hollow," and "fold" or "seam," a reference to the physical characteristics of the seed. Fresh seeds of *Acoelorrhaphe* germinate within 90 days. For further seed germination details, see *Corypha*.

Acoelorrhaphe wrightii
a-see´-lo-RAY-fee · RY-tee-eye
PLATE 3

Acoelorrhaphe wrightii grows naturally in South Florida, the Bahamas, Cuba, the Atlantic coastal plain of Central America, and the Yucatan Peninsula of Mexico, in moist or even swampy and monsoonal areas that are often flooded with fresh or brackish water The epithet honors American botanist and explorer, Charles Wright (1811–1885), best known for his botanical exploration of Cuba.

The species readily forms clumps 10–15 feet (3–4.5 m) tall and wide. The stems grow 4 inches (10 cm) per year under optimum conditions to 20 feet (6 m) tall, or sometimes 40 feet (12 m), with a diameter of 4 inches (10 cm). The oldest parts of mature stems are ringed with closely set leaf base scars, while the remaining parts are covered in a beautiful pattern of light-colored old leaf bases and dense dark brown matted fibers. The leaf crowns are spherical or nearly so and composed of 10–16 leaves, each at least 3 feet (90 cm) across on a petiole 3 feet (90 cm) long, and armed with short, vicious teeth. The leaves are semicircular or more. The blade is deeply segmented more than halfway to the petiole into many stiff, narrow, and tapering segments, each of which is again split at its tip. Leaf color is bright green above and silvery beneath. Erect inflorescences 3 feet (90 cm) long grow from and reach above the leaf crown in summer. The pendulous branches are densely packed with small, whitish, bisexual blossoms. The fruits, in drooping clusters, are ½ inch (13 mm) wide and rounded, maturing from bright orange to shiny black in midsummer to early fall.

A small form of this species occurs in western Cuba but is not in cultivation. It grows in seasonally flooded pinelands, along with *Colpothrinax wrightii*, and may be solitary trunked or few stemmed, with grayish green leaves.

This palm thrives in full sun but tolerates partial shade, is a true water lover but survives in dry soil, and does best in neutral to slightly acidic, heavy or sandy soils with lots of organic matter. It tolerates saline soil and air, and can be grown in alkaline soils if they are amended with organic matter or heavily mulched. The species is hardy in zones 9b through 11 and has been known to resprout from its roots after freezing to the ground at 15°F (−9°C). Damage to the leaves starts with any temperature below 25°F (−4°C), and the trunks are usually killed to the ground by 20°F (−7°C) or lower. Plants grown in poor or dry soil are subject to rot, especially older specimens, and manganese and potassium deficiencies can develop in plants that do not receive adequate moisture.

Because it is native to Florida, the species is often thought to handle dry conditions and is sometimes planted in unirrigated street medians. Under these conditions the palms suffer, grow slowly, and sometimes die. Propagation by division of the suckers from clumps is possible if the individual sucker is taken with roots and planted quickly.

A mature clump of this palm with its slender trunks of varying heights is the essence of the wild yet elegant tropical look. Judicious thinning of the stems in older clumps improves the look. Because the clumps are clothed from foot to head with leaves, this palm can be used as an isolated specimen in an expanse of lawn. Its best use, however, is as part of a border or swath of lower vegetation above which its beautifully elegant trunks and relatively small heads of stiff palmate leaves can punctuate the skyline. It is not a good candidate for indoor culture, but it can be grown successfully in very large conservatories and atriums given as much sunlight as it can get and good air circulation.

Acrocomia
ak-ro-KO-mee-a

Acrocomia is a genus of at least 3 solitary-trunked, pinnate-leaved, monoecious palms in tropical America. All are armed with vicious spines on all their parts. One species, *A. hassleri*, has a subterranean trunk; the others have large aboveground trunks. The genus name translates from the Greek as "highest" and "tuft of hair," an allusion to the palm's crown of prickly leaves atop a tall trunk.

Henderson et al. (1995) lumped a large number of species into *Acrocomia aculeata*, and that treatment has been followed by subsequent authors. Many hobbyists, growers, and scientists, however, do not agree with this lumping, although they readily agree that the genus has been overdescribed in the past. Bee Gunn (2004) found molecular evidence that *A. aculeata* must be divided into at least 2 species. Growers have pointed out that seedlings of *A. aculeata* have bifid leaves, while those of *A. totai* are simple and undivided; *A. aculeata* has shorter trunk spines, larger fruits and seeds, is significantly less hardy to cold, but is much faster growing when younger than *A. totai*, often surpassing the queen palm (*Syagrus romanzoffiana*) in growth speed. Resolution of this prickly taxonomic problem awaits the committed botanist(s) willing to collect and analyze these unfriendly palms from throughout their range.

Like all the genera in the tribe Cocoseae, *Acrocomia* has seeds similar to that of the coconut, with 3 "eyes" (germination pores) in a thick, hard shell (endocarp). Older seeds are viable but may take a long time to germinate. It is not unheard of for the seeds of *Acrocomia* to germinate up to 5 years after being sown. Germination tends to be sporadic, and in some cases, seeds have more than one embryo, causing up to 3 seedlings to develop, one for each "eye." Carefully removing the endocarp before sowing has been shown to speed germination greatly. Because the seeds are more prone to disease and insect attacks without their shells, they should be planted in a sterile potting medium, such as perlite.

Acrocomia aculeata
ak-ro-KO-mee-a · a-kyoo´-lee-AH-ta
PLATE 4

Acrocomia aculeata is widespread and variable from Cuba and the Caribbean Islands to southwestern Mexico and the Yucatán Peninsula, Central America, northern Colombia, most of Venezuela, and the eastern half of Brazil. It occurs in savannas and open places in monsoonal areas. Common names are macaw palm, which is also used for *Aiphanes aculeata*, and gru-gru palm. The epithet translates from the Latin as "prickly."

Mature trunks grow 15–35 feet (4.5–11 m) high and 9–16 inches (23–41 cm) in diameter, are mostly columnar and straight but occasionally show a slight-to-definite bulge near the middle. In some plants the trunks are free, or nearly so, of dead leaves and are covered in all but their oldest parts with closely set rings of black or lighter-colored needlelike spines each 1–3 inches (2.5–8 cm) long. An approximately equal number of plants retain many of the old leaves, with the bulk of the light gray to almost white trunk free of spines and smooth. The leaf crown is mostly full and rounded. The 20–30 leaves are 10–12 feet (3–3.6 m) long with mostly limp, drooping leaflets 3 feet (90 cm) long, growing from the rachis at different angles and giving the leaf a plumose aspect. The leaflets are on short spiny, hairy petioles 3 inches (8 cm) long, and the leaf rachis is very spiny. Leaf color is deep grayish green to deep pure green and even bluish green, sometimes with a grayish to silvery pubescence beneath. The inflorescences are 5 feet (1.5 m) long and grow from the leaf crown and are usually partly hidden therein. They produce separate yellowish male and female flowers in the same flower cluster and accompanied by a large woody bract (spathe) usually covered in reddish brown tomentum. The small, yellowish green to brown, rounded fruits hang in pendent clusters half hidden in the crown.

This moderately fast growing palm is always found in open, sunny areas in habitat and demands at least half a day's sun to thrive. It is drought tolerant but grows faster and looks better with adequate and regular irrigation. It seems to thrive in many soils from calcareous to rich and acidic as long as they are well drained. It is slightly tolerant to saline soil and air. Cold hardiness varies with the provenance of a given plant, but in general ranges from zones 9b through 11, although some individuals are tenderer.

Overall, this princely palm resembles a robust queen palm (*Syagrus romanzoffiana*). It is wonderful in specimen groups of 3 or more individuals of varying heights and is especially well suited to lining avenues; however, young plants pose a threat to unwary humans, especially children, because of their complete spininess. The species is not a good candidate for indoor cultivation as it demands intense light and good air circulation. In addition, its spininess makes it difficult to site in all but the largest spaces. It is occasionally grown in atriums and conservatories.

The species has many uses in its natural range. The very hard wood is sometimes used for house construction, carving, and furniture making. The pith of the trunk has been used as a source of sago. The spiny leaves are used as temporary fencing and as a source of fibers for weaving hammocks. The spiny leafstalks have been used as protective fencing. The growing point, fibrous fruit pulp, and seed kernels are sometimes eaten. Oil from the seed has been used locally as soap; and the oil and seed residue are used as livestock feed. Indigenous people of Central America used to make a potent alcoholic beverage from the sap of the tree, but the sap-collecting process requires the felling of the palm and is now rarely practiced. In Colombia, fallen flowers of the species have been used medicinally to treat respiratory tract disorders. The fruits are particularly attractive to cattle that swallow the seeds whole.

Acrocomia crispa
ak-ro-KO-mee-a · KRIS-pa
PLATES 5–7

Acrocomia crispa is endemic to Cuba and the Isle of Youth, where it grows in open savannas at low elevations. The palm was known as *Gastrococos crispa* and *Acrocomia armentalis*. The epithet is Latin for "curled" or "rippled," but the intention is obscure. The common name is Cuban belly palm, which has unfortunately been applied to another Cuban species, palmate-leaved *Colpothrinax wrightii*.

The trunk grows to 60 feet (18 m) in habitat and is light gray to almost pure white. It invariably exhibits a distinct spindle-shaped bulge (or belly) near the midpoint of the stem. The trunk is graced with closely spaced darker rings of leaf base scars and, in the younger parts, spines that were associated with the leaf bases. The numerous, slightly arching leaves are 8–10 feet (2.4–3 m) long and produce a beautiful rounded crown that looks much like that of a royal palm (*Roystonea*). The leaves of most individuals are plumose with leaflets that grow from the spiny rachis in all directions and are deep green above but silvery or grayish green beneath. The leaves of juvenile plants have spiny rachises and leaflets, but the spininess is lost as the palm matures. The inflorescences grow from the leaf crown and bear yellowish orange flowers that produce clusters of round orange fruits.

The palm is moderately fast-growing, especially after forming a few feet of trunk. It is wonderfully tolerant of calcareous soils but is also adaptable to those with a lower pH. It needs full sun from youth to old age and, while drought tolerant, grows faster and looks better with regular moisture.

Like *Roystonea regia*, this species is truly magnificent lining an avenue or other promenade, and its canopy-scape effect is one of the best. As a specimen it is always splendid and imposing as an adult. Being tolerant of such a range of soils, it is extremely valuable for adding the spice and variety to a given palm planting that everyone craves. The palm is not known to have been grown indoors and is probably a poor candidate for doing so as it requires such great amounts of light and good air circulation, not to mention space.

The bulges in the trunk have water- and nutrient-storing tissue, which is likely an adaptation to a natural habitat in which periods of drought occur. Plants grown in shade are skinnier and often have only a small belly, while those grown properly in full sun with enough room always get a nice fat belly.

Actinokentia
ak-tin'-o-KEN-tee-a

Actinokentia is a genus of 2 small solitary-trunked, pinnate-leaved, monoecious palms in New Caledonia. Although the 2 are similar in appearance, one (*A. divaricata*) has been introduced to cultivation while the other (*A. huerlimannii*) has not and is rare and threatened in the wild. The genus name combines the Greek word for "rayed" or "radiating" and *Kentia*, an out-of-date name for species of *Hydriastele*, but used here and in other compounded names to indicate generally a palm with pinnate leaves. Seeds of

Actinokentia germinate within 60–120 days. They should be harvested when fully ripe as slightly green seeds do not germinate. For further seed germination details, see *Areca*.

Actinokentia divaricata
ak-tin′-o-KEN-tee-a · di-var′-i-KAH-ta
PLATES 8–9

Actinokentia divaricata is endemic to low mountainous rain forest of central and southern New Caledonia. It is listed as threatened, although it is widespread in New Caledonia. The epithet translates from the Latin as "spreading," a reference to the branching inflorescences.

The palm grows to a maximum height of 25 feet (7.6 m) with a slender, prominently ringed gray to light brown trunk that is never more than 3 inches (8 cm) in diameter, and atop which the yellowish green to brownish green or even reddish brown crownshaft is 2 feet (60 cm) tall and scarcely thicker. The leaf crown is relatively sparse, never sporting more than 6 strongly arching leaves each 3 feet (90 cm) long. The pinkish red to reddish brown cylindrical rachis is 3 feet (90 cm) long. Widely spaced leaflets grow from the rachis at a single angle, making the leaf nearly flat. Each leaflet is 18–36 inches (45–90 cm) long, linear-oblong with a pointed apex, and matures to a glossy light to dark green with a fuzzy, often slightly waxy underside. New leaves are usually pink or red until they mature. The inflorescences grow from beneath the crownshaft, usually encircling the shaft. The blossoms are light pink to light reddish brown. The egg-shaped fruits are light to medium violet when mature and 1 inch (2.5 cm) long.

A slow grower, this palm cannot withstand drought, needs protection from the midday sun, especially in hot and dry climates, and requires a free-draining, humus-rich acidic soil. While their habitat is frost-free, little trees are reported to have withstood 30°F (−1°C) without much harm and seem adaptable even to frostless Mediterranean climes if irrigated on a regular basis.

This elegant and refined small landscape subject must be planted in an intimate setting to be effective. It has good color and is almost architectural in its appeal, with a beautiful silhouette that cries out for up-close inspection. It is perfect for patio gardens where its need for semishade and moisture can be attended to and its small-scale charm most enjoyed. It was supposedly widely grown in European greenhouses in the 19th century and is undoubtedly well suited to such treatment.

Actinorhytis
ak-tin′-o-RYT-iss

Actinorhytis is a monotypic genus of tall, slender, solitary-trunked, pinnate-leaved, monoecious palm with a prominent crownshaft. The single species is native to New Guinea and the Solomon Islands. A second species, *A. poamau*, was once recognized from the Solomon Islands, but is now placed in synonymy with *A. calapparia*. The genus name is derived from 2 Greek words meaning "ray" and "fold" and alludes to the folds in the seed's endosperm. Seeds of *Actinorhytis* germinate within 30 days. For further seed germination details, see *Areca*.

Actinorhytis calapparia
ak-tin′-o-RYT-iss · kal-a-PAHR-ee-a
PLATE 10

Actinorhytis calapparia is found in lowland rain forest from sea level to 3000 feet (900 m). It begins as an understory tree, but towers above the forest canopy at maturity. The epithet is a Latinized form of the aboriginal name for the coconut palm (Dransfield 1994). The common name is calappa palm.

The nearly white trunk grows to 40–50 feet (12–15 m) in habitat, is never more than 8 inches (20 cm) in diameter, and is closely ringed with leaf base scars. It is topped by a light green crownshaft 3 feet (90 cm) tall that is slightly bulging at its base and is scarcely thicker than the slender trunk. The base of the trunk is slightly swollen and usually bears a mass of roots anchoring the thin stem. The leaf crown is spherical but open. The leaves are remarkably beautiful: 7–10 feet (2.1–3 m) long, greatly arching, and recurved feathers of dark green with many linear leaflets 18 inches (45 cm) long arching up from the rachis at an angle that creates a V-shaped leaf. The petioles are relatively long in juvenile plants but become shorter as the tree adds height, resulting in a beautifully rounded mature crown. The inflorescences form a ring around the base of the crownshaft and are much branched with masses of cream-colored flowers. The palm, when in flower or fruit, has a thrilling visual aspect because of the extraordinarily tall crownshaft, the effect being one of a small crown beneath a large crown. The large egg-shaped fruits are 3 inches (8 cm) long and are borne in heavy, pendent masses; each one is red or reddish brown to purplish olive when ripe.

This tropical beauty is moderately cold hardy. Some specimens in central Florida have survived 30°F (−1°C) unscathed. It does not tolerate the extended periods of cool winter nights in Southern California (Geoff Stein, pers. comm.). The palm is adapted to sun or partial shade, although young plants prefer semishade or even shade. It is a true water lover and cannot withstand drought. It performs poorly in soil lacking nutrients and needs organic matter amended into calcareous soils to thrive. It grows moderately fast under optimal cultural conditions.

Among the world's most beautiful palms, it has an almost formal elegance because of the gracefully recurving, almost circular fronds, and yet the aspect is stately because of the mature height of the palm. The tree makes a beautiful silhouette. It is effective as a canopy-scape, and the nearly complete circles of the crowns of a group of individuals of varying heights against the sky are breathtaking. It should perform well in bright conservatories and atriums with enough space, good air circulation, and high levels of moisture, light, and humidity.

The seeds are reportedly used locally in much the same manner as those of *Areca catechu*, and the pulverized seed kernel has been used as a baby powder.

Adonidia
ad-o-NID-ee-a

Adonidia is, at the time of this writing, a monotypic genus of solitary-trunked, pinnate-leaved, monoecious palm; however, a second species, recently discovered on Biak Island off the coast of New Guinea, awaits a formal name and description. The known species, *A. merrillii*, was named *Veitchia merrillii* until taxonomic studies by Scott Zona indicated it should be returned to its own genus, *Adonidia*, the name by which it had been known in the early 20th century. Common names include adonidia palm, Christmas palm, Manila palm, and dwarf royal palm. The genus name translates from the Latin as "little Adonis" and alludes to the great beauty of this palm. Seeds of *Adonidia* germinate within 30 days. For further seed germination details, see *Areca*.

Adonidia merrillii
ad-o-NID-ee-a · mer-RIL-lee-eye
PLATES 11–12

Adonidia merrillii occurs on small islands off the northwest coast of Palawan (Philippines) and adjacent Sabah (Malaysia), growing in coastal forests on limestone. It has recently been discovered on Danjugan Island (Philippines). The epithet honors Elmer D. Merrill (1876–1956), botanist, plant explorer of the Asian tropics, and former director of the Arnold Arboretum.

This palm can grow to a height of 50 feet (15 m) with age and in habitat but is usually no more than 30 feet (9 m) tall in cultivation. The light to dark gray trunks are less than 1 foot (30 cm) in diameter and are stout looking in younger plants but elegantly slender appearing in taller specimens. The newer and younger parts are ringed but not strongly so and are frequently curved or crooked, especially if shaded on one side during early growth. The light green crownshaft is 1–3 feet (30–90 cm) tall and columnar with only a slight basal bulge. The leaves are 6–8 feet (1.8–2.4 m) long with a decidedly arching rachis and a relatively short petiole less than 1 foot (30 cm) long. The light to dark green leaflets grow from the rachis at an angle to create a V-shaped leaf. The ends of the leaflets are pendent, which tends to relieve the slightly formal appearance of the leaf crown. Leaflets nearest the trunk usually bear long threads from their apices because strips of tissue (called *reins*) connected the edges of leaflets in developing leaves and hang loosely from fully unfolded leaves. The leaf crown usually has 12 strongly arching leaves. The much-branched inflorescences grow from beneath the crownshaft and bear small whitish, unisexual flowers of both sexes in the same cluster. The fruits are in large grapelike clusters and when mature are brilliant scarlet. Their maturation coincides with the end of the year in the Northern Hemisphere, which has led to one of the common names, Christmas palm.

The species is very susceptible to lethal yellowing disease, limiting its landscape value in Florida and the Caribbean Basin. At Fairchild Tropical Botanic Garden in Miami, the oldest survivors are growing in a site shaded by an immense *Enterolobium* tree and it may be that the insect that transmits lethal yellowing (the leafhopper, *Myndus crudus*) prefers sunny areas (Chuck Hubbuch, pers. comm.). The tree grows moderately fast, especially when young. It also flowers and fruits with as little as 3 feet (90 cm) of trunk. It is adapted to full sun or partial shade but is not drought tolerant and needs regular, adequate moisture. It tolerates salt spray but not saline soils. It is adaptable to a range of soils with good drainage. It does not tolerate freezing temperatures and is adaptable only to zones 10b and 11, although it sometimes recovers from temperatures slightly below freezing, making it marginal in 10a.

Because of its arching leaves and often crooked trunk, the palm has an informal appearance, perfect for modern gardens. It is striking in silhouette and, if this characteristic can be emphasized, works well in groups of individuals even planted in the middle of a lawn. Nurseries often sell this palm with 3 plants to the pot, which grow to form a graceful cluster. A "golden" cultivar, with yellow-green leaves and crownshaft, and a "golden crownshaft" cultivar are popular in Southeast Asia and are becoming increasingly more available in the United States.

This palm is especially wonderful in a patio or other intimate site and is endearing in front of a contrasting background that allows its form and silhouette to be seen. It is easily grown indoors and is among the most frequently used palms in shopping malls and atriums, where it needs good light and a constantly moist medium.

Aiphanes
EYE-fay-neez

Aiphanes is composed of 24 pinnate-leaved, monoecious palms in tropical America. Most are indigenous to lower elevation forests on the eastern slopes of the Andes, but others are in the adjacent lowlands and a few in Central America and the Caribbean. Some species in Colombia and Ecuador are in great danger of extinction. The genus name translates from the Greek as "always obvious" and probably alludes to the brightly colored fruits of most species. The genus was at once known as *Martinezia*.

Most species have aboveground, solitary trunks, although a few are clustering. All the trunks are covered with spines arranged in closely set rings, and usually the leafstalks as well as the leaflets are spiny. The leaflets are often wedge shaped and have jagged apices. The inflorescences grow from among the leaf sheaths and are spikelike or branched once. The flowering branches bear both male and female flowers. The fruits are red or orange.

Few species are cultivated, perhaps because of their spininess. All of them need a rich soil and at least average, regular moisture. None are hardy to cold and most appreciate protection from the midday sun, especially in hot climates and especially when young. Seeds can take more than a year to germinate and do so sporadically. For further details of seed germination, see *Acrocomia*.

Aiphanes horrida
EYE-fay-neez · HOR-ree-da
PLATE 13

Aiphanes horrida has the widest distribution in the genus and grows naturally in the drier forests of the Andes in Venezuela, Colombia, Peru, Bolivia, western Brazil, southeastern Peru, and central Bolivia, from sea level to 5300 feet (1600 m). Common names are coyure palm and ruffle palm. Synonyms include

A. aculeata and *A. caryotifolia*. The epithet is Latin for "bristling" and "frightening," both of which apply to this palm.

The solitary trunk grows to 35 feet (11 m) tall and 6 inches (15 cm) in diameter, and is ringed in black spines. The leaf crown consists of 10–15 leaves and is wonderfully globular due to the beautifully pendent older leaves. Each linear-oblong leaf is 5–8 feet (1.5–2.4 m) long and has 50–80 leaflets that are irregularly obdeltoid and abruptly widened at the apex. Arranged in groups of 4–6, the leaflets have jagged apices and grow from different angles out of the rachis, giving the leaf a ruffled, plumose effect. There are squat black spines on the short petioles, leafstalks, and leaflets. Leaf color is deep, glossy green above and paler green beneath. The large arched, pendent inflorescences are accompanied by a spiny spathe and grow from among the leaf crown. Each small blossom is yellowish white, and the edible fruits are globose, usually less than 1 inch (2.5 cm) wide, and bright orange or bright red when mature.

This palm is fast growing when its basic requirements are met. Juvenile trees require shade, while older plants do fine in sun to partial shade. Trees of all ages require regular, adequate moisture and a rich, humus-laden, well-drained soil but flourish in calcareous soil if mulched and fertilized. The species is adaptable to zones 10b and 11, although it is infrequently successful in the warmer parts of 10a.

The tree is exceptionally beautiful as a canopy-scape. Its crown is among the most satisfying shapes in nature, being full and round but with much contrast within the orb because of the jagged, plumose leaflets. It is unsatisfactory as an isolated specimen in lawns or other open spaces, even in groups of individuals; it needs something at the bases of its trunks. It is successfully grown indoors and especially in greenhouses and conservatories if given good light, good air circulation, a fertile medium, water, and humidity. It is also often grown in containers.

Aiphanes lindeniana
EYE-fay-neez · lin-den´-ee-AH-na

Aiphanes lindeniana is endemic to mountainous rain forest of Colombia to 8600 feet (2600 m). The epithet honors Jean Linden (1817–1898), Belgian plant explorer, horticulturist, and nurseryman.

This palm is usually clustering and attains an overall height of 25 feet (7.6 m). The gray trunks are an ethereally slender 2–3 inches (5–8 cm) in diameter and covered in their younger parts with large vicious spines 4 inches (10 cm) long. The leaves are 8 feet (2.4 m) long and bear closely spaced, jagged-ended, linear wedge-shaped stiff leaflets that have prickly margins; the leaflets grow at a slight angle from the rachis to create a V-shaped leaf.

This species is among the most beautiful in the genus. Its diaphanous silhouette makes a wonderful canopy-scape. The palm is slightly susceptible to lethal yellowing disease. It flourishes in hot, tropical or nearly tropical climates and is adaptable to zones 10 and 11. It adapts to various soils, even calcareous ones, and tolerates near-drought conditions, although it grows more slowly and less robustly.

Aiphanes minima
EYE-fay-neez · MI-ni-ma
PLATE 14

Aiphanes minima is a variable, solitary-trunked, undergrowth rainforest species in Hispaniola, Puerto Rico, and the Lesser Antilles. The epithet translates from the Latin as "smallest." The common name, macaw palm, is also used for *Acrocomia aculeata*. Borchsenius and Bernal (1996) listed *Aiphanes acanthophylla*, *A. corallina*, *A. erosa*, *A. luciana*, and *A. vincentiana* under *A. minima*. A subsequent study of Lesser Antilles populations showed minor interisland separation in morphology. Populations of *A. minima* on some Caribbean islands are rare due to habitat destruction, and the threat of their loss is significant given that some populations are distinct enough in appearance to have been once considered separate species.

The trunk is 20–50 feet (6–15 m) tall and covered in closely set rings of sharp black spines in all but its oldest parts. The leaf crown is rounded and full because of the pendulous older leaves. The leaves are 6–8 feet (1.8–2.4 m) long and basically similar to those of *A. horrida*, but the leaflets are narrower, more numerous per leaf, more regularly spaced along the rachis, and less ruffled at their apices. The petiole and the rachis are covered in black spines, and leaf color is medium dark green above and much paler green beneath. Leaves of *A. minima* seedlings are spiny on both sides, while those of *A. horrida* seedlings are spiny on the underside only (Kurt Decker, pers. comm.). Older plants of both species tend to have spineless leaf upper sides (Chuck Hubbuch, pers. comm.). The inflorescences are 3–6 feet (90–180 cm) long and borne among the leaves. Blossoms are yellowish white. The rounded ½-inch (13-mm) red fruits are reportedly edible.

Adult plants do well in sun or partial shade, but juvenile plants need partial shade or mostly shade. This palm is not finicky about soil type as long as drainage is unimpeded, but a decent medium with humus is best. The tree can withstand some drought if it is not prolonged.

This exceptionally graceful and elegant palm has a full, rounded crown of pendent pinnate leaves and is more esthetically adaptable to specimen planting than *A. horrida*, especially in groups of 3 or more individuals of varying heights. It takes well to indoor cultivation if given bright light and high relative humidity.

Aiphanes ulei
EYE-fay-neez · OO-lee-eye
PLATE 15

Aiphanes ulei is indigenous to southern Colombia, eastern Ecuador, northern Peru, and western Brazil, where it occurs in mountainous rain forest to 6000 feet (1800 m). The epithet honors Ernst H. G. Ule (1854–1915), a German botanist and collector in South America.

A small, solitary-trunked species, it grows to 10–12 feet (3–3.6 m) tall, but some dwarf forms never exceed 2 feet (60 cm). The slender trunks are 1 inch (2.5 cm) in diameter and graced with rings of black spines. The sparse leaf crown has 6 leaves, each bearing black spines on its base, petiole, and rachis. The leaves are large for such a short trunk at 6 feet (1.8 m) long and give a most appealing, ruffled aspect to the palm. The leaflets are arranged in pairs along the rachis and are to 1 foot (30 cm) long but, in a pair, one is always longer than its companion. Wedge shaped with oblique, jagged apices, the leaflets are medium to deep green above and silvery green beneath, and they grow in a single flat plane from the spiny rachis.

This gem is undeservedly rare in cultivation. It is among the prettiest species in the genus and makes a perfect close-up subject, especially sited against a contrasting background.

Allagoptera
al'-la-GAHP-te-ra

Allagoptera includes 5 pinnate-leaved, monoecious palms in South America. Many have short or subterranean trunks, but one species can have a tall emergent trunk. These palms appear to be clustering because the leaf petioles are short, the leaves are numerous and full, and the growing point of the trunk turns downward at some point in the individual's history and thus is usually lower than the base of the trunk. In addition, the species with short trunks usually branch and produce more than one leaf crown. The leaves of all species consist of clusters of linear leaflets, each leaflet growing in a different plane. The spikelike, erect inflorescences grow from among the leaf bases, with the blossoms tightly packed along the apical half of the spike. The small fruits are yellowish green to brown and are crowded into club-shaped clusters.

The genus was formerly known as *Diplothemium* and *Polyandrococos*. The present genus name translates from the Greek as "change" and "wing," an obscure allusion to the plumose leaves, perhaps. Seeds of *Allagoptera* can take more than 2 years to germinate and do so sporadically. For further details of seed germination, see *Acrocomia*.

Allagoptera arenaria
al'-la-GAHP-te-ra · ar-e-NAHR-ee-a
PLATE 16

Allagoptera arenaria occurs naturally along the seacoasts of southeastern Brazil in large stands on sand dunes and in the adjacent scrublands on sandy soils. Because of development and land clearing, the species is now highly endangered. The only common name is seashore palm. The epithet is Latin and translates as "sandy," referring to the palm's habitat.

This palm reaches a height of 8 feet (2.4 m) and a width of 15 feet (4.5 m). The leaves are 4–6 feet (1.2–1.8 m) long and dense in the crown. The narrow, ribbonlike leaflets are clustered into irregularly spaced, small groups and grow from the rachis at different angles to give a plumose appearance to the leaf. Each leaflet tapers to a point and is a deep, glossy to dusky green above with a definite silvery cast beneath due to a waxy deposit. The inflorescence is a short, dense, and narrow spike of closely set small, greenish yellow flowers, the whole emerging from a paddle-shaped, persistent woody spathe. The orange, egg-shaped fruits are 1 inch (2.5 cm) long and are densely packed into semipendent narrow club-shaped clusters that are mostly hidden among the leaves. They are not only edible but also delicious in the early stages of ripening (David Witt, pers. comm.).

The species is slightly susceptible to lethal yellowing disease. The palm grows slowly, but to make up for this growth rate, it is one of the most attractive trunkless or nearly trunkless palms. It thrives in poor, sandy soil, is drought tolerant when established, and is relatively hardy to cold, being safe in zones 9 through 11 and probably marginal in 8a. It is not a good shade plant, although it grows reasonably well in semishade.

The shrub is beautiful and useful because of its high salt and drought tolerance. It may be the world's most salt-tolerant palm species. It grows in pure sand along the beach if planted above the high tide line and is useful (if slow) for dune and erosion control. It also flourishes farther inland as long as it has sun and freely draining soil.

The leaves have a beautiful shimmering quality because of the silvery backed leaflets that move in the slightest breeze. Because it needs almost full sun to thrive, its use indoors is limited, but it is commonly grown in containers outdoors and, if given enough light, should also thrive inside.

Allagoptera brevicalyx
al'-la-GAHP-te-ra · brev-i-KAY-lix
PLATE 17

Allagoptera brevicalyx is indigenous to a relatively small area of east central Brazil, mainly along the coast but also inland in dry, deciduous forest. The epithet is Latin and Greek for "short" and "calyx." The principal differences between this species and *A. arenaria* are that it is smaller, has fewer leaves, and has wider leaflets. The leaves are covered in a gray, waxy substance that renders them gray-green, and the leaflets are apically bifid. The species is probably not as hardy to cold as *A. arenaria*, but data are lacking. Otherwise, it should have the same cultural requirements.

Allagoptera campestris
al'-la-GAHP-te-ra · kam-PES-tris
PLATE 18

Allagoptera campestris grows naturally in southeastern and southern Brazil, the northeastern tip of Argentina, and extreme eastern Paraguay. The epithet is Latin for "of the fields." This small palm has a subterranean trunk and dull green plumosely pinnate leaves. The thin leaflets are arranged in whorled groups like those of the other species and are grayish green beneath. The immature fruits are reportedly edible. The palm is weedy looking when combined with larger vegetation but is effective in the landscape if planted against a background of contrasting colors and forms. It is probably hardier to cold than the above 2 species and is definitely as drought tolerant.

Allagoptera caudescens
al'-la-GAHP-te-ra · kaw-DES-senz
PLATES 19–20

Allagoptera caudescens is endemic to east central Brazil at low elevations in the Atlantic coastal forest. It was formerly known as *Polyandrococos caudescens* and *P. pectinata*. The scientific epithet means "having a trunk." It is usually a tall, stately palm, but in the northern parts of its habitat and on poor soils, it may form only short subterranean trunks, just like the other species of *Allagoptera* (Henderson et al. 1995).

The trunk attains a maximum height of 35 feet (11 m) in habitat but no more than 8 inches (20 cm) in diameter. The older stems are light brown and "stepped" because of the rough leaf base scars, but these are indistinct visually and form incomplete circles. Young trunks are almost entirely covered in fibrous, persistent leaf sheaths. There is no crownshaft, but the leaf sheaths are massive. The leaf crown is hemispherical. The leaves are 8–10 feet (2.4–3 m) long, on petioles 2 feet (60 cm) long, often with short "pseudo-spines," and are spreading and slightly arching as they mature. The leaflets of most individuals grow in a single flat plane from the rachis, but the leaflets of others (perhaps a geographical variant) are arranged in indistinct groups, the individual segments growing at slightly different angles from the leaf's rachis. The individual leaflets are 3 feet (90 cm) long, narrowly linear-lanceolate with long tapering points, and a shimmering silvery green beneath, especially when younger. This color combination of the leaf is unexpectedly demonstrated in the shaft of the unfurled new leaf, which exhibits beautiful horseshoe-shaped dark green markings on an otherwise silvery green spear. While erect before the fruits form, the infructescence becomes pendulous as the sausagelike cluster of densely packed, greenish orange to brown fruits mature.

The palm is drought tolerant when established but looks better and grows faster with adequate and regular moisture. It is not fussy about soil type, growing well on sandy and even calcareous soils, but has better color and grows faster if humus is incorporated into the medium; an organic mulch is second best. This species revels in sun, especially when past its juvenile stage. It is not hardy to cold and is adaptable only to zones 10 and 11, although some nice specimens are found in protected and warm microclimates of 9b. Its only fault is that it is not fast growing.

This species is extraordinarily beautiful at all stages and is suited, because of its form and color, to being planted even in a wall of vegetation. It is also an attractive specimen planting, even solitary and surrounded by space.

Allagoptera leucocalyx
al'-la-GAHP-te-ra · loo-ko-KAY-lix
PLATE 21

Allagoptera leucocalyx is indigenous to southern central Brazil and eastern Bolivia, where it forms large colonies in open, grassy, and sunny areas. The epithet is Greek for "yellow" and "calyx." The plant is slightly smaller than *A. arenaria* and, unlike the other *Allagoptera* species, which have plumose leaves, has stiff, almost rigid leaflets. The leaflets grow from the rachis at an angle to create a V-shaped leaf.

Ammandra
am-MAN-dra

Ammandra is a monotypic genus of rare and endangered, pinnate-leaved, dioecious palm in South America. It is closely related to and similar in appearance to *Phytelephas* species but smaller. The genus name translates from the Greek as "sand man," an allusion to the small anthers of the male flowers having the appearance of grains of sand. Seeds of *Ammandra* can take up to 3 years to germinate and do so sporadically. They can be stored for several months as long as they are soaked in water for 2–3 days before sowing.

Ammandra decasperma
am-MAN-dra · dek-ah-SPUR-ma

Ammandra decasperma is native to western Colombia in the wet coastal foothills of the Andes to 1500 feet (460 m), as well as in the intermountain valleys. The epithet is derived from Latin words meaning "10" and "seed," the maximum number of seeds in a given fruit.

The subterranean trunks of this clumping species occasionally emerge with aerial stems that are, nonetheless, recumbent and creep along the ground. The gently arching leaves are large, to 20 feet (6 m) long, and the smooth, rounded dark green petioles are 6 feet (1.8 m) long. The emerald green linear leaflets are 2–3 feet (60–90 cm) long and regularly spaced in a flat plane along the rachis. The male inflorescences are long with congested, short flowering branches, and the female inflorescences are compact. The fruits are formed in large globose heads, each with pointed projections.

The palm requires moist, warm growing conditions and a humus-rich, moist but free-draining soil. It is not adapted to full sun. A mature individual looks like a small grove of some gigantic tropical cycad species and is a magnificent sight.

Indigenous peoples reportedly eat the seeds and use the fibers of the large, tough leaf petioles to make baskets. As is the case with *Phytelephas* and *Aphandra*, the hard seeds are used as vegetable ivory to carve *objets d'art* and in the manufacture of buttons and other utilitarian products.

Aphandra
a-FAN-dra

Aphandra is a monotypic genus of solitary-trunked, pinnate-leaved, dioecious palm in South America. The genus name is a scrambled combination of *Ammandra* and *Phytelephas*, 2 closely related genera with similar inflorescences and infructescences. Seeds of *Aphandra* can take up to 3 years to germinate and do so sporadically. They can be stored for several months as long as they are soaked in water for 2–3 days before sowing.

Aphandra natalia
a-FAN-dra · na-TAH-lee-a
PLATE 22

Aphandra natalia occurs naturally in eastern Ecuador, northern Peru, and western Brazil in the rain forest on foothills of the Andes to 2640 feet (800 m). The epithet honors Natalie Uhl (1919–), Cornell University palm specialist.

This palm grows to a height of 40 feet (12 m) in habitat. All but the oldest parts of the trunk, which reaches 25–30 feet (7.6–9 m) tall, are covered in gigantic, fat, old grayish leaf bases and spine-like black leaf base fibers. The trunk is 1 foot (30 cm) in diameter when shorn of the leaf bases but 3 feet (90 cm) in diameter with them. Near the top of the trunk, the leaf base fibers are extremely long and pendent, almost ropelike, much like those of *Leopoldinia piassaba*. The leaves are 15 feet (4.5 m) long and borne on extremely long robust gray petioles that are 8 feet (2.4 m) long, making the entire leaf (with petiole) 23 feet (7 m) long. The leaves are erect, slightly arching, and never descend beneath the horizontal; indeed, they are mostly held at a 45-degree angle. The medium to dark green linear leaflets are 3 feet (90 cm) long and regularly spaced along the rachis, and grow in a single plane. The inflorescences are unusual. Male inflorescences are colossal pendent affairs 6 feet long with a peduncle 2 feet (60 cm) long at the top and short clustered secondary branches bearing spreading tufts of pure yellow blossoms. Female inflorescences are 1 foot (30 cm) long, compact tufts of wiry, yellowish sepals and petals, surrounded by tough, greenish brown, boat-shaped bracts. The fruits are in dense rounded clusters that would look like a medieval club with spikes if the large amount of hairy black fiber were removed from the clusters.

This palm is rare in cultivation outside of Ecuador, where it is grown for the long fibers near the top of the trunks that are used to make brooms and other objects. It is certainly worthy of wider cultivation in tropical climates because of its almost prehistoric appearance.

Archontophoenix
ahr-kont'-o-FEE-nix

Archontophoenix is composed of 6 solitary-trunked, pinnate-leaved, monoecious palms in warm, moist parts of eastern Australia from sea level to 4000 feet (1200 m) in elevation. All are noted for exceptionally beautiful, straight, ringed, medium to tall trunks, prominent crownshafts, and large graceful leaves. The leaflets grow from the rachis in a single plane, but the rachis twists near its middle, giving a wonderful angle to the leaf and a sense of movement even in still air. The leaflets of all but *A. cunninghamiana* have silvery undersides. The inflorescences grow in a ring, like a skirt, from the base of the crownshafts and consist of several pendent branches of small, purple, yellow, or white flowers that produce red fruits.

Cold, drought, and hot, drying winds are the worst enemies of these palms. Because the growing points are easily damaged when plants of any size are transplanted, these palms should be purchased in containers. The genus name translates from the Greek as "king" and *Phoenix*, that latter used in the general sense of a pinnate-leaved palm. All the species share the common name king palm. Seeds of *Archontophoenix* germinate within 30 days. For further seed germination details, see *Areca*.

Archontophoenix alexandrae
ahr-kont'-o-FEE-nix · a-lek-ZAN-dree
PLATE 23

Archontophoenix alexandrae is endemic to the rain forest of coastal northern and central Queensland, Australia, where it is rare in parts of its range but is overall unthreatened. Common names are Alexandra king palm, Alexandra palm, and, in Australia, the Alex palm and Alexander palm. The epithet honors Princess Alexandra of Denmark (1844–1925), consort of King Edward VII of the United Kingdom.

The palm grows to 80 feet (24 m) tall in habitat but is usually half this stature in cultivation in the continental United States. The trunk bulges slightly at its base and, above that point, is 1 foot (30 cm) in diameter; it is light brown when young, turning gray with age. The crownshaft is to 4 feet (1.2 m) tall, also slightly bulging at the base, and olive green to purplish brown. Atop it is the leaf crown 20 feet (6 m) wide and 10–12 feet (3–3.6 m) tall and seldom composed of more than 12 leaves. The leaves are 6–10 feet (1.8–3 m) long with many closely spaced, narrow and drooping, strongly ribbed leaflets 3 feet (90 cm) long and growing in one plane. Leaf color is bright grassy to deep green above and silvery or gray green beneath; new growth is often bronzy. The many-branched, pendent inflorescences encircle the trunk below the crownshaft. The small blossoms are white or nearly white, and the egg-shaped to round fruits are ½ inch (13 mm) in diameter and red when mature.

Two varieties (var. *beatricae* and var. *schizanthera*) are no longer recognized. Taxonomists now consider them one variable species. Plants labeled "*A. alexandrae* var. *beatricae*" may exhibit a pronounced "stepped" base to their trunk bases, a feature that is attractive enough to be sought by the discriminating collector.

In Mediterranean climates, this species reportedly withstands 26°F (−3°C) unscathed, but in areas subject to wet freezes even large palms die at 28°F (−2°C); thus it is basically safe only in zones 10b and 11, and marginal in 10a in wet winter climes. Like most other king palms, it is a true water lover and needs a rich, humus-laden, well-drained soil. It is subject to leaf burn in the hottest part of the day in hot or dry climates, especially when young. Relatively fast growing, it often adds 1 or more feet (30 cm) of trunk annually under good cultural conditions.

No palm is more majestic and yet graceful than a mature specimen of this species with its straight-as-an-arrow, beautifully ringed, light gray trunk and massive crown of leaves. While the tree is spectacular enough to stand alone, it looks better in groups of 3 or more individuals of varying heights. It is difficult to transplant successfully at any size. Three requirements mitigate against its use indoors: good air circulation, high relative humidity, and bright light except when young. It is successfully grown in atriums and large conservatories.

Archontophoenix cunninghamiana
ahr-kont'-o-FEE-nix · kun'-ning-ham-ee-AH-na
PLATES 24–25

Archontophoenix cunninghamiana grows naturally in the rain forest of coastal Queensland southwards into southeastern New

South Wales. It is rare in parts of its range but is overall unthreatened. Common names are piccabeen palm and bangalow palm. The epithet honors Allan Cunningham (1791–1839), a British botanist and explorer in Australia.

Mature trunks attain 60 feet (18 m) of height and 9–10 inches (23–25 cm) in diameter. The crownshaft is about 3 feet (90 cm) tall, slightly swollen at its base, and a deep olive to dark green. The leaf crown is 15 feet (4.5 m) wide and 10–12 feet (3–3.6 m) tall. The leaves are 8–10 feet (2.4–3 m) long, with many narrow grassy to deep green leaflets, each 3 feet (90 cm) long and slightly paler green beneath. The leaflets are more drooping than are those of the Alexander palm and consequently even more graceful. The newly unfurling leaves are often coppery or bronzy hued. The inflorescences are significantly larger than those of the Alexander palm, and bear lavender blossoms, one of the most attractive features of this palm. The ellipsoid to rounded pink to red fruits are ½ inch (13 mm) long and formed in pendent clusters.

Seed dealers, growers, and nurseries apply the name *A. cunninghamiana* 'Illawara' to individuals descending from palms collected in the Illawara area of the mountainous coastal region of New South Wales. This form is supposedly hardier to cold because of its southerly and elevated provenance. Alas, this has not proved to be the case in central Florida; however, the plant is certainly well adapted to the Mediterranean but nearly frostless regions of coastal California where summer heat is not great, and, according to David Witt (pers. comm.), in Orlando, Florida, this form seems to be the only one which looks good at any age when grown in the shade.

This palm grows as quickly as the Alexander palm and has the same cultural requirements. It is less heat tolerant, especially when young, but slightly more cold tolerant. Although it has survived several years in zone 9b of the continental United States, it is tender enough that it cannot be relied upon as a landscape subject in that zone, except in the drier Mediterranean climes.

The species serves the same landscape purposes as does Alexander palm but its flowers are an added bonus. Consequently, this palm should be planted where the inflorescences are easy to observe up close. Like Alexander palm, it is not recommended for use indoors, but it is successfully grown in atriums and large conservatories.

Archontophoenix maxima
ahr-kont'-o-FEE-nix · MAX-i-ma
PLATE 26

Archontophoenix maxima grows naturally west and southwest of the port of Cairns in Queensland, Australia, where it occurs in mountainous river valleys of the upper Atherton Tableland to 3500 feet (1070 m). Before John L. Dowe formally described it in 1994, it was known only as *A.* 'Walsh River'. The common name in Australia is Walsh River king palm. The epithet is Latin for "largest" and refers to the inflorescence, the largest in the genus. This palm grows as tall as piccabeen palm but has a thicker trunk. Other distinguishing features are the short to almost nonexistent petioles and the upright, rigid, scarcely arching leaves. The inflorescence bears creamy white flowers. The species is probably no hardier to cold than any other species in the genus.

Archontophoenix myolensis
ahr-kont'-o-FEE-nix · my-o-LEN-sis
PLATE 27

Archontophoenix myolensis is a rare and endangered species endemic to the mountains of northeastern Queensland west of Cairns, along streams and rivers. It is possibly the rarest species of the genus in nature. The only common name in English is the one used in Queensland, Myola king palm. The epithet is Latin for "of Myola," a district in Queensland. The tree is similar to a smaller version of *A. cunninghamiana* but has a bluer crownshaft with more pendent leaflets. It is probably one of cold-hardier species since it occurs naturally at high elevations. Despite its riverine provenance, it has proven adaptable to drier conditions in Florida.

Archontophoenix purpurea
ahr-kont'-o-FEE-nix · pur-PUR-ee-a
PLATES 28–29

Archontophoenix purpurea grows naturally in rain forest of northeastern Queensland at elevations to 4000 feet (1200 m). Before John L. Dowe and Donald Hodel described this species in 1994, it was known as *A.* 'Purple Crownshaft' or *A.* 'Mt. Lewis'. Its common name is purple king palm. The epithet is Latin for "purplish."

Mature trunks attain 50–60 feet (15–18 m) of height in habitat and 40 feet (12 m) under cultivation. They usually show diameters of 18 inches (45 cm), making this species the most massively trunked. The crownshaft is 4 feet (1.2 m) tall, slightly bulging at the base, and light violet to reddish purple. The leaf crown is no more than 20 feet (6 m) wide and 15 feet (4.5 m) tall. The leaves are 6–10 feet (1.8–3 m) long with many closely spaced, narrow leaflets 3 feet (90 cm) long and growing in one plane. Leaf color is bright grassy to deep green above, silvery or grayish green beneath. The inflorescences are pendent clusters encircling the trunk below the crownshaft. The small blossoms are purplish. The rounded 1-inch (2.5 cm) rounded, scarlet to wine-red fruits are in pendent clusters.

This species has cultural requirements similar to those of Alexandra palm; however, it seems to tolerate cooler conditions better, although it is no more tolerant of frost. It is the slowest growing species in the genus (Daryl O'Connor, pers. comm.).

Archontophoenix tuckeri
ahr-kont'-o-FEE-nix · TUK-er-eye
PLATE 30

Archontophoenix tuckeri occurs naturally in rain forest and coastal swamps of extreme northern Cape York Peninsula of Queensland, Australia, making it the most northerly occurring and most tropical species. Prior to 1994, it was known as *A.* 'Peach River' (Peach River king palm). In Australia, it is called the Iron Range or Rocky River king palm. The epithet honors the palm's original collector, Robert James Thomas Tucker (1955–1992), botanical illustrator, taxonomist, and designer of the Townsville

Palmetum. The palm is similar in appearance to the other species in the genus, especially to Alexandra palm, but is smaller. It is rare in cultivation but in habitat grows to 70 feet (21 m). The crownshaft is a typical green, but the new leaves usually are reddish bronze. The inflorescences beneath the crownshaft are held erect and bear white or creamy white flowers from which are formed large, bright red fruits.

Areca
a-REE-ka

Areca comprises about 47 pinnate-leaved, monoecious palms in India, Southeast Asia, Malaysia, Indonesia, the Philippines, New Guinea, and the Solomon Islands. Virtually all of them have some horticultural merit, although not every species is amenable to cultivation. The seeds of many species are used like that of the most famous member, the betel-nut (*A. catechu*).

The mostly solitary-trunked species are small to medium in height, although a few are tall. All have prominent crownshafts, some outstandingly colored, and all have trunks ringed with prominent leaf base scars. Leaf crowns contain few leaves, and these are arching and usually have relatively wide leaflets, the terminal 2 usually broader than the other leaflets. The inflorescences grow from beneath the crownshafts and consist of short radiating, usually stiff flowering branches bearing both male and female blossoms. The tips of the infructescence branches are persistent and bare. The fruits are often colorful.

Most species are undergrowth inhabitants of rain forest, but some (including the most famous) are adapted to growing in full sun in the open. All species are tropical and adaptable to zones 10b and 11, only occasionally being found in warm microclimates of 10a. The genus name is thought to be a Latin form of a Portuguese corruption of the aboriginal (southern India) name of one species.

Like other species in the tribe Areceae, *Areca* species are among the easiest palms to grow from seed. Most seeds in the tribe have a short viability only because they tend to germinate as soon as they ripen and fall from the tree. Seeds that are allowed to dry out completely do not germinate well, and the smaller the seed, the more likely it will dry out before sowing. Germination tends to occur quickly, but in some species can take as long as a year. Seeds shipped in a bag for several days often begin germinating before reaching their destination. Fresh seeds of *Areca* almost always germinate within 30 days and grow quickly.

Areca caliso
a-REE-ka · ka-LEE-so
PLATE 31

Areca caliso is endemic to the Philippines, where it occurs in mountainous rain forest, mainly along streams at 1000–3000 feet (300–900 m). The epithet is the aboriginal name of the species.

This solitary-trunked species grows to 25 feet (7.6 m) tall in habitat. Except in its oldest parts, the trunk is green with widely spaced, light-colored rings of leaf base scars. The dark green crownshaft is 2 feet (60 cm) tall and no wider than the trunk except at its base where it bulges slightly. The leaves are 8 feet (2.4 m) long, mostly ascending, and usually do not lie beneath the horizontal. The large leaflets are 3 feet (90 cm) long and grow from the rachis at a slight angle but not steep enough to say the leaf is V shaped. The leaflets are evenly spaced, deep green above and lighter beneath, linear-lanceolate with mostly oblique apices, and are limp enough that their tips are pendent. The fruits are yellow to red when mature.

The palm is as tender to cold as the other species and needs a humus-laden soil and constant moisture. It thrives in partial shade. It looks like *A. catechu* but is more choice for intimate sites and, because of the beautiful large leaflets, gives one of the most tropical and lush appearances of any smaller palm. It is rare in cultivation and is probably found only in botanic gardens but deserves much wider planting.

Areca camarinensis
a-REE-ka · kah-mar'-i-NEN-sis

Areca camarinensis is native to the Philippines on the island of Luzon. The epithet commemorates the province of Camarines Sur on Luzon Island, where the palm was first collected on the slopes of Mount Isarog. This solitary trunked palm can grow to 20 feet (6 m) tall and is green with widely spaced, pale leaf scars. The crownshaft is 2–3 feet (60–90 cm) long and a light green. The leaves are up to 10 feet (3 m) long, with evenly spaced leaflets in a single plane. The inflorescences are composed of stiff branches bearing small, white flowers. The clusters of ovoid fruits are reddish orange. This tropical species is intolerant of frost and looks best when protected from the midday sun, especially in hot climates. It must have moist soil rich in organic matter.

Areca catechu
a-REE-ka · KA-te-choo
PLATES 32–33

Areca catechu is naturalized in India, Southeast Asia, Malaysia, Indonesia, the Philippines, and the South Pacific Islands. The original habitat is unknown but is most likely the wet regions of peninsular Malaysia. Common names include betel palm, betel-nut palm, and areca nut palm. The epithet is a Latin form of the Malayan name for the species.

This solitary-trunked palm may attain a height of 100 feet (30 m) in habitat but is usually half that under cultivation. The trunk is 8–12 inches (20–30 cm) in diameter, with prominent whitish rings, and colored deep green except for the oldest wood, which is gray to grayish green. The full, rounded leaf crown is 12 feet (3.6 m) wide and tall. The bright green smooth crownshaft is 3 feet (90 cm) tall, with a slight bulge near its base. Mature leaves are stiffly arching, 8 feet (2.4 m) long, with 40 or more shiny, medium green leaflets which grow from the rachis at a 45-degree angle, giving the leaf a V shape. The leaflets are linear-oblong to linear-ovate, 2 feet (60 cm) long, and 4–6 inches (10–15 cm) wide with prominent grooves along the upper surface. They are jagged at their apices, and the 2 terminal leaflets are fused. The inflorescence bears male and female flowers and emerges beneath the crownshaft in spreading panicles 2 feet (60 cm) long. The small, fragrant blossoms are whitish yellow. The deep yellow to red fruits are 1–2 inches (2.5–4 cm) long, egg shaped, and borne in pendent clusters. The seed resembles a miniature coconut.

There is a beautiful cultivar with golden trunks, fruits, and crownshaft; one with a nearly white crownshaft; another with a reddish orange crownshaft; and a dwarf, stocky and compact cultivar from Thailand.

The palm flourishes in partial shade or full sun. It is a water lover and a fast grower, luxuriating in rich, humus-laden, moist but well-draining, acidic soil; poor and calcareous soils should be amended with organic material or heavy mulches. The tree is slightly tolerant to salt spray. Year-round warmth is essential for this palm, limiting its cultivation to zone 11, but it sometimes can be found in the warmest parts of 10b.

Although it is among the tallest species in the genus, the betel palm is too ethereal to be planted singly and isolated in the middle of space. Groups of at least 3 individuals of varying heights create a veritable symphony of form and color because of the splash of leaves and the gorgeous, straight dark green trunks with their widely spaced, beautiful light-colored rings. The palm is perfect as a canopy-scape. Mature trees must have a lot of head room and, since betel palm grows quickly, it is soon too large for all indoor spaces except atriums and large conservatories. Commercial nurseries grow it in small, flat pots with the seed partially exposed and market it as a miniature coconut. The seed's outward similarity to a small coconut, along with its immature first couple of leaves, makes for a beautiful tabletop plant, but it does not usually thrive for long potted this way.

Betel-nuts contain a mild narcotic, and native peoples have for millennia chewed them mixed with the leaves of one of the pepper vines (*Piper betle*) and slaked lime, and sometimes additional herbs and spices, to obtain the narcotic effect. The side effects are copious blood-red saliva, red-stained gums, and blackened teeth. The palm and this practice were made famous to the Western world by Bloody Mary in the Rodgers and Hammerstein musical and movie, *South Pacific*.

Areca concinna
a-REE-ka · kahn-SIN-na

Areca concinna is endemic to Sri Lanka in rain forest at low elevations and is endangered because of agricultural expansion. The epithet is Latin for "elegant" or "orderly." The small, sparsely clustering species has slender, elegant stems to 15 feet (4.5 m) tall with a diameter of 2 inches (5 cm) at mid height. They are emerald green in their younger parts and light tan to light gray in their older parts. The small, light green crownshaft is hardly wider than the stem and bulges slightly at the base. The spreading, slightly arching leaves are 5 feet (1.5 m) long. The medium green leaflets, each 18 inches (45 cm) long and 3 inches (8 cm) wide, grow in a single plane from the rachis, are slightly limp, heavily veined, irregularly spaced along the rachis, and variable, but mostly oblong-lanceolate with slightly jagged apices. The deep red fruits are borne in small, pendent clusters.

The palm is tender to cold and adaptable only to zones 10b and 11 and marginal in 10a. It needs constant moisture, a well-drained, slightly acidic, humus-laden soil, and partial shade, especially when young; it cannot endure full sun in hot climates. This elegant palm should be planted where its diaphanous stems and heavy-looking leaves can be seen up close. It should also be grown with its few stems at varying heights to show off the beautiful foliage and small trunks.

Areca guppyana
a-REE-ka · gup-pee-AH-na
PLATE 34

Areca guppyana grows naturally in low mountainous rain forest of Papua New Guinea and the Solomon Islands. The epithet commemorates Henry B. Guppy (1854–1926), British botanist and natural historian of the Solomon Islands.

The solitary trunk attains 10 feet (3 m) high and 3 inches (8 cm) in diameter. Like the trunks of most species in the genus, it is green with whitish rings of leaf base scars. In spite of the palm's small stature, the base of the trunk usually exhibits prominent stilt roots. The crownshaft is 1 foot (30 cm) tall, light green, and the same girth as the trunk. The few leaves (usually 6) consist of 4 or 5 widely spaced, broad, deep green, deeply plicate leaflets; the fused leaflets of the terminal pair are largest. The inflorescence is mostly erect and consists of many small, greenish white flowers of both sexes. The fruits are bright red when mature and 1½ inches (4 cm) long.

This palm does not tolerate frost. It needs protection from hot sun and a humus-laden soil that is nearly constantly moist. It must be used in an intimate setting to show up. It makes a good container plant.

Areca ipot
a-REE-ka · EE-paht
PLATE 35

Areca ipot grows naturally in the Philippines in lowland rain forest, where it is threatened because of over-collecting. The epithet is part of the aboriginal name. The solitary trunk can grow to 20 feet (6 m) tall but begins producing flowers and fruit when much smaller and looks like a diminutive betel-nut palm. The inflorescences are composed of thin, splayed branches of whitish flowers. The tight clusters of ovoid fruits are deep orange or red when mature. This species cannot tolerate frost and must be protected from the midday sun, especially in hot climates. It is a true water lover and must never dry out. In addition, it needs a rich soil that is neither heavy nor compacted.

Areca macrocalyx
a-REE-ka · mak-ro-KAY-lix
PLATES 36–37

Areca macrocalyx is indigenous to New Guinea and the Solomon Islands in dense rain forest from sea level to more than 5280 feet (1610 m). It is sometimes called dwarf betel-nut palm. The epithet is from Greek words meaning "large" and "calyx."

This species is variable, especially in leaf form. It grows to 30 feet (9 m) tall but is often smaller. The trunk is usually straight as an arrow and green in its younger parts, with widely spaced whitish rings of leaf base scars. The elongated crownshaft is usually no thicker than the trunk, even at its base, and is 4 feet (1.2 m) tall; most forms are light glaucous green but some are reddish near the base of the shaft. The mostly ascending, stiff, and unarching leaves are 6 feet (1.8 m) long with few to many deeply pleated leaflets of variable widths and spacings; some individuals have entire or almost entire unsegmented leaves. Leaf color is medium to dark green above and usually lighter green beneath. The fruits may be round or elongated but are invariably deep orange when mature and are said to be used as a betel-nut substitute.

A form of this species with brilliant red on its crownshaft seems well adapted to the climate and soil of South Florida. The species neither tolerates frost nor needs as much heat as many others. It is not drought tolerant. It luxuriates in partial shade at any age and cannot withstand full sun in hot climates.

Areca macrocarpa
a-REE-ka · mak-ro-KAHR-pa
PLATE 38

Areca macrocarpa is a rare species endemic to lowland rain forest on the island of Mindanao in the Philippines. The epithet is Greek for "large" and "fruit," an allusion to the fruits which are used for the same purposes as the betel-nut.

The stems of this clustering species grow to 20 feet (6 m) high and 6 inches (15 cm) in diameter. The crownshafts are slender and 2 feet (60 cm) tall, scarcely of greater diameter than the trunk, and deep green. The leaves are 8 feet (2.4 m) long on short petioles, strongly arching, and have closely set, stiff, erect leaflets, each 3 feet (90 cm) long and growing from the rachis at an angle which gives a distinct V shape to the leaf. The ellipsoid fruits are 3 inches (8 cm) long and red.

The species is wonderfully beautiful and, if the trunks can be kept at different heights, has one of the most thrilling silhouettes in the family. It is tropical in its requirements, needing constant warmth and moisture, partial shade, especially in hot climates, and a rich, humus-laden soil.

Areca minuta
a-REE-ka · mi-NOO-ta
PLATE 39

Areca minuta is endemic to low mountainous rain forest in Sarawak on the island of Borneo, where it occurs in the undergrowth. The epithet is Latin for "minute" or "tiny." The stems of this clustering species attain heights of 3 feet (90 cm). The crownshaft is also tiny and almost white. The leaves are 1 foot (30 cm) long and have 4–6 widely spaced, wide, glossy bright green pinnae that are ovate-elliptical with drawn-out apices; the leaflets of the terminal pair are apically obtuse and jagged. This palm must have partial shade, regular and adequate moisture, and a humus-laden soil. It makes a beautiful large groundcover and is a perfect container subject.

Areca montana
a-REE-ka · mon-TAH-na
PLATE 40

Areca montana is indigenous to Thailand, peninsular Malaysia, Sumatra, and Java in low mountainous rain forest. The species includes *A. latiloba* and *A. recurvata*. The epithet is Latin for "montane."

This small, solitary-trunked, variable species resembles *A. triandra*; indeed, some taxonomists believe it is just that. The stem reaches a height of 12 feet (3.6 m). The leaf crown is sparse with usually 6 ascending, slightly spreading leaves borne above a light bluish green crownshaft that is 18 inches (45 cm) tall and no wider than the 2-inch (5-cm) green trunk. The leaves are 4 feet (1.2 m) long on dark green petioles 1 foot (30 cm) long and bear either widely spaced, broad, 18-inch (45-cm), glossy deep green, long-tipped leaflets or regularly spaced, more numerous, narrower leaflets. The starburstlike inflorescences grow from beneath the crownshaft and consist of thin whitish branches only 6 inches (15 cm) long and bearing white flowers. The egg-shaped fruits are pink or red.

This small undergrowth palm requires protection from full midday sun, especially in hot climates. It is a true water lover and needs a rich, humus-laden, well-drained soil and a tropical climate.

Areca multifida
a-REE-ka · mul-TI-fi-da
PLATE 41

Areca multifida is endemic to low mountainous rain forest in Papua New Guinea, where it is an undergrowth subject. The epithet is Latin for "many divisions" and alludes to the numerous narrow leaflets in adult plants.

The solitary trunk grows to 12 feet (3.6 m) high, is hardly more than 1 inch (2.5 cm) in diameter, and is green in its youngest parts and tan in the older parts, with widely spaced light gray or white rings of leaf base scars. The crownshaft is 18 inches (45 cm) tall, bulges slightly at the base, and is smooth and light mint green. The leaf crown is sparse but has leaves 6 feet (1.8 m) long. The mature leaflets are unusual in the genus: they number as many as 60 per leaf, are widely spaced, narrowly linear, and 2 feet (60 cm) long with tapering apices. Juvenile plants have wider and apically jagged leaflets.

This unusual species is outstandingly beautiful and makes a stunning silhouette against contrasting vegetation. It requires partial shade, a tropical climate, regular, copious moisture, and a humus-laden soil. Even under ideal conditions, it is slow growing.

Areca novohibernica
a-REE-ka · no-vo-hi-BUR-ni-ka

Areca novohibernica occurs in the undergrowth in low-lying coastal habitats in New Britain and New Ireland, islands in the Bismarck Archipelago, itself a part of Papua New Guinea. The epithet is Latin for New Ireland.

This dwarf solitary-trunked palm has stilt roots like

A. guppyana and has been confused with that species in the past. The short trunk is green when young and light brown when mature, with widely spaced, prominent leaf base scars. It grows to about 10 feet (3 m) but begins flowering while much smaller. The crownshaft is smooth and light green. The leaves are few and short, but the broad, many-nerved segments are at once bold and graceful. The inflorescence is 10 inches (25 cm) long and erect, growing from beneath the crownshaft and bearing bright white flowers. The spherical fruits are vivid scarlet and about 1 inch (2.5 cm) in diameter.

Although not yet common in cultivation, this species deserves wider use. It needs low light, high humidity, and a rich organic soil. It is not cold tolerant.

Areca ridleyana
a-REE-ka · rid-lee-AH-na
PLATE 42

Areca ridleyana is from peninsular Malaysia in deep shade in rain forests. The epithet commemorates Henry N. Ridley (1855–1956), British botanist, the first scientific director of the Singapore Botanic Gardens, and the father of the Malaysian rubber industry. This palm produces widely spaced, pencil-thin stems in abundance from an underground stolon. The stems are no more than waist-high, each bearing about 5 small leaves, which are undivided (similar to those of *Chamaedorea metallica*) and a deep, rich green. Some forms may have pinnate leaves, but the better forms have undivided leaves. The small inflorescence is only about 4 inches (10 cm) long and bears inconspicuous, ivory-colored flowers followed by yellowish fruits. This species is not yet common in cultivation. Like many tropical understory palms, it requires constant warmth, humidity, and moisture and must be protected from direct sunlight. Its cold tolerance has not been tested but is likely low.

Areca triandra
a-REE-ka · try-AN-dra
PLATE 43

Areca triandra is a highly variable species that grows naturally in wet regions at low elevations from eastern India through Southeast Asia to Borneo. The epithet translates from the Greek as "three" and "anther."

This clustering species can have up to a dozen trunks per plant, each of them to 30 feet (9 m) tall. The trunks are green in their newer parts and grayish in their older parts, diaphanously thin and elegant, and are never more than 3 inches (8 cm) in diameter, marked with widely spaced whitish circles of leaf base scars. The smooth, green crownshaft is 3–4 feet (90–120 cm) tall and barely bulging at its base. The leaves are similar to those of betel-nut palm but usually darker green and with larger fused terminal leaflets; there are generally no more than 6 leaves per trunk. The tiny, greenish white, fragrant blossoms produce clusters of ovoid, brownish orange to reddish orange fruits 1 inch (2.5 cm) wide.

The species varies mainly in the ultimate height of the trunks and the amount of suckering; there are even some individuals with solitary trunks. One of these forms has been labeled *A. aliceae* in the past but is now regarded as a synonym of *A. triandra*.

This palm does not tolerate frost. It requires partial shade when young but, as an adult, can withstand full sun, even in hot climates. It is also a true water lover and needs a rich, humus-laden, friable soil with unimpeded drainage. Because of its suckering habit, it resprouts from its roots if frozen back in central Florida (David Witt, pers. comm.). The rooted suckers may be removed with care from the parent plant and grown on.

This relatively small but graceful, almost noble-looking palm is among the few species that can look good planted alone and surrounded by space. It is even more appealing when planted where it can be viewed up close, especially in silhouette. Some individuals form dense clumps of stems and these look better if a few of the trunks are judiciously thinned out so that the beautiful form of the others may be more apparent. The plant grows fast with proper irrigation and nutrients in a sheltered location. It makes a fine and easy large houseplant or container subject and needs a soil that never dries out and high relative humidity. It is one of the few palms for the fragrant garden. The flowers exhale a delicious lemony fragrance that is strong but not overpowering, perfuming the entire vicinity of the palm.

Areca tunku
a-REE-ka · TOON-koo
PLATES 44–45

Areca tunku is indigenous to peninsular Thailand, peninsular Malaysia, and Sumatra, where it occurs naturally in low mountainous rain forest. The epithet honors Tunku Abdul Rahman (1903–1990), the first prime minister of Malaysia. A synonym of this species is *A. befaria*.

The small solitary trunk grows up to 15 feet (4.5 m) tall and 2 inches (5 cm) in diameter. It is closely ringed, and light brown, topped by a dark green crownshaft that is 18 inches (45 cm) tall and slightly wider than the stem. The leaf crown is hemispherical. The spreading leaves are 6 feet (1.8 m) long and bear closely spaced, glossy dark green, lanceolate-acuminate, and slightly S-shaped leaflets 2 feet (60 cm) long. The fan-shaped inflorescences are borne beneath the crownshaft and consist of erect, coral-colored fleshy branches. The white, fragrant flowers produce greenish brown fruits 2 inches (5 cm) long.

Slow growing, the species needs protection from the hot sun and copious, regular moisture in a humus-laden, quickly draining soil. Its colorful inflorescence makes it one of the most decorative *Areca* species, repaying in spades any extra effort required to cultivate it.

Areca vestiaria
a-REE-ka · ves-tee-AHR-ee-a
PLATES 46–49

Areca vestiaria occurs naturally in low mountainous rain forest of eastern Indonesia, including Sulawesi and the Moluccas. It is sometimes called orange collar palm. The epithet translates from the Latin as "wrapped" or "clothed," probably a reference to the conspicuous crownshaft. A synonym is *A. langloisiana*.

This species is usually clustering but sometimes solitary trunked. Mature trunks attain a height of 20 feet (6 m) and a diameter of 3–4 inches (8–10 cm), often with distinct prop or aerial roots at the base. Clumps may attain a width of 15 feet (4.5 m) and a total height of 25 feet (7.6 m). The most striking aspect of the palm is its startlingly beautiful slender crownshaft, which is bright, shiny orange to reddish orange or even red. The leaves are similar in size and form to those of betel-nut palm with the notable exception that the petioles are extremely short and both petiole and rachis are usually an attractive orange or deep yellow. The tiny flowers are yellow to orange and are followed by small, golden-yellow to orange ovoid fruits borne in pendent clusters.

A form of this species has a maroon new leaf. By the time the new leaf fades to green, another new leaf is unfurling, giving this cultivar additional color. The crownshaft is more intensely reddish orange than that of the type. Some forms are maroon throughout their vegetative parts.

While this palm is at home in tropical conditions, it can survive and even thrive in zone 10a and seems able to tolerate extended periods of cool weather. Like most *Areca* species, it requires a rich, friable, free-draining, slightly acidic soil and does not tolerate drought. It usually looks best in partial shade but thrives in full sun except in hot climates. Propagation by division of the suckers is usually unsuccessful. *Areca vestiaria* is a plausible substitute for the more demanding *Cyrtostachys renda*.

There is hardly a more beautiful combination of form and color in the plant world, making this palm excellent for almost any landscape circumstance. It excels in close-up situations. With enough moisture, humidity, and light, it works admirably well as a houseplant or in a greenhouse.

Areca vidaliana
a-REE-ka · vee-dah'-lee-AH-na
PLATE 50

Areca vidaliana is endemic to wet forests on the island of Palawan in the Philippines, where it occurs naturally from sea level to 1000 feet (300 m). The epithet honors the collector of the original specimen, Spanish forester and botanist Sebastian Vidal (1842–1889), author of important early works on the flora of the Philippines.

This small solitary-trunked species grows to 12 feet (3.6 m) overall, with a trunk diameter of slightly more than 1 inch (2.5 cm). The slender crownshaft is 1 foot (30 cm) tall, medium green, and almost unnoticeable. The leaves are ascending and unarched, although the lower ones are spreading, and are 3–4 feet (90–120 cm) long on petioles 18 inches (45 cm) long. The regularly spaced leaflets are 1 foot (30 cm) long, glossy medium green, slightly S shaped, lanceolate overall, with the middle and lower ones long tipped and the ones near the leaf apex thicker, corrugated, and with small apical teeth corresponding to the veins. The fruits are ½ inch (13 mm) in diameter.

This choice, intimate-appearing palm needs partial shade, copious moisture, and a tropical climate.

Arenga
a-REN-ga

Arenga comprises approximately 20 pinnate-leaved, mostly monoecious palms in tropical Asia and Australia. Some are solitary trunked, others are clustering. The inflorescences are either branched once or spikelike. The trunks of almost all species are monocarpic but do not flower until they are mature (often many years), and the flowering may last for several years because the inflorescences emerge from the leaf base scars around the trunk and progress usually from top to bottom in series of 2 or 3 per year. With the clustering species, new trunks are always forming to take the places of dying (fruiting) trunks.

Some species have a slightly unkempt look and are best in large gardens where they can be appreciated from a distance. Almost all species produce remarkable amounts of "woven" black persistent fibers around their leaf bases. The flesh of the fruits of all species is poisonous and irritating to the skin.

The genus name is a Latinized form of an aboriginal name for one species. The common name of most species is sugar palm because the sap in the trunks of several is used to make sugar as well as alcoholic beverages.

Like all the genera in the tribe Caryoteae, *Arenga* is remote germinating. The stalks are 1–3 inches (2.5–8 cm) long, depending on the species, so a container at least 6 inches (15 cm) deep should be used. Fresh seeds tend to germinate quickly, but several months can elapse before the first leaf appears above the soil. Seeds should not be allowed to dry out for long. Fresh seeds of *Arenga* germinate in 60–120 days, with the first leaf appearing 30–120 days later, depending on the temperature.

Arenga australasica
a-RENG-a · aw-stra-LAY-zi-ka
PLATE 51

Arenga australasica is native to northern Queensland and the Northern Territory, Australia, in lowland forests. The epithet is Latin for "of Australasia."

This clumping species produces one or 2 stems to 30 feet (9 m) from the base(s) of which grow several smaller trunks. The tallest stems are light colored in their older parts and are strongly ringed with leaf base scars. The leaf crowns are 15 feet (4.5 m) high and wide and consist of a dozen leaves, each one 10 feet (3 m) long, including a petiole 1 foot (30 cm) long. The leaflets are narrow, stiff, pleated, dark green above and a duller and lighter shade of green beneath; they grow from the rachis at a 30-degree angle to create a V-shaped leaf. The inflorescences are large pendent panicles of yellowish male and female flowers. The black or dark purple round fruits are less than 1 inch (2.5 cm) in diameter.

This beauty is tender to cold and adaptable only to zones 10 and 11, being marginal in 10a. It is a water lover. Drought and dry air are its enemies. It needs a humus-rich, freely draining soil. While it flourishes in sun or shade, it looks better in partial shade, especially when younger. About the only fault the palm has is that it is slow growing.

Because of its height and multitrunked habit, this palm is

magnificent in the middle of a lawn. It cannot be beat as a giant hedge or barrier marking the boundary of some garden space, its lower-growing trunks creating the barrier, the taller ones a ready-made canopy-scape. This palm would only be adaptable to a large atrium or conservatory with high levels of relative humidity and light.

Arenga brevipes
a-REN-ga · BRE-vi-peez
PLATES 52–53

Arenga brevipes is indigenous to northern Borneo and the adjacent Philippine Islands, where it grows in lowland rain forest. The epithet is Latin for "short foot" and alludes to the trunks. The short stems of this sparsely clustering species are mostly subterranean, but the large leaves thrust upward to a height of 8–10 feet (2.4–3 m) on stout dark brown to nearly black petioles 2–3 feet (90 cm) long. The leaflets are short and congested near the base of the leaf, with the lowermost pointing backwards towards the petiole, but are progressively larger towards the apex where the terminal one is broad and has 3 or 4 lobes. The lower leaflets are irregularly wedge shaped with shallow lobes and undulations, and all are deep emerald green above but a startling and shimmering silvery white beneath, with a strong, darker midrib for each lobe of the leaflet. This palm is not hardy to cold and needs partial shade and a humus-laden, constantly moist but fast-draining soil.

Arenga caudata
a-REN-ga · kaw-DAH-ta
PLATES 54–55

Arenga caudata occurs naturally in southern Myanmar, Thailand, northern peninsular Malaysia, Cambodia, Vietnam, and southeastern China, where it is an undergrowth subject in monsoonal tropical forests at elevations mostly under 2500 feet (760 m). The epithet translates from the Latin as "tailed," an allusion to the tips of all the leaflets except the terminal one. It is unfortunately also called dwarf sugar palm.

This small densely clumping species shows remarkable variation of leaf morphology. The stems grow to a maximum height and width of 6–7 feet (1.8–2.1 m); they are thin, wiry, reclining, and graceful. The leaves are mostly 3 feet (90 cm) long, with slender, wiry petioles 18 inches (45 cm) long. The 4–10 leaflets range from linear to rhomboid; the nonlinear leaflets have an irregularly toothed margin and all but the terminal one have a "tail" of varying length. All forms have silvery undersides. The inflorescences, although mostly hidden by the leaves, bear creamy white, fragrant flowers of both sexes, the female ones producing pendent spikes of small, beautiful egg-shaped, brilliant red fruit. The variability of leaflet form has led to the naming of most of these configurations, often with numbers, and some with cultivar names.

This palm can withstand little cold and is adaptable only to zones 10 and 11, although protected specimens survive in 9b. It does not like drought and it needs a humus-laden, free-draining, acidic soil. Filtered sunlight is best, but the palm does reasonably well in full sun, especially in cooler but frostless Mediterranean climates. Alas, this fine landscape subject is slow growing even under ideal conditions.

It must be planted in an intimate setting to be seen, a requirement that can be difficult to achieve due to the palm's small stature. It excels as a giant groundcover in partially shaded sites and is exquisite as a border, as a specimen in a border, or alongside a path in the woods. It makes a good house plant if given bright light, regular irrigation, and good air circulation.

Arenga engleri
a-REN-ga · ENG-ler-eye
PLATES 56–57

Arenga engleri is native to Taiwan, where it grows in moist forests of the low mountains. Common names include dwarf sugar palm and Formosa palm. The epithet honors the great German taxonomist H. G. Adolf Engler (1844–1930).

This medium-sized clustering species has trunks to 10 feet (3 m) tall and clumps to 15 feet (4.5 m) tall and wide. The trunks are covered, except in their oldest parts, with a tightly adhering, woven net of black fibers, the older parts being light gray to light tan with distinct rings of leaf base scars. The leaves are 9 feet (2.7 m) long on short, stout petioles and exhibit a twist in their orientation from the top of the trunk. The leaflets are to 2 feet (60 cm) long, thick and stiff, linear (except for the apical ones), briefly lobed near the middle and irregularly jagged ended at the tips; an inward fold in the center of the leaflet gives it a slight V shape. Leaf color is medium green above but silvery beneath. The orange and yellow, fragrant flowers are borne in branched, pendent inflorescences, which grow from the leaf base scars, starting at the uppermost ones until the small, purplish round fruits mature on the lowest ones, at which point the entire stem dies. The fruits are ½ inch (13 mm) in diameter.

The species is slightly susceptible to lethal yellowing disease. It thrives in zones 9b through 11 and is even possible in 9a with occasional protection; sheltered and well-protected specimens are growing in 8b. It grows moderately fast with adequate water and light but is slow in dry soils. It thrives in clay soils and sandy ones and, even though it grows on mostly limestone soils in habitat, it luxuriates in the acidic clay soils of southeastern Texas where it is often cultivated under pine trees with a thick mulch of needles. It prefers partial or dappled shade but performs amazingly in full sun, even in hot climates if provided with enough irrigation. Suckers may be carefully removed and planted if they are rooted.

A large well-grown specimen looks good in the center of an expanse of lawn if the mass of trunks is judiciously thinned out to better reveal the appealing form of the stems. It is wonderful as a giant hedge as well as a centerpiece among lower-growing palms or other vegetation. Few palms look as good as large potted specimens, but *A. engleri* must have good light under such circumstances. It is equally at home in well-lit, airy greenhouses, atriums, or conservatories.

Arenga hastata
a-REN-ga · has-TAH-ta
PLATE 58

Arenga hastata is a small, clumping species indigenous to low mountainous rain forest of peninsular Malaysia and western Borneo. The epithet is Latin for "hastate," although the reference is obscure as no part of the leaf is shaped like an arrowhead. The palm forms clumps to 4 feet (1.2 m) high (usually less) and 5 feet (1.5 m) wide. The light-colored stems are thin and wiry, and the leaves are 2 feet (60 cm) long and pinnate, with 10 diamond-shaped, toothed leaflets; the leaflets of the terminal pair are distinctly wedge shaped with squared-off apices. The spiked inflorescences are erect and carry light orange flowers that form tiny pink or red globose fruits. The plant is tropical in its requirements and adaptable to zones 10 and 11. It needs partial shade and a moist, humus-rich, free-draining soil.

Arenga hookeriana
a-REN-ga · hoo'-ker-ree-AH-na
PLATES 59–60

Arenga hookeriana occurs naturally in southern Thailand and northern peninsular Malaysia, where it is an undergrowth subject in monsoonal tropical forests. It, or at least the sought-after form, is vulnerable in habitat and, although widespread in its distribution, is not common. The epithet honors British botanist Sir Joseph D. Hooker (1817–1911), former director of the Royal Botanic Gardens, Kew.

This small, clumping species seldom exceeds 3 feet (90 cm) high in habitat. Under cultivation and with ideal growing conditions, a clump may reach 6 feet (1.8 m) high and wide. The leaves are 1–2 feet (30–60 cm) long and 9–10 inches (23–25 cm) wide. They are variable but, in the type, are undivided and have an unusual diamond shape with deeply lobed margins, each point between the lobes ending in a short "tail." They are shallowly pleated and deep emerald green above with a beautiful silvery hue beneath. The leaf may also be truly pinnate, with 5 leaflets, only the terminal one having the above characteristics, the other 4 looking like the linear leaflets of one form of *A. caudata*. The inflorescences are seldom-branched spikes 2 feet (60 cm) long and bear orange or yellow male and female flowers. The small, round red fruits are ¼ inch (6 mm) in diameter but are sometimes heavy enough to cause the whole infructescence to nod or even droop.

This species is closely allied to *A. caudata* and hardly distinguishable from some forms of it. In fact, Henderson (2009) united the 2 species under the name *A. caudata*. Further research will be needed to determine if these palms are indeed one polymorphic species.

This beauty is adapted to zones 10 and 11. It requires at least partial shade, especially in hot climates, but grows in a range of well-drained soils. Because of its size and nearly unique leaf form, no other palm needs more judicious placement in the landscape: it cries out for an intimate site where it can be viewed up close. It is an excellent choice for a protected, partially shaded patio, as a specimen along a path, and as a large groundcover. It's hard to imagine a finer container plant. Indoors it needs good light, regular, adequate irrigation, and good air circulation.

Arenga micrantha
a-REN-ga · my-KRAN-tha

Arenga micrantha is indigenous to northeastern India, southern Bhutan, and Tibet, where it grows on steep slopes in broad-leaved subtropical mountainous forests at 4500–7000 feet (2100 m). The epithet is from 2 Greek words meaning "tiny" and "flower."

The trunks of this medium-sized sparsely clustering species grow to 10 feet (3 m) high with a diameter of 6 inches (15 cm). The leaf crown is hemispherical, with a few leaves 6–8 feet (1.9–2.4 m) long on petioles 2 feet (60 cm) long. The leaves are ascending when new but spreading when older. The leaflets are regularly arranged along the rachis, 12–18 inches (30–45 cm) long, narrowly wedge shaped, with a slightly jagged apex. They are medium dull green above with the usual silvery gray hue beneath. New leaves are V shaped with their leaflets growing from the rachis at an angle, but older leaves are flat, their leaflets growing from the rachis in a single plane. The species is unusual in that it seems to be dioecious with male and female inflorescences on separate trees.

The species was described in 1988 and is relatively new to cultivation. It is reportedly the most cold-tolerant species in the genus. It may even be adaptable to regions with hot, wet summers and dry winters. It needs a humus-rich soil.

Arenga microcarpa
a-REN-ga · my'-kro-KAHR-pa
PLATE 61

Arenga microcarpa occurs naturally in the Moluccas and on New Guinea, where it grows in mostly swampy regions. The epithet translates from the Greek as "small fruit."

The trunks of this large clumping species are relatively thin, reach 25 feet (7.6 m) tall, and are covered in their upper parts in dense black fibers, the lower parts dark green to tan and prominently ringed. The massive leaves are 12 feet (3.6 m) long and 5 feet (1.5 m) wide, with petioles 6 feet (1.8 m) long. The leaflets are 2–3 feet (60–90 cm) long and linear, with broadened but abruptly pointed tips; they are thick, tough, and evenly spaced along the rachis, which is 8 feet (2.4 m) long. Near the base of the leaves the leaflets are usually in groups and grow from the rachis at different angles, while the outer leaflets lie flat in one plane and are not arranged into groups. Leaflet color is deep green above but silvery beneath. The inflorescences are pendent panicles mostly hidden by the leaves and they carry fragrant, purplish flowers. The small, rounded fruits are red.

The palm requires tropical conditions and is adaptable only to zones 10b and 11. It is a true water lover and requires a rich soil in which it grows moderately fast. While luxuriating in full sun, it grows well in partial shade. It can be propagated most easily by separating the suckers.

This spectacular species works as an isolated specimen as well as a dominant feature of massed vegetation, to which it adds an ineluctable accent. It is unsurpassed as a gigantic hedge or barrier marking property lines or as part of a wall of great foliage, which conditions give a wonderful sense of lushness. Its visual impact is that of a large *A. engleri* specimen. It would be a candidate only for large well-lit atriums or conservatories.

Arenga obtusifolia
a-REN-ga · ahb-too′-si-FO-lee-a
PLATE 62

Arenga obtusifolia occurs naturally in southern Thailand, peninsular Malaysia, Java, and Sumatra, where it grows in rain forest along the coast and the foothills of the mountains. The epithet is Latin for "obtuse" and "foliage," a reference to the bluntly shaped leaflets.

A large, mostly clustering species, it has trunks 1 foot (30 cm) in diameter and up to 50 feet (15 m) tall; they are covered in the typical network of strong, black fibers (some of which are long and pendent) except for the oldest parts which are smooth, light colored, and prominently ringed. The leaves are among the largest in the genus, to 18 feet (5.4 m) long, on petioles 3 feet (90 cm) long. The linear leaflets to 4 feet (1.2 m) long are dark green above and silvery beneath, thick and leathery, with wavy margins and rounded but toothed apices; they sometimes grow at different angles from the rachis, especially near the base of the leaf, but are always pendent. The branched inflorescences grow from the leaf base scars on the trunk and bear many white flowers. The mature fruits are egg shaped, greenish, and 2 inches (5 cm) long.

This palm grows mostly in swampy regions and needs constant and abundant moisture, a rich soil, and sun. It does not tolerate frost and is possible only in zones 10b and 11. Under ideal conditions, it grows reasonably fast. The species is magnificent in almost any site but especially as a canopy-scape because of its usually widely separated stems. As part of a wall of foliage, it is unsurpassed. There are no data relevant to growing this palm indoors, but it requires lots of space and light.

Arenga pinnata
a-REN-ga · pin-NAH-ta
PLATE 63

Arenga pinnata has been in cultivation for so long that its exact origin is uncertain but is probably the rain forest of western Indonesia. The most popular common name is sugar palm, as the trees are tapped for the sap from which sugar, syrup, and alcoholic beverages are made. The epithet is Latin for "pinnate," which describes the leaves of all *Arenga* species.

The sugar palm is one of 3 solitary-trunked species in the genus. The trunk is 2 feet (60 cm) in diameter and 50 feet (15 m) tall, covered in its younger parts with adherent black fibers but light colored in its older parts, which usually bear the persistent leaf bases. Several fibers at each leaf base are long, stiff, and spiny. The leaves are 30 feet (9 m) long on petioles 6 feet (1.8 m) long; stiffly erect, the leaves never lie below the horizontal plane. The great leaflets are 3 feet (90 cm) long and grow at different angles from the rachis, giving the leaf a plumose effect. They are linear, slightly jagged at their apices but smooth on their margins, and are dark green above but silvery beneath. In addition, they are limp and mostly drooping, which lightens the stiff visual aspect of the erect blades. The branched inflorescences are 7 feet (2.1 m) long and bear purplish flowers with a strange aroma. The fruits are oblong, greenish yellow, and 2–3 inches (5–8 cm) long.

The species tolerates various soils, full sun to partial shade and, while a water lover, is also capable of withstanding some drought. The foliage does not tolerate temperatures below 28°F (−2°C), but the palm usually survives from temperatures in the low 20s (around −7°C), leafing out again when growth resumes. It is a fast grower if given plenty of water and a rich soil.

This palm looks good in almost any site. It should not be planted singly in areas that must be neat as the entire tree dies over a period of a year or two once it starts flowering. Otherwise it is large and spectacular enough to be sited alone surrounded by space, and it is magnificent as a large canopy-scape. It is rarely grown indoors.

Arenga porphyrocarpa
a-REN-ga · por-fy′-ro-KAHR-pa
PLATE 64

Arenga porphyrocarpa is a small, clumping species native to Java and Sumatra. The epithet is from Greek words meaning "red" or "purple" and "fruit." The palm attains a maximum height and width of 6 feet (1.8 m). It is similar to *A. caudata*, with variously lobed leaflets carried on thin, limp petioles and rachises. Leaf color is the usual green above and grayish beneath. It thrives in partial shade or full sun, likes regular, adequate moisture, and cannot endure frost.

Arenga ryukyuensis
a-REN-ga · ree-yoo′-kyoo-EN-sis

Arenga ryukyuensis is, as the epithet indicates, from the Ryukyu Islands of southern Japan. It was formerly confused with *A. engleri* but recognized as a distinct species in 2006. *Arenga ryukyuensis* is more robust with larger leaflets than *A. engleri*. The stems are up to 6 feet (1.8 m) tall and 8 inches (20 cm) in diameter. The leaves are up to 6 feet long with narrow linear leaflets that are briefly lobed only near their tips. The leaflets are strongly nerved on the upper surface; the undersides are silvery. The round ¾-inch (19-mm) fruits are orange or reddish. This palm can be used in the landscape like *A. engleri* and has similar cultural requirements. A moist, rich soil in partial shade to full sun will produce luxuriant growth.

Arenga tremula
a-REN-ga · TREM-yoo-la
PLATE 65

Arenga tremula is endemic to the Philippine Islands, where it grows in clearings in the rain forest. The epithet translates from the Latin as "trembling," an allusion to the thin leaflets.

This clumping species is similar to *A. engleri*. The trunks may attain 12 feet (3.6 m) of height and a diameter of 3–4 inches (8–10 cm). They are covered in their youngest parts with the typical netting of black fibers but are smooth, green, and prominently ringed in their older parts. The leaves are similar in size and general shape to those of *A. engleri*, but the leaflets are usually more widely spaced. They are also more limp and drooping (the longer ones even pendent) and usually a lighter green above and below. The inflorescence is a tall spike that usually rises above the mass of foliage but is sometimes pendent, and the small blossoms are greenish white to green. The rounded ½-inch (13-mm) red fruits are carried on the pendent spike.

The species is adaptable to partial shade or full sun but needs a humus-laden, moist, well-draining soil. It is not as cold tolerant as *A. engleri* and is adaptable only to zones 10 and 11, although healthy specimens are found in protected sites in 9b. Suckers may be removed and transplanted. This palm serves the same landscape uses as *A. engleri* but has an overall more delicate, fernlike appearance. The only mitigating factor for indoor use is the palm's need of abundant light.

Arenga undulatifolia
a-REN-ga · un'-dyoo-lat-i-FO-lee-a
PLATE 66

Arenga undulatifolia is found in rainforest clearings of the Philippine Islands, Borneo, and western Indonesia, where it usually grows on sloping ground in the foothills of mostly limestone mountains. The epithet translates from the Latin as "undulating" and "foliage," a reference to the wavy margins of the leaflets.

Plants in nature may be solitary-trunked or densely clustering specimens. The trunks reach 20 feet (6 m) high with total clump height of 30 feet (9 m) and a width of 20 feet (6 m) or sometimes slightly more. The stems are covered in a mat of dark fiber until nearly mature at which time the fiber falls off, leaving a smooth dark green beautifully ringed stem. The leaves are to 10 feet (3 m) long with relatively widely spaced leaflets that are 18–24 inches (45–60 cm) long, linear-oblong, generally flat or not drooping, and lobed and wavy on the margins. Leaflet color is deep dark green above and silvery green to white beneath. The leaf petiole is short and thick. The branched inflorescences grow from the leaf bases and are short and mostly hidden by the foliage. The flowers are greenish white, and the rounded 1-inch (2.5-cm) fruits are brown.

This palm is of easy culture in tropical or nearly tropical regions with sun, water, and a reasonably fertile, well-draining soil. A large species, it can serve the same general landscape use as *A. microcarpa* but is more vibrant and less somber (but no more beautiful) because of the shape of the leaflets and their silvery bottoms that lend an air of movement to a clump even in still air. It is not known to be grown indoors, but it should do well in very large atriums or conservatories with sun, high humidity, and abundant moisture.

Arenga westerhoutii
a-REN-ga · wes'-ter-HOOT-ee-eye
PLATE 67

Arenga westerhoutii occurs naturally over a wide area of the Asian tropics, from northeastern India, Thailand, Myanmar, peninsular Malaysia, Cambodia, Laos, and adjacent southern China, where it grows in clearings, often forming colonies, in the rain forest on usually limestone hills. The epithet honors J. B. Westerhout, a 19th-century Dutch assistant to the British resident councilor in the Malaysian state of Malacca and superintendent to the province of Naning (near Malacca).

This tall massive palm is one of largest species in the genus. The trunks are to 2 feet (60 cm) in diameter and covered in old leaf bases but usually lack adherent fibers; they attain heights of 60 feet (18 m), sometimes more. The leaf crown may be 20 feet (6 m) tall and 40 feet (12 m) wide. The leaves are 15–20 feet (4.5–6 m) long, flat but exhibit a twist as they grow from the trunk, are mostly ascending, and are held above the horizontal. The many narrowly lanceolate leaflets are regularly spaced along the rachis and are 3 feet (90 cm) long. Leaf color is deep, bright green, sometimes almost bluish green, above and silvery olive green to silvery brown beneath. As with *A. pinnata*, the fibers adherent to the younger leaf bases produce several long, stiff, and sharp spines. The leaves mostly hide the inflorescences, which are 2 feet (60 cm) long, branched, and proceed from near the top of the trunk downwards, growing from the leaf base scars. Separate male and female flowers are produced on the same plant. The small, deep red flowers produce round black fruits ½ inch (13 mm) in diameter.

This wonderful species cannot tolerate frost and is mostly limited to zone 11 but is marginal in 10b. It is a water lover but can withstand mild drought. The palm is not particular about soil type as long as it is well draining, but it grows faster and lusher in a rich soil.

Although it often looks disheveled and messy, this palm is potentially beautiful: it has the size and nobility of *A. pinnata* but is more graceful with its great flat and twisted, bicolored leaves. It is spectacular enough to stand isolated in space but looks better in groups of 3 or more individuals of varying heights. It also looks better when protected from strong winds.

Arenga wightii
a-REN-ga · WY-tee-eye

Arenga wightii is a moderate-sized, sparsely clumping species from the foothills of the Western Ghats of India, where it grows in monsoonal forests from elevations of 500 to 3000 feet (150–900 m). It is under threat of extinction because of agricultural expansion. The epithet honors Robert Wight (1796–1872), Scottish surgeon and botanist, who was the director of the botanic garden in Madras. The trunks grow to heights of 12 feet (3.6 m) and are densely covered in hairy fibers. The leaves are to 15 feet (4.5 m) long on petioles 3 feet (90 cm) long. The silver-backed leaflets grow from the rachis in a flat plane and have obliquely truncated apices and shallowly lobed margins. This palm is not particular about soil type as long as it is well drained. It needs a sunny site and is not hardy to cold, being adaptable to zones 10 and 11. Sap from the inflorescences is reportedly used to make an alcoholic beverage.

Asterogyne
ass'-te-ro-JY-nee

Asterogyne is composed of 5 small pinnate-leaved, monoecious palms in Central America and northern South America. All but one species are solitary trunked, and all have undivided leaves that are deeply bifid at their tips. The inflorescences are either spicate or branched once with a few rachillae. The flowering branches carry small, white or nearly white male and female blossoms, which produce red fruits. Only one species is common. The genus name is derived from the Greek words for "star" and "female," an allusion to the shape of the staminode (sterile stamens) in the female flowers. Like all the genera in the tribe Geonomateae, *Asterogyne* has small seeds that, if allowed to dry out completely, do not germinate well. Fresh seeds of Geonomateae germinate within 30–120 days, rarely longer, and seeds shipped in a bag for a couple of weeks often begin germinating before reaching their destination. Fresh seeds of *Asterogyne* almost always germinate within 45 days.

Asterogyne martiana
ass'-te-ro-JY-nee · mahr-tee-AH-na
PLATES 68–69

Asterogyne martiana has the widest natural distribution of the 5 species and is the only one that is not threatened. It ranges from southern Belize and eastern Guatemala, eastern and northern Honduras, eastern Nicaragua, Costa Rica, Panama, and western Colombia to northwestern Ecuador, where it is an undergrowth subject in low mountainous rain forest. The epithet honors the German botanist, Carl F. P. von Martius (1794–1868), author of the magnificent *Historia Naturalis Palmarum*.

The palm forms, with age, a stout, solitary trunk to 6 feet (1.8 m) high and 2 inches (5 cm) in diameter. Mature palms have up to 18 leaves, each of which is 3 feet (90 cm) long, generally wedge shaped, but with the apex deeply cleft into 2 large, pointed segments. Leaf color is emerald green above and a lighter shade beneath with a darker colored heavy midrib. New leaves are often rosy or coppery colored. The blade is also grooved with prominent transverse veins corresponding to the hidden pinnate segments. The leaves are mostly ascending from the trunk and mostly not held beneath the horizontal. The funnel-shaped crown of leaves captures falling debris and leaves, which decay and release nutrients that are taken up by the palm. Large specimens in nature have a skirt of brown dead leaves. The inflorescence is a spike with terminal branches of orange-brown flowers. The fruits are ½ inch (13 mm) long, dark purplish, and egg shaped.

This species is unusually sensitive to waterlogged soils, in which the roots often rot. It needs a humus-rich medium and partial shade. It does not tolerate frost, being adaptable only to zones 10b and 11. It also needs protection from wind if its elegant undivided leaves are to remain unsplit.

The palm demands an intimate site where its unusual elegance can be appreciated up close. Until it is old, it is wonderful along a curving shady path or as a large groundcover in partial shade. It is one of the best candidates for homes, greenhouses, atriums, or conservatories if given high relative humidity and bright light.

Asterogyne spicata
ass'-te-ro-JY-nee · spi-KAH-ta
PLATE 70

Asterogyne spicata is restricted to Guatopo National Park in northern Venezuela, where it grows in low mountainous rain forest. The epithet is Latin for "spiked," an allusion to the unbranched inflorescence. This palm is much taller than *A. martiana* but has similar vegetative characters. Its new leaves are usually reddish coppery, while its mature leaves are emerald green above and silvery beneath.

Astrocaryum
ass-tro-KAH-ree-um

Astrocaryum includes 36 very spiny, pinnate-leaved, monoecious palms in Mexico, Central America, and South America. A few are clumping species but most are solitary trunked, and a few have subterranean stems. Some have plumelike leaves because the leaflets grow from the rachis at different angles; many others have flat leaves with leaflets in one plane. All the leaflets are silvery or white beneath. The juvenile plants of many species have entire leaves, but the adults invariably have segmented leaves. The inflorescences grow from the leaf sheaths and are accompanied by a large persistent, spiny paddle-shaped bract. The flowering branches bear male and female blossoms, are usually short, and are formed at the end of long peduncles which elongate and become pendent as the fruits mature. The fruits are formed in clusters and are yellow to brown, mostly globular, and mostly spiny.

These species are of exceptional beauty at every stage of growth, have stout and vicious spines on most of their anatomy, and are difficult to handle; they are not recommended for planting in high traffic areas. Although still rare and difficult to find, they are worth the effort. Many, mostly the smaller species, are undergrowth rainforest plants and do well in shade or partial shade, while others are more adapted to the sun.

The genus name translates from the Greek as "star" and "nut," a reference to patterns on the endocarp. Two species, *A. alatum* and *A. mexicanum*, are sometimes placed in their own genus, *Hexopetion*. Seeds of *Astrocaryum* can take up to 3 years to germinate. For further details of seed germination, see *Acrocomia*.

Astrocaryum aculeatissimum
ass-tro-KAH-ree-um · a-kyoo'-lee-a-TIS-si-mum
PLATE 71

Astrocaryum aculeatissimum is indigenous to the southeastern coast of Brazil, where it grows in the rain forest and is threatened because of the felling of trees. The epithet is Latin for "very spiny."

It is a mostly clustering species, although a few plants form only one stem. The trunks grow to 30 feet (9 m) high and 6 inches

(15 cm) in diameter, and are covered in rings of black spines. The leaves are 10 feet (3 m) long and contain many linear leaflets 1 foot (30 cm) long and regularly spaced in a single plane along the spiny rachis. The pendent inflorescences are 2 feet (60 cm) long and bear creamy white blossoms. The round 1-inch (2.5-cm) fruits are brown but covered in black spines.

This species is slightly hardier to cold than most others in the genus but is still adaptable only to zones 10 and 11. It is a water lover that nevertheless needs a fast-draining, humus-laden soil. It grows in partial shade or full sun.

The palm is neither elegant nor magnificent enough to be planted in the middle of a lawn but looks good as the taller component of a wall of vegetation and as a small canopy-scape. It would be nice as a small focal point along a walk or a street were it not for its complete spininess. No data are available, but this is probably not a good candidate for a house plant and must be judiciously placed even in atriums and conservatories.

Astrocaryum alatum
ass-tro-KAH-ree-um · a-LAH-tum
PLATE 72

Astrocaryum alatum is indigenous to the rain forest of southern Nicaragua, the Caribbean coast of Costa Rica, and Panama. The epithet is Latin for "winged," but the allusion is unexplained. A synonym is *Hexopetion alatum*. This solitary-trunked palm grows to 20 feet (6 m) tall and is easily confused with *A. mexicanum*; *A. alatum* usually has slightly larger leaves and is, if possible, even more beautiful. It has the same cultural requirements as *A. mexicanum*.

Astrocaryum chambira
ass-tro-KAH-ree-um · chahm-BEE-rah

Astrocaryum chambira is indigenous to southeastern Colombia, northeastern Ecuador, northern Peru, and northwestern Brazil, where it grows in lowland rain forest and clearings therein. The epithet is the aboriginal name in Brazil. The stems of this solitary-trunked species grow to 100 feet (30 m) high in habitat but are usually 40 feet (12 m) tall in cultivation. It is rare in cultivation outside its habitat, where it is grown for the leaves from which a durable fiber is extracted. Its magnificence should certainly be more widely experienced.

Astrocaryum mexicanum
ass-tro-KAH-ree-um · mex-i-KAH-num
PLATE 73

Astrocaryum mexicanum is a solitary-trunked species from the tropical Gulf coast of Mexico, southern Belize, northern Guatemala, northern El Salvador, the northern coast of Honduras, and northernmost Nicaragua, where it grows mostly in clearings in low mountains rain forest. The epithet is Latin for "of Mexico," a reference to the palm's primary habitat. *Hexopetion mexicanum* is a synonym.

The spiny trunk grows to 20 feet (6 m) tall with a diameter of 3 inches (8 cm) and is generally free of leaf bases and indistinctly ringed. The leaf crown is 15 feet (4.5 m) tall and wide. Juvenile leaves appear simple but are pinnate, the leaflets essentially fused together unless separated by wind, each 8 feet (2.4 m) long on very spiny petioles 2 feet (60 cm) long. The leaflets of older leaves are separated and unequal in width. Their color is yellow-green to a deep grassy green, and they grow from the spiny rachis in one plane, giving a flat appearance to the leaf. The inflorescence is erect but mostly hidden among the leaves. It emerges from a rounded, dark brown, spiny large spathe. The blossoms are whitish and fragrant. The oblong brown fruits are 2 inches (5 cm) long and covered with black spines.

This palm demands a fertile, fast-draining but constantly moist soil. It does not tolerate frost but thrives in gloom when young and in partial shade or full sun when older. It grows moderately fast under ideal conditions.

This exceptional small palm is among the most beautiful in the family, especially when young. It needs a protected, partially shaded spot to keep its gorgeous young leaves from splitting and it is unsurpassed as a patio or close-up subject. Planted in groups of 3 or more individuals of varying heights there is hardly a more beautiful landscape subject. It is easy as a large houseplant or greenhouse subject but needs constant moisture and humidity.

Every part of this palm except the roots is used by indigenous peoples in the palm's habitat: the flowers and seeds for food, the leaves for thatch, and the trunks for construction.

Astrocaryum murumuru
ass-tro-KAH-ree-um · moo'-roo-MOO-roo
PLATE 74

Astrocaryum murumuru is indigenous to a vast area of Amazonia in South America, where it grows along rivers and in swamps. The epithet is an aboriginal name. A mostly solitary-trunked species, it produces an occasional clumping individual. Henderson et al. (1995) reported that some individuals have only subterranean trunks; when arborescent, the trunks can grow to 40 feet (12 m). The most distinctive characteristic of this palm is its magnificently long leaves, to 20 feet (6 m), with their evenly spaced, long, limp leaflets that are bright to deep green above but silvery or even pinkish (because of the red hairs in one variety) beneath. They are ascending and erect and form a crown shaped like a shaving brush.

This palm is tropical in its temperature requirements, is a true water lover, and needs a fertile soil. It requires partial shade when young but can endure the full tropical sun when older. It is truly magnificent because of the immense leaves, looks especially good in colonies, and lends its magnificence to any pond or stream planting.

Astrocaryum standleyanum
ass-tro-KAH-ree-um · stand'-lee-AH-num
PLATE 75

Astrocaryum standleyanum is a tall, solitary-trunked species indigenous to southeastern Costa Rica, the Caribbean coast of Panama, western Colombia, and northwestern Ecuador, where it grows in lowland rain forest. The epithet honors Paul C. Standley (1884–1963), an American botanist, collector, and author of several important floras. The trunks can reach 40 feet (12 m) high with a canopy of arching, plumose leaves each 10 feet (3 m) long. This magnificent palm does not tolerate cold and needs a wet, if not soggy, rich soil with partial shade to full sun.

Astrocaryum vulgare
ass-tro-KAH-ree-um · vul-GAH-ree

Astrocaryum vulgare occurs naturally in northeastern Brazil, French Guiana, and Surinam, where it grows in rain forest and wet savannas. The epithet is Latin for "common" and refers to the palm's distribution, not its appearance. The large and sparsely clustering species has stems to 25 feet (7.6 m) and bears long black spines arranged in beautiful dark rings. The leaves are 10–12 feet (3–3.6 m) long and among the most handsome in the genus, with their widely spaced, beautifully arching linear, dark green leaflets growing in several planes from the rachis; the visual effect is plumose but not densely so. The varying heights of the trunks in any clump give a picturesque aspect. This palm is not tolerant of cold and is adaptable only to zones 10b and 11. It is a water lover and needs sun.

Attalea
at-ta-LAY-a

Attalea is composed of nearly 70 small to large pinnate-leaved, monoecious palms mostly in South America. They inhabit both wet and semiarid climates and vary from acaulescent or subterranean-trunked species to colossally tall and mammoth-trunked forms. The trunks of most species are covered in their younger parts with old leaf bases and, when old, are remarkably straight and columnar but not so relatively thick compared to their heights.

All trunked species are solitary stemmed and have remarkably similar, long leaves with long, narrow leaflets, the ones near the bottom of the blade limp and drooping. The petioles and the rachises are robust, stout, and usually light brown, gray, or yellowish, and the unfurled spearlike new leaves are large and imposing. The leaflets have a line of brown tomentum on one of their lower margins. The leaves of all the large trunk-forming species are erect but usually arching from the midpoint of the rachis, and they rarely descend beneath the horizontal, which characteristic gives a mammoth shuttlecock appearance to the mature palms. The inflorescences are large and accompanied by even larger grooved, woody bracts. The flowers in a given inflorescence are usually either all male or mostly female with a few male blossoms. In the past, differences in the male flowers were thought significant enough to warrant placing these palms into 6 different genera: *Attalea*, *Markleya*, *Maximiliana*, *Orbignya*, *Parascheelea*, and *Scheelea*. Today, all are treated as *Attalea*.

All the species are very slow growing, especially when younger, and are adaptable to many soils as long as they are fast draining. The seeds can take over a year to germinate, and many a seed has germinated among discarded potting soil. The dwarf species need shade as juveniles, but the large species need as much sun as they can get for faster growth and better appearance. None of them are hardy to cold or adaptable outside of zones 10 and 11. The growing point of the large trunk-forming species remains below ground for several years, and in this state the giant leaves may be damaged or even killed by cold in zone 9 but the palm usually grows back with the return of sustained warmth. It is not unusual to see gigantic specimens in zone 9b (or even 9a) with leaves springing straight from the soil to 20–30 feet (6–9 m) tall. The same leaf damage and survival phenomenon mostly applies to the subterranean-trunked species as well, with the proviso that, because of their less-robust forms, the survival rate from freezes may not be as high.

None of these species are commonly cultivated in the United States, although they are among the most magnificent palms in the family. The incredible size of the larger species makes their use in small gardens impossible. They create awe when planted along an avenue, and they dominate the landscape when planted as isolated specimens, but they are nearly unparalleled as a canopy-scape.

The trunk-forming species have several human uses in their habitats: the leaves for thatch, the trunks for construction, and the fruits and seeds for fodder and oil. The genus name honors Attalus III, who was king of Pergamon, an ancient Greek city in modern-day Turkey, from 138 BC to 133 BC.

Seeds of *Attalea* can take up to 5 years to germinate and do so sporadically. Some species readily develop 2 or more seedlings from the same seed (endocarp). For further details of seed germination, see *Acrocomia*.

Attalea allenii
at-ta-LAY-a · al-LEN-ee-eye
PLATE 76

Attalea allenii occurs naturally in the lowland rain forest of the Caribbean coast of Panama and the Pacific coast of Colombia. The epithet honors Paul H. Allen (1911–1963), an American botanist and specialist in the flora of Central America and tropical fruit. This interesting species has subterranean trunks and plumose leaves 15 feet (4.5 m) long and composed of thin, narrow leaflets, except for a few near the end of the leaf that are fused at their apices. The endosperm of the seeds is almost liquid and is eaten or drunk in the same manner as the coconut palm.

Attalea amygdalina
at-ta-LAY-a · a-mig'-da-LY-na
PLATE 77

Attalea amygdalina is endemic to valleys in the Colombian Andes, where it grows in the hilly rain forest. It is endangered because of expanding agriculture. The epithet is Latin for "little almond" and refers to the shape of the edible seeds. The species never forms an aboveground trunk, but the massive ascending and erect leaves reach lengths of 20 or more feet (6 m), only their tips arching. The large rachis holds regularly spaced deep green, linear-elliptic, limp leaflets growing in a single flat plane. Each leaflet is 2–3 feet (60–90 cm) long.

Attalea butyracea
at-ta-LAY-a · byoo-te-RAY-see-a
PLATE 78

Attalea butyracea is widespread from southeastern Mexico, through Central America, into Venezuela, Colombia, eastern Ecuador, eastern Peru, western central Brazil, and northwestern Bolivia. The epithet is Greek for "like butter" and refers to the oil extracted from the seeds. This species forms trunks to 60 feet (18 m) high, the overall height of the palm nearing 100 feet (30 m). In many parts of its range, it is the most visually dominant plant in the landscape, a towering canopy-scape above the jungle or savanna. The rachis of the giant leaves is twisted to the vertical from its middle to its apex, giving an irresistible appeal of frozen movement.

Attalea cohune
at-ta-LAY-a · ko-HOO-nee
PLATE 79

Attalea cohune is indigenous to the west coast of Mexico from the tropic of Cancer southwards and eastwards along the western coast of Guatemala and the southern (Pacific) coast of El Salvador; it is also found in the states of Chiapas and Quintana Roo in Mexico, Belize, northern Guatemala, northern Honduras, and northern Nicaragua. The common name is cohune palm. The epithet is the aboriginal name for the palm. This species differs little in its vegetative appearance from *A. butyracea* and has the marvelous midpoint twist to the vertical of its great leaves. It is the most available and widely planted *Attalea* species in the United States.

Attalea crassispatha
at-ta-LAY-a · kras'-si-SPA-tha
PLATE 80

Attalea crassispatha is endemic to southwestern Haiti, where it grows on the low limestone hills of dry savannas. It is critically endangered with only a couple dozen trees existing in 1990. This palm is also interesting in a geographical sense as it is the only *Attalea* species in the Antilles. The epithet is from Greek words meaning "thick" and "spathe." The trunk of this tall species attains 60 feet (18 m) high but 1 foot (30 cm) wide. The crown is, unlike that of almost all other species in the genus, nearly round. The leaves are 10–12 feet (3–3.6 m) long and have regularly spaced leaflets growing in a single flat plane. The rachis exhibits the characteristic twist in its apical half, giving this palm a look similar to that of a straight-trunked, massive coconut.

Attalea guacuyule
at-ta-LAY-a · gwa'-koo-YOO-lay
PLATE 81

Attalea guacuyule is a massive palm native to the Pacific coast of southern Mexico. The epithet is from the plant's aboriginal name. The tree grows to 60 feet (18 m) tall, with a clean, smooth, columnar trunk. The leaves are 15 feet (4.5 m) long and have a twist to the vertical in their apical half.

Attalea humilis
at-ta-LAY-a · HYOO-mi-lis
PLATE 82

Attalea humilis is another species endemic to the eastern coast of Brazil. The epithet is Latin for "humble" or "low." The subterranean trunk seldom grows above ground and, when it does, never attains more than 3 feet (90 cm) of height. The flat leaves are usually 10–15 feet (3–4.5 m) long and numerous.

Attalea maripa
at-ta-LAY-a · mah-REE-pa
PLATE 83

Attalea maripa is native to a large area of northern South America, including southern Venezuela; eastern Colombia, Ecuador, and Peru; northern Bolivia; northeastern Brazil, including Amazonas; and the Guianas. The epithet is one of the palm's aboriginal names in Ecuador. This palm grows to 80 feet (24 m) in overall height and has wonderfully large leaves. The long pendent leaflets grow from the rachis at different angles.

Attalea phalerata
at-ta-LAY-a · fal-e-RAH-ta
PLATE 84

Attalea phalerata occurs naturally in eastern Peru, northern Bolivia, and central to northeastern Brazil. The epithet is Latinized Greek for "adorned." The trunk is among the thickest in the genus, but its overall height is 30 feet (9 m). The leaves are typical, but the leaflets may grow from the rachis in either one plane or in different planes, and the leaf itself has the typical midpoint twist to the vertical.

Attalea speciosa
at-ta-LAY-a · spee-see-O-sa
PLATE 85

Attalea speciosa has a disjunct natural distribution in southern Guyana and Suriname, in central and eastern Brazil, and in northeastern Bolivia, where it grows in savannas, clearings in the rain forest, and along rivers and streams at low elevations. The epithet is Latin for "beautiful." This palm grows to 60 feet (18 m) overall, with a straight, columnar trunk which is mostly free of leaf bases. The leaves are 15 feet (4.5 m) long and show the characteristic midpoint twist to the vertical in their apical half. The species is among the most graceful in the genus.

Bactris
BAK-tris

Bactris comprises more than 70 pinnate-leaved, monoecious palms in tropical America. They inhabit every terrestrial biome of this region except the true desert and range from subterranean or short-trunked palms to gigantic and clustering forms. Almost all are heavily armed with needlelike black spines; a few species are noteworthy for their paucity of spines. The leaflets of most species grow from different angles of the rachis to give a plumose look to the leaf. The inflorescences have clusters of rachillae that bear yellow or white flowers of both sexes.

None of the species are tolerant of freezing temperature and, while most are not fussy about soil type, all do much better with a decent soil and adequate moisture. The genus name is from the Greek and translates as "cane" or "staff," an allusion to a use of the trunks of some of the smaller species (after removal of the spines). Seeds of *Bactris* can take up to 2 years to germinate and do so sporadically. For further details of seed germination, see *Acrocomia*.

Bactris brongniartii
BAK-tris · bron-YAHR-tee-eye
PLATES 86–87

Bactris brongniartii is indigenous to a vast area of the Amazon region in northern and northwestern Brazil, the Guianas, the southern half of Venezuela, eastern Colombia, eastern Ecuador, eastern Peru, and northern Bolivia, always in low, wet areas and along streams and rivers. The epithet honors Adolphe-Théodore Brongniart (1801–1876), a French botanist.

In habitat, this densely clustering species forms walls of foliage in open spaces. The stems grow to 25 feet (7.6 m) high but are usually shorter, 2 inches (5 cm) in diameter, and tan with darker rings of spines. The leaves are 5 feet (1.5 m) long with many linear-lanceolate, light green leaflets growing from the spine-covered rachis to create a plumose leaf. Leaf spines are flattened and yellowish brown, but black at the base and tip. The purple-black fruits are edible.

This palm is nearly aquatic and makes an impression along the edge of a pond or lake and as part of a wall of other vegetation. It can grow in partial shade but reaches its full potential only in full sun.

Bactris concinna
BAK-tris · kahn-SIN-na
PLATE 88

Bactris concinna is indigenous to a large area of the western Amazon region in southeastern Colombia, eastern Ecuador, northern and eastern Peru, northern Bolivia, and western Brazil, where it grows in lowland rain forest. The epithet is Latin for "elegant" or "orderly." This clustering species has stems up to 25 feet (7.6 m) tall but usually shorter. The trunks are covered in all but their oldest parts with a dense mat of grayish leaf base fibers and rings of ½-inch (13-mm) black spines. The leaves are 4–5 feet (1.2–1.5 m) long with numerous, regularly spaced, dark green, linear leaflets growing in a flat plane from the spiny rachis. The overall appearance of the leaves is akin to those of many of the larger *Chamaedorea* species. The black fruits are 1 inch (2.5 cm) long and edible. This palm grows in sun or shade and in almost any moist to wet soil.

Bactris gasipaes
BAK-tris · GAS-i-peez
PLATES 89–90

Bactris gasipaes is unknown in the wild, but probably originated in Central America in open areas of the rain forest. It is now cultivated throughout low, wet regions of tropical America. The epithet is derived from an aboriginal name for the palm. The common name is peach palm because of the edible fruit.

This large species is found as clustering and solitary-trunked specimens. The trunks can reach heights of 50 feet (15 m) or sometimes more. They are beautifully ringed, unusually straight, and relatively slender, never more than 1 foot (30 cm) in diameter, and are variously spiny. The rounded leaf crowns are 20 feet (6 m) tall and wide. Leaves are 10 feet (3 m) long or longer and are held on very spiny petioles 1 foot (30 cm) long. The numerous leaflets are 2–3 feet (60–90 cm) long, deep green, and mostly limp and drooping. They grow in small clusters from the spiny rachis at different angles to give a rounded, plumose aspect to the leaf; the leaflets of the terminal pair are often larger than the ones beneath them. The many branched inflorescences emerge from among the leaves, from a large, spiny, and persistent bract. The blossoms are cream colored. The edible egg-shaped fruits are 2 inches (5 cm) long and turn yellow, orange, or bright red when mature.

The palm needs warmth, water, and a rich, free-draining soil. It grows fastest in full sun and cannot withstand any but the lightest frost, making it adaptable only to zones 10 and 11, with protected specimens found in 9b. The palm resprouts from the roots in central Florida if frozen to the ground and grows quickly thereafter (David Witt, pers. comm.). It can be propagated by carefully removing the suckers.

This lovely palm looks good as a single specimen, but solitary-trunked forms are always better planted in groups of 3 or more individuals of varying heights. The clumping specimens are sumptuous with their varying tiers of plumose leaves. The species needs much space, warmth, and sun to do well, and its spininess mitigates against its use as a houseplant.

Few palms are as important to the local peoples: the fruits are still a staple in the diet of many Central and South American indigenous people and the growing point is widely harvested as a vegetable. Because of its importance as a crop, this species has many local cultivars throughout tropical America, including spineless forms and forms with seedless fruits. Some cultivars produce fruits high in starch, while others produce fruits high in oils. This species is believed to be domesticated from its wild ancestor, *B. gasipaes* var. *chichagui*.

Bactris grayumi
BAK-tris · GRAY-yu-mee
PLATE 91

Bactris grayumi is indigenous to the Caribbean coastal areas of Nicaragua and Costa Rica, where it grows in lowland tropical rain forest. The epithet honors Michael H. Grayum (1949–), American botanist specializing in the plants of Central America. The leaves are similar to those of *B. militaris*, but this species is often solitary trunked and the spines on the leaves are shorter. Its inflorescence is recurved, and the fruits are orange. This palm wants the same culture as *B. militaris* and is as beautiful.

Bactris hondurensis
BAK-tris · hahn-doo-REN-sis
PLATE 92

Bactris hondurensis is a clumping and solitary-trunked species in northeastern Honduras, eastern Nicaragua, the Caribbean coasts of Costa Rica and Panama, and into western Colombia. It is found in the undergrowth of rain forest in low mountainous areas. The epithet is Latin for "of Honduras." The trunks grow to 6–8 feet (1.8–2.4 m) high and, when clustering, are never dense. The leaves are undivided, 2 feet (60 cm) long, deeply bifid apically, and light green, with grooves corresponding to the cryptic pinnate segments; occasionally the segments are free from each other, especially near the base of the leaf. The underside of the leaf is covered with white hairs. This palm must be protected from strong sun and given plenty of moisture in a friable, free-draining, and humus-laden soil. It does not tolerate frost but does well in Mediterranean climates if given enough irrigation.

Bactris major
BAK-tris · MAY-jor
PLATE 93

Bactris major has a wide distribution, from southern Mexico, through Central America, into northern and eastern Venezuela, the Guianas, central Brazil, and northeastern Bolivia, where it grows in clearings in rain forest and in open savanna regions. The epithet is Latin for "larger." The 3 varieties differ mostly in the inflorescence and female flower morphology and do not seem important to the gardener (Henderson 2000).

The trunks of this large clumping species are covered in black spines until older, when they are shed to reveal dark green smooth stems with prominent white rings. The trunks grow to 30 feet (9 m) high with a diameter of 2 inches (5 cm). Total height of the clumps is 40 feet (12 m) with a width to 30 feet (9 m). The leaves are usually 8 feet (2.4 m) long on short, spiny petioles, with 36 light green narrowly oblong leaflets growing in a single plane from the intensely spiny rachis. Each leaflet sports a prominent lighter-colored midrib and may be tapering or blunt at its end; the 2 terminal leaflets are usually broader but shorter than the others. The much-branched inflorescences emerge from spiny bracts and contain small greenish yellow blossoms. The purple to black fruits are 2 inches (5 cm) long and reportedly edible.

This extremely spiny palm is adaptable to many well-drained soils. It loves sun but grows in partial shade and needs average but regular moisture. It does not tolerate frost and is adaptable to zones 10 and 11, although marginal in 10a. It is among the finest palms for a large hedge or screen and is exquisite as free-standing clumps or as a strong, large accent in masses of other tropical vegetation. The species is not known to be grown indoors, although given enough light, humidity, and room it should succeed.

Bactris militaris
BAK-tris · mil-i-TAR-iss
PLATE 94

Bactris militaris is a clustering palm indigenous to the Pacific slope of southwestern Costa Rica, in Puntarenas province, where it grows in low and swampy regions. It is rare in the wild. The epithet is Latin for "warrior," a reference to the armed leaves. A related species from Panama, *B. neomilitaris*, is sometimes treated as a subspecies of *B. militaris*. The trunks grow to 10 feet (3 m) high and 2 inches (5 cm) in diameter in sparse clumps to 15 feet (4.5 m) wide. The unsegmented leaves are remarkable: to 10 feet (3 m) long, dark green, narrow, thick, and grooved (to correspond with the cryptic pinnae); they grow stiffly erect from the trunk and have a distinct cinnamon-hued midrib, along which are long, stiff, black widely spaced spines. The olive-shaped fruits are orange-red. This beauty is undeservedly rare in cultivation. It needs warmth, constant moisture, and protection from the hot midday sun and from desiccating, shredding winds.

Bactris plumeriana
BAK-tris · ploo-mer´-ee-AH-na
PLATE 95

Bactris plumeriana is indigenous to Hispaniola, where it grows along the margins of hilly, evergreen forests and in cleared areas. The epithet honors Charles Plumier (1646–1704), explorer and royal botanist to Louis XIV of France. In the past, this species was confused with 2 other Antillean species, namely, *B. cubensis* from Cuba, and *B. jamaicana* from Jamaica.

The mature, slender trunks of this clustering species reach 25 feet (7.6 m) high and 6 inches (15 cm) in diameter. The dark stems are made darker by the whorls of long, black spines covering all but the oldest parts. The leaves are 8 feet (2.4 m) long and covered with black spines, with about 50 leaflets growing from the spiny rachis at different angles to create a fully plumose leaf. Fruits are bright red.

The palm can form large, dense clumps and is deadly looking but also extremely beautiful. It is hard to imagine a more effective hedge subject, and the crowns are wonderful canopy-scapes.

Balaka
ba-LAH-ka

Balaka is a genus of 7 small, solitary-trunked, pinnate-leaved, monoecious palms in the island groups of Fiji and Samoa. All species have thin trunks and slender, delicate crownshafts, and all are found in the undergrowth of the rain forest. The leaflets may grow from the rachis at an angle to give a V shape to the leaves. The tips of the leaflets are irregularly jagged. The inflorescences grow from beneath the crownshafts and are open, branched twice, bearing both male and female blossoms. The small, ovoid fruits are red.

These palms are visually similar to and related to *Ptychosperma* and *Drymophloeus*. Few palm groups are more fragile looking. Unfortunately, their fragility and delicacy are real. These lovely divas resist cultivation, promptly expiring if their exacting needs are not met. All species are tender to cold and adaptable only to zones 10b and 11. They need partial shade, protection from strong winds, pure water, and a rich, moist but very fast-draining soil. They demand perfect conditions and may not be good candidates for growing indoors. The genus name is a Latinized form of the aboriginal name. Seeds of *Balaka* germinate within 30 days. For further seed germination details, see *Areca*.

Balaka diffusa
ba-LAH-ka · dif-FYOO-sa
PLATE 96

Balaka diffusa of Fiji is similar to *B. microcarpa*, although it has long been confused with *B. macrocarpa*. Most palms in cultivation under the name *B. macrocarpa* are *B. diffusa* (Hodel 2010). The leaflets are linear. Fruits are borne along the entire length of the rachillae. The epithet, meaning diffuse, alludes to the open, spreading inflorescence.

Balaka longirostris
ba-LAH-ka · lahn-jee-RAHS-tris
PLATE 97

Balaka longirostris is native to Fiji. It grows to an overall height of 25 feet (7.6 m) with an exceedingly thin trunk usually less than 2 inches (5 cm) in diameter. The crownshaft is almost the same diameter as the trunk and is olive to dusky green. The leaf crown is sparse with 3–6 ascending, erect, and partially arching leaves each 5 feet (1.5 m) long. The widely spaced light green leaflets are wedge shaped. The epithet is Latin for "long beaked," an allusion to the shape of the endocarps.

Balaka microcarpa
ba-LAH-ka · myk-ro-KAHR-pa
PLATE 98

Balaka microcarpa of Fiji can grow to 30 or more feet (9 m), and the exceptionally thin and straight trunk, 3 inches (8 cm) in diameter, gives the tree an ethereal aspect. The crownshaft is deep green and slightly bulging at its base. The leaf crown is sparse and has erect, ascending, arching leaves with short petioles. The leaflets are widely spaced, grow from the rachis at an angle that creates a V-shaped leaf, and are linear. Fruits are borne along the entire length of the rachillae and are less than ½ inch (13 mm) long. The epithet is from Greek words meaning "small" and "fruit."

Balaka seemannii
ba-LAH-ka · see-MAHN-nee-eye
PLATE 99

Balaka seemannii is endemic to Fiji. The epithet honors Berthold C. Seemann (1825–1871), a German botanist and plant explorer. This palm grows to an overall height of 20 feet (6 m) with a thin and delicate, prominently ringed trunk of 1–2 inches (2.5–5 cm) in diameter. The crownshaft is slender and dark green. The leaf crown usually consists of no more than a half-dozen glossy dark green leaves, each one 6 feet (1.8 m) long. The leaflets are wedge shaped.

Balaka tahitensis
ba-LAH-ka · tah-hee-TEN-sis

Balaka tahitensis is native to Samoa. Its epithet, meaning "of Tahiti," is a misnomer and was applied to a mislabeled specimen. This graceful beauty grows to 10 feet (3 m) tall, with linear leaflets and fruits less than 1 inch (2.5 cm) long.

Barcella
bahr-SEL-la

Barcella is a monotypic genus of pinnate-leaved, monoecious palm in the Amazonas region of Brazil, growing along the banks of the Río Negro in very sandy scrublands. The origin of the genus name is unknown but may be a Latinized form of a village (Barcelos) near where the palm was discovered (Henderson et al. 1995). Seeds of *Barcella* can take more than a year to germinate and do so sporadically. For further details of seed germination, see *Acrocomia*.

Barcella odora
bahr-SEL-la · o-DOR-a

Barcella odora is a trunkless palm with beautifully arching pinnate leaves growing directly from the ground to 6 feet (1.8 m) long. The epithet is Latin for "odorous," an allusion to the fragrant male flowers. The leaflets are 2 feet (60 cm) long, limp and pendent, medium to dark green, linear-lanceolate, with a long tapering apex, and are evenly but widely spaced along the arching rachis. The inflorescences grow from the leaf crown, are branched once and erect, and bear all-male blossoms or both male and female. The ovoid fruits are 1 inch (2.5 cm) long and turn orange when mature. This spectacularly beautiful small palm is probably not in cultivation but should be. It is related to the oil palms (*Elaeis* spp.) but has unarmed petioles. The species doubtless requires tropical conditions and copious moisture but would probably thrive in partial shade as well as full sun. It probably does not require a particularly rich medium.

Basselinia
bas-se-LIN-ee-a

Basselinia is composed of 12 monoecious species endemic to New Caledonia. Some are rare with ranges restricted to one mountain or valley. All sport crownshafts and have pinnate leaves, but the segments of some species are undivided and terminally bifid; a few species may have either pinnate or undivided leaves. In nature, the species range from 12 to 70 feet (21 m) tall and may be solitary trunked or clustering. Harold E. Moore, Jr., and Natalie Uhl divided the genus into 2 sections, one for the larger, solitary-trunked species with green crownshafts, the other for the smaller palms with colored crownshafts. The inflorescences grow from beneath the crownshafts and are branched once or twice, bearing both male and female flowers.

Almost all the species are exceptionally beautiful, and some bear colorful, intricately marked crownshafts. None are common in cultivation, and some are virtually unknown. Most are maddeningly slow-growing. The genus name honors Olivier Basselin (1403–1470), a French poet. Many New Caledonia palms seem to flower well in cultivation but rarely produce viable seeds, and this genus is no exception. Germination is difficult for *Basselinia* species, and no definitive data are available, except that germination seems to be sporadic.

Basselinia deplanchei
bas-se-LIN-ee-a · de-PLAN-shee-eye
PLATE 100

Basselinia deplanchei occurs in the undergrowth and in full sun in low, wet mountainous regions of central and southern New Caledonia. The epithet honors French botanist Émile Deplanche (1824–1875), who explored New Caledonia.

This species is among the most variable in the genus and similar in general appearance to *B. gracilis*. It is mostly clustering, but some individuals have solitary trunks. It can grow to 20 or more feet (6 m) in habitat with a slender, dark-colored, and strongly ringed trunk, atop which a small crown of mostly stiffly ascending pinnate leaves grows. Some individuals have undivided and apically bifid leaves. Leaf color is light green with prominent midveins in the undivided leaf. The segmented leaves have lanceolate leaflets, with prominent and lighter-colored multiple midribs. Each leaf is 3 feet (90 cm) long on a petiole 6 inches (15 cm) long. The crownshaft is usually swollen near its middle and red, orange, or purple with darker striations. The inflorescences grow from beneath the crownshaft and are stiffly spreading and usually the same color as the shaft. They bear tiny flowers. The small, round fruits are black when mature.

This palm needs uninterrupted and copious water in a free-draining soil. It flourishes in partial shade or full sun except in hot climates where it should be protected from the hottest midday sun. It has a modicum of cold tolerance and should do fine in frostless or nearly frostless Mediterranean climes if given enough moisture.

Basselinia favieri
bas-se-LIN-ee-a · fah-vee-AY-ree
PLATE 101

Basselinia favieri is endemic to rain forest on Mount Panié in northeastern New Caledonia, at 1000–1700 feet (300–520 m). The epithet honors Joseph Favier, who assisted Harold E. Moore in collecting the type.

The mature stem of this solitary-trunked species attains a height of 30 or more feet (9 m). It is dark tan, with darker rings of leaf base scars, and is topped by a light green to light greenish gray crownshaft that is 3 feet (90 cm) long and slightly swollen at its base. The leaves are to 9 feet (2.7 m) long, ascending but gracefully arching, and bear many linear-acuminate emerald green leaflets each 4–5 feet (1.2–1.5 m) long. In newer leaves, the leaflets grow from the rachis at an angle that gives a slight V shape to the leaf. Older leaflets are limper and allow the leaf to assume an almost flat cross-section. The large spidery, many-branched rose-colored inflorescences grow from beneath the crownshaft. The small fruits are round and black when mature.

This palm needs constant moisture in a fast-draining, humus-laden soil. It is not hardy to cold but reportedly withstands light frosts undamaged and is adaptable to zones 10 and 11. It needs partial shade when young and always in hot climates. It is beautiful enough to serve well as anything but an isolated specimen. In groups of 3 or more individuals of varying heights, it is magnificent, and its choice form makes it a perfect patio subject.

Basselinia glabrata
bas-se-LIN-ee-a · glah-BRAH-ta
PLATE 102

Basselinia glabrata comes from wet forests of New Caledonia, frequently along stream banks. The epithet is from Latin and translates as "smooth" or "hairless," a reference to the glabrous crownshaft. This species was formerly known as *Alloschmidia glabrata*.

This palm grows to a height of 40 feet (12 m) in habitat. The trunk is no more than 6 inches (15 cm) in diameter and is light gray, except for the newest growth that is tan, and indistinctly ringed. The beautiful olive green crownshaft is 2 feet (60 cm) tall and not much wider than the trunk. The leaves are borne on short petioles, are 6 feet (1.8 m) long, and have 24 pairs of slightly S-shaped, light green leaflets, each of which is 2 feet (60 cm) long, and all of which grow from the whitish rachis at a slight angle, resulting in an almost flat leaf. The species exhibits an unusual flowering habit with several inflorescences in various stages of maturity on the upper part of the trunk, commencing beneath the crownshaft. Each inflorescence bears both male and female flowers and consists of thin, ropelike, grayish green and pendent branches with tiny greenish blossoms. The elliptical fruits are 4 inches (10 cm) long and become black when mature.

The palm needs protection from the midday sun, especially in hot climates. It requires a moist, fast-draining soil. It is probably intolerant of frost, but no data are available to confirm this suspicion. This slender, graceful but slow-growing palm is elegance personified. It would seem best planted where its comeliness can be seen against a contrasting background, such as in a patio or as a tableau of specimens towering above lower-growing vegetation under high trees.

Basselinia gracilis
bas-se-LIN-ee-a · GRAH-si-lis
PLATES 103–104

Basselinia gracilis occurs naturally on the island of New Caledonia, where it grows in rain forest at elevations from sea level to 5280 feet (1610 m). The epithet is Latin for "graceful."

Most individuals are clustering. The trunk varies from 5 to 20 feet (1.5–6 m) tall and from 1 to 4 inches (2.5–10 cm) in diameter but is always smooth, free of leaf bases and fibers, and distinctly ringed. The color ranges from gray or green to brown and black. The crownshafts are 4–18 inches (10–45 cm) tall and they range from green to pink, orange, red, purple, brown, or black. The leaves may be entire and deeply bifid at the apex, with strong and lighter-colored veins, or completely pinnately segmented; the latter condition is associated with leaflets that may be few, wide, oblong, and S shaped with long tips or may be many, linear, and slightly S shaped. The segmented leaves are always spreading but never arching, and the leaf crown may be elongated. Inflorescences grow from beneath the crownshaft and are red to brown or purple, with stringy branches. The round ½-inch (13-mm) fruits are deep red to black when mature.

Some of the palm's vegetative characters are so inconstant that it is nearly impossible for the nonbotanist to realize that one species is being discussed. Interestingly, the leaf forms and trunk heights are related to elevation in the palm's habitat: those forms with segmented leaves and the tallest trunks are found at the lowest elevations, with a continuum of gradual diminution of leaf segmentation and trunk height at higher elevations.

The most important cultural requirement is regular, copious water. The species does not tolerate full sun in hot climates and usually thrives in partial shade in all but cool and nearly frostless Mediterranean climes, where it reportedly withstands light frosts unscathed. The little trees are slow growing but are adaptable to several soils as long as they are well drained and not too acidic. All forms of this species are eminently suited to intimate sites and are perfect as small to medium-sized patio subjects or, because of their rich colorings, contrasting components of other vegetation.

Basselinia humboldtiana
bas-se-LIN-ee-a · hum-bolt′-tee-AH-na
PLATE 105

Basselinia humboldtiana occurs naturally in wet cloud forest at 2500–3000 feet (760–900 m) on New Caledonia. The epithet refers to Mt. Humboldt in the palm's habitat.

This solitary-trunked palm grows to an overall height of 30 feet (9 m) with a uniform trunk diameter of 4 inches (10 cm). The slender trunks are usually light tan and show closely set darker rings of leaf base scars. The crownshaft is 18 inches (45 cm) tall, mostly cylindrical, slightly wider than the trunk, but sometimes bulging near the base of the shaft. It is covered in a dense light brown or whitish felt. The leaf crown is tight and compact because of the short petioles, which hold mostly ascending but slightly spreading light green leaves, each 5 feet (1.5 m) long and strongly arching but never descending beneath the horizontal plane. The leaflets are uniformly closely spaced, lanceolate, and slightly S shaped, growing from the rachis at an angle that gives a V-shaped leaf. The species is not known in cultivation outside of a few botanical gardens and is reportedly slow growing.

Basselinia pancheri
bas-se-LIN-ee-a · pahn-SHER-ee
PLATES 106–107

Basselinia pancheri occurs naturally in rain forest from sea level to 3000 feet (900 m) on New Caledonia. The epithet honors Jean Armand Isidore Pancher (1814–1877), a French botanist who collected plants on the island.

This variable species is usually found as a solitary-trunked specimen. The stem can attain a height of 25 feet (7.6 m) with a diameter of 4 inches (10 cm) and is mostly green with widely spaced brown rings of leaf base scars, especially in the younger parts; it ages to light gray. The brown to purplish brown crownshafts are 18 inches (45 cm) tall and bulging at their bases. The leaf crown varies in size and shape and may be almost tiny or relatively large, with leaves 1–4 feet (30 cm to 1.2 m) long. They may be pinnately segmented or entire and apically bifid, the segmented ones with leaflets that are generally wide and S shaped. In all cases they are dark green above, paler green beneath, and thick and leathery with deeply corrugated veins.

This attractive species warrants wider use in gardens. It is easily grown in tropical or nearly tropical climates if given copious, regular moisture and a well-drained, humus-rich soil that is not too acidic. It flourishes in partial shade and full sun except in hot, dry climates.

Basselinia tomentosa
bas-se-LIN-ee-a · to-men-TO-sa
PLATE 108

Basselinia tomentosa occurs in wet forests on ridge tops at 3000 feet (900 m) in southern central New Caledonia. The epithet is Latin for "tomentose" and refers to the feltlike, hairy covering on the crownshaft.

This solitary-trunked species grows to 50 feet (15 m) tall and 10 inches (25 cm) in diameter. The gray trunk is marked with closely set, ridged rings of leaf base scars. The crownshaft is 3 feet (90 cm) tall, wider than the trunk, nearly uniformly cylindrical, and covered with a gray or purplish gray tomentum. The leaf crown is dense but describes less than a semicircle because of the stiffly ascending leaves that exhibit little arch. They are 6–7 feet (1.8–2.1 m) long on short, almost nonexistent petioles. The leaflets are 3 feet (90 cm) long, evenly spaced, deep emerald green, stiff, rigid, and lanceolate. They grow from the rachis at a steep angle that creates a distinctly V-shaped leaf. The spidery inflorescences are deep yellow or golden brown.

Basselinia velutina
bas-se-LIN-ee-a · vel-loo-TEE-na
PLATE 109

Basselinia velutina occurs in wet forests at 1200–5280 feet (370–1610 m) in northeastern New Caledonia. The epithet is Latin for "velvety" and refers to the tomentum on the crownshaft. The stems of this solitary-trunked species are tan and grow to 30 feet (9 m). The crownshaft is 3 feet (90 cm) tall and is gray, reddish, or brown because of its feltlike covering. The leaf crown is full, dense, and rounded. The greatly arching, almost recurved leaves are 8–9 feet (2.4–2.7 m) long with many narrow, rigid, long-tipped, medium green leaflets, each 3–4 feet (90–120 cm) long and growing from the rachis at an angle that creates a V-shaped leaf. The form and silhouette are beautiful and, were it not so slow growing, this palm would probably be more widely cultivated. It is reportedly frost resistant.

Basselinia vestita
bas-se-LIN-ee-a · ves-TEE-ta
PLATE 110

Basselinia vestita occurs in the undergrowth of wet forest at 3000 feet (900 m) in central New Caledonia. The epithet is Latin for "clothed," an allusion to the feltlike hairs on the crownshaft and leaf underside.

This small, clustering species attains 10 feet (3 m) maximum height. The little trunks are 1 inch (2.5 cm) in diameter and dark colored with indistinct lighter-colored rings of leaf base scars. The crownshafts are 6 inches (15 cm) tall, bronzy orange to almost red, with dark, curved striations of purplish black feltlike tomentum; they are narrow at their bases but bulging near their summits. The leaf crown is a semicircle of 6 tightly packed leaves less than 1 foot (30 cm) long on short, almost nonexistent petioles. The 4–6 leaflets per leaf are so congested along the short rachis that the leaf looks almost palmate; indeed, some leaves are unsegmented except for their deeply bifid apices. The leaves are a deep, almost bluish green above and are pleated with deep veins, but are a gorgeous silvery purple beneath because of their feltlike covering.

This little palm is near perfect for an intimate site where it can be seen up close. It is, alas, almost completely unknown in cultivation.

Beccariophoenix
be-kahr′-ee-o-FEE-nix

Beccariophoenix is a genus of solitary-trunked, pinnate-leaved, monoecious palms of Madagascar. The name is derived from the surname of great Florentine palm taxonomist Odoardo Beccari (1843–1920) and *Phoenix*, which when used in combined forms signifies a pinnate-leafed palm (rather than the date palm, specifically). Some growers have coined the moniker of window palm for this genus, which would be apt were the name not already applied to most *Reinhardtia* species. It is sometimes called giant windowpane palm because on young plants the leaflets nearest the petiole are partially separated at their bases, creating narrow "windows" on that part of the blade. Seed germination is fairly quick for this member of the tribe Cocoseae. Seeds germinate at the same time and within 60–90 days or occasionally longer.

Beccariophoenix alfredii
be-kähr′-ee-o-FEE-nix · al-FRE-dee-eye
PLATES 111–112

Described in 2007, *B. alfredii* is found in the seasonally dry interior of Madagascar at an elevation of 3400 feet (1046 m), growing in sandy soils along streams and watercourses. The specific epithet honors Alfred Razafindratsira, the Malagasy nurseryman who first brought attention to this remarkable palm.

The trunk is up to 40 feet (12 m) tall and 1 foot (30 cm) in diameter, grayish brown and clean, with closely spaced leaf scars. The lower leaves of the crown do not hang below the horizontal, and the leaves twist so the leaf blade is turned on edge. Each leaf is a shiny medium green, up to 14 feet (4.3 m) long, and borne on a short petiole. The leaflets are regularly arranged in one plane. The inflorescences are borne below the leaves on short peduncles not exceeding 5 inches (13 cm) long. The inflorescence has a horse-tail shape, bearing a cluster of 30–50 rachillae on which are borne the yellowish white male and female flowers. The flattened-spherical fruits are purplish red when ripe.

This palm is new to horticulture but probably is at least as hardy to cold as the better-known *B. madagascariensis*; it should do well in zones 10b and 11 and probably even in 10a. If we take our cue from its natural habitat, we should site this palm in sandy soils with abundant moisture. It should have some shade when young but can take full sun as an adult. Like its sister species, it makes a fine accent specimen, grown on its own where its stature and size will inspire awe.

Beccariophoenix madagascariensis
be-kahr′-ee-o-FEE-nix · mad-a-gas′-kar-ee-EN-sis
PLATES 113–114

Beccariophoenix madagascariensis occurs naturally in low, mountainous rain forest, where it grows in sandy soil. It is critically threatened because of development, expansion of slash-and-burn agriculture, and the felling of trees for their edible "hearts of palm." Were it not for the wide dissemination of seed in the 1990s, the species would be almost extinct. The epithet is Latin for "of Madagascar."

The trunk attains a maximum height of 40 feet (12 m) and a diameter of 1 or more feet (30 cm) in habitat. It is covered, except in its oldest parts, with large, persistent leaf bases and tightly woven brown fibers. The grass green leaves are 15 feet (4.5 m) long and grow almost directly from the top of the trunk with little or no petiole, although there is a "false petiole" of old fibers adherent to the robust rachis. The leaflets are regularly arranged along the rachis, linear-lanceolate, undulating on their margins, 3 feet (90 cm) long in mature palms, and usually shortly bifid at their apices. They are bright yellow-green to deep green above and grayish or silvery beneath. In young plants most of the leaflets remain unsegmented, but the leaflets nearest the petiole are partially separated. The leaflets of mature palms are pendent, whereas those of young plants are stiff. John Dransfield described the inflorescences thus: "When this tree is in bud (and it often is) there seem to be torpedoes poking out of the crown; these are the extraordinary peduncular bracts at the tips of long peduncles" (Dransfield and Beentje 1995). When the bracts open, the much more prosaic flowering branches, 2 feet (60 cm) long, emerge with bright yellow blossoms. The fruits are 1 inch (2.5 cm) long, ovoid with pointed ends, and purplish brown.

This palm is not as tender to cold as was first believed; it thrives in zones 10b and 11 and is marginal in 10a. It grows in sandy soils with abundant moisture and does well in partial shade, especially when young, but can endure sun even at that age. Both species of *Beccariophoenix* are exceptional even as isolated specimens, especially when younger and exhibiting their "leaf windows." When mature, they are magnificent: their great leaves resemble those of the largest *Attalea* species, and their crowns are among the most impressive canopy-scapes. Young plants do fine in a greenhouse with enough light and moisture, and older plants would seem to be good candidates for large conservatories and atriums.

Bentinckia
ben-TINK-ee-a

Bentinckia includes 2 tall, graceful, pinnate-leaved, monoecious palms in India and the Nicobar Islands. They are similar to *Archontophoenix* species and have prominent crownshafts beneath their leaf crowns. The inflorescences grow from nodes beneath the crownshaft and consist of spreading branches bearing both male and female blossoms. The fruits are red to purplish black when mature. The genus name honors Lord William Henry Cavendish Bentinck (1774–1839), British statesman and Governor-General of India. Seeds of *Bentinckia* germinate within 60 days. For further details of seed germination, see *Areca*.

Bentinckia condapanna
ben-TINK-ee-a · kahn-da-PAN-na
PLATE 115

Bentinckia condapanna is endemic to southern peninsular India in the evergreen forests of the Western Ghat and Palni Hills, where its growing point is so relished by elephants and humans alike that the plant now exists only in the most inaccessible places. The epithet is the Malayan name for the species.

This palm grows to an overall height of 40 feet (12 m) with a slender, smooth trunk no more than 7 inches (18 cm) in diameter. The wax-covered, deep green to tan cylindrical crownshaft is hardly wider than the trunk but is 3 feet (90 cm) tall. The sparse leaf crown holds 6–8 leaves at a time, none of which usually lie beneath the horizontal plane. The leaves are 5–6 feet (1.5–1.8 m) long, spreading, and arching, with many stiff, dark green, apically bifid leaflets, each 2 feet (60 cm) long, growing from the rachis at an angle that gives a V-shaped leaf. The inflorescences have a pinkish cast when new and bear beautiful deep red ovoid fruits that are 3 inches (8 cm) long.

This species is rare in cultivation, which is a great shame as its silhouette is among the world's most beautiful. It should be much more widely propagated, not only for ornamental purposes but also because of its threatened existence in habitat.

Bentinckia nicobarica
ben-TINK-ee-a · ni-ko-BAH-ri-ka
PLATES 116–117

Bentinckia nicobarica is endemic to the Nicobar Islands in the Bay of Bengal, where it grows on low hills in rain forest near the coast. It is threatened because of the lamentably rapid clearing of the forests and the harvesting of the palm's growing point, considered a delicacy. The epithet means "of the Nicobar Islands."

The smooth, light brown to deep gray trunk grows to 50 feet (15 m) tall and is ringed with circles of whitish leaf scars. The sparse but beautifully rounded leaf crown is 12 feet (3.6 m) wide and tall. The slender, columnar crownshaft bulges slightly at its base, is 6 feet (1.8 m) tall, smooth, and light grayish green. The deep green leaves are 6–8 feet (1.8–2.4 m) long on short petioles, with many linear leaflets growing from the rachis in one plane; the rachis itself usually has a slight twist which gives a special appeal to the leaf, similar to the leaves of *Archontophoenix* species. The leaflets are 18–24 inches (45–60 cm) long, slightly limp and drooping. Although the rachis arches and curves gracefully, the leaf crown is relatively sparse and most of the living leaves do not descend below the horizontal plane. The many-branched, pendent inflorescences grow in a ring from beneath the large crownshaft. The round ½-inch (13-mm) fruits are red to black.

This lovely species needs a nearly tropical climate, being adaptable only to zones 10b and 11. It is a water lover but not a swamp dweller and needs a humus-rich, free-draining soil. It is fast growing in partial shade and even faster in full sun. It is similar in form to *Archontophoenix cunninghamiana* and serves approximately the same landscape use, but has a more rounded leaf crown (the leaves descend below the horizontal) and is even prettier than *A. cunninghamiana* as a silhouette. The tree is dramatic in groups of 3 or more individuals of varying heights.

Bismarckia
bis-MAR-kee-a

Bismarckia is a monotypic genus of large palmate-leaved, dioecious palm. The name honors the German chancellor, Otto E. L. von Bismarck (1815–1898). Like all the genera in the tribe Borasseae, *Bismarckia* is remote germinating. Fresh seeds of Borasseae generally begin to germinate within 1–4 months but can take up to a year or more. Depending on warmth, a month or more can elapse between the time the germination stalk first emerges to when the first leaf emerges from the ground. Seeds of *Bismarckia* can take 1 to 18 months to germinate. The germination stalk can go down as far as 12 inches (30 cm), so a container at least 16–18 inches (41–45 cm) deep should be used.

Bismarckia nobilis
bis-MAR-kee-a · NO-bi-lis
PLATES 118–119

Bismarckia nobilis is endemic to northern and western Madagascar in open grasslands. The epithet is Latin for "noble." The common name is Bismarck palm.

The gray to tan or brown trunk is slightly swollen at its base, 12–18 inches (30–45 cm) in diameter, and at most 80 feet (24 m) tall for old specimens in habitat but half that for cultivated individuals. It is free of leaf bases for most of its height and is ringed with closely set grooves, the indentations of former leaf bases. The full, rounded to oblong leaf crown is 25 feet (7.6 m) tall and 20 feet (6 m) wide. Each leaf is carried on a petiole 6–8 feet (1.8–2.4 m) long with a few short spines along its margins and covered with a white waxy substance and patches of cinnamon-colored scales. The leaves are costapalmate and exhibit a wedge-shaped hastula on the upper side at the juncture of blade and petiole. To 10 feet (3 m) in diameter, the nearly round leaves are divided to one-third the width of the blade into 20 or more stiff, tapering segments and are colored yellowish green to grayish green or even bluish gray, often with a thin red margin. Long, pale fibers hang from between the leaf segments (fibers are absent in the similar-looking *Latania loddigesii*). The long, pendent inflorescences bear tentacle-like branches with tiny brownish flowers. The fruits are 1½-inch (4-cm) brown spheres.

Bismarck palm grows fastest in full sun and has better leaf color there. It needs a perfectly draining medium but is not particular about the actual components. It is more cold hardy than the horticulturally similar *Latania loddigesii*. Its hardiness to cold is partially dependent on the color of its leaves: bluer color indicates a more hardy leaf—but note that this is only the leaf; the growing points (trunks and roots) are unaffected by leaf color. In general, the blue leaves are safe in zones 9b through 11, while the greener (especially the pale green forms) are more tender to cold, zone 10a being usually the limit. Seedlings of the bluish- or silvery-leaved forms are easily identified by their purplish or maroon young leaves, whereas the plants destined to remain green are green in their seedling stage as well (David Witt, pers. comm.). The palm is drought tolerant but grows faster and bigger with regular, adequate moisture. It is difficult to transplant plants when young and, as such, cannot be done other than with container-grown specimens. Older plants should be root pruned a few months before the scheduled move, and even then great care must be taken to not break the delicate growing point by jostling the plant during the move.

Because of its dimensions and beauty, Bismarck palm is wonderful as an isolated specimen surrounded by space. It is even more glorious in groups of 3 or more individuals of varying heights. This is a tree for large landscapes and it looks crowded in small gardens. Nothing but banyans and other giant trees are more imposing in the landscape. Although rarely grown indoors, it could be tried in large, sunny spaces with good air circulation.

Borassodendron
bo-ras'-so-DEN-drahn

Borassodendron is composed of 2 solitary-trunked, palmate-leaved, dioecious palms in southern Thailand, peninsular Malaysia, and Borneo. They are closely related to *Borassus* species in details of inflorescences, flowers, and fruits but grow in rain forest rather than in savanna. Whereas *Borassus* leaves are similar to other palmate-leaved genera like *Sabal* and *Washingtonia*, *Borassodendron* leaves are palmate with deep splits between the segments, have a delicate appearance, and are more similar to *Itaya* and *Chelyocarpus*. The inflorescences grow from among the leaf bases, the males a long thick, pendent branch with secondary branches of wormlike catkins of flowers, the females unbranched and spikelike with large blossoms bunched at the apex, each flower subtended by a bract. The hard, glossy fruits are large and rounded.

Both species are slow growing even under optimum conditions. The genus name combines *Borassus* with the Greek word for "tree," a reference to the affinity of this genus to *Borassus*. Seeds generally germinate within 60 days but can take as long as 120 days. Seeds that are allowed to dry out before sowing do not germinate as well as those kept slightly moist. The germination stalk can go down 8 inches (20 cm), so a container at least 12 inches (30 cm) deep should be used. For further details of seed germination, see *Bismarckia*.

Borassodendron borneense
bo-ras'-so-DEN-drahn · bor-nee-EN-see
PLATE 120

Borassodendron borneense is endemic to the rain forest of Borneo. The epithet is Latin for "of Borneo." The palm is similar to the much more widely grown *B. machadonis* but is more ethereal looking and not as robust. It is undeservedly rare in cultivation.

Borassodendron machadonis
bo-ras'-so-DEN-drahn · mah-cha-DO-nis
PLATES 121–122

Borassodendron machadonis occurs naturally in southwestern Thailand and in northern peninsular Malaysia, where it grows in the wet forests of low, limestone mountains. It is nowhere abundant and is under threat in much of its range because of the harvesting of its edible growing point. The epithet honors A. D. Machado, a late 19th-century naturalist and ethnologist of Perak, Malaysia.

The trunk attains a height of 50 feet (15 m) in habitat but a diameter of less than 1 foot (30 cm); it bears conspicuous raised rings in its older parts and is dark gray to brown. The leaves are carried on stout, unarmed petioles, which are 6 feet (1.8 m) long and have hard, sharp, knifelike edges. The leaf is usually circular, and the segments are generally deeply cleft to the juncture of the petiole. The essentially palmate blade is 8 feet (2.4 m) wide, and the leaflets, which can number 60 or more, are typically deep green, linear, and apically blunt. Female flowers have a very strong, fruity fragrance that is more powerful than pleasant. The fruits are formed in tight clusters of large, purplish black globes.

This palm has a modicum of cold tolerance and thrives in zones 10b and 11, occasionally in sheltered locations in 10a. It is not particular about soil type as long as it is well drained. It needs regular, adequate moisture, protection from wind, and partial shade to full sun. Cold, dry winds are inimical to its appearance; indeed, strong winds of any temperature or moisture content tatter the leaves.

When mature, the tree has a relatively thin trunk that makes it a remarkably beautiful specimen. When young, the tree is adapted to close-up situations where its form can be appreciated. This species is a good candidate for large, well-lit atriums and conservatories with high relative humidity.

Borassus
bo-RAS-sus

Borassus comprises 5 large palmate-leaved, dioecious palms in monsoonal parts of tropical and southern subtropical Africa, Madagascar, southern Pakistan, coastal India, Southeast Asia, Malaysia, Indonesia, and New Guinea. These massive palms have large costapalmate leaves on thick, spiny petioles, and each leaf has a large hastula on both surfaces at the juncture of petiole and blade. Only 2 species are in general cultivation; others are found in botanical gardens. All the species have many human uses: the trunks in construction, the growing point as a vegetable, the sap for alcoholic beverages, and the leaves for making woven materials. The genus name is derived from the Greek word for the immature flower spike of the date palm, but its application to these palms is unclear.

Seeds generally germinate within 60 days but can take as long as 120 days. Seeds that are allowed to dry out before sowing do not germinate as well as those kept slightly moist. The germination stalk can go down 18 inches (45 cm), so a container at least 24 inches (60 cm) deep should be used. For further details of seed germination, see *Bismarckia*.

Borassus aethiopum
bo-RAS-sus · ee-thee-O-pee-um
PLATES 123–124

Borassus aethiopum occurs naturally in open, monsoonal woods and savannas of tropical Africa and northern Madagascar. The epithet is Latin for "of Ethiopia." *Borassus sambiranensis* is a synonym. The only common name is palmyra palm, but this moniker also applies to *B. flabellifer* and the whole genus.

Mature palms can attain a trunk height of 80 feet (24 m) and a trunk diameter of 3 feet (90 cm). The stems bulge slightly at their bases and usually near the middle of the trunk, and they usually have complete or incomplete sets of adherent leaf bases, sometimes at several separate areas. The leaf crown is 20 feet (6 m) wide and tall. The leaves are 10–12 feet (3–3.6 m) in diameter on stout petioles 6 feet (1.8 m) long. Near the trunk the petioles are brown and toward the blade they are yellowish (but black or yellow-brown in juveniles); they bear large recurved teeth along the margin. The nearly circular blade is stiff with non-drooping, tapering segments, which are separate for more than half their length. Leaf color is deep green on both sides. The heavy rounded or flattened fruits are 6–8 inches (15–20 cm) in diameter, orange to yellowish brown, and similar in size to a coconut, each one containing 1–3 large seeds.

The palm is not hardy outside of zones 10 and 11 but is marginal in warm microclimates of 9b. It is drought tolerant but looks better and grows faster with regular, adequate moisture. It is not particular about soil type as long as it is well drained, but it must have sun. This massive palm is not for small gardens. Mature trees are a dominant part of any landscape and are spectacular from a distance. As a canopy-scape, this palm is superb and tropical looking. It is not a good choice for indoors, even in atriums and conservatories.

Borassus flabellifer
bo-RAS-sus · fla-BEL-li-fer
PLATE 125

Borassus flabellifer appears indigenous to open savannas in India, southern China, Southeast Asia, and Indonesia. It is widely cultivated and its current distribution may be the result of human activity in the distant past. Common names are palmyra palm, toddy palm, and lontar palm. The epithet is Latin for "fan-bearing," an allusion to the palmate leaf.

Mature palms can attain a trunk height of 60 feet (18 m) in cultivation and a trunk diameter of 3 feet (90 cm). The stems are smooth, cylindrical, and dark gray, once the dead leaves have dropped off. The leaf crown is 25 feet (7.6 m) wide and tall. The leaves are 8–10 feet (2.4–3 m) in diameter on stout yellow petioles 6 feet (1.8 m) long, which bear large, black, irregular teeth along their margins. The blade is nearly circular and is dramatically costapalmate. The drooping, tapering segments extend to more than halfway through the depth of the blade. Leaf color is deep green on both sides, although the leaves of young plants often exhibit a decidedly bluish cast. The heavy, roundish fruits are 6–8 inches (15–20 cm) in diameter, blackish, and edible.

Toddy palm is slightly susceptible to lethal yellowing disease and is not hardy to cold outside of zones 10 and 11. It is drought tolerant when established but grows faster and looks better with regular irrigation. It is slow growing when young but moderately fast as it matures. It is adaptable to most well-drained soils. This massive palm is not for small gardens. Mature trees are a dominant part of any landscape and are spectacular from a distance. They are sometimes planted when young as if they would never grow to their natural dimensions. As a canopy-scape, this palm is superb and tropical looking. It is not recommended for indoor cultivation.

Not only the fruit but every other part of this palm has been used for millennia by the indigenous peoples of tropical Asia to whom the plant is secondary in importance only to the coconut; and it is of primary importance for some inland peoples where the coconut does not usually grow. It has been used for building materials, writing material (the leaves), baskets and mats, cordage, food, beverage, and pollen for beekeepers.

Borassus madagascariensis
bo-RAS-sus · mad-a-gas'-kar-ee-EN-sis
PLATE 126

Borassus madagascariensis is endemic to western Madagascar. The epithet means "of Madagascar." Like its sister species, it is a massive palm, with a bulging trunk and large leaves. The green or yellow-green petioles are well armed with fine, irregular, black teeth. The leaves are deep green on both surfaces. Each fruit is massive, 10–14 inches (25–36 cm) long, yellowish green, and tapers to a rounded point. While new to cultivation, this palm appears to thrive under the same conditions as *B. aethiopum* and *B. flabellifer*.

Brahea
BRA-hee-ya

Brahea consists of 11 palmate-leaved, hermaphroditic palms in Mexico and Central America, mostly in the drier regions. Most species are solitary trunked, but a few are clustering and one has a subterranean trunk. The weakly costapalmate leaves are stiff and usually not pure green, often bluish gray to bluish green. The leaves have a hastula at the juncture of blade and petiole, and one of the most salient characteristics of the genus is the size of the inflorescence, which often greatly exceeds that of the leaf crown. Ironically, the flowers are very small, even by palm standards, but are borne in groups along the fuzzy rachillae. The larger, arborescent species are similar to each other in appearance and similar to *Washingtonia robusta* in the landscape, although not as tall at maturity; the smaller, clustering species are unusual, especially in the color of their leaves.

There is a real need for detailed study of this genus. *Brahea* palms are probably one of the worst choices for indoor cultivation. They invariably need sun and good air circulation and are adapted to semiarid open situations. The genus name honors the Danish astronomer Tycho Brahe (1546–1601). Seeds germinate within 120 days and should not be allowed to dry out before sowing. For further details of seed germination, see *Corypha*.

Brahea aculeata
BRA-hee-ya · a-kyoo-lee-AH-ta
PLATE 127

Brahea aculeata is endemic to the desert of northwestern Mexico at low elevations. It is sometimes called hesper palm or Sinaloa hesper palm. The epithet is Latin for "prickly," but the species is no pricklier than any other.

The stem of this solitary-trunked palm grows to 15 feet (4.5 m) high and 8 inches (20 cm) in diameter. It is usually covered in old, dark gray leaf bases but, when clean, is cinnamon colored. The leaf crown is sparse, and the semicircular light green leaves are held on spiny petioles 2 feet (60 cm) long. The length of the inflorescences does not exceed the leaf crown. The flowers are white, and the fruits are round and shiny black.

This species is one of the most drought resistant in the genus and must have full sun from youth to old age. It prefers a calcareous soil but grows well in many very well drained soils. It is adaptable to zones 9 through 11 and is marginal in 8b but not in areas subject to wet freezes in winter. This airy, up-close subject looks good mixed with cactus or succulent plantings. Its size precludes its use as a specimen standing alone, but it looks good in groups of 3 or more trees of varying heights.

Brahea armata
BRA-hee-ya · ahr-MAH-ta
PLATE 128

Brahea armata occurs naturally in Baja California and the Sonoran Desert of northwestern Mexico, where it grows in canyons (where water is available) from sea level to 5200 feet (1580 m) in elevation. Common names are blue hesper palm and Mexican blue palm. The epithet is Latin for "armed."

The massive trunk reaches a height of 50 or more feet (15 m) and a diameter of 18 inches (45 cm); it usually has a slightly bulging base and closely set, indistinct rings, and is usually free of leaf bases. The leaf crown is full, rounded, and 15–18 feet (4.5–5.4 m) wide and 20 feet (6 m) tall. The circular leaves are distinctly bluish green, sometimes an icy blue, almost never pure green. The blade is 6 feet (1.8 m) wide and divided into as many as 50 tapering segments, each extending half way to the center of the blade and stiff to drooping on its ends. The petioles are 4–5 feet (1.2–1.5 m) long and have margins armed with vicious teeth. The inflorescences grow from among the leaves and are long, arching panicles of cream-colored flowers. They extend well beyond the leaf crown and, in younger palms, sometimes reach the ground. The round 1-inch (2.5-cm) fruits are black when mature.

Blue hesper palm demands full sun and a soil with perfect drainage. It does best in slightly calcareous soils but grows in any medium that is not very acidic. It is drought tolerant but grows and looks better when not stressed for long periods with drought. Growth rate is slow to moderately slow under any conditions. The palm is adaptable to zones 8b through 11 but is marginal in areas subject to wet freezes in winter. It is beautiful enough to be often planted as a single specimen, but it excels as a loud accent in a mass of vegetation where its bluish starburst of leaves lends an almost unmatched fillip to other forms and colors. It is stunningly spectacular in bloom.

Brahea brandegeei
BRA-hee-ya · bran-DEE-jee-eye
PLATE 129

Brahea brandegeei occurs naturally in low mountains of southern Baja California and the Sonoran desert area of northwestern Mexico, where it grows in canyons. Common names are San José palm, San José fan palm, and San José hesper palm. The epithet honors Townshend S. Brandegee (1843–1925), pioneering California botanist and the original collector of the species.

Trunks attain a height of 30–35 feet (9–11 m) and a diameter of 1 foot (30 cm). The stems usually retain most of the leaf bases, and these can be decorative. The leaf crown is 10 feet (3 m) wide and 8 feet (2.4 m) tall. The circular 3-feet (90-cm) leaves are held on petioles 3 feet (90 cm) long. Leaf color is yellow-green to deep green above but waxy grayish green beneath. The inflorescence is 4–5 feet (1.2–1.5 m) long and usually shorter than the leaf crown. The rounded 1-inch (2.5-cm) fruits are usually yellow and reportedly edible when ripe.

This species is adaptable only to zones 9 through 11. It is drought tolerant once established but grows more slowly and less robustly when stressed by drought. It is adaptable to various free-draining soils and does not need as much alkalinity as does *B. armata*. San José palm is distinguished by its trunks that are usually covered (except for the oldest parts) in a crisscrossing pattern of leaf bases.

Brahea calcarea
BRA-hee-ya · cal-CAH-ree-a
PLATE 130

Brahea calcarea is indigenous to the western coastal mountain areas of Mexico, from Mazatlán to Guatemala, where it grows in dry, open woods. The epithet is Latin for "limestone," which indicates that this species tolerates alkaline soil. A synonym is *B. nitida*.

The trunks attain a height of 30 feet (9 m) and a diameter of 1 foot (30 cm) in habitat. The leaf crown is dense with leaves 3 feet (90 cm) wide borne on unarmed petioles 2 feet (60 cm) long. The leaves are circular or nearly so, a deep, shiny green above and a lovely chalky white beneath. The great inflorescences far exceed the leaf crown and are similar to those of *B. armata*. The ripe fruits are yellow or orange.

The palm is adaptable to zones 9b through 11 in areas subject to wet freezes in winter but is probably a good candidate for 8b in the drier climes. It needs more water than many other species and grows on various well-drained soils including calcareous ones. It thrives in full sun or partial shade. This species is similar to and has the same landscaping uses as *B. armata*.

Brahea decumbens
BRA-hee-ya · dee-KUM-benz
PLATE 131

Brahea decumbens is endemic to the foothills of the Sierra Madre Oriental in northeastern Mexico, where it grows on limestone formations in scrublands. The epithet is Latin for "decumbent" or "reclining."

Trunks are no more than 6 feet (1.8 m) high and are usually prostrate to some extent. In addition, the palm is suckering and with time forms extensive thickets in habitat, but in cultivation a mature clump is no more than 15 feet (4.5 m) wide and 8 feet (2.4 m) tall. The hemispherical to nearly circular leaves are 2–3 feet (60–90 cm) wide. The leaflets are stiff. Juvenile leaves are green, but older plants have nearly blue leaves. They are held on petioles 2 feet (60 cm) long that show tiny teeth near their bases. The inflorescence panicles are shorter than the leaf crown. The fruits are unknown.

The species is drought tolerant and must have excellent drainage in a calcareous soil, although it seems to do well in any medium that is not very acidic. In dry climates it is hardy in zones 8 through 11 but is probably not as hardy to cold in areas subject to wet freezes in winter. Although rare, this small palm is desirable as a tall groundcover or as an accent in the cactus and succulent garden. The nearly unique color of its mature leaves is unexcelled.

Brahea dulcis
BRA-hee-ya · DOOL-sis
PLATE 132

Brahea dulcis occurs naturally in the foothills of the mountains of eastern and western Mexico, extreme southern Mexico, Guatemala, southern Belize, the mountains of El Salvador, and Honduras, and is the most widespread species in the genus. It grows in dry, open woods at 1000–5280 feet (300–1600 m) in elevation. Its common name is rock palm. The epithet is Latin for "sweet," a reference to the edible fruit.

The species is mostly a solitary-trunked palm, but some individuals are found clustering. Trunks attain 20 feet (6 m) of height and 9–10 inches (23–25 cm) in diameter; they are generally free of leaf bases and deep gray to brown with closely set indistinct rings. The leaf crown is 10 feet (3 m) wide and 10–12 feet (3–3.6 m) tall. The leaves are usually 3 feet (90 cm) in diameter, hemispherical, a dull, yellowish green or sometimes bluish green, and are held on spiny petioles 2 feet (60 cm) long. The leaflets are stiff. The inflorescences are hairy panicles 7 feet (2.1 m) long that extend beyond the leaves. The globular ½-inch (13-mm) fruits are light brown to greenish brown when mature and are reportedly edible and sweet.

As might be expected for a wide-ranging species, *B. dulcis* is indifferent to soil type as long as it is well drained and not too acidic. It is drought tolerant when established but grows faster and looks better with regular irrigation. It thrives in full sun in any climate and, when young, enjoys partial shade. Depending on the provenance, the palm is hardy in zones 9 through 11; individuals from the northeastern part of its range have more cold tolerance.

Except for the lack of blue coloration, the palm has the same landscape uses as does *B. armata*. When young and grown in partial shade, the leaves are exceptionally large and tropical looking. All parts of the palm are used in habitat: the trunks for construction, the leaves for thatch, and the fruits for food.

Henderson et al. (1995) combined several species under *B. dulcis*, but there is still some disagreement over the disposition of some names. *Brahea salvadorensis*, formerly included in *B. dulcis* by Henderson et al. (1995), is now recognized as a distinct species. The taxonomic confusion stems from the fact that this genus is poorly known. The palms are not adequately represented in museum reference collections. Moreover, the juvenile forms of this species are variable across wide geographic areas—as they indeed are in many palm species—and, in the past, separation of species was determined on this sort of phenomenon. Then there is the problem of regional variation. Further botanical studies are needed.

Brahea edulis
BRA-hee-ya · ED-yoo-lis
PLATE 133

Brahea edulis is endemic to the island of Guadalupe, off the northwestern coast of Baja California, where it grows in splendid isolation. The species has been rescued from the brink of extinction by the Mexican government, which removed the feral goats from the island. The goats were eating seeds and seedlings and threatened the survival of this palm, as well as the other unique plants of Guadalupe, destroying the fragile island ecosystem. The common name is Guadalupe palm. The epithet is Latin for "edible."

Trunks grow to 35 feet (11 m) high with a diameter of 14 inches (36 cm). They are deep brown to dark gray with closely set almost indistinct rings and are generally free of leaf bases. The leaf crown is 12 feet (3.6 m) wide and 10 feet (3 m) tall. The leaves are hemispherical, 4–6 feet (1.2–1.8 m) wide with tapering and slightly pendulous leaflets, each of which extends halfway into the depth of the blade. Their color is light to deep green on both sides. The leaf petioles are 5 feet (1.5 m) long and often free of spines. The inflorescences are panicles 4 feet (1.2 m) long and do not exceed the radius of the leaf crown. The 1-inch (2.5-cm) round fruits are black when mature and are edible.

Guadalupe palm needs full sun and thrives on poor, dry and rocky soils, although it has proven amenable to cultivation and supplemental irrigation, as long as the soil has unimpeded drainage. This palm is not hardy outside of zones 9 through 11 in areas subject to wet freezes in winter but is found in warm parts of zone 8 in drier climes. Except for the lack of blue coloration, the palm has the same landscape uses as *B. armata*. It is one of the least adaptable palms for indoor use.

Brahea moorei
BRA-hee-ya · MOR-ee-eye

Brahea moorei grows in the Sierra Madre Oriental of northeastern Mexico in dry, open woods at an elevation of 5280 feet (1610 m). Powder palm, a moniker coined by Yucca Do Nursery in Texas, is not yet in wide use but aptly describes the color of the leaf underside. The epithet honors the great palm taxonomist Harold E. Moore, Jr. (1917–1980) of Cornell University.

The species has underground trunks; some individuals are suckering but most are single stemmed. The leaves are nearly circular and are held on smooth and spineless petioles 2 feet (60 cm) long. The deeply divided leaf blade is light, shiny green above and, in mature plants, chalky white beneath. The inflorescence exceeds the leaf crown.

This species is among the hardiest in the genus, being adaptable in zones 8 through 11. It does not like cold and wet but, if planted high and dry in such climates, usually comes through fine. It is not particular about soil as long as it is well drained; it even tolerates a relatively high pH medium. It seems to thrive in full sun and partial shade.

Brassiophoenix
bras'-see-o-FEE-nix

Brassiophoenix comprises 2 pinnate-leaved, monoecious palms endemic to Papua New Guinea. The outwardly similar species are distinguished only by their disjunct geographical ranges and by the shapes of their endocarps and seeds.

These small palms grow to a maximum overall height of 30 feet (9 m). The slender trunks are 3 inches (8 cm) in diameter at most, green and ringed with darker-colored leaf base scars in their younger parts and light gray in their older parts. The uniformly cylindrical crownshafts are 2 feet (60 cm) tall and scarcely wider than the trunks, a light green to silvery green. The sparse, open leaf crowns are usually rounded because of the ascending and spreading leaves, which are stiff and only slightly arching on short, almost nonexistent petioles. The leaves are 6 feet (1.8 m) long with regularly and widely spaced leaflets 1 foot (30 cm) long. The unusual leaflets are diamond shaped with the point of attachment to the rachis a narrow point. They are a uniform deep green and deeply 3-lobed, the center lobe extended and with undulate margins, the other 2 lobes shorter and jagged on their apices. The inflorescences grow from beneath the crownshaft and are branched twice with ropelike branches bearing flowers of both sexes. The ellipsoid to round fruits are 1 inch (2.5 cm) long and orange when mature.

These palms need a frost-free climate and copious, regular water in well-drained, humus-laden soil. They luxuriate in partial shade and must be protected from wind, especially in hot and dry climates, as the leaflets are easily tattered and spoiled. They are neither fast growing nor terribly slow growing. The genus has been in cultivation for several years but not widely so, and this is a mystery as the species are unusual and tropical looking. The species most common in cultivation is *B. drymophloeoides* (Zona and Essig 1999).

The genus name is derived from the surname of Leonard J. Brass (1900–1971), Australian plant explorer and the original collector of the genus in New Guinea, and *Phoenix*, meaning a generalized pinnate-leaf palm. Seeds of *Brassiophoenix* germinate within 30 days. For further details of seed germination, see *Areca*.

Brassiophoenix drymophloeoides
bras'-see-o-FEE-nix · dry-mo'-flee-O-i-deez
PLATE 134

Brassiophoenix drymophloeoides occurs naturally in the southeastern peninsular region of Papua New Guinea. The epithet is Greek for "similar to *Drymophloeus*." The endocarp of this species has 5 ridges.

Brassiophoenix schumannii
bras'-see-o-FEE-nix · shoo-MAH-nee-eye

Brassiophoenix schumannii is found in northwestern Papua New Guinea. The epithet honors Karl M. Schumann (1851–1904), German botanist and original collector of the species. The endocarp of this species has 9 ridges.

Burretiokentia
bur-ret'-ee-o-KEN-tee-a

Burretiokentia is composed of 5 solitary-trunked, pinnate-leaved, monoecious palms endemic to New Caledonia. Three of them were described by Donald Hodel and Jean-Christophe Pintaud in 1998. The species are known for their strikingly beautiful trunks and large flat leaves. The trunks are distinctly ringed and dark green in their younger parts. The inflorescences grow from beneath the crownshafts and are dense with single ropelike branches bearing both male and female flowers.

The species are relatively fast growing for New Caledonian palms and are usually trouble-free and pest-free. They are among the most beautiful tropical-looking palms in the world. The genus name is a combination of 2 words, one of which honors Karl E. Maximilian Burret (1883–1964), German botanist and palm specialist, and *Kentia*, an illegitimate name for *Hydriastele*, which when used in combined forms, refers to any graceful, pinnate-leaf palm. Seeds of *Burretiokentia* germinate within 90 days. For further details of seed germination, see *Areca*.

Burretiokentia dumasii
bur-ret'-ee-o-KEN-tee-a · doo-MAH-see-eye
PLATE 135

Burretiokentia dumasii occurs in mountainous rain forest of central and western New Caledonia at an elevation of 2000 feet (610 m). It is rare in habitat and in cultivation. The epithet honors Marc Dumas, founder of Association Chambeyronia, an organization to promote and protect the endemic palms of New Caledonia. The species attains an overall maximum height of 40 feet (12 m) in habitat, with a beautiful deep green trunk that is 5 inches (13 cm) in diameter and ringed with widely spaced whitish leaf base scars. The leaf sheaths do not form a true shaft but are broad and a dirty silvery green. The leaf crown is open and dome shaped with spreading leaves 6–7 feet (1.8–2.1 m) long on petioles 1 foot (30 cm) long. The leaves are mostly straight and flat, the lanceolate deep green leaflets, 2–3 feet (60–90 cm) long, growing from the rachis in a single plane. The newly opening leaves are usually deep cherry or wine red.

Burretiokentia grandiflora
bur-ret'-ee-o-KEN-tee-a · gran-di-FLOR-a
PLATE 136

Burretiokentia grandiflora occurs in mountainous rain forest in southeastern New Caledonia at 600–2900 feet (180–880 m). The epithet is Latin for "large flowered," an allusion to the inflorescences. The palm is similar in general appearance and size to *B. dumasii*, but the inflorescences are much larger at 3–4 feet (90–120 cm) wide, the wider leaflets are slightly more widely spaced, and the newly expanding leaves are green and not reddish; it is nevertheless as beautiful as *B. dumasii*. This rare and relatively newly described species is still uncommon in cultivation.

Burretiokentia hapala
bur-ret'-ee-o-KEN-tee-a · ha-PAH-la
PLATE 137

Burretiokentia hapala grows at elevations from sea level to 1600 feet (490 m) in rain forest in northern New Caledonia. The epithet is Greek for "soft" and alludes to the woolly inflorescences.

The trunk attains a height of 30 feet (9 m) and a diameter of 6 inches (15 cm); it is emerald green in all but its oldest parts and is strongly ringed with widely spaced tan leaf base scars. The tan crownshaft is 3 feet (90 cm) tall and is often loose, the upper and mostly separated leaf bases rising above the tightly closed part. The leaf crown is almost spherical with spreading leaves 7–8 feet (2.1–2.4 m) long on short petioles usually less than 2 inches (5 cm) long; there is often a slight twist to the rachis of older leaves. The medium green leaflets are 3 feet (90 cm) long, linear-elliptic with long apices, and are regularly and closely spaced along the rachis, from which they grow in a single flat plane. The inflorescences grow from beneath the crownshaft and consist of pendent, light brown, and densely woolly branches 1 foot (30 cm) long. There may be as many as 6 or more inflorescences at a time growing from several nodes (leaf scar rings) near the top of the trunk. The ovoid reddish brown fruits are ½ inch (13 mm) in diameter.

Although it does not tolerate drought, this palm does tolerate the lightest frosts. It needs a rich, humus-laden, free-draining soil, and it luxuriates in partial shade when young but adapts to full sun in all but the hottest and driest climes when older. It is fast growing. The distinctly ringed, almost pure green trunks and the large leaves are more than exciting in the landscape and are as tropical looking as the palm world has to offer. This species is a marvelous candidate for up-close and intimate plantings, where the curiously furry inflorescence can be enjoyed, as well as specimen groupings. When young, the leaf crown is especially beautiful.

Burretiokentia koghiensis
bur-ret'-ee-o-KEN-tee-a · ko-gee-EN-sis
PLATE 138

Burretiokentia koghiensis occurs naturally in low mountainous rain forest in southeastern New Caledonia, where it is endangered. The epithet is Latin for "of Koghi," a mountain in the palm's habitat. This species and *B. vieillardii* are the tallest in the genus, growing to an overall maximum height of 60 feet (18 m). *Burretiokentia koghiensis* differs from *B. vieillardii* in having (1) erect, ascending leaves that form a shuttlecock-like leaf crown, (2) a loosely formed crownshaft covered in white tomentum, and (3) reddish new leaves which sometimes retain a purplish cast when older. It is still rare in cultivation.

Burretiokentia vieillardii
bur-ret'-ee-o-KEN-tee-a · vee-YAHR-dee-eye
PLATES 139–140

Burretiokentia vieillardii grows in mountainous rain forest of New Caledonia at 1000–4000 feet (300–1200 m). The epithet honors Eugène Vieillard (1819–1896), a French botanist who collected in New Caledonia.

The trunk grows to 60 feet (18 m) tall in habitat, is green in the younger parts and light to dark brown in the older parts, has a diameter of 6–7 inches (15–18 cm), and is prominently ringed with lighter-colored leaf base scars. The crownshaft, the most distinct in the genus, is 3 feet (90 cm) tall, much bulged at the base, and coppery green to reddish brown with darker striations. The leaf crown is the most rounded in the genus, as the older leaves are spreading and slightly arching, 8 feet (2.4 m) long on petioles 8 inches (20 cm) long. The deep green leaflets are 3 feet (90 cm) long, regularly spaced, and linear-elliptic with drawn-out apices, growing from the rachis in a single flat plane. The inflorescences consist of orange-yellow smooth ropelike, spreading or pendent branches 2 feet (60 cm) long that produce ellipsoid dark red fruits ½ inch (13 mm) long.

The palm needs copious, regular moisture and partial shade when young. It is not tolerant of freezing temperatures but can succeed in cool but frost-free Mediterranean climates if enough irrigation can be supplied. It also needs a humus-laden, moist soil that is freely draining, but it tolerates slightly acidic to slightly alkaline conditions. It is fast growing. This is the most beautiful species in the genus with its leaf crown of ascending but arching leaves, plump crownshaft, and slender and distinctly ringed trunk. Its silhouette is stunning and, planted in groups, is incomparably attractive.

Butia
byoo-TEE-a

Butia includes 18 mostly solitary-trunked, pinnate-leaved, monoecious palms in relatively dry grasslands and savannas of South America. A few species have subterranean trunks; the others have aboveground trunks, which are mostly covered with persistent leaf bases. The petioles usually bear teeth, unlike the related *Syagrus*. The inflorescences grow from the leaf crown and are branched once or are spikelike, bearing both male and female flowers, and are accompanied by a large, mostly persistent, smooth woody bract that is often tomentose. Some species are the world's cold-hardiest pinnate-leaved palms.

Because the plants freely hybridize in cultivation, many of them are nearly impossible to attach to a specific name; moreover, even the unadulterated species with solitary aboveground trunks are similar enough to make identification difficult. Horticulturists have capitalized on this genus' fortunate ability of hybridizing with other related genera, such as *Jubaea*, *Parajubaea*, and *Syagrus*, producing interesting intergeneric hybrids that are much sought after by collectors. The genus name is derived from a Portuguese corruption of an aboriginal term meaning "spiny." Seeds can take up to 3 years to germinate and do so sporadically. Often 2 or 3 seedlings develop from the same seed. For further details of seed germination, see *Acrocomia*.

Butia archeri
byoo-TEE-a · AR-cher-eye
PLATE 141

Butia archeri has a small natural range in southeastern interior Brazil in the states of Minas Gerais, São Paulo, and Goiás, where it grows in sandy savannas and open pastures. The epithet honors W. Andrew Archer (1894–1973), American botanist and collector of the original specimen.

The solitary trunks infrequently emerge above ground and then grow no more than 3 feet (90 cm) high and are often recumbent. The leaves are 3 feet (90 cm) long, have short, smooth petioles, and are silvery-grayish green, with leaflets arching from the rachis at a 45-degree angle to give a V-shaped leaf. The woody bract of the inflorescence is smooth. Overall, the appearance is that of a small and trunkless *B. capitata*.

The species demands a sunny position, is drought tolerant, and would seem appropriate for planting at the base of taller palms or other vegetation or lining a sunny pathway. Its hardiness to cold is uncertain, but it is probably a candidate for zones 9 through 11; because of its underground trunk it will doubtless come back even if the leaves are killed by cold.

Butia campicola
byoo-TEE-a · kam-pi-KO-la
PLATE 142

Butia campicola has a natural range in southeastern Paraguay and adjacent Brazil, where it grows in open, sandy savannas and grasslands and is in grave danger of extinction because of agricultural expansion. Its slender leaflets actually mimic grass. The epithet is Latin for "inhabitant of the field." This species is similar horticulturally to *B. archeri* and has the same cultural requirements and landscape uses. The main differences between it and *B. archeri* are that the trunks are sometimes clustering but are hidden underground, the inflorescence is unbranched, and the bract is tomentose.

Butia capitata
byoo-TEE-a · kap-i-TAH-ta
PLATES 143–144

Butia capitata occurs naturally on sandy soils in the states of Bahia, Goiás, and Minas Gerais, Brazil. The epithet is Latin for "with a head," an allusion to the shape of the leaf crown. Common names are pindo palm, jelly palm, and wine palm, the latter unfortunate as it is also the English vernacular for *Jubaea chilensis*.

The stems of this solitary-trunked species can measure 11 inches (28 cm) in diameter and 13 feet (4 m) tall, although cultivated individuals may be taller. The stems are usually covered with old leaf bases unless they are of great age and height, in which case the leaf base scars are incomplete circles of ridges. The leaves are 4–7 feet (1.2–2.1 m) long on short petioles armed with flat, rigid fibers. Many narrow, thin leaflets grow from the rachis at a 45-degree angle, giving the leaf a distinct V shape. The leaf rachis has a pronounced and beautiful arch, sometimes recurving as much as 180 degrees, creating a perfectly spherical crown. Leaf color varies from yellowish green to silvery green to almost gray or bluish gray. The inflorescence is many branched but relatively short and accompanied by a large, smooth, spoon-shaped woody bract. The flowers are yellow to orange. The spindle- or top-shaped yellow to deep orange fruits are 1 inch (2.5 cm) long and are formed in pendent clusters; they are edible and sweet.

Pindo palm is not fussy about soil type as long as it is loose, sandy, and freely draining. It needs full sun for robust growth and good color and, although drought tolerant, likes moisture. It is one of the hardiest pinnate-leaved palms with a trunk, being adaptable to zones 8 through 11 in drier climates.

This outstanding landscape subject creates a tropical effect with its pleasing, stiff yet gracefully arching, much recurved, bluish or silvery green feathery fronds and the light to dark gray trunk with adhering leaf bases. Nothing stands out quite like it against the darker greens of other vegetation. The palm is beautiful enough to stand alone as a specimen plant. It is seldom used indoors because it requires intense light and good air circulation.

Butia eriospatha
byoo-TEE-a · er'-ee-o-SPA-tha
PLATE 145

Butia eriospatha is indigenous to Argentina and extreme southeastern Brazil and grows in open, mostly deciduous forests and grasslands. The epithet is Greek for "woolly spathe." The species is similar to *B. capitata* and differs mainly in details of its inflorescence, the most salient of which is the beautiful light brown tomentum that covers the emerging inflorescence bract. In addition, the leaves are never as silvery or bluish as are those of *B. capitata*. The 2 species have the same cultural requirements and landscaping uses. It may be even more cold hardy than *B. capitata*.

Butia microspadix
byoo-TEE-a · my-kro-SPAY-dix
PLATE 146

Butia microspadix is native to southeastern Brazil, where it grows in open grasslands. This is another of several elusive species of *Butia* that hide in plain sight among grasses (Noblick 2006). They are grass mimics. The epithet is Greek for "tiny" and "spathe." The trunk remains underground. Except for the densely woolly spathe of the inflorescence, the species is similar to *B. archeri* and *B. campicola*.

Butia odorata
byoo-TEE-a · o-do-RAH-ta
PLATES 147–150

Butia odorata is native to southernmost Brazil and adjacent Uruguay. It has long been treated as a variety of *B. capitata*, but Noblick (in Lorenzi et al. 2010) made the case that this southern variant should be treated as a distinct species. Because of its southern provenance, it is believed to be more cold hardy than true *B. capitata*. Noblick stated that most of the palms in cultivation outside Brazil under the name *B. capitata* are likely to be *B. odorata*. The epithet means "fragrant" in Latin and refers to the flowers.

The species is nearly identical to (and often confused with) *B. capitata*. It differs mainly in details of fruits and seeds. The fruits of *B. odorata* are 1 inch (2.5 cm) long, ovoid or globose, yellowish orange to reddish orange or purple, whereas, the fruits of *B. capitata* are larger and shaped like a spindle or top. For culture and uses in the landscape, see *B. capitata*.

In cultivation, *B. odorata* (misidentified as *B. capitata* until recently) has hybridized with *Syagrus romanzoffiana* to form one of the most beautiful palm hybrids, ×*Butiagrus nabonnandii* [byoo-ty-AG-rus · nah-bo-NAN-dee-eye] (see plate 149). The epithet honors a French nurseryman and hybridizer, Paul Nabonnand (1860–1937), who first produced the cross. In the United States, this hybrid is usually called the mule palm because its seeds are generally sterile. Its appearance varies depending on which species is the male or female parent and whether the individual palm has been further back-crossed. Individuals favoring the *Syagrus* parent are especially desirable as they are faster growing and absolutely the most tropical looking palm for nontropical areas. In southern France, the palm is reported to withstand temperatures in the upper teens (about −6°C) without damage.

To create a species that looks like *Jubaea chilensis* but is more tolerant of summer rainfall, *Butia odorata* has been crossed with it to give ×*Jubautia* [joo-BOW-tee-ah] (see plate 150). The hybrid is intermediate between the parents. It is not commonly cultivated.

Butia paraguayensis
byoo-TEE-a · pah'-ra-gwah-YEN-sis
PLATES 151–152

Butia paraguayensis occurs naturally in eastern and southern Paraguay, southern Brazil, northeastern Argentina, and northern Uruguay, where it grows in open, sunny, and very sandy areas. The epithet is Latin for "of Paraguay." This species is similar to *B. yatay* and may be a variant of it.

While *B. paraguayensis* is often a trunkless palm in habitat, almost all the individuals cultivated in the United States have robust trunks about the height of *B. capitata*. The leaf bases of arboreal specimens seem to be more persistent on the trunks of *B. paraguayensis* than on trunks of *B. capitata*, and the leaves are even more arched and recurved, almost always bending back to the trunk itself. Leaf color is never as silvery as it is in *B. capitata* but is, at most, only glaucous. The fruits are ovoid, greenish yellow or purple. This palm is a fine substitute for contrast for the ubiquitous *B. capitata* and is pleasing to contemplate up close. The 2 species have identical cultural requirements and cold hardiness.

Butia purpurascens
byoo-TEE-a · pur-pur-RAS-senz
PLATES 153–154

Butia purpurascens is native to the state of Goiás in northeastern Brazil. The epithet is Latin for "purplish." The species is similar to *B. capitata* but lacks teeth on its petioles and has purplish inflorescences and flowers. The fruit is dark purple or almost black. The species is probably the least hardy to cold in the genus.

Butia yatay
byoo-TEE-a · yah-TY
PLATE 155

Butia yatay occurs in southernmost Brazil, northeastern Argentina, and northern Uruguay. The epithet is the aboriginal name in Uruguay. It is the tallest growing species in the genus and has the longest leaves. These are held on long petioles armed with teeth on their margins. The crown is more open than that of *B. capitata* and the leaves are not as strongly arched, but the leaf color is similar. Cultural requirements of the 2 species are nearly identical, but *B. yatay* is probably even more tolerant of cold. It has the most tropical look in the genus because of its large, loose crown and superior height. It should be planted much more widely than it is at present.

Calamus
KAL-a-mus

Calamus consists of more than 375 mostly climbing, pinnate-leaved, dioecious palms from tropical Africa eastwards through the South Pacific; they reach their pinnacle of diversity in the forests of Indonesia and Malaysia. There are both clustering and solitary-trunked species, and a few have subterranean stems. The climbing species do not have a distinct, compact leaf crown; rather the leaves grow widely separated from one another along the stem. The lower portion of the stem becomes spineless with age as the old leaves fall away, because, in reality, the spines are borne on the leaf bases, not on the stem itself.

Many species climb trees and other vegetation to reach an opening in the forest canopy. All species are spiny, and the climbers ascend by the spines that attach themselves to supports. In addition, the climbers have specialized organs called *cirri* (singular *cirrus*) that are modified and extended leaf rachises. Alternatively, they have *flagella* (singular *flagellum*) that are long, whiplike, sterile inflorescences—organs unique to the genus *Calamus*. Both types of organ have backward-pointing spines (usually in pairs) or clawlike clusters of hooks that aid in the climbing process. In almost all cases, these specialized climbing organs are formed only when the palm has attained some age and, more importantly, has an aboveground stem. The inflorescences are also whiplike, long, and sparsely branched. The male and female inflorescences are similar in appearance except for the small flowers. The small, ellipsoid fruits are usually orange, red, or brown but also white, and are covered in distinct, overlapping scales. In some species, the fruits are edible.

Most species are tropical or nearly tropical in their cultural requirements and, because juvenile plants are undergrowth subjects, they need partial shade when young; ironically, most do not climb if kept in dense shade. Indeed many of these climbers simply remain as small, clustering palms on the ground for years until enough light reaches them; and many do not flower until they can climb high enough to reach the forest canopy and full sun. In addition, almost all need copious moisture.

Calamus is the largest genus in the palm family, and most of the species are little known. The few species in cultivation are mainly found in the collections of enthusiasts and botanical gardens because of their great spininess and ultimate size. Yet many species are of outstanding beauty. They are best placed where they can clamber up specimen trees so that their beautiful leaves may be seen without trampling through other vegetation or endangering one's person. Most of the species are easily grown in containers, and their stems may be repeatedly cut back to limit their height and to initiate compact clustering. Alternatively, most can be kept in dense shade, which prevents them from initiating climbing.

The genus name is a Latinized form of a Greek word meaning "reed" or "cane." All species are called *rattans*, a Malayan term referring to the use of the pliable stems for making furniture and other articles. Like other species of the Calameae, *Calamus* species are among the easiest palms to grow from seed. Most seeds of Calameae have a short viability only because they tend to germinate as soon as they ripen and fall from the tree. Seeds shipped in a plastic bag from a foreign country can begin germinating before reaching their destination. Seeds that are allowed to dry out completely do not germinate well, and the smaller the seed, the more likely it will dry out before sowing. Seeds of most *Calamus* species germinate within 30 days.

Calamus australis
KAL-a-mus · aw-STRAH-lis
PLATE 156

Calamus australis is a clustering and climbing species endemic to the rain forest of eastern Queensland, Australia, where it can climb to heights of 80 or more feet (24 m). The epithet is Latin for "southern," in this case the southern continent. The palm is called wait-a-while and lawyer cane in Australia. The very spiny stems are usually 1 inch (2.5 cm) in diameter, and the leaves are 6 feet (1.8 m) long. The light green leaflets are evenly spaced along the rachis in a single plane and are 1 foot (30 cm) long and linear-acuminate. Flagella 12 feet (3.6 m) long are produced from the spiny leaf axils. Fruits are white. This climber produces one of the most tropical-looking vistas if it can be planted so that its layers of beautifully feathery leaves can be seen.

Calamus caryotoides
KAL-a-mus · kar-ee-o-TO-i-deez
PLATE 157

Calamus caryotoides is a small, clustering and climbing species endemic to northeastern Australia, where it grows in the rain forest and in drier forests and clearings. The epithet is from the Greek meaning "similar to *Caryota*," although in our opinion, the leaflets are not that similar. The palm is called wait-a-bit (or wait-a-while) and fishtail lawyer cane in Australia.

This palm is capable of making large clumps with up to 100 stems. It usually climbs no higher than 40 feet (12 m), mainly by flagella 3 feet (90 cm) long, but in garden sites, it is generally much smaller. The stems are relatively thin, ½ inch (13 mm) in diameter, pliable, and not particularly spiny. The leaves are 1 foot (30 cm) long and have 2–12 pinnae, each one glossy dark green, linear to wedge shaped, and jagged at the apex. The yellowish brown fruits are formed in long, pendent clusters.

This garden-friendly species of *Calamus* is not ferociously spiny and thus is not as difficult to work with as its spinier kin. As a diminutive curiosity, it should be more widely grown.

Calamus ciliaris
KAL-a-mus · si-lee-AH-ris
PLATE 158

Calamus ciliaris is native to low montane forests of western Java and western Sumatra. The epithet is Latin for "fringed," an allusion to the soft bristles on the margins of the leaflets. This species is clustering or solitary with stems to 30 feet (9 m) tall and up to 4 inches (10 cm) in diameter, but usually smaller in cultivation. The leaves are about 6 feet (1.8 m) long and bear cirri. The rachis and leaf sheath are covered in pale, straw-colored spines. The leaflets, which are evenly arranged, have soft, yellowish bristles along their margins and main veins, and are slender, linear, and medium green on both sides. This species is very decorative when young, with an elegant and graceful appearance. Mature specimens impart the tropical look that only rattans can give.

Calamus erectus
KAL-a-mus · ee-REK-tus
PLATE 159

Calamus erectus is widespread from eastern India to southern China and south to Thailand. It occurs in seasonally dry forest, in hillsides, from lowlands to over 5000 feet (1520 m) in elevation. This is one of the non-climbing *Calamus* species, and as such, it bears neither flagella nor cirri. It does, however, have plenty of sharp spines. The epithet is Latin for "erect" and describes the salient feature of its growth habit.

The clustering stems grow up to 15 feet (4.5 m) tall and are 1½ inches (4 cm) in diameter. They are stiff and non-flexible. The leaves are 9–15 feet (2.7–4.5 m) long, with regularly arranged leaflets. One of the most distinctive features of this species is the combs or collars of long, yellow spines that ring the petioles at frequent intervals. The flower stalk is long, and the large greenish brown fruits are up to 2 inches (5 cm) long and 1 inch (2.5 cm) wide. This species is occasionally cultivated in botanical gardens and specialist collections.

Calamus javensis
KAL-a-mus · jah-VEN-sis
PLATE 160

Calamus javensis is indigenous to southern Thailand, peninsular Malaysia, Java, Borneo, and Sumatra, where it grows in rain forest. The epithet is Latin for "of Java." This species is among the smallest, most delicate-looking species in the genus. Although it is a clustering climber, its stems are never longer than 30 feet (9 m) and are less than ½ inch (13 mm) in diameter. It climbs by flagella 3 feet (90 cm) long at the ends of the leaves, which are 2 feet (60 cm) long. The dark green elliptic-acuminate leaflets are 6 inches (15 cm) long and unusual in that the lowest pair is usually backwards pointing, and the leaflets of the terminal pair are fused. The fruits are whitish. Only rarely seen, this species deserves to be better known. Its small size makes it more amenable to garden situations.

Calamus moti
KAL-a-mus · MO-tee
PLATE 161

Calamus moti is endemic to northeastern Australia, where it grows in the rain forest. The epithet is an aboriginal name for the palm. The common name in Australia is yellow wait-a-while. The stems of this large, clustering, climbing, spiny palm reach 70 feet (21 m) long. The clumps have many stems, and a single plant can appear like a colony. The recurved leaves are 6–10 feet (1.8–3 m) long and have deep green, evenly spaced, long, linear leaflets; they much resemble the leaves of *Howea forsteriana*. The plant hangs onto its supports by 1-inch (2.5-cm) yellow spines on its stems and by whiplike flagella 12 feet (3.6 m) long. The spines on the leaf sheaths are arranged in combs. This beautiful species takes readily to container culture with heavy pruning. Individual stems can be removed when they become large and problematic.

Calamus viminalis
KAL-a-mus · vi-mi-NAH-lis
PLATE 162

Calamus viminalis is indigenous from India to southern China to Southeast Asia and Malaysia and Indonesia. As one might expect from such a wide-ranging palm, the species is highly variable and includes several previously recognized varieties and species. It grows in lowland forest margins and scrubby vegetation in full sun; rarely, it can be found inside evergreen forests. The epithet is Latin for a long, flexible willow shoot suitable for weaving. This large clustering palm produces leaf sheaths with long, yellow spines that are reddish near their tips (as if having already drawn blood). The leaves are 3–5 feet (90–150 cm) long, with leaflets clustered and in many planes. The fruits are whitish. Although the attractively plumose leaves and colorful leaf sheath spines recommend this palm for cultivation, it is suitable only for large gardens, where it can climb and form a thicket.

Calyptrocalyx
ka-lip'-tro-KAY-lix

Calyptrocalyx comprises 26 mostly small, mostly clustering, pinnate-leaved, monoecious palms in the rain forest of New Guinea, with one exception from the Moluccas (see *C. spicatus* below). The inflorescences are unbranched spikes of tiny male and female flowers, and the fruits are orange, red, or black. Most of the plants grown today are relatively new to cultivation, and many of them entered cultivation with provisional names provided by collectors. Some species have undivided leaves with bifid apices; most newly opened leaves are orange, red, or purple. Almost all these palms are undergrowth subjects, a few rising to or above the forest canopy.

All are tender to cold and adaptable only to zones 10b and 11. All of them need a rich, humus-laden soil that is moist but fast draining, and none are particularly fond of calcareous soil. All are adaptable to partial shade and a few to full sun. Most need protection from strong winds lest their beautiful leaves become marred; dry winds, especially when cold, are inimical and often fatal. All the smaller species make excellent container subjects if given enough water and a fast-draining medium, and all of them are beautiful along a path or as a border to contrasting vegetation.

As with *Dypsis*, there exist many plants in cultivation that are referred to by their aboriginal common name or by their collection locality. These are very attractive palms, sometimes exhibiting especially colorful new leaves or unusual leaflet shapes. Some of these are simply superior clones of known species; others may represent undescribed species. Additional study will be needed to sort out the identification of these interesting palms. The genus name is from Greek words meaning "covered calyx." The genus *Paralinospadix* is now incorporated into *Calyptrocalyx*. Seeds of *Calyptrocalyx* germinate within 45 days. For further details of seed germination, see *Areca*.

Calyptrocalyx albertisianus
ka-lip'-tro-KAY-lix · al-ber-tis'-see-AH-nus
PLATE 163

Calyptrocalyx albertisianus is a large, beautiful, solitary-trunked or infrequently clustering species, growing ultimately to heights of 25 feet (7.6 m). The epithet honors Luigi d'Albertis (1841–1901), Italian naturalist and explorer of New Guinea.

The trunk is usually dark tan, less than 1 foot (30 cm) in diameter in mature specimens, and ringed with cream-colored leaf base scars. The leaves are 10 feet (3 m) long with spreading, glossy deep green elliptic-acuminate leaflets, each 18 inches (45 cm) long and growing in a flat plane from the rachis. The new leaves unfold a beautiful cherry to maroon color that lasts for 2 weeks before it turns to brownish orange and finally deep green. The fruits are red, and the endosperm is deeply ruminate.

The species grows reasonably fast once past the seedling stage and with good soil and abundant moisture. It is tolerant of partial shade to full sun, except in the hottest climates where it needs protection from the midday sun.

Calyptrocalyx arfakiensis
ka-lip'-tro-KAY-lix · ahr-fah'-kee-EN-sis
PLATE 164

Calyptrocalyx arfakiensis is a small, solitary-trunked species with undivided leaves that are apically bifid. The epithet is Latin for "of Arfak," a mountain range in West Papua Province, Indonesia. The species is choice because of the slender, elegant stem that measures 6 feet (1.8 m) high and ½ inch (13 mm) in diameter when mature. The leaves are a delicious, glossy deep green with corrugations corresponding to the fused pinnae. The fruits are black, and the endosperm is ruminate.

Calyptrocalyx awa
ka-lip'-tro-KAY-lix · AH-wah
PLATE 165

Calyptrocalyx awa is a clustering palm with bifid or irregularly segmented leaves. It grows up to 6 feet (1.8 m) tall in lowland rain forest. The foliage of seedlings has a metallic luster that disappears as the plants mature. The new leaves emerge orange to bronze on juveniles and adults. The flowers have a rosy tinge to their petals and are borne on an inflorescence of 2 spikes. The round orange fruits are highly decorative but less than ½ inch (13 mm) in diameter, and the endosperm is homogeneous. The species epithet is the aboriginal name for this palm, although it is also applied to another palm species.

Calyptrocalyx doxanthus
ka-lip'-tro-KAY-lix · dahx-AN-thus
PLATE 166

Calyptrocalyx doxanthus is a small, clustering species to 6 feet (1.8 m) high with thin, delicate stems. The leaves are 3 feet (90 cm) long with 8–10 pairs of 6-inch (15-cm) elliptic-acuminate leaflets growing in a flat plane from the rachis. New leaf color is pink to deep rose, but the color settles down to a subdued green with darker veins. The undersides of the mature leaves are buff colored. The long tapering tip of each leaflet is especially decorative. The inflorescence is a solitary spike. Fruit color is red, and the endosperm is homogeneous. The epithet is from 2 Greek words meaning "glory" or "praise" and "flower," an allusion to the unique morphology of the flower. This species was introduced into horticulture by Greg Hambali.

Calyptrocalyx elegans
ka-lip'-tro-KAY-lix · EL-e-ganz

Calyptrocalyx elegans is a small, clustering species to 15 feet (4.5 m) high. The clumps are not dense, and the leaves are either mostly unsegmented and apically bifid or completely pinnate, the latter type of leaf looking much like that of a *Chamaedorea* species, with limp, linear light green regularly spaced leaflets. The inflorescence is a solitary spike. The fruits are red, and the endosperm is ruminate. This species is sought after because of the beautiful dark red newly unfurled leaves. The epithet is Latin for "elegant."

Calyptrocalyx forbesii
ka-lip'-tro-KAY-lix · FORB-zee-eye
PLATE 167

Calyptrocalyx forbesii is among the most beautiful and graceful species in the genus. It is native to the Central and Milne Bay Provinces of Papua New Guinea. The epithet honors the collector of the original specimen, Henry O. Forbes (1851–1932), Scottish naturalist and explorer of New Guinea.

This large clustering palm produces several elegantly slender, light-colored, ringed trunks that can attain heights of 20 or more feet (6 m). The elegance continues with the ascending and beautifully arching pinnate leaves, which are 8 feet (2.4 m) long, with many dark green, linear-acuminate, 18-inch (45-cm) leaflets that grow from the rachis in a single plane. The new leaf color is dark reddish brown. The inflorescence has 1–4 spikes. The fruits are orange or red, and the endosperm is homogeneous.

A clump of this species is as graceful as an areca palm (*Dypsis lutescens*) and, like the latter, benefits visually if some of the densely packed stems are judiciously cut away so that the form of the others is more readily apparent.

Calyptrocalyx hollrungii
ka-lip'-tro-KAY-lix · hol-RUNG-ee-eye
PLATES 168–169

Calyptrocalyx hollrungii is a small species to 8 feet (2.4 m) high that forms dense and sometimes large clumps. The leaves are 18 inches (45 cm) long, undivided and apically bifid, medium green, with deep corrugations corresponding to the fused pinnae; some individuals have segmented pinnae of varying widths. The new leaves are a deep maroon or wine color. The small, ellipsoid fruits are orange or red, and the endosperm is homogeneous. This species has been cultivated longer than most others have, but it was known as *Paralinospadix hollrungii*. The epithet honors Max U. Hollrung (1858–1937), a German chemist and plant pathologist, who co-wrote one of the important early floras of New Guinea.

Calyptrocalyx lauterbachianus
ka-lip'-tro-KAY-lix · lah-tur-bah-kee-AH-nus

Calyptrocalyx lauterbachianus is named for Carl A. G. Lauterbach (1864–1937), German naturalist, explorer, and director of the German New Guinea Company. It is a medium-sized solitary or sometimes clustering palm. Its leaves have leaflets grouped together along the rachis and arrayed in different planes. Several long green spikelike inflorescences emerge from each node on the mature stem. The fruits are elongate and slightly curved, red, up to 1½ inches (4 cm) long and 1 inch (2.5 cm) wide. The endosperm is ruminate. Many plants cultivated under this name appear to be of another species.

Calyptrocalyx leptostachys
ka-lip'-tro-KAY-lix · lep-to-STAY-kees

Calyptrocalyx leptostachys is known from Mount Yule, Papua New Guinea. The epithet means "slender spike" and describes the inflorescence. The stem of this solitary palm reaches 6 feet (1.8 m) tall but is less than 1 inch (2.5 cm) in diameter. The leaves are irregularly pinnate. Fruits are small, globose to ellipsoid, and the endosperm is ruminate. This species is still poorly known, but recent introductions allow it to take its place among the many delightful species of this genus.

Calyptrocalyx micholitzii
ka-lip'-tro-KAY-lix · mee-ko-LIT-zee-eye
PLATES 170–171

Calyptrocalyx micholitzii is a small solitary-trunked species with undivided deep green leaves that are apically bifid. The palm grows to 2 feet (60 cm) high and is a true miniature. The new leaves are purplish orange and often show an attractive mottling of dark and lighter green when mature. The epithet honors Wilhelm Micholitz (1852–1932), a collector of insects and plants, mostly orchids and mosses, in New Guinea.

Calyptrocalyx pachystachys
ka-lip'-tro-KAY-lix · pa-kee-STAY-kees
PLATES 172–173

Calyptrocalyx pachystachys is a small, solitary-trunked species with 6 feet (1.8 m) of slender stem. The gorgeous leaves are 3 feet (90 cm) long and graced with elliptic-acuminate to obovate-acuminate, glossy dark green "fat" leaflets that often have a slight S shape. The newly unfurled leaflets are a beautiful cherry red. A variegated cultivar looks more chlorotic than beautiful. The epithet is from Greek words meaning "thick spike," a reference to the inflorescence.

Calyptrocalyx pauciflorus
ka-lip'-tro-KAY-lix · paw-si-FLO-rus
PLATE 174

Calyptrocalyx pauciflorus is a small, solitary-trunked or clustering species with a stem that usually remains underground. The beautiful leaves are 2 feet (60 cm) long and either undivided except for the bifid apex or irregularly divided. They are deep green, wide at the apex, with a lighter-colored midrib and corrugations corresponding to the fused pinnae. The inflorescence has one or 2 spikes. The fruits are red, and the endosperm is ruminate. The epithet is Latin for "few flowered."

Calyptrocalyx polyphyllus
ka-lip'-tro-KAY-lix · pah-lee-FYL-lus
PLATE 175

Calyptrocalyx polyphyllus is a beautiful clustering species with slender trunks to 10 feet (3 m) tall. The clumps are neither large nor dense, which makes them more elegantly handsome. The leaves are 3–5 feet (90–150 cm) long with many elliptical, long-acuminate, glossy deep green, 8-inch (20-cm) leaflets in mature plants. New growth is typically colored maroon to cherry red. The male flowers are pleasantly fragrant. The fruits are red, and the endosperm is ruminate. The epithet is 2 Greek words meaning "many" and "leaf."

Calyptrocalyx spicatus
ka-lip'-tro-KAY-lix · spi-KAH-tus
PLATE 176

Calyptrocalyx spicatus is atypical in its origin, being the only species indigenous to the island of Ambon in the Moluccas. The epithet is Latin for "spiked" and refers to the long unbranched inflorescence. The solitary trunk is atypically large for the genus, growing to 40 or more feet (12 m) tall, and is light tan, beautifully ringed with darker leaf base scars, and typically straight as an arrow and slender, no more than 10 inches (25 cm) in diameter. The leaves are sumptuous, 12 feet (3.6 m) long, and pinnately divided into many narrow, lanceolate, pointed, deep green, pendent leaflets, each 2 feet (60 cm) long. Some people compare the leaves to those of *Howea forsteriana*, but others see them as much more like *Hydriastele costata*. The leaves are glorious and sparse, with never more than 12 per leaf crown. The fruits are globose to ellipsoid, to 2½ inches (6 cm) long, and red or orange. The endosperm is ruminate.

Calyptrogyne
ka-lip'-tro-JY-nee

Calyptrogyne is composed of 17 small, solitary-trunked, pinnate-leaved, monoecious palms in tropical America, all undergrowth subjects in rain forest. Most species are outwardly similar to *Geonoma* and *Asterogyne* species but are not as well represented in cultivation. One wonders why, as they are at least as ornamental as those species. The genus is most closely related to *Calyptronoma* in the technical details of its inflorescences and flowers but not in its vegetative characters. The inflorescences are mostly simple erect spikes held well above the leaf crowns, the flowers being pollinated by bats. The genus name is derived from 2 Greek words meaning "cap" or "lid" and "female," a reference to the corolla of the female flower, which falls away like a lid as the flower opens. Seeds germinate within 60 days. For further details of seed germination, see *Asterogyne*.

Calyptrogyne ghiesbreghtiana
ka-lip'-tro-JY-nee · gees-brek'-tee-AH-na
PLATES 177–178

Calyptrogyne ghiesbreghtiana is the most widespread species in the genus and the only one in cultivation. It occurs from the Mexican state of Chiapas, southwards and eastwards through Central America into Costa Rica missing El Salvador, where it grows in rain forest from sea level to 4800 feet (1460 m). The epithet honors Auguste B. Ghiesbreght (1810–1893), a Belgian plant collector in Mexico. The species is sometimes called rat's tail palm because of the thin inflorescences, with flowers that smell of garlic.

The stem is subterranean, and the leaves are 3–4 feet (90–120 cm) long, pinnately divided into dark green, deeply ribbed segments of varying widths, growing from a stout and lighter-colored rachis. More rarely, the leaves are undivided and deeply bifid apically. New growth is usually colored a deep wine or maroon. Four subspecies are recognized, based on technical details of the inflorescence bracts and fruit size.

This palm is elegant and, because of the color of its new leaves, is widely sought after. Although some populations are found at high elevations, the species is sensitive to frost and, even under canopy, can be damaged in zone 10a. It wants regular, adequate moisture, partial shade, and a humus-rich soil.

Calyptronoma
ka-lip'-tro-NO-ma

Calyptronoma consists of 3 solitary-trunked, pinnate-leaved, monoecious palms in the Greater Antilles. All are denizens of low-lying, wet areas. The inflorescences grow from beneath the leaf crown and consist of a spray of short, congested flowering branches on a stout peduncle. The small fruits are reddish purple when mature. Some authors place these species in the genus *Calyptrogyne*. They are wonderfully beautiful feather palms that resemble short coconuts. In the garden, they are interchangeable, differing only in the technical details of their flowers and fruits.

All 3 species are water lovers and drought intolerant. They do best on a rich, slightly acidic soil. The genus name was unexplained but was likely compounded from the Greek word meaning "lid" or "cap" and *Geonoma*, a closely related genus. As in *Calyptrogyne*, the petals of the female flower are joined to form a cap that falls away as the flower opens. Seeds of *Calyptronoma* germinate within 60 days. For further details of seed germination, see *Asterogyne*.

Calyptronoma occidentalis
ka-lip'-tro-NO-ma · ox-i-den-TAL-is
PLATES 179–180

Calyptronoma occidentalis is found only on Jamaica, in the swampy margins of lakes and streams. The epithet is Latin for "western." The palm grows to a maximum height in habitat of 40 feet (12 m) and is much like *C. rivalis* in appearance.

Calyptronoma plumeriana
ka-lip'-tro-NO-ma · ploo-mer'-ee-AH-na
PLATE 181

Calyptronoma plumeriana is indigenous to Cuba and Hispaniola, where it grows in wet forests and along streams. The epithet honors Frenchman Charles Plumier (1646–1704), one of the first botanists to explore the West Indies. The palm grows to 30 feet (9 m) tall and is much like *C. rivalis* in appearance. In the palm's native haunts, the male flowers are eaten for their sweet nectar.

Calyptronoma rivalis
ka-lip'-tro-NO-ma · ri-VAL-is
PLATE 182

Calyptronoma rivalis occurs naturally on Hispaniola in wet areas along streams. It also occurs on Puerto Rico, where it is exceedingly rare and threatened with extinction. The epithet translates from the Latin as "of the river."

The palm attains a maximum height in habitat of 50–60 feet (15–18 m). The light to dark brown or grayish trunk is 1 foot (30 cm) in diameter and covered in its younger parts with much brown fiber and in its older parts by closely set rings. The leaves are 10 feet (3 m) long on unarmed petioles 2–3 feet (60–90 cm) long, seldom lie below the horizontal, and fall quickly after dying. The bright green leaflets are 2 feet (60 m) long, regularly spaced along the rachis, and are pendent. The branched inflorescences have a thick peduncle and bear tiny white male and female flowers. The small fruits are pea shaped.

This palm needs sun except as a seedling, when it prefers partial shade. It is tender to cold and adaptable only to zones 10b and 11, although some protected specimens can be found in 10a. It is fast growing once past the seedling stage and if given enough sun and moisture. The species is beautiful up close and used as a canopyscape or part of a massive wall of vegetation. Its elegant silhouette is used to advantage against a contrasting background of differently colored and textured foliage or even man-made structures.

Carpentaria
kahr-pen-TAHR-ee-a

Carpentaria is a monotypic genus of tall pinnate-leaved, monoecious palm. The name honors a place near the palm's habitat, the Gulf of Carpentaria in northern Australia. Seeds of *Carpentaria* can take 30 days to one year to germinate. In Florida, seeds germinated really well in late spring following the season in which they were sown. For further details of seed germination, see *Areca*.

Carpentaria acuminata
kahr-pen-TAHR-ee-a · a-kyoo'-mi-NAH-ta
PLATE 183

Carpentaria acuminata is endemic to the Northern Territory, Australia, where it grows along rivers and streams in the rain forest at low elevations. The epithet is Latin for "acuminate" or "pointed," an inexact allusion to the shape of the mature leaflets. The common name is Carpentaria palm.

The solitary trunk grows 50–55 feet (15–17 m) high and 8 inches (20 cm) in diameter, is light gray to almost white, and has widely spaced, prominent rings of the same color. The deep green crownshaft is smooth and 3–5 feet (90–150 cm) tall with a slightly bulging base. The spherical leaf crown is 12 feet (3.6 m) wide and tall, and the gracefully arching leaves are 10–12 feet (3–3.6 m) long on short petioles and have numerous linear leaflets, growing from the rachis at a slight angle to create a V-shaped blade. Each leaflet is 2 feet (60 cm) long, limp, pendent at its apex, and tapering towards its apex. Leaf color is deep emerald green above with a decidedly bluish cast beneath. The inflorescences originate beneath the crownshaft, are 3–4 feet (90–120 cm) long, much branched, and pendent. They form a skirt around the base of the crownshaft and bear white flowers. The globose ½-inch (13-mm) fruits are scarlet when ripe.

This palm needs lots of water; its enemies are drought and cold. It thrives in a rich soil and is among the few palms that do not demand perfect drainage. It tolerates partial shade when young but needs full sun as it gets older. The species grows quite fast with warmth, sun, and adequate moisture. Its cultivation is limited to tropical or nearly tropical regions, and it is adaptable only to zones

10b and 11, although some mature specimens can be found in protected spots in 10a.

With its relatively diaphanous and straight trunk, Carpentaria palm is elegance personified, and yet it is noble in size and appearance. It is beautiful as an isolated specimen surrounded by space but is even more attractive in groups of 3 or more individuals of varying heights, and it is stunning as a canopy-scape. The large clusters of red fruits are highly ornamental. It is not known to be grown indoors, but it should be adaptable to large conservatories or atriums with lots of light.

Carpoxylon
kahr-PAHX-i-lahn

Carpoxylon is a monotypic genus of solitary-trunked, pinnate-leaved, monoecious palm. The name is derived from Greek words meaning "fruit" and "wood." Seeds germinate within 90 days. For further details of seed germination, see *Areca*.

Carpoxylon macrospermum
kahr-PAHX-i-lahn · mak-ro-SPER-mum
PLATE 184

Carpoxylon macrospermum is endemic to the Vanuatu islands in low-elevation rain forest. It is exceedingly rare and was thought to be extinct, known only from descriptions of its fruits and seeds, until the discovery of the trees themselves in 1987 by Australian botanist John L. Dowe, who almost accidentally found it in cultivation on the island of Espíritu Santo. A governmental organization (Profitable Environmental Protection Project) was established in Vanuatu to try to protect and increase the remaining populations of this critically endangered species. The epithet is derived from Greek words meaning "large seed."

The mature trunks reach a height of 90 feet (27 m) in habitat with a swollen base and are 1 foot (30 cm) in diameter otherwise, deep green in their younger parts, gray in their older parts, and closely ringed with slightly raised, whitish leaf base scars all over. The crownshaft is 4–5 feet (1.2–1.5 m) tall, straight, columnar, slightly bulging at its base, and light green or light tan. The beautifully recurved leaves are 12 feet (3.6 m) long on short petioles usually less than 1 foot (30 cm) long. The leaflets range from 1 foot (30 cm) long at the base of the leaf to 4 feet (1.2 m) long in the middle and as short as 8 inches (20 cm) at the apex. They are olive green above, silvery green beneath, thick, and leathery, and grow at a 45-degree angle from the grooved rachis to create a V-shaped leaf. The large infructescences are borne at the base of the crownshaft, and the large red egg-shaped fruits are 2 inches (5 cm) long.

This slow-growing palm seems to thrive in partial shade to full sun in nearly tropical regions. It needs copious, regular moisture. Dowe (1989a) reported that the growing point of the palm is sometimes harvested. This species has one of the most beautiful forms in the palm family and its silhouette is alluring with the much arched, recurving leaves and relatively thin trunk. There could be no more beautiful canopy-scape. The palm is undoubtedly suited to large greenhouses, atriums, and conservatories with bright light, warmth, moisture, and humidity.

Caryota
kar-ee-O-ta

Caryota comprises about 13 solitary-trunked or clustering, monoecious, monocarpic palms in tropical Asia, India, Malaysia, Indonesia, the Philippines, New Guinea, the Solomon Islands, the Vanuatu islands, and northeastern Australia. It is the only palm genus with bipinnate (doubly pinnate) leaves, which makes all the species immediately recognizable, whether solitary trunked or clustering. They are called fishtail palms because of their wedge-shaped, jagged-ended leaflets.

Most species have elongated leaf crowns in which the leaves grow not only from the top of the trunk but also down its length, sometimes considerably so. The trunks die off after they finish flowering and the fruits mature. This phenomenon is not a great problem for the clumping types, as there are always new trunks to fill the void; but, for the single-trunked species, a logistical problem is created as these species are almost invariably of large size. The beginning of flowering does not, however, signal the immediate death of a trunk because the flowering process may take several years to complete, the inflorescences forming at each leaf node, from the top down. While each palm produces both male and female blossoms, almost always in separate inflorescences, the male flowers tend to produce viable pollen when the pistils of the female flowers are not receptive; thus seed is not always forthcoming from a single, isolated individual.

All species are relatively fast growing when past the juvenile stage and with commencement of trunk production. All are indigenous to forests and need rich, humus-laden soils. The fruits have calcium oxalate crystals, which are irritating to the skin and mucous membranes. In spite of this poisonous aspect of the fruit, most species have edible terminal buds and the sweet sap exuding from incised inflorescences is used to make sugar or liquor. In the past, the pith of the larger trunks was processed to produce sago.

Because herbarium specimens are difficult to create, the names of some species are based on inadequate scraps that are difficult to interpret and match to living plants. Hence, taxonomic confusion still exists in this genus. The name is from a Greek word that means "nut," referring, one assumes, to the smooth seed within the fruit. Seeds germinate after 30–90 days, depending on how fresh they are. For further details, see *Arenga*.

Caryota cumingii
kar-ee-O-ta · koo-MIN-jee-eye

Caryota cumingii is a solitary-trunked species from the rain forest of the Philippines. The epithet commemorates British naturalist Hugh Cuming (1791–1865), who collected extensively in the Philippines. The palm grows to 30 feet (9 m) tall with a stout, pale, prominently ringed, straight trunk. The leaves are 10 feet (3 m) long and 6 feet (1.8 m) wide, in 2 or 3 closely set tiers atop the trunks; the apices of the leaves are slightly pendulous as are the secondary rachises, giving a fine, ruffled appearance to the blade. The species is tropical in its cultural requirements and adaptable only to zones 10b and 11. It needs copious water, a rich but well-draining soil, and full sun or partial shade. Because of its gorgeous trunk, it makes a fine specimen or canopy-scape.

Caryota kiriwongensis
kar-ee-O-ta · kee'-ree-won-GEN-sis

Caryota kiriwongensis is the leviathan of the genus, its massive single trunk towering to 100 feet (30 m) tall. It is a rare palm, endemic to high elevation rain forest of one small area in Thailand. The epithet means "of Kiriwong."

The trunk is about 2 feet (60 cm) in diameter and slightly swollen in the middle. The leaves are borne in a tight, distinct crown at the apex of the trunk, giving this palm the aspect of a gigantic tree fern. The leaves are about 25 feet (7.6 m) long, and the inflorescences are about 5 feet (1.5 m) long and pendulous. The mature fruits are purplish or reddish, and the endosperm is homogeneous, an unusual condition in *Caryota* and shared only with *C. ophiopellis* and *C. zebrina*. The first leaf of the seedling is pinnate, rather than two-lobed as is typical for the genus.

Causing a sensation when it was brought to the attention of palm enthusiasts in the late 1990s, this species is still not widespread in cultivation. Indeed, only very large gardens can accommodate this palm of Brobdingnagian proportions. It is a specimen plant, but one's pulse quickens at the mental image of a formal avenue lined with this palm. Its cold hardiness is unknown.

Caryota maxima
kar-ee-O-ta · MAX-i-ma
PLATE 185

Caryota maxima has a large natural distribution from southern China, Myanmar, Thailand, peninsular Malaysia, Laos, Vietnam, Sumatra, and Java, where it grows in low mountainous rain forest. Common names are giant fishtail palm and black palm. The Latin epithet means "largest." This species includes *C. ochlandra*.

In habitat, the solitary trunk can reach a height of almost 100 feet (30 m), but does not approach the gigantic proportions of *C. kiriwongensis*, and is 1 foot (30 cm) in diameter and gray to tan. The crown is elongated, and triangular leaves are 10–20 feet (3–6 m) long on a petiole 4 feet (1.2 m) long, mostly stiff and flat but usually pendent at their apices. The leaflets are narrowly triangular to almost linear, have a jagged end with a long point on one side, are 1 foot (30 cm) long or sometimes longer, and are generally pendulous. The many-branched, pendulous inflorescences are 6 feet (1.8 m) long and bear 1-inch (2.5-cm), elliptical, pink to red fruits in pendent clusters 8 feet (2.4 m) long.

This water lover is adaptable only to zones 10b and 11, although a few good specimens are found in favorable microclimates of 10a. It needs a rich, humus-laden soil but can endure the full sun of any but the hottest climates, even when young.

Caryota maxima is closely allied visually and genetically to *C. rumphiana*. Because of the elongated leaf crown, it has as much in common visually with narrow-crowned dicot trees as it does with other palms. It is almost unbelievably spectacular and tropical looking in groups of 3 or more individuals of varying heights to give a great tiered effect to the massive fernlike leaves. It is also sufficiently large and sensational to stand isolated, although it looks better either in groups of individuals of varying heights or as a canopy-scape. It can be grown indoors in a truly monstrous atrium or conservatory with lots of light and moisture.

Caryota mitis
kar-ee-O-ta · MY-tis
PLATE 186

Caryota mitis is widespread and indigenous to open, cleared areas in the rain forest of the Philippines, southern China, the Andaman and Nicobar Islands, Thailand, peninsular Malaysia, Southeast Asia, Sumatra, Borneo, Java, and Sulawesi. It is one of the few clustering species in the genus. The epithet is Latin for "mild" or "gentle," but the meaning is unclear. The only common English name is the somewhat trivial clustering fishtail palm.

Clumps of this palm can be 20 feet (6 m) tall and 15 feet (4.5 m) wide with densely packed stems to 12 feet (3.6 m) tall. The leaves are 8 feet (2.4 m) long and triangular on petioles 3 feet (90 cm) long. The leaflets are 6 inches (15 cm) long, narrowly triangular with usually 3 lobes, the terminal one being drawn out into a long point. There are but few leaves atop any trunk but, because the clumps are dense, the visual effect is a wall of leaves. The inflorescences are 2 feet (60 cm) long and pendent, with many tiny purplish flowers. The rounded ½-inch (13-mm) fruits are deep reddish black when mature.

Caryota mitis is slightly susceptible to lethal yellowing disease. It is tropical in its climate requirements and adaptable only to zones 10b and 11, although nice specimens are found in favorable microclimates of 10a. It needs abundant water and a rich, humus-laden, free-draining soil, and thrives in full sun or partial shade. The plants are fast growing under ideal conditions. Propagation is via seed and rooted division of the suckers, with the former method being easier. Often, seedlings volunteer beneath the parent tree. This species is most beautiful as a large element in masses of other vegetation, but is effective standing alone if sited as a focal point. It is adaptable to indoor cultivation if given enough light and space.

Caryota monostachya
kar-ee-O-ta · mo-no-STAY-kee-a
PLATE 187

Caryota monostachya is native to southern China and Vietnam. The epithet is from the Greek meaning "one spike" and refers to the unbranched inflorescence, a unique condition in the genus.

The clumps of clustering trunks can be 10 feet (3 m) tall and 5 feet (1.5 m) wide. Individual stems are slender, less than 2 inches (5 cm) in diameter, with leaves borne throughout the length of the stem and forming a dense, bushy screen. The inflorescences may be as long as 3 feet (90 cm) and are unbranched, although rarely an inflorescence may have 2 or 3 flower-bearing branches. Flowers are reddish purple, and the fruits are 1¼ inches (3 cm) in diameter, purplish when mature.

Not yet common outside botanical gardens, this species has the habit of a shrub, recommending it for smaller gardens. It makes an excellent screen in a site that has partial shade, abundant moisture, and loamy soil. Its cold hardiness is not yet known, but accessions from higher elevations and latitudes may be more cold hardy than those from lowland rain forest.

Caryota no
kar-ee-O-ta · NO
PLATE 188

Caryota no is endemic to the rain forest of Borneo, where it is endangered because the trees are felled for their edible growing points. The epithet is the aboriginal name for this palm. The only common English name seems to be giant fishtail palm.

The gray to almost white solitary trunks attain 75 feet (23 m) of height and are 12–18 inches (30–45 cm) in diameter except for a bulge in the middle that is 2 feet (60 cm) in diameter. The stems are beautifully ringed on their older parts, the rings widely spaced, and the upper portions of the trunks are covered in giant leaf bases. The leaf crown is 25 feet (7.6 m) wide and 20 feet (6 m) tall. The triangular leaves are 15 feet (4.5 m) long and 12 feet (3.6 m) wide, erect, and mostly ascending; while the rachis is unbending, the subsidiary stalks are pendulous, giving the palm a lush, tropical look. The narrowly triangular leaflets are 1 foot (30 cm) long, irregularly toothed on their apices, and slightly pendent. The pendent inflorescences are 6 feet (1.8 m) long, and the fruits are black when mature.

This palm is not hardy to cold and is adaptable only to zones 10b and 11. It needs a lot of water and a rich, humus-laden, well-drained soil. It grows well in full sun or partial shade when young. Startlingly beautiful, this species looks as much like a colossal tree fern as a palm. It is unsurpassed when planted in groups of 3 or more individuals of varying heights to create a stunning tiered, fern effect, and is even effective as an isolated specimen. Because it needs immense space, only the largest atriums or conservatories would be able to accommodate this species.

Caryota obtusa
kar-ee-O-ta · ob-TU-sa
PLATE 189

Caryota obtusa is indigenous to wet mountainous forests in southern China, northern Laos, and northeastern Thailand, where it grows usually in colonies from elevations of 4000–5280 feet (1200–1610 m). A synonym is *C. gigas*. There is no widely accepted common English name, but mountain fishtail and giant fishtail are sometimes used. One has to wonder about the need for a common name, so short and sweet is the epithet, which is Latin for "obtuse" and refers to the rounded tips of the individual leaflets.

In nature, the trunks of this enormous species attain heights of more than 100 feet (30 m), are tan and columnar, swollen in the middle, and are usually supported by a cone of stilt roots at their bases. The canopy is sparse with leaves which grow only from the top of the trunk. The triangular leaves are enormous, 20 feet (6 m) long and half as wide. The leaf apex is pendent as are the secondary rachises. The inflorescence is among the largest for a flowering plant; it is much branched, pendent, and to 20 feet (6 m) long. The red fruits are globose.

The species is moderately cold hardy but not to the degree claimed by some growers. It has endured temperatures as low as 30°F (−1°C) with little or no damage and is a good palm for the cooler tropical or nearly tropical regions, such as coastal Southern California, if it receives enough moisture. It needs a rich, humus-laden soil and copious moisture, and is adapted to full sun in all but the hottest regions.

One can only think of a tall and massive tree fern when considering the landscape potential of this palm. It is large and beautiful enough to be a specimen plant, especially in groups of 3 or more individuals of varying heights. Its spindle-shaped trunks recall Doric columns, even more so when planted in formal designs. As a canopy-scape nothing could be more impressive. Its incredibly beautiful leaf silhouette is absolutely magical against the sky. It can be grown indoors in a truly monstrous atrium or conservatory with lots of light and moisture.

Caryota ophiopellis
kar-ee-O-ta · o'-fee-o-PEL-lis
PLATE 190

Caryota ophiopellis is a rare solitary-trunked palm from Vanuatu, where it grows in lowland rain forest near the coast. It was described by John L. Dowe in 1996. The common name is snakeskin palm, which is a literal translation of the Greek epithet and refers to the distinctive mottled pattern on the leaf sheaths and petioles.

The palm grows to 20 feet (6 m) high, and the trunk is 18 inches (45 cm) in diameter, white with wide, dark rings of leaf base scars spaced 1 foot (30 cm) apart. The leaf crown is sparse and the leaves 10 feet (3 m) long, with deep green leaflets regularly spaced along the secondary rachises; the leaflets have a smooth margin on one side and are shaped like an irregular triangle, with the shorter sides jagged. The leaf sheath and petioles are deep green to almost black and irregularly circled with white or pinkish bands of tomentum, which extend to the rachis and secondary rachises. The round ½-inch (13-mm) fruits are deep crimson when ripe. The stigmatic scar on the fruit is not truly apical, as it is in the similar *C. zebrina*.

This undergrowth palm is not good in full sun or in calcareous soils but can thrive with regular water. It is adaptable only to zones 10b and 11.

Caryota rumphiana
kar-ee-O-ta · rum-fee-AH-na
PLATE 191

Caryota rumphiana is a variable species occurring naturally in open, cleared areas of the rain forest of the Philippines, eastern Indonesia, New Guinea, the Solomon Islands, and northern Queensland, Australia. The epithet honors German botanist Georg E. Rumphius (ca. 1627–1702), who resided in eastern Indonesia, where he wrote *Herbarium Amboinense*, a seminal work in tropical Asian botany.

Mature trunks attain 60 feet (18 m) of height and 12–18 inches (30–45 cm) in diameter, and are deep tan to light gray, with beautiful, widely set rings on the older parts. The leaf crown is large and extended, 20–40 feet (6–12 m) high and 20 feet (6 m) wide. The triangular leaves are 20 feet (6 m) long and 15 feet (4.5 m) wide on petioles 3 feet (90 cm) long. The leaflets are linearly triangular to almost linear, have jagged ends, are 18 inches (45 cm) long, and are generally pendulous. The pendulous inflorescences are 6 feet (1.8 m) long, and many branched, with pinkish purple flowers. The rounded ½-inch (13-mm) fruits are black when mature.

The palm is slightly susceptible to lethal yellow disease. It does not tolerate cold and is possible only in zones 10b and 11. It is a true water lover and needs a humus-rich soil as well as full sun when older. The species is fast growing, which is good because the palm rapidly reaches full size, but is also bad because it can start flowering and therefore dying when it is only 15 years old. The landscaping attributes of this majestic palm are the same as those of *Caryota no*. Indoors it would need an extraordinarily large space and good light.

Caryota urens
kar-ee-O-ta · YOO-renz
PLATE 192

Caryota urens occurs naturally in open, cleared areas of the rain forest of Sri Lanka and India through southern China, Nepal, Myanmar, and Malaysia. The epithet is Latin for "stinging," a reference to the irritating juice in the fruits. This species seems to have more common English names than any other, although none are that apt: solitary fishtail palm, toddy palm, jaggery palm, and wine palm.

Mature trunks grow to 40 feet (12 m) high and about 1 foot (30 cm) in diameter, and are light to dark gray with widely spaced rings on the older parts. The leaf crown is extended, 15–20 feet (4.5–6 m) tall and 20 feet (6 m) wide. The leaves are generally 12 feet (3.6 m) long and widely triangular on stout petioles 2 feet (60 cm) long. The leaflets are narrowly triangular and jagged on one of the wide ends, drawn out into a point on the other; they are usually 1 foot (30 cm) long and are a grassy to glossy deep green. The inflorescences are large, much branched and pendulous, to 10 feet (3 m) long, with whitish flowers. The round ½-inch (13-mm) fruits are red.

The species is adapted only to zones 10b and 11 and is marginal in 10a. It needs abundant, regular water and a well-drained humus-rich soil. It is a moderately fast grower in partial shade to full sun. The sap in the trunks is tapped to make sugar, alcoholic beverages, and sago. The trunks are also used for construction.

Caryota zebrina
kar-ee-O-ta · ze-BRY-na
PLATES 193–194

Caryota zebrina was described in 2000 by Dransfield et al. as a species with striped sheaths, petioles, and rachises. It has been a collector's item and quite the rage. The epithet is Latin for "zebra-like." The common name is zebra fishtail palm.

This solitary-trunked species is endemic to Indonesian New Guinea, where it grows in the Cyclops Mountains at an elevation of 2500–4600 feet (760–1400 m). The species has a remarkable similarity to *C. ophiopellis*: the petioles of both are striped. Otherwise the trunk of *C. zebrina* can attain 50 feet (15 m) of height, and the leaflets are more diamond shaped and glossier than those of *C. ophiopellis*. The secondary rachises have few leaflets, while the apical one is much larger than the others and also heavily red-veined. Both species have a homogeneous endosperm, an unusual condition in *Caryota*. The fruit of *C. zebrina* has an apical stigmatic scar, whereas the scar is displaced from the apex of the fruit in *C. ophiopellis*.

The species is adapted to zones 10b and 11 and is marginal in 10a. It thrives with abundant, regular water and a rich, well-draining soil high in organic matter. *Caryota zebrina* is easier to grow on alkaline soils than is *C. ophiopellis*. It is a moderately fast grower in partial shade. Both species are best used where their strikingly patterned petioles can be appreciated up close.

Ceratolobus
se-ra´-to-LO-bus

Ceratolobus is a genus of 6 densely clustering, spiny, climbing, pinnate-leaved, dioecious palms in Thailand, peninsular Malaysia, Sumatra, Borneo, and Java, where they grow in low mountainous rain forest, often forming nearly impenetrable thickets.

The palms climb mainly by their spiny, flexible stems sprawling and leaning against supports; they do not have flagella but they do exhibit short cirri. None of the species reach the giant dimensions of some *Calamus* and *Laccosperma* species and other rattans. The leaflets may be diamond shaped or linear, but new growth is usually tinged pinkish. The unusual inflorescence is enclosed by a single large, usually pendent and persistent bract that opens partially to allow pollinators access to the flowers.

None of the species are commonly in cultivation, and the genus is doubtless exceedingly rare in enthusiasts' collections and in botanical gardens. The genus name comes from Greek words meaning "horn" and "pod" and refers to the unusual inflorescences. Seeds are not generally available, so little information is known about their germination.

Ceratolobus glaucescens
se-ra´-to-LO-bus · glaw-SES-senz
PLATE 195

Ceratolobus glaucescens is native to lowland rain forest in Thailand and western Java, where it is exceedingly rare. In cultivation, this species is not seen outside of botanical gardens. The epithet is Latin for "slightly glaucous." The sheaths, petioles, and leaf rachises are covered with spines. Leaflets are in 6–8 pairs and are diamond shaped with an elongated drip-tip. They are beautifully chalky white on the underside. The endosperm is homogeneous. It is strictly tropical but takes readily to cultivation if given adequate moisture and warmth.

Ceroxylon
se-RAHX-i-lahn

Ceroxylon includes about a dozen tall, solitary-trunked, pinnate-leaved, dioecious palms in rain forest of the Andes. The genus comprises not only the tallest palms in the world (in fact, they are the world's tallest monocots) but also those growing at the highest elevations. Some of these species are beyond spectacular in appearance. The straight, columnar trunks are heavily ringed and are covered thinly or thickly with wax in all but their oldest parts. The leaf scars (rings) on the trunk are distinctive in that they are wavy and uneven; they do not encircle the trunk in a straight, level line. There is no true crownshaft, but the leaf bases are large and sheathing and sometimes give the appearance of a crownshaft. The large inflorescences are obvious even from a distance because they project well beyond the leaf crown or hang below it, as do the large clusters of red or orange fruit.

All species are difficult to grow outside their native habitats and are impossible in hot, humid climes in which the night temperatures do not drop much; they are most at home in cool, moist climates and, while the species from high elevations are frost tolerant, they do not tolerate heat. All species are extremely slow growing, often taking many years to develop a trunk. The vernacular name wax palm or Andes wax palm is applied to most species. The genus name is derived from Greek words meaning "wax" and "wood."

Like other species of the tribe Ceroxyleae, *Ceroxylon* species are among the easiest to grow from seed, and seeds germinate quickly. Seeds allowed to dry out completely before sowing do not germinate as well as those kept slightly moist, and the smaller the seed, the more likely it will dry out before sowing. Seeds shipped in a bag for several days often begin germinating before reaching their destination. Fresh seeds of *Ceroxylon* generally germinate within 60 days but can continue sporadically for several months.

Ceroxylon alpinum
se-RAHX-i-lahn · al-PY-num
PLATE 196

Ceroxylon alpinum occurs naturally in 6 disjunct regions of the central Andes in Venezuela, Colombia, and Ecuador, where it grows in wet forests at 4500–6500 feet (1370–1980 m). Because of land clearing, the species is almost extinct. The epithet is Latin for "alpine." Two subspecies are recognized: **C. alpinum subsp. alpinum**, from Venezuela and Colombia, and **C. alpinum subsp. ecuadorense** [ek-qua-dor-EN-see], endemic to Ecuador. The trunk grows to 65 feet (20 m) tall, is tan to almost white with widely spaced darker rings, and is covered in a thin layer of wax. The leaves are 8–10 feet (2.4–3 m) long with many 18-inch (45-cm) pendent leaflets that grow in a single plane and are evenly spaced along the rachis; they are bluish green above but gray beneath. The inflorescences are pendulous and hang against the leaf bases.

Ceroxylon amazonicum
se-RAHX-i-lahn · am-a-ZAH-ni-kum
PLATE 197

Ceroxylon amazonicum is endemic to southeastern Ecuador, where it grows in the rain forest at 2640–3800 feet (800–1160 m), the lowest elevations of any *Ceroxylon*. The epithet is Latin for "of Amazon." This is one of the smaller species with whitish trunks attaining 60 feet (18 m). The beautiful leaves resemble those of the coconut palm (*Cocos nucifera*) and exhibit a twist to the apical half of the rachis which causes the leaflets in that part of the leaf to be held vertically; they are deep green above and have a gray, yellowish, or whitish waxy coating beneath.

Ceroxylon ceriferum
se-RAHX-i-lon · se-RI-fe-rum
PLATE 198

Ceroxylon ceriferum is indigenous to northern Colombia and Venezuela, where it grows in the rain forest at 6000–9000 feet (1800–2700 m). The epithet is Latin for "wax-bearing." The tan trunks reach heights of 75 feet (23 m). The leaves are similar to those of *C. alpinum* and *C. amazonicum* except that the rachises are much more ascending or erect, especially when young, and the leaf crown is never fully rounded.

Ceroxylon parvifrons
se-RAHX-i-lahn · PAHR-vi-frahnz
PLATE 199

Ceroxylon parvifrons occurs naturally in the wet Andes of Venezuela, Colombia, Ecuador, Peru, and Bolivia at 6000–10,000 feet (1800–3000 m), the greatest naturally occurring elevation of any palm. The epithet is from Latin words meaning "small frond." The species is one of the smaller ones with a tan trunk to 40 feet (12 m) tall. The smallish sharply ascending, arching leaves are unusual for the genus in having stiff, bluish green leaflets. These grow from the rachis at an angle which gives the leaf a V shape. The leaf crown is sparse and never forms more than a semicircle; the leaf bases are large and prominent and almost form a crownshaft.

Ceroxylon parvum
se-RAHX-i-lahn · PAHR-vum
PLATE 200

Ceroxylon parvum occurs naturally in 2 disjunct regions of the Andes: the eastern slopes in Ecuador and also in southern Peru and western Bolivia, where it grows in rain forest and the clearings therein from 4500 to 5700 feet (1370–1740 m). The epithet is Latin for "small." This species is the smallest in the genus, the trunk attaining only a maximum of 30 feet (9 m) in habitat, with a diameter of 4–7 inches (10–18 cm). The leaf crown is globular and the individual leaves, each 6 feet (1.8 m) long, bear narrow leaflets which grow at several angles off the rachis to create a plumose leaf.

Ceroxylon quindiuense
se-RAHX-i-lahn · keen-dee'-oo-EN-see
PLATE 201

Ceroxylon quindiuense is endemic to the Andes of Colombia, where it grows at 6000–10,000 feet (1800–3000 m) under wet conditions. It is the national tree of Colombia. The epithet is Latin for "of Quindío," a region in the western Andes.

This species is the tallest palm (and tallest monocot) in the world, the straight, columnar trunks growing to 200 feet (60 m) high. They are white, due to a wide thick coating of wax, and bear beautiful dark and widely spaced rings in their younger parts. They have a stiff, shuttle-cock crown of leaves as juveniles, but the crown becomes more hemispherical once the palms reach reproductive maturity. The leaves are 15–18 feet (4.5–5.4 m) long, with pendent dark green leaflets 2 feet (60 cm) long and bearing a thick white or yellowish indument underneath.

There is no more spectacular palm tree but, even if one has a favorable climate, one would not live long enough to see such grandeur on one's property. The good news is that even young individuals are massive and splendid with their great ascending leaves.

Ceroxylon ventricosum
se-RAHX-i-lahn · ven-tri-KO-sum
PLATE 202

Ceroxylon ventricosum is indigenous to the Andes of Ecuador and southern Colombia at 6000–10,000 feet (1800–3000 m) in wet forests. The epithet is Latin for "bulged" or "swollen." This species is similar in general appearance to *C. quindiuense*, differing mainly in its height, leaf form, and trunk size. The trunks grow to half the height of *C. quindiuense* and seem more massive because they are as thick or even thicker; they are also usually slightly swollen or bulging near their middles, are white or nearly so, and bear beautiful undulating dark rings of leaf base scars. The leaf crown is possibly the most rounded in the genus, and the leaves are sumptuous because they are plumose, almost like a royal palm (*Roystonea*). One is tempted to call this the alpine royal. It is as beautiful as the royal palm.

Ceroxylon vogelianum
se-RAHX-i-lahn · vo-gu-lee-AH-num
PLATE 203

Ceroxylon vogelianum has the largest natural range in the genus and is indigenous to the Andes in Venezuela, Colombia, Ecuador, Peru, and western Bolivia, where it grows in the rain forest at 6,000–10,000 feet (1800–3000 m). The epithet honors Julius R. T. Vogel (1812–1841), a German botanist and specialist in South American legumes. It is one of the smaller species, to 35 feet (11 m) overall, with relatively thin, indistinctly ringed trunks that are brown in their older parts and light green in their younger parts. The leaves are the most plumose in the genus because the bluish green leaflets grow in clusters and from all directions around the straight, stiffly upright rachis. The inflorescence is on a long stalk that arches gracefully below the crown.

Chamaedorea
ka-mee-DOR-ee-a

Chamaedorea consists of more than 100 mostly small to quite small, pinnate-leaved, dioecious palms in tropical America, with most species in Mexico and Central America. Many are solitary trunked, a few are clustering, some are truly trunkless, and one is a climber. Some species exhibit true crownshafts, others have rudimentary crownshafts, but most have no crownshaft. All the trunked species have prominent rings on their stems and, because of this characteristic, are usually referred to as bamboo palms.

Many species exhibit startling variation of leaf detail between young and adult individuals. Provenance is also an important consideration. The leaves are either pinnate or undividedly pinnate, and all have an even number of leaflets. Some species have the unusual characteristic (for palms) of producing adventitious roots on their stems, which allow them to be air-layered, in effect shortening excessively tall, leggy plants. Unfortunately, the lower part of the stem of the original plant will not sprout anew and invariably dies.

The inflorescences are either male or female and grow from stem nodes among the leaves or beneath the leaf crown. Their form ranges from simple spikes to severally branched, and the tiny flowers are often fragrant. In many species, the green branches of the inflorescence turn bright orange as the fruits mature and contrast nicely with the shiny black fruits. Almost all species have fruits with calcium oxalate crystals that are irritating to sensitive skin.

Almost all the species hail from mountainous rain forests or cloud forests and are undergrowth subjects, which allows them to thrive in shade or partial shade. (In the following descriptions, only the cold hardiness of the individual species is mentioned unless the particular palm's other cultural requirements are different from the majority.) Few large palm genera have as many cultivated representatives as does this one. Virtually all species make excellent subjects for indoors or greenhouses, atriums, and conservatories. The genus name is derived from Greek words that mean "on the ground" and "gift," in reference to their typically small stature and great beauty. Happening upon them in the forest would indeed be like finding a gift on the ground.

Seeds can take 90 days to a year or more to germinate and do so in a fairly short time frame. They can be stored for a few months but are best planted soon after harvest. For further details of seed germination, see *Hyophorbe*.

Chamaedorea adscendens
ka-mee-DOR-ee-a · ad-SEN-denz
PLATE 204

Chamaedorea adscendens is a small, rare solitary-trunked species indigenous to northern Guatemala and southern Belize, where it grows in rain forest up to 2000 feet (610 m). The epithet is Latin for "ascending," referring to the upright stem. The trunks grow to 6 feet (1.8 m) high. The leaves, mostly divided pinnately in mature individuals but bifid in some individuals, are 12 inches (30 cm) long. The obovate leaflets have rounded ends, and are thick and leathery, a distinct bluish green, with a velvety appearance that adds even more to the unusual color and sheen of the leaves. This little beauty is not hardy to cold but, because of its small size, is easily protected from occasional frosts and is adaptable to zones 10b and 11 and marginal in 10a.

Chamaedorea allenii
ka-mee-DOR-ee-a · al-LEN-ee-eye

Chamaedorea allenii is a solitary-trunked species from Colombia, Panama, and perhaps also Costa Rica, where it grows in montane rain forests. The epithet honors Paul H. Allen (1911–1963), American botanist specializing in Central American plants, palm and orchid enthusiast, and charter member of the International Palm Society.

This palm grows to 6 feet (1.8 m) tall on a green prominently ringed stem ½ inch (13 mm) in diameter. The leaves of seedlings and young plants may be bifid, but those of adult plants are pinnate and medium green. The most distinctive feature of this species is the inflorescence: in both sexes it is spicate (rarely forked), pendulous from the end of a long stalk, densely covered with short-lived, yellow-orange flowers. The crowded spike looks like a slender ear of corn (maize). The fruits are black when ripe. Not reliably cold hardy outside of zone 10a, this species makes an excellent subject for conservatory or greenhouse culture.

Chamaedorea amabilis
ka-mee-O-ee-a · a-MAH-bi-lis
PLATE 205

Chamaedorea amabilis occurs naturally in the cloud forests of Costa Rica and Panama, where it is nearly extinct because of land clearing and overzealous collecting. The epithet is Latin for "lovely." The solitary stems are never more than 6 feet (1.8 m) tall or ½ inch (13 mm) in diameter, and are green and indistinctly ringed. The obovate leaves consist of fused pinnae and are deeply bifid apically. They are bright green, and have toothed margins and S-shaped ridges corresponding to the margins of the conjoined pinnae. This species is a deliciously beautiful little thing for undergrowth planting, needing copious moisture and a rich, well-drained soil. It is not hardy to cold and is adaptable only to zones 10b and 11, although its diminutive size makes it possible with protection in 10a. It struggles to survive in hot, tropical climates.

Chamaedorea arenbergiana
ka-mee-DOR-ee-a · ar'-en-bur-gee-AH-na

Chamaedorea arenbergiana is indigenous to southern Mexico, central Guatemala, and northeastern Honduras, where it grows in low mountainous rain forest. The epithet commemorates Luis Englebert, 6th Duke of Arenberg (1750–1820), Belgian patron of science, from whose garden the palm was first described.

The solitary trunk sometimes attains a height of 10 feet (3 m) but is only 1 inch (2.5 cm) in diameter, light to dark green, with long internodes and gorgeous rings of golden-brown leaf base scars. The leaves are 5–6 feet (1.5–1.8 m) long with handsome, widely spaced glossy dark green ovate-lanceolate leaflets, 2 feet (60 cm) long and 6 inches (15 cm) wide, each with a long drip tip; they are more closely spaced and more definitely S shaped in juvenile plants.

This palm is one of the choicest "bigger" species in the genus and is irresistibly magnificent when planted in an intimate site. It cannot tolerate frost and needs a humus-laden soil, lots of water, and partial shade. It does well in a frostless Mediterranean climate.

Chamaedorea benziei
ka-mee-DOR-ee-a · BEN-zee-eye

Chamaedorea benziei comes from the Pacific slope of Chiapas, Mexico, at over 5000 feet (1520 m). The epithet, bestowed by Donald Hodel in 1992, honors palm enthusiast James Benzie of Orange, California. The solitary stem, 1–1½ inches (2.5–4 cm) in diameter, reaches 15 feet (4.5 m) tall. The leaves are 4 feet (1.2 m) long, with 40–44 linear leaflets, which are thick and durable and covered with a fine waxy "bloom." The much-branched inflorescence, borne among the leaves on a long peduncle, is erect and spreading. This species has been in continuous cultivation in Southern California since the mid-20th century, suggesting that it is amenable to cultivation. It would be stunning in an intimate grouping in a garden or interior-scape.

Chamaedorea brachypoda
ka-mee-DOR-ee-a · bra-kee-PO-da
PLATE 206

Chamaedorea brachypoda is a clustering species indigenous to low mountainous rain forest in northeastern Honduras, where it is now threatened with extinction because of agricultural expansion and collection of plants for the nursery trade. The epithet is derived from Greek words meaning "short" and "foot," an allusion to the rhizomes. The separate clumps can be large, wide, and dense with stems to 5–6 feet (1.5–1.8 m) tall. Atop the stems reside the mostly undivided and apically bifid leaves, 18 inches (45 cm) long, obdeltoid, dark green, and heavily ribbed. The lower leaves are sometimes partially segmented. This palm does not tolerate cold and is adaptable only to zones 10b and 11, although with protection it does nicely in microclimates of 10a. It needs constant moisture, partial to nearly full shade, and a rich, slightly acidic soil. This species makes a wonderful groundcover as each plant spreads far by creeping rootstocks.

Chamaedorea cataractarum
ka-mee-DOR-ee-a · kat'-a-rak-TAH-rum
PLATE 207

Chamaedorea cataractarum occurs naturally in the Mexican states of Oaxaca, Chiapas, and Tabasco, where it grows along rivers and streams in the rain forest. The common names are cat palm, cataract palm, and cascade palm, all of which are reasonable alternatives to the somewhat clumsy Latin epithet which translates as "of the cataracts." In nature, this palm is a rheophyte, thriving in fast-moving water currents.

The clustering trunks, which are usually not visible because they are prostrate, are unusual in being forked near their bases. Clumps can be 6 feet (1.8 m) high and 10 feet (3 m) wide. The leaves are 4 feet (1.2 m) long with many narrowly lanceolate, dark green shiny leaflets 1 foot (30 cm) long growing in a flat plane along both sides of the rachis. The inflorescences are 18 inches (45 cm) long, much branched, and grow near the ground from the leaf nodes, bearing small, yellow blossoms. The round ½-inch (13-mm) fruits are borne in pendulous clusters.

This palm is adaptable to zones 10 and 11 and in favorable microclimates of 9b. It thrives in full tropical sun if given regular water. In the landscape it is used as a clump of contrasting foliage within a larger mass of vegetation or as a tall groundcover. When used as a screen or hedge, it has the unfortunate characteristic of trapping the old, dead leaves among the green leaves, so one must be vigilant in removing dead leaves to keep the hedge looking its best. Planted in a heavy pot and plunged into a pond, this palm can even be grown as a semiaquatic plant, mimicking its natural habitat along streams and river banks. It is delicate and ferny in appearance and lends a beautiful gracefulness to any site in which it is planted. Unlike most species, it needs so much water and humidity that it is a less-than-ideal houseplant.

Chamaedorea correae
ka-mee-DOR-ee-a · kor-RAY-ee
PLATE 208

Chamaedorea correae is endemic to Panama, where it grows in high-elevation cloud forest. The epithet honors Mireya Correa, a Panamanian botanist. The very slender, solitary stem is procumbent, meaning that it is not entirely self-supporting and frequently leans on other plants or scrambles along the ground, rooting at the nodes. Consequently, the plant is a good candidate for air-layering when it becomes too leggy. The leaves are bifid, leathery, and dull green, with toothed margins, although some populations exhibit leaves that divide into a few pinnae. This palm is an elegant subject for greenhouses or conservatories, where high humidity and soil moisture can be maintained without interruption. It requires a soil high in organic matter. So beautiful is this palm that it is worth any effort to meet its horticultural requirements. Its cold tolerance is unknown.

Chamaedorea costaricana
ka-mee-DOR-ee-a · kost-a-ree-KAH-na
PLATE 209

Chamaedorea costaricana occurs in southern Mexico and throughout Central America except for Belize, growing in lowland to mid-elevation rain forest. Plant size in habitat is related to rainfall amounts: plants in areas of abundant rainfall (usually at the higher elevations of the palm's range) are larger in all aspects than their counterparts in areas of more seasonal and less total rainfall (Hodel 1992). The epithet is Latin for "of Costa Rica." One common name is the Costa Rican bamboo palm.

Mature clumps of this clustering species are 6–10 feet (1.8–3 m) wide and 12–25 feet (3.6–7.6 m) tall, with mature trunks to 20 feet (6 m) high and 1 inch (2.5 cm) in diameter. The dark green stems have light green rings. The crownshaft is indistinct. The leaves are 4 feet (1.2 m) long and beautifully arching on thin petioles 2 feet (60 cm) long. The many, slightly pendent leaflets are linear-lanceolate, 8 inches (20 cm) long, olive green to deep green, and thin. They arise from the rachis in one plane, giving a flat appearance to the blade. From below the leaves emerge the many thin, pendent branches, each 3 feet (90 cm) long and carrying small, yellow to light orange flowers. The round black fruits are usually less than ½ inch in diameter (13 mm).

This palm is adaptable to zones 10 and 11 and is occasionally found in favorable microclimates of 9b. It often regrows new canes if the old ones are frozen to the ground and the ground itself does not freeze. It does well in a frostless or nearly frostless Mediterranean climate.

Few plants are as robustly graceful as a well-grown clump of this species. It is one of the finest landscape subjects for shade or partial shade, and is simply superb as a patio specimen or as a large accent in masses of vegetation. With many stems per clump, and all at varying heights, the trunks are usually obscured by a wall of foliage; such clumps can be made much more beautiful if judiciously thinned to allow the lovely trunks to show.

Chamaedorea dammeriana
ka-mee-DOR-ee-a · dahm-mer-ee-AH-na

Chamaedorea dammeriana is found at mid-elevations in Nicaragua, Costa Rica, and Panama. The epithet honors German botanist C. L. Udo Dammer (1860–1920). *Chamaedorea chazdonii* is a synonym of this species. The solitary stem is 3–6 feet (90–180 cm) tall, with magnificent pinnately divided leaves, 12 inches (30 cm) long. The pinnae are S shaped and somewhat "cupped" (with margins turned up) or "hooded" (with margins turned down), and the apical pinnae often are wider than the others. A well-grown plant can hold a dozen or more glossy green leaves at one time. The inflorescence is unbranched or divided into 2 or 3 branches. The shiny black fruits are held on orange infructescences, making this an extremely attractive species. One can scarcely imagine a more beautiful small palm for an intimate garden setting or a conservatory. It requires a rich growing medium and adequate moisture at all times.

Chamaedorea deckeriana
ka-mee-DOR-ee-a · dek'-e-ree-AH-na
PLATE 210

Chamaedorea deckeriana originates in Costa Rica and Panama, where it grows in rain forest. According to Hodel (1992), the epithet "honors 19th-century German botanist G. H. Decker, from whose garden the type originated." The solitary trunk is 1 inch (2.5 cm) thick and sometimes attains a height of 8 feet (2.4 m) but is usually shorter; it is medium green and bears widely spaced dark rings of leaf base scars. The leaves are 2 feet (60 cm) long, undivided except for the bifid apex, borne on petioles 2 feet (60 cm) long, and are dark and show corrugations that correspond to the fused pinnae. This palm languishes and even dies in calcareous soils and needs a humus-laden, acidic medium.

Chamaedorea deneversiana
ka-mee-DOR-ee-a · den-e-ver'-see-AH-na
PLATE 211

Chamaedorea deneversiana comes from Panama and Ecuador but has not been found in intervening Colombia, despite several decades of active palm exploration of that country. It may occur in Costa Rica, but additional collections are needed. It is exceedingly rare. The epithet honors American botanist Greg de Nevers (1955–), specialist in Central American palms and aroids. This solitary slender stem, up to 9 feet (2.7 m) tall, often leans and develops roots from the nodes. It bears only a few leaves in its crown, but the leaves are about 1 foot (30 cm) long, undivided except for the bifid apex. They resemble the more common *C. metallica*, and often have a similar, but less pronounced, bluish metallic sheen. The inflorescence, with several pedant branches, is borne below the leafy crown. The fruits are black. Not yet common in cultivation, this species is nevertheless a handsome palm worthy of a place in any collection where it can be sheltered from wind and assured constant moisture.

Chamaedorea elatior
ka-mee-DOR-ee-a · ee-LAY-tee-or
PLATES 212–213

Chamaedorea elatior is a climbing palm of Mexico, Guatemala, and Honduras, at low to middle elevations. Only this species, the rattans (*Calamus* and its kin), *Desmoncus*, and 2 species of *Dypsis* in Madagascar have the ability to climb. The epithet is Latin for "taller," presumably in reference to the stature of this species, which because it is a climber, is taller than other species of *Chamaedorea*.

The trunk is usually solitary, although clustering individuals are known, and is slender to about 60 feet (18 m) long. The leaves are up to 5 feet (1.5 m) long, arching and regularly pinnate, with medium green, lax pinnae. This species climbs by a remarkable adaptation of the leaf tip. The stiff, short, and backwards-pointing (toward the base of the leaf) apical pinnae function like grappling hooks, allowing the palm to hook onto neighboring trees. The leaves do not have the true cirrus of certain *Calamus* species, but the distinctive pinnae of *C. elatior* leave no doubt as to their function. The inflorescence is well branched and held close to the stem. The staminate flowers have a resinous fragrance.

This species has been cultivated in European greenhouses and conservatories since the mid-19th century. It has been long cultivated in the United States as well, although usually only as a curiosity.

Chamaedorea elegans
ka-mee-DOR-ee-a · EL-e-ganz
PLATE 214

Chamaedorea elegans grows naturally in the rain forest and cloud forests of Mexico, Guatemala, and Belize from sea level to 4500 feet (1370 m) in elevation. It is exceedingly rare in much of its original habitat (primarily Mexico) because of overcollecting. The epithet is Latin for "elegant." Common names are parlor palm and neanthe bella, the latter an illegitimate binomial that was never scientifically described or recognized.

The solitary stems seldom exceed 6 feet (1.8 m) tall but can reach 10 feet (3 m), and are ½–1 inch (13–25 mm) in diameter, light green with closely set dark green rings. The leaf crown is never more than 5 feet (1.5 m) wide and 3 feet (90 cm) high. The leaves are 18 inches to 3 feet (45–90 cm) long on thin petioles 6–12 inches (15–30 cm) long, erect but gracefully arching and seldom lying below the horizontal. The leaflets are narrowly ovate to linear, 8 inches (20 cm) long, and light to deep green; they grow from the rachis in a flat plane. The inflorescences are erect and branched, growing from among the leaves, and usually are longer than the crown. The tiny flowers are bright yellow, and the stems of the cluster turn orange to red after the flowers fade. The ¼-inch (6-mm) round black fruits are in erect clusters.

This palm is adaptable only to zones 10b and 11 and is marginal in 10a. Because the stems of older plants form adventitious roots, the palm may be air-layered, but the old stems do not produce leaves after being cut.

No species more deserves its descriptive epithet than this lovely little palm: it is delicate and elegant from youth unto old age. It is unexcelled as an accent in shady borders and, if expense is not a problem, makes one of the most beautiful tall groundcovers. Few things are more striking than an area devoted to this palm with small plants on the perimeter and larger plants towards the center of the mass. It is worth plunging pots of this palm into shady borders during the frostless seasons in almost any zone for the great beauty they lend to such areas. Parlor palm is probably the world's most widely grown indoor palm.

Chamaedorea ernesti-augusti
ka-mee-DOR-ee-a · er-nest'-ee-aw-GUS-tee
PLATE 215

Chamaedorea ernesti-augusti occurs naturally in lowland rain forest of southern Mexico, northern Guatemala, western Belize, and northern Honduras. The epithet commemorates Ernest August I (1771–1851), king of Hanover from 1837 until his death. The epithet has long been spelled "*ernesti-augustii*" but nomenclatural authorities now regard that spelling as incorrect.

Mature stems of this solitary-trunked species sometimes attain a height of 10 feet (3 m) but are usually 6 feet (1.8 m) with a diameter rarely more than ½ inch (13 mm). They are light to dark green with closely set rings of a much lighter color. The leaf crown is small and compact, 2 feet (60 cm) wide and 12–18 inches (30–45 cm) tall. The leaves are pinnate but undivided, held on thin petioles 4 inches (10 cm) long, and are broadly obovate with bifid apices very deeply cleft into 2 fishtail-like wedges. Leaf color is light to deep green, and the glossy surface has longitudinal grooves corresponding to the fused leaflets. The inflorescence is an erect spike arising from the leaf bases, with tiny white to yellow unisexual flowers. The spikes are significantly taller than the leaves and turn orange or red after the flowers fade. The egg-shaped fruits are ½ inch (13 mm) long and black.

This palm is not hardy to cold and is adapted only to zones 10b and 11 and marginal in 10a. Because stems of older plants form adventitious roots, the palm may be air-layered, but the old stems do not put forth leaves after being cut. This little palm is attractive as a single-specimen accent in masses of vegetation of a border; however, it is almost unsightly planted alone and surrounded by space, although it makes a nice silhouette. In groups of 3 or more individuals of varying heights, it makes a lovely tableau of leaves and stems.

Chamaedorea fragrans
ka-mee-DOR-ee-a · FRAY-granz
PLATE 216

Chamaedorea fragrans is indigenous to the eastern foothills of the Peruvian Andes, where it grows in rain forest to 2500 feet (760 m). The epithet is Latin for "fragrant" and refers to the flowers. This clustering species makes large, dense clumps to 15 feet (4.5 m) wide and looks as much like a small bamboo as it does a palm. The green stems grow to 10 feet (3 m) tall with widely spaced internodes. The dark green leaves are undivided except for the deep apical cleft, which is so pronounced that one must look closely to tell that the 2 resulting segments are not separate leaves. This species is excellent in a shady garden or in a large pot, as an indoor palm.

Chamaedorea frondosa
ka-mee-DOR-ee-a · frahn-DO-sa
PLATE 217

Chamaedorea frondosa is endemic to Honduras, where it grows in cool, wet cloud forests. The epithet is from the Latin for "leafy" in reference to the full crown of leaves. Less than 3 feet (90 cm) tall, this small, understory palm has entire, apically bifid leaves about 20 inches (50 cm) long. There may be as many as 15–20 leaves in the crown, and they are deep green and strongly nerved, with jagged margins. The inflorescence is erect and either unbranched or divided into 2–5 branchlets. The green flowers give way to black fruits on an orange infructescence. This lovely palm is perfect for a moist, shady garden or conservatory, not too hot in the summer but never freezing in the winter. Its dwarf stature and leafy crown give it a bushy appearance not seen in other *Chamaedorea* palms.

Chamaedorea geonomiformis
ka-mee-DOR-ee-a · jee-o-no'-mi-FOR-mis
PLATE 218

Chamaedorea geonomiformis occurs naturally in 2 regions of tropical America, the first being southern Mexico, central Guatemala, southern Belize, and northeastern Honduras, and the second in southwestern Costa Rica. It grows in rain forest from sea level to 3000 feet (900 m). The epithet means "in the form of *Geonoma*," another genus of tropical American palms. Individuals from the northern and southern extremes of the distribution have smaller leaves, so some taxonomists separate these diminutive forms as *C. tenella*.

The palm grows to 6 feet (1.8 m) high and has leaves that are 1 foot (30 cm) long, glossy green, and entire, with corrugations corresponding to the fused pinnae. The nerves are not prominent. The leaves are deeply bifid apically and are thin. The inflorescence is unbranched or sparsely branched. Those of the male are pendulous, while those of the female are erect. This beauty is not adapted to sun and needs regular moisture. It is adaptable to zones 10 and 11 and is easy to grow.

Chamaedorea glaucifolia
ka-mee-DOR-ee-a · glaw-si-FO-lee-a
PLATE 219

Chamaedorea glaucifolia is endemic to the state of Chiapas in Mexico, where it grows on limestone hills in the rain forest at 1600–3200 feet (490–980 m). It is in danger of extinction because of land clearing for coffee plantations. The epithet is Latin for "glaucous foliage."

The thin solitary stems grow to 12 feet (3.6 m) high, and the leaves are 6 feet (1.8 m) long and mostly ascending. The petiole is 1 foot (30 cm) long, and it and the rachis are usually covered in a waxy white bloom that is easily rubbed away to reveal a dark green surface. The 100 or more leaflets are narrow, gray-green and irregularly arranged in clusters along the rachis from which they also grow at different angles to give a diffuse plumose effect. The male flowers are fragrant.

In spite of its relatively tall stature, this palm needs to be sited where its sheer leaves can be appreciated: a strongly contrasting background is recommended. It is easy to grow and tolerant of lighter shade than other species of *Chamaedorea*.

Chamaedorea graminifolia
ka-mee-DOR-ee-a · gra-mi'-ni-FO-lee-a
PLATE 220

Chamaedorea graminifolia has a disjunct distribution in southeastern Mexico, central Guatemala, and southernmost Belize, but also in northwestern Costa Rica, where it grows over limestone from low elevations up to 2300 feet (700 m). The epithet is from Latin meaning "grassy foliage" and alludes to the narrow, grass-like leaflets.

This clustering species has well-separated stems to 10 feet (3 m) tall. The leaves are widely spaced from one another, leaving well-spaced leaf scars looking like nodes on a bamboo stem. The leaves are 4–5 feet (1.2–1.5 m) long and are beautifully arching on petioles 10 inches (25 cm) long. The evenly and widely spaced, slightly pendent leaflets, 50–80 per leaf, are 1 foot (30 cm) long or longer, deep green on both surfaces, and linear-lanceolate. The male flowers are fragrant. This species has a beautiful silhouette and needs to be planted against a wall or contrasting foliage where its delicate form can be seen.

Chamaedorea hooperiana
ka-mee-DOR-ee-a · hoo-per'-ee-AH-na
PLATE 221

Chamaedorea hooperiana is endemic the state of Veracruz in Mexico, where it grows at 3000–5000 feet (900–1520 m) in the last remaining fragments of tropical rain forest in North America. The epithet honors Louis Hooper, who grew this species for years in his La Habra, California, garden before it received a scientific name.

The trunks of this clustering palm are 12 feet (3.6 m) long. The clumps are to 12 feet or more (3.6 m) wide, and the stems bend gracefully outwards from the center of the mass. They are very dark green and ringed with light brown leaf base scars. Their upper parts are covered with the brown, persistent leaf sheaths. The leaf crown is sparse with leaves 3 feet (90 cm) long on petioles 1 foot (30 cm) long. The many leaflets are evenly spaced in a single flat plane, each one linear-elliptic. The inflorescences grow from among the leaves with coral-colored branches and bear small, round green fruits which mature to jet black, giving a festive combination of hues.

There is no more beautiful species in the genus, and it is glorious sited almost anywhere in partial shade. Alas, it is not cold tolerant and needs good soil that is continuously moist and yet well drained; however, the palm is well adapted to a frostless Mediterranean climate.

Chamaedorea klotzschiana
ka-mee-DOR-ee-a · klaht-skee-AH-na

Chamaedorea klotzschiana is endemic to the rain forest of the Mexican state of Veracruz, where it is in danger of extinction because of land clearing and overcollecting for the nursery trade. The epithet honors German botanist Johann F. Klotzsch (1805–1860).

The slender, dark green trunk grows to 12 feet (3.6 m) tall and is distinctly ringed with raised, nearly white leaf base scars. The crownshaft is small, plump, and endearing. The leaves are 3 feet (90 cm) long and unusual because of the distinctive arrangement of the leaflets, which grow in irregularly spaced groups along the rachis and, within each group, grow at slightly different angles from the rachis. Each leaflet is linear-elliptic, glossy medium to deep green, strongly nerved, and long tapering apically. The coral-colored infructescences produce shiny black little round fruits. This palm is not hardy to cold and is at home only in zones 10b and 11. It is a little gem, perfect in a site where it can be viewed up close, especially against a contrasting background.

Chamaedorea linearis
ka-mee-DOR-ee-a · lin-ee-AHR-iss

Chamaedorea linearis is indigenous to eastern Venezuela, northern and western Colombia, western Ecuador, the western slope of the Peruvian Andes, and west central Bolivia, where it grows in rain forest from sea level to 9000 feet (2700 m) in elevation. The epithet is Latin for "linear," a reference to the leaflets.

The solitary trunk attains a height of 30 feet (9 m) and is never more than 3 inches (8 cm) in diameter; it is yellow-green to pure yellow in its older parts and has widely spaced brown rings of leaf base scars and a small cone of aerial roots at its base. A crownshaft is obvious in most individuals. The leaves are 10 feet (3 m) long, with many slightly limp, linear but wide and pointed, bright green leaflets 2 feet (60 cm) long and growing in a nearly flat plane from the rachis. Several inflorescences grow from each node around the upper part of the trunk and have numerous, pendulous branches bearing white flowers. The fruits are round and bright red when mature. The palm's wide natural distribution results in a natural variability among individual plants, mainly in leaf size, number of leaflets, and trunk height. Some individuals even show fused pinnae and a bifid leaf.

This robust palm is adaptable to zones 10 and 11 and is an excellent choice for cool but frostless Mediterranean climates. It grows faster than most other species, especially with adequate moisture and a good soil. It is among the most beautiful and the largest *Chamaedorea* species. The yellow trunk, white flowers, and red fruits make it one of the most colorful species. It is good as a specimen planted alone but even better in groves of individuals of varying heights under a canopy.

Chamaedorea metallica
ka-mee-DOR-ee-a · me-TAL-li-ka
PLATES 222–223

Chamaedorea metallica has a small natural range in the rain forest of southern Mexico and is near extinction because of overcollecting for the nursery trade. The epithet is Latin for "metallic," a reference to the color and sheen of the leaves. The palm is sometimes called miniature fishtail palm and metallic palm.

The solitary trunk sometimes grows to 8 feet (2.4 m) high but 3–4 feet (90–12 cm) is more common, with a diameter of ½ inch (13 mm). The stem is dark green and ringed. The leaf crown is 3 feet (90 cm) wide and 2 feet (60 cm) tall. The leaves are usually 1 foot (30 cm) long, undivided, and pinnate, but some plants have 3–8 S-shaped leaflets of unequal widths. Unsegmented leaves usually exhibit a slight to pronounced "cupped" form, the margins of the blade higher than the center or rachis. The leaf is obovate but, like that of *C. ernesti-augusti*, has a bifid apex that is deeply cleft into 2 fishtail-shaped wedges. The leaves are carried on petioles 2–6 inches (5–15 cm) long and are held half-erect, never descending below the horizontal plane. Leaf color is a remarkable deep blue-green with a distinct metallic sheen, especially on the upper leaf surface. Although the leaves are not usually variegated, the grooves corresponding to the fused leaflets in the undivided parts of the blade may create a two-toned hue because of the different levels of light reflection; some plants show a definite difference in the color of the leaflets' veins. Male flowers are borne on thin pendulous branched inflorescences, females on erect spikes. The flowers are deep yellow to bright orange. The round ½-inch (13-mm) black fruits are carried on erect, orange spikes.

This palm is not hardy to cold, being adaptable only to zones 10b and 11 and marginal in 10a. It thrives in shade and is not adapted to full sun. It may be air-layered due to the production of adventitious roots along the older stems; the original stem dies. Because of its form and unique coloration, this little palm is unexcelled in the landscape. It is beautiful in silhouette and, planted in groups of individuals of varying heights, makes a shimmering mass of beauty. Planted en masse, it is one of the most exquisite large groundcovers.

Chamaedorea microspadix
ka-mee-DOR-ee-a · my-kro-SPAY-dix
PLATES 224–225

Chamaedorea microspadix is endemic to the foothills of the Sierra Madre Oriental in central eastern Mexico, where it grows in forest, often over limestone. The epithet is from Greek words meaning "small" and "flower stalk," which description seems inaccurate as the inflorescences are not that small. A common name is hardy bamboo palm.

This clustering species makes clumps to 12 feet (3.6 m) tall and 8 feet (2.4 m) wide. The individual trunks are to 8 feet (2.4 m) tall but ½ inch (13 mm) in diameter and so relatively slender that they often recline because of the weight of the leaves. The stems are dark green with widely spaced whitish rings. The leaves are 2 feet (60 cm) long, with widely spaced, narrowly elliptical leaflets, each 8–10 inches (20–25 cm) long and S shaped; the terminal 2 leaflets are slightly wider than the others, and all grow from the rachis in a flat plane, although the leaf is beautifully arching and its leaflets are slightly pendulous at their ends. Leaf color is usually dark green above, with a soft, almost satiny feel, and silvery beneath. The inflorescences are erect spikes of tiny yellowish white flowers; they grow from beneath the leaf crown. The rounded ½-inch (13-mm) fruits are deep orange to scarlet and are borne in drooping clusters. Female plants bear fruit several times per year, usually in the warmer months.

This is one of the most cold hardy species in the genus. It takes a temperature in the mid-20s Fahrenheit (around −4°C) without damage, and its trunks usually survive temperatures in the low 20s (around −7°C). Colder temperatures are likely to cut it to the ground, but it almost always grows new trunks and it has reportedly resprouted from temperatures in the low teens (−11°C).

It is a wonderful landscape subject. The clumps are dense, and the leaves on each thin trunk are widely spaced and cover a good part of the stem's length. The palm is used as a screen or an accent in borders and other massed vegetation. It has a remarkable architectural look when its trunks are thinned (so that each one is visible) and pruned (so that they are of varying heights); its silhouette under such conditions is extraordinarily pleasing.

Chamaedorea nationsiana
ka-mee-DOR-ee-a · nay-shun-see-AH-na

Chamaedorea nationsiana is found on limestone soils on the Atlantic slope of Guatemala. The epithet honors biologist James D. Nations, who has worked tirelessly to conserve forests in Central America. The solitary stem is 1 inch (2.5 cm) in diameter and up to 10 feet (3 m) tall with an attractive crown of leaves. Each leaf is 5 feet (1.5 m) long and bears up to 22 evenly spaced pinnae up to 20 inches (50 cm) long and 3½ inches (9 cm) wide. The male inflorescences are unbranched and pendulous; several are borne at each node. The densely flowered female spike is erect and one to a node. The thickened female inflorescence turns orange and holds a crowded cluster of shiny black fruits. This medium-sized *Chamaedorea* is similar to *C. arenbergiana* (which see) and would be at home in a shady garden with moist, neutral soil high in organic matter.

Chamaedorea neurochlamys
ka-mee-DOR-ee-a · nyoo'-ro-KLA-mees
PLATE 226

Chamaedorea neurochlamys is from southern Mexico, Guatemala, Belize, and Honduras. It grows in lowland wet forest, sometimes over limestone. The epithet is from the Greek meaning "nerved covering," but the reference is obscure. Henderson et al. (1995) included this species within their *C. pinnatifrons*.

The single, slender trunk attains 12 feet (3.6 m) tall, but in garden settings is considerably shorter. The leaves are about 3 feet (90 cm) long, held at a 45-degree angle from the trunk. The leaflets, a rich, glossy green, are arrayed more or less regularly along the rachis, but pinnae at the middle of the rachis are more widely spaced than those at either the base or the tip of the leaf. The most distinctive feature of this palm is the white apex of the leaf sheath, close to the petiole. In related species (*C. pinnatifrons*), the leaf sheath is evenly green. Fruits are vaguely kidney shaped, bright red during development but brownish when fully mature. Another brilliant choice for a small shady garden, this palm needs protection from cold, drought, and other cruelties of the modern world.

Chamaedorea oblongata
ka-mee-DOR-ee-a · ahb'-lon-GAH-ta
PLATE 227

Chamaedorea oblongata occurs naturally in southern Mexico, northern Guatemala, western Belize, central Honduras, and north central Nicaragua, where it grows in lowland rain forest. The epithet is Latin for "oblong" and refers to the leaflets, the fruits, or both. The solitary trunk reaches 10 feet (3 m) tall with a diameter of scarcely 1 inch (2.5 cm). The ascending leaves are 2 feet (60 cm) long on petioles 1 foot (30 cm) long. The leaflets are always thick, leathery, and glossy; they are 1 foot (30 cm) long, ovate, with a long, tapering point, and may be closely set or widely spaced.

Chamaedorea oreophila
ka-mee-DOR-ee-a · o-ree-o-FI-la
PLATE 228

Chamaedorea oreophila is from Veracruz and Oaxaca in Mexico, on limestone soils, at sites 3300–5000 feet (1000–1520 m) in elevation. The name, which is Greek for "mountain loving," refers to the elevations favored by this palm. This small palm grows no more than 10 feet (3 m) tall and bears a tight crown of leaves, each with a regular array of narrow, deep green pinnae. The most distinctive features of this species are found in the inflorescences, which are erect among the leaves. Males bear several inflorescences at each node, and the rachilla is pendulous. In females, the rachillae are erect. In both sexes the flowers are crowded onto the unbranched rachillae. Fruits are red. The cold hardiness of this palm is untested, but probably not great.

Chamaedorea pinnatifrons
ka-mee-DOR-ee-a · pin-NA-ti-frahnz
PLATE 229

Chamaedorea pinnatifrons has one of the widest areas of distribution in the genus, from southern Mexico, Belize, and Central America into Venezuela, Colombia, Ecuador, Peru, Brazil, and Bolivia. It grows in wet forests from sea level to 8000 feet (2440 m). The epithet is Latin for "pinnate leaf," a characteristic common to the whole genus.

This solitary-trunked species grows to 12 feet (3.6 m) high. There is usually a loose, indistinct crownshaft, above which the leaves, 4–6 feet (1.2–1.8 m) long, form a sparse and often elongated crown. Each leaf has 8 pairs of wide S-shaped to lanceolate, tapering bright green leaflets that are thin but crinkly; rare individuals have leaves with fused leaflets and bifid apices. The inflorescences are much branched, and the round fruits are red or black when mature. Morphological variety is to be expected in such a widespread species and, in this one, it is the form and size of the leaves.

The species is adaptable to zones 10 and 11 and does well in nearly frostless Mediterranean climates. It is a picturesque palm that should be planted where it can be viewed up close. It is truly delicious as a miniature canopy-scape, especially if it can be sited against a contrasting background.

Chamaedorea plumosa
ka-mee-DOR-ee-a · ploo-MO-sa
PLATE 230

Chamaedorea plumosa is endemic to the Mexican state of Chiapas, where it grows in deciduous forests from 2000 to 4000 feet (610–1200 m) in elevation. The epithet translates from the Latin as "plumose." This solitary-trunked species grows to 10 feet (3 m) tall and 1½ inches (4 cm) in diameter, with a loose, indistinct crownshaft. The leaves are almost unique for the genus because of the thin, grasslike leaflets that grow in clusters irregularly spaced around the rachis. The palm is not hardy to cold and is adaptable only to zones 10 and 11 but is adaptable to full sun (David Witt, pers. comm.).

Chamaedorea pochutlensis
ka-mee-DOR-ee-a · po'-choot-LEN-sis

Chamaedorea pochutlensis is a clustering species indigenous to the rain forest and evergreen forests of the Sierra Madre Occidental in western Mexico from the southern half of the state of Nayarit down to the Isthmus of Tehuantepec. The epithet is Latin for "of Pochutla," a town in the state of Oaxaca near where the species was first collected. The species is so similar in appearance and floral details to *C. costaricana* that Henderson (1995) wrote that it "is the northern counterpart of and similar to *C. costaricana*." The stems of *C. pochutlensis*, however, are not as tall as those of *C. costaricana*. Both species are among the most beautiful in the genus and have the same cultural requirements.

Chamaedorea pumila
ka-mee-DOR-ee-a · POO-mi-la
PLATE 231

Chamaedorea pumila is a small, solitary-trunked species from rain forest distributed from Costa Rica to Colombia. It is in danger of extinction because of collecting for the nursery trade. The epithet is Latin for "dwarf." Synonyms include *C. sullivaniorum* and *C. minima*.

The short trunk is usually less than 18 inches (45 cm) tall, decumbent, and not readily apparent. The leaves are 1 foot (30 cm) long on short petioles to 6 inches (15 cm) long and are undivided and generally oblong, broken only by an apical cleft 4 inches (10 cm) long. The thick, leathery blade is deep green with a velvety sheen and bears distinct ribs corresponding to the fused pinnate segments; its margin is toothed. There is often a suffusion of darker hues from the midrib, and the refraction of light along the ribs sometimes results in a reddish or purplish tint.

This little gem does not tolerate cold and is adaptable only to zones 10b and 11 and in 10a with protection. It is also intolerant of full sun in any climate and needs constant moisture in a free-draining, humus-laden soil.

Chamaedorea pygmaea
ka-mee-DOR-ee-a · pig-MAY-ah

Chamaedorea pygmaea occurs at high-elevation cloud forests in Colombia, Panama, and Costa Rica. The irresistible epithet means, of course, "pygmy."

This little palm has little or no aboveground stem. Its leaves seem to arise directly out of the ground and are about 1 foot (30 cm) long. The leaf may be undivided with a bifid tip, or divided into several leaflets. The lower margin of each leaflet merges smoothly and gradually with the rachis, a condition termed "decurrent." The inflorescence also seems to arise from the ground; the females are unbranched or forked, the males are many branched. Fruits are black.

This palm has a long history of cultivation in Europe but then disappeared. Fortunately, it is making a comeback. A charming, diminutive palm, it is perfect for container or tub culture in a greenhouse, or as a low groundcover in a shady garden. Mass plantings of this species, forming drifts beneath taller shrubs and palms, are breathtaking.

Chamaedorea radicalis
ka-mee-DOR-ee-a · ra-di-KAL-iss
PLATES 232–233

Chamaedorea radicalis is endemic to the foothills of the Sierra Madre Oriental in northeastern, Mexico, where it grows in oak forests. The epithet translates from the Latin as "from the root," an allusion to the inflorescences which, in the most common form of the palm (trunkless), grow from the base of the plant.

The palm occasionally produces a light green, ringed stem as tall as 12 feet (3.6 m) and 1 inch (2.5 cm) or slightly more in diameter. These trunked individuals usually exhibit a distinct crownshaft. The leaf crown is 5 feet (1.5 m) tall and 6 feet (1.8 m) wide.

The leaves are unusually broad for a *Chamaedorea* species, 3 or more feet (90 cm) long and 2 or more feet (60 cm) wide, and are held on thin petioles 6 inches (15 cm) long. The linear-lanceolate leaflets are 1 foot (30 cm) long and grow from the rachis in not quite a single plane, giving a slight V shape to the leaf. Their color is medium to dark green but they are slightly glossy. The leaf crown is sparse, open, and more fernlike than that of most other *Chamaedorea* species; there are seldom more than 6 leaves in a crown. The inflorescences grow from beneath the leaves and are 4 feet (1.2 m) long and erect to arching. The small, unisexual flowers are pale yellow to orange. The round ½-inch (13-mm) fruits are densely clustered and are orange to red when mature.

This species is among the cold hardiest species in the genus. It endures temperatures in the mid-20s Fahrenheit (around −4°C) without leaf damage and regrows from temperatures in the mid-teens (around −9°C). It has one of the most beautiful leaf shapes in the genus. Plants with little or no apparent trunk often look like clustering palms, especially when there are several leaves in a crown, and they make choice large groundcovers. Those with apparent trunks are uncommonly graceful and tropical looking and are simply superb in groups of 3 or more individuals of varying heights. Single-trunked specimens are not good isolated in space, but they are beautiful as canopy-scapes.

Chamaedorea sartorii
ka-mee-DOR-ee-a · sar-TOR-ree-eye
PLATE 234

Chamaedorea sartorii is distributed in wet forest, often over limestone, in southern Mexico and in Honduras. The species was named in honor of Carl C. Sartorius (1796–1872), a German planter living in Veracruz, Mexico, as well as an artist, author, and collector of plants for Berlin's botanical garden.

This single-stemmed species has a full crown of shiny, regularly pinnate leaves. Each dark green leaflet is 8–16 inches (20–41 cm) long and 1–3 inches (2.5–8 cm) broad, somewhat S shaped and tapering to a long, drawn-out tip. The female inflorescences are especially attractive, borne well below the crown on slender, arching, bright red stalks. The red female flowers are followed by shiny black fruits.

Delicate and slender, this species is an excellent choice for an intimate garden setting, where its jaunty red inflorescence stalks can brighten a shady spot. Like most species in the genus, it requires an organic substrate and plentiful moisture. It is unfussy about soil pH, thriving on both acid and alkaline soils.

Chamaedorea seifrizii
ka-mee-DOR-ee-a · see-FRIT-zee-eye
PLATES 235–236

Chamaedorea seifrizii is indigenous to the Yucatán and adjacent areas in Mexico, all of Belize, northern Guatemala, and northeastern Honduras, where it grows in lowland monsoonal forest. This species is the one most often referred to as bamboo palm; another moniker is reed palm. The epithet honors American botanist William Seifriz (1888–1955), who first collected the species near the ruins of Chichen Itza.

The trunks of this clumping species grow to 10 feet (3 m) high and are usually less than 1 inch (2.5 cm) in diameter. They are light to dark green and indistinctly ringed, with an elongated but indistinct crownshaft. A mature clump can reach 12 feet (3.6 m) high and 6 feet (1.8 m) wide. The leaves are 3 feet (90 cm) long and composed of several mostly linear-lanceolate leaflets, the terminal 2 of which are often wider and shorter than the others. Leaflets are broadly lanceolate to narrow and linear but usually 8 inches (20 cm) long, sometimes straight and stiff looking but also often (for the wider ones) soft and half-pendulous. The leaves are carried on petioles 2 inches (5 cm) long. The inflorescences are short and erect with fragrant, tiny greenish yellow flowers densely packed along the few branches. The ¼-inch (6-mm) rounded fruits are black.

This species is variable in leaf form. Some plants formerly classified as *C. erumpens* have a pair of unusually wide and fat leaflets. Some forms have narrow, almost grasslike leaflets, while others are wider, looser, and even pendent.

The palm is tender to cold and adaptable only to zones 10b and 11 and is marginal in 10a. The clumps usually have a stiff but elegant aspect that is best as an accent or silhouette. Unpruned, the stems are densely clustered and usually provide a wall of leaves from top to bottom of the clump; if judiciously thinned the beauty of the silhouette is greatly enhanced. It is outstanding as an indoor plant.

Chamaedorea stolonifera
ka-mee-DOR-ee-a · sto-lo-NI-fe-ra
PLATE 237

Chamaedorea stolonifera is endemic to the Mexican state of Chiapas, where it grows in rain forest from elevations of 2000 to 2640 feet (610–800 m). It is highly endangered as its habitat is being cleared primarily for coffee growing and was thought to be extinct until David Besst rediscovered it in the 1980s. The epithet is Latin for "stolon-bearing."

The clustering trunks sprout from creeping stolons near the surface of the soil, creating single clumps to 30 feet (9 m) across. The stems can grow to 6 feet (1.8 m) high but are ¼ inch (6 mm) in diameter; they are a solid deep green with widely spaced nodes that each produce a single leaf. The few leaves extend downwards from the top of the trunk for several nodes like those of a bamboo or other types of grass. They are each 1 foot (30 cm) long on thin petioles 2 inches (5 cm) long. The blade is simple, very deeply bifid apically, and medium to deep green.

This little palm makes one of the best groundcovers for a shady or partially shady site, needing only lots of moisture and a nearly tropical climate. It thrives on slightly alkaline soils but adapts to slightly acidic ones also and, while it doesn't grow terribly fast, can nevertheless cover a lot of space in a year's time. It is also a superb indoor plant.

Chamaedorea tenerrima
ka-mee-DOR-ee-a · ten-er-REE-ma
PLATE 238

Chamaedorea tenerrima is an uncommon species endemic to high-elevation rain forest in Guatemala. The epithet means "very delicate or slender," apparently in reference to its stature.

The stem of this palm is short, less than 6 feet (1.8 m) tall, and thin, often leaning under the weight of its own crown of leaves. The leaves are the most unusual and distinctive in the entire genus: about 18 inches (45 cm) long, they have curiously, backward-pointing, S shaped leaflets. The largest segments on the leaf are the terminal pair, which is both longer and wider than the others. They are stiffly spread 180 degrees, giving the entire leaf an awkward, improbable appearance.

The back-swept leaflets are unique within the genus, and the color photo of this species that appeared on the cover of *Principes* in 1991 caused a sensation among enthusiasts, who wanted this unusual palm in their collections. While still uncommon in cultivation, it responds readily to a moist, shady site rich in organic matter and protected from wind and cold. Its cold hardiness remains untested.

Chamaedorea tepejilote
ka-mee-DOR-ee-a · tay-pay'-hee-LO-tay
PLATES 239–240

Chamaedorea tepejilote is widespread from the rain forest of southern Mexico, through Central America, and into extreme northwestern Colombia. The epithet is the aboriginal name for the palm in El Salvador.

This species is usually found in nature as a solitary-trunked tree, but there are many clustering individuals. Mature trunks may be 10–22 feet (3–6.7 m) tall and ½–3 or more inches (13 mm to 8 cm) in diameter; most individuals have a loose, indistinct crownshaft. Even when very tall, the stems are straight and noble looking. They are light to medium green and prominently ringed. The leaf crown is sparse and 10 feet (3 m) wide but 3 feet (90 cm) tall. The leaves are beautifully arching, 5–6 feet (1.5–1.8 m) long, on short thin petioles that are no more than 18 inches (45 cm) long and usually less than 12 inches (30 cm). The leaflets are 2 feet (60 cm) long, lanceolate with long tapering ends, and S shaped. They are light to dark green and mostly pendent from the rachis, although they grow in only one plane. The inflorescences are 2 feet (60 cm) long with pendent branches arising from beneath the crownshaft and carrying yellow flowers; the male inflorescences are thinner than the females. The rounded ½-inch (13-mm) fruits are black. A species with such a widespread natural range must be expected to be variable in form; in this case the variation is in the size of the leaves and the shape and number of leaflets.

The immature inflorescences of this species are edible and sold fresh in Central American markets. Specific cultivars are grown for their larger, tastier inflorescences. Specialty grocery stores in the United States carry the pickled version, which is an acquired taste. This palm does not tolerate cold and is adaptable only to zones 10b and 11 with some specimens found in favorable microclimates of 10a. Because of its large leaves and thick trunk, it is not the most graceful species in the genus. Nevertheless, it is unparalleled as a canopy-scape, and it is stunning in groups of 3 or more individuals of varying heights.

Chamaedorea tuerckheimii
ka-mee-DOR-ee-a · turk-HY-mee-eye
PLATE 241

Chamaedorea tuerckheimii occurs naturally in the premontane rain forest of southern Mexico and northeastern Guatemala. The species is nearly extinct in the wild due to collection for the nursery trade. The epithet honors Hans von Türckheim (1853–1920), coffee planter, naturalist, and German consul in Guatemala, who first brought this species to the attention of botanists in Berlin's botanical garden. Common names are potato chip palm and ruffled palm, the latter often applied to *Aiphanes horrida*.

The solitary trunk is usually under 3 feet (90 cm) tall overall, ¼ inch (6 mm) in diameter, and often recumbent. The leaf crown is compact because of the short petioles and in many specimens is almost a rosette. The leaves are 7 inches (18 cm) long, 3 inches (8 cm) wide, undivided, shortly bifid apically, and obovate, with shallowly toothed margins; they are strongly grooved above and beneath and exhibit a prominent, lighter-colored midrib. Leaf color is light to dark green above and grayish green beneath; the grooves are so deep that they often create highlights and shadows in a way that the blade seems to be variously hued. The inflorescences are sparsely branched and bear white blossoms that yield small, egg-shaped black fruits.

This little palm requires consistently high humidity and pure water. If these demands are not met, the edges and tips of its leaves will be spoiled by browning. This diminutive diva is not hardy to cold and is adaptable only to zones 10b and 11, being marginal in 10a. It is small enough to be easily protected in even colder climates, but again, high humidity and mineral-free water are an absolute must. This is not an easy palm to grow well.

Chamaedorea warscewiczii
ka-mee-DOR-ee-a · wahr-se-WIK-zee-eye
PLATE 242

Chamaedorea warscewiczii is indigenous to mountainous rain forest in Costa Rica and Panama. The epithet honors Józef Warszewicz (1812–1866), Polish plant collector in Central America, who introduced this species to cultivation. The solitary trunk grows to 12 feet (3.6 m) high and 1 inch (2.5 cm) in diameter. There is no true crownshaft, but the leaf sheaths are prominent and persistent. The leaf crown is extended along the upper part of the trunk with leaves 2 feet (60 cm) long on petioles 1 foot (30 cm) long. The widely spaced leaflets are broadly ovate with long drip tips and are glossy deep emerald green on both surfaces; the apical pair of leaflets is usually significantly wider than the others. The inflorescences grow from the extended leaf crown and consist of long, erect peduncles at the tips of which are the long, pendent, spidery ropelike yellow-green branches. The round ½-inch (13-mm) fruits are black.

Chamaedorea woodsoniana
ka-mee-DOR-ee-a · wood-so'-nee-AH-na
PLATE 243

Chamaedorea woodsoniana has a large natural range from southern Mexico, through Central America, and into Colombia, where it grows in montane rain forest. The epithet honors American plant collector Robert E. Woodson (1904–1963), one-time curator of the herbarium of the Missouri Botanical Garden. This species grows to 40 feet (12 m) high with a green and distinctly ringed trunk to 4 inches (10 cm) in diameter, atop which sits an attractive green crownshaft that is usually swollen at its base. The leaves are 3–4 feet (90–120 cm) long on petioles 1 foot (30 cm) long. The linear leaflets grow from the rachis at an angle to form a V-shaped leaf; they are deep green above and slightly paler beneath. The inflorescences are much branched and coral red, as are the pendulous clusters of fruit. This beautiful palm is adaptable only to zones 10 and 11.

Chamaerops
ka-ME-rahps

Chamaerops is a monotypic genus of palmate-leaved, dioecious palm from dry parts of the Mediterranean region. It is usually a clumping species, but some individuals remain naturally solitary trunked. It has the distinction of being the world's most northerly occurring palm species. The genus name is derived from Greek words meaning "on the ground" and "shrub," an allusion to the stature of the plant. Fresh seeds germinate within 90 days. Seeds that are allowed to dry out completely and then rehydrated may take 9 months or longer to germinate, and the percentage of seeds that germinate is not as high as that of fresh seed. For further details of seed germination, see *Corypha*.

Chamaerops humilis
ka-ME-rahps · HY00-mi-lis
PLATES 244–247

Chamaerops humilis occurs naturally on the Atlantic coasts of Spain, Portugal, and Morocco; the Mediterranean coasts of Spain, France, Italy, and the islands of Sardinia, Sicily, and Malta; and along the North African coast in Algeria, Libya, and Tunisia. It grows mostly in coastal areas, including rocky foothills near the sea, but also extends up into the Atlas Mountains of Morocco to elevations slightly above 5280 feet (1610 m). The epithet is Latin for "humble" or "small." The common names are dwarf palm, European fan palm, and Mediterranean fan palm.

The trunks vary from less than 1 foot (30 cm) to 20 feet (6 m) high and, when mature, usually have a diameter of 1 foot (30 cm) and are covered in all except the oldest parts with a dense mat of dark gray to dark brown or almost black fibers from the old leaf bases, which create the effect of large, dark, tightly packed scales. On a large plant, the leaf crown can be 10 feet (3 m) wide and tall. Mature clumps are 25 feet (7.6 m) tall and 30 feet (9 m) wide, with full rounded crowns, the dead leaves adherent only near the base of the crown. Leaves are 3 feet (90 cm) wide, a little more than half-circular, with several dozen narrow, stiff, and tapering segments which are free for two-thirds or more of their length. The stiff petioles are 2–5 feet (60–150 cm) long with vicious forward-pointing spines. Leaf color ranges from deep grassy green through shades of grayish green and bluish green. The small, erect panicles of yellow flowers grow from the leaf crown and are 6 inches (15 cm) long; they are mostly hidden by the leaves. The plants are usually dioecious, but some individuals bear both male and female blossoms. The round ½-inch (13-mm) fruits are yellow-orange or brown and are formed in slightly pendulous clusters.

This slow-growing, drought-tolerant species succeeds in several well-drained soils. It withstands temperatures to 15°F (−9°C), especially in dry climates; suckering individuals survive even lower temperatures but lose most of their aboveground growth and take a long while to grow back, while solitary-trunked individuals are usually killed outright. This palm is not at home in moist, tropical climates where its growth is slower and where it fails to develop its full set of characteristics. Transplanting of rooted suckers is occasionally successful.

Single-trunked plants are best used as a canopy-scape; they look unnatural as isolated specimens, although a sort of architectural effect is obtained from them, especially as outdoor tub subjects. Clustering plants are picturesque enough to be planted as isolated specimens or as formal architectural specimens in large raised planters. Old clumps create a tableau that is the essence of luxuriance and the semiarid tropical look. As a silhouette against a large wall or differently colored foliage, nothing is better. This palm is not a good choice for growing indoors, as the plants tend to decay without good air circulation and bright light.

A complete, illustrated account of the many cultivars and geographical variants of this species is not available but much needed. Some garden forms have few leaflets on wedge-shaped leaves; others have many leaflets that spread more than 360 degrees. Other forms differ in the posture of the trunk(s), spininess of the petiole, color of the leaf, or size and shape of the fruits. Numerous varieties have been named in the past, but today only one naturally occurring variety with waxy, silvery leaves is given recognition: *C. humilis* var. *argentata* [ar-jen-TAH-ta]. The epithet is Latin for "silvered," as in most individuals, the leaves are silver on both surfaces. This variety seems to be slower growing than the type but also may be hardier to cold and drought. A synonym is *C. humilis* var. *cerifera*.

Chamaerops humilis 'Vulcano' (see plate 247) is widely available in Europe. It is more compact than the standard wild species, with a dense head of leaves borne on short, less spiny petioles. It is slow-growing. Its compactness makes it an ideal container plant. It takes its name from Vulcano, a tiny island off the northern coast of Sicily, where it was supposedly found.

Chambeyronia
shahm-bay-RO-nee-a

Chambeyronia comprises 2 solitary-trunked, pinnate-leaved, monoecious palms in New Caledonia. One is rare in habitat and reportedly difficult in cultivation. The other is gaining popularity because of its unique beauty. The genus name honors Charles-Marie-Léon Chambeyron (1827–1891), a French naval officer who mapped much of the coast of New Caledonia and assisted botanist Eugène Vieillard in his palm collecting in the island. Seeds of *Chambeyronia* germinate within 90 days. For further details of seed germination, see *Areca*.

Chambeyronia lepidota
shahm-bay-RO-nee-a · le-pi-DO-ta
PLATE 248

Chambeyronia lepidota is endemic to wet mountainous forests in northeastern New Caledonia in soils derived from weathered schist. The epithet is Latin for "scaly," a reference to the surface of the large crownshaft.

The brown, indistinctly ringed trunk grows to 35 feet (11 m) high and 6 inches (15 cm) in diameter. The remarkable crownshaft is 3 feet (90 cm) tall, fat, and distinctly bulging at the base, and covered in a chocolate-red feltlike tomentum with lighter striations. The leaf crown is mostly a semicircle, and the stiffly arching leaves at 6 feet (1.8 m) long are reminiscent of those of *Hedyscepe*; they have many dark green linear-lanceolate leaflets growing from the reddish brown rachis at an angle that gives a V-shaped leaf. The red fruits are produced on inflorescences that are the color and texture of the crownshaft.

Because of the unusual soil on which it grows in habitat, this palm is difficult to establish in cultivation and few individuals exist outside their native haunts. It doubtless needs copious moisture and probably partial shade when young. It is not tolerant of frost but would probably do well in a cool but nearly frost-free Mediterranean climate. It may be intolerant of calcium.

The tree has an unusually beautiful architectural aspect and would be stunning as a small canopy-scape or an intimate close-up subject.

Chambeyronia macrocarpa
shahm-bay-RO-nee-a · mak-ro-KAR-pa
PLATES 249–251

Chambeyronia macrocarpa is endemic to rain forest on the island of New Caledonia at 2000–3000 feet (610–900 m). The epithet is Greek for "large fruit." Several growers have coined the moniker red-leaf palm because most specimens have a colored new leaf; alas, many pinnate-leaved palms have this characteristic. The names red feather palm and flame-thrower palm are also gaining acceptance.

The trunk, which grows to at least 40 feet (12 m) tall and has a diameter of 10 inches (25 cm), is gray to greenish gray with widely spaced light-colored rings. The deep emerald green crownshaft is 3–4 feet (90–120 cm) high and bulging at its base. The circular leaf crown is 8 feet (2.4 m) wide and tall and has 8–10 leaves, which are 12 feet (3.6 m) long and wonderfully arching up, down, and back towards the trunks. Each of the widely spaced linear-oblong leaflets is 3–5 feet (90–150 cm) long, tapers to a point, and grows from the rachis in a single plane. The petiole is usually less than 1 foot (30 cm) long, and the leathery leaflets are heavily veined with a prominent midrib and 2 lateral veins, giving an almost pleated look to each pinna. Mature leaves are deep bluish green, but newly emerging leaves are often highly colored in shades of carmine red, orange, scarlet, purplish red, or bronzy purple, the color changing daily as the leaf matures. The inflorescences are erect panicles 1 foot (30 cm) long that emerge from beneath the crownshaft and carry green and white large flowers. The fruits are 2 inches (5 cm) wide and round or slightly egg shaped, and deep red when mature, hanging on pendent branches.

The species is a water lover and needs a rich, humus-laden, well-drained soil. It is relatively fast growing but not hardy outside of zones 10 and 11 without protection. It relishes partial shade, especially when young or in hot climates. It can tolerate full sun only where soil moisture is adequate and consistent.

Besides having gorgeously colored leaves, this palm has an even more incredibly beautiful form: a tall and straight, light-colored trunk topped by a perfectly round crown of a few large and gloriously arching leaves with their widely spaced and wide leaflets. It should receive a favored spot in any planting and is unparalleled for up-close viewing. It looks best as a canopy-scape but is so slow growing that it will probably be first planted as an isolated specimen, and it is extraordinarily beautiful enough to be esthetically pleasing as such. The species is well adapted to cultivation under glass with abundant moisture, a well-draining medium, and lots of light.

The variability of the species in nature mostly corresponds with particular geographical locations in New Caledonia. Pintaud (2000) pointed out that there are very tall forms to 90 or more feet (27 m) in the southern part of the island, a form with a green- and yellow-striped crownshaft (known as the "watermelon" type; see plate 251), a form with a yellow crownshaft (the "Hookeri" type; see plate 250), a form which never has red new growth, and a form that is especially stout and robust and has a crownshaft covered with white tomentum. These forms are not given formal, botanical recognition.

Chelyocarpus
che´-lee-o-KAHR-pus

Chelyocarpus is composed of 4 solitary-trunked and clustering hermaphroditic palms in lowland rain forest of South America. The trunks are smooth and clean except for their newest parts. All species have palmate or minimally costapalmate leaves divided into 2 halves and split nearly to the petiole into pleated and wedge-shaped segments; the overall appearance of a leaf is like a pinwheel with 40 or more marginal indentations. Because of their long petioles, the 2 taller species have sumptuously beautiful leaf crowns with the "pinwheels" gloriously displayed. The inflorescences grow from the leaf sheaths and are branched once or twice. The flowers are creamy white, and all but one species is pleasantly fragrant. The attractive fruits are green, brown, or yellowish and succulent looking.

These palms are beautiful because of their large leaves. All are tender to cold and adaptable only to zones 10b and 11. They need copious water and a rich, humus-laden, well-drained soil. They luxuriate in partial shade (especially when young) or full sun. The species are still unaccountably rare in cultivation. The genus name is from Greek words meaning "turtle" and "fruit," a reference to the corky, cracked surface of the fruits of *C. ulei*, the most widespread species. Seeds of *Chelyocarpus* germinate within 45 days and should not be allowed to dry out before sowing. For further details of seed germination, see *Corypha*.

Chelyocarpus chuco
che'-lee-o-KAHR-pus · CHOO-ko
PLATE 252

Chelyocarpus chuco is found in the Amazon regions of extreme northern Bolivia and northern Brazil, where it grows in swamps and seasonally inundated forests. The epithet is one of the aboriginal names for the palm. This primarily clustering species has stems to 60 feet (18 m) high. It sports a sparse crown of circular leaves 5 feet (1.5 m) in diameter and with 40 or more segments; leaf color is deep green on both sides. The leaf segments are the most uniform in size of the genus. The species is slightly susceptible to lethal yellowing disease.

Chelyocarpus ulei
che'-lee-o-KAHR-pus · OO-lee-eye
PLATE 253

Chelyocarpus ulei is indigenous to western Ecuador, southeastern Colombia, and Peru, where it occurs in the lowland rain forest of the Amazon region. The epithet honors Ernst H. G. Ule (1854–1915), a German botanist and collector in South America. This palm grows to 20 feet (6 m) high and has beautiful leaves 4–5 feet (1.2–1.5 m) wide with 6–12 segments cleft to the petiole. Leaf color is deep green above and silvery or even bluish green beneath because of the white tomentum. The leaves give the typical and thrilling pinwheel effect. The blossoms are not pleasantly fragrant to most people; Harold E. Moore, Jr. (1972) described the odor as that of fishmeal or burning rubber.

Chuniophoenix
choo'-nee-o-FEE-nix

Chuniophoenix includes 2 (possibly 3) clustering, hermaphroditic palms in China and Vietnam. It is among the few genera of palmate-leaved palms whose leaves do not have hastulas. The leaves are deeply divided almost to the petiole into segments of varying widths and numbers. The inflorescences grow from the leaf bases and are branched once or twice. The important character for ornamental plantings is the deep orange to red fruits, which, alas, are mostly hidden by the leaf crown.

These palms are amazingly hardy for their tropical origins; all species have withstood temperatures in the mid- and upper 20s Fahrenheit (−4 to −2°C) in central Florida without damage. They are still uncommon in cultivation, although one cannot imagine why. The genus name is derived from the surname of Chen Huanyong, formerly Anglicized as Woon-Young Chun (1890–1971), botanist and founder of China's Institute of Agriculture and Forestry, and *Phoenix*, the date palm (but used in the general sense of "palm" in compound names). Seeds of *Chuniophoenix* germinate with 60 days and should not be allowed to dry out before sowing. For further details of seed germination, see *Corypha*.

Chuniophoenix hainanensis
choo'-nee-o-FEE-nix · hy-nah-NEN-sis
PLATE 254

Chuniophoenix hainanensis is indigenous to the island of Hainan, off the southern coast of China, and the adjacent southern mainland. It grows in low mountainous evergreen forest, where it is highly endangered because of expanding agriculture. The epithet is Latin for "of Hainan."

Although this palm is a robust-looking clumper, the clumps are not that wide. The trunks grow to 10 feet (3 m) high and 6 inches (15 cm) in diameter. The leaves are 4 feet (1.2 m) wide on petioles 3 feet (90 cm) long. The many segments are roughly equally sized and deep grayish green; they are free almost to the petiole and are mostly shallowly bifid at their apices. The inflorescence is not noticeable because it grows from the center of the leaf crown and is short. What is usually noticeable is the unpleasant odor of the maroon blossoms it bears. The fruits are 1 inch (2.5 cm) wide, deep red when mature, and more readily visible in larger specimens than smaller ones. The endosperm is ruminate. The species seems at home on calcareous soils but needs supplemental water in drought. It also relishes partial shade but grows well in full sun except in the hottest climates.

Chuniophoenix humilis
choo'-nee-o-FEE-nix · HYOO-mi-lis
PLATE 255

Chuniophoenix humilis is recognized by Henderson (2009) as endemic to Hainan. The epithet is Latin for "humble" or "small." The species is scarcely distinguished from *C. nana* in having slightly more "hooded" leaf segments, that is, the leaflet margins are turned down, especially near the tip, making an inverted cup of the tip of the segment. Zona (1998) considered *C. humilis* synonymous with *C. nana*.

Chuniophoenix nana
choo'-nee-o-FEE-nix · NA-na
PLATE 256

Chuniophoenix nana is indigenous to northern Vietnam, where it grows in low mountainous rain forest. The epithet is Latin for "diminutive." This species grows to 3–4 feet (90–120 cm) high. It looks like a small clump of lady palm (*Rhapis excelsa*) with its tightly clustering, slender canes and similar leaves. The circular leaves have 6 widely separated, flat segments, which are thin, glossy medium green, and distinctly ribbed. The flowers are white and sweetly scented. The red fruits each contain a single seed with a homogeneous endosperm. This tiny undergrowth species does not like full sun, especially in hot climates. It also prefers a slightly acidic, humus-laden soil, turning unpleasantly yellow-green when the soil is too alkaline. The palm is not drought tolerant.

Clinosperma
kly-no-SPUR-ma

Clinosperma consists of 4 solitary-trunked, pinnate-leaved, monoecious palms confined to New Caledonia. The genus name is derived from Greek words meaning "slanted" and "seed." This genus now includes the genera *Brongniartikentia* and *Lavoixia*. Seeds of *Clinosperma* are not generally available. They can take 60–120 days to germinate. For further details of seed germination, see *Areca*.

Clinosperma bracteale
kly-no-SPUR-ma · brak-tee-AH-lee
PLATE 257

Clinosperma bracteale is endemic to rain forest of southern and central New Caledonia from sea level to 3800 feet (1160 m). The epithet is Latin for "with bracts," referring to the inflorescence.

The stem attains a height near 50 feet (15 m) and a uniform diameter of 4 inches (10 cm); it is light gray to light tan and shows distinct darker rings of leaf base scars except in its oldest parts. The unusual and loosely formed crownshaft is 2 feet (60 cm) tall, often white, sometimes greenish and white, because of the waxy covering of small scales, but it may also be mauve or light brown to dark brown due to the amount of chocolate tomentum. Its most unusual feature, however, is its almost triangular shape, as it usually bulges near the base but on one side only. The leaf crown is sparse and open with spreading light green leaves that seldom lie beneath the horizontal. The petioles are about 1 foot (30 cm) long and hold leaves 5 feet (1.5 m) long that arch only near their tips. The stiff, linear-lanceolate leaflets are regularly spaced along the rachis, are 3 feet (90 cm) long, and grow from the rachis at a slight angle to give a shallow V shape to the leaf. Beautiful once-branched inflorescences grow from beneath the crownshaft. The flowering branches are knobby and deep wine red maturing to medium red; they look like an undersea branching coral. The tiny yellowish white flowers of both sexes, followed by the shiny round green fruits that mature to black, add to the general colorfulness.

This palm needs copious, regular moisture but is not fussy about soil type as long as it is well drained and not too calcareous. It luxuriates in partial shade or full sun except in hot, especially dry climes. In spite of its tall stature, this beauty has a miniature look about it that is neat, orderly, and appealing. It's hard to misplace it in the landscape. It's also hard to find it for sale, and it is, alas, slow growing, especially when young.

Clinosperma lanuginosa
kly-no-SPUR-ma · la-noo'-ji-NO-sa
PLATE 258

Clinosperma lanuginosa occurs naturally in mountainous rain forest of northeastern New Caledonia from elevations of 2200 to 4000 feet (670–1200 m). The epithet is Latin for "woolly" and refers to the tomentose crownshaft. The species was once known as *Brongniartikentia lanuginosa*.

This palm attains a maximum overall height of 20 feet (6 m) with a dark green or dark brown trunk 4 inches (10 cm) thick and bearing distinctly indented rings of leaf base scars. The crownshaft is usually tight, distinct, and 18 inches (45 cm) tall, covered in ruddy brown tomentum, and slightly swollen at its base. The leaf crown is sparse, and the leaves are short and ascending, never falling beneath the horizontal, usually 5 feet (1.5 m) long on petioles 2 feet (60 cm) long. The thin but stiff lanceolate-acuminate leaflets are closely and regularly spaced and grow from the rachis at almost a 45-degree angle to give a V-shaped leaf. The fruits are ½ inch (13 mm) long, almond shaped, and an attractive red. Again, this species is uncommon in cultivation but worth seeking out and trying in any reasonably moist, protected spot in the garden.

Clinostigma
klyn-o-STIG-ma

Clinostigma comprises 11 solitary-trunked, pinnate-leaved, monoecious palms in the South Pacific islands. These mostly tall species inhabit rain forest from sea level to elevations above 5000 feet (1520 m). All have prominent crownshafts and leaves with usually pendent leaflets, and most have stilt roots at the bases of their trunks. The inflorescences grow from beneath the crownshafts, borne on a stocky, short peduncle, are much branched, and bear unisexual flowers of both sexes.

These species are still rare in cultivation, especially in the Western Hemisphere, but are among the most beautiful things the natural world has to offer, and all are more than worthy of cultivation in tropical climates. None are frost tolerant and, while some can adapt to the temperatures of a frostless Mediterranean climate, they need such abundant, regular moisture as well as such high humidity that they are seldom grown there. They do not like an alkaline soil, which limits their use in South Florida, unless the soil is amended with humus or the trees are mulched with organic material. Since they start out as undergrowth plants, they relish shade when young and are not adapted to hot and dry climates at any age. None of these species are fast growing but neither are they terribly slow. The species from higher elevations are at their prime in regions similar to the wetter parts of the Hawaiian Islands.

The genus name is from Latin words meaning "bent" and "stigma," an allusion, perhaps, to the eccentric position of the stigmatic remains on the mature fruit. Seeds of *Clinostigma* generally germinate within 45 days. Long-term viability is not good. For further details of seed germination, see *Areca*.

Clinostigma exorrhizum
klyn-o-STIG-ma · ex-o-RY-zum
PLATE 259

Clinostigma exorrhizum is endemic to Fiji, where it grows in mountainous rain forest at 800–4000 feet (240–1200 m). The epithet is from Greek words meaning "outside of" and "root," an allusion to the prominent stilt roots.

This species grows to a maximum overall height of 70 feet (21 m) with a diameter of 1 foot (30 cm). It anchors itself to the wet slopes on which it grows by prickly stilt roots at the base of its trunk that may be 10 feet (3 m) tall. The crownshaft is 6 feet (1.8 m) tall, light green to light bluish green, of slightly greater diameter than the trunk, and usually uniform in diameter. The spreading leaves are slightly but beautifully arching, 15 feet (4.5 m) long, and borne on petioles 2 feet (60 cm) long; the rachises do not usually fall below the horizontal, but the great pendulous leaflets create nevertheless an almost globular crown. The regularly spaced deep green leaflets are 3 feet (90 cm) long, linear-lanceolate, long-tipped, and pendulous from the rachis. The small, ovoid fruits are red or pink when mature. No palm has a more beautiful leaf crown than this one.

Clinostigma harlandii
klyn-o-STIG-ma · hahr-LAN-dee-eye
PLATE 260

Clinostigma harlandii is endemic to the Vanuatu islands, where it grows in rain forest at 1300–4500 feet (400–1370 m). The epithet honors A. E. Harland, the palm's 19th-century original collector. The trunk attains a height of 80 feet (24 m) and a diameter of 1 foot (30 cm), is green in its youngest parts with closely set whitish rings of leaf base scars, and is light gray in the older parts. A cone of stilt roots 6 feet (1.8 m) tall supports it. The cylindrical crownshaft is 4 feet (1.2 m) tall, bulges at the base where it is light yellowish or even orange glaucous green to pure emerald green, but above that is scarcely wider than the trunk and is light green or almost white near its summit. The leaf crown and leaves are similar to those of *C. exorrhizum* but lighter colored. The tiny round fruits are deep red when mature.

Clinostigma ponapense
klyn-o-STIG-ma · po-na-PEN-see
PLATE 261

Clinostigma ponapense is endemic to Ponapei, one of the Caroline Islands east of the Philippines, where it grows in mountainous rain and cloud forest. The epithet is Latin for "of Ponapei."

The trunk is mostly light grayish green except for the oldest parts that are gray. It attains a height of 60 feet (18 m) in habitat and is beautifully ringed with brownish leaf base scars in its younger parts. A spreading mass of stilt roots 4–5 feet (1.2–1.5 m) tall supports the trunk. The crownshaft is 6 feet (1.8 m) tall, slightly bulging at its base, and a uniform mint green. The leaf crown is sparse and hemispherical, with no more than 10 leaves at any time. The leaves are 15 feet (4.5 m) long, spreading, and arching at their tips. The many evenly spaced glossy dark green leaflets are 4–5 feet (1.2–1.5 m) long, linear-lanceolate, S shaped, and, as in most of the other species in the genus, pendent in trees past the seedling stage, making them one of the world's most beautiful palm leaves.

The species is extraordinarily picturesque and, in silhouette, looks like a gigantic tree fern. It does not tolerate drought or freezing temperatures and is hardy in zones 10 and 11.

Clinostigma samoense
klyn-o-STIG-ma · sah-mo-EN-see
PLATE 262

Clinostigma samoense is endemic to Samoa, where it grows in low mountainous rain forest. The epithet is Latin for "of Samoa." The tan, heavily ringed trunk attains a height of 50 feet (15 m) and a diameter of 1 foot (30 cm). The leaf crown is among the most beautiful in the palm world: it is semirounded and slightly more than hemispherical, with spreading, ascending leaves that are 15–20 feet (4.5–6 m) long and slightly arched at their tips. The many leaflets are a uniform deep green, regularly spaced along the rachis, 3–4 feet (90–120 cm) long, linear-lanceolate, and pendent. There is no more beautiful palm species.

Clinostigma savoryanum
klyn-o-STIG-ma · sa-vor'-ee-AH-num
PLATE 263

Clinostigma savoryanum is endemic to the Bonin Archipelago (Ogasawara Islands) southeast of the main Japanese islands, where it grows in low mountainous rain forest. The epithet honors Nathaniel Savory (1794–1874), an adventurer from Massachusetts and founder of one of the first colonies in the archipelago.

This species is among the smaller ones in the genus. The trunk attains a height of 40 feet (12 m), is tan in its older parts, deep green in the younger parts, and relatively slender at 10 inches (25 cm) in diameter. The crownshaft is 4–5 feet (1.2–1.5 m) tall, bulges slightly at its base, and is a light glaucous mint green to almost white. The leaves are 10 feet (3 m) long on short petioles and form a slightly more-than-hemispherical crown; they are typically ascending, spreading, and slightly arching, with narrowly lanceolate pendent leaflets 3 feet (90 cm) long spaced regularly along the rachis. While the habitat of the species is mostly extratropical geographically, it is climatically tropical, and the palm is as tender to cold as are the other species in the genus.

Coccothrinax
ko'-ko-THRY-nax

Coccothrinax is a genus of about 50 small to medium, palmate-leaved, hermaphroditic palms. All the species are indigenous to the Caribbean Basin, South Florida, and the Yucatán Peninsula of Mexico. They include solitary-trunked as well as clustering species, whose trunks are covered in the younger parts with long burlaplike or spinelike fibers. The leaves are mostly circular and usually silvery beneath. All the species have creamy white flowers borne on inflorescences that either exceed the leaves in length or curve downward below the leaves, depending on the species. Fruits are usually purple-black, but some species have white fruits. The species are difficult to identify as seedlings or juveniles. Preliminary research by Carl E. Lewis at Fairchild Tropical Botanic Garden has shown that some species form interspecific hybrids in cultivation whereas others do not.

All species are drought tolerant and of unusual beauty, their only drawback being that they are slow growing. They are relatively cold hardy, most of them being adaptable to zones 9b through 11, but they need abundant heat in the summer. Almost all species tolerate salinity in the soil and air. If not grown directly in the ground, all species need disproportionately large containers, even as small seedlings, as they dislike crowded roots. Many of the Antillean species grow in soils derived from limestone and are at home when cultivated in alkaline soils. Some of the Cuban species are from unusual soils that are acidic, high in heavy metals and lacking in calcium; they do not thrive in limey soils.

The genus name comes from the Greek word for "berry" and *Thrinax*, a related genus that these palms resemble. The species are all called thatch palms for their use in making thatched roofs. Many species are used in broom-making. Fresh seeds of *Coccothrinax* germinate within 90 days. Seeds that are allowed to dry out completely and then rehydrated may take 9 months or longer to germinate. For further details of seed germination, see *Corypha*.

Coccothrinax alexandri
ko′-ko-THRY-nax · al-ex-AN-dree
PLATE 264

Coccothrinax alexandri is endemic to the far eastern tip of Cuba, where it grows along the coast, on limestone outcrops. Although it has a limited range, it does not appear to be endangered unless at some time this part of the coast begins to be developed or is threatened by rising sea level. The epithet honors Alejandro López, a resident of Maisí, Cuba, who assisted in collecting this palm.

This species is the tallest in the genus. The solitary gray trunk can reach more than 60 feet (18 m) tall and 6 inches (15 cm) in diameter in habitat, although perhaps less in cultivation. The leaf crown of 15 leaves can measure 7 feet (2.1 m) across. Leaves are orbicular, 4 feet (1.2 m) in diameter, and deep green on top while light silvery underneath. The tips of the segments are stiff. Inflorescences are borne among the lower leaves and bear spherical ¼-inch (6-mm) reddish black fruit.

Two subspecies are recognized: **C. alexandri subsp. alexandri** grows on the north side of the eastern tip of Cuba, while **C. alexandri subsp. nitida** [NI-ti-da] grows on the south side and has shinier leaves. The northern subspecies was introduced into cultivation in the year 2000, when seeds found their way to Florida and elsewhere. It is a fairly vigorous grower, rapid by *Coccothrinax* standards, and prefers sunny areas with well-drained, alkaline soils. Its cold hardiness is unknown, but it could be assumed it would be hardy only to zones 10 and 11, like many others in this genus.

The elegance of this tall statuesque beauty shows off best against a solid background or the open sky. Of course, an odd numbered grouping of staggered heights makes a gorgeous canopy-scape.

Coccothrinax argentata
ko′-ko-THRY-nax · ahr-jen-TAH-ta
PLATE 265

Coccothrinax argentata occurs naturally in South Florida and the Bahamas, where it grows in open woods, pinelands, and coastal dunes. The epithet is Latin for "silvered." Common names are silver thatch palm, silver palm, silvertop palm, and Florida silver palm.

The solitary trunk may grow to 20 feet (6 m) with a diameter of 5 inches (13 cm), or may stay under 3 feet (90 cm), depending on the provenance and genetic make-up of the individual; plants from the Florida Keys grow taller than those from the Florida mainland. The upper part of the stem is covered with matted dark brown fiber that often hangs in wide patches from the leaf axils. The leaf crown is no more than 6 feet (1.8 m) wide and 8 feet (2.4 m) tall. The circular to half-circular leaves are 3 feet (90 cm) in diameter on petioles 2 feet (60 cm) long and have deep lanceolate segments that are pendulous at their tips, a shiny deep green above and a silvery to a metallic brownish green beneath. The branched inflorescences are 2 feet (60 cm) long and borne among the leaves. The rounded ½-inch (13-mm) fruits are black when mature.

In the lower Florida Keys, this species sometimes forms suckers, but the young shoots never grow very much and are suppressed by the dominant, main stem. In these same Keys populations, botanists have found hybrids between this species and co-occurring *Leucothrinax morrisii*. The hybrid, in appearance somewhat intermediate between the 2 parents, has not been introduced into cultivation.

Like most other *Coccothrinax* species, this one relishes full sun and is not particular about soil type, thriving under limey conditions. Being drought tolerant, it needs a quickly draining medium. It is adaptable to zones 9b through 11 and, while sometimes briefly surviving in colder regions, grows slowly enough that its long-term survival in the landscape is doubtful, especially in wetter climates, where fungus attacks can be fatal.

There is something alluring about the beautiful rounded, glossy dark green and silvery leaves of almost all *Coccothrinax* palms, and this one is no exception. It looks especially good in groups of 3 or more individuals of varying heights and is superb in a patio or courtyard where it can be seen up close. This species, like all the others, is superb as a canopy-scape, and few landscape subjects are as beautiful and as immune to saline conditions as it is, making it a perfect candidate for the seashore. It is not known to be grown indoors and would seem to be a poor candidate for such.

Coccothrinax argentea
ko′-ko-THRY-nax · ahr-JEN-tee-a
PLATE 266

Coccothrinax argentea is endemic to Hispaniola, where it grows in grasslands and low-elevation pinelands. The epithet is Latin for "silver," an allusion to the leaf underside. Common names are Hispaniolan silver thatch palm and Dominican silver thatch palm.

The trunks of this mostly solitary-trunked species can attain a height of 30 feet (9 m) but are more commonly 20 feet (6 m) tall. They are 4 inches (10 cm) in diameter, and the younger parts are covered with a mat of tightly woven dark gray to dark brown or black fibers from the old leafstalks. Total height of a solitary-trunked specimen is 35 feet (11 m) with a crown width to 10 feet (3 m) and height of 8–10 feet (2.4–3 m). The perfectly circular 3-foot (90-cm) leaves are divided nearly to the center of the blade, with many narrowly lanceolate, tapering segments. The segments are slightly pendent at their tips and are a deep, shiny green above with a striking silvery hue beneath. The leaves are carried on petioles 3–4 feet (90–120 cm) long, which gives an open, graceful appearance to the crown. The branched inflorescence is 1 foot (30 cm) long and hangs below the leaf crown. The round ½-inch (13-mm) fruits are black when mature.

The species grows well in most well-draining soils, even calcareous ones. It is not hardy to cold and is adapted to zones 10 and 11. Full sun is important even for seedlings. This is among the most graceful palms, and it is taller than most other *Coccothrinax* species. Its leaf crown is perhaps the most beautiful in the genus, providing breathtaking little starburst effects when the tree is planted en masse, and scintillating in the breeze to show off the silvery undersides. A single plant works well in an intimate space and is superb as a canopy-scape. It is not known to be grown indoors and would probably be difficult to do so.

Coccothrinax "azul"
ko'-ko-THRY-nax · a-ZOOL
PLATE 267

Coccothrinax "azul" is an unidentified (and perhaps unnamed) species from central Cuba, where it grows in serpentine soils with *C. clarensis*. The name *azul*, Spanish for "blue," refers to the leaves, which are bluish silver above and whitish silver below. Many palms that grow in serpentine soils show unusual bluish tendencies, but no other *Coccothrinax* species has leaves that are this blue, which is as intense as that of the blue forms of *Bismarckia*. This species may be a blue form of *C. macroglossa*, but additional taxonomic research is needed to confirm this identification.

The palm grows slowly to a height of perhaps 20 feet (6 m) with a trunk diameter of no more than 5 inches (13 cm). The fiber is particularly stiff and needlelike. The leaves are V shaped and not as stiff as those of some other species. Whether this palm becomes a separate species or is lumped with another, it appears to have several distinct taxonomic characters. It would be a stunning addition to the landscape even though it would take years to grow from seed to a size where it could be really appreciated.

Coccothrinax barbadensis
ko'-ko-THRY-nax · bahr-ba-DEN-sis
PLATE 268

Coccothrinax barbadensis occurs naturally in the Lesser Antilles, Barbados, Trinidad, and Tobago, and northernmost Venezuela, where it grows on calcareous soils in scrubland not far from the coast. It is sometimes called silver palm. The epithet translates from the Latin as "of Barbados."

The solitary trunk reaches 40 feet (12 m) high, making it one of the taller species in the genus, although the trunk is never more than 6 inches (15 cm) in diameter and thus very delicate looking. The leaf crown is 6 feet (1.8 m) wide and 5 feet (1.5 m) tall. The leaves are 3 feet (90 cm) wide, circular, with the narrowly lanceolate, tapering segments reaching only halfway to the center of the blade; they are pendulous at their ends. Leaf color is deep, shiny olive green to emerald green above but an almost chromium-like silvery and shiny hue beneath. Inflorescences are 2 feet (60 cm) long and pendent, drooping slightly below the leaf crown. The round ½-inch (13-mm) fruits are black when mature. The palm has the same cultural requirements as *C. argentea*. It is not hardy to cold outside of zones 10 and 11, although some protected specimens exist in 9b.

Coccothrinax borhidiana
ko'-ko-THRY-nax · bor-HEE-dee-ah-na
PLATE 269

Coccothrinax borhidiana is an endangered species endemic to Matanzas, Cuba, where it grows in scrublands near the ocean. The epithet honors Hungarian botanist Atilla Borhidi (1932–), a specialist in the plants of Cuba.

The solitary trunk grows, with great age, to a maximum height of 15 feet (4.5 m) in habitat and is usually clothed in a heavy skirt of dead leaves. The leaves are borne on very short petioles, and the leaf crown is tight and dense and, from any distance, causes the palm to look much like a yucca. Each leaf is almost circular with stiff linear, pointed segments that are a deep to bluish green above and a lighter hue beneath. From the side, the leaves appear to be in tight tiers. The inflorescences rise well above the leaf crown and become pendulous when the round, purplish black fruits mature.

This palm is hardy in zones 10 and 11, needs full sun, and is drought tolerant and somewhat accepting of salinity. It grows in well-drained, alkaline soils near the sea. With its tight leaf crown and small stature, it is probably the most distinctive and unusual species in the genus. It would not look out of place in a cactus or succulent garden. This little gem should be set near a path or other site where it can be seen up close. Although it grows in sparse stands in open areas in its native haunts, it does not lend itself to specimen planting in an open space.

Coccothrinax boschiana
ko'-ko-THRY-nax · bah-shee-AH-na
PLATES 270–271

Coccothrinax boschiana is endemic to a single, dry forest location in southern Dominican Republic, where it grows on steep limestone cliffs overlooking the ocean. The epithet honors Juan Emilio Bosch (1909–2001), author, historian, and first democratically chosen president of the Dominican Republic.

The slender solitary trunk can grow to 35 or more feet (11 m). It is gray and covered in ornamental coarse leaf bases near the top. The crown measures 4–5 feet (1.2–1.5 m) across and consists of 12 semiorbicular leaves held stiffly upright. The leaf is khaki green on top and silver underneath. It appears slightly folded lengthwise and is held in a manner that shows off its handsome underside. Inflorescences emerge from below the crown, producing a dark purplish fruit that holds a tiny spherical seed less than ⅛ inch (3 mm) in diameter.

The palm is a slow grower and requires well-drained alkaline soil to do best. It is not cold hardy, being best suited for zones 10 and 11. It has outstanding salt tolerance. This species has to be one of the most ornamental in the genus. The mix of tan-colored leaf base fiber with its intricate pattern and tightly held crown showing off the green and whitish coloration of the leaves is simply gorgeous. Being a slender palm and more delicate in appearance than other species of the genus, it is an ideal candidate to be planted in groupings of staggered heights that will attract all who come upon it.

Coccothrinax clarensis
ko'-ko-THRY-nax · klah-REN-sis
PLATE 272

Coccothrinax clarensis is endemic to central Cuba, growing on riverbanks in what is otherwise a dry part of Cuba. The epithet is Latin for "of Santa Clara," a reference to the city near which the first examples of this species were collected.

The solitary trunk grows to 15 feet (4.5 m) tall and 2–3 inches (5–8 cm) in diameter. The petioles are about 18 inches (45 cm) long, giving a compact appearance to the crown. The leaves are 3 feet (90 cm) wide, circular, with many narrowly lanceolate, rigid segments that are free for the majority of their length. The leaf is green above, but paler beneath due to the dense covering of silvery scales. The long inflorescence extends well beyond the leaf crown. The round ½-inch (13-mm) fruits are black.

This species requires access to soil moisture. Although its native soils are serpentine, nevertheless, it has proven adaptable to a variety of well-drained soils. It is not hardy to cold.

Coccothrinax crinita
ko'-ko-THRY-nax · kri-NEE-ta
PLATES 273–274

Coccothrinax crinita is endemic to western Cuba, where it grows in seasonally wet grasslands and hillsides. It is threatened because of its few numbers and restricted range. The epithet is Latin for "hairy." Common names are old man palm, old man thatch palm, and mat palm.

The solitary trunk attains a height of 25 feet (7.6 m) with great age and a diameter of 8 inches (20 cm); if the trunk retains its fibrous covering, it appears to be of greater width. The most distinctive characteristic is the shag of long, pale brown fibers that envelops the trunks; up close the fibers look like a loosely combed, brown wig. Only old trunks that are subject to wind or fire lose this covering. The rounded leaf crown is 8 feet (2.4 m) wide and tall. The leaves are to 5 feet (1.5 m) wide and circular, and are carried on petioles 4 feet (1.2 m) long. The segments are free for three-fourths of their length and are narrowly lanceolate and rigid. Leaf color is dark, shiny green above and silvery green beneath. Flower clusters are 5 feet (1.5 m) long and extend beneath the leaf crown. The rounded, 1-inch (2.5-cm) fruits are lavender when mature.

Taxonomists recognize 2 subspecies based on the length of the leaf base fibers. The one with longer fibers, **C. crinita subsp. crinita**, is horticulturally more desirable and more common in cultivation. The short-haired subspecies, **C. crinita subsp. brevicrinis** [bre-vi-CREE-nis] is passed over by some growers, as it lacks the visually arresting, long fibers. Nevertheless, the short fibers cover the trunk in an attractive way.

The palm has the same cultural requirements as *C. clarensis*, *C. boschiana*, and *C. borhidiana*, with the possible exception that it is not as drought tolerant. It also seems to be more cold tolerant than the other species and is grown successfully in zones 9b through 11.

Because of its unusual trunk, this species is not as easy to place in the landscape as are the other *Coccothrinax* species. It is nice in small groups of individuals of varying heights, and can be a conversation piece when planted where it can be enjoyed up close. The trunk should not be obscured by other plants nor should it be sprayed directly by irrigation sprinklers. This palm is not known to be grown indoors.

Coccothrinax ekmanii
ko'-ko-THRY-nax · ek-MAH-nee-eye
PLATE 275

Coccothrinax ekmanii occurs naturally in southwestern Dominican Republic, where it grows on limestone hills in scrublands near the coast. The epithet honors the intrepid Swedish plant explorer and botanist Erik Ekman (1883–1931), who collected extensively in Hispaniola. Henderson et al. (1995) included *C. munizii* of Cuba in this species, but the 2 are now generally regarded as distinct.

This species is one of the tallest solitary-trunked species in the genus, attaining a maximum height of 50 feet (15 m) in habitat. The leaves are wedge shaped, with narrow, stiff segments that are deep green above, distinctly silvery hued beneath. The leaf sheath has coarse, stiff, spine-tipped fibers, reminiscent of the leaf sheath of *Zombia antillarum*. The inflorescences are short, heavy, and pendent, hanging beneath the leaf crown. The fruits are pale and slightly warty.

This palm is hardy only to zones 10 and 11. It thrives in alkaline soil, needs full sun, and is drought tolerant once established. It is exceptionally slow growing. Its size makes it suitable for specimen plantings in groups of 3 or more individuals of varying heights. It also is a wonderful canopy-scape.

Coccothrinax gracilis
ko'-ko-THRY-nax · GRA-si-lis
PLATE 276

Coccothrinax gracilis is endemic to Hispaniola, where it grows in open coastal areas on limestone near sea level. The epithet is Latin for "graceful." The palm grows to 30 feet (9 m) in its native haunts. It has leaves that are circular and bright green above but silvery beneath. The inflorescences are short and curved down beneath the leaves. This palm has an airy appearance because of its height and relatively thin trunks. It is well suited as a canopy-scape or even in specimen groups of individuals of varying heights. It thrives in full sun in any well-drained soil, but is not hardy to cold and is only good in zones 10 and 11.

Coccothrinax gundlachii
ko'-ko-THRY-nax · goont-LAHK-ee-eye
PLATE 277

Coccothrinax gundlachii is endemic to central and eastern Cuba, where it grows on alkaline soil in open savannas or scrublands near the coast. It is rare in cultivation. The epithet honors German-born naturalist Johannes C. Gundlach (1810–1896), one of the foremost students of Cuban natural history. Henderson et al. (1995) synonymized *C. camagueyensis* and *C. clarensis* here, but these 3 species are now regarded as distinct.

In habitat, the solitary trunk grows to 30 feet (9 m) high and 8 inches (20 cm) in diameter. It is covered in all but its oldest parts with light gray, closely adhering, spine-tipped fibers forming a beautiful woven pattern. The sparse leaf crown is hemispherical or nearly spherical and contains 8–10 leaves, which are 3–4 feet (90–120 cm) wide, circular or nearly so, with deeply cleft, long, and narrow rigid segments. Leaf color is glossy deep green above and lighter green beneath. The inflorescences are long and mostly erect, extending above the leaf crown. This is a tall, stately species.

Coccothrinax hiorami
ko'-ko-THRY-nax · hee-o-RAH-mee
PLATE 278

Coccothrinax hiorami is endemic to eastern Cuba, where it grows in arid, calcareous soil. The epithet honors Brother Hioram, né Jean Frange Langorce (1875–1936), collector of Cuban plants. *Coccothrinax guatanamensis* was synonymized under this species by Henderson et al. (1995), but the 2 species are now believed to be distinct. The sturdy trunk grows to 30 feet (9 m) tall, and the leaves tend to be deeply divided with leaflets somewhat lax. The fiber is open and soft and, with the remnants of the old leaf bases, forms a distinct pattern on the trunk. The inflorescences are about as long as the leaves. This palm does best in full sun on well-drained alkaline soils. The drooping segment tips give it a somewhat informal look in the garden.

Coccothrinax litoralis
ko'-ko-THRY-nax · li-to-RAL-is
PLATE 279

Coccothrinax litoralis is endemic to Cuba, where it grows in sand on both the north and south coasts. It is a common sight on the northern keys in particular. The epithet refers to the littoral (shoreline) forests where it is found.

This species is closely related to *C. argentata* but differs in being far more robust. It has a smooth solitary light gray trunk to about 20 feet (6 m) tall and 6 inches (15 cm) in diameter. The crown holds 10–12 leaves and measures 6–7 feet (1.8–2.1 m) across. The leaves are a deep dark green on top with a contrasting light silver coloration underneath; each measures about 3 feet (90 cm) across and is supported by a petiole 2 feet (60 cm) long. The shape is half circular with deeply divided segments that are pendulous at the tips. Branched inflorescences are also 2 feet (60 cm) long and emerge from among the leaves. The fruits are dark purple to black.

This palm is far faster growing than *C. argentata* but not nearly as cold hardy, being adaptable to zones 10 or 11. It prefers alkaline soils with a full sun exposure and is quite salt tolerant. An elegant palm such as this demands a landscape situation where its silhouette can be appreciated against the sky or blank background. Like most *Coccothrinax* species, it is ideally suited to be planted in odd-numbered groupings of 3 or more.

Coccothrinax miraguama
ko'-ko-THRY-nax · meer-a-GWAH-ma
PLATES 280–281

Coccothrinax miraguama is indigenous to Cuba, where it grows in savannas, open woods, and near the coast. The epithet recalls the common name used for this species in Cuba, "miraguano."

When mature, the solitary trunk reaches a height of 40 or more feet (12 m) in habitat but under cultivation grows no more than 30 feet (9 m) with a diameter of 6 inches (15 cm). An exceptionally beautiful woven mesh design of old leaf fibers and narrowly triangular leaf bases covers younger parts of the stems, while older parts are bare and show closely set rings. The leaf crown is more rounded than that of most other species of *Coccothrinax* and is a maximum of 12 feet (3.6 m) wide and 8 feet (2.4 m) tall. The leaves are circular or nearly so, to 5 feet (1.5 m) in diameter on petioles 3–4 feet (90–120 cm) long. The linear-lanceolate segments are free halfway to the center of the blade and are rigid with a stiff tip. Leaf color is shiny deep green to almost bluish green above and silvery gray or grayish green beneath. The inflorescences are up to 3 feet (90 cm) long, branched and hanging beneath the leaf crown. The rounded ½-inch (13-mm) red fruits mature to deep purple or black.

Taxonomists recognize 4 subspecies, which differ in details of the leaf sheath, number of leaf segments, stamen size and number, and other technical details. The type, ***C. miraguama* subsp. *miraguama***, is described above. ***Coccothrinax miraguama* subsp. *roseocarpa*** [ro-see-o-CAHR-pa] has reddish purple fruit at maturity and is found in an inland area of Matanzas, Cuba. ***Coccothrinax miraguama* subsp. *arenicola*** [ah-re-NI-ko-la] grows in seasonally flooded, sandy savannahs in southwestern Cuba and the Isle of Youth. ***Coccothrinax miraguama* subsp. *havanensis*** [ha-va-NEN-sis] is native to the vicinity of Havana. The differences among the subspecies are poorly understood and have little bearing on their landscape use.

This palm is hardy only in zones 10b and 11 and is marginal in 10a. It needs full sun and grows in most well-drained soils, including calcareous ones. This species is generally considered the most beautiful in the genus. Its attractions are the wonderful crown of separated, round starburst-shaped leaves with their silvery undersides, the beautiful design of the trunk fibers, and the colorful fruit. The palm is exceptionally beautiful in groups of 3 or more individuals of varying heights and is unexcelled as a canopy-scape. It is among the best close-up landscape subjects or silhouettes for patios, courtyards, and other intimate sites. It is not known to be grown indoors and is probably not a good candidate for such.

Coccothrinax moaensis
ko'-ko-THRY-nax · mo-ah-EN-sis
PLATE 282

Coccothrinax moaensis is endemic to the heavy metal-laden soils in the vicinity of Moa, the city in eastern Cuba for which this palm is named. It grows amid the hard-leaved scrub vegetation that is typical of these ultramaphic (serpentine) soils. The small, tough leaves of the surrounding vegetation suggest dry chaparral, but in fact, the region around Moa is one of the wettest places in Cuba. Henderson et al. (1995) did not accept this species, believing it to be a form of *C. miraguama* specialized for growing on serpentine soils, but these 2 species are now regarded as distinct.

This improbable and extraordinary palm is remarkable. The trunk attains a height of 15 feet (4.5 m) and 3 inches (8 cm) in diameter after many decades, perhaps even centuries. It retains a thick covering of old leaf sheaths that disintegrate into a burlaplike covering, which in this species is not spiny. The leaves are rigid pinwheels of dark green above, silvery below. Each leaf has 18–22 widely splayed segments. The inflorescence is curved and hangs below the leaves. Each palm bears only 4 or 5 leaves in the crown, so the effect is almost comical and artificial. Unfortunately, this species is one of the slowest *Coccothrinax*. It needs sandy, well-drained, acidic soil and total protection from cold.

Coccothrinax montana
ko'-ko-THRY-nax · mon-TAH-na
PLATE 283

Coccothrinax montana is a rainforest species endemic to the island of Hispaniola, where it grows in the Dominican Republic and perhaps in Haiti in the central mountain range. The epithet refers to the palm's montane habitat of 5000–6000 feet (1520–1800 m) elevation.

The solitary trunk grows to 40 or more feet (12 m) in habitat and is topped with a crown 5–6 feet (1.5–1.8 m) in diameter bearing 8–10 leaves. In cultivation, it will likely grow no more than 20 feet (6 m) tall. Leaf bases have extremely coarse fiber wrapping around the upper trunk and forming a striking crisscross pattern. The stiffly held, orbicular leaf is 24–30 inches (60–76 cm) in diameter and is supported by a petiole 18–24 inches (45–60 cm) long. Leaf color is deep khaki green on top with some silvery scales on the underside. Deep lanceolate segments have a distinctive V shape running their entire length. This palm is one of 2 *Coccothrinax* species (the other being *C. torrida*) that bears white fruits rather than the usual purple to black fruits.

This very slow-growing species is not cold hardy and should be planted only in zones 10 or 11. While an emergent palm in habitat, it does quite well in full sun even when young and should be planted in a well-drained soil. Patience will be rewarded in time with an eminently elegant landscape specimen.

Coccothrinax pauciramosa
ko'-ko-THRY-nax · paw'-see-ra-MO-sa

Coccothrinax pauciramosa is endemic to eastern Cuba, where it grows in open, serpentine savannas and on limestone hills at low elevations. The epithet is Latin for "few branches," a reference to the inflorescences. The slender solitary trunk can attain a maximum height of 40 feet (12 m) but is usually much shorter and is covered in all but its oldest parts with dense woven dark fibers, some of which are long and spiny. The leaf crown is sparse, each leaf nearly circular with stiff segments; the blade is whitish on its lower surface. The inflorescence extends above and beyond the leaves. This beauty needs full sun and a well-draining soil. It is hardy only to zones 10 and 11.

Coccothrinax pseudorigida
ko'-ko-THRY-nax · su-do-RID-ji-da
PLATE 284

Coccothrinax pseudorigida is endemic to eastern Cuba, where it grows in open savannas. The epithet is compounded from the Greek *pseudo* ("false") and *rigida*, in reference to *C. rigida*, with which it has been confused. Henderson et al. (1995) believed this to be a synonym of *C. pauciramosa*, but currently both species are regarded as distinct. The solitary trunk to 15 feet (4.5 m) tall is completely covered by the old leaf sheaths, which have rigid spines almost ¼ inch (6 mm) wide. The effect is very reminiscent of *Zombia antillarum*. Each leaf has about 18 stiff segments, which are pale gray on the underside. The crown of leaves is spherical, and the inflorescence extends above and beyond the crown. This species is hardy in zones 10 and 11. It needs full sun and sandy, acidic soil.

Coccothrinax salvatoris
ko'-ko-THRY-nax · sal-va-TOR-is
PLATE 285

Coccothrinax salvatoris is endemic to eastern Cuba, where it grows along the coast on low limestone hills adjacent to salt marshes. The epithet honors Salvador Rionda (1888–1961), general superintendent of an important sugar mill in pre-revolutionary Cuba. The solitary trunk reaches 20 feet (6 m) high with much thick, coarse, and spiny fiber on all but the oldest parts. The leaves are circular with silvery undersides and have stiff segments. The inflorescence is curved and hangs below the leaves.

Coccothrinax scoparia
ko'-ko-THRY-nax · sko-PAH-ree-a
PLATE 286

Coccothrinax scoparia is endemic to Hispaniola, where it most commonly occurs in pine woodlands in cool, montane regions of the southern Dominican Republic. The epithet is Latin for "broom shaped," which presumably refers to the leaf but also recalls that the leaves of this species (and many others in the genus) are used for making brooms. Henderson et al. (1995) did not accept this species, believing it to be a form of *C. miraguama* of Cuba, but these 2 species are now regarded as distinct.

This handsome species grows to about 30 feet (9 m) tall, and the trunk, which is usually covered with soft burlaplike leaf sheaths, is about 6 inches (15 cm) in diameter. The leaves, green above and silvery below, have stiff, inflexible segments that spread in a full circle. The inflorescence is short and hangs below the leaves. The fruits are small and black. This palm is easy to grow and unfussy about soils as long as they are well drained. It may have some tolerance to cold, but unfortunately, the limits of its hardiness have not been tested.

Coccothrinax spissa
ko'-ko-THRY-nax · SPIS-sa
PLATE 287

Coccothrinax spissa is an endangered palm confined to a small area of southern Dominican Republic, where it grows in savannas and cleared forest areas. The epithet is Latin for "thick," an allusion to the swollen trunk.

The solitary trunk attains a height of 20–30 feet (6–9 m), is light colored, usually free of leaf bases, smooth, robust, and usually but not always showing a slight to pronounced bulge near the middle of the stem. The leaf crown is full and the leaves are semicircular, with deep but limp, pointed, light to deep green segments that exhibit a silvery sheen beneath. The inflorescences curve down and hang below the leaves. They are short and branched. The round ½-inch (13-mm) fruits are deep purple to nearly black.

This striking palm is hardy only to zones 10 and 11, being marginal in 10a. It is drought tolerant and not particular about soil type as long as it is well drained. It is slow growing even in full or nearly full sun.

Cocos
KO-kos

Cocos is a monotypic genus of solitary-trunked, pinnate-leaved, monoecious palm now found on all tropical seashores. The name is a Portuguese corruption of an aboriginal name meaning "monkey," an allusion to the 3 facelike depressions at one end of the nut. Seeds can take more than a year to germinate and do so sporadically. Coconuts are usually sown without removing the husk. For further details of seed germination, see *Acrocomia*.

Cocos nucifera
KO-kos · noo-SI-fe-ra
PLATES 288–289

Cocos nucifera has been cultivated for so long and was carried to tropical regions at such an early date that its exact places of origin are obscured but are probably the Indo-Malayan region of Southeast Asia. The common name is coconut palm. The epithet is Latin for "nut-bearing."

Mature trunks can reach a height of 90–100 feet (27–30 m). The base of the stem is always swollen and its diameter above that point is nearly constant at 1 foot (30 cm). The trunks of large palms lean or curve and, especially near the shore, are often prostrate and contorted. They are picturesquely ringed with crescent-shaped leaf base scars, and some individuals exhibit a beautiful color pattern of light gray or almost white trunk with darker scars.

The leaf crown is usually round, full, and to 30 feet (9 m) wide and tall. The leaves are 20 feet (6 m) long on stout yellowish petioles 3–4 feet (90–120 cm) long. The leaflets are to 3 feet (90 cm) long and narrowly lanceolate, and arise in one plane from the rachis. Young leaves have stiff leaflets, but older ones tend to be pendulous. The younger leaves also tend to have a twist of the rachis, which renders the leaflets vertical for the apical half of the blade. Leaf color varies according to environmental conditions and the variety of the palm, ranging from deep, shining green to yellowish green.

The palm starts blooming and fruiting at an early age, often with no more than 3 feet (90 cm) of trunk growth. The inflorescences appear in panicles 5 feet long (1.5 m) below the crown of leaves and are accompanied by a large boat-shaped woody spathe, which is usually deciduous before the fruits form. The flowers are tightly packed on the flowering branches, and their color ranges from white to almost yellow. The coconut is 1 foot (30 cm) long, 3-sided, and, when fully mature, bright yellow to green or reddish brown, depending on the cultivar and the age of the fruit. When fully dry, the husk of the fruit turns a pale gray-brown.

Coconuts are exceedingly handsome on the tree, hanging in great clusters below the crown of leaves; and, yes, they can be fatal if they fall from any height onto one's head, although this is a very rare phenomenon. The edible nut must be liberated from its hard, brown stone (endocarp) before consumption. The coconut is one of the world's most economically important palms, second only to African oil palm (*Elaeis guineensis*). Every part of the tree is used and some of the uses, like copra (the dried endosperm from which coconut oil is expressed) and the edible nuts, constitute major industries in the Philippines, Indonesia, Trinidad and Tobago, Jamaica, and Mexico. The trunks are used for construction and the leaves for thatch and for weaving many utensils.

This species is susceptible to lethal yellowing disease. It is drought tolerant when established and extremely salt tolerant. It survives in dry, poor, and limey soil but looks and grows much better with good soil and adequate water. It grows moderately fast under optimal conditions, especially when younger, but is almost intolerant of cold and is adaptable only to zones 10b and 11.

The coconut palm is the icon of the tropics and tropical beauty. While it is commonplace in tropical areas, it is yet the most beautiful tree on earth. One of the most picturesque palm tableaux is created by planting a number of coconuts of varying heights in a group; siting them no more than 10 feet (3 m) from each other encourages them to naturally bend towards the optimum lighting conditions for each tree, and the result is a natural-appearing island. The coconut palm is unsurpassed as a canopy-scape, and its silhouette is both stunning and a universally recognized emblem of the tropics. This palm is difficult to maintain in any enclosure, as it needs good air circulation and bright light. In spite of this, it is often successfully grown for years in large atriums and conservatories.

Many cultivated forms exist, several of which are readily available. ***Cocos nucifera* 'Tall Jamaican'** (synonym 'Jamaican Tall'), one of the tallest growing and reportedly the most tender to cold, is very susceptible to lethal yellowing disease, as is ***C. nucifera* 'Panama Tall'**. Three varieties are less susceptible to lethal yellowing: ***C. nucifera* 'Green Malayan Dwarf'**, to 60 feet (18 m) high with deep green leaves; ***C. nucifera* 'Golden Malayan Dwarf'**, also to 60 feet (18 m) high but with beautiful yellow-green leaves, dazzling golden fruit, and golden to almost orange petioles and rachises (these cultivars are "dwarf" only in the sense that they bear fruit while still quite short); and the outstandingly spectacular ***C. nucifera* 'Maypan'**, tall growing and robust with leaves that are longer and broader than those of the type. Lethal yellowing, once established in an area, such as South Florida, will never go away. Even the resistant varieties have some susceptibility to the disease, especially as they grow older and taller. Nevertheless, the coconut is such a beautiful and useful palm that many people will always want to have it in their gardens.

Colpothrinax
kol′-po-THRY-nax

Colpothrinax is composed of 3 palmate-leaved, hermaphroditic palms in tropical America. The leaves are costapalmate but, unlike those of many *Sabal* species, not so much that the costa is evident and curves downward. The inflorescences are much branched, do not extend beyond the leaf crown, and bear bisexual flowers. The fruits are black when mature. All the species are slow growing and not cold hardy, being adaptable only to zones 10 and 11. Despite horticultural myth, these palms do not like calcareous soils and dry conditions, although *C. wrightii* is able to withstand seasonal droughts such as those that occur in its native habitat.

The genus name combines the Greek word meaning "swollen," referring to the trunks of the Cuban species, and *Thrinax*, a similar genus. In reality, *Colpothrinax* is much more closely related to *Pritchardia*. Seeds of *Colpothrinax* can take more than 2 years to germinate, even when fresh, and do so sporadically. Do not give up on any seeds during that time. For further details of seed germination, see *Corypha*.

Colpothrinax cookii
kol′-po-THRY-nax · KOOK-ee-eye
PLATES 290–291

Colpothrinax cookii is indigenous to 4 disjunct and small areas in central Guatemala, southern Belize, Costa Rica, and Panama, where it grows in montane rain forest. The epithet honors Orator F. Cook (1867–1949), an American botanist. This palm grows to an overall height of 40 feet (12 m) with a trunk diameter of 1 foot (30 cm). The newer parts of the stem are usually covered in a tangled mass of long fibers. The circular 3-foot (90-cm) leaves are borne on petioles 4 feet (1.2 m) long. The linear segments are yellowish green above and grayish green to almost white beneath, free almost to the petiole, and pendulous at their apices. This palm has exceptionally beautiful leaves and is slow growing enough that it may be used as a close-up subject in partial shade or full sun. Older trees make specimen plantings when spaced in groups of 3 or more individuals of varying heights.

Colpothrinax wrightii
kol′-po-THRY-nax · RY-tee-eye
PLATE 292

Colpothrinax wrightii is endemic to western Cuba and the adjacent Isle of Youth, where it grows in open, seasonally flooded pinelands. The epithet honors American botanist, Charles H. Wright (1811–1885), famed for his explorations of the American Southwest, Mexico, and Cuba. Common names are Cuban bottle palm, Cuban belly palm, Cuban barrel palm, and, in Cuba, *barrigona*.

The solitary trunk can attain a height of 30 feet (9 m) high with age. It almost invariably has a distinct bulge near or beneath the middle of the stem; a few individuals in habitat even have a second bulge often twice the diameter of the trunk above and beneath the "belly." It is assumed that this swelling stores water for the palm in times of drought; cultivated specimens which receive regular and ample amounts of moisture usually have less-pronounced bulges. The leaves are similar in size and form to those of *C. cookii* but have generally stiffer segments and are a darker, almost bluish green above.

The species does not take kindly to calcareous soils, preferring instead red clays and quartz sands. It grows faster and looks better with regular, ample moisture. It also needs full sun from youth unto old age but seems more tolerant of cold than does *C. cookii*, although it is still tender. The palm is as beautiful as and more majestic than most *Pritchardia* species; it also looks from a distance like a stout, robust palmetto (*Sabal* species). It is probably at its best as a specimen. This species is not known to be grown indoors and is probably a poor candidate for such as it requires much sunlight.

Copernicia
ko-pur-NEE-see-a

Copernicia includes 28 species and several putative natural hybrids of mostly solitary-trunked, palmate- or slightly costapalmate-leaved, hermaphroditic palms of Cuba, where most species are endemic, as well as Hispaniola and South America. Usually a significant skirt of dead leaves hangs on the trunks, and living leaves are often covered with wax. The leaves are held on usually short, sometimes almost nonexistent, very spiny petioles, and a hastula, often large and spiny-edged, is present at the juncture of petiole and leaf blade. The much-branched inflorescences grow from the leaf crown and are often longer than the leaves. They bear small, bisexual blossoms that produce rounded blackish fruits.

All species except *C. alba* and *C. prunifera* are slow to exceedingly slow growing, especially when young. A popular misconception is that *Copernicia* species like truly dry conditions. In habitat, most grow in lowland conditions that are seasonally flooded. Even during the dry season they generally receive occasional rain but can tolerate a couple of months without rain because the water table is high. The soils they grow in may also have some bearing on their moisture requirements as well: *C. baileyana*, *C. fallaensis*, *C. glabrescens*, and *C. hospita* grow in alkaline, red clays that hold water; other species such as *C. brittonorum*, *C. gigas*, *C. macroglossa*, and *C. rigida* grow in coastal sand close to mangrove swamps that stay wet at all times and are very salt tolerant. While copernicias can tolerate dry conditions, they grow best with regular, adequate moisture. Regular feedings with a good palm fertilizer are sufficient, and none of the species except perhaps *C. cowellii* requires any additional trace elements. In South Florida and similar climates, copernicias have proven cold hardy, only minor leaf burn occurring on specimens when temperatures reach the upper to mid-20s Fahrenheit (around −2 to −4°C) during the infrequent cold snaps.

A few of the Cuban species apparently hybridize in some areas of the island, often making it difficult to determine their parentage. Some of the most incredibly beautiful specimens can be found in these habitats. The genus name honors the great Polish astronomer and polymath Mikołaj Kopernik (1473–1543), better known as Copernicus. Seeds of *Copernicia* germinate within 30 days and should not be allowed to dry out before sowing. For further details of seed germination, see *Corypha*.

Copernicia alba
ko-pur-NEE-see-a · AHL-ba
PLATE 293

Copernicia alba occurs naturally in eastern Bolivia, Paraguay, northern Argentina, and southeastern Brazil, where it grows in seasonally flooded savannas. The epithet is Latin for "white," an allusion to the waxy white undersides of the leaves. The species is commonly called caranday palm.

This palm can attain a height of 100 feet (30 m) in habitat. The trunks of large specimens are less than 1 foot (30 cm) in diameter and retain many old leaf bases in their younger parts; the older parts are mostly smooth, sometimes indistinctly ringed, and light gray to almost pure white. The leaves are 3 feet (90 cm) wide, borne on long petioles, and are deeply divided into linear, pointed segments that are grayish green to pure green above, silvery green beneath, and glaucous and waxy on both surfaces. The length of the inflorescence exceeds the leaf crown, and the fruits are egg shaped and black when mature.

The species extends farther south than any other species in the genus and is the hardiest to cold. It is adaptable to zones 9b through 11, and protected specimens are occasionally found even in 9a. It adapts to a range of soils, including calcareous and heavy clays and, while drought tolerant once established, can endure flooded soil for a short time. It is among the fastest growing species in the genus and is the tallest growing. It needs full sun.

This palm is noble, yet graceful. Its dense leaf crown is beautiful up close and from afar, and it works wonderfully as a canopy-scape. It is probably most picturesque in groups of 3 or more individuals of varying heights, simulating its colonial habit in the wild where great stands of the palm occur. It is not known to be grown indoors, but could be given a sunny and airy site in a conservatory or atrium.

Copernicia baileyana
ko-pur-NEE-see-a · bay-lee-AH-na
PLATE 294

Copernicia baileyana is endemic to Cuba, where it grows in savannas and open woodlands. The epithet honors American horticulturist and palm taxonomist, Liberty Hyde Bailey (1858–1954). The common name is Bailey fan palm.

Mature trunks attain a height of 60 feet (18 m) in habitat but are usually no more than 40 feet (12 m) under cultivation, with a diameter of about 2 feet (60 cm). The trunks are usually of even diameter from within 1 foot (30 cm) of the ground to the leaf crown but sometimes show a slight swelling in the middle. They are remarkably smooth below the few, pendent, dead leaves and are pale gray to nearly white. The leaf crown is usually rounded, 15–20 feet (4.5–6 m) wide and tall, dense and fully packed, with stiff segments. Leaves are 5 feet (1.5 m) wide and nearly circular, with many stiff, narrowly lanceolate segments that are free for one-third of the way to the center of the blade. The leaves are held on stout petioles 4 feet (1.2 m) long. Leaf color is light to deep green above and lighter grayish green beneath, where the leaf is covered in a waxy bloom. The much-branched inflorescence panicles are 7 feet (2.1 m) long, curving down from the center of the leaf crown. The rounded brown to black fruits are less than 1 inch (2.5 cm) in diameter.

This species is not hardy to cold and is limited to zones 10 and 11. Although drought tolerant when established, it grows faster and looks better with regular, adequate moisture. It is not particular about soil type as long as it is well drained, but it needs full sun, especially when past the juvenile stage.

The first thing one notices about mature specimens of this magnificent palm is the massive trunks, like stout, unadorned columns carved from marble. The second thing noticed is the dense crown of large deep green rounded leaves with their almost startling halo effect of stiff, comblike segments. The palm is eminently suited to specimen planting in large areas and to planting in formal rows, forming a living colonnade reminiscent of the great Hypostyle Hall of the ancient Egyptian temple of Karnak. Would that earlier generations of landscapers planted this species widely so that we could now enjoy these magnificent palms in public spaces. The palm is not known to have been grown indoors and is probably a poor candidate for such.

Copernicia berteroana
ko-pur-NEE-see-a · ber'-te-ro-AH-na
PLATE 295

Copernicia berteroana is endemic to seasonally dry parts of Hispaniola. The epithet honors Italian plant collector and botanist, Carlo L. G. Bertero (1789–1831).

The palm grows to a height of 30 feet (9 m) in habitat. The slender, brownish gray trunk is 8–9 inches (20–23 cm) in diameter, and mostly smooth and free of leaf bases. The shiny circular leaves are 3 feet (90 cm) in diameter, a pure, grassy green on both surfaces. The leaf segments are free for two-thirds of their length and pendulous, at least in older specimens. The inflorescences are slightly longer than the diameter of the leaf crown, and the mature fruits are black.

This species is drought tolerant but looks better and grows faster with regular, adequate moisture. It is not hardy to cold and is adapted only to zones 10 and 11. It needs full sun from youth to old age but tolerates light shade when young. It is not fussy about soil type as long as it is well drained. It is one of the faster growing species but could never be considered fast.

In the landscape this palm is wonderful as a canopy-scape or even isolated in expanses of lawn if planted in groups of 3 or more individuals of varying heights. It resembles the much taller and faster-growing Mexican fan palm (*Washingtonia robusta*).

Copernicia brittonorum
ko-pur-NEE-see-a · brit'-to-NO-rum
PLATE 296

Copernicia brittonorum is endemic to western Cuba, where it grows in savanna, open dry woodland, and almost on the beach.

It is critically endangered. The epithet means "of the Brittons" and honors husband and wife botanists Nathaniel L. Britton (1859–1934) and Elizabeth G. Britton (1858–1934), both affiliated with the New York Botanical Garden. This species is similar to *C. hospita*, differing mainly in the much longer inflorescences and in its leaves that are always circular, green, and with segments whose apical ends are deeply divided and pendent.

Copernicia cowellii
ko-pur-NEE-see-a · ko-WEL-ee-eye
PLATES 297–298

Copernicia cowellii is endemic to eastern Cuba, where it grows in dry pine scrub and open savannas. It is rare and endangered. The epithet commemorates the original co-collector of the species (with N. L. and E. G. Britton), American botanist John F. Cowell (1852–1915).

The trunk attains 6 feet (1.8 m) of height but is seldom seen because of the adhering dead leaves. The overall height of old individuals is 8 feet (2.4 m), making it the smallest species in the genus. The leaf crown is dense and compact, causing the plant to look like a yucca. In fact, this palm has one of the most unusual looks in the genus. The leaves are circular but do not readily appear so because they are held on short petioles and are packed tightly in the crown. Each thick, rigid leaf is 3 feet (90 cm) in diameter and deep olive green above but a waxy bluish gray beneath. The inflorescences are much longer than the leaf crown and bear round, black fruits.

This palm is adaptable only to zones 10 and 11 and is marginal in 10a. It needs full sun and is drought tolerant. In habitat, it grows on serpentine soils, which are high in magnesium, iron, and nickel, and are generally toxic to plants. The species has only been in cultivation for a few years, and it will be interesting to see how it adapts to cultivation outside its habitat.

Although the palm is not large enough to be planted as a specimen, a group of them, combined with caudiciform oddities, such as *Pachypodium* spp. or even *Hyophorbe lagenicaulis* (which see) would be spectacular. It works well as a border for larger vegetation but must not be combined with water-loving plants. It also looks good in a cactus or succulent garden as a focal point. Displayed against a contrasting background, it is incredibly effective, as it is planted along a wide, sunny path where it can be viewed up close.

Copernicia curtissii
ko-pur-NEE-see-a · kur-TIS-see-eye
PLATE 299

Copernicia curtissii is endemic to Cuba in western Pinar del Río province, as well as the Isle of Youth to the south. It grows as a green form near the coast and as a whitish silver form inland on open savannahs often prone to some flooding during the wet season. Stems are used in construction because of their strength and durability. The epithet honors American botanist Allen Hiram Curtiss (1845–1907), who collected the type specimen.

This solitary-trunked or clustering palm is similar in look and stature to *C. hospita*. Stems can reach nearly 20 feet (6 m) in height with a diameter of 10 inches (25 cm) and are smooth in their older parts while holding persistent leaf bases near the crown. The leaf crown has a diameter of 15 or more feet (4.5 m) and is made up of as many as 20 nearly orbicular leaves. Petioles 4 feet (1.2 m) long support leaves 5–6 feet (1.5–1.8 m) long with 70–80 segments. The inflorescence is 10 feet (3 m) long and extends far beyond the leaves, producing golden, spherical fruits about ¾ inch (19 mm) in diameter.

This species tolerates drought but thrives best in an open, sunny area with regular watering. It is easily adaptable to zones 10 and 11 and will tolerate some light cold snaps at or just below freezing. The clustering form will usually have a central main stem with 2–4 smaller stems rising around it. The silver form is especially eye-catching and, with several smaller stems surrounding a main stem, can be a major focal point in a landscape.

Copernicia ekmanii
ko-pur-NEE-see-a · ek-MAH-nee-eye
PLATE 300

Copernicia ekmanii is endemic to northern Haiti, where it grows near the coast. It is rare and threatened with extinction because of its small range. The epithet commemorates the great Swedish botanist Erik Ekman (1883–1931), who collected extensively in Hispaniola.

The palm attains a maximum height of 15 feet (4.5 m) in habitat, with trunks that are usually smooth and free of old leaves, but under cultivation the trunk can grow to 30 feet (9 m). The circular 3-feet (90-cm) leaves are a waxy bluish gray on both surfaces and are borne on spiny petioles 2–3 feet (60–90 cm) long. The segment tips are stiff and fine. The inflorescences project well beyond the leaf crown. The small, ovoid fruits are a shiny black.

This species is drought tolerant but not hardy to cold and is adaptable only to zones 10 and 11. It requires a well-drained soil and full sun. Because of its leaf color, it is wonderful against a background of contrasting colors and textures.

Copernicia fallaensis
ko-pur-NEE-see-a · fah-lah-EN-sis
PLATE 301

Copernicia fallaensis is endemic to Cuba and is known from a single location, where it is threatened with extinction. The epithet is Latin and translates as "of Falla," a small town in northern central Cuba near the palm's habitat. This species was treated as a synonym of *C. baileyana* by Henderson et al. (1995), but is now regarded as a separate species. It is the largest species in the genus and similar in appearance to *C. baileyana* but larger. The trunks attain a height of 60 feet (18 m) and a diameter of 3 feet (90 cm). The leaves, in contradistinction to those of *C. baileyana*, are always silvery blue-green, diamond shaped (not circular) in outline, and 8 feet (2.4 m) wide.

Copernicia gigas
ko-pur-NEE-see-a · GY-gas
PLATE 302

Copernicia gigas is endemic to southeastern coastal Cuba in open woodlands, savannas, and salt marshes. The epithet is Latin for "gigantic," an apt appellation. *Copernicia vespertilionum* was synonymized here by some authorities, but it is now regarded as a hybrid with *C. rigida*. This palm is similar in all respects to *C. baileyana* except for its leaf size and shape: its leaves are slightly larger with much larger leaf bases and much longer petioles, and they are shaped like a wedge or a triangle. There are also differences between the 2 species in the bracts of the inflorescence. In a landscape design, however, they could be interchangeable. Both species, along with *C. fallaensis*, are magnificent palms for large garden settings.

Copernicia glabrescens
ko-pur-NEE-see-a · gla-BRES-senz
PLATE 303

Copernicia glabrescens is endemic to western Cuba, where it grows in open savannas at low elevations. The epithet is Latin for "becoming smooth or hairless." This species is similar to *C. hospita*, except that it clusters with time, always has pure green leaves, and has longer inflorescences that greatly exceed the leaves. The few differences of floral details are important only to taxonomists. The thin trunks can grow to 30 feet (9 m) in habitat and are covered in old leaf bases and petiole parts. The species has the same cold tolerance and general cultural needs as *C. hospita* but, in clustering individuals, is an even more desirable garden subject.

Copernicia hospita
ko-pur-NEE-see-a · HAHS-pi-ta
PLATE 304

Copernicia hospita is widespread in Cuba, in savannas and open woodlands. The epithet is Latin for "harboring" and is probably an allusion to the large shag of dead leaves adhering to the trunk that often harbors rodents, bats, birds, and other creatures.

Mature trunks grow 20–25 feet (6–7.6 m) high with a diameter of 1 foot (30 cm) and are smooth on the older parts. The leaf crown is 18 feet (5.4 m) wide and tall. The leaves are 5–7 feet (1.5–2.1 m) wide, circular to wedge shaped, with many rigid segments that are free for one-third of their length, becoming slightly pendulous at their ends when they mature. Leaf color is grayish green to bluish green to almost pure white or silver, and both surfaces are covered in a waxy bloom; there are tiny teeth along the margins of each segment. The petiole is short but extends into the leaf blade, making it costapalmate, and there is a distinct hastula on the top of the juncture of leaf and petiole. The leaf crown is densely packed, with leaves that almost overlap. The inflorescence is a panicle 6 feet (1.8 m) long that extends beyond the leaves and bears small, brownish yellow flowers. The round ½-inch (13-mm) fruits are black.

This species is adaptable to zones 10 and 11, needs sun, and is drought tolerant but looks and grows better with regular moisture. The color and form of the leaves, especially in young plants, make this small palm choice for patio or courtyard settings. It is also beautiful in groups of 3 or more individuals of varying heights. It is not known to be grown indoors but is probably adaptable to large and sunny conservatories.

Copernicia macroglossa
ko-pur-NEE-see-a · mak-ro-GLAHS-sa
PLATE 305

Copernicia macroglossa is endemic to northwestern and central Cuba, where it grows in savannas and in salt marshes near the coast. The common name is Cuban petticoat palm. The epithet is 2 Greek words meaning "large" and "tongue," an allusion to the very large flaplike hastula at the junction of leaf blade and petiole.

Mature trunks attain a height of 15 feet (4.5 m) and are 8 inches (20 cm) in diameter but, especially in younger plants, are covered with a shag of dead leaves as wide as the leaf crown, which is 12–15 feet (3.6–4.5 m) wide and tall. The crown is unusually round and full in young palms, and the leaves are so densely packed that the palm closely resembles a haystack with green foliage at the top. As the palm gets older and taller, the lower skirt of leaves will fall off, revealing a smooth, slender stem. By that time, the palm resembles a "lollipop" tree in a child's drawing, or one of Dr. Seuss' trufula trees as illustrated in *The Lorax*. Leaves are 5–7 feet (1.5–2.1 m) wide, wedge shaped or half circular, and are carried on short petioles that extend only a little distance into the blade. The many, stiff, narrowly lanceolate segments are free for two-thirds their length and are armed along their margins with tiny teeth. Leaf color is light to deep green above and grayish green and waxy beneath. There is a distinct and relatively long hastula protruding from the top of the leaf at the juncture of blade and petiole. The inflorescence is a panicle at least 6 feet (1.8 m) long; it extends beyond the leaf crown and bears short, stout branchlets crowded with brownish yellow blossoms. The rounded ½-inch (13-mm) fruits are black.

This species is adaptable to zones 10 and 11 and is marginal in 9b. It needs sun and a well-draining soil. It is one of the most widely cultivated species in the genus and justly so, as it is dramatic in the landscape. It may be used as a large accent in a wide border and, even when it grows up (which takes a long while), its height is not too great to remain visually "in place." Its character makes it adaptable to the cactus or succulent garden, and a group of these palms is stunning if the individuals are of differing heights. It is not known to be grown in enclosures and is probably a poor candidate for such.

Copernicia prunifera
ko-pur-NEE-see-a · proo-NI-fe-ra
PLATE 306

Copernicia prunifera is endemic to northeastern Brazil, where it grows in low-lying monsoonal areas, especially along rivers and lakes. The common name is carnauba wax palm. The epithet translates from the Latin as "bearing waxy powder," referring to the white powdery wax covering the leaves.

Mature trunks grow to a height of 50 feet (15 m) in their native haunts but are usually no more than 35 feet (11 m) under cultivation. They are 10 inches (25 cm) in diameter and often retain the spirally arranged knobby leaf bases, mostly on the lower parts of the trunks rather than the upper portions as is the case with most other palm species. The full, rounded leaf crown is 15 feet (4.5 m) wide and open. The circular 5-foot (1.5-m) leaves are carried on petioles 3 feet (90 cm) long with strong teeth on their margins. The many linear, rigid segments are conjoined for half their length. Leaf color is deep yellow-green to blue-green, and both surfaces are covered in a gloss of wax, especially the lower surface. The inflorescence is a narrow panicle 7 feet (2.1 m) long that extends well beyond the leaves and bears brownish yellow flowers. The round 1-inch (2.5-cm) fruits are brown to black.

This species has unusual hardiness to cold considering its tropical origins. David Witt (pers. comm.) reported that it seems hardy in Orlando, Florida (zone 9b/10a). It is slow growing but significantly faster than most other *Copernicia* species except for *C. alba*. It needs sun and appreciates regular, adequate moisture, although it is drought tolerant when established. It seems to thrive on almost any soil type except the quite acidic or quite alkaline.

This moderately sized palm is excellent in a large patio or courtyard, where the form of its trunk and leaf crown may be appreciated. It looks especially wonderful in groups of 3 or more individuals of different heights, and its silhouette is extraordinarily beautiful. It is probably a better candidate than most other copernicias for atriums or conservatories. The wax from the leaves, carnauba wax, is a valuable commodity much used in cosmetics, food, and pharmaceuticals. Refineries in Brazil still process this wax.

Copernicia rigida
ko-pur-NEE-see-a · RI-ji-da
PLATES 307–308

Copernicia rigida is endemic to eastern Cuba, where it grows in savannas, open pinelands, and serpentine scrub near the coast. The epithet is Latin for "rigid," an allusion to the leaf segments.

The palm grows to 50 feet (15 m) high in habitat. The trunk is covered near its summit with a skirt of dead leaves in all but its oldest parts, and it tends to have the shape of a lollipop when older. The stiff, erect, narrow, and wedge-shaped leaves are medium green above and waxy and grayish underneath, with small prickles along the margins of the rigid, linear segments. The petiole extends into the blade to form a prominent costa. The inflorescences extend beyond the leaf crown and bear small, brownish flowers, which produce small, round black fruits.

This species is not hardy to cold and does not survive unprotected outside of zones 10 and 11. It is drought tolerant and needs full sun and a well-drained soil. It is most unusual looking, especially when smaller, as it usually has a long leaf crown of erect, rigid leaves. Because of its slow growth, it can be sited as a curiosity or specimen for a long time; in old age, it is magnificent as a canopy-scape, the crown then usually globular atop the stout, straight trunk. This palm is probably not a good candidate for indoors, but no data are available.

Copernicia tectorum
ko-pur-NEE-see-a · tek-TOR-um
PLATE 309

Copernicia tectorum is indigenous to northern Colombia, northwestern Venezuela, and the Netherlands Antilles, where it grows in monsoonal savannas (*llanos*), often forming great colonies. The epithet is Latin for "of the roofs," an allusion to the use of the leaves as thatch.

The palm grows to 40 feet (12 m) in habitat. The trunks are 1 foot (30 cm) in diameter and are covered in an attractive pattern of old leaf bases except in their older parts which are smooth and light to dark gray. The circular 3-foot (90-cm) leaves are carried on well-armed petioles 5 feet (1.5 m) long. Leaf color is medium to deep green on both surfaces. The inflorescences are as long as the diameter of the leaf crown and are not as apparent as in other *Copernicia* species, where the panicles extend beyond the crown. The brown egg-shaped fruits are 1½ inches (4 cm) long.

This species is tender to cold and adaptable only to zones 10 and 11. It thrives in heavy clay, sandy, and calcareous soils. It can withstand waterlogged soil as well as drought conditions but grows faster and looks better with regular moisture. It needs full sun. This palm looks a lot like a *Livistona* species from a distance and is excellent as a canopy-scape or as a specimen. It is not known to be grown indoors but can probably adapt to a sunny, airy conservatory.

Copernicia ×*textilis*
ko-pur-NEE-see-a · tex-TIL-iss
PLATE 310

Copernicia ×*textilis* is another Cuban palm found in open woodlands. Although originally described as a species (*C. textilis*), it is believed by some authors to be a natural hybrid between *C. hospita* and *C. baileyana*. The epithet is Latin for "textile" and refers to the use of the leaves for materials such as thatch and weaving of handicrafts.

The solitary trunk is sometimes slighted bulging in the middle, and is up to 30 feet (9 m) tall and 20 inches (50 cm) in diameter. The crown of leaves is rounded and 18–20 feet (5.4–6 m) across. The circular leaves, about 4 feet (1.2 m) in diameter, are borne on stout, well-armed petioles. The segments are stiff and erect, shiny green above and waxy white on their lower surface. The highly branched inflorescences extend well beyond the crown and bear many, thin branchlets. This palm looks like a smaller, less formal version of *C. baileyana* and, as such, would be a handsome addition to any informal landscape.

Corypha
ko-RY-fa

Corypha consists of 6 gigantic solitary-trunked, palmate-leaved, hermaphroditic palms in tropical Asia, Malaysia, Indonesia, and Australia. The monocarpic plants die after producing enormous terminal inflorescences. In a garden setting, the very large dead palms can pose a major logistical problem, but they are a boon for cavity-nesting birds. Were it not for the spectacular beauty of the leaves, especially when young, and the many years that it takes for the palms to flower, they would be garden outcasts. Even so, they are best relegated to estate-sized gardens, parks, botanical gardens, and wherever the death of the trunks does not pose a threat to adjacent gardens or structures.

The palms grow slowly when young but faster with the advent of trunk formation. Only 2 species are commonly planted. Both dominate the landscape. The genus name is Greek for "summit," an allusion to the giant terminal inflorescence. *Corypha* is remote germinating. The germination stalk ranges from 1 to 3 inches (2.5–8 cm), depending on the species, so a deep container at least 6 inches (15 cm) deep should be used. Fresh seeds of *Corypha* tend to germinate quickly within 120 days.

Corypha umbraculifera
ko-RY-fa · um-brak'-yoo-LI-fe-ra
PLATES 311–313

Corypha umbraculifera is of unknown origin because of its long history of cultivation and is always found in association with human habitation; it probably originated in the monsoonal plains and open forests of southern India and Sri Lanka. The epithet is Latin for "shade-bearing," an allusion to the giant leaves and immense crown. While not strictly an English name, talipot palm is a commonly used name.

Mature trees grow to 90 feet (27 m) high with a trunk diameter to 3 feet (90 cm). The stems are mostly free of leaf bases and ringed when older, but younger parts have many immense leaf bases. The leaf crown is 40 feet (12 m) wide and tall when young but half that size when mature. The plants take many years to form a trunk and grow slowly (but are nevertheless massive) until trunk formation is initiated. The leaves are 20 feet (6 m) wide, circular but folded into a trough, and are carried on stout petioles 10 feet (3 m) long and armed with black teeth. The leaves are costapalmate with the petiole projecting several feet into the blade, resulting in a deep V shape. There is a large hastula on the blade at the juncture of leaf and petiole. The segments are stiff, tough, and 3 inches (8 cm) wide and 7 feet (2.1 m) long. Leaf color is light to deep green on both sides. The terminal panicle is 30 feet (9 m) tall with great branches that make it 40 feet (12 m) wide. This is the largest inflorescence in the Plant Kingdom. The flowers are creamy white to almost yellow and number in the millions, creating a Christmas-treelike affair with enormous plumose branches. The round 2-inch (5-cm) green to brown fruits create a show almost as spectacular as the flowers.

The palm is tender to cold and adaptable only to zones 10b and 11, although specimens are found in microclimates of 10a. It needs average, regular moisture and, although drought tolerant when established, does not attain its spectacular proportions if constantly deprived of moisture. It is adaptable to most soils with good drainage but looks its best with adequately fertile soil.

This colossal palm is not for small gardens. Its leaf crown is as large as some small houses. Its massive proportions make it marvelous as a specimen, but it is simply breathtaking if room can be found to plant groups of 3 or more individuals of varying heights, as in Flamengo Park, Rio de Janeiro, Brazil; such practice also ensures the perpetuation of its landscape role. Its slow growth when young and before it forms a trunk allows it to be used as a tropical-appearing giant shrub and accent for many years. The life span of a single tree is from 30 to 80 years. Only as an immature plant could the palm be considered for a large atrium or conservatory, and it would need lots of light.

Talipot palm has many non-ornamental uses. The trunks are felled for the starchy pith and the sap from which alcoholic beverages are made; the leaves are used for thatch and for making many household items; and the growing point is harvested as a vegetable.

Corypha utan
ko-RY-fa · OO-than
PLATES 314–315

Corypha utan is native to a vast area from northeastern India through Southeast Asia and northwards to the Philippine islands and south to northern Australia. The epithet is one of the aboriginal names. The species is similar to *C. umbraculifera* but smaller in all its parts, has spiraling leaf bases, and has much more deeply segmented leaves. It is also less hardy than *C. umbraculifera*. It is slightly susceptible to lethal yellowing disease.

Cryosophila
kry'-o-SAH-fi-la

Cryosophila comprises 10 palmate-leaved, hermaphroditic palms in Mexico, Central America, and northwestern Colombia, some of them very rare in habitat. The genus is instantly recognizable by the prominent, often branched spines on the trunk, especially its lower half, which are, in fact, aerial roots that harden into spine-like projections from the trunk. They usually root if they reach the soil. The circular leaves are borne on long petioles and usually are silvery on their undersides. As in the genus *Chelyocarpus*, the leaves are divided into 2 halves, which are split into numerous segments. The 2 main divisions are not overly obvious in most cases to the untrained eye as all the segments are close together. The short, congested, and much-branched inflorescences are mostly hidden below the leaf crown by large overlapping bracts; they bear small, white, bisexual flowers. The fruits, when mature, are either white or green.

These palms generally grow on limestone soils, and most species are similar and difficult to distinguish one from another. All the species are intolerant of freezing temperatures and need nearly constant moisture. The genus was formerly named *Acanthorrhiza*, meaning "spiny root." The current genus name was not explained when it was coined. It is from Greek words meaning either "cold-loving" or "bleak-loving." Either derivation must have been a botanical joke, as the palms are inhabitants of warm, dense tropical forests. The common English name, rootspine palm, applies to the whole genus. Seeds of *Cryosophila* germinate within 30 days and should not be allowed to dry out before sowing. For further details of seed germination, see *Corypha*.

Cryosophila guagara
kry'-o-SAH-fi-la · gwah-GAH-ra
PLATE 316

Cryosophila guagara is native to the Pacific coastal rain forest of Costa Rica and adjacent Panama. The epithet is an aboriginal name. This species differs from *C. stauracantha* in having a long, pendulous inflorescence with large, persistent bracts and in having more numerous and generally longer root spines.

Cryosophila nana
kry'-o-SAH-fi-la · NA-na
PLATE 317

Cryosophila nana is indigenous to the western coast of Mexico from Mazatlán to almost the border of Guatemala, where it grows in mountainous, semiarid deciduous forests, and pine and oak woodlands. The epithet is Latin for "small," referring to the overall stature of this palm. This species differs from *C. stauracantha* in its shorter stature, its leaves that are green on both surfaces, and in not needing constant moisture. It is the only palm we know of that can be rooted from a cutting: the entire crown and some length of stem can be cut and rooted like a cutting. Of course, the decapitated base will die.

Cryosophila stauracantha
kry'-o-SAH-fi-la · sto-ra-KANTH-a
PLATES 318–319

Cryosophila stauracantha occurs naturally in lowland, ever-wet to seasonally dry forest of southern Mexico, northern Guatemala, and Belize. The epithet is from Greek and may be translated as "spiny pole," in reference to the spiny trunk of this palm. The palm is called give-and-take in Belize (Henderson et al. 1995).

Mature trunks may attain a height of 30 feet (9 m) with a diameter of 4–5 inches (10–13 cm). They are mostly covered, and densely so, with gray to almost white short root spines, although the upper parts often are free of the roots, which either do not form at these levels or have been worn away. The sparse but beautifully rounded leaf crown is 15 feet (4.5 m) wide and tall. The perfectly circular 6-feet (1.8-m) leaves are carried on delicate, slender petioles 6 feet (1.8 m) long. The leaf segments are clustered in groups of 2–4 and extend halfway to the center of the blade. Leaf color is dark green above and grayish green to gray beneath. There is a small but distinct hastula at the junction of leaf blade and petiole. The inflorescence is 2–3 feet (60–90 cm) long and grows from among the leaves, bearing small, white flowers. The round 1-inch (2.5-cm) fruits are green to white.

The species is tropical in its requirements and is adaptable only to zones 10b and 11, although sometimes found in favorable microclimates in 10a. It is a true water lover. It thrives in a rich soil but is adaptable to those that are sandy if fertilized and mulched with compost; it prefers slightly alkaline to slightly acidic soils. It likes partial shade, especially when young, but can adapt to full sun when older.

Because of the shape and form of the leaves dancing on their slender stalks and because of the delicate and thin trunk, there is no more graceful palm. It has one of the most beautiful silhouettes of any plant and nothing in the landscape sings more alluringly than a canopy-scape of this palm. It is perfect also in a courtyard or patio, and groups of 3 or more individuals of varying heights create a veritable oratorio of grace and movement. It is not known to be grown indoors, although there would seem to be no reason why it couldn't be if given enough light, space, and moisture.

Cryosophila warscewiczii
kry'-o-SAH-fi-la · wahrs-se-WIK-zee-eye
PLATE 320

Cryosophila warscewiczii is native to the Caribbean coastal rain forest of Costa Rica and northern Panama. The epithet honors Józef Warszewicz (1812–1866), Polish plant collector in Central America. This species is taller than *C. stauracantha* and is slightly susceptible to lethal yellow disease. The leaves are less divided than those of other species, and the fruits are larger.

Cryosophila williamsii
kry'-o-SAH-fi-la · WIL-yam-zee-eye
PLATE 321

Cryosophila williamsii is a rare and endangered species indigenous to northeastern Honduras, where it grows in low mountainous rain forest. The epithet honors Louis O. Williams (1908–1991), an American botanist and specialist in the flora of Central America. This species is shorter than *C. stauracantha* and has a short inflorescence that is held close to the trunk, just below the leaves.

Cyphokentia
sy-fo-KEN-tee-a

Cyphokentia is a genus of 2 species of solitary-trunked, pinnate-leaved, monoecious palms found in the moist forests of New Caledonia. The genus name is formed from a Greek word meaning "bump" and *Kentia*, an out-of-date name for species of *Hydriastele* but used here and in other compounded names to indicate generally a palm with pinnate leaves. The name refers to the bump on the side of the fruit that corresponds to the stigmatic remains. Very few seeds of *Cyphokentia* have been available, and they have been difficult to germinate, as have seeds of most New Caledonian species. For further details of seed germination, see *Areca*.

Cyphokentia cerifera
sy-fo-KEN-tee-a · ser-RI-fer-a
PLATE 322

Cyphokentia cerifera occurs naturally in mountainous rain forest from sea level to 2500 feet (760 m). The epithet means "wax bearing" and refers to the wax on the crownshaft. The palm was formerly known as *Moratia cerifera*.

The solitary trunk grows slowly to 60 feet (18 m) with a diameter of 6–7 inches (15–18 cm) in habitat. It is light to dark brown with distinct but closely spaced, ridged rings of leaf scars in its upper parts. The crownshaft is 2 feet (60 cm) tall, cylindrical, but slightly bulged at its base and slightly tapering above that point. It is deep orange but covered with a thick layer of pure white wax, which, unless removed, makes the shaft the same color. The leaf crown is open and nearly spherical because of the few arched and recurving leaves, each one 7 feet (2.1 m) long and borne on a short orange to yellow petiole less than 1 foot (30 cm) long. The leaflets grow from the rachis at an angle that creates a V-shaped leaf. They are a deep yellowish to pure emerald green and are linear-lanceolate, long tapering, and stiff. The inflorescences, which grow from beneath the crownshaft and are 3 feet (90 cm) wide, are large, spreading grayish green flowering branches that have a gigantic spidery appearance. The round ½-inch (13-mm) fruits are reddish.

This is a stunningly beautiful palm. Its white crownshaft and round crown with recurving leaves create a tableau that seems the essence of what a palm should look like. If only it were easier and faster growing! The trees are so slow growing as to be impractical for the average person, which accounts for this palm's rarity in cultivation.

Cyphokentia macrostachya
sy-fo-KEN-tee-a · ma-kro-STAK-ee-a
PLATE 323

Cyphokentia macrostachya is endemic to New Caledonia, where it grows in mountainous rain forest at elevations from sea level to 3000 feet (900 m). The epithet is formed from Greek words meaning "large" and "spike," an allusion to the inflorescences.

In habitat, the light gray or tan trunk attains a height of 50 feet (15 m) but is 6 inches (15 cm) in diameter except at its expanded base. The elegant cylindrical crownshaft is 3 feet (90 cm) tall, slightly wider than the trunk, and is light gray to almost pure white. The leaf crown is sparse, and the leaves, on short petioles 2 feet (60 cm) long, are 10 feet (3 m) long and beautifully recurved, with evenly and widely spaced light green stiff and narrowly lanceolate leaflets 4 feet (1.2 m) long that grow from the rachis at an angle to give a V-shaped leaf. The inflorescences grow in a circle from nodes beneath the crownshaft and are thin, ropelike, and pendulous branches 3 feet (90 cm) long bearing both male and female flowers. The fruits are ½ inch (13 mm) long, egg shaped, and bright red when mature.

This beautiful palm is rare in cultivation and slow growing. It needs partial shade when young, a frost-free climate, and constant moisture in a rich, humus-laden soil. Its silhouette is as beautiful as that of any palm. It makes a perfect canopy-scape but is so slow growing as to be planted for future generations.

Cyphophoenix
sy-fo-FEE-nix

Cyphophoenix is composed of 4 solitary-trunked, pinnate-leaved, monoecious palms in New Caledonia. The inflorescences grow from beneath the crownshaft and are large, spreading, multi-branched, and succulent; they bear both male and female blossoms, which produce ellipsoid red or reddish brown fruits ½ inch (13 mm) long. The genus name is formed from Greek words for "bump" and *Phoenix*, the date palm (but used in the sense of "palm," as in many palm names that are compounded with *Phoenix*), and probably alludes to the prominent remains of the stigma (the "bump") on the apex of the fruit. Fresh seeds of *Cyphophoenix* germinate sporadically in 45–120 days. Time of harvest of the fruits is critical. The percentage of seeds that germinate has never been good and *may* be the result of using seeds that are not fully ripe. For further seed germination details, see *Areca*.

Cyphophoenix alba
sy-fo-FEE-nix · AHL-ba
PLATES 324–325

Cyphophoenix alba is endemic to New Caledonia, where it grows in rain forest from sea level to 2000 feet (610 m). The epithet is Latin for "white" and refers to the bloom (waxy coating) on the crownshaft and inflorescences. This species was once known as *Veillonia alba*.

The solitary trunk grows to a maximum height in habitat of 50 feet (15 m). It is 6 inches (15 cm) in diameter and light tan to light gray or almost white in its older parts but green in its youngest parts; there is a white waxy coating on the younger parts and widely spaced, deeply indented darker brown rings of leaf scars on all but the oldest parts. The loosely formed crownshaft is bulging at its base, 4 feet (1.2 m) tall, and waxy white but overlain, especially at its top, with a reddish brown feltlike tomentum. The leaf crown is sparse but full and rounded, and the leaves are large in comparison to the height and width of the trunk. They are 10 feet (3 m) long, including a petiole 2 feet (60 cm) long. The thick, heavy leaflets are 4 feet (1.2 m) long and almost 3 inches (8 cm) wide at their midpoint and are linear-lanceolate, long tipped, and bright to deep green above and beneath. The much-branched, ropelike inflorescences are pinkish coral when new and covered with white waxy scales when in flower, giving a distinctly glaucous rosy hue. The ovoid fruits are reddish to greenish brown when mature and, as they grow, the inflorescence changes from a glaucous white to a grayish green infructescence.

There is hardly a more colorful and attractive palm species. It is like a gigantic candy cane. It is still rare in cultivation and is reportedly slow growing. It needs regular, copious water, a fast-draining nonalkaline soil, partial shade in hot climates, and a tropical or nearly tropical climate. It reportedly does well in a cool but frostless Mediterranean clime if given enough water.

Cyphophoenix elegans
sy-fo-FEE-nix · EL-e-ganz
PLATES 326–327

Cyphophoenix elegans is endemic to northeastern New Caledonia, where it grows in low mountainous rain forest. The epithet is Latin for "elegant."

The trunk attains a height of 50 feet (15 m) in habitat and is 6 inches (15 cm) in diameter, light green in its younger parts, and a light to deep gray in its older parts, with distinct whitish rings of leaf scars. The bulbous crownshaft is 2–3 feet (60–90 cm) high, markedly swollen at its base, and olive green to light silvery green. The leaf crown is less than a semicircle because of the short-petioled, ascending but also beautifully arching and recurved leaves

6 feet (1.8 m) long. The leaflets are 3 feet (90 cm) long, a deep yellow-green on both sides, linear-lanceolate, and stiffly ascending, growing from the rachis at an angle which gives the leaf a V shape.

The tree is relatively fast growing for a New Caledonian palm, needing only year-round warmth and copious, regular moisture in a free-draining, humus-laden soil. Its silhouette is almost breathtakingly beautiful and makes one of the finest canopy-scapes.

Cyphophoenix fulcita
sy-fo-FEE-nix · ful-SEE-ta
PLATES 328–329

Cyphophoenix fulcita is endemic to southern New Caledonia, where it grows in very wet, low, montane rain forest. The epithet is Latin for "propped up" and refers to the stilt roots. The species was formerly known as *Campecarpus fulcitus*.

In habitat, the palm grows to a height of 50 feet (15 m), with a smooth, distinctly ringed green trunk that is 5 inches (13 cm) in diameter and supported on a massive cone of densely packed stilt roots 6 feet (1.8 m) tall. The greatly bulging crownshaft is 3 feet (90 cm) tall and dull green except the top half, which is covered in a pinkish white feltlike tomentum. The leaf crown is hemispherical, with leaves 8–10 feet (2.4–3 m) long. The stiff, linear leaflets are 4 feet (1.2 m) long and regularly spaced, growing from the rachis at an angle to give a slight V shape to the leaf. The inflorescences grow from nodes encircling the trunk beneath the crownshaft and consist of ropelike, pendent dingy green branches that bear tiny male and female blossoms. The egg-shaped cinnamon-colored fruits are curved and ½ inch (13 mm) long.

The species is very rare in cultivation. It is surely intolerant of cold and needs a humus-laden soil and partial or full shade when young. Mature palms are beautiful, almost the essence of palminess, with their elegantly straight, green and smooth, distinctly ringed trunks, cone of prop roots, and handsome leaves, but individuals are so slow growing, especially when young, that planting the species should be looked on as a gift for future generations.

Cyphophoenix nucele
sy-fo-FEE-nix · noo-SEL-ee
PLATE 330

Cyphophoenix nucele is endemic to the small island of Lifou in the Loyalty Islands, east of New Caledonia, where it grows at low elevations near the coast on limestone outcrops. It is rare and endangered, with fewer than 100 individuals in habitat. The epithet is the aboriginal name, which translates as "coconut palm" and "sling," because at one time the fruits were used by islanders as projectiles for hunting birds.

The trunk attains a height of 50 feet (15 m) in habitat and is slightly more than 6 inches (15 cm) in diameter. It is light green in its younger parts and tan in its older parts with distinct white rings of leaf base scars. The crownshaft is 2 feet (60 cm) high, barely thicker than the trunk itself, and is almost pure white because of the dense, feltlike tomentum that covers it. The erect, ascending, deep emerald green leaves are 6 feet (1.8 m) long on short petioles. The crown usually has 8 leaves and is shaped like a shaving brush. The linear-lanceolate, long-tipped, stiff leaflets are 2½ feet (75 cm) long and grow almost in a single plane from the rachis.

The species is still rare in cultivation but has the same cultural requirements as *C. elegans* with one exception: it is one of the few New Caledonia palms that is tolerant of limey soils.

Cyphosperma
sy-fo-SPUR-ma

Cyphosperma includes 4 solitary-trunked, pinnate-leaved, monoecious palms from the rain forests of Fiji, the Vanuatu islands, and New Caledonia. The inflorescences emerge among the leaves and persist as the leaves fall away, so that eventually they are found below the leaf crown. All species are wonderfully tropical looking and exotic. The genus name is derived from Greek words meaning "bump" and "seed," an allusion to the shape of the seeds. Very few seeds have been available, and virtually no data are known. Seed viability is most likely quite short. Bottom heat as well as proper time of harvest seems crucial. For further details of seed germination, see *Areca*.

Cyphosperma balansae
sy-fo-SPUR-ma · ba-LAHN-see
PLATES 331–332

Cyphosperma balansae is endemic to New Caledonia, where it grows in mountainous rain forest from 1300 to 3200 feet (400–980 m). The epithet honors Benedict Balansa (1825–1891), a French botanist and plant explorer in South America, Southeast Asia, and Turkey.

The trunk attains a height of 50 feet (15 m) in habitat and is 8 inches (20 cm) in diameter. It is deep green in its younger parts but chocolate to nearly black in its older parts, with beautifully distinct light yellow rings of leaf scars, making this one of the most beautiful stems of any palm. The sheaths are 2 feet (60 cm) long and prominent but do not form a crownshaft; they are the color of the trunk at their bases with light green summits. The leaf crown is shaped like a shaving brush because of the erect, ascending leaves, each 8–10 feet (2.4–3 m) long on a short petiole generally less than 1 foot (30 cm) long. The light green stiff leaflets are 3 feet (90 cm) long, linear-lanceolate with acuminate tips, and grow from the rachis at an angle that creates a V-shaped leaf. The inflorescences are extraordinarily long at 7–8 feet (2.1–2.4 m) and consist of a sturdy peduncle 6 feet (1.8 m) long with many apical reddish brown branches bearing both male and female flowers. The round ½-inch (13-mm) fruits are borne in large pendent clusters and are deep red when mature.

This species is one of the most strangely beautiful palms and is more tropical looking than it actually is with its incredibly exotic trunk. It is still rare in cultivation but is reportedly easily and quickly grown if given ample, regular moisture and a humus-rich soil. It seems to thrive in partial shade when young and is adaptable to full sun in all but hot, dry climates. It needs, alas, a nearly tropical climate and is not frost tolerant. Seedlings with entire, unsegmented leaves are much sought after (Hodel and Pintaud 1998).

Cyphosperma tanga
sy-fo-SPUR-ma · TAHN-ga
PLATE 333

Cyphosperma tanga is endemic to the island of Viti Levu in Fiji, where it grows in low, mountainous rain forest. The epithet is the aboriginal name for the species.

The trunk attains a height of 20 feet (6 m) in habitat and a diameter of 6 inches (15 cm). It is chocolate brown in all but the oldest parts and is graced with wavy yellowish rings of leaf base scars. The leaf crown is in the form of a shaving brush because of the stiffly ascending and erect leaves, which are 6 feet (1.8 m) long, deep green above and paler green beneath. They are borne on short light green petioles, are wedge shaped, and are undivided except at their apices where there are irregularly sized segments. The blade shows deep corrugations above and below corresponding to the fused segments. The inflorescence consists of an erect peduncle 6 feet (1.8 m) long with an apex that terminates in a spray of thin branches 2 feet (60 cm) long bearing greenish yellow blossoms. The ellipsoid fruits are ½ inch (13 mm) long and yellow.

This species is undeservedly rare outside of botanical gardens. It is exotic and tropical looking and is on a par with *Phoenicophorium* and *Verschaffeltia* in beauty. It needs constant warmth and moisture, a humus-laden, slightly acidic soil, and full sun once past the seedling stage.

Cyphosperma trichospadix
sy-fo-SPUR-ma · try-ko-SPAY-dix
PLATE 334

Cyphosperma trichospadix is endemic to the islands of Vanua Levu and Taveuni in Fiji, where it grows in mountainous rain forest and cloud forest at 2000–4000 feet (610–1200 m). The epithet is derived from Greek words meaning "hair" and "spadix," an allusion to the hairiness of the inflorescence.

The trunk grows to a maximum height of 18 feet (5.4 m) in habitat with a diameter of 3 inches (8 cm). The stem is a beautiful smooth light green in all but the oldest parts where it is a light tan with indistinct lighter-colored rings of leaf base scars. This is the only species in the genus to form a true crownshaft. The crownshaft is 3 feet (90 cm) tall, scarcely wider at any point than the trunk itself, and a beautiful silvery green. The leaf crown is open and almost hemispherical with only a few ascending but gracefully arching leaves 6–7 feet (1.8–2.1 m) long on bright green petioles 4 feet (1.2 m) long. The lanceolate-acuminate leaflets are 2 feet (60 cm) long, deep green on both sides, and widely spaced, growing from the rachis in a nearly single flat plane.

This is the most typical-looking species in the genus because of its open, full leaf crown and colored trunk. It is nevertheless of great beauty. It is extremely rare in cultivation but not deservedly so and should do well in all but the hottest and driest nearly tropical climes if given substantial, regular moisture, partial shade, especially when young, and a free-draining, humus-laden soil.

Cyrtostachys
sir'-to-STAK-ees

Cyrtostachys consists of 7 clumping and solitary-trunked, pinnate-leaved, monoecious palms in Thailand, Malaysia, Indonesia, New Guinea, and the South Pacific Islands. The spreading inflorescences form beneath the crownshaft and are much branched. The flowers are borne in little pits on the branches of the inflorescence. Only 2 species are cultivated outside of tropical botanical gardens but, if the others are at all comparable in beauty, then this is one of the most badly neglected palm genera for horticulture. All species need tropical growing conditions and abundant moisture.

The genus name is derived from Greek words meaning "curved" and "spike," referring to the inflorescence. The taxonomy of the genus was recently revised by Heatubun et al. (2009). Seeds generally germinate within 90 days but can take 190 days or sometimes longer. If the tiny seed is allowed to dry out completely before sowing, it does not germinate. Seeds should be harvested when fully ripe. For further details of seed germination, see *Areca*.

Cyrtostachys elegans
sir'-to-STAK-ees · EL-e-ganz
PLATE 335

Cyrtostachys elegans is indigenous to swampy regions of rain forest in Indonesian New Guinea. The epithet is Latin for "elegant." Sparsely suckering, this palm produces slender stems to 50 feet (15 m) tall. The leaves are gracefully curved and bear pendulous, evenly arranged leaflets. The fruits are ½ inch (13 mm) long, ellipsoid or curved (banana shaped), and black. This palm's cultural requirements are the same as those for *C. renda*.

Cyrtostachys glauca
sir'-to-STAK-ees · GLAW-ka

Cyrtostachys glauca is a sparsely clustering species from low, mountainous, rain forest in Papua New Guinea. The epithet is Latin for "glaucous" and refers to the crownshaft and undersides of the leaflets.

The tallest trunks attain a maximum height of 30 feet (9 m) but usually number only 2 or 3, most of the suckers being much shorter. The younger parts of the stems are a beautiful green with prominent rings, the older parts a light gray. The prominent crownshafts are 3 feet (90 cm) tall, glaucous blue-green, and smooth. The bright green leaves are 10–12 feet (3–3.6 m) long and flat or nearly flat because the leaflets grow from the rachis in one plane. The slender petioles are 2–3 feet (60–90 cm) long. The pits on the inflorescence branches are shallow, and the fruits are similar to those of *C. loriae*.

Although not as demanding of moisture as is *C. renda*, this species is a water lover and needs sun and a rich soil. It grows faster than *C. renda* but is as tender to cold. It is a wonderfully picturesque and beautiful palm because of the form of the clumps with their levels of leaves and heights of trunks. It looks best in a wall of vegetation, where it also serves as a marvelous canopy-scape.

Cyrtostachys loriae
sir'-to-STAK-ees · LOR-ee-eye

Cyrtostachys loriae is a widespread species from the rain forest of New Guinea and the Solomon Islands. It was named for Lamberto Loria (1855–1913), Italian explorer and ethnographer best known for his work in New Guinea. *Cyrtostachys kisu, C. brassii,* and *C. peekeliana* are now regarded as synonyms.

The beautifully slender, dark green trunk has prominent and widely spaced white rings of leaf base scars; it is solitary and grows straight as an arrow to 90 feet (27 m) high. The leaves are 6–8 feet (1.8–2.4 m) long on nearly nonexistent petioles, forming a spherical crown; they are medium to deep green, and the pendent leaflets are neatly and regularly arranged. The inflorescence is said to be "more robust" than in other species. The fruits are ½ inch (13 mm) long, ellipsoid to curved, and black.

This extremely elegant palm is highly garden-worthy. It is less demanding of water than *C. renda*, but it is strictly tropical, requiring absolute protection from drying winds and cold temperatures.

Cyrtostachys renda
sir'-to-STAK-ees · REN-da
PLATES 336–337

Cyrtostachys renda is native to lowland rain forest and peat swamps in Thailand, Sumatra, Malaysia, and Borneo. The epithet is an aboriginal name for the palm. Common names are sealing-wax palm and lipstick palm. *Cyrtostachys lakka* is a synonym.

The trunks of this clustering species grow to 60 feet (18 m) high in habitat but under cultivation are usually no more than half that with diameters of only 2–3 inches (5–8 cm). Mature clumps may attain a total height of 35 feet (11 m) and a width of 20 feet (6 m). The younger parts of the stems are green, but the older portions are light brown to gray or sometimes almost white and are beautifully ringed. The crownshaft is 3 feet (90 cm) long, smooth, and slender, slightly bulging at the base, and a brilliant reddish orange to vivid scarlet. The leaf crown is 10 feet (3 m) wide and 6 feet (1.8 m) tall. The leaves are 4–5 feet (1.2–1.5 m) long on short petioles less than 1 foot (30 cm) long. The petiole and rachis are the color of the crownshaft. The narrowly elliptical, tapering leaflets are 1 foot (30 cm) long, narrowly grooved and stiff, growing from the rachis at an angle that creates a V shape to the blade. Leaf color is light to deep green above and a distinctly lighter grayish green beneath. There are usually only 6–8 leaves on any given trunk, and they almost never descend below the horizontal plane. The beautiful red inflorescences grow from beneath the crownshaft and are short, sparsely branched with whiplike branches. They bear small, greenish white or greenish yellow flowers. The round ½-inch (13-mm) fruits are black.

This palm needs abundant, constant moisture and constantly tropical temperatures. It is marginal in zone 10b. While it readily adapts to full sun when mature, it likes partial shade when young. Cold or dry winds are inimical to it. The richer the soil, the better, and this palm must not come in contact with limestone as alkalinity is lethal. This species can be cultivated as an emergent aquatic or semiaquatic plant, growing at the margin of ponds or in pots submerged or partially submerged in water. There are forms with light green as well as orange or yellowish brown crownshafts, and several named cultivars including 'Ruby' and 'Theodora Buhler'.

This species is one of the most beautiful and sought after; every part of the plant is choice. The slender green trunks are heartbreakingly beautiful and so thin and lissome as to often bend as if they were the long stalks of big blossoms in a bouquet. The clumps are usually open enough that the trunks do not need to be thinned, and the younger stems add incredibly beautiful tiers of leaves from top to bottom; however, it is the magnetic attraction of the long, red crownshafts that makes this species irresistible. It is hard to imagine a site in which it would not be the center of attraction, but it looks exceptionally beautiful when made a part of other vegetation. The palm does well in large, humid, and constantly warm enclosures with good light.

Daemonorops
dee-MON-o-rahps

Daemonorops is a large and diverse genus of 115 spiny, mostly climbing, dioecious palms from India eastwards to New Guinea and the Philippines, primarily in peninsular Malaysia, Sumatra, and Borneo. Some species are solitary trunked, but most are clustering. A few are dwarf in stature; many others are gigantic and high climbing. A few of the trunkless species have no specialized climbing organ, but most others develop a cirrus at the end of the leaf rachis. In a few species the stems die after flowering and fruiting. The stems of all species are spiny, and several species have sets of interlocking spines that form "ant galleries" in which the insects make their homes, a symbiotic relationship between palm and insect.

The inflorescences grow from the leaf axils and are usually accompanied by large boat-shaped and podlike, erect or pendent, spiny woody bracts that open by splitting lengthwise to reveal much-branched but short stalks of male or female blossoms. The fruits are scaly, 1 inch (2.5 cm) wide, usually globose, and light brown to dark brown to reddish brown or black. The "pods" often have a sharply pointed beak that may be as long as the part of the bract that holds the inflorescences themselves.

All species require a tropical or nearly tropical climate and copious, regular moisture, but almost all flourish in full sun or partial shade. The high-climbing species may be pruned to remain in bounds. Like other rattan species, the juvenile plants are much smaller, are less viciously spiny, and make wonderful rosettes of pinnate leaves, especially for containers. The genus name is derived from Greek words meaning "demon" and "shrub," an allusion to devilish spines on the plants. Seeds germinate within 30 days. For further details of seed germination, see *Calamus*.

Daemonorops angustifolia
dee-MON-o-rahps · an-gu'-sti-FO-lee-a

Daemonorops angustifolia is indigenous to peninsular Malaysia, peninsular Thailand, Sumatra, and Borneo, where it grows mostly in clearings and along banks of streams and rivers. In Malaysia, it is called *rotan ayer* or water rattan. The epithet is Latin for "narrow leaf" and refers to the shape of the leaflets.

This species grows from sea level to 2640 feet (800 m). It is a giant clustering climber, the beautiful stems with their long internodes growing to 100 feet (30 m) long but only 1 inch (2.5 cm) or slightly more in diameter. The spreading leaves are 6–8 feet (1.8–2.4 m) long and bear numerous narrowly lanceolate leaflets that are semiglossy, semipendent, 1 foot (30 cm) long, and growing from the rachis in a flat plane, making for an extremely handsome leaf. The cirrus is another 3 feet (90 cm) long and bears widely spaced grapnel-like backward-pointing pairs of spines 2 inches (5 cm) long. The woody inflorescences are 18 inches (45 cm) long and erect.

Daemonorops curranii
dee-MON-o-rahps · kur-RAN-ee-eye
PLATE 338

Daemonorops curranii is endemic to Palawan Island in the Philippines, where it grows in dense rain forest at low elevations. The epithet honors Hugh M. Curran (1875–1960), American forestry officer and collector in the Philippines. This small, clustering and short climbing species has flat leaves that are 6 feet (1.8 m) long and terminate in a short cirrus. It is a beautiful clumper that is easily kept in bounds and in its rosette stage by pruning. It makes an incredibly nice large groundcover or hedge or barrier planting in shade or sun.

Daemonorops jenkinsiana
dee-MON-o-rahps · jen-kin'-see-AH-na
PLATE 339

Daemonorops jenkinsiana is indigenous to northeastern India, Bhutan, Myanmar, and northern Thailand, where it grows in low mountainous rain forest. The epithet commemorates Major Francis Jenkins (1793–1866), the original collector. The stems of this large clustering climber grow to lengths of 60 or more feet (18 m). The spreading leaves are 6–8 feet (1.8–2.4 m) long, with leaflets 18 inches (45 cm) long, closely and regularly spaced, linear and pendent. This extremely spiny but gorgeous species is best planted in outlying areas of a garden where its marvelous leaves can be appreciated.

Daemonorops melanochaetes
dee-MON-o-rahps · mel'-a-no-KEE-teez
PLATE 340

Daemonorops melanochaetes is indigenous to peninsular Thailand, peninsular Malaysia, Sumatra, and Java, where it grows in rain forest from sea level to 1000 feet (300 m), often forming large colonies. The epithet is from Latin and Greek words meaning "black" and "bristle" and alludes to the long spines that cover the sheaths, petioles, and leaf rachises. The densely clustering stems grow to 100 feet (30 m) long, the leaves are 10 feet (3 m) long on petioles 2–3 feet (60–90 cm) long, and the terminal cirrus is 3 feet (90 cm) long. The many narrow, dark green, pendent leaflets are 2 feet (60 cm) long and regularly spaced along the rachis, growing in a single plane. This palm is easily kept at a manageable height and, as such, makes a beautiful large component in other vegetation.

Daemonorops ochrolepis
dee-MON-o-rahps · o-kro-LEP-iss

Daemonorops ochrolepis is endemic to the Philippines, where it grows abundantly from sea level to 3500 feet (1070 m) in primary rain forests. The epithet is Greek for "ochre scale" and refers to the fruit color. This large clustering climber is armed with dense brown spines on the leaf sheaths, petioles, and leaf rachises. The stems are 1 inch (2.5 cm) thick and widely collected for making rattan furniture and other handicrafts. Narrow dark green leaflets grow in a single plane along the leaf that is 8 feet (2.4 m) long including the petiole and ends in a long dangling cirrus. Light yellow fruit are borne in clusters along an infructescence 5 feet (1.5 m) long. This palm makes an attractive element of other tall vegetation where its beauty can be appreciated both near and far.

Deckenia
de-KEN-ee-a

Deckenia is a monotypic genus of spiny, solitary-trunked, pinnate-leaved, monoecious palm in the Seychelles. The name honors a German explorer of Africa, Madagascar, and the Mascarene Islands, Karl K. von der Decken (1833–1865). Seeds of *Deckenia* germinate within 60 days. For further details of seed germination, see *Areca*.

Deckenia nobilis
de-KEN-ee-a · NO-bi-lis
PLATES 341–343

Deckenia nobilis grows in rain forest on hills, slopes, and low mountains, usually within sight of the coast but also to 2000 feet (610 m). In some areas, it forms large, dense colonies, whose individuals litter the ground so that no other vegetation, including their own seedlings, can grow. The species is endangered because of habitat destruction and the harvesting of its growing point as a vegetable. The epithet is Latin for "noble" and refers to the size and appearance of the species.

The large trunk can attain a height of 100 or more feet (30 m) but a diameter of no more than 18 inches (45 cm). It is straight as an arrow, light gray or light tan, and beautifully and distinctly ringed with darker leaf base scars. The stem sports a smooth, cylindrical crownshaft that is 3–4 feet (90–120 cm) long, bulging at its base, and usually a light green but sometimes bluish green or purplish. Young trees are invariably covered on their trunks, crownshafts, and petioles with golden spines 2–3 inches (5–8 cm) long, but these gradually disappear as the palm ages and, usually by the time the trunk is 12 feet (3.6 m) tall, most are gone. The leaf crown is large and rounded, and the leaves are 10–12 feet (3–3.6 m) long. The linear-lanceolate, long-tipped, deep green, limp, and half-pendent leaflets are 3–4 feet (90–120 cm) long. The leaf rachis usually exhibits a twist near its midpoint so that the apical half of the leaf is oriented in a vertical plane, much like that of *Archontophoenix* and *Cocos* (coconut) species. The bracts of the inflorescences are spiny no matter what the age of the palm, and these buds grow from nodes beneath the crownshaft. The inflorescences are sprays of yellow branches bearing both male and female blossoms. The small, ovoid fruits are black.

This large and magnificent species is one of the most beautiful palms but is very tender to cold and is marginal even in zone 10b. It also needs constant moisture, a fast-draining, humus-laden soil, full sun when past the seedling stage, and constantly high humidity. It is fast growing when past the seedling stage and up until the trunk attains 20 feet (6 m) of height. There is no finer subject for tropical climates.

Desmoncus
dez-MON-cus

Desmoncus is composed of about 7 variable, spiny, pinnate-leaved, monoecious palms in tropical America. All but one are climbers and use cirri to get a footing on other vegetation, rocks, or whatever. The stems of these species are mostly lacking spines, but all have extensions of the leaf bases that are like tough sleeves extending from one petiole or node to the next above. These sheaths, technically called *ochreas*, are invariably spiny. Unlike the leaves of nonclimbing palms, the leaves of these climbers extend from the top of the stem to its bottom in some instances, but more rarely are found on the top third or half of the stem. The inflorescences are elongated with few or many short flowering branches bearing either male and female blossoms, or often only male flowers. These palms produce attractive clusters of red or orange rounded fruits.

The few species in cultivation grow best with a rich soil, copious water, and partial shade or full sun. All species are indigenous to lowland rain forest, and none tolerate cold. The genus name comes from Greek words meaning "band" and "hook" and refers to the hooks along the climbing organ at the end of the leaf. Seeds can take more than a year to germinate and do so sporadically. For further details of seed germination, see *Acrocomia*.

Desmoncus giganteus
dez-MON-cus · jy-gan-TEE-us

Desmoncus giganteus is indigenous to southern Colombia, eastern Ecuador, northern Peru, and extreme western central Brazil, where it grows in lowland rain forest. The epithet is Latin for "gigantic," an allusion to the palm's imposing size. The stems grow to 80 feet (24 m) long, with leaves confined to the upper parts thereof. The leaves are soft and limp and, therefore, pendent, and look, from any distance, as much like leaves of a *Dracaena* species or a large aroid than those of a palm. Each leaflet is 2 feet (60 cm) long and 3 inches (8 cm) wide, linearly ovate-lanceolate, glossy deep green above, and waxy, whitish green beneath. Straight black spines cover the leaf rachis and the ochreas. This species is most likely not in cultivation but is so interesting and even attractive that it needs to be. The stems could be cut back regularly, as is done with several *Calamus* species, to keep the handsome leaves at eye level.

Desmoncus orthacanthos
dez-MON-cus · or-tha-KAN-thos
PLATES 344–345

Desmoncus orthacanthos is the most far ranging species of the genus, from southeastern Mexico southwards through Central America and most of the northern half of South America east of the Andes, from sea level to elevations near 3000 feet (900 m). It is found mostly along rivers, in cleared areas of the rain forest, and in coastal forests. The epithet is from Greek words meaning "straight spine."

The stems of this clustering species attain lengths of 40 feet (12 m) but a diameter of 1 inch (2.5 cm). The leaves are 3–6 feet (90–180 cm) long and grow only on the upper part of the stem. The leaflets vary in size, shape, and arrangement and may grow in clusters or singly, be regularly or irregularly spaced, bear short spines or not, and be linear-elliptic or ovate; they may be straight and stiff, or limp and curling. Two invariable characteristics are the rachis, which is erect, stiff, and straight as an arrow, and its attendant spines, which are straight, stiff, black, and 2 inches (5 cm) long. The stems are used in the palm's habitats to weave baskets.

Desmoncus polyacanthos
dez-MON-cus · pah-lee-a-KAN-thos
PLATE 346

Desmoncus polyacanthos is indigenous to a large area of northern South America east of the Andes, where it grows in almost every type of habitat below 3000 feet (900 m). The epithet is derived from Greek words meaning "many" and "spine," an apt allusion to the prickly stems, petioles, and rachises. The stems grow to maximum lengths of 50 feet (15 m), clambering over and hooking onto other vegetation by the cirri. The leaves are 6 feet (1.8 m) long with evenly spaced long dark green elliptic and wavy margined leaflets 1 foot (30 cm) long.

Dictyocaryum
dik´-tee-o-KAHR-ee-um

Dictyocaryum includes 3 large, mostly solitary-trunked, pinnate-leaved, monoecious palms in the rainy cloud forests of the mountains of northern South America, although one species is also found at lower elevations. The 3 have large cones of spiny stilt roots at the bases of their trunks, enormous plumose leaves held on short petioles in a sparse crown, tall crownshafts, and unusually large inflorescences growing from beneath the shafts. The inflorescences are branched once, the flowering branches pendent in a curtainlike fashion from the stiff and horizontally held peduncle. The globose 1-inch (2.5-cm) fruits are yellow when ripe.

These species are related to and closely resemble *Iriartea* species. All require abundant, regular amounts of moisture and a noncalcareous soil. They are not frost tolerant, but the ones from high elevations should do well in a frostless but cool, Mediterranean climate if given enough water. The trunks are used for construction in the palm's habitat. The genus name is from Greek words meaning "net" and "nut," an allusion to the markings on the seed coat. Seeds germinate within 120 days and should not be allowed to dry out before sowing. For further details of seed germination, see *Iriartea*.

Dictyocaryum lamarckianum
dik'-tee-o-KAHR-ee-um · la-mark'-ee-AH-num
PLATE 347

Dictyocaryum lamarckianum is the most widespread species, occurring in the rain forest of easternmost Panama, the high Andes of Colombia, Ecuador, Peru, and into western Bolivia at 3000–6000 feet (900–1800 m), where it often grows in immense colonies on mountain slopes. The epithet honors French naturalist Jean B. P. M. de Lamarck (1744–1829).

The trunk grows to 70 feet (21 m) high in habitat but slightly more than 1 foot (30 cm) in diameter and sits atop a massive cone of spiny stilt roots. A slight bulge is usually visible near the middle of the stem, which is light gray to white, with distinct and widely spaced darker rings of leaf base scars in its younger parts. The smooth, light green crownshaft is 6–8 feet (1.8–2.4 m) tall and bulged at its base but otherwise hardly exceeds the girth of the trunk. The sparse crown never has more than 6 leaves, each 12–16 feet (3.6–4.9 m) long and on a short petiole which from any distance (like the ground) is mostly invisible. The leaves are erect and slightly arching near their apices, never lie beneath the horizontal, but are plumose with linear leaflets 2 feet (60 cm) long that are deep green above and silvery glaucous beneath. The immense inflorescences burst out from large upwardly curving horn-shaped spathes beneath the crownshafts and are yellowish white with unisexual blossoms of both sexes.

This species is a true water lover. It also needs a humus-rich, slightly acidic soil; these requirements keep it from being grown in hot, nearly tropical climates like South Florida where the night temperatures remain high in the summer and where the soil is mostly alkaline. The palm flourishes in partial shade when young but, when older, needs lots of light. It withstands temperature near freezing for short periods but is also killed by short periods of freezing temperatures.

This palm is truly magnificent and unparalleled as a specimen tree. It looks best in groups of 3 or more individuals of varying heights. Its skyline silhouette is second to none and matches the *Iriartea* and *Wettinia* species. Its only drawback is that it is not particularly fast growing.

Dictyosperma
dik'-tee-o-SPUR-ma

Dictyosperma is a monotypic genus of solitary-trunked, pinnate-leaved, monoecious palm. The name is from Greek words meaning "net" and "seed," an allusion to the marking on the seed coat. Seeds can take 60–90 days to germinate. For further details of seed germination, see *Areca*.

Dictyosperma album
dik'-tee-o-SPUR-ma · AL-bum
PLATES 348–349

Dictyosperma album is endemic to the Mascarene Islands, where it grows in coastal forests. It is in danger of extinction because of the felling of trees for their edible growing points; for one of its naturally occurring varieties, only one individual remains in habitat. The epithet is Latin for "white," an allusion to the most common color of the crownshaft. This palm reputedly resists strong winds by losing its leaves. Common names are hurricane palm and princess palm.

The mature trunk grows to 30 feet (9 m) tall, sometimes 40 feet (12 m), and is 6 inches (15 cm) in diameter except for its swollen base. The stem is colored light to dark gray or light to dark brown and is closely set with rings. The distinctive crownshaft, which is partially hidden by the rounded crown of leaves, is 4 feet (1.2 m) tall, bulbous at its base, light green to gray or almost white (and sometimes with a reddish hue), and usually smooth but sometimes covered in a short felt. The beautifully full leaf crown is 15 feet (4.5 m) wide and 10 feet (3 m) tall. The leaves are 8–12 feet (2.4–3.6 m) long on petioles usually less than 1 foot (30 cm) long. The rachis of young leaves has a beautiful rounded arch and is twisted to 90 degrees for about the last third of its length, resulting in the leaf plane being almost vertical from this point on. The narrowly lanceolate leaflets are 2–3 feet (60–90 cm) long, taper to a point, and grow from the rachis in a single flat plane. The new leaves are slow to unfold their leaflets and remain in a great thick, needlelike shoot before unfurling. The deep green leaflets have a prominent midrib and are usually drooping on older leaves. Young plants often have a reddish petiole, but the color is lost when the plant is older. The inflorescences are 2 feet (60 cm) long and usually form a ring around the trunk below the crownshaft. They carry creamy white to yellow to dark red, single-sexed, small, fragrant flowers. The ovoid fruits are ½ inch (13 mm) wide and deep purple to black

This species encompasses 3 separate varieties, all in danger of extinction in habitat. The 2 cultivated varieties are separated according to color of male flower buds and the presence of persistent reins. *Dictyosperma album* var. *album* has yellowish buds, and its reins quickly fall away as the leaf matures. *Dictyosperma album* var. *conjugatum* [kon-ju-GAY-tum], from Latin for "conjoined," is shorter and has dark maroon buds and persistent reins.

This species is slightly susceptible to lethal yellowing disease. It also is a water lover and, while it survives in poor, dry soil, its growth is so changed that it is scarcely recognizable to specimens grown under optimum conditions. It is adapted to full sun but grows readily in partial sun. It needs a good, well-draining soil. It is nearly tropical in its temperature requirements and is adapted to zones 10b and 11, although some nice specimens are to be found in favorable microclimates of 10a. The palm grows moderately slow to moderately fast, depending on the conditions.

The species is visually similar to *Archontophoenix cunninghamiana* and serves generally the same landscape uses, but it seems slightly heavier and nobler in appearance. While good as an isolated specimen surrounded by space, the palm is incredibly tropical and lush looking when planted in groves or in groups of 3 or more individuals of varying heights. It is seldom grown indoors but is adaptable to large sunny conservatories with good air circulation.

Dransfieldia
drans-FEEL-dee-a

Dransfieldia is a monotypic genus of understory, clustering, and monoecious palm from western New Guinea in lowland rain forest. The genus name honors John Dransfield (1945–), palm taxonomist extraordinaire. Seeds can take 60 days to germinate. For further details of seed germination, see *Areca*.

Dransfieldia micrantha
drans-FEEL-dee-a · my-KRAN-tha
PLATE 350

Dransfieldia micrantha has slender, ringed stems and a sparse crown of leaves. The leaflets are regularly arranged in a single plane, and each leaflet has prominent longitudinal ridges that run the length of the major veins. The new leaf is often pinkish. The inflorescence is borne beneath the crownshaft and is purplish. The flowers are reddish purple with white stamens. The shiny black ellipsoid fruits are about ½ inch (13 mm) long. The endosperm is ruminate.

This species resembles a small, clustering *Ptychosperma* species and can be used as such in the landscape. It should be planted in an intimate setting where its colorful foliage and inflorescences can be appreciated. It is hardy only in Zones 10 and 11. It seems not to be fussy about soil type, as long as it has abundant moisture and good drainage. The epithet is Latin for "small flower."

Drymophloeus
dry-MO-flee-us

Drymophloeus comprises 7 solitary-trunked, pinnate-leaved, monoecious, stilt-rooted palms in the Moluccas, New Guinea, the Solomon Islands, and Samoa. These lovely, mostly small palms have straight, beautifully ringed trunks, elegant crownshafts, and variously incised leaflets; the terminal pair of leaflets is often united. The inflorescences are small and sparsely branched, and the fruits are orange or red.

All the species are water lovers, require protection from the midday sun, especially when young, and need a humus-laden, free-draining soil. They are intolerant of cold and are adaptable to zones 10b and 11. The meaning of the genus name was unexplained and controversial. Zona (1999) suggested that it comes from Greek words meaning "forest" and "marsh reed," an allusion to the thin, reedy stems of these forest dwellers. Seeds germinate within 30 days. For further details of seed germination, see *Areca*.

Drymophloeus hentyi
dry-MO-flee-us · HENT-ee-eye
PLATES 351–352

Drymophloeus hentyi is endemic to rain forest on the island of New Britain. The epithet honors E. E. (Ted) Henty (1915–2002), Australian botanist and curator of the herbarium at Lae, Papua New Guinea. The results of molecular analyses place this species in *Ponapea*, where it will soon be reclassified. The species grows to 15 feet (4.5 m) and has a beautiful medium green crownshaft. The recurving leaves have wedge-shaped leaflets with oblique, jagged apices.

Drymophloeus litigiosus
dry-MO-flee-us · li-tij´-ee-O-sus
PLATE 353

Drymophloeus litigiosus is indigenous to the Moluccas and northwestern New Guinea, where it occurs in the undergrowth of rain forest. The epithet is Latin for "contentious," and may highlight the variability of the species and the taxonomic difficulties caused by this variation. A synonym for this species is *D. beguinii*. This choice little palm grows to 12 feet (3.6 m) high from a thin trunk. The light green crownshaft bulges at its midpoint in a sort of spindle shape. The leaves have regularly spaced leaflets growing in a flat plane and apices that are usually jagged; the leaflets of the terminal pair are larger than the others.

Drymophloeus oliviformis
dry-MO-flee-us · o-liv´-i-FOR-mis
PLATE 354

Drymophloeus oliviformis occurs naturally in the Moluccas and West Papua Province, Indonesia, in the understory of lowland rain forest. The epithet is Latin for "in the form of an olive" and alludes to the shape of the fruit. The species is similar in most respects to *D. litigiosus* except that the terminal pair of leaflets is united to form a single large segment that may or may not be cleft. David Fairchild (1943) described the leaflets as rubbery when he found the palm on Amboina in 1940.

Drymophloeus pachycladus
dry-MO-flee-us · pak-ee-KLAY-dus
PLATE 355

Drymophloeus pachycladus is indigenous to the Solomon Islands in the South Pacific, where it grows in low mountainous rain forest. The epithet is derived from Greek words meaning "thick" and "branched" and refers to the inflorescence. The results of molecular analyses indicate that this species should be classified in *Veitchia*. It is one of the larger species of *Drymophloeus* with trunks growing as tall as 20 feet (6 m). The prominent crownshaft is 3 feet (90 cm) tall, olive green, and slightly bulging at its midpoint. The leaves are 6–8 feet (1.8–2.4 m) long, beautifully arching, with widely spaced, large, and wide, wedge-shaped leaflets that are variously incised and lobed apically. This palm is perfection as a small canopy-scape and, because of its near-ravishing form, even works as a specimen palm, albeit on a small scale.

Drymophloeus subdistichus
dry-MO-flee-us · sub-DIS-ti-kus
PLATE 356

Drymophloeus subdistichus is endemic to the Solomon Islands in the South Pacific, where it grows in rain forest at elevations below 1000 feet (300 m). The epithet is from Latin words meaning "almost two-ranked," an allusion to the arrangement of the flowers in the inflorescences. The results of molecular analyses indicate that this species should be classified in *Veitchia*.

This palm is one of the larger species in the genus. The light brown to grayish trunk attains a height of 30 or more feet (9 m) and a diameter of less than 1 foot (30 cm), with closely set darker rings of leaf base scars. The stem grows atop a small cone of stilt roots 10 inches (25 cm) tall and with a diameter scarcely wider than the base of the actual trunk. The crownshaft is 4 feet (1.2 m) tall and a beautiful silvery green except at its top, which is suffused with black or dark purple. The leaves are 7–8 feet (2.1–2.4 m) long on short petioles and are ascending but arched from their midpoints. The light to deep green leaflets are 1 foot (30 cm) long, 2–3 inches (5–8 cm) wide, oblanceolate, and obliquely jagged at their apices; they grow from the rachis at an angle which gives a slight V shape to the leaf.

Because of its more than delightful overall form, this palm is striking enough to be used as a specimen, even surrounded by space; in groups of 3 or more individuals of varying heights, it is a magnificent canopy-scape. Few forms are lovelier.

Dypsis
DIP-sis

Dypsis is composed of more than 150 massive to tiny solitary-trunked or clustering, pinnate-leaved, monoecious palms growing in open savannas as well as dry and wet forests throughout Madagascar, the Comoro Islands, and Pemba Island in the far western Indian Ocean. Almost all the species have a crownshaft and many have plumose leaves. The inflorescences are formed mostly from within the leaf crown but sometimes beneath it, and the flowers are invariably unisexual, with both sexes present. The fruits are mostly brightly colored, but in some species are black or brown.

The groundbreaking work of John Dransfield and Henk Beentje in Madagascar united *Chrysalidocarpus*, *Macrophloga*, *Neodypsis*, *Neophloga*, *Phloga*, and *Vonitra* into a single genus. As *Dypsis*, is the oldest name for these palms, it is the one that must be used. The etymology of the genus name is obscure.

Collectors have been at work in the palms' native haunts for a century, and many of the species have been in cultivation for almost that long. This phenomenon has led to at least some species described from cultivation well before they were found in the wild. Alas, cultivation may be the only way of preserving some of Madagascar's palms, as most of the island's present ecosystems are in a state of irreversible degradation.

Seeds of *Dypsis* germinate within 90 days and do so in a short time. Seeds allowed to dry out completely before sowing do not germinate as well as those kept slightly moist. For further details of seed germination, see *Areca*.

Dypsis albofarinosa
DIP-sis · al'-bo-fahr-ee-NO-sah
PLATE 357

Dypsis albofarinosa was described from cultivated plants in Hawaii and only recently found in the wild in Madagascar. The epithet Latin for "white powder" and refers to the thick indument covering the upper stems and crownshafts.

This clustering species has greenish stems with white indument on their youngest parts and prominent leaf scar rings. The overall height is 15 feet (4.5 m) with clumps reaching 10 feet (3 m) wide. The species bears a strong resemblance to *D. baronii* and *D. onilahensis*, but is easily distinguished by the thick white indument covering the crownshaft and by the longer leaf petioles that give the leaves an even more graceful appearance than its counterparts.

The striking white coloration of the upper stems, offset by the deep green leaves, make this palm a dramatic landscape subject in any garden or as a patio container specimen. This charmer grows easily in neutral to acidic soils when not subject to freezing temperatures.

Dypsis ambositrae
DIP-sis · ahm-bo-SI-tree
PLATE 358

Dypsis ambositrae is endemic to the central highland of Madagascar, where it grows in forest at 4200–4800 feet (1280–1460 m). It is in danger of extinction. The epithet is Latin for "of Ambositra," a town in the palm's habitat.

This sparsely clustering species has stems to 20 feet (6 m) tall with a diameter of 4–5 inches (10–13 cm). They are light brown or gray in their older parts but green and distinctly ringed with whitish leaf base scars in the younger parts. The grayish green and glaucous crownshaft is 3 feet (90 cm) tall and hardly wider than the trunk. The leaf crown is round and dense with leaves 6 feet (1.8 m) long on grayish green petioles 1 foot (30 cm) long; the leaves are beautifully arching and recurved. The leaflets grow from the rachis at an angle of almost 90 degrees to create a V-shaped leaf; they are 3 feet (90 cm) long and erect, except near their apices where they are also shortly bifid.

This beauty is reportedly easy of cultivation and has some hardiness to cold.

Dypsis ankaizinensis
DIP-sis · an'-kah-ee-zee-NEN-sis

Dypsis ankaizinensis is known only from Mount Tsaratanana in eastern Madagascar, where it is thought to be quite common. It grows between 4200 and 6000 feet (1280–1800 m) either on ridge tops or in forest depressions. The forest form has much taller trunks. The epithet means "of Ankaizina," the name for the southern foothills of Mount Tsaratanana.

The trunk of this large solitary species can reach 45 feet (14 m) in height and 15 inches (38 cm) in diameter. Immense leaves begin from a leaf sheath that is covered in dense brown scales extending up along the 5-inch (13-cm) petiole. The inflorescences emerge from among the leaf sheaths and bear elliptical fruit 1 inch (2.5 cm) long and about ¾ inch (19 mm) wide. *Dypsis tsaravoasira* and *D. hovomantsina* appear to be closely related.

As of this writing, this palm is not known to be in cultivation and has not been visited in habitat for over 85 years. If ever cultivated, it would definitely be a magnificent specimen for a large landscape. It is included here because of confusion over a clustering palm with this name that has found its way into numerous gardens and is most likely a form of *D. madagascariensis*.

Dypsis arenarum
DIP-sis · ar-e-NAR-um
PLATES 359–360

Dypsis arenarum grows in sand near the coast of eastern Madagascar by fresh water rivers and clearings. It is similar in appearance to *D. lutescens*, which grows in the same vicinity and is also critically endangered because of development and fires. The epithet is Latin for "of the sands." The description of this clustering species is virtually identical to that of *D. lutescens*. Differences are in its longer leaf petioles, smaller number of leaflets (28–30 on each side as opposed to 44–59 in *D. lutescens*), inflorescence branching only twice instead of 3 times, and smaller fruit.

This palm is not nearly as common as *D. lutescens* in cultivation, but requires the same growing conditions and is equally sensitive to cold. It has somewhat brighter green leaves that show off dramatically against a contrasting background, and its densely clumping habit makes an impenetrable screening plant. It should also perform well as a container plant for indoor or outdoor patio areas.

Dypsis baronii
DIP-sis · ba-RO-nee-eye
PLATE 361

Dypsis baronii is endemic to northern, central, and eastern Madagascar, where it grows on steep slopes in wet forest from 2500 to 4500 feet (760–1370 m). The edible growing point is still harvested. The epithet honors Richard Baron (1847–1907), who collected the original specimen.

This clustering species is rarely found with solitary trunks. Mature stems attain a height of 25 feet (7.6 m) and a diameter of 6 inches (15 cm). They are deep green on all but their oldest parts and are encircled with widely spaced white or gray rings of leaf base scars. The crownshaft is 1 foot (30 cm) tall, light green to yellow-green, slightly bulging at the base or sometimes the middle, and waxy and smooth. Some individuals are very colorful, with chalky white crownshafts and burgundy-red petioles. The sparse leaf crown usually has 6 leaves, each 6 feet (1.8 m) long on a petiole 1 foot (30 cm) long which, when newly expanded, is often red. The leaves are erect but arching, and the stiff, deep green leaflets are 1 foot (30 cm) long and grow from the rachis at an angle that gives the leaf a V shape. The much-branched inflorescences may grow from atop or beneath the crownshaft. They are coral-colored and bear brown or whitish single-sexed small blossoms. The globose ½-inch (13-mm) fruits are bright yellow.

This palm is better adapted to nearly frostless Mediterranean climates than to those that are constantly warm or tropical. It needs a lot of moisture and is not hardy to cold, being adaptable only to zones 10 and 11. It loves a rich, humus-laden, well-drained soil and flourishes in full sun or partial shade. There is no more graceful landscape subject. The palm should be planted against a contrasting background so that its diaphanous form can be appreciated. It is superb as a canopy-scape and nothing could be choicer than a clump of this palm in a patio or courtyard. It is not known to be grown indoors but is probably adaptable to such conditions.

Dypsis beentjei
DIP-sis · BENT-jee-eye
PLATE 362

Dypsis beentjei is an understory lowland rainforest species from the eastern coast of Madagascar. It is endangered, known from only a single population of perhaps 30 plants in waterlogged flats adjacent to streams on ultramaphic rock. The epithet honors Royal Botanic Gardens, Kew, botanist Henk Beentje (1951–), who has been instrumental in sorting out the palms of Madagascar.

This clustering species has subterranean stems with only the crowns of bifid leaves above ground. Up to 9 leaves are held erect from each crown with the pale brown petioles measuring 20–24 inches (50–60 cm) and the solid bifid leaf another 24 inches (60 cm) long. The pale creamy colored rachis sets off the deep green foliage. Bright red fruits just complete what is simply a gorgeous palm.

Constant humidity and moisture in a warm setting with no chance of freezing would be the ideal conditions for this little beauty. An intimate shady spot in the garden where its festive coloration could be enjoyed up close would be a perfect location.

Dypsis bejofo
DIP-sis · beh-JO-fo
PLATE 363

Dypsis bejofo is, in its habitat of the rain forest of northeastern Madagascar, an endangered species because of its few numbers. The epithet is the aboriginal name for the palm. In habitat, the mature trunks reach 80 feet (24 m) tall with a diameter of 1 foot (30 cm) or slightly more. The crownshaft is 4–5 feet (1.2–1.5 m) tall, slightly bulging at its base, and glaucous bluish green or almost greenish white. The leaf crown is sparse, but the leaves are massive, each as long as 20 feet (1.8 m), with linear deep green leaflets 3–4 feet (90–120 cm) long growing from the rachis at all angles to create a plumose leaf. This palm may or may not be in cultivation. Many plants labeled as such do not seem to fit the scientific description.

Dypsis bosseri
DIP-sis · BOS-ser-eye
PLATE 364

Dypsis bosseri is endemic to the eastern coast of Madagascar where it is known only from a single collection. Since that collection, its lowland forest habitat has been destroyed and it may now be extinct in the wild. The epithet honors Jean Bosser (1922–), collector of the type specimen of this species.

The slender brown solitary stem is marked by prominent rings left by old leaves. The overall height of the palm reaches 8 feet (2.4 m). The top of the short creamy crownshaft is covered in blackish scales. The deep green leaf begins immediately from the crownshaft with no petiole present. When younger, leaves tend to be bifid, but eventually they develop 4 or 5 leaflets on each side of the rachis with the terminal leaflets remaining bifid.

While distinctive by itself, an odd-numbered grouping of 3 or more individuals would make this handsome palm stand out in a shady landscape. It is doubtful that it has great cold hardiness. It does best in acidic soil.

Dypsis cabadae
DIP-sis · ka-BAH-dee
PLATE 365

Dypsis cabadae is endemic to the Mayotte islands, part of the Comoros archipelago, but the species was first identified from plants cultivated in a garden in Cuba. The epithet honors Emilio Cabada, a physician in Cienfuegos, Cuba, who first cultivated the palm. The only common English name seems to be cabada palm.

Mature trunks of this large clustering species grow to 25 feet (7.6 m) high but are less than 6 inches (15 cm) in diameter. They are olive green to deep dull green and are graced with widely spaced light gray to almost white rings. Leaf crowns are 10–15 feet (4.5 m) wide but 8 feet (2.4 m) tall. Total height of a mature clump is 35 feet (11 m) and the width is 20–25 feet (6–7.6 m). The green to almost silvery gray crownshaft is 3–5 feet (90–150 cm) tall, smooth, and slightly bulging at its base. The leaves are 8–10 feet (2.4–3 m) long on petioles 1 foot (30 cm) long. The rachis arches stiffly upward and results in a crown of half-erect leaves, none of which descend below the horizontal plane. The leaflets grow at an angle of 20–30 degrees, creating a slight V shape to the blade. They are bright medium green, 2 feet (60 cm) long, and pendent at the tip. The inflorescences, 5 feet (1.5 m) long, are pendent panicles of small, yellow flowers. The rounded ½-inch (13-mm) red fruits are borne in pendent clusters.

The species is slightly susceptible to lethal yellowing disease. The palm is tender to cold and adaptable only to zones 10b and 11, although mature specimens are found in the most favorable microclimates of 10a. The tree is a water lover but can endure average amounts of moisture. It looks better in full sun but is adaptable to partial shade. It never makes a dense clump and often takes several years to start suckering, a phenomenon that emphasizes the extraordinary beauty of the trunks.

The palm's moderate size and gorgeous trunks make it one of the most successful tropical or subtropical large patio or courtyard landscape subjects. It is wonderfully attractive in silhouette against a large wall of vegetation of contrasting color or form. As a canopy-scape, it is unsurpassed. The only landscape use to which it is not suited is as an isolated specimen surrounded by a large space. This palm is successfully grown in conservatories and atriums.

Dypsis carlsmithii
DIP-sis · karl-SMITH-ee-eye
PLATE 366

Dypsis carlsmithii was described in 2002 from a cultivated palm growing in the former garden of Donn Carlsmith (1929–2003), on the island of Hawaii. The species has since been found in the wild in Madagascar.

Formerly known informally as *Dypsis* "Stumpy," this massive solitary species has a stem 20 inches (50 cm) in diameter and as much as 20 feet (6 m) tall. The crownshaft is over 4 feet (1.2 m) long, green at the bottom to reddish brown at the top, and often covered with whitish wax. The leaves are 10 feet (3 m) long, curved and slightly twisted near the end. Regularly spaced leaflets are 3 feet (90 cm) long and deep green creating a distinct V shape to the leaf. The inflorescence, also 10 feet (3 m) long, emerges from among the lower leaves and produces small, black fruit.

While an excellent grower in the well-drained acidic volcanic soil of Hawaii, this palm does fine in Southern California and will tolerate the hot nights of South Florida albeit growing more slowly there. It seems to tolerate the occasional cold temperature to near freezing, but is not likely to appreciate much colder.

A canopy palm of such massive proportions needs a grand landscape to show off its attributes fully. A grouping of individuals forming a majestic silhouette against the sky would be awe-inspiring when seen from afar and simply humbling up close. It is a palm that needs plenty of space to be appreciated and should not be crowded among other trees or palms unless it can rise well above them.

Dypsis catatiana
DIP-sis · ka-tah'-tee-AH-na
PLATE 367

Dypsis catatiana is endemic to northern and eastern Madagascar, where it grows in rain forest from sea level to more than 6000 feet (1800 m). The epithet commemorates Louis Domingue Maria Catat (1859–1933), the original collector of the species.

The green stem of this solitary-trunked, small undergrowth palm attains a height of 4 feet (1.2 m) and a diameter of ½ inch (13 mm). It forms no true crownshaft, but the sheaths are prominent. The leaves are glossy deep green above but paler and usually with a brownish hue beneath. They are up to 4 feet (1.2 m) long and are usually unsegmented and apically bifid, although many specimens exhibit segmented leaves with 3–5 pairs of variably sized slightly S-shaped pinnae.

This choice little thing must be protected from the midday sun, especially in hot climates. It probably luxuriates in frostless Mediterranean climes if given constant moisture, partial shade, and a humus-laden, quickly draining soil. No data are available for cold hardiness, but the species is probably sensitive to frost.

Dypsis confusa
DIP-sis · kahn-FYOO-sa

Dypsis confusa is a rare palm endemic to Masoala, Mananara, and Betampona in northeastern Madagascar, where it grows in lowland rain forests and peat swamps near sea level to 900 feet (270 m). The epithet comes from the fact that this species had long been confused with other similar species.

This solitary or clustering species produces stems to 21 feet (6.4 m) tall and 1½ inches (4 cm) in diameter supporting a crown of 5–8 leaves. The crownshaft is pale green and covered with reddish scales. Leaves are nearly 4 feet (1.2 m) long, including the petiole. Leaflets number 11–28 on each side of the rachis and are arranged in groups of 2–4 in mostly a single plane. The inflorescence grows from among the leaves and produces an orange-red ellipsoid fruit less than ½ inch (10 mm) long.

Like many other *Dypsis* species, this one is fairly new to cultivation and it is still being tested as to its needs in the landscape. It should be an attractive addition to a shady locale, where it could possibly emerge into more direct sunlight as it ages.

Dypsis coriacea
DIP-sis · kor-ee-AY-see-a
PLATE 368

Dypsis coriacea is a lowland rainforest species endemic to northeastern Madagascar where it grows at 600–1200 feet (180–370 m) on steep banks near streams or on ridge tops. The epithet is Latin for "leathery" and refers to the leaf's texture. This species can be solitary or clustering and is similar in appearance to *D. catatiana* differing with its shiny thick green leaves and stiff inflorescence. Stems grow only to 6 feet (1.8 m) tall and support a crown of 6–9 leaves. The leaves are entirely bifid and about 18 inches (45 cm) long. The emergent new leaf is a deep red. An equally choice small addition for an intimate garden location as *D. catatiana*, it should have the same cultural requirements. This palm is also new to cultivation and it will take time to see how it adapts to various climes throughout the world.

Dypsis crinita
DIP-sis · kri-NEE-ta
PLATES 369–370

Dypsis crinita is a sparsely clustering species from northern Madagascar, where it grows in mountainous rain forest along banks of streams and rivers at elevations from 600 to 800 feet (180–240 m). The epithet is Latin for "hairy," an allusion to the persistent leaf base fibers.

The species is sometimes found as a solitary-trunked specimen in nature; the trunks also are sometimes branched above ground. They are dark brown and grow to 40 feet (12 m) tall in their native haunts, but are not this large in cultivation. The most distinctive feature of the species is the great amount of persistent light brown to cinnamon-colored leaf fibers on the younger parts of the stems, but the most beautiful feature of the species is its leaves. These are an elongated ellipse in shape, usually 6–8 feet (1.8–2.4 m) long on petioles 3 feet (90 cm) long, with regularly arranged, glossy medium to dark green, limp, narrowly elliptic-lanceolate leaflets growing from the rachis in a single flat plane. The rachises invariably twist 90 degrees in their apical halves to give the leaf and its leaflets a vertical orientation for half their length. The visual effect is reminiscent of a large cycad or *Phoenix roebelenii* but even lovelier. The new leaves are usually a beautiful orange brown to pinkish brown shortly after they unfold.

This palm is a true water lover, almost an aquatic. It needs a rich, humus-laden soil with pH that is not too high. It does not tolerate full sun in hot climates and is best in partial shade in any clime. While not hardy to cold, the palm has survived temperatures in the upper 20s Fahrenheit (around −2°C) in central Florida, under canopy. Nothing could be more beautiful as part of a wall of vegetation and as an accent or focal point therein.

Dypsis decaryi
DIP-sis · de-KAHR-ee-eye
PLATES 371–372

Dypsis decaryi is endemic to a small area of extreme southeastern Madagascar, where rain forest gives way to spiny forest. It grows in dry hillside forest on poor soil. The epithet honors the original collector, French botanist Raymond Decary (1891–1973). The common name is triangle palm, and it is sometimes still called by its synonym, *Neodypsis decaryi*.

This solitary-trunked species grows to 35 feet (11 m) in habitat and 25 feet (7.6 m) under cultivation. The trunks are free of leaf bases beneath the pseudo-crownshaft and are dark gray with closely set rings of leaf base scars. The stems are stocky and to 20 inches (50 cm) in diameter. The leaf crown is 15 feet (4.5 m) wide and tall. The leaves grow from the trunk in 3 distinct vertical rows, and the wide, overlapping leaf bases of the living leaves form a large triangle at the top of the trunk that is 3 feet (90 cm) tall and 2 feet (60 cm) wide. The leaves are 10 feet (3 m) long and stiffly erect with only the tips pendent. The leaf petiole is short, 1 foot (30 cm) long, its base expanded into a large, plump, and broadly triangular sheath. It is covered when young in a fine, felty, rust-colored tomentum that is soon shed to reveal a chalky bloom on the older bases. The leaflets are 2 feet (60 cm) long and grow at a steep angle from the rachis to give a definite V shape to the leaf. Only the tips of leaflets near the tip of the blade are pendent. The lower leaflets are extended at their tips into long, thin reins. Leaf color is grayish green to bluish green. The dead leaves tend to fall cleanly from the trunk, giving a distinctive inverted triangle to the leaf crown. The inflorescence grows from among the lower leaves, is much branched and 4–5 feet (1.2–1.5 m) long, and bears small, yellow unisexual flowers of both sexes. The ovoid fruits are 1 inch (2.5 cm) long and greenish yellow to white.

The species is slightly susceptible to lethal yellowing disease. It is safe only in zones 10 and 11, although it is found in 9b as a protected individual. It wants full sun but endures partial shade. It is drought tolerant but grows better with regular moisture. It needs a well-draining soil but is not particular as to type.

This is a sensational landscape subject whose great fountain sweep of grayish leaves is seen to best advantage against a dark green background. One of the most dramatically beautiful placements is at Leu Botanical Gardens in Orlando, Florida, where the palm is planted in a widened circle in the middle of a broad walkway and surrounded on all sides by other palms and cycads. This setting not only allows a close-up view of its form but also singles it out from its surroundings, and the color of its leaves is striking against the darker greens around it. The silhouette of the triangle palm is almost amazingly decorative. The palm can work well in almost any landscape but should not be crowded into a small space where it looks quite uncomfortable. It is not generally grown indoors, but it succeeds given lots of light and good air circulation.

Dypsis decipiens
DIP-sis · de-SIP-ee-enz
PLATE 373

Dypsis decipiens is endemic to the plateau region of Madagascar, where it grows in exposed, rocky sites at 4000–6000 feet (1200–1800 m) and is endangered because of habitat destruction due to agriculture and wildfires. The epithet is Latin for "deceiving," a term usually applied to plants that look remarkably like other species. This palm is so distinctive that "in this case, it is a particularly inappropriate name" (Dransfield and Beentje 1995).

The species is mostly found in nature as a solitary-trunked individual, but individuals with 2 or more stems are known. The mature trunk can attain a height of 60 feet (18 m) in habitat and is encircled in its younger parts with wide, dark rings. The dark gray stem is massive and columnar, resembling those of royal palms (*Roystonea*) but even more so those of gigantic specimens of *Hyophorbe*. The stem usually bulges in the middle, but may also have a distinct, often grotesque bulge only near the base of the column; with age the upper stem invariably tapers. The crownshaft is 2 feet (60 cm) tall, a light, waxy green, sometimes almost white, and often scarcely thicker than the adjacent trunk, although many individuals show a slight bulge near the base of the column. The arching leaves are carried on short petioles less than 1 foot (30 cm) long. The stiff leaflets are usually deep green on both surfaces but sometimes are bluish and glaucous. They grow from the rachis at a steep angle to give the leaf a V shape, but this phenomenon is obscured because of the short petioles and the arrangement of the leaflets in plumose groups. The short but much branched inflorescences grow from beneath the crownshaft and bear many small, yellow blossoms of both sexes. The globose ½-inch (13-mm) fruits are yellow.

This species seems temperamental in its climatic requirements, flourishing in nearly frostless Mediterranean climates but languishing in hot, moist tropical or subtropical climes. The trees reportedly are immune to light frosts and are drought tolerant but slow growing. They seem to be not particular about soil type as long as it is well drained. From all reports, they need full sun when past the seedling stage. David Witt (pers. comm.) reported that the species is a "tillering" palm that tries to pull itself downwards as a seedling and therefore needs to be planted high to counteract this tendency in wet climates.

This imposing palm works even as an isolated specimen, surrounded by space. It is more pleasing in groups of 3 or more individuals of varying heights; and it is visually arresting as a canopy-scape. It is not known to be grown indoors and may not be amenable to such conditions.

Dypsis dransfieldii
DIP-sis · dranz-FEEL-dee-eye
PLATE 374

Dypsis dransfieldii is endemic to the Masoala Peninsula in northeastern Madagascar, where it grows in coastal white sand forests at nearly sea level. It is endangered since it is found in only one site and that site is considered fragile. The epithet honors John Dransfield (1945–), the foremost modern palm botanist to study the palms of Madagascar.

This clustering palm has 3–5 stems that grow to 18 feet (5.4 m) tall and are as much as 1½ inches (4 cm) in diameter. The crown is comprised of 6–12 leaves. The leaf sheath is covered with thick reddish-brown fibers reminiscent of *D. crinita* and *D. fibrosa*. Deep green leaves can be more than 5 feet (1.5 m) long with 33 or 34 regularly spaced leaflets on each side of the rachis in a single plane or ever so slightly pendulous. The inflorescence is 8 feet (2.4 m) long and grows from among the leaf sheaths, projecting well beyond the leaves. The ovoid warty fruit is about ¾ inch (19 mm) long. This is another *Dypsis* species that is just making its way into cultivation. Its cultural needs are still not well understood.

Dypsis faneva
DIP-sis · fah-NEE-vah
PLATE 375

Dypsis faneva hails from northeastern Madagascar where it is endemic to coastal or lowland rain forests from near sea level to 900 feet (270 m). It is considered endangered with fewer than 50 plants left in habitat. The epithet means "flag" in Malagasy and refers to the leaf's large, terminal leaflets.

The brown stems of this clustering palm grow to 18 feet (5.4 m) tall with a diameter to 2 inches (5 cm). Some short stilt roots emerge from the stem bases. The crownshafts are a pale yellowish green with scattered brown scales. The leaf crown holds 8 or 9 arching leaves that are as much as 40 inches (102 cm) long including the petiole, if present. Leaflets number 8–21 on each side of the rachis and are held stiffly in a single plane. The terminal pair of leaflets is enlarged and forms a V-shaped segment. The spreading inflorescence emerges from among the leaves branching twice.

The manner in which the palm holds its leaves is somewhat reminiscent of some clustering species of *Pinanga*, and thus this species can be used in a similar way in the landscape. It has a distinctive bearing that is quite attractive but perhaps not as a main focal point. Seeing its leaves emerging from the background will give the viewer pause to take a second look to appreciate this palm's unique beauty. This species is fairly new to cultivation and growers have yet to test its range of hardiness in various climes other than Hawaii, where so many *Dypsis* species do so well.

Dypsis fasciculata
DIP-sis · fa-sik-yoo-LAHT-ah

Dypsis fasciculata is endemic to the white sand lowland forests of northeastern Madagascar from near sea level to 700 feet (210 m). It is considered vulnerable due to fires that often occur in its habitat. The epithet refers to the grouped (fascicled) leaflets along the leaf rachis.

This solitary or clustering species has green stems to 18 feet (5.4 m) tall and 1½ inches (4 cm) thick supporting a leaf crown of about 8 leaves. The crownshaft is 10 inches (25 cm) long and covered in reddish-brown scales. Leaves up to 4 feet (1.2 m) long bear 11–23 drip-tip leaflets on each side of the rachis held in groups of 2–6. The inflorescence emerges from among the leaves and is shorter than the leaves, branching twice.

The cultural conditions for this species are not yet well known as it is new to cultivation. It does not grow as well in its nutrient-poor natural habitat as it does when fed and watered in cultivation. With good soil and fertilizer, it should be a handsome addition to the landscape. Its cold hardiness is unknown at this time.

Dypsis fibrosa
DIP-sis · fy-BRO-sa

PLATES 376–377

Dypsis fibrosa is endemic to mountainous rain forest of eastern Madagascar at elevations from sea level to 2600 feet (790 m). The epithet is Latin for "fibrous."

The trunks of this sparsely clustering species attain heights of 20 feet (6 m) in nature and are often branched above ground. The stems are usually covered with a mass of long, deep brown fibers, which in the youngest parts are especially long and are pendent. The leaves are 6 feet (1.8 m) long, with widely but regularly spaced dark green, narrow, linear-elliptic, and S-shaped leaflets growing in a flat plane. The leaf rachis, like that of *D. crinita*, is twisted in its apical half to give a vertical orientation to the leaf and the leaflets from that point on. New leaves are usually light red or pink.

The species does not tolerate cold, is adaptable only to zones 10b and 11, and is marginal in 10a. It is a water lover and likes a humus-laden soil. It thrives in partial shade or full sun except in the hottest climates. It is robust looking enough to be used as a specimen, a contrasting component of other vegetation, or a close-up subject.

Dypsis heteromorpha
DIP-sis · het'-e-ro-MOR-fa

PLATES 378–379

Dypsis heteromorpha is endemic to the Tsaratanana, Marojejy, and Anjanaharibe areas of northern Madagascar. It grows in montane rain forest at 4000–6500 feet (1200–1980 m). The epithet means "having different shapes or forms" and refers to the leaves, which may be entire (in juveniles) or regularly pinnate.

This clustering palm generally has 3–6 stems but occasionally is solitary. Stems can reach a height of 35 feet (11 m) with a diameter up to 5 inches (13 cm). The leaf crown holds 10 leaves, which are tristichous (in 3 ranks) in habit. The leaf sheath is waxy white at the bottom changing to a reddish brown at the base of the petiole and extending up into the petiole. The arching leaf has regularly spaced leaflets that are held in a V shape. The pendulous inflorescence emerges below the leaves and produces an ellipsoid fruit nearly 1 inch (2.5 cm) long. The endosperm is ruminate. *Dypsis baronii* is closely related to this species. Its cultural requirements should be quite similar as well.

Dypsis hiarakae
DIP-sis · hy-ah-RAH-kee

Dypsis hiarakae is endemic to northern Madagascar where it grows at 750–1800 feet (230–550 m) elevation in submontane rain forests. It is considered rare, as only a few hundred individuals exist in habitat. The epithet refers to a village within its range on the west coast of Masoala. The stem of this solitary palm grows to 18 feet (5.4 m) tall with a 1-inch (2.5-cm) diameter and supports a crown of 7–9 leaves. Leaves are arching and 3 feet (90 cm) long including the petiole. The 9–14 leaflets are spaced almost regularly along the rachis or in groups of 2 or 3 on each side of it. The inflorescence emerges from below the leaves and produces a small, ellipsoid red fruit less than ⅓ inch (8 mm) long. The stems are used to make blowpipes.

Dypsis hildebrandtii
DIP-sis · hil-de-BRAND-tee-eye

PLATES 380–381

Dypsis hildebrandtii is endemic to montane rain forests of eastern Madagascar, growing mainly between 2000 and 3000 feet (610–900 m) elevation and occasionally down to 1000 feet (300 m). The epithet honors German plant collector Johannes Hildebrandt (1847–1881), who perished in Madagascar while collecting for Berlin's botanical garden.

This small, clustering or solitary-trunked species has stems to only 6 feet (1.8 m) tall and a leaf crown holding 4–10 leaves. The light green crownshaft is lightly covered in reddish-brown scales. Petioles are short or nonexistent, and the bright green leaves are entirely bifid or made up of 2 or occasionally 4 leaflets. The newly emergent leaf is a deep red. Inflorescences emerge from among the leaves and produce a small, ellipsoid red fruit less than ½ inch (13 mm) long.

This delightful palm can be a small focal point in an intimate space beneath other palms or trees, where it can be enjoyed up close. A soil with a humusy top layer would be ideal. Because the species comes from higher elevations, it would not do well in hot Florida summers but rather a tropical humid clime where temperatures get down in the mid-60s Fahrenheit (around 18°C) at night.

Dypsis hovomantsina
DIP-sis · ho'-vo-mant-SEE-na
PLATE 382

Dypsis hovomantsina is endemic to the lowland rain forest of northeastern Madagascar, where it is rare and threatened because of agricultural expansion. The epithet is the aboriginal name, which means "stinking," a reference to the smelly but edible growing point.

The solitary stem attains a height of 40 feet (12 m) and a diameter of 1 foot (30 cm); it is a light cinnamon color in its older parts, the younger parts a grayish green with widely spaced rings of brown leaf base scars. The bulging crownshaft is 3 feet (90 cm) high and is a beautiful waxy white except near its summit where it has a reddish-brown feltlike tomentum. The leaf crown is hemispherical with spreading, barely arching leaves 10 feet (3 m) long on thick petioles 18 inches (45 cm) long. The narrow, long dark green leaflets grow at all angles from the rachis to give a wiry plumose effect.

This odd-looking palm is not always beautiful or tidy in appearance but it is fascinating to contemplate because of the unusually colored crownshaft and the large wiry leaves. It is tender to cold and needs copious, regular moisture in a humus-laden soil.

Dypsis lanceolata
DIP-sis · lan'-see-o-LAH-ta
PLATE 383

Dypsis lanceolata is endemic to the elevated rain forest of the Comoro Islands, east of Mozambique and northwest of Madagascar, where it grows at 1000–3000 feet (300–900 m). The epithet is Latin for "lanceolate," in reference to the shape of the leaves.

The small, mostly clustering trunks attain a maximum height of 20 feet (6 m), are olive green in all but their oldest parts, and bear widely spaced, darker rings of leaf base scars. The elongated crownshaft is light silvery green or even white and is covered in tiny whitish scales. The leaves on older palms are 8 feet (2.4 m) long on short petioles. The wide, glossy light green, tapering, pointed leaflets grow from the rachis at an angle when young but from nearly the same plane when older.

This exceptionally beautiful species resembles *D. cabadae* but is smaller. It needs abundant, nearly constant moisture, a tropical climate, full sun to partial shade, and a moist, well-drained, rich soil.

Dypsis lantzeana
DIP-sis · lant-zee-AH-na
PLATE 384

Dypsis lantzeana is endemic to northeastern Madagascar, particularly around the Bay of Antongil, in lowland rain forest up to 1000 feet (300 m) in elevation. While a common palm in its habitat, it is considered vulnerable due to farming. The epithet is the Latinized form of the surname of the explorer who first collected this species in 1871.

Stems of this solitary or clustering species can reach a height of 12 feet (3.6 m) and are less than 1 inch (2.5 cm) in diameter. The short crownshaft is pale green or ivory and covered in dark scales. The leaf crown has 6–15 leaves, each 18 inches (45 cm) long on a petiole that can be up to 4 inches (10 cm) long or entirely absent. The leaves are either entirely bifid or have 3–9 leaflets of irregular width and spacing on both sides of the rachis. The newly unfurling leaf has a red tinge. The inflorescence is 20 inches (50 cm) long, emerges from the older leaf bases, and produces a tiny red elliptical fruit less than ⅓ inch (8 mm) long.

This delightfully attractive palm makes a picturesque container plant or charming landscape specimen in a shady location of the garden. The colorful combination of bright green foliage with emergent reddish leaf and red fruit is a treat to the beholder. The palm appreciates a soil with a humusy top layer and consistent watering to show off best. It does not tolerate freezing temperatures well.

Dypsis lastelliana
DIP-sis · las-tel'-ee-AH-na
PLATES 385–386

Dypsis lastelliana is endemic to northwestern and northeastern low, mountainous rain forest and lowland wet forests of Madagascar. The epithet honors the original collector, M. de Lastellé. Common names are teddy bear palm and redneck palm.

The solitary trunk, which can grow to 65 feet (20 m) in habitat but is usually half that under cultivation, is 10–15 inches (25–38 cm) in diameter and has phenomenal color in its younger parts: the stem is deep brown to reddish brown with wide, closely spaced pure white rings of leaf base scars. The rounded leaf crown is the glory of the palm; 12 feet (3.6 m) wide and 8 feet (2.4 m) tall. The crownshaft is loosely formed and 2 feet (60 cm) tall, but has an outstanding color of deep orange-brown to reddish brown and is felty to the touch. The leaves are 10 feet (3 m) long, erect, and slightly arching but some fall beneath the horizontal. The short petioles are never more than 1 foot (30 cm) long. The many narrowly lanceolate leaflets are 18 inches (45 cm) long and grow from the rachis in a flat plane but are pendent and glossy deep green on both sides. The leaf crown is not dense, and the old leaves fall cleanly, leaving only a half crown. The inflorescences are borne from among the lower leaves and are 2 feet (60 cm) long and much branched. They bear small, whitish, unisexual flowers. The ovoid fruits are 1 inch (2.5 cm) long and orange.

This species has been confused with *D. leptocheilos* (see following). Almost all growers in Florida still think they are growing *D. lastelliana*, although there are exceedingly few representatives of it in that state; most so-named individuals are actually *D. leptocheilos*.

Dypsis leptocheilos
DIP-sis · lep-to-KY-los
PLATES 387–388

Dypsis leptocheilos is endemic to Madagascar but precisely where seems unclear. It was first described from a garden in Tahiti, and because it was thought to be a cultivar of *D. lastelliana*, it was named 'Darianii'. It is now known in the trade as the teddy bear palm. The epithet *leptocheilos* is from Greek words meaning "thin" and "lip" and refers to the thin bracts on the rachillae.

This species is generally similar to *D. lastelliana* but has a lighter-colored, even looser crownshaft, is shorter, and has leaves 12 feet (3.6 m) long that are elongated ellipses with flatter, much more spreading leaflets. The leaf bases bear a reddish brown indument of rough, loose scales. The leaves are as beautiful as those of *D. lastelliana*. Horticulturally the difference is that *D. leptocheilos* endures cooler temperatures at or slightly below freezing and does not require as much moisture or as good a soil as *D. lastelliana*; furthermore, it thrives in South Florida and coastal Southern California on alkaline soils.

The species is tender to cold and adaptable only to zones 10b and 11. It needs regular moisture and a humus-laden, well-drained, slightly acidic soil. It is a "tillering" palm that tries to pull itself downwards as a seedling (David Witt, pers. comm.) and therefore needs to be planted high to counteract this tendency in wet climates.

This beautiful palm does not work well as an isolated specimen surrounded by space; it needs to be in groups of 3 or more individuals of varying heights or used as a canopy-scape. Old, tall specimens look much like a straight-trunked coconut from any distance. This palm needs space and good light to do well indoors and is only suited to large greenhouses, atriums, or conservatories.

Dypsis louvelii
DIP-sis · loo-VEL-ee-eye
PLATE 389

Dypsis louvelii is endemic to the rain forest of central eastern Madagascar at 1000–3500 feet (300–1070 m). The epithet honors M. Louvel (1877–1957), a French forester and plant collector in Madagascar. This small undergrowth palm has a thin trunk to 3 feet (90 cm) high in nature. The leaf crown is sparse, but the deep, dark green leaves are beautiful, each one is 3 feet (90 cm) long, linear wedge shaped, entire and undivided except for the deep single apical cleft, and heavily ribbed throughout. The new growth has a beautiful orange or reddish brown hue. The slightly branched inflorescences are long and form beautiful scarlet fruits, which cause the infructescence to become pendent. This palm does not tolerate frost but succeeds in cooler but frostless Mediterranean climates. It needs partial shade in all climates and a constantly moist but well-draining soil.

Dypsis lutescens
DIP-sis · loo-TES-senz
PLATE 390

Dypsis lutescens is endemic to sandy riverbanks and clearings in the wet forests of eastern Madagascar. Because of its small distribution in isolated areas and the development of agriculture there, the species is critically threatened in the wild, but ironically, it is one of the commonest palms in cultivation. The epithet is Latin for "becoming yellow," a reference to the petioles. In its long history under cultivation, the palm has acquired several common names, including areca palm, yellow bamboo palm, butterfly palm, and golden cane palm.

The trunks of this clustering species can attain a height of 25 feet (7.6 m) with a diameter of 2–3 inches (5–8 cm). The younger parts are deep green to yellow or even orange, depending on how much light they receive; the yellow color results from high light levels. The rings are relatively closely set. The leaf crown is 12 feet (3.6 m) wide and 10 feet (3 m) tall. A mature clump can reach a height of 35 feet (11 m) and a width of 20 feet (6 m). The crownshaft is 3–4 feet (90–120 cm) long, grayish green to almost silvery gray, and slightly bulging at its base. The leaves are 6–8 feet (1.8–2.4 m) long, narrowly ovoid, and beautifully arched. They are carried on lissome petioles 2 feet (60 cm) long, and the thin, narrowly lanceolate yellow-green to dark green leaflets, also 2 feet (60 cm) long, grow from the rachis at a 40-degree angle, creating a distinct V shape to the blade. The petiole and rachis are light green to almost orange, depending on the amount of light the plant receives, and the lightweight leaf moves in the slightest breeze. There are only 6–8 leaves in a single crown, but their great arches give a full visual effect. The flower stalks are pendent and branched and grow among the crownshafts. The flowers are yellow. The ovoid fruits are 1 inch (2.5 cm) long and yellow-orange.

The palm is tender to cold and is restricted to zones 10b and 11, although its fast growth and clumping habit allow it to be widely grown in 10a with occasional damage. It loves water and needs sun to look its best but is not particular about soil type as long as it is well drained. Rooted suckers are easily transplanted but may sulk for a while after being planted.

Several interesting cultivars have been developed from this species in Thailand, including a weeping form, a dwarf form, and a strange "strict" form, with stiffly erect leaves and leaflets. None of these cultivars seem to be available outside Thailand.

Few plants are as breathtakingly graceful as a well-grown clump of this palm. It is so common in tropical areas that it is often overlooked, especially by palm collectors, in favor of less common and often harder-to-grow palms, most of which are not as beautiful. There is no more handsome screen than one made of mature clumps of this species planted 10 feet (3 m) apart; its wall of leaves from top to bottom creates a world unto itself, one of unexcelled grace and frozen movement. As a patio or courtyard subject, it is impossible to beat. It even looks good in the middle of a lawn surrounded by space, and as a major accent among other masses of vegetation it is more than eye-catching. Like many other clustering palms, the clumps are more beautiful, especially as a silhouette, if judiciously and occasionally thinned so that the trunks and their lovely forms may be better seen. This is one of the commonest indoor subjects. It needs high light levels to look good. It is also subject to spider mite infestation in low humidity situations.

Dypsis madagascariensis
DIP-sis · mad-a-gas'-kar-ee-EN-sis
PLATES 391–393

Dypsis madagascariensis is endemic to northern Madagascar, where it grows in dry, open areas of semideciduous forest as well as in rain forest. There are both clustering and solitary-trunked individuals. The solitary forms are known as **D. madagascariensis var. lucubensis** [loo-coo-BEN-sis]. This species is endangered because of habitat destruction and harvesting of its edible growing point. The epithet is Latin for "of Madagascar."

Mature trunks reach a height of 25 feet (7.6 m) and a diameter of 6 inches (15 cm). They are slightly swollen at their bases, and gray to greenish (especially the younger parts) with prominent and widely spaced rings. The leaf crown is 20 feet (6 m) wide and 15 feet (4.5 m) tall. The clustering forms produce few trunks, and they are not dense, attaining a total width of 30 feet (9 m). The crownshaft is only about 1 foot (30 cm) tall, smooth, light green, and thicker at the top than at the bottom. The stiffly arching leaves are 10 feet (3 m) long and held on petioles 6–8 inches (15–20 cm) long. The dark green leaflets are 2 feet (60 cm) long and grow from the rachis in small groups, emanating from the stem at different angles to give a plumose appearance to the blade. The arrangement of the leaflets, their width, and even their color vary: some plants have plumose leaves with narrow leaflets, others have leaflets that spring from the rachis at a 40-degree angle, giving a V shape to the blade. The inflorescences are densely branched, grow from beneath the crownshaft, and bear yellow flowers. The ovoid fruits are ½ inch (13 mm) long and black.

This species is intolerant of cold and is adaptable only to zones 10b and 11. It relishes full sun and is drought tolerant but looks and grows much better with regular, adequate moisture. It seems not particular about soil type as long as it is well drained. It has more in common visually with *D. cabadae* than it does with *D. lutescens*, but its trunks are less attractive. It is (or should be) used in the landscape as is *D. cabadae*; it can, however, look better as an isolated specimen. It is difficult to maintain indoors in any but the brightest and airiest of large glasshouses.

Dypsis malcomberi
DIP-sis · mal-KAHM-bur-eye
PLATES 394–395

Dypsis malcomberi is a rare and endangered species endemic to southeastern Madagascar, where it grows in wet forest at 1000–2600 feet (300–790 m). The epithet honors botanist Simon Malcomber (1967–), a member of the Dransfield-Beentje expeditions to Madagascar.

The large solitary trunk is distinctly ringed and attains a height of 80 feet (24 m) in habitat. The mint green, wax-covered crownshaft is 5–6 feet (1.5–1.8 m) tall and slightly bulging at its base, mostly cylindrical otherwise. The leaf crown is sparse and hemispherical with spreading and slightly arching, deep green, plumose leaves 10–12 feet (3–3.6 m) long.

This beautiful species looks from a distance like an elegant version of *Syagrus romanzoffiana* and should serve the same landscaping purposes, except that it is tropical in its requirements. It grows fast with regular, adequate moisture in a humus-laden, well-drained soil and withstands full tropical sun once past the juvenile stage.

Dypsis mananjarensis
DIP-sis · ma-nan'-ja-REN-sis
PLATES 396–397

Dypsis mananjarensis is endemic to eastern Madagascar, where it grows in dry and wet habitats at low elevations. The epithet is Latin for "of Mananjary," a region of the palm's habitat. Growers refer to it this palm as *Dypsis* "Mealy Bug."

The solitary trunk is very straight-growing and attains a height of 80 feet (24 m) in habitat and a diameter of 1 foot (30 cm). It is ringed with whitish widely spaced leaf base scars and is greenish in its youngest parts but gray in its older parts. The glaucous crownshaft is 4 feet (1.2 m) tall, bulging at its base, and varies from silvery green to yellow or orange-yellow because of large waxy scales that cover it; the scales give the palm a palpable gritty appearance, which looks like a severe infestation of mealy bugs and has given rise to the common name. The sparse leaf crown is hemispherical or less and consists of leaves 10 feet (3 m) long that are spreading, unarching, and arranged in 3 ranks. The short petioles are covered in the same scales as the crownshaft. The many dark green narrowly lanceolate leaflets are 3–4 feet (90–120 cm) long and grow from the scaly rachis in groups and from different angles to give a plumose effect.

This palm is an impressive one when mature and a fast, robust grower until mature. It is not frost tolerant but should grow well in a frostless Mediterranean climate as it does not need copious moisture or humidity.

Dypsis marojejyi
DIP-sis · mah-ro-JAY-zhee-eye
PLATE 398

Dypsis marojejyi is endemic to the Marojejy area in northeastern Madagascar, as its epithet indicates. Although the area is protected, the conservation status of this species is considered vulnerable. The palm grows from 2000 to 3300 feet (610–1000 m) in submontane rain forest. Growers call it *Dypsis* "Mad Fox," short for Madagascar foxtail.

This gorgeous solitary palm has a trunk to 18 feet (5.4 m) tall and 1 foot (30 cm) in diameter atop stilt roots. The leaf sheaths are densely covered with attractive rusty-brown hairs. Dark green leaves are plumose with groups of 3–6 leaflets regularly spaced along the rachis that can be 12 feet (3.6 m) long. New leaves emerge in shades of red, pink or bronze. The inflorescence is 3 feet (90 cm) long, emerges from among the leaves, and produces yellowish-green ovoid fruit that is 1 inch (2.5 cm) long and ¾ inch (19 mm) wide.

With its brownish-red leaf bases and dark green plumose leaves, this exquisite palm is an immediate eye-catcher that becomes a

Dypsis nauseosa
DIP-sis · naw-see-O-sa
PLATE 399

Dypsis nauseosa is endemic to coastal Madagascar in the former Fianarantsoa Province where it grows from 150 to 600 feet (46–180 m) elevation in possibly dry forest. It is critically endangered due to forest cutting. The epithet refers to the palm heart, which is bitter and supposedly poisonous. The wood is used for roofing beams and the outer part of the trunk is used for floor boards.

The solitary pale brown trunk rises to as much as 45 feet (14 m) tall and 1 foot (30 cm) in diameter and is swollen at the base. The leaf crown has 12 or 13 leaves, straight, up to 12 feet (3.6 m) long, with nearly pendulous leaflets that are regularly spaced on the rachis. The large inflorescence can be 7 feet (2.1 m) long and emerges from among the lower leaves.

This large canopy palm needs an open space to show off its stiff silhouette. As with many *Dypsis* species, this one is relatively new to cultivation and large specimens are yet to be seen. Due to its habitat, it should tolerate some dry conditions and may be a palm better suited for a California or Mediterranean-type climate.

Dypsis nodifera
DIP-sis · no-DI-fe-ra
PLATE 400

Dypsis nodifera is a widespread species endemic to a large portion of eastern Madagascar. It occurs in moist rain forests from near sea level to 3000 feet (900 m). The epithet means "bearing nodes," which likely refers to the bumps or nodules on the mature stem. The species is virtually identical to *D. pinnatifrons* with the main taxonomic difference being in the male flowers and in the ruminate endosperm in the seed. As this palm grows often in the same habitat as *D. pinnatifrons*, its cultural requirements are the same.

Dypsis onilahensis
DIP-sis · o-ni'-la-HEN-sis
PLATE 401

Dypsis onilahensis occurs in small pockets of evergreen forest throughout Madagascar. The epithet is Latin for "of Onilahy," a river in southern Madagascar. The species is closely related to, similar in appearance to, and has the same cultural requirements as *D. baronii*. Its powdery white crownshaft is a delight to behold.

Dypsis ovobontsira
DIP-sis · o-vo-bahnt-SIR-ah

Dypsis ovobontsira is endemic to the Mananara Biosphere Reserve in northeastern Madagascar. This rainforest species grows at 800 feet (240 m) elevation in serpentine soils that have a deep humusy layer. With fewer than 10 plants left in a single site, its conservation status is considered critical. The epithet is the aboriginal (Betsimisaraka) name for this species. The large solitary trunk grows to 30 feet (9 m) tall and 5 inches (13 cm) in diameter and holds a crown of 6 leaves. The leaf sheath is green with a covering of dense brown and white scales. The leaf petiole is also covered with white scales that become sparse along the rachis. Arching leaves are over 8 feet (2.4 m) long and spiral. This palm is new to cultivation and should make an elegant addition to tropical and perhaps subtropical landscapes. Little is known of the conditions it will tolerate as of this writing.

Dypsis pachyramea
DIP-sis · pak-ee-RA-mi-a
PLATE 402

Dypsis pachyramea is a small, clustering species endemic to the Masoala Peninsula in northeastern Madagascar. It is known from 2 locations and is considered vulnerable. Its habitat is lowland rain forest up to 1200 feet (370 m) elevation where it grows on steep slopes as well as in the bottoms of valleys. The epithet means "thick branches." This is one of the smallest *Dypsis* species reaching an overall height of 2 feet (60 cm). The crown is made up of 8 leaves that are narrowly bifid, deeply plicate, and less than 18 inches (45 cm) long. Bright red fruit are produced on small inflorescences hidden beneath the leaves. The palm requires shade and a soil with a humusy top layer. The stems tend to recline and take root, making this small palm an ideal groundcover especially if planted in multiples under a canopy of large and mid-story vegetation.

Dypsis paludosa
DIP-sis · pa-loo-DO-sa
PLATE 403

Dypsis paludosa is endemic to northeastern Madagascar where it grows in coastal peat swamp forests. The habitat is threatened making the palm vulnerable even though it is quite widespread within its range. The epithet means "marsh dwelling" and refers to the palm's habitat.

This clustering species is closely related to *D. procera*. Stems can reach 18 feet (5.4 m) tall and 2 inches (5 cm) in diameter, and are gray-brown near the base changing to green below the leaves. The crown has 9–12 leaves, and the crownshaft is covered in reddish-brown scales. The leaves are 3–4 feet (90–120 cm) long, emerge directly without a petiole, and have 2–13 leaflets of irregular width and length when the leaf is not entirely bifid. The inflorescence can be 4 feet (1.2 m) long and produces elliptical fruit about ½ inch (13 mm) long and ⅓ inch (8 mm) wide.

The ideal growing condition for this palm is an acidic humusy soil that is never allowed to dry out completely. A shady spot under a high canopy with contrasting foliage will show off its charming attributes to the fullest. It would also be delightful as a container plant for the patio.

Dypsis pembana
DIP-sis · pem-BAH-na
PLATES 404–405

Dypsis pembana is a rare clustering (sometimes solitary) species endemic to the island of Pemba off the coast of Africa, where the borders of Kenya and Tanzania meet. It grows in a single lowland rain forest. The epithet is Latin for "Pemban." The palm grows to 40 feet (12 m), with a light brown or gray trunk that is straight and encircled with widely spaced whitish rings of leaf base scars. The plump crownshaft is 1 foot (30 cm) tall, olive green, and bulging at its base. The leaves are 6 feet (1.8 m) long on short petioles, and have deep green, linear, tapering leaflets that grow from the rachis at a steep angle to create a V-shaped leaf. The species does not tolerate cold and is adaptable only to zones 10b and 11. It needs abundant, nearly constant moisture and a rich but well-draining soil. It thrives in full sun to partial shade.

Dypsis pilulifera
DIP-sis · pil-yoo-LIF-e-ra
PLATES 406–407

Dypsis pilulifera is endemic to northern and eastern Madagascar, where it grows in rain forest at a moderate elevation of 2600 feet (790 m) and is endangered because of being felled for its edible growing point. The epithet is Latin for "bearing little globes," a reference to the fruits.

The solitary trunk can grow nearly 100 feet (30 m) tall but is scarcely more than 1 foot (30 cm) in diameter. It is brown or dark gray in the older parts but greenish in the younger parts and distinctly ringed with brown leaf base scars in all parts. The crownshaft is 3–5 feet (90–150 cm) tall, bulging at its base, and green, brown, or yellowish according to the age of most sheaths. Well-colored individuals have been grown as **D. 'Orange Crush'**. The leaf crown is sparse and slightly more than hemispherical. The leaves are 12–15 feet (3.6–4.5 m) long, spreading and slightly arching, and borne on short petioles. The pendent, deep green leaflets are 3–4 feet (90–120 cm) long and grow in widely spaced groups.

This palm is one of the most attractive palms in Madagascar and is beautiful in silhouette. It is fast growing if given a decent soil, regular, adequate moisture, and a tropical climate.

Dypsis pinnatifrons
DIP-sis · pin-NAT-i-frahnz
PLATE 408

Dypsis pinnatifrons is endemic to eastern Madagascar, where it occurs in the undergrowth layer of rain forest at elevations from sea level to 3000 feet (900 m). The epithet is a combination of Latin words meaning "pinnate" and "leaf."

The solitary trunk can attain 40 feet (12 m) of height in its native haunts, is never more than 6 inches (15 cm) in diameter, and is as straight as an arrow. All but the oldest parts of the trunk are light to dark green with distinct and widely spaced white encircling rings of leaf base scars. The crownshaft is swollen at its base and light green to nearly white. The leaf crown is elongated, and the leaves are widely spaced along the upper part of the stem. Each leaf is held on a short petiole and has widely spaced, S-shaped or obovate, medium green, tapering and long-tipped leaflets arranged in widely spaced groups, except near the apex of the leaf where they are usually more closely set. The newly unfurled leaves are usually purplish red or pink. The inflorescences grow from the top of the crownshaft and consist of many thin, light yellow branches carrying tiny flowers. The small fruits are brown when mature.

The species is not tolerant of cold and needs a site in partial shade, with abundant, regular moisture and a rich, fast-draining acidic soil. It is one of the most beautiful plants in the genus.

Dypsis plumosa
DIP-sis · ploo-MO-sa
PLATE 409

Dypsis plumosa was described from cultivated plants in Hawaii and California. The original seed found its way from Madagascar into cultivation in the early 1990s, but the species has not yet been rediscovered in Madagascar. The epithet refers to the leaves, which are plumose.

The solitary trunk attains a height of at least 8 feet (2.4 m) but cultivated palms are more likely much taller when mature. The trunk is green at the top changing to brown in the older parts with distinct rings and a diameter of up to 10 inches (30 cm). The crownshaft is covered with whitish to grayish brown tomentum. The crown holds 7–10 leaves to 8½ feet (2.6 m) long including the petiole. About 120 pairs of narrow leaflets are irregularly spaced along the rachis and radiate out at different angles, giving a plumose effect to the leaf. The ends of the leaflets have a droopy, weeping habit. Inflorescences emerge below the leaves, are about 3 feet (90 cm) long, and eventually bear yellowish to brownish fruit measuring about ¾ inch (19 mm) long and about ½ inch (13 mm) wide.

Although this species has not been long in cultivation, it has found its way into many gardens throughout the world. It has a moderate growth rate and tolerates acidic to alkaline soils that drain well. Short-term temperatures to freezing or slightly below do not seem to cause damage and it has done well in tropical climates as well as mild Mediterranean or California climes.

The semblance of this palm to *Syagrus romanzoffiana* has earned it the name of Madagascar queen palm. It has almost a wistful elegance that looks best against a solid background or in odd-numbered groupings.

Dypsis prestoniana
DIP-sis · pres-to'-nee-AH-na
PLATE 410

Dypsis prestoniana is endemic to central eastern and southeastern Madagascar, where it grows in low mountainous rain forest. The epithet honors Paul Preston (1948–), president of McDonald's Restaurants Limited (UK), "which sponsored the four-year Palms of Madagascar fellowship" (Dransfield and Beentje 1995).

The straight, solitary trunk attains a height of 35 feet (11 m) in habitat, with a diameter of slightly more than 1 foot (30 cm). It is light grayish green to dark gray and has distinct dark gray rings of leaf base scars in all but its older parts. There is no true closed crownshaft, but the leaf sheaths form a loose inverted pyramid that is 4 feet (1.2 m) high and green, often with brownish maroon tomentum near the summit of the mass and sometimes covering most of it. The leaf crown is usually less than hemispherical because of the erect leaves, each 7 feet (2.1 m) long, borne on a short petiole, and gracefully arching near the apex. The dark green leaflets are 4 feet (1.2 m) long and grow from the rachis in groups and at different angles and are mostly erect but have arching and pendent apices; the overall effect is that of a partially plumose V-shaped leaf.

This beautiful, robust-appearing palm is still rare in cultivation. All that is currently known is that it seems slow and need lots of water and a tropical or nearly tropical climate.

Dypsis procera
DIP-sis · pro-SER-a
PLATE 411

Dypsis procera is a sparsely clustering, variable species endemic to the low mountainous rain forest of northeastern Madagascar. The epithet is Latin for "tall." The trunks can grow to 15 feet (4.5 m) high and are usually less than 1 inch (2.5 cm) in diameter; because of the long internodes, they are elegantly reminiscent of bamboo stems. The leaves are 4 feet (1.2 m) long and range from entire but apically bifid to several broad leaflets to pinnate in segmentation but are always deep green. This palm is intolerant of cold and drought. It needs a rich, moist, fast-draining soil.

Dypsis psammophila
DIP-sis · sam-mo-FI-la
PLATE 412

Dypsis psammophila is endemic to eastern Madagascar where it grows in white sand coastal forests at near sea level. Its conservation status is critical due to habitat destruction with fewer than 100 plants left in the wild. The epithet is Greek for "sand loving," referring to its habitat. Closely related to *D. lutescens*, this clustering palm is smaller overall, has more slender stems, lacks scales on the leaflet underside, and has an inflorescence branched twice. Its cultural requirements are virtually identical to *D. lutescens* and it should have the same cold sensitivity. Because of its somewhat smaller appearance, it could be a daintier substitute.

Dypsis remotiflora
DIP-sis · ree-mo-ti-FLOR-a
PLATE 413

Dypsis remotiflora is endemic to southeastern Madagascar where it grows in lowland rain forest. Presumed extinct because of forest destruction, it was recently rediscovered and has entered into cultivation. The epithet refers to the flowers being distant from one another on the inflorescence.

The solitary or clustering stems reach a height of 3 feet (90 cm) and are less than ½ inch (13 mm) in diameter. The small crownshaft is pale yellowish green and covered in dark brown scales. Leaves reach 16 inches (41 cm) long and are entirely bifid when the palm is young. As the palm ages, 2–4 leaflets develop on each side of the rachis. The new emergent leaf is a dark red.

Another *Dypsis* species new to cultivation, this diminutive beauty could act almost like a groundcover in a shady locale in the landscape, but perhaps would look best if only 1–3 plants were placed together so its attractiveness would not be lost in the crowd. It would look superlative against a blank background such as mulch so its form and grace could be appreciated.

At this time, its cultural needs are still unknown, but it might be assumed this palm needs a soil with a humusy top layer, medium shade, and moist conditions to look its best. It is doubtful it would have much cold tolerance if any.

Dypsis rivularis
DIP-sis · riv-yoo-LAR-iss
PLATES 414–415

Dypsis rivularis is a rare, solitary-trunked, stilt-rooted species endemic to northwestern Madagascar, where it grows along streams and rivers. The epithet is Latin for "of rivers" and refers to the palm's habitat.

The trunk is usually 6–8 feet (1.8–2.4 m) tall but in habitat may reach 15 feet (4.5 m). The crownshaft is loose and ill defined in younger plants but often distinct and red at its base in older trees, at which age it is usually deep olive green to brownish green. The leaves are 6–8 feet (1.8–2.4 m) long with a gracefully arching rachis. According to John Dransfield (Dransfield and Beentje 1995), they appear "untidy" up close because the medium to dark green, slightly S-shaped, linear-lanceolate, and pendent leaflets tend to grow in irregular groups along the rachis; the leaflets of some leaves are almost regularly arranged. From any distance this "untidy" effect is not noticeable and, especially when planted in groups, the palm has a lovely, graceful appearance. It is another true water lover and is intolerant of alkaline soils. It does not tolerate frost and is adaptable only to zones 10b and 11 and probably marginal in 10a.

Dypsis robusta
DIP-sis · ro-BUS-tah
PLATE 416

Dypsis robusta was described from a cultivated palm in Jeff Marcus' garden in Hawaii. Its origin is most assuredly Madagascar, but it has not been found in habitat. The epithet refers to the robust nature of this species in both size and growth.

The massive solitary trunk has a diameter of nearly 16 inches (41 cm) at eye level, flaring to over 2 feet (60 cm) at ground level. It is gray with pronounced leaf scar rings and presumably will get quite tall, but currently the tallest specimen is just over 30 feet (9 m). The crown holds 10–14 very large dark green leaves supported by leaf sheaths that are covered in white indument and brown scales. The regularly spaced leaflets hang from the rachis to create a ∧-shaped leaf. The leaf tends to twist, giving the appearance that it is plumose although it is not.

Such a mammoth canopy palm needs space away from other vegetation. It demands attention because of its sheer size and is best viewed from afar so its full grandeur can be appreciated.

Dypsis saintelucei
DIP-sis · saynt-LOO-see-eye
PLATE 417

Dypsis saintelucei is a solitary (rarely clustering) species endemic to extreme southeastern Madagascar, where it grows in coastal forest on sandy soil at elevations near sea level. It is rare and critically threatened because of its few numbers and exploitation for the leaves, which are used to make lobster traps. The epithet is a Latinized form of Sainte-Luce, the area in which the species is native.

The trunk grows to 30 feet (9 m) in habitat with a diameter of 6 inches (15 cm) and is usually straight as an arrow in cultivation. It is distinctly ringed with undulate, whitish leaf base scars and is brown or gray except for the youngest parts, which are covered with wax. The outstanding crownshaft is light green, almost white or bluish white in many specimens, and is waxy, 1–2 inches (2.5–5 cm) wider than the trunk, nearly cylindrical, and 2 feet (60 cm) tall. The leaf crown has the form of a shaving brush because of the erect leaves, each 6–7 feet (1.8–2.1 m) long. The leaves are held on short petioles, arch apically, and grow in 3 ranks, but this fact is usually obscured visually because of their density in the crown. The medium to dark green leaflets are stiff and grow from the rachis at an angle to create a V-shaped leaf.

This outstandingly attractive palm, with visual affinities to both *D. cabadae* and *D. madagascariensis*, is reportedly easy of cultivation and seems to do well in a frostless Mediterranean climate. It needs moderate moisture and full sun except for hot climates. It is not tolerant of freezing temperatures.

Dypsis sanctaemariae
DIP-sis · sahnk-tee-MAHR-ee-eye
PLATE 418

Dypsis sanctaemariae is a small, clustering species endemic to the island of Sainte-Marie, off the northeastern coast of Madagascar, in the Indian Ocean. It occurs as an undergrowth subject in windswept coastal forest near sea level, and is rare and critically endangered because of its few numbers and real estate development. The epithet is Latin for "of Sainte-Marie."

The stems grow to 3–4 feet (90–120 cm) high and ½ inch (13 mm) in diameter. There is no crownshaft, but the persistent leaf bases clasp the stem and usually have a reddish patch of color near the beginning of the rachis (there is no true petiole). The leaf crown is extended. The leaves are 4 feet (1.2 m) long, either entire and very deeply bifid or divided into a few wide, stiff segments, and are medium to deep green on both surfaces.

This attractive little species is still rare in cultivation. It seems to be intolerant of cold, needs partial shade and regular, adequate moisture, but is not fussy about soil type as long as it is well drained.

Dypsis scottiana
DIP-sis · skot-tee-AH-nah
PLATE 419

Dypsis scottiana is endemic to the white sand forests of southeastern Madagascar where it grows from near sea level by the coast to 1500 feet (460 m). Its conservation status is considered vulnerable. The epithet honors British botanist George Francis Scott Elliot (1862–1934), who collected in Madagascar.

This clustering species has light gray stems to 12 feet (3.6 m) tall and less than 1 inch (2.5 cm) in diameter with each stem holding 4–7 leaves. The crownshaft is densely covered in reddish scales, which turn brown with age. Leaves can be as long as 2 feet (60 cm) including the short petiole. Shiny green leaflets are arranged in groupings of 2–8 along the rachis. The inflorescence is up to 22 inches (56 cm) long and produces small, red fruit.

A palm that does well in tropical moist conditions such as Hawaii, this species is still being tested elsewhere to find out what climate tolerances it can handle. It is unlikely to be cold hardy below freezing. It should tolerate some dryness and poorer soils, although it should excel in a nutrient-rich medium and consistent watering.

Dypsis simianensis
DIP-sis · sim-ee-ah-NEN-sis
PLATE 420

Dypsis simianensis is a clustering palm endemic to southeastern Madagascar where it grows from lowland rain forest at 200 feet (60 m) to semidry, rocky forest at 800 feet (240 m). It is endangered with perhaps 40 plants left in habitat. The epithet is derived from Simianona River, along which this palm was first collected.

This palm grows in clumps of 2–12 stems that are up to 6 feet (1.8 m) tall and less than ½ inch (13 mm) in diameter. The crown has 5–15 leaves. Leaf sheaths are a pale yellowish-brown and sprinkled with reddish scales. The narrow leaves are entirely bifid reaching a length of 16 inches (41 cm) including the petiole. The new emergent leaf is a deep red changing to a dark green after a few days.

Still another of fairly new understory *Dypsis* to cultivation, this species should become an appealing addition to tropical landscapes. As it is grown under different cultural conditions, more will be learned about its tolerances to soil, shade, sun, and cold.

Dypsis sp. 'Mayotte'
PLATE 421

Dypsis sp. 'Mayotte' is said to originate from Mayotte, a cluster of small islands in the Indian Ocean, between northwestern Madagascar and the African mainland. The material in cultivation has not been critically examined and identified, so the provisional cultivar name 'Mayotte' is still used.

This medium-sized solitary or clustering species has a trunk about 20 feet (6 m) high and 4–5 inches (10–13 cm) in diameter. The elongated, silvery crownshaft is slightly thicker than the trunk. The leaves are 6–8 feet (1.8–2.4 m) long, have short petioles, and bear regularly arranged, evenly spaced, slender, green leaflets that grow from the rachis at a V-shaped angle when young but in nearly the same plane when older. The tip of the rachis is delicately recurved, giving the crown a supremely graceful aspect. This fast-growing species resembles *D. pembana* but the leaflets are not curled. It needs a tropical climate, full sun to partial shade, and a moist, well-drained, rich soil.

Dypsis sp. 'Pink Crownshaft'
PLATES 422–423

Dypsis sp. 'Pink Crownshaft' is undoubtedly endemic to some part of Madagascar. It is just one of several plants awaiting decisions from the taxonomic community on whether they represent new species or merely forms of known species. In the 1980s and 1990s much seed emerged from Madagascar identified only with location or Malagasy names. Once grown, some of these are turning into palms that do not fit the mold of known species. With the destruction of habitats in Madagascar, there is a good possibility some of these unknowns are now extinct in the wild. This particular palm has been seeding for several years now in Australia and Hawaii. Also known as *Neophloga* 'Pink Crownshaft', it has some distinctive features that may well warrant its being named a bona fide species.

This solitary-trunked palm has many similarities to *D. pinnatifrons* and *D. nodifera*. The grouped leaflets are generally wider, giving the leaves a fuller, more plumose look. The new emergent leaf is a deep red that generally stays tinged red until the next emergent leaf unfolds. The crownshaft has a noticeably pink coloration, and the inflorescence is a showy bright pink.

Cultural conditions are identical to *D. pinnatifrons* although it seems slightly more cold hardy and tolerant of soils that are not terribly acidic. It seems capable of tolerating full tropical sun, but its leaves will be a deeper green when placed in partial shade. Like a lady of high society, this is the palm world's answer to dripping extravagance. No matter where placed in the garden, she will draw gawking attention and, when gathered with others of her species, will cause jaws to drop.

Dypsis tokoravina
DIP-sis · to-ko-RAH-vee-na
PLATES 424–425

Dypsis tokoravina is endemic to Maroantsetra and Mananara in northeastern Madagascar where it grows along swamp edges in lowland rain forest at 1300 feet (400 m). It is heavily endangered with perhaps 20 plants left in a habitat threatened by an increase in agriculture. The epithet comes from an aboriginal (Betsimisaraka) word that translates to "group leaf" and refers to the grouped pinnae.

This is a massive solitary species with a trunk to 60 feet (18 m) tall and 2 feet (60 cm) in diameter at its base supporting a crown of 10–14 glossy deep green leaves. The swollen leaf sheath is grayish-brown and barely forms a crownshaft. Leaves are 9 feet (2.7 m) long including the petiole. On each side of the rachis are 80–110 leaflets in irregular groupings that radiate out in 3–8 planes giving a plumose look to the leaf. The inflorescence emerges from amongst the leaves and can be 9 feet (2.7 m) long producing obovoid fruit not quite 1 inch (2.5 cm) long.

Cultural conditions are still being studied for this species to see how it will tolerate various climates. It is growing very well in tropical areas such as Hawaii and Australia, but its cold tolerance still needs to be tested. This large canopy palm looks best out in the open where its dark green foliage can be admired. When planted in odd numbered groupings of 3 or more, it will form a splendid silhouette against the sky.

Dypsis tsaratananensis
DIP-sis · sah'-ra-ta-nah-NEN-sis

Dypsis tsaratananensis is endemic to Mount Tsaratanana in northern Madagascar where it grows at 3000–5000 feet (900–1520 m). It has not been seen in the wild by any botanist in over 75 years and is not in cultivation. Palms with this name have been introduced into gardens are of an unknown species that bear no semblance of the actual species.

Dypsis tsaravoasira
DIP-sis · sah'-ra-vo-a-SEE-ra
PLATE 426

Dypsis tsaravoasira is a large rare, solitary-trunked species endemic to northeastern Madagascar, where it grows in rain forest at elevations up to 3400 feet (1040 m). It is under threat of extinction because of its few remaining individuals and agricultural expansion. The epithet is the palm's aboriginal name.

In habitat, the trunk grows to a height of 80 feet (24 m) and a diameter of 18 inches (45 cm). It is smooth, straight and, in its younger parts, green and distinctly ringed with lighter-colored leaf base scars. The cylindrical crownshaft is 4–5 feet (1.2–1.5 m) tall, a few inches wider than the trunk beneath it, and yellowish green. The leaf crown resembles a giant shuttlecock because of the erect leaves 10 feet (3 m) long on short petioles. The leaves are beautifully arching and recurved in the apical third of their length. The stiff leaflets are 4 feet (1.2 m) long and grow from the rachis at a slight angle from the horizontal.

This great palm has one of the most beautiful silhouettes in the world. Alas, it is almost unknown in cultivation and no data are available on its cultural requirements, although it cannot be immune to frost and doubtless requires a humus-laden soil and copious moisture.

Dypsis utilis
DIP-sis · YOO-ti-lis
PLATE 427

Dypsis utilis is a clustering or solitary-trunked, branching species endemic to eastern Madagascar, where it grows in low mountainous rain forest usually along streams and rivers. The epithet is Latin for "useful" and alludes to the long, tough fibers from the leaf bases, which are used for rope production and were profitably exported in the past.

The stems grow to 50 feet (15 m) tall and 1 foot (30 cm) in diameter. They often branch dichotomously well above the ground but usually only twice per stem, and the branches grow parallel and close together. The trunks are light gray or light brown and exhibit many pendent, thin, brown leaf base fibers near the leaf crowns but form no true crownshafts. The crowns are hemispherical, with many densely packed leaves 10–12 feet (3–3.6 m) long on petioles 2–3 feet (60–90 cm) long. The leaves usually exhibit a twist of the rachis at the midpoint so that the leaflets are aligned vertically from that point to the apex of the leaf. The leaflets are 2–3 feet (60–90 cm) long, linear-lanceolate, deep green, usually pinkish orange when new, and are slightly limp when old.

The species needs constant and copious water and is not hardy to freezing temperatures. It seems to not be fussy about soil type but needs full sun once past the juvenile stage.

Elaeis
e-LAY-is

Elaeis consists of 2 large solitary-trunked, palmate-leaved, monoecious palms, one in Central and South America, the other in Africa. The trunk of the American species creeps along the ground (or sometimes underneath it) for a time when young and then turns upright. The species are not cultivated much as ornamentals, although they are worthy of being so. The genus name is based on the Greek word for the olive tree, an allusion to the valuable oil extracted from the fruits of both species. Seeds can take 2 years or longer to germinate and do so sporadically. For further details of seed germination, see *Acrocomia*.

Elaeis guineensis
e-LAY-is · gin-ee-EN-sis
PLATES 428–429

Elaeis guineensis is native to western and central tropical Africa, usually in cleared or open spaces and along streams and rivers and sometimes in swampy areas. It is known as the African oil palm. The epithet is Latin for "of Guinea," a country in West Africa.

A mature trunk reaches a height of 60 feet (18 m) or sometimes more in habitat but, under cultivation, is often no more than 40 feet (12 m). The diameter of the stem is 2 feet (60 cm), and older parts are covered in rings of knobby leaf base scars, which create a chiseled look, while the younger parts are usually densely covered in tough wedge-shaped leaf bases. The full, rounded leaf crown is 25 feet (7.6 m) wide and 20 feet (6 m) tall. The leaves are 15 feet (4.5 m) long and erect but gently arching near their ends. The stout green petioles are 3–5 feet (90–150 cm) long and armed with long, fibrous spines. The deep green leaflets are 3 feet (90 cm) long, limp and drooping, and grow from the rachis in small clusters at slightly different angles, giving a plumose effect. The inflorescences usually consist of single-sexed flowers, but both male and female flowers occur on the same tree, and a few inflorescences have both sexes in the same cluster. The inflorescences are 1 foot (30 cm) long and wide and grow from among the lower leaves, carrying densely packed small, whitish blossoms. The ovoid fruits are 2 inches (5 cm) long and densely packed into clusters 1 foot (30 cm) wide. They turn black when mature.

A rich, humus-laden soil is best, although this palm can adjust to other soils. It luxuriates in full sun but adapts to partial shade. It is not hardy to cold and is suitable only to zones 10 and 11. It is moderately slow growing but grows faster with ample moisture and a rich soil.

The African oil palm is one of the few palms that looks decent as an isolated specimen, so massive and spectacular is it; it is not for a small space, however. It looks better planted in groups of 3 or more individuals of varying heights. Young palms with little trunk do not show the characteristics of older plants with some trunk; their leaves are stiff and not as graceful. This palm has been called the poor man's Canary Island date palm (*Phoenix canariensis*) as it is significantly cheaper to obtain, grows faster, has the same visual impact, and is not subject to lethal yellowing disease; it is, of course, good only for tropical or nearly tropical regions. This palm is not known to be grown indoors but, given enough space and high light levels, it should be possible to do so.

The oil extracted from the seeds is a valuable commodity in the international market. Vast plantations in Latin America, West Africa, and Southeast Asia are devoted to its cultivation. The oil is used in the manufacture of many products, such as margarine, detergent, and soap, and is used as a fine lubricating agent in industry.

Elaeis oleifera
e-LAY-is · o-lee-I-fe-ra
PLATE 430

Elaeis oleifera occurs naturally in 2 disjunct regions of tropical America: the first is primarily in Central America (Honduras, Nicaragua, Costa Rica, Panama, and northern Colombia), and the second in South America (northern Peru, Ecuador, the northern Amazon region of Brazil, Surinam, and French Guiana). It grows in wet, low-lying areas near streams, rivers, and lakes, and was probably introduced eons ago into Brazil by the native peoples. The epithet is Latin for "oil-bearing." The common name is American oil palm.

A mature trunk is prostrate near the bases and to 20 feet (6 m) long and 1 foot (30 cm) in diameter; it is covered in old leaf bases. The leaves are 12 feet (3.6 m) long, erect but beautifully arching, on spiny petioles 4–5 feet (1.2–1.5 m) long. The deep green, linear leaflets are 2 feet (60 cm) long and grow from the rachis in one plane that would create a flat leaf were they not mostly pendent from the rachis. The inflorescences grow in tight, compact, and rounded clusters from among the leaves and bear either male or female flowers. The egg-shaped fruits are ½ inch (13 mm) long, remain in tight clusters, and are yellow, orange, or red, often all 3 colors.

Because the lower portion of the trunk is prostrate, the entire trunk forms aerial roots, and the base of the stem often dies and rots away while the upper portion is erect, the species can actually move its location to a small extent. It seems to have the potential of being immortal unless destroyed by mechanical factors, because the stem avoids the stresses of great vertical height, which might cause other aerial, solitary-trunked palms to die.

This species is more tender to cold than its African relative and is adaptable only to zones 10b and 11. It needs copious water, as it is nearly aquatic, and a soil that incorporates some humus. It grows in full sun or partial shade but is always slow growing. This unusual palm is beautiful as a part of a wall of vegetation, from which its beautiful leaves with their drooping leaflets stand out in a tropical-looking contrast. Its half-prostrate manner prevents it from being well used in most other situations.

Eleiodoxa
e-lee'-o-DAHX-a

Eleiodoxa is a monotypic genus of clustering, pinnate-leaved, dioecious palm in Thailand, peninsular Malaysia, and eastwards to Borneo and Sumatra. The name is from Greek words meaning "swamp" and "glory," an allusion to the size, beauty, and natural habitat of the species. Seeds germinate within 45 days. For further details of seed germination, see *Calamus*.

Eleiodoxa conferta
e-lee'-o-DAHX-a · kahn-FER-ta

Eleiodoxa conferta occurs naturally in lowland tropical swamp forests, often forming extensive colonies. The epithet is Latin for "congested" and refers to the inflorescences.

This densely clustering species has subterranean stems. The leaves arise directly from the ground and are each 12 feet (3.6 m) long, stiffly ascending, and barely arching, on deep green petioles 10 feet (3 m) long. The leaf sheath near the base of each petiole bears partial rings of spines that are 2–3 inches (5–8 cm) long and black when new but age to almost white. The regularly spaced leaflets are 5 feet (1.5 m) long, linear-lanceolate, with obliquely terminated apices, medium to dark green, and margined with short spines; they grow in an almost flat plane from the rachis. The leaves look overall similar to those of a large *Arenga* species. The inflorescences grow from the underground trunks and are compacted short spikes of male or female flowers. The fruits on female plants are in compact clusters and are small, scaly, and reddish brown. The underground stem which produced the inflorescence dies off after the fruits mature but, since the species is so densely clustering, the dying leaves are soon replaced by those from another underground trunk.

The species must have copious water and a rich acidic soil, and it is intolerant of frost or even cold, being adaptable only to zone 11 and marginal in 10b. The palm is tropical looking and its large leaves make a stunning contrast with other vegetation when it is incorporated into same. Its size and clustering habit lend themselves to the palm's being sited as a specimen plant. The fleshy seed coat of this species is pickled and eaten in Borneo.

Eremospatha
e-ree'-mo-SPAY-tha

Eremospatha is a genus of 10 pinnate-leaved, clustering, hermaphroditic, spiny, climbing palms in wet tropical Africa. Most species are large and high climbing with relatively slender stems. Juvenile leaves are entire except for the bifid apex and are borne on petioles, while the adult leaves are pinnately segmented, lack a petiole, and form cirri at their apices. Spines are confined to the leaf rachis, the margins of the leaflets, and the cirri. The inflorescences grow from the leaf axils, are relatively short, and sparsely branched, and bear bisexual flowers that produce small, oblong, scaly fruits.

None of the species seem to be in cultivation, and the genus is not well known even to most taxonomists. All species are tropical in their temperature requirements and all need copious, regular moisture. The genus name is derived from Greek words meaning "lack of" and "spathe" and alludes to the inconspicuous bracts in the inflorescence. Seeds are not generally available, so germination information is nonexistent.

Eugeissona
yoo-ji-SO-na

Eugeissona includes 6 clustering, spiny, polygamous palms in peninsular Malaysia, Thailand, and Borneo. There are both trunk-forming and trunkless species. The species forming aerial stems usually have basal stilt roots supporting the trunks and are unusual in that the junction of aerial roots and trunks forms a basket in which much leaf litter can collect, forming an individualized compost pile. The trunks, petioles, and rachises bear spines, which, on the petioles, can be large. The large leaves are borne on long petioles. The inflorescences are erect and spikelike, and bear large male and bisexual flowers; the inflorescence stem and bracts turn woody and persistent as the ovoid scaly fruits form. The stems flower terminally and die after the fruits ripen.

Some growers consider trunk-forming species like *E. tristis* untidy because of the masses of stilt roots and leaf litter; others find the large feathery leaves to be gorgeous, and when incorporated into other vegetation, the ungainly trunks are hardly noticeable. They are, however, weedy in habitat, growing in disturbed forest. All the species are tropical and adaptable only to zones 10b and 11. The genus name is from 2 Greek words meaning "good" and "roof," an allusion to the use of the leaves of some species for thatching roofs. Seeds germinate within 60 days and must be sown immediately as long-term viability is nil. For further details of seed germination, see *Calamus*.

Eugeissona insignis
yoo-ji-SO-na · in-SIG-nis
PLATE 431

Eugeissona insignis is endemic to the rain forest in Sarawak on the island of Borneo. The epithet is Latin for "remarkable." It is another trunkless or nearly trunkless species. The great leaves are rigidly ascending and slightly arching when new but later are spreading and gently arching. They are 20 feet (6 m) long. The regularly spaced, thin leaflets are 3 feet (90 cm) long and pendulous, creating one of the most beautiful leaves of any palm.

Eugeissona tristis
yoo-ji-SO-na · TRIS-tis
PLATE 432

Eugeissona tristis occurs naturally in peninsular Malaysia and southern Thailand, where it grows in low, mountainous rain forest or in low, swampy areas but usually on sloping ground. The Latin epithet translates literally as "sad" and, by extension, as "dull" or "colorless," an allusion to the drab brown and purple inflorescences. In habitat, this trunkless or almost trunkless species often forms large and impenetrable thickets of petioles and leaves. The leaves are 20 feet (6 m) long and bear regularly spaced leaflets 3 feet (90 cm) long that grow from the rachis in a nearly flat plane. This species is considered by many to be "untidy" because, in habitat, the palms usually have several dead leaves. The landscaping solution to this dishevelment is to limit the size of the individual clumps and therefore make it much easier to remove the unsightly deceased leaves. The living leaves are as beautifully magnificent as those of any species in the genus.

Euterpe
yoo-TUR-pee

Euterpe comprises 7 solitary-trunked or clustering, pinnate-leaved, monoecious palms in tropical American rain forest. The inflorescences grow from beneath the crownshaft and are sparsely branched, covered in short hairs, and whitish; they bear both male and female blossoms, which produce small, round black fruits. All the species have tall, slender crownshafts that may be green, yellow, or orange, and gorgeous pinnate leaves with pendulous, widely spaced long leaflets. They are without exception intolerant of frost and drought. Few tropical plants are as wonderfully graceful as these palms: their long drooping leaflets are comblike against the sky and create a constant visual effect of movement even in still air.

According to Henderson et al. (1995), the "genus is considered by many people to contain the most beautiful American palms" and yet only 3 of the species are in cultivation. In Florida, one reason for their rarity is that, in the parts of the state with winters warm enough to grow them, the soil is usually not to their liking: they need an acidic or slightly acidic medium.

The genus name honors one of the Greek Muses and translates as "good delight." Many species once known under the name of *Euterpe* are now segregated into the closely related *Prestoea*. Fresh seeds germinate within 45 days.

Euterpe catinga
yoo-TUR-pee · ka-TIN-gah
PLATE 433

Euterpe catinga is a native of the western Amazon region of Brazil, Colombia, Peru, and Venezuela as well as the Guiana Highlands of Brazil, Guyana, and Venezuela extending into the Andes in Ecuador and Peru. Its habitat ranges from poorly drained, acidic, white sand soils below 1000 feet (300 m) to wet upland forest patches of 3300–5500 elevation (1000–1680 m). Uses include the leaves for thatching, stems for construction, and the fruit to make açaí juice. The epithet refers to the caatinga, the deciduous scrub forest of the Río Negro in the Amazon Basin.

The gray stems are solitary or clustering with heights to nearly 50 feet (15 m) by 6 inches (15 cm) in diameter supporting crowns of 6–11 leaves. The crownshaft is orange with a mass of black fibers at the top. Leaves are 7–8 feet (2.1–2.4 m) long, including the petiole, and the rachis is covered in dark scales. Regularly spaced leaflets arranged in a single plane number 38–75 on each side of the rachis. The inflorescence emerges from below the crownshaft and produces a globose ½-inch (13-mm) fruit that is purple-black or brown when mature.

Two varieties are recognized: **E. catinga var. catinga** comes from the lower elevations in the western Amazon to the higher elevations of the Guiana Highlands; while **E. catinga var. roraimae** [ro-RY-mee] comes from the Guiana Highlands and extends into the Andes. Variety *roraimae* is distinguished by having much denser scales on the petiole and rachis as well as leathery leaves.

This palm is not cold hardy and prefers moist, acidic soils to grow well. Nighttime lows should not be higher than the mid-60s Fahrenheit (around 18°C). If the proper cultural conditions can be met, the orange crownshaft and bright green leaves of this beauty will create a focal point particularly where the palm can be gazed upon against the open sky or a contrasting background.

Euterpe edulis
yoo-TUR-pee · ED-yoo-lis
PLATES 434–435

Euterpe edulis occurs naturally in the Atlantic coastal forests of eastern Brazil, southwards to extreme southeastern Paraguay and extreme northeastern Argentina. It is nearly extinct because of the felling of trees for their edible growing points. The epithet is Latin for "edible," a reference to the growing point (heart-of-palm).

The species is mostly a solitary-trunked palm but is rarely found clustering. Mature trunks grow to 30 feet (9 m) high and 6 inches (15 cm) in diameter. They are light gray to light brown, graced with widely set darker rings of leaf base scars. The crownshafts are 3 feet (90 cm) tall and 6 inches (15 cm) in diameter, deep green but often showing a suffusion of orange or red. The leaf crown is 15 feet (4.5 m) wide and 10 feet (3 m) tall. The light to medium green leaves are 10 feet (3 m) long with unarching stiff rachises and drooping narrowly lanceolate, widely spaced and tapering leaflets 2–3 feet (60–90 cm) long on short petioles 1 foot (30 cm) long. Few if any of these elegant leaves descend below the horizontal plane. The inflorescences consist of many thin, furry branches, growing from beneath the crownshaft, bearing tiny whitish flowers. The unopened inflorescence is reportedly edible. The round edible fruits are ½ inch (13 mm) in diameter and dark purple to black when mature.

The species does not tolerate cold and is adaptable only to zones 10b and 11, although a few specimens are found in protected sites in 10a. It needs constant moisture and a rich acidic soil. It thrives in partial shade to full sun except in tropical desert regions where it needs lots of irrigation and protection from the hottest midday sun. The palm is moderately fast growing under optimal conditions.

This palm looks like a delicate and straight-trunked coconut. It is extraordinarily beautiful planted in groups of 3 or more individuals of varying heights. It looks good as a solitary individual only when used as a canopy-scape. It is eminently adaptable to growing under glass if provided with ample moisture, humidity, light, and space.

Euterpe oleracea
yoo-TUR-pee · o-ler-RAY-see-a
PLATES 436–437

Euterpe oleracea is indigenous to 2 disjunct regions of northern South America, one being western and northern central Colombia and extreme northern Ecuador, and the other being coastal areas of northeastern Venezuela, the island of Trinidad, the Guianas, and northeastern Brazil. It grows in low, wet areas along rivers and streams. The palm is usually called the assaí [a-sah-EE] palm in English-speaking countries, from the Portuguese name (*açaí*) for a popular drink made from the fruits in the Amazon region. The epithet is a Latin word that means "pertaining to a vegetable garden" and alludes to the palm's edible terminal bud (heart-of-palm).

The species is clustering but often grows only a solitary trunk until that stem (almost) reaches its total height; then it produces suckers. The palm produces pneumatophores under swampy conditions in its habitat. Mature clumps may be 50 feet (15 m) wide and 90 feet (27 m) tall, while mature trunks attain a height of 80 feet (24 m) in habitat but are usually no more than 60 feet (18 m) under cultivation. They are never more than 6 inches (15 cm) in diameter and are green on their younger parts and light gray to light brown on the older parts. The crownshaft is 3–4 feet (90–120 cm) tall, smooth, and slender, and almost 6 inches (15 cm) in diameter throughout. Its color is usually a light to medium green, but some plants have crownshafts tinged with yellow, brown, or even purple. Leaf crowns are 15 feet (4.5 m) wide and 10 feet (3 m) tall. The leaves are 8–12 feet (2.4–3.6 m) long, with straight, stiff rachises along which grow the narrow, widely spaced light to dark green pendulous leaflets, each 3 feet (90 cm) long. The leaves seldom descend beneath the horizontal and, like all the other species of *Euterpe*, fall cleanly from the trunk after dying and leave a neat crown. The inflorescences are 3 feet (90 cm) long, growing from beneath the crownshaft, and are many branched; the branches are furry and bear tiny white flowers. The rounded 1-inch (2.5 cm) fruits are dark purple when mature.

This species has the same cultural requirements as *E. edulis* but needs even more moisture. There is no more beautiful and spectacular palm. It combines grace and nobility like few other things in nature do, and mature clumps are beautiful enough to stand isolated but look better as canopy-scapes. This palm is eminently adaptable to cultivation under glass if enough space, humidity, moisture, and light can be provided.

Euterpe precatoria
yoo-TUR-pee · pre-ka-TOR-ee-a
PLATES 438–440

Euterpe precatoria has the widest distribution in the genus: from eastern Belize, eastern Guatemala, all the rest of Central America, into Colombia, eastern Ecuador, eastern Peru, northern Bolivia, north central Brazil, and the eastern Guianas. In the mountain regions of its range it occurs to elevations of more than 5300 feet (1620 m), in wet forests, while in the lower areas of its range it grows along rivers, in swamps, and along lakes, often in areas that are seasonally inundated. The epithet is derived from the Latin word for "to pray," a reference to the fact that Jesuit missionaries once used the seeds of this palm to manufacture rosaries. A common name seems to be mountain cabbage palm.

In nature, the palm is usually found as a solitary-trunked specimen, only rarely as a clustering individual. The mature trunks are less than 1 foot (30 cm) in diameter but can attain heights of 60 feet (18 m). They are gray with widely, regularly spaced dark rings and a cone of stilt roots at the base. The crownshaft is 6 feet (1.8 m) tall, bright green, smooth, and barely wider in diameter than the trunk. The leaf crown is sparse and open with erect leaves whose rachises arch and droop at their apices. The leaves can be 15–20 feet (4.5–6 m) long and are carried on relatively short petioles. The pendulous, medium green leaflets are numerous, regularly spaced, linear and tapering, and as long as 3 feet (90 cm). The inflorescences form a large, encircling skirt beneath the crownshaft and consist of many branches carrying bright yellow blossoms. The round 1-inch (2.5 cm) fruits are purplish black when mature.

A naturally occurring variety, **E. precatoria var. longivaginata** [lon'-jee-va-ji-NAH-ta], occurs in elevated areas; it has leaves with nearly flat and nonpendulous leaflets and generally shorter trunks. The word *longivaginata* is Latin for "long-sheathed." The type, *E. precatoria* var. *precatoria*, hails from lower elevations and has taller and always solitary trunks and larger inflorescences.

The species needs constant, abundant moisture, a humus-rich, acidic soil, and partial to full sun. It is not cold tolerant and succeeds only in zones 10 and 11 but can adapt to cool but frostless Mediterranean climates if enough moisture is provided. There is no more beautiful species of palm. It has every virtue of the family except for bright coloration. It is fetching enough to be planted in almost any site, including as an isolated specimen, but looks its best when used as a canopy-scape or mixed in a wall of contrasting vegetation. The palm is adaptable to large conservatories or atriums if enough light, space, and moisture can be provided.

Gaussia
GOW-see-a

Gaussia is composed of 5 medium-sized, solitary-trunked, pinnate-leaved, monoecious palms in the West Indies and Central America. All have trunks with swollen bases, tapering tops, sparse leaf crowns, and cones of roots at their bases; they form indistinct, loose crownshafts. The inflorescences grow from the leaf sheaths and consist of a stiff, elongated peduncle at the end of which radiate the much-branched flowering branches bearing both male and female blossoms. The fruits are red when ripe.

These palms would probably be more popular if they did not retain their dead leaf rachises long after the leaflets have fallen. The species are mostly found on mogotes (limestone hills). They are intolerant of real cold and are adaptable only to zones 10 and 11, although some specimens are found in protected areas of 9b.

The genus name honors the German mathematician and physicist Karl Friedrich Gauss (1777–1855). Seeds can take 90 days to a year to germinate and then germinate rather quickly. They can be stored for a few months but are best planted soon after harvest. For further details of seed germination, see *Hyophorbe*.

Gaussia attenuata
GOW-see-a · at-ten′-yoo-AH-ta
PLATE 441

Gaussia attenuata is endemic in western Puerto Rico and the Dominican Republic on steep hillsides of the monsoonal limestone plains. It is rare and in danger of extinction because of its few numbers. The epithet is Latin for "attenuated," a reference to the narrow, tapered apex of a mature trunk.

With time, the trunk can attain a height of 50 feet (15 m) and a diameter of 1 foot (30 cm). It tapers dramatically near the light green crownshaft so that, in comparison with the rest of the trunk, sometimes seems almost grotesquely thin. The few leaves are 6 feet (1.8 m) long on petioles 8 inches (20 cm) long. The deep green, wide lanceolate leaflets are 2 feet (60 cm) long and grow at an angle from the rachis, giving a V-shaped leaf.

This palm is drought tolerant and slow growing. It needs full sun and a free-draining alkaline soil. The species is slightly susceptible to lethal yellowing disease.

Gaussia gomez-pompae
GOW-see-a · go-mez-POM-pee
PLATE 442

Gaussia gomez-pompae is endemic to Mexico in the states of Oaxaca, Tabasco, and Veracruz. It grows on steep limestone slopes and outcroppings and is considered vulnerable due to habitat destruction. The epithet honors Mexican botanist Arturo Gomez-Pompa (1934–).

The smooth light gray trunk is up to 40 feet (12 m) tall with a diameter of 12 inches (30 cm) and supports a crown of up to 10 leaves. The arching dark green leaves are 8–10 feet (2.4–3 m) long with regular spaced leaflets. The inflorescence emerges below the leaves and produces a globose orangish-red fruit about ½ inch (13 mm) across.

This palm is ideally suited to alkaline well-drained soil and full sun. It tolerates drought well but cannot stand being kept too wet. Very short periods of freezing temperatures down to 30°F (−1°C) do not seem to adversely affect it. By itself, the palm has a graceful appeal, but the ideal placement for this palm would be on a raised bed as a grouping of 3 or more of staggered heights.

Gaussia maya
GOW-see-a · MAH-ya
PLATES 443–444

Gaussia maya occurs in scrublands on rocky, limestone outcrops and hills in Belize, the adjacent Mexican state of Quintana Roo, and in the Petén region of Guatemala. The epithet refers to the ancient Maya people who lived in the palm's habitat. The common name is Maya palm.

Mature trunks attain a height of 60 feet (18 m) and are enlarged at their bases, the rest of the stem columnar and of unvarying diameter, light gray or light brown, with widely spaced, whitish rings of leaf base scars. The trunks are never more than 1 foot (30 cm) in diameter above their bases, and usually less, which characteristic often sees them leaning gracefully because of the weight of the leaf crown. The crownshaft is smooth and deep green. The leaves are erect from the trunk but gracefully arching, 8–10 feet (2.4–3 m) long on short petioles. The narrow light to deep green leaflets are 18 inches (45 cm) long and grow in groups at different angles from the rachis to give the leaf a plumose effect. The species has an interesting method of flowering: the inflorescences arise from the leaf base scars on the trunk and stop developing until several others form above them, at which point the first (lowest) inflorescence commences with flowering. This process results in 12 or more branched inflorescences on the trunk at one time, spaced spirally around the upper part of the stem, and in various stages of flowering and fruiting.

The species is possibly more tolerant of cold than others in the genus and has been successfully used in Orlando, Florida (David Witt, pers. comm.). It also likes more moisture than the other species, and the soil need not be as alkaline. This is one of the most beautiful species in the genus and certainly the most "typical" looking. It is excellent as a canopy-scape and equally wonderful as a specimen grouping of 3 or more individuals of varying heights. It is probably the only *Gaussia* species adaptable to indoor cultivation; it needs lots of light and good air circulation.

Gaussia princeps
GOW-see-a · PRIN-seps
PLATES 445–446

Gaussia princeps is endemic to extreme western Cuba, where it grows in monsoonal forest areas on limestone hills. The epithet is Latin for "prince," an allusion to the lofty habit. The mature trunks attain a height of 30 feet (9 m); they are light gray or even white, with greatly expanded bases, the remainder of the stem tapering to the crownshaft. The palm is similar to *G. attenuata* but twice as tall and faster growing. It is extremely nice as a canopy-scape.

Gaussia spirituana
GOW-see-a · spi′-ri-too-AH-na
PLATE 447

Gaussia spirituana is endemic to central Cuba where it grows on limestone cliffs in monsoonal savannas. The Latin epithet is based on the name of one area in the palm's habitat, Sancti Spiritus. The palm is somewhat grotesque appearing because of its greatly swollen trunk that tapers in maturity to a slim crownshaft. Its maximum height is 18 feet (5.4 m), and it grows slowly.

Geonoma
jee-o-NO-ma

Geonoma includes about 50 mostly small, solitary-trunked or clustering, pinnate-leaved, monoecious palms in tropical America;

however, a recent reexamination of the genus recognized 140 species and subspecies (Henderson 2011). Most are found in the undergrowth of wet regions from sea level to 10,000 feet (3000 m). A few are tall and rise above the canopy. Almost all species have extremely hard and durable trunks that are used for construction by local people.

Some of the larger species have loose or rudimentary crownshafts similar in many respects to *Chamaedorea* and *Calyptrogyne*. As in the genus *Chamaedorea*, some species exhibit a greater variation of leaf detail among their individual forms than between 2 species in the genus. Within a given species, leaves may be undivided or divided and leaflet form may vary greatly. Many species have colorful newly emergent leaves from light pink to bright red to deep maroon. Most species do not form a true crownshaft. Furthermore, their similarities to species of the other 2 genera have led to some erroneous labeling in the nursery trade. The inflorescences are as variable as the leaves and may be much branched or simple spikes; they may grow from among the leaves or beneath the leaf crown and crownshaft. They bear flowers of both sexes, and the female blossoms produce fruits that are blue or black when mature.

Almost all these species are at least as beautiful as those in *Chamaedorea* but are not nearly so well represented in cultivation; the larger and often more beautiful species of *Geonoma* are especially rare in gardens, most likely because of their remote habitats and slow growth habit. Only species from high elevations can withstand frost, but several others perform well in a cool, frostless Mediterranean climate. Their most important requirement is constant moisture and, for most, constantly high humidity. Few are adaptable to limey soils, and all need one that is fast draining. The genus name is from the Greek for "colonist" and refers to the habit of many species, which often grow in large colonies. Seeds germinate within 120 days. For further details of seed germination, see *Asterogyne*.

Geonoma congesta
jee-o-NO-ma · kahn-JES-ta
PLATE 448

Geonoma congesta is a mostly clustering undergrowth species indigenous to northeastern Honduras, eastern coastal Nicaragua, Costa Rica, and Panama. It is found in low mountainous rain forest. The epithet is Latin for "congested" and alludes to the dense arrangement of flowers on the inflorescence, which is 2 feet (60 cm) long, including the peduncle.

The stems grow to 20 feet (6 m) high but are 1 inch (2.5 cm) in diameter. They are green in their younger parts but light brown in their older parts. In habitat, they are sometimes reclining on the ground and rooting at the leaf base scars. The leaf crown is full but not usually rounded as most leaves are erect or horizontally spreading. Each one is 5 feet (1.5 m) long in mature individuals and nearly that size in juveniles, borne on a wiry petiole that is 1 foot (30 cm) long and has sharp margins. They are occasionally found entire with deeply bifid apices, but mostly are segmented into 6–10 pinnae of varying widths, each deeply grooved and apically tapering.

Because of its size, this is one of the most attractive species in the genus. It is neither fast growing nor hardy to cold, being adaptable to zones 10b and 11, but is lovely at all stages of growth.

Geonoma cuneata
jee-o-NO-ma · kyoo-nee-AH-ta

Geonoma cuneata is indigenous to extreme southern Nicaragua, Costa Rica, Panama, central western Colombia, and western Ecuador, where it occurs in the undergrowth of low mountainous rain forest to 3800 feet (1160 m). The epithet is Latin for "cuneate," an allusion to the wedge-shaped leaf.

The palm is found mostly solitary trunked but occasionally as a clustering specimen. The stem is often short or even subterranean but sometimes is aerial and as tall as 5–6 feet (1.5–1.8 m). The leaf crown is full and nearly rounded, with glossy leaves that are 3–4 feet (90–120 cm) long, obovate, and either entire and deeply bifid apically, or segmented into 6–14 broad, slightly falcate, acuminate and deeply grooved pinnae; the leaflets of the terminal pair are usually much wider than the rest.

This species is among the more variable species in the genus. The most recent taxonomic treatment (Henderson 2011) recognized 9 subspecies, which are distinguished on technical details of the leaf and inflorescence.

This beautiful little palm is suitable for underplanting in partial shade. It is adaptable to zones 10 and 11 and seems to grow well in a frostless Mediterranean climate if given enough moisture.

Geonoma densa
jee-o-NO-ma · DEN-sa
PLATE 449

Geonoma densa is a solitary-trunked, sometimes suckering species with a disjunct distribution in northern South America: in northwestern Venezuela; in northwestern and in north central Colombia, on both sides of the Andes; in extreme southern Ecuador and adjacent northwestern Peru; and in extreme southeastern Peru and adjacent west central Bolivia. In all cases, it is an undergrowth subject in mountainous rain forest or cloud forests at 4800–8000 feet (1460–2440 m). The epithet is Latin for "dense" and alludes to the short, almost congested, yellowish green branches. The newest monograph of *Geonoma* (Henderson 2011) treated this species as a synonym of *G. undata* subsp. *undata*.

The trunk grows to a height of 20 feet (6 m) in habitat, is 1–3 inches (2.5–8 cm) in diameter, and exhibits beautiful undulating rings of leaf base scars. While no true crownshaft is formed, the leaf sheaths are persistent and the inflorescences are borne at the nodes beneath them. The leaf crown is mostly hemispherical, the leaves 4–5 feet (1.2–1.5 m) long on hard, slender petioles 8–12 inches (20–30 cm) long. The many regularly, widely spaced, lanceolate to ovate-acuminate leaflets are glossy dark green, leathery, and pleated; new growth is usually bronzy pink.

This beauty is not completely adaptable to hot tropical regions like South Florida, although it usually survives there. It readily grows in frostless or nearly frostless Mediterranean climes if provided with regular, copious moisture.

Geonoma deversa
jee-o-NO-ma · dee-VER-sa
PLATE 450

Geonoma deversa has one of the most widespread distributions in the genus, from southern Belize through northern South America, where it grows mostly in low mountainous rain forest, but occasionally is found up to 3800 feet (1160 m). The epithet, from the Latin meaning "swept or turned away," alludes to the S-shaped leaflets. Henderson (2011) recognized 4 subspecies that differ in technical characteristics that make little difference to growers.

The stems of this clustering species attain heights of 12 feet (3.6 m) in habitat but are usually less than 1 inch (2.5 cm) in diameter. The leaf crown is extended along the upper part of the stem, and the leaves generally spreading and arching. As is to be expected, the leaf form of such a widespread species is variable but, in general, leaves are 2 feet (60 cm) long. They may be undivided except for the bifid apex but are usually segmented into 4–8 fat, wide, S-shaped, widely spaced, and long-tipped pinnae; the new growth is deep rosy bronze. The species does not tolerate frost but is otherwise easy to grow if given regular, generous moisture, partial shade, and a humus-laden soil.

Geonoma epetiolata
jee-o-NO-ma · ee-pee′-tee-o-LAH-ta
PLATE 451

Geonoma epetiolata is indigenous to the Caribbean coasts of Costa Rica and Panama, where it is an undergrowth subject in low mountainous rain forest. The epithet is from 2 Latin words meaning "without a petiole." The stem of this small, solitary-trunked species grows to 5 feet (1.5 m) in habitat. The leaves are undivided with a bifid apex, are held on short or nonexistent petioles, and are erect in the crown. Each leaf is to 2 feet (60 cm) long and obovate, the apical points S shaped. New growth is rosy colored and mottled yellow-green, and the undersides of the leaves are red. This palm is one of the most spectacularly colorful palms in the Americas. Everyone who sees it wants to grow it, but few can. It is one of the most horticulturally demanding palms, intolerant of frost, low humidity, hard water, and drying winds, among other things.

Geonoma ferruginea
jee-o-NO-ma · fer-roo-JI-nee-ah
PLATE 452

Geonoma ferruginea is an understory palm found on Nicaragua and Costa Rica in pre-montane to montane rain forests from 1300 to 4000 feet (400–1200 m). The epithet is from Latin meaning "rust colored," which probably refers to the color of the stems.

The light brown stems of this clustering palm can attain a height of 12 feet (3.6 m) with a diameter of 1 inch (2.5 cm), but are often much shorter and at times form rather large clumps. Leaves can be 30 inches (76 cm) long with usually 3 leaflets and sometimes as many as 15 on either side of the rachis. Leaflets are sickle shaped and regularly spaced. The inflorescence emerges from below the leaves. The small, globose black fruits are only ⅓ inch (8 mm) in diameter.

Being from a higher elevation, this palm does not adapt well to Florida's hot summers. It prefers nighttime lows in the mid to low 60s Fahrenheit (around 16–18°C) along with ample moisture and a soil with a humusy top layer.

Geonoma interrupta
jee-o-NO-ma · in-ter-RUP-ta
PLATE 453

Geonoma interrupta is the most widespread species in the genus, occurring from southeastern Mexico into northern South America, as well as in the Caribbean Basin in Haiti and the Lesser Antilles, although some populations, including the Antillean ones, were recently separated as *G. pinnatifrons* [pin-NA-ti-frahnz] by Henderson (2011). *Geonoma interrupta* grows mostly in lowland rain forest and swamps but is also found to 4500 feet (1370 m). Vegetatively, it is among the most variable species. The Latin epithet refers to the pinnately divided ("interrupted") form of the leaf.

This species is found both as a solitary-trunked and as a clustering palm in habitat and is one of the largest in the genus, with stems to 30 feet (9 m) high and 4 inches (10 cm) in diameter. The leaves are never undivided in specimens past the seedling stage, but the variation in leaf form between young plants and (nearly) mature individuals is large, the younger ones having as few as 2 wide S-shaped leaflets. Older plants usually bear leaves with 12–40 differently sized leaflets, all glossy deep green and generally S shaped with long tips and distinct grooves corresponding to the leaflet veins.

This palm is certainly one of the most beautiful in the genus, and a large clump is a veritable symphony of leaf form from the ground upwards. It does well in frostless Mediterranean climes if given enough irrigation.

Geonoma longivaginata
jee-o-NO-ma · lon′-jee-va-ji-NAH-ta
PLATE 454

Geonoma longivaginata is an undergrowth, sparsely clustering species in low mountainous rain forest of southern Nicaragua, Costa Rica, and Panama. The epithet is Latin for "long sheathed." The stems grow to heights of 12–15 feet (3.6–4.5 m) and are usually less than 1 inch (2.5 cm) in diameter. The leaf crown is sparse and hemispherical with leaves 3 feet (90 cm) long that bear glossy deep green leaflets when mature but a beautiful coppery or wine color when new. They are regularly spaced, 8 inches (20 cm) long, and slightly S shaped; the leaflets of the terminal pair are much wider than the others.

Geonoma schottiana
jee-o-NO-ma · shot-tee-AH-na
PLATE 455

Geonoma schottiana is endemic to eastern Brazil, where it grows in coastal forests as well as dense second-growth forests at 1200–5000 feet (370–1520 m). The epithet honors German botanist Heinrich Wilhelm Schott (1794–1865).

This solitary or clustering species has stems to 12 feet (3.6 m) tall and 1½ inches (4 cm) in diameter. The hemispherical crown holds 7–12 leaves, each 3 feet (90 cm) long with numerous narrow leaflets irregularly spaced along the rachis. Each leaflet has 3 conspicuous veins on the upper surface. The inflorescence emerges from below or among the lowest leaves and produces a globose black fruit that is almost ½ inch (13 mm) across.

Though not adaptable to South Florida's hot summers, this is a delightful addition to frost free Mediterranean climates given ample water, a shady environment, and a soil with a humusy top layer. A fine specimen by itself, it excels in a grouping of 3 or more.

Geonoma undata
jee-o-NO-ma · un-DHA-ta
PLATE 456

Geonoma undata is a large solitary-trunked species with the second widest distribution in the genus, from Mexico into northern South America and on the islands of Dominica, Guadeloupe, and Martinique in the Lesser Antilles, always growing in montane rain forest at 4500–7700 feet (1370–2350 m). The epithet is Latin for "wavy," a reference to the undulate folds of the leaflets.

The trunk attains a height of 30 feet (9 m) and a diameter of 4 inches (10 cm) in habitat. The loose reddish brown crownshaft bulges near its midpoint and is cylindrical and 2 feet (60 cm) high. The leaf crown is full and slightly more than hemispherical with leaves 6–8 feet (1.8–2.4 m) long that bear many sickle-shaped, deeply corrugated leaflets that are 2 feet (60 cm) long, usually of differing widths and may or may not grow from the rachis at different angles but almost always have pendent tips. The large inflorescences grow from beneath the crownshaft.

This beautiful species is still rare in cultivation, and few data are known about its requirements. It would probably be a good candidate for frostless or nearly frostless Mediterranean climates.

Guihaia
gwee-HY-ya

Guihaia consists of 2 small, clustering, palmate-leaved, dioecious palms in southern China and northern Vietnam, where they grow in crevices of steep limestone hills and cliffs. The inflorescences of both sexes grow from the leaf crown and are erect and much branched, the females producing small, ellipsoid bluish black fruits. The genus name is the ancient name of one of the Chinese provinces to which the palms are native. Seeds of *Guihaia* germinate within 120 days and should not be allowed to dry out before sowing. Seeds are very small and dry out easily. For further details of seed germination, see *Corypha*.

Guihaia argyrata
gwee-HY-ya · ahr-jee-RAH-ta
PLATE 457

Guihaia argyrata occurs naturally on limestone cliffs in extreme southern China and northern Vietnam. The epithet is Greek for "silvery." Grant Stephenson, a palm broker in Houston, Texas, has coined the apt vernacular names Vietnamese silver-backed fan palm and Chinese needle palm.

Mature trunks grow slowly to 3 feet (90 cm). They are covered in brown fibers, and those near the top of the stem are spinelike, reminiscent of the crown of *Rhapidophyllum*. Mature clumps probably do not exceed 6 feet (1.8 m) high and wide. The usually circular 3-foot (90-cm) leaves have deep divisions extending almost to the petiole, which is 3–4 feet (90-120 cm) long. A small, dark hump-shaped hastula sits atop the center of the leaf at the juncture of petiole and blade. The narrow segments are light to deep green above and are shallowly once pleated, the fold being reduplicate rather than induplicate, which is the usual condition for fan palms. There is a distinct and lighter-colored vein atop the pleat of the upper surface, and the undersides of the leaf are a beautiful and shiny light tan overlain with a bright and tangible white to silvery sheen, the result of small, silvery scales. The fruits are ¼ inch (6 mm) wide.

The species is probably hardy in zones 8b through 11, possibly even hardier. It has survived unscathed a temperature of 22°F (−6°C) in containers in Houston, Texas (Grant Stephenson, pers. comm.). It is drought tolerant but grows better with regular, adequate moisture. It thrives on calcareous soils but seems to like various well-drained soils. It luxuriates in partial shade to full sun. Plants at Fairchild Tropical Botanic Garden are smaller, have fewer leaf segments, and have more pleats per segment when grown in sun than those grown in partial shade (Dransfield and Zona 1998). The plants are slow growing no matter what the conditions.

This palm is suited to sites where its color and form can be readily appreciated. It works wonderfully as an accent with other contrasting vegetation but is not suited to specimen planting, isolated with surrounding space. It is not known to be grown indoors but probably could be if given enough light and good air circulation.

Guihaia grossifibrosa
gwee-HY-ya · gros'-si-fy-BRO-sa
PLATE 458

Guihaia grossifibrosa is indigenous to the same general region as *G. argyrata*. The epithet is from Latin words meaning "fat" and "fibrous," an allusion to the thick fibers of the leaf bases. It differs from the above species in being slightly taller, lacking "needles" in the crown but having more leaf base fibers of greater length that are adherent to the petioles, and in having leaves with thinner segments that are free all the way to the petiole and are of only a slightly paler color on the lower surfaces and are not silvery hued.

Hedyscepe
he-di-SEE-pee

Hedyscepe is a monotypic genus of solitary-trunked, pinnate-leaved, monoecious palm on Lord Howe Island. The genus name is derived from Greek words meaning "pleasant" and "shade," although one is hard put to understand how the canopy would ever provide much shade or how that amenity might be needed in such a cool habitat. Seeds generally germinate within 90 days but can take longer. Seeds allowed to dry out completely before sowing do not germinate as well as those kept slightly moist. For further details of seed germination, see *Areca*.

Hedyscepe canterburyana
he-di-O-pee · kant'-er-bur-ee-AH-na
PLATE 459

Hedyscepe canterburyana grows in mountainous moist forest from 1000 to 2400 feet (300–730 m) in elevation. It occurs mostly on cliffs and ridges near the coast where the rainfall is high and there is often cloud cover. In its habitat it is called umbrella palm or big mountain palm. The epithet honors Sir John Manners-Sutton (1814–1877), third Viscount of Canterbury, governor of the state of Victoria in Australia.

The robust and stout light brown to dark gray trunk attains a height of 35 feet (11 m) in habitat, a diameter of 1 or more feet (30 cm), and has closely set darker brown rings of leaf base scars. The crownshaft is 2 feet (60 cm) tall, slightly thicker than the trunk, cylindrical, and light green to bluish gray-green. The leaf crown is hemispherical or slightly less, with erect, slightly arching leaves, each 8–9 feet (2.4–2.7 m) long on a short petiole. The leaflets are 1 foot (30 cm) long, lanceolate, stiff, and rigid, and grow from the rachis at an angle that creates a V-shaped leaf. They are deep green above but paler green beneath. The inflorescences grow from nodes beneath the crownshaft and consist of rigid, fleshy spreading branches bearing male and female flowers. The fruits are obovoid or oblong, 2 inches (5 cm) long, and dark red.

This palm requires a moist Mediterranean or cool tropical climate, such as the mountains of Hawaii, northern New Zealand, and southern coastal Australia. It is not adapted to hot climates, where it only languishes and slowly dies, nor does it stand much cold. It also needs a humus-laden, moist but-fast draining soil. Even under optimum conditions it is slow growing.

This species is related genetically and visually to the genus *Rhopalostylis*; it is architectural and formal appearing and lends itself to sites where the strong form or silhouette can be seen against the sky or contrasting foliage or structures. It looks good as a specimen planting, even surrounded by space in certain formal sites, but is better in groups of 3 or more individuals of varying heights.

Hemithrinax
he-mi-THRY-nax

Hemithrinax consists of 3 solitary-trunked, palmate-leaved, hermaphroditic palms endemic to Cuba. This genus, first described in 1883, was included within *Thrinax* for many decades, until DNA studies showed that the species were deserving of their own genus once more (Lewis and Zona 2008). Morphologically, *Thrinax*, *Leucothrinax*, and *Hemithrinax* are quite similar with one visual difference being conspicuous leaf cross veins in the first 2 genera and inconspicuous veins in the latter. The genus name compounds the Greek suffix for "half" with *Thrinax*, a related genus. It was, perhaps, an allusion to the mistaken notion that *Hemithrinax* has half as many stamens as *Thrinax*.

The species form relatively small leaf crowns of circular leaves. The leaf bases are split where attached to the trunk. The inflorescences grow from the leaf crown and bear white, bisexual flowers. The globular fruits are bright white. Each species is confined to its own unique, small habitat. All species are extremely slow growing and do not adapt well to soils other than what exists in their native haunts. None are particularly hardy to cold, being adaptable to zones 10 and 11 but marginal in 9b.

Fresh seeds germinate within 90 days. Seeds that are allowed to dry out and then rehydrated may take 9 months or longer to germinate. For further details of seed germination, see *Corypha*.

Hemithrinax ekmaniana
he-mi-THRY-nax · ek-MAH-nee-ah-nah
PLATES 460–461

Hemithrinax ekmaniana is endemic to 3 small mogotes (limestone hills) near the coast of northeastern Cuba, where it is an extremely rare and endangered species. The epithet honors Swedish plant explorer and botanist Erik Ekman (1883–1931). This species is one of the world's most unusual palms and looks more like a large yucca than a palm.

The plant never attains a height greater than 15 feet (4.5 m), even in habitat, and its slender trunk, 2 inches (5 cm) in diameter, is hidden in the youngest parts (just beneath the living leaf crown) by a dense, rounded shag of dead leaves, which gives to older individuals the appearance of a big lollipop. The leaves are held on petioles 4 inches (10 cm) long and grow so close together in the crown that it is impossible to discern the outline of any individual leaf without extracting it from the palm's crown; in truth each one is three-quarters of a circle. The segments are rigid, taper to a stiff, sharp point and are medium to deep green above but silvery green to almost silvery blue beneath. The inflorescences are long enough to exceed the foliage crown and are white with yellowish white, bisexual flowers that produce round ½-inch (13-mm) white fruits.

This extremely slow growing species only entered cultivation outside Cuba in the mid-1990s when a few seeds made their way out and were dispersed to several botanical gardens throughout the world. A particularly nice grouping is now planted at Montgomery Botanical Center in Miami, Florida, and although still small, these palms have begun to fruit. In 2006, more seed reached Florida and other parts of the world, and now this palm's widespread cultivation is assured. It will become a choice landscape candidate for gardens with well-drained alkaline soil. While a darling specimen on its own, an odd-numbered grouping is a simply captivating sight to behold. It would not look out of place growing with cacti and succulents.

Hemithrinax rivularis
he-mi-THRY-nax · riv-yoo-LAR-iss
PLATE 462

Hemithrinax rivularis is endemic to a small area of northeastern Cuba, where it grows along the coast in salt marshes and along the streams of sparsely wooded low savannas in mostly serpentine soils. The marble-sized fruits are harvested and fed to pigs. The epithet is Latin for "of the river."

This small species grows to a maximum height of 20–25 feet (6–7.6 m) in habitat, with a trunk only 3 inches (8 cm) in diameter. There is often a small area of shag or pendent dead leaves beneath the small, globular and dense leaf crown; otherwise, the trunks are smooth and almost white. The leaves are 3 feet (90 cm) wide, deep green above and a lighter, dusky green beneath. The leaf segments are stiff, nonpendent at their apices, and taper to a sharp point. The upper surface of the leaves is very glossy; no other palm has shinier leaves.

This palm looks much like a small *Copernicia* species. It is extremely rare in cultivation and is reportedly not easy to grow because of its adaptation to the serpentine soil of its habitat. A few of these slow growers are beginning to make their way into collections. We hope to learn more about its horticultural requirements as this species is more widely grown.

Heterospathe
het'-er-o-SPAY-thee

Heterospathe comprises about 40 small to large, solitary-trunked or clustering, pinnate-leaved, monoecious palms in the Philippines, Micronesia, New Guinea, eastern Indonesia, the Solomon Islands, the Vanuatu islands, and Fiji. The genus name is derived from Greek words meaning "varied" and "spathe," an allusion to the 2 dissimilar bracts enclosing the inflorescence buds.

There are small undergrowth and large emergent species. None of them form a crownshaft, but many have colored new growth. The inflorescences are formed in the leaf crown but are often found beneath it by the time the single-sexed blossoms open. They consist of several short, usually white or yellowish branches and are enclosed by 2 large, leathery bracts when new, one bract usually persisting until the fruits are formed. The fruits are round or oblong and orange to red.

All species are tender to cold and are only possible in zones 10b and 11, although at least one (*H. delicatula*) hails from the mountains of New Guinea and may be adaptable to frostless Mediterranean climes. These palms invariably need constant moisture and a slightly acidic, humus-laden, fast-draining soil. Only one species is common in cultivation in the United States, but all would seem more than worthy of growing; they are much more widely grown in Australia but are still not common.

Seeds of *Heterospathe* germinate within 60 days. They should be harvested when fully ripe and even then may not germinate well. Some batches of seeds germinate far better than others for no apparent reason even though the same techniques are used. For further details of seed germination, see *Areca*.

Heterospathe cagayanensis
het'-er-o-SPAY-thee · ka-gah'-ya-NEN-sis
PLATE 463

Heterospathe cagayanensis is endemic to the Cagayan Islands in the southern Philippines, where it occurs in the undergrowth of dense rain forest at low elevations. The epithet is Latin for "of Cagayan." The palm is found both as an acaulescent and as a trunked species, the latter form slowly growing to 20 feet (6 m). The beautifully arching leaves are 6 feet (1.8 m) long on petioles only 2 feet (60 cm) long. The widely spaced, narrow, deep green, S-shaped, narrowly lanceolate, long-tipped leaflets are also 2 feet (60 cm) long. This lovely species is not for the full sun of hot climates. It needs constant, abundant moisture and a humus-laden soil.

Heterospathe delicatula
het'-er-o-SPAY-thee · de-li-ka-TOO-la
PLATE 464

Heterospathe delicatula is endemic to extreme eastern Papua New Guinea, where it grows in mountainous rain forest at 3000 to more than 5300 feet (900–1580 m). The epithet translates from the Latin as "diminutively delicate" and refers to the small stature of the palm. This species has a subterranean trunk with ascending, erect leaves 5–6 feet (1.5–1.8 m) long on petioles 18–24 inches (45–60 cm) long. The deep green leaflets are regularly and widely spaced along the rachis, each one 6–10 inches (15–25 cm) long, linear-lanceolate, limp, and pendent, and a beautifully brilliant cherry to wine color when new. The inflorescences are colorful with their reddish brown bracts and branches and their brownish and purple flowers. The fruits are bright red.

Heterospathe elata
het'-er-o-SPAY-thee · ee-LAH-ta
PLATE 465

Heterospathe elata is widespread in the Philippines, Micronesia, and the Moluccas in rain forest at low elevations, where it can be an undergrowth subject for years before usually ending up as an emergent canopy tree. The epithet is Latin for "tall." The common name is sagisi palm.

The mature stem of this solitary-trunked species grows to 50 feet (15 m) high with a diameter of 8–12 inches (20–30 cm). It is smooth and gray with wide and widely spaced rings of leaf base scars on the younger parts and a swollen base. The leaf crown is 15 feet (4.5 m) wide and tall, full, and nearly round. The leaves are 6–10 feet (1.8–3 m) long on petioles 2 feet (60 cm) long with limp, dark green narrowly lanceolate and tapering leaflets that are 2–3 feet (60–90 cm) long and pendent with age. They grow from the rachis in a flat plane, and the rachis is usually twisted to 90 degrees at its midpoints, resulting in the distal half of the blade being almost vertical in alignment. New leaves of younger plants are an unusual shade of pinkish bronze. The panicles of the inflorescences are 4 feet (1.2 m) long and much branched, and grow from among the leaves, bearing many small, whitish single-sexed blossoms of both sexes. The rounded ⅓-inch (8-mm) fruits ripen from green to white and finally to red and are borne in pendent clusters hanging beneath the leaf crown.

This species does not tolerate frost and is adaptable only to zones 10b and 11. It also needs partial shade when young, full sun when older, a humus-laden soil or a permanent organic mulch, and regular, adequate moisture. Its growth rate is slow when young and until it forms an aboveground trunk, but moderate to fast afterwards; the young stems usually sit or grow slowly sideways for a few years before they start growth upwards. The leaves of this species are unexpectedly fragile and easily broken by strong winds. Older plants form a dense mat of surface roots that can impede the growth of nearby plants, especially those that grown from tubers or bulbs, so care should be taken when siting this palm or underplanting it.

This palm is grace and elegance personified. Its leaves are reminiscent of those of the coconut, if not as large, and it is possibly even more attractive than the coconut when it is young as the leaf crown is often extended, with leaves arching out from the trunk at some distance from the palm's growing point. The species is beautiful enough to stand as an isolated specimen but looks better in groups of 3 or more individuals of varying heights. It has a silhouette almost as beautiful as that of the coconut and is stunning as a canopy-scape; against a background of contrasting foliage, it is most delightful. In short, it is hard to misplace it in the landscape. It is possible indoors in a large space with good light and high relative humidity.

Heterospathe elmeri
het′-er-o-SPAY-thee · EL-mer-eye
PLATE 466

Heterospathe elmeri is endemic to Camiguin Island, north of the main island of Luzon, in the Philippines, where it grows in dense rain forest at low elevations. The epithet honors Adolph D. E. Elmer (1870–1942), an American botanist who collected extensively in the Philippines. This solitary-trunked species grows to an overall height of 25 feet (7.6 m). The mostly erect leaves are 10–12 feet (3–3.6 m) long and have numerous narrowly linear-lanceolate dark green, limp, and pendent leaflets that are 2 feet (60 cm) long and regularly spaced. This beautiful species is extremely slow growing. The seeds are reportedly used as a betel-nut substitute in habitat, and the growing point is edible.

Heterospathe glauca
het′-er-o-SPAY-thee · GLAW-ka
PLATE 467

Heterospathe glauca is endemic to the island of Bacan in the Moluccas, where it grows in dense rain forest at low elevations. The epithet is Latin for "glaucous" and refers to the leaf bases or leaf sheaths.

This solitary-trunked species grows to an overall height of 30 feet (9 m). There is no true crownshaft, but the silvery green leaf bases are prominent and result in a cylindrical pseudo-crownshaft. The leaf crown is hemispherical or slightly more rounded and composed of 10–12 spreading leaves on short, stout petioles; the leaves are 10 feet (3 m) long, mostly straight but arch near the apex. The regularly spaced deep green, broadly lanceolate leaflets have irregularly truncated apices and are 2 feet (60 cm) long or more; they are limp and pendent when mature.

This species is one of the most attractive species in the genus and, although rare in cultivation in the United States, should do well in calcareous soils if kept constantly moist and heavily mulched with organic material.

Heterospathe intermedia
het′-er-o-SPAY-thee · in-ter-MEE-dee-a

Heterospathe intermedia is endemic to the montane rain forests of Luzon and Mindanao in the eastern Philippines. The epithet is Latin for "intermediate" and refers to the palm's appearance, intermediate between 2 other species.

This moderately tall species has a solitary trunk to 35 feet (11 m). Leaf sheaths are grayish green with the leaf petiole and rachis a glossy yellowish-green. Each leaf is gracefully recurved. The deep green leaflets are arranged in a single plane and somewhat leathery. The newly emerging leaf is a deep red color. The fruits are a rich red.

This palm is not particularly cold hardy and is best suited for zones 10 and 11. It wants a shady to partly shady location where it can eventually grow up through the canopy. The soil should be neutral to acidic with a humusy top layer and should be regularly watered.

Heterospathe longipes
het′-er-o-SPAY-thee · LON-ji-peez
PLATE 468

Heterospathe longipes occurs in Fiji on the islands of Taveuni and Vanua Levu, where it grows in low mountainous rain forest. The epithet means "long foot," and refers to the long petioles. This species was formerly known as *Alsmithia longipes*.

Mature trunks attain a height of 12 feet (3.6 m), a diameter of 3 inches (8 cm), and are light brown with closely set rings of leaf base scars. The leaves are 8 feet (2.4 m) long on petioles 2–2½ feet (60–75 cm) long; they are erect, never descending below the horizontal, and are extended down the trunk to create an elongated crown. New leaves are beautiful rose or red for 2 weeks. The regularly spaced, linear-lanceolate leaflets are 2 feet (60 cm) long, each with a prominent and lighter-colored midrib. The leaflets grow from the rachis in almost a flat plane and are deep green above and lighter green beneath. The inflorescences grow from among the leaves and are 4 feet (1.2 m) long, with several thin, spidery, tan branches, which bear tiny male and female flowers. The egg-shaped fruits are 1 inch (2.5 cm) long and turn red when mature.

This palm is barely in cultivation and little data are available, but it is adaptable only to zone 11 and is marginal in 10b. It needs constant moisture, high humidity, partial shade, and a soil that is not too poor.

Heterospathe minor
het′-er-o-SPAY-thee · MY-nor
PLATE 469

Heterospathe minor is endemic to the Solomon Islands, where it grows in low mountainous rain forest. The epithet is Latin for "smaller," in reference to its fruits. This solitary-trunked species grows to an overall height of 20 feet (6 m). The leaves of older palms are borne on exceptional petioles 6 feet (1.8 m) long and are themselves no more than 8 feet (2.4 m) long and usually nearer 6 feet (1.8 m). The leaves form a crown that is less than hemispherical because of their ascending nature. The numerous leaflets are regularly spaced along the rachis, medium to deep green, 3 feet (90 cm) long, narrowly lanceolate, long tipped, and slightly S shaped; they grow from the rachis at an angle which creates a V-shaped leaf and are a beautiful wine or rosy bronze when new.

Heterospathe negrosensis
het'-er-o-SPAY-thee · neg-ro-SEN-sis
PLATE 470

Heterospathe negrosensis is endemic to the Philippines, where it grows in lowland rain forest. The epithet is Latin for "of Negros," a large island in the central Philippines. The solitary trunk is 8 feet (2.4 m) tall but the leaves are 15 feet (4.5 m) long, including the petioles which are 4–5 feet (1.2–1.5 m) long. The numerous leaflets are deep green, 3–4 feet (90–120 cm) long, narrowly lanceolate, limp, and partially pendent.

Heterospathe philippinensis
het'-er-o-SPAY-thee · fi-li-pee-NEN-sis
PLATE 471

Heterospathe philippinensis is endemic but widespread in mountainous rain forest in the Philippines, where it grows to 5300 feet (1620 m). The epithet is Latin for "of the Philippines." This variable species has both solitary-trunked and clustering individuals. The slender stems grow to a height of 6 feet (1.8 m), and the leaf crown is hemispherical or less. The leaves are 4–5 feet (1.2–1.5 m) long and have deep green leaflets 1 foot (30 cm) long that grow from the rachis at an angle to create a slightly V-shaped leaf.

Heterospathe phillipsii
het'-er-o-SPAY-thee · fi'-LIP-see-eye

Heterospathe phillipsii is a rare and endangered species endemic to low mountainous rain forest on the island of Viti Levu in Fiji, and was described in 1997. It is threatened because of logging activities. The epithet honors Richard (Dick) H. Phillips (1923–1999), horticulturist and collector of palms in Fiji.

The solitary trunk attains a height of 40 feet (12 m) and a diameter of 6–7 inches (15–18 cm). The nearly spherical leaf crown has sumptuously beautiful, spreading, dark green leaves 13–14 feet (4–4.3 m) long on petioles 12–18 inches (30–45 cm) long. The dark green leaflets are 2½ feet (75 cm) long, wide-lanceolate, closely and evenly spaced along the rachis, and limp and semipendent. The small, oblong fruits are bright red when mature.

This species is among the most beautiful species in the genus. It is almost as tall as *H. elata* and its leaves are even more attractive. The palm is faster growing than most other species but, alas, is no hardier to cold.

Heterospathe salomonensis
het'-er-o-SPAY-thee · sa'-lo-mo-NEN-sis

Heterospathe salomonensis is endemic to the Solomon Islands, where it grows in low mountainous rain forest. The epithet is Latin for "of Solomon." This species is similar in stature and appearance to *H. elata* but is slower growing.

Heterospathe scitula
het'-er-o-SPAY-thee · SI-tyoo-la
PLATE 472

Heterospathe scitula is a critically endangered palm endemic to Luzon Island in the Philippines. It grows in lowland rain forest from 300 to 600 feet (90–180 m). The epithet is Latin for "elegant" or "neat" and refers to the palm's appearance.

This clustering species has small, thin stems to 6 feet (1.8 m) tall with a diameter about ½ inch (13 mm). The crown holds 8 arching leaves, each about 20 inches (50 cm) long including the petiole. The leaf sheath is covered in brown scales and does not form a crownshaft. Leaflets number 2–10 on each side of the rachis and are rather widely spaced in a single plane. The newly emerging leaf is bronzy red in color. Inflorescences emerge from below the leaves and produce a glossy red, spherical fruit ⅓ inch (8 mm) in diameter.

This small eye-catching beauty is ideally suited for a shady intimate landscape location, where it can be appreciated close up. It is not cold hardy, being suited only for zones 10 and 11. It will do best in acidic soils with a humusy top layer and needs regular watering.

Heterospathe woodfordiana
het'-er-o-SPAY-thee · wood-ford'-ee-AH-na
PLATES 473–474

Heterospathe woodfordiana is endemic to Santa Isabel Island in the Solomon Islands, where it grows in low mountainous rain forest. The epithet honors Charles M. Woodford (1852–1927), a British naturalist, author and Resident Commissioner of the Solomon Islands. The solitary brown trunk grows to 12 feet (3.6 m) tall and 3–4 inches (8–10 cm) in diameter. The hemispherical leaf crown holds 10 leaves that are each 5 feet (1.5 m) long. The regularly spaced, linear-elliptic, long-tipped, soft leaflets are pendent at their tips. New growth is a beautiful maroon, and the small fruits are deep red.

Howea
HOW-ee-a

Howea comprises 2 solitary-trunked, pinnate-leaved, monoecious palms on Lord Howe Island off the central eastern coast of Australia. The unusual inflorescences are unbranched pendent spikes 6 feet (1.8 m) long, growing from the lower crown and carrying tiny creamy white, unisexual flowers of both sexes. The species are slightly susceptible to lethal yellowing disease. The genus name is a Latinized form of the name of the palm's island home. Seeds of *Howea* differ from that of other genera of Areceae in that they germinate sporadically from 30 days to as long as 2 years. Germination is best when seeds are fresh and fully ripe before harvesting. For further details of seed germination, see *Areca*.

Howea belmoreana
HOW-ee-a · bel-mor'-ee-AH-na
PLATE 475

Howea belmoreana is endemic to Lord Howe Island, where it grows in hilly, moist forests. Common names are sentry palm and kentia palm. The epithet honors Somerset Richard Lowry-Corry (1835–1913), 4th Earl Belmore, who was governor of New South Wales.

Mature trunks attain heights of 40 feet (12 m) in habitat with a diameter of 6 inches (15 cm). They are light brown to gray with slightly swollen bases and closely set darker leaf scar rings. The leaf crown is 10 feet (3 m) wide and 8 feet (2.4 m) tall, full, and rounded. The leaves are 7–10 feet (2.1–3 m) long, strongly arching and forming almost a semicircle. They are held on strongly arching petioles 3 feet (90 cm) long. The dark green leaflets are 3–4 feet (90–120 cm) long and grow from the rachis at a 45-degree angle, creating a strong V shape to the leaf. The fruits are 2 inches (5 cm) long, ovoid, and brown to reddish brown when ripe.

The palm needs average but regular moisture and is adaptable to zones 10 and 11. It is slow growing at all stages but thrives in partial shade, especially in hot climates; indeed, it is difficult to maintain in hot or tropical climates. It is magnificent in nearly frostless Mediterranean climates.

Because of its strongly arching and trough-shaped leaves, the palm has unusual grace and a striking architectural aspect. It is unsurpassed for a silhouette display against the sky as a canopyscape, or large expanses of walls or different background landscape colors. It is accommodating in this respect because of its slow growth, retaining whatever proportions for a relatively long time. It is exceptionally beautiful in groups of 3 or more individuals of varying heights, and yet is distinctive and appealing enough as an isolated specimen. This is one of the most durable indoor plants, especially when young, adapting to low light as well as some neglect.

Howea forsteriana
HOW-ee-a · for-ster′-ee-AH-na
PLATE 476

Howea forsteriana is endemic to low, coastal elevations on Lord Howe Island. The epithet honors William Forster (1818–1882), an Australian politician and poet.

Mature trunks reach heights of 50–60 feet (15–18 m) but are usually half that under cultivation. The stem is 6 inches (15 cm) in diameter, swollen at the base, and gray to tan, with closely spaced, grooved rings of leaf base scars. The leaf crown of mature palms is usually 20 feet (6 m) wide and 12–15 feet (3.6–4.5 m) tall; it is extended down the trunk but never quite rounded. The leaves are 8–12 feet (2.4–3.6 m) long on petioles 4–5 feet (1.2–1.5 m) long. The rachis arches beautifully, and the many dark green, narrowly lanceolate and tapering, limp leaflets are 2–3 feet (60–90 cm) long and pendent. There is much tightly woven brown fiber associated with the bases of the leafstalks and it is persistent for each leaf until the leaf dies and falls. The ellipsoid fruits are 2 inches (5 cm) long and orange or red when mature.

The species needs average but regular moisture. It is adaptable to a range of free-draining soils. It grows in zones 10 and 11; older palms sometimes withstand a temperature of 28°F (−2°C) unharmed, but anything colder is trouble and temperatures in the low or mid-20s (around −7 to −4°C) can be fatal.

Younger plants have an astoundingly graceful and attractive fountainlike aspect, and there is likely nothing more elegant in all nature than these wonderful curves accented by the great drooping leaflets. The palm should be sited where there are no other prima donna plants to compete with its visual perfection except more of its kind. It is one of the few tall palms that are beautiful enough to stand alone as an isolated specimen, but it is even more choice in groups of 3 or more individuals of varying heights, and its silhouette is very beautiful. It is the second most (if not the most) popular indoor palm and has been used for more than 150 years as an interiorscape subject. It is a divine diva, thriving and proffering her great beauty in spite of abuse and bad lighting, gracing every imaginable chamber from saloons and cheap hotel lobbies to spacious presidential ballrooms.

Hydriastele
hy-dree-AS-ste-lee

Hydriastele is composed of nearly 50 mostly clustering, pinnate-leaved, monoecious palms in northern Australia, New Guinea, and the adjacent Bismarck Archipelago. The species form crownshafts, below which the green stem often has dark blotches or spots. The leaf crowns are sparse, especially in clustering forms, but often elongated, with distinctive, relatively short leaves. The leaflets are mostly irregularly wedge shaped with jagged apices; the leaflets of the terminal pair are larger than the rest, and a few species, such as *H. rheophytica*, have typically dissected, linear leaflets. The inflorescences grow from beneath the crownshafts, are branched once, and look like large whisk brooms; they produce pendent clusters of small, red, purple, or black fruits.

Most of these beauties have been in cultivation for many years, but none are common, especially in the United States. All are intolerant of cold and adaptable only to zones 10b and 11. They are mostly of easy culture and seem not fussy about soil type, needing mainly copious, regular moisture. They seem to do best when given partial shade in their youth.

The genus now includes all the species previously placed in *Siphokentia*, *Gulubia*, and *Gronophyllum*. The genus name is from 2 Greek words, one for a water nymph, the other meaning "column," the allusion, supposedly, to the size of the palms and the swampy habitats of some species.

Seeds of *Hydriastele* germinate within 60 days. They should be harvested when fully ripe and even then may not germinate well. Some batches of seed germinate far better than others for no apparent reason even though the same techniques are used. Slightly green seeds do not germinate at all. For further details of seed germination, see *Areca*.

Hydriastele affinis
hy-dree-AS-ste-lee · AF-fi-nis
PLATE 477

Hydriastele affinis is endemic to the New Guinea rain forest. The epithet is Latin for "related" or "similar" and probably refers to other species in the genus with similar traits, such as *H. pinangoides*.

The green stems of this clustering species grow to at least 12 feet (3.6 m) tall with a 3-inch (8-cm) diameter and have broadly spaced rings. The ascending leaf crown holds 8 or 9 deep green leaves. Leaflets are fishtail shaped and clustered in groupings of 2–6 along the rachis. The newly emergent leaf is pink or cream in color. A short inflorescence about 12 inches (30 cm) long produces a bright red ovoid fruit.

This palm requires a moist shady location with acidic soil that has a humusy top layer. It is not cold hardy, being best suited for zones 10 or 11. Given proper conditions, it is a fast grower, becoming a moderate-sized specimen rather quickly. With its broad fishtail leaflets, pinkish new leaves, and red fruit, this exceedingly attractive palm will stand out in any landscape if given a spot with contrasting foliage around it.

Hydriastele beguinii
hy-dree-AS-ste-lee · be-GWIN-ee-eye
PLATE 478

Hydriastele beguinii occurs naturally on the Moluccan island of Halmahera and adjacent islands, where it grows in lowland rain forest. The epithet honors Victor Beguin (1886–1943), a Dutch botanist and collector in the Moluccas and Indonesian New Guinea. This species was formerly called *Siphokentia beguinii*.

The solitary trunk attains a maximum height of 25 feet (7.6 m) and is 2–3 inches (5–8 cm) in diameter. It is dark green in its younger parts and is heavily ringed with wide, tan leaf base scars. The crownshaft is 2 feet (60 cm) tall, not much thicker than the trunk, cylindrical, sometimes slightly swollen at its midpoint, and bluish or olive green. The hemispherical leaf crown often is elongated and consists of 6–8 spreading, arching leaves 4–6 feet (1.2–1.8 m) long. The leaflets grow in a flat plane off the rachis, are limp and semipendent, and are irregularly shaped, except for the leaflets of the terminal pair, which are always the broadest. The leaflets range from linear-lanceolate to rhomboid, are glossy deep green above with duller matte color beneath, heavily ribbed with jagged tips, and held on short petioles. The short inflorescences resemble whisk brooms. The oblong fruits are ¾ inch (19 mm) long and deep red when mature. With its variously shaped leaflets, this beautiful small palm adds great visual interest to plantings among other vegetation.

Hydriastele costata
hy-dree-AS-ste-lee · kos-TAH-ta
PLATE 479

Hydriastele costata is indigenous to the Aru Islands of Indonesia, the island of New Guinea, the Bismarck Archipelago, and northeastern Queensland, Australia, where it grows in lowland rain forest and swamps, often forming large stands. The epithet is Latin for "ribbed," an allusion to the fruits. The species was formerly known as *Gulubia costata*.

The solitary trunk of this very tall, magnificent species attains a height of 100 feet (30 m) in habitat, with a diameter of 2 feet (60 cm). The crownshaft is 4–5 feet (1.2–1.5 m) tall, slightly bulging at its base, and medium to deep green often with purplish hues near its base and a suffusion of brown near its summit. The sparse leaf crown is hemispherical or slightly more so. This species is one of 2 with straight, unarching rachises, and the leaves are 12–15 feet (3.6–4.5 m) long on petioles 1–2 feet (30–60 cm) long. The many lanceolate, medium to dark green leaflets are pendent, 3–4 feet (90–120 cm) long, and give to each leaf a sort of curtain appearance. Each leaflet is 2 inches (5 cm) wide at its base and very deeply bifid at its apex, which characteristic adds to the comblike appearance of the curtain of segments. The individual inflorescences are not extraordinarily large, but they produce ellipsoid bluish gray, white-striped ("ribbed") fruits 3 inches (8 cm) long.

The species is nearly tropical in its temperature requirements and adaptable only to zones 10b and 11. It needs constant, copious moisture and sun when past the juvenile stage. In spite of its preference for a humus-laden soil, the species is thriving at Fairchild Tropical Botanic Garden in Miami, Florida, where its site in the rainforest area is heavily mulched and watered. It is slow growing while a seedling, a stage in which the mostly undivided leaves look much like those of an *Areca* species; once past this stage, however, it is fast.

This exquisite species stands out in the landscape as hardly any other specimen of vegetation can. It is incomparably spectacular because of its great stiff, curtainlike leaves and, as a canopy-scape, is matchless. Groups of 3 or more individuals of varying heights create an irresistible tableau of tropical beauty.

Hydriastele cylindrocarpa
hy-dree-AS-ste-lee · si-lin´-dro-KAHRP-a
PLATE 480

Hydriastele cylindrocarpa is indigenous to the Vanuatu islands and the Solomon Islands, where it grows in mountainous rain forest at 300–3200 feet (90–980 m). The epithet is Latin for "cylindrical-fruited," an allusion to the shape of the small, yellow to orange fruits. This species was formerly known as *Gulubia cylindrocarpa*.

This species is one of the tallest palms in the genus with mature trunks attaining heights of nearly 100 feet (30 m) but diameters of usually less than 1 foot (30 cm). The stems are light gray to almost white, have closely spaced rings of leaf base scars, and are basally expanded. The light or silvery green crownshaft is 3 feet (90 cm) tall, slightly bulged at its base, and cylindrical above. The rounded leaf crown contains 15–18 greatly arching, recurved leaves 7 feet (2.1 m) long on petioles 1 foot (30 cm) long. The linear-lanceolate leaflets are 3 feet (90 cm) long, olive green above and grayish green beneath. They are stiff and shortly bifid apically and grow from the rachis at an angle that gives a V-shaped leaf.

Before they grow above the forest canopy, young palms have a short cone of stilt roots and usually undivided leaves, a characteristic that leads to the palm's being often misidentified at this stage as a *Pelagodoxa* species. Mature palms are enchantingly beautiful, especially in silhouette.

Hydriastele dransfieldii
hy-dree-AS-ste-lee · dranz-FEEL-dee-eye
PLATE 481

Hydriastele dransfieldii is endemic to Biak Island off the northwestern coast of New Guinea, where it grows in lowland rain forest. The epithet honors British palm taxonomist John Dransfield (1945–), of the Royal Botanic Gardens, Kew. This species was once called *Siphokentia dransfieldii*. The species was described in 2000 and rapidly achieved popularity in cultivation because of its charming aspect. It is similar to *S. beguinii* except for details of the inflorescence and its much stiffer leaflets which are also of more uniform sizes and shapes: obliquely wedge shaped with truncated, jagged apices. The plant is also reported to be more robust under cultivation and has slightly larger but similarly shaped and colored fruits.

Hydriastele flabellata
hy-dree-AS-ste-lee · fla-bel-LAH-tah

Hydriastele flabellata is indigenous to West Papua, Indonesia, on the island of New Guinea, where it grows at about 1400 feet (430 m) in dense, very humid rain forest at the base of limestone outcrops. The stems are used for arrow shafts and fish spear handles. The epithet is Latin for "fan shaped" and refers to the leaf blade.

Clustering stems can be 6 feet (1.8 m) in height and ½ inch (13 mm) in diameter with a crown of 7 leaves. The crownshaft is green with light speckling. Leaves are 10–12 inches (25–30 cm) long and are usually undivided or have just 2 leaflets on each side of the rachis with jagged margins. The spicate inflorescence is 4–6 inches (10–15 cm) long and produces an obovoid, pointed fruit that is less than ½ inch (13 mm) long and bright red.

This delightful understory charmer is perfect for a cozy spot where it can be viewed up close. It prefers neutral or acidic soil with a humusy top layer and will not tolerate any freezing temperatures.

Hydriastele hombronii
hy-dree-AS-ste-lee · hom-BRO-nee-eye

Hydriastele hombronii is endemic to the Solomon Islands, where it is widespread in low mountainous rain forest. The epithet honors Jacques B. Hombron (1800?–1852), a French botanist, surgeon and collector in the islands. The species was formerly called *Gulubia hombronii*.

The trunks of mature palms grow to 60 feet (18 m) tall in habitat with diameters of 10–12 inches (25–30 cm) and swollen bases. The crownshaft is 2–3 feet (60–90 cm) tall, almost cylindrical, and scarcely thicker than the trunk. It is olive green with a suffusion of orange at its summit. The leaf crown is hemispherical to almost spherical and is dense, with as many as 20 strongly arching leaves 6 feet (1.8 m) long on petioles less than 1 foot (30 cm) long. Juvenile leaves are usually entire and unsegmented. The light green leaflets are 18–24 inches (45–60 cm) long, linear-lanceolate, and shortly bifid apically. They grow from the rachis at a 45-degree angle, creating a V-shaped leaf. The small, ellipsoid fruit are dull red.

The species has a beautiful silhouette and makes an exciting and lovely canopy-scape. It is tender to cold and needs constant moisture but isn't fussy about soil type.

Hydriastele kasesa
hy-dree-AS-ste-lee · ka-SAY-sa
PLATE 482

Hydriastele kasesa is endemic to the Bismarck Archipelago of Papua New Guinea, where it occurs in lowland rain forest. The epithet is the aboriginal name of the palm. This densely clustering species grows to a maximum height of 15 feet (4.5 m). The slender stems are topped by slender, light green crownshafts 2 feet (60 cm) tall, above which grow the sparse crowns of leaves 3 feet (90 cm) long. The leaflets are widely spaced along the rachis, each one 1 foot (30 cm) long, irregularly wedge shaped, with an obliquely truncated, jagged apex; the leaflets of the terminal pair are usually united and much wider than the others. This palm looks better if some of the stems are judiciously cut out so that the others may be more readily seen.

Hydriastele ledermanniana
hy-dree-AS-ste-lee · lee'-der-man-nee-AH-na
PLATE 483

Hydriastele ledermanniana is endemic to north central Papua New Guinea, where it occurs in the undergrowth of low mountainous rain forest. The epithet honors Carl L. Ledermann (1875–1958), Swiss horticulturist and botanist who collected plants in New Guinea. This species was long cultivated as *Gronophyllum ledermanniana*.

The stem of this moderate-sized solitary-trunked species grows to 30–40 feet (9–12 m) but is usually less than 3 inches (8 cm) in diameter. The crownshaft is 2 feet (60 cm) long, slightly bulging at its base, and is otherwise cylindrical and a beautiful silvery green to nearly white. The spreading leaves are 8 feet (2.4 m) long, on petioles 18 inches (45 cm) long, and form a crown that is slightly more than hemispherical. The wedge-shaped, pendent leaflets are widely spaced in groups and are 1 foot (30 cm) long with jagged apices and beautiful silvery undersides.

This species is another one that is undeservedly rare in cultivation. It is reportedly tropical in its temperature requirements. It seems to need constant, copious moisture and a humus-laden, free-draining soil in partial shade.

Hydriastele longispatha
hy-dree-AS-ste-lee · lon-jee-SPAY-tha
PLATE 484

Hydriastele longispatha is endemic to north central West Papua Province, Indonesia, where it grows in mountainous rain forest at 500–4600 feet (150–1400 m). The epithet is Latin for "long" and "spathe" and alludes to the primary leathery bract that is 4 feet (1.2 m) long and initially covers the inflorescence before the latter expands and the former falls away.

The trunk attains a height of 80 or more feet (24 m) in habitat and is less than 1 foot (30 cm) in diameter. It is light gray to nearly white, smooth, almost shiny, straight as an arrow and is ringed with closely set darker leaf base scars. The crownshaft is 3–4 feet (90–120 cm) tall, slightly thicker than the trunk, cylindrical, smooth and glaucous, light green at its base and silvery green, almost white at its summit. The leaf crown is rounded with 18 strongly arching, recurved leaves 8 feet (2.4 m) long on petioles 10 inches (25 cm) long. The linear-lanceolate leaflets are 3 feet (90 cm) long, stiff and erect, growing from the rachis at a 45-degree angle to create a V-shaped leaf. The leaflets are light green or olive green and are deeply bifid and slightly pendent at their apices. The round ½-inch (13-mm) fruits are shiny, bright red.

The species is still undeservedly rare in cultivation. It should do well in almost any frost-free climate if given constant, copious moisture and a humus-laden soil. It is arguably the most beautiful species in a genus that is rife with beauty. Its thrilling silhouette is comparable to that of *Actinorhytis calapparia*. A group of 3 or more individuals of varying heights is among the most exquisite sights the palm family has to offer.

Hydriastele macrospadix
hy-dree-AS-ste-lee · mak-ro-SPAY-dix
PLATE 485

Hydriastele macrospadix is endemic to the Solomon Islands, where it grows in mountainous rain forest at elevations from sea level to 3000 feet (900 m). The epithet is from Greek words meaning "large" and "inflorescence," although the inflorescence is no larger than those of most other species. It was once called *Gulubia macrospadix*.

The tan, slender, tapering trunk grows to 40–50 feet (12–15 m) high but is 10 inches (25 cm) in diameter at its midpoint. The crownshaft is at least 3 feet (90 cm) tall and light to medium green except near its top where it is suffused with a purplish black hue. The leaf crown is spherical. The leaves are 6–7 feet (1.8–2.1 m) long, borne on petioles 1 foot (30 cm) long, and are beautifully arching and recurved at their apices. The dark green leaflets are 2 feet (60 cm) long, evenly and widely spaced, and grow from the rachis at a 45-degree angle, giving to the leaf a V shape; each one is broadly lanceolate with an obliquely cut apex that terminates in a pendent tip. The ellipsoid fruits are crimson and ½ inch (13 mm) long.

Hydriastele microcarpa
hy-dree-AS-ste-lee · myk-ro-KAHR-pa
PLATE 486

Hydriastele microcarpa is indigenous to the Moluccas and Ceram between Sulawesi and New Guinea, where it grows in low mountainous rain forest. The epithet is from Greek words meaning "small" and "fruit," an allusion to the clusters of small, round red berrylike fruits that are reportedly used as a betel-nut substitute in habitat. A synonym is *Gronophyllum microcarpum*.

The solitary trunk attains a height of 30 feet (9 m). The crownshaft is 2 feet (60 cm) tall, slightly bulging at its base and slightly tapering to its summit, and an even dark green. The leaf crown is slightly less than hemispherical with ascending, slightly arching leaves 6–8 feet (1.8–2.4 m) long. The many stiff and erect, dark green leaflets grow from the rachis at a 45-degree angle to give a V-shaped leaf. The species is tropical in its requirements and needs abundant moisture. It thrives in partial shade or full sun except in hot, dry climes.

Hydriastele microspadix
hy-dree-AS-ste-lee · myk-ro-SPAY-dix
PLATE 487

Hydriastele microspadix is endemic to northern Papua New Guinea, where it grows in lowland rain forest. The epithet is from Greek words meaning "tiny" and "spadix." It is a densely clustering species growing to an overall height of 15–20 feet (4.5–6 m). The crownshaft is 2–3 feet (60–90 cm) tall, usually slightly bulging near its midpoint and nearly white except near its summit where it is suffused with greenish brown. The leaves are 5 feet (1.5 m) long and bear regularly and widely spaced, linear-oblong leaflets 2–3 feet (60–90 cm) long with obliquely truncated, jagged apices; the leaflets of the terminal pair are larger than the others, often significantly so, and all the leaflets are medium to deep green above but duller and paler green beneath.

Hydriastele montana
hy-dree-AS-ste-lee · mon-TAH-na

Hydriastele montana is endemic to Arfak, a mountain range in West Papua Province, Indonesia, where it grows in rain forest at moderate elevations. The epithet is Latin for "mountainous" and refers to the palm's habitat. The species was formerly known as *Gronophyllum montanum*.

The stems of this sparsely clustering species grow to 30 feet (9 m) tall and 2 inches (5 cm) in diameter. The silvery green to nearly white crownshafts are 2–3 feet (60–90 cm) tall and scarcely wider than the trunk itself. The leaf crown is sparse and less than hemispherical with stiffly ascending leaves 4–5 feet (1.2–1.5 m) long on petioles 1 foot (30 cm) long. The few leaflets occur in widely spaced groups and are linearly to broadly wedge shaped, with jagged apices; the leaflets of the terminal pair are larger than the others. New growth is rosy to reddish.

This species is almost nonexistent in cultivation, and little is known about its requirements. It probably needs a nearly tropical climate, great and regular moisture, partial shade, and a humus-laden soil.

Hydriastele palauensis
hy-dree-AS-ste-lee · pa-low-EN-sis
PLATE 488

Hydriastele palauensis is a rare and endangered species endemic to the Palau islands east of the southern Philippines, where it grows in wet forest at low elevations on limestone soils. It is called Rock Island palm in Palau, and the epithet is Latin for "of Palau." This species was previously known as *Gulubia palauensis*. The solitary trunk attains a height of 60 feet (18 m) in habitat with a diameter of 6 inches (15 cm). The smooth crownshaft is 2 feet (60 cm) tall, scarcely thicker than the trunk, cylindrical, and light green. The leaf crown is nearly spherical, with small, arching leaves 3 feet (90 cm) long on petioles 10 inches (25 cm) long. The stiff, linear-lanceolate leaflets are 2 feet (60 cm) long and deeply bifid apically, and grow from the rachis at a 45-degree angle to give a V-shaped leaf. The species, like the rest in the genus, has a beautiful silhouette but seems not to be in cultivation.

Hydriastele pinangoides
hy-dree-AS-ste-lee · pi-nan-GO-i-deez
PLATES 489–490

Hydriastele pinangoides is endemic to northern West Papua Province, Indonesia, where it occurs in the undergrowth of low mountainous rain forest. The epithet translates as "similar to *Pinanga*," but the similarity is difficult to see. A synonym is *Gronophyllum pinangoides*.

The slender stems of this densely clustering species grow to 20 feet (6 m) high and less than 1 inch (2.5 cm) in diameter. The light green crownshafts are scarcely wider than the stems, and the leaf crowns are extended along the upper parts of the stems. The leaves are 3–4 feet (90–120 cm) long including the petioles, which are 18 inches (45 cm) long. The 4–10 leaflets grow in oppositely placed pairs along the rachis, each one rhomboidal or obliquely wedge shaped, glossy dark green, with a jagged apex and a generally undulating surface. The leaflets of the terminal pair are united to some extent, and the new growth is usually deep rose or wine red.

The species is as tender as any of the rest, needs partial shade, especially in hot climates, copious, regular moisture, and a humus-laden soil. It is probably the most widely grown species in the genus, although it is still not that common. It should be planted in an intimate site where the individual leaflets may be seen up close.

Hydriastele ramsayi
hy-dree-AS-ste-lee · RAM-zay-eye

Hydriastele ramsayi is a tall, solitary-trunked species endemic to Arnhem Land, the extreme northern region of the Northern Territory, Australia, where it grows in low, swampy areas on sandy soils, often in large stands. It is called northern kentia palm in Australia. It was formerly known as *Gulubia ramsayi*. The epithet commemorates, Edward P. Ramsay (1842–1916), naturalist and curator of the Australian Museum.

The solitary trunk attains a height of 50 feet (15 m) in habitat and is light gray to nearly white and often bulged at its midpoint. It is also distinctly ringed with closely set leaf base scars. The crownshaft is light green to nearly white and 3 feet (90 cm) high. It is usually bulging at its base and tapering to its summit, although many individuals have a cylindrical shaft that is about as thick as the trunk itself. The leaf crown is wonderfully full and round with greatly arching and recurved leaves 8–10 feet (2.4–3 m) long on short to nearly nonexistent petioles. The numerous stiff leaflets grow from the rachis at a 45-degree angle to create a distinctly V-shaped leaf. They are linear-lanceolate, long tipped, and slightly pendent at their apices, and range from deep olive green to more commonly bluish and glaucous green.

This species has great visual affinities to another of the world's most beautiful palm species, *Actinorhytis calapparia*. Its trunk, however, is much more robust, particularly in cultivation. It does not tolerate cold, needs constant, copious moisture, and flourishes in full sun. It is reportedly difficult to grow when young, and the roots are brittle and difficult to transplant. Once established, however, this palm grows fast and robustly.

Hydriastele rheophytica
hy-dree-AS-ste-lee · ree-o-FI-ti-kah
PLATE 491

Hydriastele rheophytica is native to the upper tributaries of the Taritatu (formerly Idenburg) River in Papua Province, Indonesia, and in the West Sepik Province of Papua New Guinea. It grows up to 2500 feet (760 m) elevation along rocky river banks where it is often inundated. The epithet refers to its being a rheophyte, a plant that lives in fast-moving water currents.

Up to 30 stems can make up a clump, which can be as much as 18 feet (5.4 m) tall. The stems are 1 inch (2.5 cm) thick, quite pliable, and generally lean or at times are decumbent. There are 3–12 leaves in the crown and the crownshaft is dark green with dense crustose scales falling off over time. Leaves can be 4 feet (1.2 m) long with 18–32 widely spaced leaflets on each side of the rachis in a single plane. The inflorescence emerges below the crownshaft producing beautiful pinkish-cream flowers that in turn produce tiny globose red fruit only ¼ inch (6 mm) in diameter.

While this species has been cultivated for a time in Australia and Hawaii, its cultivation tolerances are still not well known. It is a water lover so does well where kept constantly moist and it prefers sun or light shade, but other than that, it is not well known how much cold it can tolerate or if it likes high heat. Regardless, it is worth a try as it can be a rather charismatic subject in that wet area of the landscape where other plants find difficulty in growing.

Hydriastele rostrata
hy-dree-AS-ste-lee · ro-STRAH-ta
PLATE 492

Hydriastele rostrata is endemic to eastern and central Papua New Guinea, where it grows in lowland rain forest. The epithet is Latin for "beaked" and refers to the shape of the fruit.

The stems of this densely clustering, large species grow to heights of 30 or more feet (9 m). The crownshaft is 3 feet (90 cm) tall, barely thicker than the trunk, not conspicuously bulged at any point, cylindrical, and medium to deep green. The leaves are 6 feet (1.8 m) long on wiry petioles 2 feet (60 cm) long, and the leaflets are spaced along the rachis in groups and are of varying widths but are all narrowly or widely wedge shaped with obliquely truncated, jagged apices.

This impressive palm looks better if a few of the stems in a clump are judiciously cut out so that the interesting forms and silhouettes of the stems and leaflets may be better seen. It is rare in cultivation and this fact is almost unfathomable as the palm is robust and easy of culture in a tropical or nearly tropical climate.

Hydriastele valida
hy-dree-AS-ste-lee · VAL-ee-da
PLATE 493

Hydriastele valida is endemic to northwestern Papua New Guinea, where it grows in low mountainous rain forest. The epithet is Latin for "valid" or "true," a reference to the strength and robustness of the foliage (Essig 1982). The species was originally described as *Gulubia valida*.

The solitary trunk attains a height of 40–50 feet (12–15 m) in habitat and is usually less than 1 foot (30 cm) in diameter. The crownshaft is usually 3 feet (90 cm) tall, light green, slightly thicker than the trunk, and cylindrical. The leaf crown is dense and nearly spherical with 20 or more slightly arching leaves 7 feet (2.1 m) long on stout petioles 1 foot (30 cm) long. The stiff lanceolate leaflets are 2–3 feet (60–90 cm) long and grow from the rachis at an angle that creates a V-shaped leaf. The ellipsoid fruits are dark red to dark purple and ½ inch (13 mm) long.

This extremely attractive species has an exceptionally beautiful leaf crown and an arresting silhouette. It is rare in cultivation and should be much more widely planted in tropical or nearly tropical regions.

Hydriastele vitiensis
hy-dree-AS-ste-lee · vee-tee-EN-sis
PLATE 494

Hydriastele vitiensis is endemic to Fiji, where it grows in low mountainous rain forest. The epithet means "of Fiji." It was described in 1982 as *Gulubia microcarpa*, but when placed in *Hydriastele*, the species had to be given a new name because *H. microcarpa* was already in use for a species from the Moluccas (see above).

This tall, slender-trunked species grows to 80 feet (24 m) high in habitat with a trunk diameter that is usually less than 1 foot (30 cm). The light green crownshaft is barely thicker than the trunk, 2½ feet (75 cm) tall, and cylindrical. The leaf crown is hemispherical and consists of 16 or 17 arching leaves 7 feet (2.1 m) long on petioles 1 foot (30 cm) long. The stiff leaflets are 3 feet (90 cm) long and grow from the rachis at an angle that makes a V-shaped leaf. The small, cylindrical fruits are gray or white when ripe.

The species is still rare in cultivation but needs a humus-rich soil, partial shade when young, full sun when older, and a frostless climate. The palm has a beautiful silhouette and is stunningly attractive in groups of 3 or more individuals of varying heights.

Hydriastele wendlandiana
hy-dree-AS-ste-lee · wend-lan'-dee-AH-na
PLATE 495

Hydriastele wendlandiana is endemic to Australia in northern and northeastern Queensland and the Northern Territory, where it grows in swampy lowland rain forest. The epithet honors Hermann Wendland (1825–1903), German botanist and palm specialist. The 2 common names in Australia honor places in the Northern Territory: Florence Falls palm or Latrum palm (after a small river by that name).

This sparsely clumping species is occasionally found as a solitary-trunked individual. The stems attain a height of 60 feet (18 m), are usually only 6 inches (15 cm) in diameter, and are light gray to almost white, straight and distinctly ringed in their younger parts with darker rings of leaf base scars. The crownshafts are relatively short considering the heights of the trunks; they are 2 feet (60 cm) tall, swollen near their midpoints, and light silvery green to whitish blue-green. The leaves are 6 feet (1.8 m) long and bear yellowish green to deep green, wedge-shaped leaflets of varying widths. Wide leaflets are interspersed with narrow ones in no particular pattern, the wider leaflets always fewer in number, and all are shortly or even minutely toothed at their apices; however, the leaflets of the terminal pair are considerably shorter and wider than most of the others.

This species is easy of culture in tropical or nearly tropical climates if given regular, adequate moisture. It is not suited to calcareous soils and must be heavily mulched (or the soil greatly amended) in such sites.

Hyophorbe
hy-o-FOR-bee

Hyophorbe includes 5 solitary-trunked, pinnate-leaved, monoecious palms endemic to the Mascarene Islands of the Indian Ocean, where they are threatened with extinction. Several species have unusual swollen trunks and all have prominent, colorful crownshafts and distinctive upward-pointing horn-shaped inflorescence buds beneath the crownshafts. The much-branched, spreading inflorescences emerge from these buds, bearing male and female blossoms.

Two species are commonly planted in tropical or nearly tropical climates; a third (*H. indica*) is common; the fourth (*H. vaughanii*) is rare in its habitat and in cultivation; and the fifth (*H. amaricaulis*) is, in habitat, reduced to one individual which does not produce viable seeds. The genus name is Greek for "pig" and "food," the allusion probably to the fruits being at one time used for fodder.

Like other genera in the tribe Chamaedoreeae, *Hyophorbe* has seeds that germinate either all at one time or sporadically for up to a year or more. Seeds of Chamaedoreeae tend to be viable longer than those of many other palm species but are best planted soon after harvest. Seeds allowed to dry out completely before sowing do not germinate as well as those kept slightly moist, and the smaller the seed, the more likely it will dry out before sowing. Fresh seeds of *Hyophorbe* germinate in 60 days to a year or more and do so sporadically.

Hyophorbe indica
hy-o-FOR-bee · IN-di-ka
PLATES 496–497

Hyophorbe indica is endemic to Réunion Island, where it grows in moist lowland forest. The epithet is Latin for "Indian," a reference to the Indian Ocean and, by extension, Réunion.

The light gray to light tan trunk attains a height of 30 feet (9 m) and a diameter of 8 inches (20 cm), and is beautifully ringed with widely spaced dark leaf base scars. The crownshaft is usually distinctly bulging at its base, 2–3 feet (60–90 cm) tall, and variable in color, especially in younger individuals: from light green to deep dark green to chocolate or reddish brown. The leaf crown is sparse but hemispherical or even rounded, with 6 leaves 6–8 feet (2.4 m) long on petioles 2 feet (60 cm) long; the petiole and rachis color is usually the same as the crownshaft in young specimens and may be light to dark green or brownish red. The leaves are gracefully arching and bear numerous, regularly spaced, light green to dark green, linear-lanceolate leaflets that, in younger palms, grow from the rachis in an almost flat and single plane but, in older specimens, are more rigid and grow at angles that create a V-shaped leaf.

The species is fast growing and generally carefree in tropical or nearly tropical climates. It prefers a humus-laden, slightly acidic soil but is adaptable to other soils as long as they are fast draining. It is more shade tolerant than the other species in the genus, doing well in the partial shade of the understory. It makes an attractive canopy-scape and is more than beautiful in groups of 3 or more trees of varying heights.

Hyophorbe lagenicaulis
hy-o-FOR-bee · la-je′-ni-KAW-lis
PLATES 498–499

Hyophorbe lagenicaulis is endemic to coastal savannas and hilly forests of the Mascarene Islands, where it is critically endangered. The epithet is derived from Greek words meaning "bottle" and "stem." The common name is bottle palm.

The mature trunk sometimes attains a height of 20 feet (6 m) and is light gray to almost white with closely set rings around all but the oldest parts, which are smooth. The shape of the trunk is unusual among palms: it is thick, 2 feet (60 cm) in diameter, and is a column for two-thirds of its height and then tapers abruptly to a bottle-neck form until it reaches the crownshaft, which is 2–3 feet (60–90 cm) tall, mint green, and smooth and waxy, with an enlarged base. The leaf crown is 8 feet (2.4 m) wide and tall with usually 6 leaves. The leaves are 6–12 feet (1.8–3.6 m) long and much arching, to the point that they form semicircles. The petiole is short, 10 inches (25 cm) long, and stout at its base. The deep green, linear-lanceolate, tapering leaflets are 2 feet (60 cm) long and grow from the rachis at an angle of 45 degrees or more, lending a deep V shape to the leaf. The inflorescences grow in a whorl around the base of the crownshaft and are upward-pointing, horn-shaped green buds 3 feet (90 cm) long and composed of several bracts which fall, one by one, to emit the much-branched greenish yellow panicles that bear small, white, unisexual flowers. The rounded 1-inch (2.5-cm) fruits are brown when mature.

This palm is tender to cold and adaptable only to zones 10b and 11. It needs full sun, average but regular moisture, and a humus-rich, sandy, well-drained soil. The crown of leaves is always beautiful and dramatically architectural. The trunks are an acquired taste; some think them dumpy, insolent, and grotesque, while others think them shapely and dramatically architectural. In any case, they could never be called "graceful." Where they are planted would seem to make all the difference as to whether or not they are pleasing in the landscape. Rather than trying to hide their unusual form, it should be accented. They are especially attractive when planted with succulents and other pachycaulous (thick-trunked) plants, such as species of *Pachypodium*, *Adenium*, and *Beaucarnea*. Given enough space and light, the palm makes a remarkably attractive indoor plant.

Hyophorbe verschaffeltii
hy-o-FOR-bee · ver-sha-FEL-tee-eye
PLATE 500

Hyophorbe verschaffeltii is endemic to coastal savannas and hilly forests of the Mascarene Islands, where it is critically endangered. The epithet honors Belgian nurseryman, botanist, and author Ambroise Verschaffelt (1825–1886). The common name is spindle palm.

The mature trunk grows to 25 feet (7.6 m) high with a diameter from slightly less than 1 foot (30 cm) to 18 inches (45 cm). The vernacular name well describes the overall shape of the stems: they are spindle shaped with the greatest diameter anywhere from the middle of the trunk to beneath the crownshaft. The crownshaft is 2–3 feet (60–90 cm) tall, mint green to powdery blue, and smooth and waxy with an enlarged base. The leaf crown is 10 feet (3 m) wide and tall. The leaves are 6–10 feet (1.8–3 m) long and arched almost exactly like those of the above species. The leaflets generally grow from the rachis at a 45-degree angle, giving the leaf blade a V shape. The leaflets are more widely spaced than those of *H. lagenicaulis*. There are usually 6 leaves in the leaf crown but, because of their arch, the crown is quite round. The inflorescence is similar to that of *H. lagenicaulis*.

This species is slightly susceptible to lethal yellowing disease. It is slightly more tolerant of cold than *H. lagenicaulis*, but it is still marginal in zone 10a. Given enough space and high light levels, the spindle palm makes a remarkably attractive indoor plant.

Hyospathe
hy-o-SPAY-thee

Hyospathe consists of 4–6 small, pinnate-leaved, monoecious palms in tropical American rain forest. The inflorescences grow from beneath the leaf crown, are branched once, small, and whitish, and look like little starbursts. They bear male and female blossoms and turn a beautiful crimson as the small, purplish-black fruits mature. Because the leaves of both species vary greatly in size and segmentation, botanists have described as many as 17 different species, but nowadays, only a few variable species are recognized. The palms are rare in cultivation and generally limited to specialty collectors and a few botanical gardens. They are not tolerant of freezing temperatures and need copious, regular moisture with partial shade and a humus-laden soil. The genus name means "hog" and "spathe" and is a literal translation into Greek of the Brazilian aboriginal name. Seeds germinate within 60 days.

Hyospathe elegans
hy-o-SPAY-thee · EL-e-ganz
PLATE 501

Hyospathe elegans is widely distributed from Costa Rica, through Panama and into most of northern South America, where it grows under canopy from sea level to 6000 feet (1800 m). The epithet is Latin for "elegant."

The mostly solitary-trunked species is sometimes found as a clustering specimen with slender trunks. The single-trunked specimens may attain a height of 20 feet (6 m) with a diameter of 1 inch (2.5 cm). They form an elongated but skinny tan crownshaft of leaf sheaths. Leaf form and size are extremely varied. Larger plants usually have leaves 3–4 feet (90–120 cm) long with 4 wide to 24 narrow, regularly spaced, sickle-shaped pinnae, while smaller plants often have undivided leaves 8–20 inches (20–50 cm) long. The wiry petioles are 1 foot (30 cm) long and the leaf blades, whether divided or not, are beautifully corrugated with raised veins. The larger individuals with segmented leaves are beautiful and are reminiscent of large, robust *Chamaedorea* species, while the smaller individuals resemble *Asterogyne* species.

Hyphaene
hy-FEE-nee

Hyphaene comprises about 8 palmate-leaved, dioecious palms in southern and eastern Africa, Madagascar, Arabia, and western India. Most species are clustering but some are solitary-trunked. A few are small, but most are moderate to large palms. They grow in dry savannas, along rivers and streams, on the margins of forests, and often in coastal regions inland of the dunes, but always in open, sunny, and mostly hot areas where groundwater is available to the thirsty roots. They are mostly lowland occupants, except for *H. compressa*.

The genus is unusual in having species with trunks that naturally and dichotomously branch. The larger species with clustering and branching stems almost always end up with only one main trunk because of the great diameter of the leaf crown which pushes over the less-dominant trunks. Highly branched specimens with dense crowns are vulnerable to wind damage. The stems are covered in Y-shaped leaf bases in their younger parts, but these bases eventually fall away to leave a trunk that is closely ringed. The leaf crown is usually hemispherical, but several dead leaves hang beneath the crown to create the visual effect of a spherical crown. The leaves are held on petioles with marginal forward-pointing thorns, exhibit a distinct hastula on the underside of the juncture of petiole and blade, and are extremely costapalmate with a strong and curving costa (rib) that arches downwards, much like in *Sabal* species, to create an almost pinnate-looking leaf. Leaf color ranges from deep green to silvery green to almost blue in some species, and the blades usually have a thin, glaucous covering of wax on both surfaces. The leaf segments are always stiff and tough and usually have several threadlike segments (again, as in many *Sabal* species) between them.

The short inflorescences grow from the leaf crown, are branched once, and bear tiny creamy white or yellow male or female flowers. The large, leathery fruits are usually pear-shaped or oblong, and are constricted in the middle, often looking like large cashew fruits; they are invariably orange to deep red to dark reddish brown. They usually take a year (sometimes two) to ripen on the tree. The flesh of the fruits of most species is edible, although fibrous and not delicious, and is often fed to livestock, while the sap from the trunks is sometimes fermented to make an alcoholic toddy. The seeds of most species are extremely hard and were, in the past, used as vegetable ivory and carved into *objets d'art*.

All species are drought tolerant and, in their native haunts, usually are found on poor soils. They grow and look better if supplied with regular, adequate moisture and a decent soil. They are tolerant of salinity in the soil and air, especially the coastal species. They are sun lovers from youth unto old age, and most of them die if planted under canopy. None are hardy to cold but most are adaptable to zones 10 and 11, especially in drier climates, and many survive as small and shrublike in 9b. All are slow growing.

The genus name is derived from the Greek word for "weave" and alludes to the fibers in the flesh of the fruits. All species are generally referred to as doum palms or simply doum, the word being a French transliteration of an Arabic word for one of the species.

Seeds generally germinate within 90 days but can take 180 days or longer. The germination stalk can go down 12 inches (30 cm), so the container in which seeds are sown should be at least 16 inches (41 cm) deep. For further details of seed germination, see *Bismarckia*.

Hyphaene compressa
hy-FEE-nee · kom-PRES-sa
PLATE 502

Hyphaene compressa is indigenous to Somalia, Kenya, Tanzania, and Mozambique, where it grows near the coasts and extends inland along streams and rivers to 4500 feet (1370 m). The epithet is Latin for "compressed" and refers to the shape of the fruit, which is flattened on 2 sides.

The trunk is mostly solitary, to 60 feet (18 m) tall in habitat, but some specimens produce 2–4 main trunks. The trunks branch as many as 4 or even 5 times with age and, until they are old, are covered in gray leaf bases. Leaf color is usually deep green but may also be silvery green or even bluish green. The fruits are mostly pear shaped or oblong and deep orange or light brown; they smell like gingerbread when broken apart.

This species is one of the tallest, most massive species in the genus. Old, mature palms are incredibly beautiful and interesting with their tall, branched trunks atop which sit the relatively small leaf crowns; few palm species are as picturesque.

Hyphaene coriacea
hy-FEE-nee · ko-ree-AY-see-a
PLATE 503

Hyphaene coriacea occurs naturally in Somalia, Kenya, Tanzania, Mozambique, northeastern South Africa, and eastern Madagascar, where it grows mostly along the coasts, even among the dunes. The epithet is Latin for "coriaceous" and alludes to the leathery fruits.

In habitat, this species is often shrubby and almost nondescript because of the nearly sterile soils it grows in; but in cultivation with proper care, it can become one of the most attractive species in the genus, with trunks 30 feet (9 m) tall. It is mostly a clustering species but can also be found in habitat as a solitary-trunked individual; the trunks do not normally branch. The fruits are oblong, constricted in the middle, and deep orange or light chestnut brown. The species seems unusually cold tolerant. It is now found in many locations of central Florida, zone 9b (David Witt, pers. comm.).

Hyphaene dichotoma
hy-FEE-nee · dy-KO-to-ma
PLATE 504

Hyphaene dichotoma is endemic to western India, where it grows in open savannas at low elevations and is endangered because of land clearing. The epithet refers to the branching habit of the species. The older trunks of this mostly clustering species dichotomously branch 3 or 4 times. The leaf crowns are large and globular, and the leaves are deep green to slightly silvery green, nearly orbicular if flattened, and 3 or slightly more feet (90 cm) across. The shiny, leathery fruits are pear shaped and light to deep orange-brown. This massive species is as beautiful and as picturesque as *H. compressa* when it is old, but it grows slowly, and small plants are not likely to assume the branching habit for many years. Reports indicate this species is unusually hardy to cold and may be possible as a tree in drier regions of zone 9b.

Hyphaene petersiana
hy-FEE-nee · pee-ter-see-AH-na
PLATE 505

Hyphaene petersiana is indigenous to Tanzania, Congo, Angola, Namibia, Botswana, and Zimbabwe, where it grows in open savannas and along streams and rivers in the desert areas. The epithet honors German explorer and zoologist Wilhelm Peters (1815–1883).

This unusual species is usually found as a solitary-trunked palm and is always nonbranching. The trunk attains a height of 60 or more feet (18 m) in habitat and is often slightly swollen near or slightly above its middle. The leaf crown is spherical and consists of grayish green, very deeply costapalmate leaves 6 feet (1.8 m) wide on petioles 4 feet (1.2 m) long.

The silhouette of this palm is exceptional because of its spherical outline and its airiness due to the exceptionally long petioles that allow the leaves to be seen; it is one of the most beautiful canopy-scapes. It is faster growing than most other species in the genus, especially if provided with moisture and a decent soil.

The species is among the most useful to native peoples who use every part of it: the trunks for construction; the leaves for an extensive basket weaving industry; the sweet flesh of the fruits for food; and the hard seeds as a vegetable ivory for making utensils and *objets d'art*.

Hyphaene thebaica
hy-FEE-nee · thee-BAY-i-ka
PLATES 506–508

Hyphaene thebaica is indigenous to northern and northeastern Africa, where it grows near the coasts in open savannas and along streams and rivers inland, always at low elevations. The epithet is Latin for "of Thebes," an ancient city on the Nile River. A common name, gingerbread palm, refers to the taste of the fruits.

The species is sparsely clustering and freely branching, even at a relatively young age. It is similar in general appearance to *H. compressa* and *H. dichotoma* but does not grow as tall, although it is almost as massive in its old age because of its repeated and dense branching. Individual trunks never attain more than 50 feet (15 m) of height and are usually 30 feet (9 m) or less. The leaves and leaf crown are smaller than the aforementioned 2 species and are a deep green to grayish or silvery green. The fruits are 3 inches (8 cm) long, pear shaped, and light brown. This palm is exceptionally picturesque and appealing at all ages. It is a fabled species with a long record of cultivation and use in ancient Egypt.

Iguanura
ig-wah-NOO-ra

Iguanura is composed of about 32 delicately beautiful, small, clustering and solitary-trunked, pinnate-leaved, monoecious palms in peninsular Thailand, peninsular Malaysia, Sumatra, and Borneo. All are undergrowth plants in rain forest. The leaves are either entire with the segments united and apically bifid, or with segments divided; the margins of the leaflets are always toothed, and the newly unfurled leaves of many species are pinkish, reddish, purplish, or bronzy for an unusually long time. The leaves last a long time also, so that in habitat, epiphytes, algae, and fungi often find homes on them. The inflorescences grow from the leaf crown in all species except *I. bicornis*, which has a rudimentary crownshaft, and are elongated and mostly spikelike, but some are branched.

All are tender to cold and adaptable only to zones 10b and 11; they need warmth year-round, even at night, and thus are not suitable for frostless Mediterranean climates. They are water lovers, need a slightly acidic, humus-laden soil, and cannot take the full sun of hot climates. All species are eminently well adapted to greenhouse or container culture.

The genus name is coined from the New World Spanish word for "lizard" (*iguana*, which itself is from a South American aboriginal word) and the Greek word for "tail," alluding to the whiplike, scaly inflorescences of some species. Seeds of *Iguanura* germinate within 60 days. For further details of seed germination, see *Areca*.

Iguanura bicornis
ig-wah-NOO-ra · by-KOR-nis
PLATE 509

Iguanura bicornis is indigenous to peninsular Thailand and peninsular Malaysia, where it grows in low mountainous rain forest. The epithet is Latin for "two horned" and refers to the shape of the fruits.

The clustering stems attain heights of 8–9 feet (2.4–2.7 m) with a slender, loose, light green crownshaft to 1 foot (30 cm) in height. The leaf crown is sparse. The leaves never lie beneath the horizontal, are 18 inches (45 cm) long, and are borne on slender petioles 1 foot (30 cm) long. The few leaflets are irregularly spaced along the rachis, each 8–10 inches (20–25 cm) long, irregularly trapezoidal, and S shaped; the apex is wider than the point of attachment, obliquely squared, and jaggedly toothed. New growth is a beautiful purplish bronze. The inflorescences grow from beneath the crownshaft and are branched twice, thin and spreading; they bear both male and female flowers. The egg-shaped red fruits are ½ inch (13 mm) long and have 2 small lobes at one end.

Iguanura elegans
ig-wah-NOO-ra · EL-e-ganz
PLATE 510

Iguanura elegans is endemic to Sarawak on the island of Borneo, where it grows in lowland rain forest. The epithet is Latin for "elegant." The leaves of this sparsely clustering species are either entire and undivided with the usual bifid apex, or divided into irregularly spaced, wide, slightly S-shaped segments. Both forms have deeply corrugated blades or segments, and the new growth is a beautiful rosy bronze. An exceptionally beautiful form has leaves variegated in cream and shades of green, the new growth also rose colored.

Iguanura geonomiformis
ig-wah-NOO-ra · jee'-o-no-mi-FOR-mis
PLATE 511

Iguanura geonomiformis is endemic to peninsular Malaysia, where it occurs in the undergrowth of low mountainous rain forest. The epithet is Latin for "in the form of *Geonoma*," an unrelated tropical American genus of palms with many vegetative similarities to *Iguanura*. This densely clustering species has thin stems never more than 3 feet (90 cm) high. The ascending leaves are 3 feet (90 cm) long and may be undivided with deeply bifid apices, or divided into a few wide segments with even larger and wider apical segments, or exhibit fine and regular pinnate segmentation.

Iguanura palmuncula
ig-wah-NOO-ra · pahl-MUNK-yoo-la
PLATE 512

Iguanura palmuncula is endemic to Borneo, where it grows in low mountainous rain forest. The epithet is Latin for "little palm." The clustering stems are no more than 6 feet (1.8 m) tall. The leaves are 2–3 feet (60–90 cm) long, a uniform emerald green, and either segmented into broad S-shaped leaflets or entire and apically bifid.

Iguanura polymorpha
ig-wah-NOO-ra · pah-lee-MOR-fa
PLATES 513–514

Iguanura polymorpha is indigenous to peninsular Thailand, Malaysia, and Borneo, where it grows in low mountainous rain forest. The epithet is from Greek words meaning "many forms," although the species is not as vegetatively polymorphous as *I. geonomiformis*. *Iguanura speciosa* is a synonym. The stems of this densely clustering species attain heights of 10 feet (3 m). The sparse leaf crowns have leaves 3 feet (90 cm) long on petioles 1 foot (30 cm) long. The leaflets are similar in shape and size to those of *I. bicornis*, and the plant overall is similar in appearance except for the fruits.

Iguanura tenuis
ig-wah-NOOR-a · TEN-yoo-iss

Iguanura tenuis is endemic to peninsular Thailand, where it grows in lowland rain forest up to 1800 feet (550 m). The epithet is Latin for "thin" and refers to the very slender rachillae of the inflorescence. Two varieties are recognized: *I. tenuis* var. *tenuis* has clustering stems and *I. tenuis* var. *khaosokensis* [kow-so-KEN-sis] has solitary stems. The stems grow to 4–5 feet (1.2–1.5 m) tall with a diameter of less than 1 inch (2.5 cm). The leaf crown has 7 leaves, each 20 inches long including the petiole. On each side of the rachis are 4–5 broad leaflets. From among the leaves emerge inflorescences 3 feet (90 cm) long, which in turn produce ovoid, pinkish fruit ½ inch (13 mm) long and ¼ inch (6 mm) wide.

Iguanura wallichiana
ig-wah-NOO-ra · wah-lik'-ee-AH-na
PLATES 515–516

Iguanura wallichiana is among the most widespread species in the genus. The epithet honors Nathaniel Wallich (1786–1854), superintendent of the botanical gardens in Calcutta, India. This palm is among the most variable in vegetative characters, and 2 naturally occurring forms have been identified, both clustering.

Iguanura wallichiana var. *wallichiana* is indigenous to peninsular Thailand, peninsular Malaysia, and Sumatra, where it grows in low mountainous rain forest. The stems grow to heights of 9 feet (2.7 m) and the leaves are 4–5 feet (1.2–1.5 m) long. The latter are spreading and slightly arching, segmented into broad, corrugated, S-shaped, regularly spaced leaflets, which are a most beautiful rosy bronze when new.

Iguanura wallichiana var. *major* [MAY-jor] is endemic to peninsular Malaysia, where it grows in low mountainous rain forest. This form is usually not as large as the type and has undivided stiff and ascending obovate, apically bifid and deeply corrugated leaves that are 2½ feet (75 cm) long and, like the type, are a beautiful purplish or rosy bronze when newly opened. This variety is most sought after in horticulture; in fact, it was so avidly collected and grown in greenhouses and conservatories in Victorian England that it is still endangered in habitat.

Iriartea
ir'-ee-AHR-tee-a

Iriartea is a monotypic genus of large solitary-trunked, pinnate-leaved, monoecious palm in tropical America. The name honors Don Bernardo de Iriarte (1735–1814), a Spanish (Canary Islands) politician and benefactor of explorations to the New World. Like other species in the tribe Iriarteeae, *Iriartea* species are among the easiest palms to grow from seed. If they are allowed to dry out completely, seeds do not germinate well, and the smaller the seed, the more likely it will dry out before sowing. Germination of Iriarteeae tends to occur within 30–180 days depending on the species. Fresh seeds of *Iriartea* almost always germinate within 60 days.

Iriartea deltoidea
ir'-ee-AHR-tee-a · del-TOY-dee-a
PLATES 517–519

Iriartea deltoidea occurs naturally in extreme southeastern Nicaragua, through Costa Rica and Panama, and into western and southern Colombia, southern Venezuela, northern Ecuador, eastern Peru, northern Bolivia, and the western Amazon region of Brazil. It grows from sea level to 4200 feet (1280 m), always in rain forest and almost always on the slopes of the hills and mountains where, in some localities, the species forms great colonies. The epithet refers to the more-or-less triangular leaflets.

The trunk attains a height of 80 feet (24 m) in habitat and is 1 foot (30 cm) in diameter, light gray to white in older individuals. The presence of a midpoint bulge seems associated with the elevation at which the palm grows: lowland individuals have the swelling, highland individuals do not. The base of the stem is, in older palms, supported by a dense cone 4–6 feet (1.2–1.8 m) tall of nearly black stilt roots bearing short, prickly thorns; young and nonemergent individuals seldom exhibit the prop roots. These roots themselves often branch near their bases, in much the same manner as do the prop roots of the red mangrove (*Rhizophora mangle*).

The crownshaft is a light grayish green, 3–4 feet (90–120 cm) tall, usually bulging at its base but may also, at that point, be the same diameter as the trunk and then expanded near its summit. The leaf crown is sparse with 6 leaves but is usually spherical because of the great spreading leaves, each of which is held on a petiole 12–15 feet (3.6–4.5 m) long and is stiff and straight except near its apex where it is slightly arched. The leaflets are narrowly wedge shaped, 2–3 feet (60–90 cm) long, of varying widths, and deep emerald green on both surfaces, with obliquely truncated, jagged apices. They grow from the rachis at different angles to create a plumose, rounded leaf. The leaflets of younger trees are exceptional in having a different form: they are immense, 2–3 feet (60–90 cm) long, flat, lustrous, and irregularly fan shaped or wedge shaped with jagged apices, growing from the rachis in a single, nearly flat plane. It is only when the young trees grow above the canopy into the full sunlight that the much narrower, plumosely arranged leaflets are formed, possibly as an adaptation to the light intensity above the forest cover.

The inflorescences are formed beneath the crownshaft. The primary bracts are remarkable, forming before opening, downward-curving, narrowly horn-shaped buds 10 feet (3 m) long and as many as 16 overlapping woody, hirsute bracts. The inflorescences consist of many cream-colored pendent branches 2–3 feet (60–90 cm) long, bearing both male and female flowers. The round 1-inch (2.5-cm) yellowish green fruits mature to bluish black.

The species has many uses in habitat: the trunks are used for house construction, blowguns, and spears (young individuals), and, hollowed out, for canoes; the inside of the leaf sheaths (crownshafts) is administered to women in childbirth to ease the rigors of labor; the leaves are used for thatch; the growing point is edible and eagerly eaten by the native people, as are the seeds.

Mature trees are among the most magnificent palms. The great plumose leaves are incomparably beautiful and create an exceptionally alluring silhouette and canopy-scape. Alas, the species does not tolerate frost, and needs copious, constant moisture and a humus-rich, constantly moist, and yet fast-draining soil. It appreciates partial shade as a juvenile but adapts to full sun even then and only comes into its own with full exposure once past its early youth.

Iriartella
ir'-ee-ahr-TEL-la

Iriartella consists of 2 clustering, pinnate-leaved, monoecious palms in tropical America. They are undergrowth species in low mountainous rain forest, often forming colonies. They send up either stems from the creeping rootstocks (rhizomes) or adventitious shoots from original stems at a point well above the ground. The slender, reedy stems are supported atop a small mass of short stilt roots. The loosely formed crownshafts are covered in a short, prickly tomentum that can be irritating to the skin. The green once-branched inflorescences bear green flowers of both sexes. The small, ellipsoid fruits are orange or red. Neither species seems to be in cultivation outside of a few botanical gardens, but both are attractive and worthy of being grown. They are intolerant of frost and need regular, adequate moisture as well as partial shade. The genus name translates as "little *Iriartea*" and alludes to the similarity of the juvenile leaves of *Iriartea* and the mature leaves of *Iriartella*. Seeds germinate within 60 days and should not be allowed to dry out completely before sowing. For further details of seed germination, see *Iriartea*.

Iriartella setigera
ir'-ee-ahr-TEL-la · se-ti-JE-ra
PLATE 520

Iriartea setigera is indigenous to southwestern Guyana, southern Venezuela, extreme eastern Colombia, and northwestern Brazil. The epithet is Latin for "bristle-bearing" and alludes to the short hairs on the leaf sheaths.

The stems, which grow to heights of 35 feet (11 m) but are no more than 2 inches (5 cm) in diameter, were formerly used in habitat for making blowguns; they are deep green and have widely spaced distinct gray rings of leaf base scars. The little crownshafts are barely thicker than the stems, a bright, light green, cylindrical, and covered in a short, stiff tomentum that is irritating to the skin. The leaf crown is sparse and hemispherical with 6 spreading leaves 4–5 feet (1.2–1.5 m) long on wiry petioles 1–2 feet (30–60 cm) long. The 8–10 leaflets are large considering the length of the leaf, light to medium green on both surfaces, obliquely wedge shaped or diamond shaped, with an undulating surface and jagged apices; the leaflets of the terminal pair are usually wedge shaped and significantly broader than the others.

Because of the light color of the leaves and crownshaft as well as the shape of the leaflets, this palm would make a stunning contrast with other vegetation or palms under a high canopy.

Iriartella stenocarpa
ir′-ee-ahr-TEL-la · ste-no-KAHR-pa

Iriartella stenocarpa occurs naturally in extreme southeastern Colombia, eastern Peru, and adjacent extreme western Brazil, where it grows in low mountainous rain forest. The epithet is from Greek words meaning "slender" and "fruit." This species differs from *I. setigera* in its smaller stature, smaller and lower (in elevation) natural distribution, smaller clumps, and in its leaves that are sometimes entire and unsegmented but usually with 3–5 segments shaped like those of *I. setigera*.

Itaya
ee-TY-ya

Itaya is a rare monotypic genus of palmate-leaved, hermaphroditic palm in northern South America, where it is in great danger of extinction because of land clearing. It is similar taxonomically and visually to *Chelyocarpus* but is readily distinguished by the petiole bases, which are split in mature leaves. The inflorescences are much branched but shorter than the leaf crown and consist of flowering branches bearing small, white, bisexual flowers that produce round greenish fruits. The genus name is that of a river in Peru near where the palm was discovered. Seeds of *Itaya* germinate within 60 days and should not be allowed to dry out before sowing. For further details of seed germination, see *Corypha*.

Itaya amicorum
ee-TY-ya · ah-mee-KO-rum
PLATES 521–522

Itaya amicorum is indigenous to northeastern Peru, southeastern Colombia, and extreme northwestern Brazil, where it occurs in the undergrowth and along rivers and streams in lowland tropical rain forest. The epithet is Latin for "of friends" and was bestowed by Harold E. Moore, Jr. (1972), because "of the spirit of the program under which I first encountered the species and for my associates in Peru."

The little palm attains a total height of 15 feet (4.5 m) with a clean trunk that is 3 inches (8 cm) in diameter. The leaves are the reason one lusts after this beauty; they are circular, 6 feet (1.8 m) in diameter, with wedge-shaped segments that are apically toothed and divided to the petiole. The 10–16 segments are deeply pleated and light, shiny green above and silvery green or even white beneath. The leaves are reminiscent of those of the larger *Licuala* species and are held on petioles 7–8 feet (2.1–2.4 m) long, giving the leaf crown the most beautiful aspect of a globular constellation of giant pinwheels spinning in the breeze against the sky.

This species is tender to cold and is adaptable only in zones 10b and 11. It needs a rich soil with unimpeded drainage and nearly constant moisture, although it thrives in partial shade or full sun when older. It makes one of the world's most beautiful canopy-scapes.

Johannesteijsmannia
yo-hahn′-nes-tezh-MAH-nee-a

Johannesteijsmannia comprises 4 undivided but palmate-leaved, hermaphroditic palms in southern Thailand, peninsular Malaysia, Sumatra, and western Borneo. These are undergrowth palms in tropical rain forest, where they occur on slopes and ridges, never in swampy areas. One species has a short trunk; the others form little or no aboveground stem, the leaves growing directly from the ground in immense rosettes that trap falling leaves and debris, which in turn decay in place and feed the palm.

The genus has the third largest undivided leaves in the palm family; only *Manicaria* of the American tropics and *Marojejya* of Madagascar have larger undivided leaves and, in both cases, the leaves are usually split or segmented by the wind. The diamond-shaped or lanceolate leaves are held on long petioles, and the strongly pleated blades exhibit a distinct midrib or extension of the petiole through the blade. Both the petiole and the bottom margin of the leaves bear tiny thorns, and the blade apex is toothed or jaggedly incised, corresponding to the corrugations on both surfaces of the blade. The leaves appear to be more pinnate than palmate, mainly because of the relative narrowness and length of the blades. That they are basically palmate is proved by the small hastula on the undersides of young leaves before they are fully expanded; this hastula is usually worn off by the time the leaf is unfurled.

The relatively short inflorescences are partially hidden in the leaf crowns by the size of the leaves and, in habitat, by leaf litter and other debris. They are enclosed in bud by several felty bracts that persist after the short, curved cream-colored branches of the inflorescences emerge. The bisexual flowers smell bad but produce light brown, globose fruits with corky, conical projections.

These palm species are almost maniacally sought after by gardeners and collectors in suitable climates. Drawings and photographs seldom convey their extraordinary size and elegance. Amazingly, although they are adaptable only to tropical or nearly tropical climates, they nevertheless withstand temperatures at or near freezing for short periods. They resent cold, dry winds and demand partial shade especially when young. They need a humus-laden, friable soil that is constantly moist but quickly draining, and they are difficult to maintain without constantly high relative humidity.

The genus is called joey palms, a shortened form of the scientific name, which honors Johannes Teijsmann (1808–1882), Dutch gardener and botanist at the Bogor (then Buitenzorg) botanical garden in Indonesia.

Seeds of *Johannesteijsmannia* germinate within 45 days and should not be allowed to dry out before sowing. The shell often cracks and breaks off, but this seems to cause little problem with germination and may even enhance it. If cracked, the shell should be removed so water does not become trapped between it and the seed and cause disease. For further details of seed germination, see *Corypha*.

Johannesteijsmannia altifrons
yo-hahn′-nes-tezh-MAH-nee-a · AL-ti-frahnz
PLATE 523

Johannesteijsmannia altifrons has the widest natural distribution of the genus and is found in Sumatra, peninsular Malaysia, southern Thailand, and western Borneo, where it grows in mountainous rain forest at 1000–3000 feet (300–900 m). It is threatened in Malaysia because of forest destruction and the gathering of the leaves for thatch. The epithet is Latin for "tall frond" and alludes to the centermost leaf in a rosette that may reach as high as 20 feet (6 m). The species is sometimes called diamond joey.

The petioles are 6–10 feet (1.8–3 m) long and are armed with tiny sawlike teeth, which are also found on the lower margins of the younger leaf blades. The diamond-shaped blades are 10 feet (3 m) long in older plants and 6 feet (1.8 m) wide at their broadest points. They are light to medium green on both surfaces and are held erect and slightly spreading, but the older ones near the margins of the rosettes are usually spreading horizontally and pendent. There may be as many as 24 giant leaves in a single rosette. This species is the most widely cultivated in the genus, and for good reason: it is supremely attractive.

Johannesteijsmannia lanceolata
yo-hahn′-nes-tezh-MAH-nee-a · lan-see-o-LAH-ta
PLATE 524

Johannesteijsmannia lanceolata is endemic to southeastern peninsular Malaysia, where it grows in low mountainous rain forest and is highly endangered. The epithet is Latin for "lanceolate," referring to the shape of the leaves. The species differs from *J. altifrons* in having linear leaves 7–8 feet (2.1–2.4 m) long and only 1 foot (30 cm) wide on petioles 3 feet (90 cm) long.

Johannesteijsmannia magnifica
yo-hahn′-nes-tezh-MAH-nee-a · mag-NI-fi-ka
PLATES 525–526

Johannesteijsmannia magnifica is endemic to peninsular Malaysia, where it grows in low mountainous rain forest and is endangered. The epithet is Latin for "magnificent." This species is similar to *J. altifrons*, but the leaf underside has a gorgeous, almost shimmering, white tomentum, which often has a bluish hue and creates an unbelievably beautiful leaf.

Johannesteijsmannia perakensis
yo-hahn′-nes-tezh-MAH-nee-a · per-a-KEN-sis
PLATE 527

Johannesteijsmannia perakensis is endemic to central peninsular Malaysia, where it grows in low mountainous rain forest. The epithet is Latin for "of Perak," a province in the palm's habitat. This palm forms an aboveground stem, which grows to 12 feet (3.6 m) high. The leaves are similar in appearance and dimensions to those of *J. altifrons*, slightly narrower. A mature or nearly mature specimen is truly glorious, especially when incorporated into other vegetation, but the trunk takes many years to grow very high.

Juania
WHAH-nee-a

Juania is a rare and endangered monotypic genus of pinnate-leaved, dioecious palm endemic to the Juan Fernández group of islands in the eastern Pacific, 350 miles (560 km) west of Valparaiso, Chile. It is threatened by disturbance to the delicate and fragile ecosystem of its island home. The genus name is taken from the name of the islands. Seeds germinate within 90 days. For further details of seed germination, see *Ceroxylon*.

Juania australis
WHAH-nee-a · aw-STRAH-lis
PLATE 528

Juania australis grows in moist, cool forests on steep hills and ridges at 600–2500 feet (180–760 m). The epithet is Latin for "southern."

The solitary trunk attains a height of 50 feet (15 m) in habitat and is 1 foot (30 cm) in diameter. It is light to dark gray in all but its youngest parts and is ringed with light colored, closely spaced leaf base scars. There is no crownshaft but, in young plants, the leaf sheaths are persistent and a deep olive green. The leaf crown is hemispherical to nearly rounded, with spreading and slightly arching leaves 6–8 feet (1.8–2.4 m) long on short petioles. The medium to dark green leaflets are 2 feet (60 cm) long, stiff, linear-lanceolate, and shortly bifid at their tips. They grow from the rachis at an angle, giving a slight V shape to the leaf. The inflorescences grow from the leaf sheaths and are accompanied by long, woody bracts.

The species is closely related to *Ceroxylon* species. It seems nearly impossible to grow, and only a few individuals are growing outside of the palm's island home. It has been successfully grown in coastal mainland Chile, Ireland, and San Francisco. In New Zealand and other cool Mediterranean climates, it seems to last only a while. Small palms in southern Europe were killed by temperatures of 25–29°F (−4 to −2°C) and do not last a season in climates that have hot summer temperatures with high nighttime temperatures. It is probably the most demanding and difficult palm to grow.

Jubaea
joo-BAY-a

Jubaea is a monotypic genus of massive, solitary-trunked, pinnate-leaved, monoecious palm. The name supposedly honors Juba II (ca. 25 BC–ca. 23 AD), king of the ancient kingdom Mauretania, an area corresponding to western Algeria and northern Morocco. Seeds can take more than a year to germinate and do so sporadically. For further details of seed germination, see *Acrocomia*.

Jubaea chilensis
joo-BAY-a · chi-LEN-sis
PLATES 529–530

Jubaea chilensis is endemic to a small area in central Chile, where it grows in the scrub and low forests at low elevations. The epithet is Latin for "of Chile." The common name is wine palm, Chilean palm, or Chilean wine palm.

Mature trunks grow to 80 or more feet (24 m) in habitat with a diameter of 6 feet (1.8 m), making them some of the thickest palm trunks. They sometimes show a slight bulge near the middle and are usually dark gray, retaining their rings of diamond-shaped leaf base scars even on the oldest parts of the stems. The leaf crown is 25 feet (7.6 m) wide and 15–20 feet (4.5–6 m) tall. The leaves are 8–12 feet (2.4–3.6 m) long on short petioles that are generally less than 1 foot (30 cm) long. On some individuals, the leaves are stiff and nearly erect, but in others they are gracefully spreading, with the narrow, stiff leaflets 2 feet (60 cm) long and growing from the rachis in one plane to give a flat shape to the blade. Leaf color is a dull green above and a lighter, often grayish green beneath. The leaf crown may be hemispherical or almost completely rounded. The inflorescences are branched once and have large paddle-shaped woody, persistent, felt-covered bracts. The flowering branches are usually 4 feet (1.2 m) long and erect, bearing small, dirty purple unisexual blossoms of both sexes. The round 1-inch (2.5-cm) fruits are yellow to orange, and hang in pendent clusters mostly hidden by the leaves.

The species is drought tolerant when established but grows more quickly and looks better with average but regular moisture. It is hardy to cold and is adaptable to zones 9 through 11 in areas subject to wet freezes in winter, and to zones 8 through 11 in drier, Mediterranean climes. It is not recommended for hot, humid climates like those found in Florida and along the Gulf coast; although it is perfectly hardy in most parts of these areas, it grows slowly, is never as large or beautiful, and tends to die out because of year-round high heat and humidity and because of the little difference between day and night temperatures. In cooler climates like those of Southern California, the Mediterranean, South Africa, and areas of Australia, it is often the most magnificent palm grown. It is not particular about soil type as long as it is fast draining. It needs as much sun as possible.

Few gardens are large enough to accommodate a mature Chilean palm, but since it grows so slowly, it is usually planted without regard to its ultimate size. It is one of the few palms that look good in straight rows lining a drive or large avenue or flanking an entrance. There is something massively "Egyptian" about the thick, massive trunks that is reminiscent of the columns of the ancient temple of Karnak. This palm is architectural and yet among the most magnificent subject for a canopy-scape. It is not known to be grown indoors and is not recommended for such situations.

This palm is felled for its sap, which is boiled down into syrup. The practice has decimated the remaining stands, many of which are now protected. The seeds are edible and delicious, tasting much like coconut, and are harvested commercially.

Jubaeopsis
joo-bay-AHP-sis

Jubaeopsis is a rare monotypic genus of clustering, pinnate-leaved, monoecious palm. The name means "similar to *Jubaea*," although the 2 genera are neither visually nor evolutionarily close. Seeds can take more than a year to germinate and do so sporadically. For further details of seed germination, see *Acrocomia*.

Jubaeopsis caffra
joo-bay-AHP-sis · KAF-fra
PLATE 531

Jubaeopsis caffra grows at low elevations along 3 rivers near the coast of Pondoland in northeastern South Africa. The species is sometimes called Pondoland palm. It was once known as kaffir palm, a moniker that should be eschewed as it is a derogatory term for a person of color and comes from an Arabic word meaning "infidel" or "heathen." The epithet is a Latinized form of the word "kaffir."

The stems of this sparsely clumping species are thick and robust looking. They grow to heights of 20 feet (6 m) in habitat and are covered in all but their oldest parts with persistent, woody leaf bases. The leaf crowns are, at most, hemispherical and often look like a shaving brush due to the large, erect leaves. The leaves are 12–15 feet (3.6–4.5 m) long, gracefully arching from their midpoints, and usually have a twist to the rachis from the midpoint on, resulting in a vertical alignment of the leaflets in the apical half. The stout petioles are 3–4 feet (90–120 cm) long and may be green, yellow, or deep orange. The leaflets of new leaves and those of juvenile plants are stiff and grow from the rachis at an angle that gives a V-shaped leaf, while those of older leaves in older plants are often nearly pendent. The leaflets are light to medium green, 3 feet (90 cm) long, lanceolate to narrowly lanceolate, with obliquely truncated and shortly bifid apices, and are regularly spaced along the rachis. The cream-colored inflorescences are 3–4 feet (90–120 cm) long; they grow out of the leaf bases and emerge from 2 bracts, the upper of which is large, woody, and persistent. They carry flowers of both sexes and produce round 1-inch (2.5-cm) yellow fruits.

This palm seems to need sun even from an early age and, while not fussy about soil type, does best with a fast-draining, fertile one. It languishes if not provided with regular, adequate moisture. It would seem to be a perfect candidate for Mediterranean climates but is not so good in tropical or nearly tropical regions, where night temperatures in summer are high. It tolerates frost moderately and is adaptable to zones 10 and 11.

The species is slow growing: its trunks take many years to grow to any size. Even when young, however, it is a beautiful clumper because of the great arching leaves, and it gives a wonderful contrast when incorporated into other vegetation. It is even beautiful as a specimen surrounded by space, especially when older. It is a close relative of *Cocos* (coconut), and much has been written about the similarities between the 2 genera. While the leaves may be similar, there is no way the ponderous-looking and clumping trunks of *Jubaeopsis* could be mistaken for those of the coconut; they are almost unique.

Kentiopsis
kent-ee-AHP-sis

Kentiopsis is composed of 4 large solitary-trunked, pinnate-leaved, monoecious palms in New Caledonia. All but one species is endangered. The inflorescences are formed beneath the crownshafts and consist of reddish or purplish much-branched sprays of waxy, furry white blossoms. The fruits are egg shaped, 1 inch (2.5 cm) long, and bright red or dull purple when mature. All the species are intolerant of cold and adaptable only to zones 10b and 11. They are slow growing. The genus name translates as "similar to *Kentia*," and likens this genus to *Howea forsteriana*, which was once called *Kentia forsteriana*.

Seeds germinate within 30 days for *Kentiopsis piersoniorum*, 120 days for *K. oliviformis*, and sporadically over 2 years for *K. magnifica*. There is no standard rule for the species in this genus. Seeds allowed to dry out completely before sowing do not germinate as well as those kept slightly moist. Seeds should be harvested when fully ripe as slightly green seeds do not germinate well. For further details of seed germination, see *Areca*.

Kentiopsis magnifica
kent-ee-AHP-sis · mag-NIF-i-ka
PLATE 532

Kentiopsis magnifica is endemic to northern New Caledonia, where it grows in colonies in mountainous rain forest at 1000–2000 feet (300–610 m). The epithet is Latin for "magnificent." A synonym is *Mackeea magnifica*. The trunk can attain a height of 80 feet (24 m) and is straight, columnar, less than 1 foot (30 cm) in diameter, and strongly ringed, except in the oldest parts. The crownshaft is 3–4 feet (90–120 cm) tall, slightly wider than the stem, of uniform width, and a beautiful bluish green to purplish black. The leaves are 8 feet (2.4 m) long on short petioles and exhibit a vertical twist from their midpoints to their apices; they often are a handsome reddish brown or cherry red when newly unfurled. The deep green, leathery leaflets are regularly spaced along the rachis in a single flat plane and are linear-acuminate and 2–3 feet (60–90 cm) long.

Kentiopsis oliviformis
kent-ee-AHP-sis · o-liv'-i-FOR-mis
PLATE 533

Kentiopsis oliviformis is endemic to central New Caledonia, where it grows in colonies in wet forests and valleys and drier hillsides under 1000 feet (300 m) elevation. It is in danger of extinction because of agricultural expansion. The epithet is Latin for "in the form of an olive," a reference to the fruits.

This palm is the tallest one native to New Caledonia, with trunks reaching to almost 100 feet (30 m) tall and 1 foot (30 cm) in diameter. The crownshaft is 3 feet (90 cm) tall, scarcely wider than the trunk, and dull purplish to brownish green. The leaves are 10–12 feet (3–3.6 m) long, sharply ascending and, except when dying, never lie beneath the horizontal. The deep green, linear-acuminate leaflets are 2–3 feet (60–90 cm) long and grow in a single flat plane from the rachis.

The species seems adaptable to most soils, even alkaline and calcareous ones, and it is becoming popular in South Florida. The palm is of exceptional beauty, both as a juvenile and as a mature specimen. It is majestic enough to be an isolated specimen but looks best in groups of 3 or more individuals of varying heights. It is one of the grandest canopy-scapes.

Kentiopsis piersoniorum
kent-ee-AHP-sis · peer-sahn-ee-OR-um
PLATE 534

Kentiopsis piersoniorum is endemic on the steep, wet slopes of Mount Panié in New Caledonia, at elevations between 1000 and 3000 feet (300–900 m). The epithet honors the Pierson family of New Caledonia, who helped Jean-Christophe Pintaud and Donald Hodel explore the palm flora of the island.

The trunk can reach a height of 50 feet (15 m), atop which resides the purplish gray, waxy crownshaft which is 3 feet (90 cm) tall and has a diameter slightly greater than the trunk and does not vary much along its length. The beautifully arching, light to deep grayish green leaves are 8 feet (2.4 m) long with regularly spaced, stiff, linear-acuminate leaflets growing from the rachis at a steep angle to give the leaf a V shape.

This species is the only one in the genus that is not critically endangered in habitat. It is unmatched as a beautiful canopy-scape with its globular crown of recurved leaves. Specimen groups of 3 or more individuals of varying heights are equally esthetically satisfying.

Kentiopsis pyriformis
kent-ee-AHP-sis · py-ri-FOR-mis
PLATE 535

Kentiopsis pyriformis is endemic to southeastern New Caledonia, where it grows in rain forest and is much endangered. The epithet is from Latin words meaning "in the form of a pear" and refers to the fruit shape. The palm grows to 70 feet (21 m) overall. Its crownshaft is purplish or coppery colored, and the sparse leaf crown consists of arching, deep green leaves 8 feet (24 m) long with sharply ascending stiff leaflets. This species is still rare in cultivation.

Kerriodoxa
ker'-ree-o-DOX-a

Kerriodoxa is a monotypic genus of solitary-trunked, palmate-leaved, dioecious palm. It is sometimes called white elephant palm. The genus name is a combination of 2 words, one of which honors botanist Arthur F. G. Kerr (1877–1942), the original collector, and the other of which is Greek for "glory." Seeds of *Kerriodoxa* germinate within 30 days and should not be allowed to dry out before sowing. For further details of seed germination, see *Corypha*.

Kerriodoxa elegans
ker'-ree-o-DOX-a · EL-e-ganz
PLATES 536–537

Kerriodoxa elegans is endemic to the low mountainous and wet forests of Phuket, an island off the west coast of peninsular Thailand. The epithet is Latin for "elegant."

Mature trunks grow to 15 feet (4.5 m) high with a diameter of 5–8 inches (13–20 cm). They are generally light brown and smooth and show closely set rings on the older parts; the younger portions are covered with narrowly triangular leaf bases. The leaf crown is 10–12 feet (3–3.6 m) wide but usually 8–9 feet (2.4–2.7 m) tall. The circular leaves are 6–8 feet (1.8–2.4 m) wide, with the tapering narrowly lanceolate, once-pleated segments and divisions extending to a third or almost halfway through the blade. The stout, black petioles are 3 feet (90 cm) long, and the leaf segments are slightly pendent at their ends. A small hastula sits atop the blade near its center. Leaf color is light to medium green above, but the undersurface is covered in a short but dense white tomentum. The male inflorescences are much branched and about 1 foot (30 cm) long; the females are about the same size but more sparingly branched. Both are white and carry small, whitish flowers. The round 2-inch (5-cm) fruits are creamy yellow when ripe.

This palm has proven amazingly hardy to cold considering its tropical origins; it is adaptable to zones 10 and 11 and is found in favorable microclimates of 9b. It loves moisture, a rich, well-drained soil, and partial shade, although large palms can adapt to full sun. Constantly high relative humidity seems to be important to the palm's health; the only problem with the species in southern and central Florida comes in the dry season (winter) when the leaves can look ragged, and if not watered every day, the plant can die.

The great, spreading leaves are astonishingly handsome and spectacular, especially in young plants with little trunk. There are no more beautiful palmate leaves; the nearest comparison to them is young leaves of *Saribus rotundifolius* or some of the larger-leaved species of *Pritchardia*. The leaves should be protected from damaging high winds.

This palm is perfect for a patio or courtyard. It is also a beautiful accent in a bed of lower vegetation. Planted in groups of 3 or more individuals of varying heights, it creates a tableau of near rapturous beauty and, if one has a hillside, nothing could be more splendid than having these wonderful leaves seen from above or below. Seed is now widely distributed. It is not known to be grown indoors, but there should be no reason why it wouldn't succeed with high light and humidity. It would also seem to be one of the choicest palms for containers outdoors.

Korthalsia
kor-THAL-see-a
PLATE 538

Korthalsia consists of about 26 clustering, climbing, pinnate-leaved, monoecious palms in lowland wet forests of the Andaman and Nicobar Islands, Myanmar, Thailand, Laos, Vietnam, Cambodia, southeastern China, Sumatra, Indonesia, and eastwards to New Guinea and the Philippines. The genus name honors Pieter Willem Korthals (1807–1892), Dutch botanist and explorer, who collected in Indonesia.

The mostly large, high-climbing species are the only rattans with aerial stems that branch. The stems are covered in all but the oldest parts in leaf bases clothed with long, needlelike spines, which are also found (in much smaller sizes) on the petioles and rachises of climbing individuals. The leaves of juvenile plants are unlike those of older plants and are entire, usually linear-lanceolate, with distinct petioles and toothed margins, and are undivided except for the bifid apex in some species. Once the plants begin to climb, the pinnate leaf is formed with an extension of the rachis into a barbed cirrus, allowing stems to hook onto other vegetation in the plant's scramble upwards. The leaflets of climbing individuals are linear, wedge shaped or diamond shaped, and deep green above but silvery or silvery brown beneath; they grow in a roughly flat plane from the rachis until the stem reaches sunlight, at which time they usually become pendent.

Another distinction of the genus is that the leaflets of most species are borne on short stalks, which grow directly from the rachis. Several species have persistent, much swollen leaf sheaths in which large biting ants reside who, when disturbed, first click their large and hard mandibles together as a warning and then rush out to attack whatever occasions the disturbance.

The large inflorescences grow from the ends of the stems, not from the stem's nodes, and, after flowering and fruiting, the stem dies back to a living branch at some distance below the inflorescence. Each inflorescence consists of stout, robust primary branches that branch once again, the long, felt-covered, wormlike, and pendent flower stalks growing from the second set of branches and bearing bisexual blossoms. The fruits are round or egg shaped, brown, orange-brown, or reddish brown, and are covered with large, overlapping scales.

After they begin to climb, these plants are viciously spiny in their younger parts, and this is most likely why they are so rare in cultivation. If space can be found for them, however, they are beautiful in a wall of vegetation where they create a lovely contrast to their surroundings. The juvenile, nonclimbing plants would make great container subjects. The plants require a tropical climate and copious moisture, and, if kept shaded and occasionally pruned, will keep the unsegmented, friendly leaf form for a while. The stems of many species are used in habitat to make rope, baskets, and other utensils, but have no commercial value for making rattan furniture because they cannot be polished. Seeds germinate within 30 days. For further details of seed germination, see *Calamus*.

Korthalsia scortechinii
kor-THAL-see-a · skor-te-KEE-nee-eye
PLATE 539

Korthalsia scortechinii is widespread in forests up to 3000 feet (900 m) in Thailand, Singapore, and peninsular Malaysia. The name honors Benedetto Scortechini (1845–1886), Italian priest and botanist who collected plants in Malaysia and Australia. It is a high-climbing rattan with stems about ½ inch (1.5 cm) thick. The stems are completely covered by the inflated leaf sheaths, which are covered with short spines. In the palm's natural habitat, ants nest inside the inflated sheaths and defend the palm against all attackers, including botanists. The leaflets are narrow, somber dark green. The species is not known to be in cultivation and would likely be challenging outside the wet tropics.

Laccospadix
lak-ko-SPAY-dix

Laccospadix is a monotypic genus of pinnate-leaved, monoecious palm from Atherton Tableland in Australia. The genus name is from Greek words meaning "pit" and "spadix," an allusion to the flowers being formed in depressions on the branches of the inflorescences. In Australia, the species is called Atherton palm and misty mountain palm. Seeds of *Laccospadix* germinate within 60 days. For further details of seed germination, see *Areca*.

Laccospadix australasicus
lak-ko-SPAY-dix · aus-tra-LAY-si-kus
PLATES 540–541

Laccospadix australasicus is endemic to rain forest in northeastern Queensland, Australia, where it occurs in the undergrowth at 2500–4500 feet (760–1370 m). The epithet is Latin for "Australian."

The species is found mostly as a clustering palm, with stems to 10–12 feet (3–3.6 m) tall and no more than 2 inches (5 cm) in diameter. If solitary, the trunk usually grows to 25 feet (7.6 m) tall and 4 inches (10 cm) in diameter. Whether clustered or solitary, the stems are deep green, almost black, in their younger parts and deep tan or brown in their older parts. They are distinctly ringed with lighter leaf base scars and are covered with the persistent leaf bases and short fibers in their younger parts. The leaf crown is never more than hemispherical, even in solitary-trunked forms, because the leaves are ascending, mostly erect, and gracefully arching. They are 6 feet (1.8 m) long on stout, green petioles 3 feet (90 cm) long. The leaflets are linear-elliptical, long-tipped, light to medium green on both surfaces, and widely and evenly spaced along the rachis in a single flat plane. New growth is sometimes a beautiful wine to deep bronzy red, and there is some evidence that the solitary-trunked specimens more often produce the colored growth than do the clustering individuals. The inflorescences are 3–4 feet (90–120 cm) long, grow from the leaf crown, and are remarkable single, cream-colored, and pendent unbranched spikes that bear tiny male and female flowers. The oblong, juicy fruits grow directly on the spike, are ½ inch (13 mm) long, and are bright cherry red when ripe. The endosperm is ruminate.

This palm needs partial shade at all ages. It does not grow well in hot or tropical climates with warm nighttime temperatures. It wants a moist, relatively cool but frost-free climate with no extreme fluctuations in temperature. It is not well adapted to calcareous soils but needs a friable, humus-laden medium that is constantly moist but not soggy. This species is suitable for coastal Southern California and similar climates. It is an elegant and graceful small palm. The clustering forms are a veritable symphony of lovely leaf forms among other vegetation, and the solitary-trunked form is scintillating when surrounded by space and a short groundcover, all under canopy. One of the most beautiful sitings is in the Rain Forest Pyramid at Moody Gardens in Galveston, Texas, where a large, curving trunked solitary specimen leans out over a pathway, its beautifully ringed stem and lovely, ethereal leaf crown often with colored new growth.

Laccosperma
lak-ko-SPUR-ma

Laccosperma comprises 6 clustering, spiny, climbing, pinnate-leaved, hermaphroditic palms in rainforest swamps of tropical West Africa. They are mostly large high-climbing rattan palms with large leaves. The stems (actually the adherent leaf sheaths) are invariably spiny, as are the petioles and rachises. The leaflets also have small barbs and are linear, usually regularly spaced along the spiny rachis, and mostly pendent. The ends of the rachises are formed into long cirri with pairs of backward-pointing spines. The inflorescences form at the ends of the stems and are usually tall and branched twice, with the flowering branches hanging down from the secondary branches like a curtain. After the bisexual flowers form fruits, the stems die.

These species are almost completely unknown to horticulture outside of their native haunts. The leaves are large, striking, and even beautiful, but the plants are so spiny that their use in gardens is limited. The genus name is from Greek words meaning "pit" and "seed" and calls to mind the irregular pits and bumps on the surface of the seeds of some species. Seeds are not generally available, so germination information is nonexistent.

Lanonia
la-NO-nee-a

Lanonia was described in 2011 to accommodate 8 species of dioecious, understory palmate-leaved palms from Vietnam, southern China and Hainan, and Java (Indonesia). The species were all formerly placed in *Licuala*, and strongly resemble it. The new genus was discovered during a molecular study, when much to the surprise of the researchers and other palm taxonomists, a group of mostly Vietnamese "*Licuala*" was shown to be distinct from true *Licuala*. *Lanonia* differs from that genus in having unisexual flowers and a small glandlike structure at the tip of the costa, on the underside of the leaf. The genus name is taken from the Vietnamese common name for these palms, *la non*, meaning "hat palm."

Seeds take 60–90 days to germinate and do so sporadically. Ideally, seeds should not be allowed to dry out prior to sowing. For further details of seed germination, see *Corypha*.

Lanonia dasyantha
la-NO-nee-a · da-zee-AN-tha
PLATE 542

Lanonia dasyantha is indigenous to extreme southeastern China and Vietnam, where it occurs in the undergrowth of low mountainous rain forest. The epithet is Latin for "hairy flower," a reference not to the flower itself, but to the rachillae, which are densely covered with rusty brown hairs.

With time the sparsely clustering species forms short, stout trunks 3 feet (90 cm) high. The segmented leaves, 18 inches (45 cm) wide, are held on thin petioles 1–2 feet (30–60 cm) long. The leaf outline is circular, and the blade consists of 9 or 10 segments, each with an obliquely truncated and toothed apex. The middle segments are the broadest, the ones adjacent to them progressively more narrow and slightly shorter as they approach the basal portion of the leaf. The upper surface of the leaves is a lustrous medium to dark green mottled with lighter green; the undersides a lighter green.

This is a gem of a palm, its beautiful, little round leaves as pretty in a delicate way as any small *Licuala*. It is adaptable to zones 10 and 11, and even in warm microclimates of 9b.

Latania
la-TAN-ee-a

Latania is composed of 3 large palmate-leaved, solitary-trunked, dioecious palms endemic to the Mascarene Islands, where they are now nearly extinct. They are similar in general appearance, especially as adult plants, and are suspected of hybridizing when grown near each other, so that it may be impossible to assign a specific name to such seedlings. All the species, especially when older, resemble *Bismarckia nobilis*. The salient differences are the greatly flaring trunk base of *Latania*, the leaves that lack fibers among the segments, and the sculptured endocarp (stone). In contrast, *Bismarckia* has a stouter, taller trunk, larger and less costapalmate leaves that retain fibers among their segments, and a smooth endocarp.

Latania species are slightly susceptible to lethal yellowing disease and are much less cold hardy than *Bismarckia*. They are adaptable only to zones 10b and 11, although mature specimens are found in favorable microclimates of 10a. These drought-tolerant palms grow faster and look better with average but regular moisture. They are not fussy about soil type as long as its drainage is unimpeded; they even flourish in calcareous soils.

These palms are spectacular because of their size and color, and are amenable to almost any landscape situation other than small spaces. Their exceptionally straight trunks and rounded, dense crowns make them unexcelled as canopy-scapes. They look good as isolated specimens but are at their best in groups of 3 or more palms of varying heights. They are not known to be grown indoors, but there would seem to be no good reason why they cannot be, given enough light and space. The genus name is a Latinized form of the vernacular name, *latanier*.

Seeds can take up to 18 months to germinate and do so sporadically. The germination stalk can go down 3 inches (8 cm), so the container in which seeds are sown should be at least 6 inches (15 cm) deep. For further details on seed germination, see *Bismarckia*.

Latania loddigesii
la-TAN-ee-a · lo-di-GEE-zee-eye
PLATE 543

Latania loddigesii is endemic to savannas and open woodlands of Mauritius Island, where it has been wiped out by development and agricultural expansion. Wild palms still survive on Round Island and other small islets. The epithet honors British nurseryman George Loddiges (1784–1846), whose nursery and arboretum greatly influenced 19th-century European horticulture. The common name is blue latan palm.

Mature trunks grow to at least 35 feet (11 m) high and are 10 inches (25 cm) in diameter, swollen at the base, and closely set with the rings of deeply indented leaf base scars. The stem is usually deep gray but may be almost black (or sometimes dark brown), and the leaf crown is full, rounded, and 12 feet (3.6 m) wide and tall. The leaves are 6–8 feet (1.8–2.4 m) wide and costapalmate, with the petiole protruding so far into the leaf that the blade is folded and usually does not spread to its full width. The petiole is 4–6 feet (1.2–1.8 m) long and is armed with small teeth along its lower margin when young, and covered in a white tomentum which does not rub off with age; it is heavily tinged with red along its margins in young palms. The lanceolate leaf segments are stiff and armed with fine, tiny teeth along the margins; they are never pendent, and they are free for half their length. Leaf color is bluish gray-green to silvery blue. The inflorescences are shorter than the leaves but sometimes 6 feet (1.8 m) long. The male inflorescences are shorter than the females and are composed of several branches that bear small, yellowish brown blossoms, while the female inflorescences are longer single stems with flowers of the same color. The brown plum-shaped fruits are 3 inches (8 cm) long and 1 inch (2.5 cm) wide.

Latania lontaroides
la-TAN-ee-a · lon-ta-RO-i-deez
PLATES 544–545

Latania lontaroides, commonly known as red latan palm, is endemic to Réunion Island. The epithet is Greek for "similar to *Lontarus*," an illegitimate name for *Borassus*. This species differs from *L. loddigesii* in having distinctly red petioles and red-margined, red-veined stiff leaflets when young; adult leaf color is never silvery blue but rather a deep, dull green above with grayish green beneath. Because young plants of both species have similar leaf and petiole, many growers believe they have this species when they most likely have the much more common *L. loddigesii*.

Latania verschaffeltii
la-TAN-ee-a · ver-sha-FEL-tee-eye
PLATE 546

Latania verschaffeltii, commonly known as yellow latan palm, is endemic to Rodrigues Island, where it is critically endangered. The epithet honors Belgian horticulturist, nurseryman, and author Ambroise Verschaffelt (1825–1886). Young plants have bright yellow petioles covered in white tomentum. Mature plants differ from

L. loddigesii in having much thinner, much less stiff leaves that are never silvery or bluish but rather yellowish to deep green on both surfaces.

Lemurophoenix
lee-moo′-ro-FEE-nix

Lemurophoenix is a monotypic genus of large pinnate-leaved, monoecious palm from Madagascar. The genus name is a combination of "lemur," the endemic primate emblematic of the island, and *Phoenix*, the date palm, but in combination meaning simply "a pinnate palm." According to Dransfield (1994), the aboriginal name for the palm in Madagascar translates as "the palm of the red-tufted lemur." Seeds have not been easy to germinate, but when they do, they take 6 months or perhaps a bit longer. Seeds allowed to dry out completely before sowing do not germinate as well as those kept slightly moist. For further details of seed germination, see *Areca*.

Lemurophoenix halleuxii
lee-moo′-ro-FEE-nix · ha-LOO-zee-eye
PLATE 547

Lemurophoenix halleuxii is endemic to northeastern Madagascar, where it grows on steep slopes in rain forest at 600–1400 feet (180–430 m). It is critically endangered, with only a handful of small, scattered populations known to exist. The epithet honors Dominique Halleux, who assisted in bringing this species to the attention of John Dransfield.

The solitary trunk attains a height of 60 feet (18 m) in habitat and a diameter of 18 inches (45 cm). It is light to dark gray and distinctly ringed with darker leaf base scars. The crownshaft is 4–5 feet (1.2–1.5 m) tall, slightly thicker than the trunk, cylindrical, and light to pinkish gray. The leaf crown is hemispherical with leaves 12 feet (3.6 m) long on stout petioles 1 foot (30 cm) long. The erect to spreading leaves are mostly completely straight. The deep green linear-lanceolate leaflets are 3 feet (90 cm) long and regularly spaced along the rachis; they are wine colored when new, at least in younger plants. The inflorescences grow from beneath the pinkish crownshaft and consist of many branches, each 6 feet (1.8 m) long on a large peduncle 3 feet (90 cm) long. The branches bear both male and female flowers, and the globose 1-inch (2.5-cm) fruits are pale brown with an usual corky warty surface.

When immature, this truly massive palm has the look of a gigantic *Ravenea rivularis*. It is grand and beautiful at all stages but is reportedly not easy to grow, needing ultra-tropical conditions with constant warmth and copious, regular moisture. While quickly forming an impressive leaf crown, this palm takes a long while to begin forming a trunk. Its great beauty would seem to justify whatever it takes to grow it in a tropical climate or in a large conservatory.

Leopoldinia
lee′-o-pol-DIN-ee-a

Leopoldinia includes 2 small, pinnate-leaved, monoecious palms in the Amazon region of South America, where they grow along rivers and in lowlands that are seasonally flooded for months at a time. Both are characterized by much fiber among their leaf bases, fibers that in one species obscure the trunk. The inflorescences are much branched, brown, and hairy, and bear tiny unisexual flowers that produce, in one species, unique flattened fruits. Neither species seems to be in cultivation, but there seems no reason why they could not be, if given a tropical, wet climate.

The genus name honors Carolina Josephina Augusta Leopoldina of Habsburg (1792–1873), whose father, Maxmillian I Joseph, king of Bavaria, sponsored an expedition to Brazil where Carl F. P. von Martius discovered and later described this genus. Seeds of *Leopoldinia* are not generally available, and the seeds that were tried were dried out and did not germinate. Seeds should probably be kept moist prior to sowing. For further details of seed germination, see *Areca*.

Leopoldinia piassaba
lee′-o-pol-DIN-ee-a · pee-a-SAH-ba
PLATE 548

Leopoldinia piassaba is indigenous to Brazil, Colombia, and Venezuela, where it grows in forests, never in full sun. The epithet is one of the aboriginal names for the tough and long leaf base fibers, which have many uses by the native peoples, including commercial production of brushes, brooms, and baskets.

The solitary stem grows to 15 feet (4.5 m) tall and is 6 inches (15 cm) in diameter but looks as though it is much thicker, 2–3 feet (60–90 cm) thick, because of the large mass of long, hanging brown or reddish brown cordlike fibers that cover it. The leaves are 12–15 feet (3.6–4.5 m) long on long petioles and have widely and regularly spaced, narrowly lanceolate, stiff, light green leaflets 2–3 feet (60–90 cm) long and growing in a single plane; they have undulate margins and are often apically bifid. The flattened kidney-shaped fruits are edible and used locally to make a beverage. This impressive palm looks larger than it is, especially up close, and would seem to be a perfect landscape candidate for a spectacular specimen or small palm grove in a jungle setting.

Lepidocaryum
lep′-i-do-KAR-ee-um

Lepidocaryum is a monotypic genus of clustering, palmate-leaved, dioecious palm in South America. The name is from Greek words meaning "scale" and "nut," an allusion to the scaly fruits. Seeds germinate within 45 days. For further details of seed germination, see *Calamus*.

Lepidocaryum tenue
lep′-i-do-KAR-ee-um · TEN-yoo-ee

Lepidocaryum tenue is indigenous to a large area of the Amazon region in western Brazil and adjacent regions of Venezuela, Colombia, and Peru, where it occurs in the undergrowth of lowland rain forest. The epithet is Latin for "thin" or "slender" and alludes to the stem.

The species usually forms a large colony by its rambling

rhizome. The stems grow as tall as 12 feet (3.6 m), are 1 inch (2.5 cm) in diameter, and are beautifully ringed with grayish leaf base scars. The leaves are held on wiry green petioles 2 feet (60 cm) long and are divided to the petioles into 2 parts; each half is split into 2–11 narrowly lanceolate segments that are 18–30 inches (45–76 cm) long and ½–3 inches (13 mm to 8 cm) wide. Each segment is deeply ribbed and lustrous medium green, and bears tiny white prickles along its margins as well as on the veins or ribs. There is no distinct hastula, only (usually) a slight swelling on the top of the juncture of petiole and blade. The leaf is usually less than hemispherical. The inflorescences consist of short branches at the ends of peduncles 2 feet (60 cm) long that are pendent in fruit. The female inflorescences produce attractive oblong to nearly round 1-inch (2.5-cm) fruits that are covered in orange to reddish brown overlapping scales.

As many as 9 species were formerly recognized, but it seems obvious, now that the palms have become better known to taxonomists, that these are manifestations of one variable species. Henderson et al. (1995) recognized 3 naturally occurring varieties according to the number of leaf segments, number of inflorescence branches, and size and shape of the fruits.

These delicately attractive little things have leaves reminiscent of *Rhapis* species (lady palm) and would make truly beautiful large groundcovers; however, they are almost completely absent from cultivation, probably due to their need of a tropical climate, partial shade, copious, regular moisture, and a rich, humus-laden soil that is always moist but freely draining and not limey. The leaves are widely used for thatch in the Amazon region.

Lepidorrhachis
lep′-i-do-RAY-kis

Lepidorrhachis is a monotypic genus of solitary-trunked, pinnate-leaved, monoecious palm endemic to Lord Howe Island. It is called little mountain palm in Australia. The name is from Greek words meaning "scale" and "rachis," in reference to the scales on the sheaths, rachis, and pinnae of the leaves. Seeds of *Lepidorrhachis* germinate within 60 days. For further details of seed germination, see *Areca*.

Lepidorrhachis mooreana
lep′-i-do-RAY-kis · mor-ee-AH-na
PLATE 549

Lepidorrhachis mooreana is restricted to the summits of Mounts Gower and Lidgbird in windswept, cool but frostless, constantly moist cloud forests above 2460 feet (750 m) in elevation. It has one of the most restricted natural ranges of any palm. The epithet honors Charles Moore (1820–1905), botanist, horticulturist, and director of the Royal Botanic Gardens in Sydney, Australia.

The trunk grows to 10–12 feet (3–3.6 m) in habitat and is often shorter. It is only 6 inches (15 cm) in diameter but, because of the small height, looks thicker and more robust. It is succulent light green in its younger parts and usually light gray in its older parts, always with closely spaced rings of leaf base scars. The thick, succulent-looking leaf sheaths form an open crownshaft that is loose, split, and light green, often with a pinkish buff hue. The stout, short petioles hold leaves 5 feet (1.5 m) long. The ascending but arching leaves create a less-than-hemispherical leaf crown that is more like a shaving brush. The leaflets are 2 feet (60 cm) long, stiff and leathery but not rigid, medium to dark green, and lanceolate, and grow from the rachis at an angle that creates a V-shaped leaf; older leaves have semipendent leaflets. The inflorescences grow from nodes beneath the loose crownshaft and are short, succulent, densely branched, and spreading. The inflorescences are unusual in that they bear either male or female flowers, and a mature palm produces both kinds of inflorescences. The small, round ¼-inch (6-mm) fruits are red when mature.

The species has visual affinities to another Lord Howe endemic, *Hedyscepe*, but the genera are unrelated. In fact, botanists have yet to determine which genera are most closely related to *Lepidorrhachis*.

This species is among the most alluring palm species, and one made possibly even more alluring because there are so few regions in which it thrives. It requires a cool, moist, nearly frost-free climate, such as that in San Francisco, California, or parts of New Zealand, or perhaps high elevations of Hawaii. It is impossible in warm tropical or hot climates, especially those with warm nighttime temperatures. It grows in poor but constantly moist, free-draining soil. This palm is reportedly extremely slow growing but has endured temperatures in the upper 20s Fahrenheit (−2°C), but these temperatures were without drying winds, which are also inimical to it.

Leucothrinax
loo′-ko-THRY-nax

Leucothrinax is a monotypic genus of solitary-trunked, palmate-leaved, hermaphroditic palm found in Florida and the Caribbean, where it grows on limestone soils. The genus was described in 2008, based on molecular and morphological characters that separate it from *Thrinax*, *Coccothrinax*, *Hemithrinax*, and *Zombia*. The genus name combines the Greek word for "white" and the generic name *Thrinax*, in reference to the whitish color of the leaves and the palm's resemblance to (and long nomenclatural exile in) *Thrinax*. Fresh seeds germinate within 90 days. Seeds that are allowed to dry out completely and then rehydrated may take 9 months or longer to germinate. For further details of seed germination, see *Corypha*.

Leucothrinax morrisii
loo′-ko-THRY-nax · mor-RIS-ee-eye
PLATE 550

Leucothrinax morrisii is indigenous to the Bahamas, Florida Keys, western Cuba, Navassa Island, Puerto Rico, and the Lesser Antilles islands of Anguilla and Barbuda, where it grows in open deciduous forests at low elevations and along the coasts. It is called Key thatch palm, silver thatch palm, brittle thatch palm, and peaberry palm in the United States and broom palm and buffalo-top in the Lesser Antilles. It was formerly known as *Thrinax morrisii*. The epithet honors Sir Daniel Morris (1844–1933), Imperial

Commissioner of Agriculture in the West Indies and assistant director at the Royal Botanic Gardens, Kew.

The trunk grows to a height of 30 feet (9 m) in habitat, with a diameter of 8 inches (20 cm). It usually has a small area of crisscrossing leaf bases and fibers beneath the leaf crown; otherwise, the stem is clean and relatively smooth. The leaves vary in shape and color according to the age and habitat of the palm: younger palms tend to have leaves that are not complete circles and that also are more bluish green than mature individuals, which are almost always circular. The undersides of the leaves are always a lighter green than the upper surfaces. Individual leaves are relatively small and usually 2 feet (60 cm) wide on long petioles 3 feet (90 cm) long. The soft linear-lanceolate leaf segments are free almost to the petiole and normally do not grow in a single plane. Both the inflorescences and the clusters of round ⅓-inch (8-mm) white fruits are almost hidden in the leaf crown.

The species is adaptable to zones 9b through 11; isolated specimens thrive in warm microclimates (sometimes with protection) in 9a. The rounded, airy crown is beautiful as a canopy-scape, but the palm is equally satisfactory in an intimate site where its little starburst-shaped leaves dance in a breeze. This palm can be grown indoors with a great amount of light.

Licuala

li-ku-AH-la

Licuala is a genus of about 100 small to medium, solitary-trunked or clustering, palmate-leaved, hermaphroditic palms in India, Myanmar, southeastern China, Thailand, Southeast Asia, Sumatra, Malaysia, Indonesia, New Guinea, Borneo, the Philippines, northeastern Australia, the Solomon Islands, and the Vanuatu islands, where they are mostly undergrowth subjects in rain forest. The center of natural distribution of the genus is Borneo. In some forests, a single species dominates the undergrowth, leaving room for few other species to grow. A few *Licuala* species grow in exposed sites, especially along riverbanks and in swampy areas.

All the species have circular or diamond-shaped leaves. Several have undivided leaves, but most have deeply (but sometimes unevenly) segmented leaves, and all have deeply corrugated leaf blades. All the species except the largest make excellent container subjects. Most species have armed petioles and all have leaf sheaths with much fiber that is usually adherent to the stem for varying periods. A few species with segmented leaves have the unusual characteristic of leaf blades bearing a petiolule, which is the petiole-like stalk that bears a leaflet or leaf segment. The central segment of the leaf may be borne on a petiolule. The inflorescences grow from the crown and may be branched or spikelike, short or long. They bear small, yellowish-green flowers, which produce rounded red fruits. Some species from New Guinea have large sculptured endocarps.

Recent DNA studies show that several, mostly Vietnamese species formerly treated in *Licuala* do not belong in this genus. These species are now placed in the new genus *Lanonia* (which see).

Several species whose leaves have smaller segments are difficult to distinguish, especially when not in flower and fruit, but many are among the most desirable ornamental palms. The smaller ones cry out for intimate settings in which their individual beauties may be best appreciated. None are hardy to cold, although a few can tolerate occasional sharp drops in temperature, and fewer still withstand frost. They invariably need regular, adequate moisture, a well-drained soil, and partial shade. In the following text, no mention of cold hardiness is made unless the species tolerates frost, and no mention is made of exposure requirements unless the particular species has different needs from the majority. The genus name is derived from a Moluccan vernacular name for one of the species.

Seeds can take a couple of months to 3 years to germinate and do so sporadically. Ideally, seeds should not be allowed to dry out prior to sowing. Dried seed can be rehydrated, although it will take longer to germinate. Seeds of *Licuala* are viable longer than that of some genera, but they are still best planted soon after harvesting. For further details of seed germination, see *Corypha*.

Licuala beccariana

li-ku-AH-la · bek-kahr'-ee-AH-na
PLATE 551

Licuala beccariana is a solitary-trunked species endemic to low, mountainous rain forest of Papua New Guinea, where it occurs in the undergrowth. The epithet honors the great Italian botanist Odoardo Beccari (1843–1920). The trunk grows to 12–15 feet (3.6–4.5 m) high and is slender and fiber covered. The leaves are deeply segmented on thin petioles 3–4 feet (90–120 cm) long. Each leaf is slightly more than hemispherical and has 6 or fewer, widely separated, dark green segments, which are narrowly wedge shaped, deeply corrugated, and obliquely truncated and jagged at the apex; one segment is usually much broader than the rest. The endocarps are distinctively ridged and furrowed. This species is certainly among the most desirable for the garden because of its size and delicately segmented large leaves on long petioles that seem to shimmer in the slightest breeze.

Licuala bintulensis

li-ku-AH-la · bin-too-LEN-sis
PLATE 552

Licuala bintulensis is a solitary-trunked species indigenous to swampy coastal areas of rain forest in Sarawak on the island of Borneo. The epithet is Latin for "of Bintulu," a region of Sarawak. This lovely palm grows to 10 feet (3 m) tall and that height mainly due to its long, thin leaf petioles. The trunk itself is usually very short. The beautiful deep bluish green, deeply divided, circular leaves have 6–9 distinctly wedge-shaped, deeply corrugated leaflets. The apex of each leaflet is often slanted and always toothed, with the indentations corresponding to the corrugations of the segment itself.

Licuala cabalionii

li-ku-AH-la · kah'-bahl-YON-ee-eye

Licuala cabalionii is a rare, solitary-trunked species endemic to Vanuatu, where it occurs in rain forest from sea level to 800 feet

(240 m). The epithet honors Pierre Cabalion (1947–), a French ethnobotanist in Vanuatu.

The trunk grows to a height of 15 feet (4.5 m) in habitat with a diameter of 3 inches (8 cm) and is smooth and light gray or light brown except near its summit where it holds old leaf bases with attendant fibers. The leaf crown is breathtakingly beautiful and open, with leaves 3 feet (90 cm) wide on thin petioles that are 8 feet (2.4 m) long and armed with short teeth at the leaf base. The circular leaves are divided to the petiole into 12 pleated, glossy deep green segments, the center one much wider than the others. Each segment is soft and slightly pendent at its apex. The round ½-inch (13-mm) fruits are deep yellow or orange.

This species is one of the most beautiful species in the genus, with an incomparable silhouette. It is among the most tropical in its requirements and possible only in zones 10b and 11. It grows rapidly with copious, regular moisture and a humus-laden soil in partial shade or full sun except in the hottest, driest climates.

Licuala cordata
li-ku-AH-la · kor-DAH-ta
PLATES 553–554

Licuala cordata is a small, solitary-trunked species indigenous to the low mountainous rain forest in Sarawak. The epithet is Latin for "heartlike," a reference to the shape of the leaf blade. The palm grows to 3–4 feet (90–120 cm) high on a short, slender trunk that is sometimes not apparent. The leaves are circular or almost so, 18–24 inches (45–60 cm) across, sometimes with one or more shallow points and often undulate; the leaf base is usually lobed. The entire blade is deeply corrugated and a glossy light to emerald green. Occasionally specimens have leaves divided into 6 wide, wedge-shaped segments, but plants with undivided leaves are, by far, more desirable.

Licuala distans
li-ku-AH-la · DIS-tanz
PLATE 555

Licuala distans is a solitary-trunked species endemic to peninsular Thailand, where it grows in low mountainous rain forest. The epithet is Latin for "distant" in the sense of "widely separated" and alludes to the long inflorescence that extends beyond the crown of leaves.

Mature trunks attain a height of 10–12 feet (3–3.6 m). The deep green leaves are 4 feet (1.2 m) wide on stout, armed petioles 6–7 feet (1.8–2.1 m) long. The circular leaf blade is 4 feet (1.2 m) across and deeply divided into many linear wedge-shaped, corrugated segments. The tip of each segment is oblique and deeply indented, with the indentations corresponding to the corrugations of the segment. The branched inflorescences are 5–7 feet (1.5–2.1 m) long and extend up and out of the leaf crown. The globose fruits are deep red.

This palm is simply magnificent and should be sited where its large pinwheel-like leaves can be best displayed. In spite of its relatively large size, it does not relish full sun, especially in hot climates.

Licuala fordiana
li-ku-AH-la · for-dee-AH-na

Licuala fordiana is native to southeastern China and Vietnam. It was named for Charles Ford (1844–1927), first superintendent of the botanical garden in Hong Kong. This small, clustering palm produces very short stems. The long, slender petioles support round leaves about 3 feet (90 cm) across. The blade is divided into 15–20, very narrowly wedge-shaped segments, all about the same size. The inflorescence is erect and slightly shorter than the leaves, but elongates as fruits develop. The rachillae are hairy, and the globose red fruits are about ¼ inch (6 mm) in diameter.

Licuala glabra
li-ku-AH-la · GLA-bra
PLATES 556–557

Licuala glabra is a small, solitary-trunked species in southern peninsular Thailand and northern peninsular Malaysia, where it occurs in the undergrowth of mountainous rain forest to 4000 feet (1200 m), often in immense colonies; it also grows in lowland rain forest. The epithet is Latin for "glabrous" and alludes to the smooth inflorescences. Mature trunks can grow to 6 feet (1.8 m) high, and the overall height of the palm can be 12 feet (3.6 m). The leaves are held on long, thin petioles and are circular or nearly so, with wedge-shaped segments. *Licuala glabra* var. *selangorensis* [se'-lan-go-REN-sis] has fewer, generally broader segments (especially the middle ones), which are not as uniformly shaped as those of the type.

Licuala grandis
li-ku-AH-la · GRAN-dis
PLATES 558–559

Licuala grandis is indigenous to rain forest in Vanuatu and the Solomon Islands. It is sometimes called the ruffled fan palm. The epithet is Latin for "grand" or "spectacular."

The solitary trunk of mature trees seldom grows to more than 10 feet (3 m) tall and 3 inches (8 cm) in diameter. It is usually covered in tightly woven brown fibers from the leaf bases, which are narrowly triangular; old trunks are usually gray to almost white and nearly smooth but exhibit closely set semicircular rings of leaf base scars. The leaf crown is 8 feet (2.4 m) wide and 6 feet (1.8 m) tall. The undivided, pleated leaves are 3 feet (90 cm) wide on heavy petioles that are 3 feet (90 cm) long and marked with small, curved teeth on the margins of their lower parts. The leaf is semicircular to broadly wedge shaped or diamond shaped, and the leaf margins are deeply toothed, the indentations as well as the pleats in the blade approximately matching the fused segments. Leaf color is medium to deep, shiny green on both sides. The leaves are mostly erect, the older ones pendent, and they usually do not create a fully rounded crown. The blade is usually wavy and undulating, especially along its margins. There are more than 12 leaves per palm (often as many as 20) on relatively long petioles; the visual effect is that of being densely packed into the leaf crown. The flower stalks are 6 inches (15 cm) long and grow from among the leaves. They are sparsely branched and bear small, yellowish white flowers. The rounded ½-inch (13-mm) fruits are bright red when ripe.

The species does not tolerate cold and is possible only in zones 10b and 11. It needs copious, constant humidity and moisture in a fast-draining, humus-laden soil. It is eminently adapted to partial shade, especially when young, but older plants can take full sun, except in the hottest climates. These small palms are among the choicest in the world of horticulture. They have an exquisite elegance matched by few other palms. This palm is probably the last one the gardener should consider planting as an isolated specimen; its penchant for partial shade adds to its usefulness and need to be a part of other vegetation. The leaf crowns are stunning when planted in groups of 3 or more individuals of varying heights and as accents in masses of other vegetation. As a patio or courtyard subject this palm is perfect anywhere its beauty may be enjoyed up close. The palm is unexcelled for indoors and excellent as container specimens outdoors. It is often purchased as a pot of 3 or more small plants, which should be separated to allow each palm's leafy crown to develop fully. This magnificent species needs good light but not full sun, a rich soil, regular feedings (fish emulsion is good), and should never be allowed to dry out.

Licuala lauterbachii
li-ku-AH-la · low-ter-BAH-kee-eye
PLATE 560

Licuala lauterbachii is indigenous to the undergrowth of rain forest in Papua New Guinea and the Solomon Islands. The epithet honors Carl A. G. Lauterbach (1864–1937), German naturalist and explorer in New Guinea.

The mature solitary trunk grows to 20 feet (6 m) high and less than 6 inches (15 cm) in diameter. It is mostly free of fibers and leaf bases, light gray to almost white, and indistinctly grooved with leaf base scars. The leaf crown is 12 feet (3.6 m) tall and wide. The leaves are 3–5 feet (90–150 cm) wide, circular to semicircular to oval, and composed of 15–30 segments, each of which is wedge shaped, squared at its apex where it is also pleated and toothed corresponding to the fused segments. The segment divisions extend to or almost to the petiole, and the visual effect is that of a large green pinwheel, made more dazzling because of the offset and not circular leaf. The thin, graceful petiole is 4–5 feet (1.2–1.5 m) long with short marginal spines on its lower portions and gives to the leaf crown a full and nearly rounded aspect. Leaf color is grayish to medium to deep green on both surfaces. The inflorescences are 3 feet (90 cm) long, mostly erect, and sparsely branched, with small, whitish blossoms. The round 1-inch (2.5-cm) orange to red fruits are borne in half-pendent clusters.

This species does not tolerate cold and is adaptable only to zones 10b and 11. It needs an abundance of moisture and a rich, well-draining soil. Young plants should not be subjected to the full sun of hot summer climates. The leaf crown provides one of the most beautiful silhouettes in nature. Its large pinwheel-like blades dance on their delicately long stalks and move in the slightest breeze. It is superbly attractive as a patio or courtyard subject and, in groups of 3 or more individuals of varying heights, creates a veritable symphony of forms. It is wonderful as a large potted palm, indoors or out, asking only for bright light, constant moisture, and a rich, friable medium.

Licuala longipes
li-ku-AH-la · LON-ji-peez
PLATE 561

Licuala longipes is indigenous to low mountainous rain forest of southern Myanmar and peninsular Malaysia. The epithet is Latin for "long" and "foot," referring to the length of the petioles. The solitary stem is usually underground or very short. The leaves are to 7 feet (2.1 m) wide, a deep and lustrous green, and divided almost to the petiole into 12 giant wedge-shaped segments. The apical segments are 1 foot (30 cm) wide, while the smaller basal segments are 4–5 inches (10–13 cm) wide. Leaf outline is a giant oval to almost a circle, and the petioles may be 8 feet (2.4 m) long. The inflorescence is much shorter than the leaves, compact, and much branched. This species is an amazing sight, so large are the leaves and so tall are the stalks. It is perhaps the grandest and most beautiful trunkless species. It seems to be among the more tender (to cold) species in the genus.

Licuala malajana
li-ku-AH-la · mah-lay-AH-nah
PLATE 562

Licuala malajana grows in Thailand and peninsular Malaysia in forests on slopes, ridges, and hillsides. The species is morphologically variable over its large range. The epithet is Latin for "Malayan."

This species is either solitary or clustering, with or without an aboveground stem. When clustering, 1 or 2 stems are dominant. The main stems can reach 10 feet (3 m) tall but are only about 1 inch (2.5 cm) in diameter. The crown bears 11–25 leaves, each on a slender, finely spiny petiole up to 6 feet (1.8 m) long. The leaf blade is broadly circular, 2–3 feet (60–90 cm) wide, divided into 5–33, glossy equal-sized segments. The central segment is wider than the lateral segments. The inflorescence is shorter than the leaves. The smooth, globose ½-inch (13-mm) fruits are dull orange when ripe.

This is yet another charming *Licuala* species for an unsurpassed tropical look in a moist, well-drained site in partial shade. It is not cold hardy outside zones 10 and 11.

Licuala mattanensis
li-ku-AH-la · mat-ta-NEN-sis
PLATES 563–564

Licuala mattanensis is endemic to Sarawak on the island of Borneo, where it grows in lowland rain forest, usually in acidic peaty soil. The epithet is Latin for "of Matang," a region of Sarawak.

The solitary trunk takes a long time to form and reaches a maximum height of 3 feet (90 cm). The leaves are 2 feet (60 cm) wide and often smaller, hemispherical or slightly more rounded, and consist of 12 linear wedge-shaped segments. The species has 3 traits that are not typical of the genus: unarmed petioles; large, elongated, slightly curved fruits; and beautifully variegated forms. The most sought after cultivar is **L. mattanensis** 'Mapu' (see plate 564), which has light yellow tessellation and is often referred to as *L. "mapu."* Upon seeing this palm, grown men have gone weak in the knees. It is that breathtaking. This is one of the more tender (to cold) species in the genus.

Licuala olivifera
li-ku-AH-la · o-li-VIF-e-ra

Licuala olivifera is indigenous to lowland rain forest and swamps of peninsular Malaysia and Sarawak. The epithet is Latin for "olive-bearing," an allusion to the size and shape of the fruits, but could equally well apply to the color of the leaf segments. This solitary-trunked species has a short, often unapparent trunk. The circular leaves are to 4 feet (1.2 m) in diameter on petioles 3 feet (90 cm) long and have 24 (or sometimes more) narrow wedge-shaped, deeply ribbed segments. The uppermost segments are wide and have squared apices, with progressively narrower and shorter segments on both sides towards the base of the circle. The matte, almost olive green leaf is among the most beautiful in the genus.

Licuala orbicularis
li-ku-AH-la · or-bik'-yoo-LAR-iss

PLATE 565

Licuala orbicularis is an undergrowth inhabitant of rain forest in Sarawak, where it is nearly extinct because of clearing of the rain forest. The epithet is Latin for "orbicular," an allusion to the leaf shape.

The species is solitary trunked or trunkless with the stem never emerging above the ground. The total height of the plant is 6–8 feet (1.8–2.4 m). The leaves are to 5 feet (1.5 m) wide and 3 feet (90 cm) long and shaped like a squared-off fan, although some specimens have orbicular leaves. The segments are fused, and the blade is slightly indented along its margins, corresponding to the fused segments that are deeply pleated within the blade. A prominent midvein in the middle of the blade extends through its entire length, and the leaf margins are undulating and wavy. The dark green, shiny leaves are carried on elegantly thin, long petioles to 6 feet (1.8 m) long or more and 1 inch (2.5 cm) thick at the base. The lower part of the petiole is margined with small teeth. This little palm seldom blooms in cultivation, but the inflorescence is 3 feet (90 cm) long and pendent, and bears small, whitish flowers. The round ½-inch (13-mm) fruits are red when mature and are borne in narrow, pendent clusters.

This palm does not tolerate cold and is adaptable only to zones 10b and 11. It needs constant warmth, moisture, and humidity but is no good in full sun in tropical climes. It also needs a humus-rich soil with unimpeded drainage.

This astonishingly beautiful accent plant for shady beds has great shiny, rounded, and corrugated blades that add an almost unique elegance and charm to tropical sites that are difficult to landscape. It should be grown with plants whose leaves are of a different shape and texture, like ferns and small, pinnate-leaved palms such as *Chamaedorea elegans*, lest the other leaf forms detract from its exceptional grace and elegance. This palm is a near perfect choice indoors if given enough water and a rich soil.

Licuala paludosa
li-ku-AH-la · pa-loo-DO-sa

PLATE 566

Licuala paludosa occurs naturally in peninsular Malaysia, Thailand, Sumatra, and Vietnam, where it grows in swampy areas at low elevations. The epithet is Latin for "swampy."

This sparsely clustering species has stems to 12–15 feet (3.6–4.5 m) tall with a diameter of 1 inch (2.5 cm); the overall height of the palm may exceed 20 feet (6 m). The trunks are mostly smooth and free of leaf bases except for the youngest parts. The circular 3-foot (90-cm) leaves are held on slender, spiny petioles 18 inches (45 cm) long; they are similar to those of *L. spinosa* but generally larger, a darker green, and the segments are more varied in width with 1 or 2 quite wide.

The species is more beautiful than *L. spinosa* because of its leaves and its less dense clumps. In fact, it is arguably the most beautiful divided-leaf species in the genus. Compared with *L. spinosa*, it is, alas, more tender to cold, needs more water and a better soil that is on the acidic side, but grows as fast. It grows in partial shade or full sun.

Licuala parviflora
li-ku-AH-la · pahr-vi-FLOR-a

Licuala parviflora is endemic to northern Papua New Guinea, where it grows in low mountainous rain forest. The epithet is Latin for "small flowered." The solitary stem slowly attains a height of 6 feet (1.8 m). The leaves are held on petioles 2 feet (60 cm) long, are 2–3 feet (60–90 cm) wide, and are semicircular or slightly more rounded. They bear 15–20 segments, which are diaphanously thin and nearly the same length and width. The leaf crown is also sparse. The whole aspect of this little gem is fairylike and utterly charming. It is tropical in its requirements and needs copious, regular moisture.

Licuala peekelii
li-ku-AH-la · pee-KEL-ee-eye

PLATE 567

Licuala peekelii is endemic to the Bismarck Archipelago, where it is widespread in lowland forests. In the past, it has been confused with *L. lauterbachii*, to which it bears some resemblance. It is named for Gerhard Peekel (1876–1949), a German-born Catholic missionary and botanist, who wrote an early, unpublished account of the archipelago's flora.

The stems of this clustering palm reach 10–15 feet (3–4.5 m) tall. The leaves are held on slender petioles 3 feet (90 cm) long and bearing teeth at their base. The semicircular blade is about 3 feet (90 cm) wide and split into approximately a dozen wedge-shaped segments of unequal widths. The tip of each segment bears short teeth corresponding to the tips of the joined leaflets. The inflorescence extends well beyond the leaves. The globose fruits are ½ inch (13 mm) in diameter.

This elegant beauty forms a compact plant, perfect for an intimate, sheltered spot in the garden. It looks great as a container specimen, too. It needs plenty of water and shade.

Licuala peltata
li-ku-AH-la · pel-TAH-ta
PLATES 568–569

Licuala peltata occurs naturally from northeastern India, Bhutan, Bangladesh, the Andaman and Nicobar Islands, through Myanmar, Thailand, and into peninsular Malaysia, where it grows in low, mountainous rain forest. It is threatened with extinction in India and Bangladesh. The epithet is Latin for "peltate," describing a circular leaf attached to its petiole at the center of the leaf blade, as is the case with sacred lotus and nasturtium leaves. Although the leaf of *L. peltata* is not truly peltate, its larger, circular shape suggests a peltate leaf.

The trunk of this mostly solitary-trunked, rarely clustering species slowly grows to 20 feet (6 m) high. The leaves are 6 feet (1.8 m) wide on petioles that may be 12 feet (3.6 m) long but are more often half that length. This great length creates a large, open, and nearly rounded leaf crown. The leaf blade is hemispherical or slightly greater and often is slightly diamond shaped; it normally is divided into 15–25 wedge-shaped, lustrous, deep green segments of nearly equal length and width except for the most apical ones. All have deeply toothed apices. The inflorescences are erect spikes 10–12 feet (3–3.6 m) long from which pendent flowering branches grow. Flowers are yellowish. The rounded ½-inch (13-mm) fruits are deep red.

Licuala peltata **var.** *sumawongii* [soo-ma-WONG-ee-eye], sometimes erroneously referred to as *L. elegans*, a different species, has entire, undivided leaf blades, which are so relatively thin that they often fold to some extent because of their own weight. This variety grows naturally only in southern Thailand and peninsular Malaysia, where it may now be extinct. It was named for Watana Sumawong, a Thai collector who introduced it to cultivation. It looks much like *L. grandis* and is much sought after by gardeners and collectors almost to the exclusion of the segmented, typical form, **L. peltata var. peltata**. Some people find it difficult to believe the 2 forms can be the same species, but if they could see the Sumawong form in a windy and exposed site, they would see the resemblance, as the leaves usually become variously segmented under these conditions. Both varieties are equally beautiful.

This magnificently beautiful, tropical-looking species seems surprisingly hardy to cold, having withstood unscathed temperatures slightly below freezing. It is also adaptable to a range of soils including slightly alkaline ones, as long as they are fast draining. In addition, the species can tolerate a modicum of drought but looks and grows much better if provided regular, adequate moisture. Protection from drying wind will prevent the leaves from being disfigured. The palm responds well to a deep organic mulch in South Florida and luxuriates in partial shade or nearly full sun once past the juvenile stage, always being a healthier and darker green with some shade.

This palm is so attractive that it is hard to misplace in the garden if given enough space; it should not be crowded into tight spaces with other vegetation. It should be planted in a wind-protected site as its leaf blades are relatively thin and become easily tattered, but, even then, it is unusually beautiful. It looks its best as a specimen planting under a high canopy with 3 or more individuals of varying heights in a single group; in this situation, it is a symphony of shimmering leaf form. Once past the juvenile stage, this palm is unusually fast growing for a *Licuala* species. Indoors it needs a large space and high light and humidity.

Licuala petiolulata
li-ku-AH-la · pee′-tee-o-loo-LAH-ta
PLATE 570

Licuala petiolulata is endemic to low mountainous rain forest in Sarawak. The epithet is Latin for "having a petiolule" and alludes to the "little stalks," the petiolules, of some leaf segments. This solitary-trunked species has a short stem that is often not apparent, but it may eventually attain 2–3 feet (60–90 cm) high. The small, nearly perfectly circular leaves, up to 2½ feet (75 cm) wide, are held on thin petioles 3–4 feet (90–120 cm) long; they are as beautiful as any in the genus. The blade consists of 15–20 elegantly thin, wedge-shaped segments divided to the petiole, each with an obliquely truncated, jagged apex. The central segments are usually wider than the others, and their basal portions are narrow and stalklike, looking like short petiolules. Their color is a medium, almost olive to deep green on both surfaces. The species does not tolerate cold.

Licuala platydactyla
li-ku-AH-la · plat-ee-DAK-ti-la
PLATE 571

Licuala platydactyla is endemic to northern Papua New Guinea, where it grows in lowland rain forest. The epithet is from Greek words meaning "flat" and "fingered," an allusion to the leaf segments. The solitary trunk grows to 10 feet (3 m) high in habitat. The leaves are held on stout petioles 3 feet (90 cm) long or sometimes longer. The leaf blade mostly consists of 3 or 4 wedge-shaped segments. The central segment is much broader than the others in mature or nearly mature plants and is usually deeply bifid. All segments have obliquely truncated, jagged apices and are a medium to deep green on both surfaces. The inflorescence is shorter than the leaves; individual flowers are borne on short stalks.

Licuala poonsakii
li-ku-AH-la · poon-SAH-key-eye
PLATE 572

Licuala poonsakii comes from southeastern Thailand. It is named after Poonsak Vatcharakorn, a plant collector employed by Nong Nooch Tropical Botanical Garden, Thailand.

The trunk is either clustering and up to 3 feet (90 cm) tall or solitary and then much taller, about 2 inches (5 cm) in diameter. The circular leaf blades are divided into about 10 wedge-shaped segments, which are truncate and toothed at their apices. The middle segment is the largest, 3–5 inches (8–13 cm) broad at the apex, with the other segments decreasing in size toward the base of the blade. The slender petioles are 3 feet (90 cm) long and bear teeth near their bases. The inflorescences are a little longer than the leaves and bear just a few rachillae, which are held parallel to the main axis of the inflorescence. The globose fruits are about ¼ inch (6 mm) across.

This delicate palm is a good choice for an intimate, well-protected, and warm part of the garden. It is requires shade and well-drained, humus-rich soil.

Licuala ramsayi
li-ku-AH-la · RAM-say-eye
PLATES 573–574

Licuala ramsayi is endemic to low, swampy rain forest of northeastern Queensland, Australia. The epithet commemorates the original collector, Edward P. Ramsay (1842–1916), naturalist and curator of the Australian Museum.

This solitary-trunked species starts as an undergrowth subject but, in maturity, reaches and usually overtops the tree canopy. It sometimes attains a height of 60 feet (18 m) in habitat but is usually no more than 30–40 feet (9-12 m) tall under cultivation. The diameter is never more than 8 inches (20 cm), and the almost-white trunk is smooth and free of leaf bases except for the youngest parts. The leaf crown is 18 feet (5.4 m) wide and tall. The leaves are 5–6 feet (1.5–1.8 m) wide and nearly circular (often shaped like a scallop's shell) but are divided into 10 or more wedge-shaped segments that are squared off on their jagged ends. The segments vary in width, and the divisions between them extend to the petiole, giving a visual effect like that of a pinwheel. Leaf color is light to medium green on both sides, and each segment is deeply pleated, the pleats corresponding to fused segments. The elegantly thin petioles are 6 feet (1.8 m) long and armed with short teeth along the lower margin. The leaf crown usually has 12 leaves. The inflorescences are unusually long, usually spilling out of the leaf crown; they are much branched and bear small, white flowers. The round 1-inch (2.5-cm) fruits are red.

Two varieties of this species are recognized; both are in cultivation. **Licuala ramsayi var. ramsayi** has conspicuously armed petioles and bears flowers in small clusters along the rachillae. **Licuala ramsayi var. tuckeri** [TUK-er-eye] has unarmed or sparsely armed petioles, and the flowers are solitary; the varietal name honors Australian Robert Tucker (1955–1992), horticulturist and specialist in palms and Pandanaceae.

The species does not tolerate cold and is adaptable only to zones 10b and 11 (at least as the magnificence it should be). It needs constant moisture and humidity, as well as a rich soil. It usually survives on drier soils but is always stunted and less colorful. It luxuriates in partial shade, but older plants can take full sun. This is the tallest, most spectacular, and most beautiful species in the genus. Its leaves form incredibly beautiful silhouettes and dance in the slightest breeze. The palm looks its best among or near masses of other darker vegetation against which its light-colored trunks and great pinwheel-like leaves stand out. It is unequalled as a canopy-scape. It is a good candidate for growing indoors when young, or when planted in large conservatories, if it can be given enough light and moisture.

Licuala sallehana
li-ku-AH-la · sahl'-le-HAH-na
PLATE 575

Licuala sallehana is a rare, densely clustering species endemic to peninsular Malaysia, where it grows as an undergrowth subject in lowland rain forest. It was described in 1997. The epithet honors Salleh Mohammad Nor (1940–), former director of the Forest Research Institute Malaysia.

The thin reedlike stems grow in clusters to a height of 4 feet (1.2 m) and are covered with a fine, almost woven net of fibers and leaf bases. The leaves are unusual in the genus as they are undivided, lanceolate to an elongated diamond shape, 2½ feet (75 cm) long and 8 inches (20 cm) wide near their apices. They are deep green and deeply grooved on long petioles armed with triangular teeth on their lower halves. The inflorescences are short and produce small, round fruits that are black when mature.

Two varieties are recognized: **L. sallehana var. sallehana** has entire leaves, and **L. sallehana var. incisifolia** [in-sy'-si-FO-lee-a] has divided leaves and is slightly less desirable for gardens. The overall aspect of this extraordinarily beautiful and choice palm is that of a small, clustering *Johannesteijsmannia* species.

Licuala sarawakensis
li-ku-AH-la · sahr'-a-wah-KEN-sis
PLATE 576

Licuala sarawakensis is endemic to lowland rain forest in Sarawak on the island of Borneo. The epithet is Latin for "of Sarawak." This small, solitary-trunked species has a short stem that is usually unapparent. The circular leaves are 18 inches (45 cm) wide and consist of 6–8 wedge-shaped segments of varying widths. The segments of adult plants are of mixed widths without a pattern to their variation in breadth, wider segments adjacent to narrower ones; juvenile plants usually exhibit a central segment that is significantly broader than the others. Leaf color is medium to light green on both surfaces. This species is unusually attractive because of its circular leaves and diminutive size. It is exquisite as a groundcover in tropical climates.

Licuala scortechinii
li-ku-AH-la · skor-te-KEE-nee-eye
PLATE 577

Licuala scortechinii is native to wet, lowland forests of Thailand and peninsular Malaysia. It is named for Benedetto Scortechini (1845–1886), Italian priest and botanist who collected plants in Australia and Malaysia. *Licuala delicata* is a synonym.

This species generally has a solitary trunk. The slender stem is less than 3 feet (90 cm) tall and 1 inch (2.5 cm) in diameter. It will produce a few weak suckers in time. The slender petiole is 6–12 inches (15–30 cm) long, with a few prominent but irregularly scattered teeth. The small leaves are only 12–14 inches (30–36 cm) across. The 4 or 5 segments of the leaf are wedge shaped and truncate. The middle segment is widest, the segments becoming smaller and shorter toward the sides. The long inflorescence is branched only at its tip and is a little longer than the leaves. The bracts of the inflorescence bear a rusty brown, fuzzy tomentum.

One can scarcely imagine a more delicate, diminutive *Licuala*. The species is still not common in cultivation, but if a number of plants could be had, they would make a lovely underplanting in a shady, moist spot in a zone 10 or 11 garden. Alternatively, this species would be ideal in a decorative container in a greenhouse or conservatory. It requires abundant moisture in the soil and in the air, protection from drying winds, and shelter from cold.

Licuala spinosa
li-ku-AH-la · spi-NO-sa
PLATES 578–579

Licuala spinosa has one of the largest distributions in the genus and is indigenous to the coastal plains of the Nicobar and Andaman Islands, Myanmar, Vietnam, Thailand, peninsular Malaysia, Sumatra, Java, Borneo, western Indonesia, and the Philippines. The epithet is Latin for "spiny." The species is sometimes called spiny licuala and in Australia is known as mangrove fan palm.

The mature trunks of this densely clustering species grow to 15 feet (4.5 m) high and are covered with fibers and leaf bases in their upper parts but are almost smooth and light colored on the lower portions. A clump can measure at least 15 feet (4.5 m) wide and 20 feet (6 m) tall. The leaves are almost completely circular, 2 or more feet (60 cm) wide, and consist of 10–15 wedge-shaped segments, which are strongly pleated and obliquely squared on their ends. The apex of each segment is deeply toothed, the indentations corresponding to the pleats in the blade. Each segment is approximately the same width but not always the same length, which phenomenon creates the visual aspect of a starburst as well as that of a pinwheel. The leaves are carried on thin and delicate petioles 3–4 feet (90–120 cm) long with stout curved teeth on their margins. The inflorescences are 3–8 feet (90–240 cm) long and sparsely branched; they bear small, yellowish white flowers. The round ¼-inch (6-mm) red fruits are in short pendent clusters.

This species is hardier to cold than most others and is adaptable only to zones 10 and 11. It is a true water lover and, while it needs partial shade when young, can readily adapt to full sun when older. It needs a well-drained soil and abundant, constant moisture. This palm is a splendid hedge subject. It also is perfect as a large accent among other vegetation, in which case it looks better if a few of the trunks (especially those of the same height) are judiciously pruned out so that the wonderful silhouette of the leaves is more readily apparent. Like almost all other species in the genus, it is a superb patio or courtyard subject. It is wonderful indoors but needs more light than most other species.

Licuala triphylla
li-ku-AH-la · try-FIL-la
PLATE 580

Licuala triphylla occurs naturally in Thailand, peninsular Malaysia, and western Borneo, where it grows in low mountainous rain forest. The epithet is Latin for "three leaves," a general reference to the sparse segmentation of the leaf. Among the several synonyms for this species are *L. pygmaea*, *L. stenophylla*, and *L. filiformis*. This solitary-trunked species has a subterranean stem. The leaves are held on petioles 2–3 feet (60–90 cm) long and are hemispherical or slightly more so. They consist of 3–5 widely divergent wedge-shaped segments, the middle one much wider than the others with a truncated, jagged apex. The other 2–4 segments are obliquely truncated and more pronouncedly toothed at their apices.

Linospadix
ly-no-SPAY-dix

Linospadix comprises 7 small, clustering or solitary-trunked, pinnate-leaved, monoecious palms in New Guinea and northern and eastern Australia. They are undergrowth palms. A given species may have leaves that vary in size, shape, and number of segments. All the species have a long, spikelike inflorescence bearing both male and female flowers. In fruit, they are pendent strings or clusters of red berrylike fruits. All the species are intolerant of drought and need shade to partial shade especially in hot climates. They also need a cool but frost-free climate and a soil that is acidic, humus laden, moist, and well drained. The genus name is from Greek words meaning "thread" and "spadix." Seeds germinate within 60 days. For further details of seed germination, see *Areca*.

Linospadix microcaryus
ly-no-SPAY-dix · my-kro-KAR-ree-us
PLATE 581

Linospadix microcaryus is endemic to northeastern Queensland, where it grows in rain forest from sea level to 5000 feet (1520 m). The epithet is from Greek words meaning "tiny nut."

It is a sparsely clustering species whose stems grow to 10 feet (3 m) high but are usually ½ inch (13 mm) in diameter. They are a deep green and have distinct whitish rings of leaf base scars. The loose pseudo-crownshaft is silvery green to almost white or brownish. The petioles of new leaves have a whitish indument. The leaves and the leaf crowns vary in form and appearance. The former may be undivided and deeply bifid at their apices with a prominent midrib, in which case the leaf crown is usually of a tufted appearance, or the leaves may be divided into as many as 20 irregularly or regularly spaced linear pinnae, in which case the leaf crown is open and airy. In either case, the leaves are deep green on both surfaces. The fruits are round.

Linospadix minor
ly-no-SPAY-dix · MY-nor
PLATE 582

Linospadix minor is endemic to northeastern Queensland, Australia, where it grows in rain forest from elevations near sea level to 3800 feet (1160 m). The epithet is Latin for "smaller" and betrays the fact that, when it was first discovered, it was thought to be a small species of *Areca*. This is the most variable species in the genus: its stem heights range from 2 to 20 feet (60 cm to 6 m);

clumps may be sparse or dense; and the leaves may be short or long and have few separate pinnae or many separate leaflets. The petioles are green. The fruits are elongated.

Linospadix monostachyos
ly-no-SPAY-dix · mo-no-STAY-kee-os
PLATE 583

Linospadix monostachyos is endemic to Australia in the states of Queensland and New South Wales in dense and wet rain forest at 500–800 feet (150–1160 m). The epithet is Latin and Greek for "single spike," although the species forms more than a single inflorescence at a time. The common name in Australia is walking-stick palm.

The solitary trunk grows to a height of 12 feet (3.6 m) with a width of 1 inch (2.5 cm) and is light to dark green with closely spaced darker rings of leaf base scars. The dense leaf crown is hemispherical to three-quarters hemispherical, and the leaves are 2–3 feet (60–90 cm) long and beautifully arching on petioles 1 foot (30 cm) long. The 10–20 deep green leaflets are of variable widths and spacing, but the terminal pair are always broader and have obliquely truncated, jagged apices, whereas the other leaflets are more linear and long tipped.

This species is by far the most widely grown and the most beautiful Australian species, making a perfect little palm of exquisite proportions. It is also the most cold-tolerant species in the genus, withstanding unscathed temperatures slightly below freezing. It does not, however, endure regions with hot summers.

Linospadix palmerianus
ly-no-SPAY-dix · pal-mer-ree-AY-nus
PLATE 584

Linospadix palmerianus is native to the rain forest of Queensland, Australia, at 1000–5000 feet (300–1520 m). The epithet honors Edward Palmer (ca. 1840–1899), author, legislator, and amateur anthropologist.

The clustering stems are 3–15 feet (90 cm to 4.5 m) tall and less than 1 inch (2.5 cm) in diameter. The leaves are 18–24 inches (45–60 cm) long and are either simple and bifid, or evenly divided into lanceolate segments, about 10–12 per side. The terminal pair are broader than the others and have truncate tips. The tips of the lateral leaflets are slender and pointed. The leaflets are dark green and glossy on the upper surface, lighter green below. The long, slender inflorescences bear small, white flowers followed by red or yellow elongate fruits.

A delicate palm that would grace any garden, this species forms an open cluster of stems that do not require thinning to look their best. It is supremely elegant and tropical.

Livistona
liv-i-STO-na

Livistona is composed of 27 mostly large to massive solitary-trunked, palmate-leaved, hermaphroditic (or possibly dioecious) palms that are widespread in the Old World tropics and subtropics, from Somalia, Djibouti, the extreme southern Arabian Peninsula, eastern India, Bangladesh, Myanmar, southern China, and Indochina to the Philippines, New Guinea, Australia, Indonesia, and Malaysia. The genus is well represented in Australia, where there are several species in deserts (with access to underground water), as well as mesic and monsoonal regions.

A few species are found naturally with branching stems due to injury of the growing point. Most species have leaf segments whose ends are pendulous when the plants are older but often rigid or shallow when young. The petiole is never split where it joins the trunk, as it is in *Sabal*. The leaves are costapalmate, but the phenomenon is indistinct in many species, especially in younger individuals. All species have a fairly distinct hastula on top of the juncture of petiole and blade. The inflorescences grow from among the leaf crown, are branched twice or more, and bear bisexual flowers that produce blue, green, black, or brown fruits.

Recent comparative studies of DNA have shown that *Livistona*, as we once knew it, had to be split into 2 genera and that the old genus *Saribus* had to be resurrected. Those species with 3-pronged inflorescences, including the familiar *L. rotundifolia*, are now treated in *Saribus*.

Most *Livistona* species, and especially those from Australia, are unusually hardy to cold considering their mostly tropical origins, even on that continent. The genus name honors Patrick Murray (1634?–1671), Baron of Livingston(e), a Scottish student of natural history, whose plant collections formed the basis for those of the Royal Botanic Garden, Edinburgh.

Seeds of *Livistona* generally germinate within 90 days unless they are allowed to dry out before sowing. Dried seeds take longer to germinate and do not germinate as well as those kept slightly moist. For further details of seed germination, see *Corypha*.

Livistona australis
liv-i-STO-na · aw-STRAH-lis
PLATE 585

Livistona australis is endemic to the east coast of Australia from central Queensland to the state of Victoria, where it grows in hilly as well as swampy forests and the edges of low, mountainous rain forest in the northern parts of its range. It is the world's southernmost *Livistona* species and the second most southerly occurring palm species, the honor of first place going to *Rhopalostylis sapida* in New Zealand. The epithet is Latin for "southern" and refers to the southern continent of Australia. Common names for this species are Australian cabbage palm or cabbage tree.

The mature solitary trunk is straight as an arrow and can grow to heights of 75 feet (23 m) in habitat but is generally no more than half that under cultivation and is about 1 foot (30 cm) in diameter. While dead leaves adhere to the trunk for a while, the trunk is mostly clear of leaf bases and is light to dark brown or light to dark gray with closely set deep rings of leaf base scars and several vertical fissures. The round leaf crown is 15 feet (4.5 m) tall and wide. The leaves are held on petioles 6 feet (1.8 m) long that have spines along their margins. The leaves are 3–4 feet (90–120 cm) wide and semicircular to almost circular. The blade is divided for about half to two-thirds its length into many segments, and these are split to half their length. The segment tips are pendulous. The leaf is glossy deep to grayish green. The sparsely branched inflorescences are 4 feet (1.2 m) long and bear small, white blossoms. The black fruits are ½ inch (13 mm) across.

Some populations of this species show slight morphological differences and have, in the past, been referred to as *Livistona* 'Eungella Range' or *L.* 'Paluma Range'. The differences are so slight as to warrant no taxonomic status (Dowe 2009, 2010).

This relatively hardy palm is adaptable to zones 9 through 11, although it is marginal in areas of 9a subject to wet freezes in winter. It grows in a range of soils but is indigenous to regions with fertile soil and plenty of year-round moisture, under which conditions it looks its best. It needs sun from youth to old age. The Australian cabbage palm has nobility and, because of the pendent leaf segments, gracefulness. Its straight columnar trunk invites its use as an avenue tree, planted single file along one or both sides. While this practice works if the palms are planted far enough apart so that their large leaf crowns do not conflict with each other, the fan palm looks even better in groups of 3 or more individuals of varying heights. It is near perfection as a canopy-scape. It needs a lot of space and nearly full sun but is sometimes grown indoors as a juvenile plant and is often grown in large atriums or conservatories.

Livistona benthamii
liv-i-STO-na · ben-THAM-ee-eye
PLATE 586

Livistona benthamii is indigenous to southern Papua New Guinea, and northern Queensland and the Northern Territory in Australia, where it grows in lowland rain forest, especially along the edges of lagoons, swamps, and rivers. The epithet honors George Bentham (1800–1884), one of the greatest British taxonomists of the 19th century.

The trunk is tan or cinnamon colored in its younger parts but deep gray in its older parts. It attains a height of 50 feet (15 m) in habitat but is 6 inches (15 cm) in diameter. It is usually covered in its lower parts with persistent petiole stalks which, even with time and weathering, are 2 feet (60 cm) long and give to the trunk a decidedly spiny if not grotesque look unless they are removed mechanically or by hurricane winds. Once the petiole stubs are gone, the stems usually exhibit picturesque knobs similar to those in several *Phoenix* species; these are the old, persistent, lignified leaf bases. The open leaf crown is nearly spherical in older trees but is elongated in younger individuals, the leaves growing from an extended area of the upper trunk. The petiole is 6 feet (1.8 m) long and bears perpendicular spines at its base. The leaves are 4 feet (1.2 m) wide with deep green pendent segments. The divisions between segments extend three-quarters into the blade. The leaves of seedlings and young individuals are circular with narrow segments divided nearly to the petiole, giving them the appearances of small papyrus plants. The fruits are purple-black.

This exceptionally attractive species is fast growing if given a rich, moist soil and is beautiful at all stages. It requires some shade when young, and even as an adult, it looks best in light shade. It is tolerant of saline conditions but, alas, not high winds if its appearance is important. It is adaptable to zones 9 through 11 and, while it survives in Mediterranean climes, it grows slowly.

Livistona carinensis
liv-i-STO-na · kar-i-NEN-sis
PLATE 587

Livistona carinensis is a geographic outlier in the genus. It comes from 3 widely scattered populations in oases of Djibouti, northern Somalia, and eastern Yemen. The epithet is Latin for "of Carin," the oasis in Somalia from which the species was first identified. Until the late 1980s, the species was near extinction, as it was being felled for its durable trunks. In addition, habitat is being converted to subsistence agriculture. The species was formerly known as *Wissmannia carinensis*.

In habitat, the solitary trunk grows to 100 feet (30 m) high and 15 inches (38 cm) in diameter, has a heavy, swollen base, and produces aerial roots near the ground and closely spaced rings of leaf base scars near the summit. The leaf crown is spherical or elongated and is open because of the long leaf petioles. The weakly costapalmate, semicircular leaves are 3 feet (90 cm) wide. The petioles are 4 feet (1.2 m) long and, on their undersides, are orange to yellow and have backwards-curving teeth along their margins, similar in shape and viciousness to those on the petiole margins of *L. saribus*. The segments are free for three-quarters of their length and are stiff, linear-lanceolate, and a deep green on both surfaces. Inflorescences are 7 feet (2.1 m) long and usually extend beyond the leaf crown. The tiny round brown fruits are less than 1 inch (2.5 cm) in diameter.

This is a beautiful species, especially when it is older. Its leaf crown is exceptionally nice and open, and the long petioles allow the leaves to be readily seen, even from a distance. It also seems easy, if slow, to grow and is adaptable to zones 10 and 11 and is marginal in drier climates of 9b. It is salt tolerant and drought tolerant (for short periods). It is adaptable to full sun, even in its youth.

Livistona chinensis
liv-i-STO-na · chi-NEN-sis
PLATE 588

Livistona chinensis occurs naturally in open woodlands on various, often sandy, soils in the southern Japanese islands as far north as the southeastern part of Shikoku, in Taiwan, where it is endangered, and in Hainan Island, China. The epithet is Latin for "of China." The common name is Chinese fan palm.

The mature stems grow slowly to 45 feet (14 m) tall with a diameter of 1 foot (30 cm). They are a deep brown to reddish brown when young (due to the adherence of leaf bases) and gray with indistinct closely set rings of leaf base scars in older palms. The leaf crown is densely packed with 40–60 leaves, usually full and almost round, and 18 feet (5.4 m) wide and 20 feet (6 m) tall. The leaves are 6 feet (1.8 m) wide on petioles 6 feet (1.8 m) long. In younger palms the petioles have teeth along their margins. Leaf shape is nearly circular, sometimes diamond shaped, and the segments are free for about half their length. The segments are split about 5 inches (13 cm), and these smaller segments are pendulous in older palms but more erect in juvenile plants. Leaf color is bright green to glossy olive green. The much-branched inflorescences are 6 feet (1.8 m) long and bear small, whitish flowers. The round or elongated 1-inch (2.5-cm) fruits are glossy greenish blue to pale green when mature.

This species is slightly susceptible to lethal yellowing disease. It is among the hardiest species in the genus and is adaptable to zones 9 through 11 and in favorable microclimates of 8b, especially those with drier climates. It withstands drought but looks its best and grows its fastest only with regular, adequate moisture. It is adaptable to and looks wonderful in partial shade as a juvenile but needs full sun to attain its ultimate height. It is not fussy about soil type as long as it is freely draining.

Livistona chinensis and *L. australis* are difficult to distinguish from each other at a distance, especially when past their juvenile stages. The salient differences between them are that the Australian cabbage palm has a taller trunk with much more distinct rings of leaf base scars, leaves that are more of a dusky green, and leaf segments that are usually longer and even more pendulous, at least when the palm is older. The Chinese fan palm has the same landscape uses as its Australian counterpart but is more choice as a young plant because of its wide, nearly circular leaves; when grown in partial shade, the leaves stay rounded and the segments remain shallower and a deeper green for much longer. It is well adapted to indoor cultivation, especially when young, but needs a great amount of space and light when older.

Livistona concinna
liv-i-STO-na · kon-SIN-na
PLATE 589

Livistona concinna is endemic to Queensland, Australia, where it is known as the Cooktown or Kennedy River Livistona or Cooktown fan palm. It occurs at low elevations, in seasonally flooded forests, river banks, and mangrove margins. Its distinctiveness had been appreciated even before the species was formally named in 2001. The epithet is Latin for "neat" and "orderly" and refers to the arrangement of the leaves in the crown.

The solitary trunk grows to nearly 100 feet (30 m) tall and about 1 foot (30 cm) in diameter. The petioles are 4–10 feet (1.2–3 m) long, green throughout, and armed with curved black spines. The leaf blade is 6 feet (1.8 m) long, medium green above, lighter green below, and glossy on both sides. The conspicuously drooping segments are free for about 60 percent of their length, and each segment is cleft for about 40 percent of its length. The globose fruits are shiny black at maturity.

This graceful species grows lustily if given a rich, moist soil. It is tolerant of saline conditions but not of cold. Like many coastal palms from Queensland, it is tolerant of high winds. It is for zones 10 and 11.

Livistona decora
liv-i-STO-na · de-KO-ra
PLATE 590

Livistona decora is endemic to the eastern Queensland coastal region of Australia, where it grows in open woods or along the edges of forests and in swampy areas near the coast. The epithet is Latin for "decorative." Common names in Australia are weeping cabbage palm and ribbon fan palm. This species was previously known as *L. decipiens*.

Mature trunks grow to 50 feet (15 m) high with a diameter of 10 inches (25 cm). They are light brown and, except in young plants, covered in closely set deep rings of leaf base scars. The leaf crown is especially rounded and is densely packed with 40–60 leaves. The leaves are 9 feet (2.7 m) wide but are so deeply costapalmate that they are almost V shaped with much of the diameter thus foreshortened and the general leaf a linear diamond shape; the visual length of a given leaf is therefore around 6 feet (1.8 m). The leaf segment divisions extend almost to the intrusive petiole and are divided for half their lengths into finer, pendent segments, resulting in a decidedly weeping visual effect from the curtains of segments that hang vertically for 3 or more feet (90 cm). The petiole is 6 feet (1.8 m) long and has short teeth on its margin. Leaf color is deep green to almost bluish green above and glaucous grayish green beneath. The much-branched inflorescences are up to 9 feet (2.7 m) long and bear small, bright yellow blossoms. The round ½-inch (13-mm) fruits are shiny black when mature.

This hardy palm is adaptable to zones 9 through 11 and sometimes 8b in the drier winter climates. It is among the more cold-tolerant species in the genus, surviving brief exposure to temperatures as low as 20°F (−7°C). Its soil preference is wide as long as the medium is freely draining. This palm is drought tolerant when established but looks better and grows faster with regular, adequate moisture. It is adapted to full sun from youth to old age. It is more graceful than many *Livistona* species because of the weeping curtain effect of the leaves that look almost pinnate from a distance and are delicate appearing. The leaf crown provides one of the most beautiful silhouettes and its canopy-scape is almost spellbinding. The tree may be planted as a single specimen, even surrounded by space, but is much more attractive in groups. It needs space and high light levels indoors and thus is not a good subject long term.

Livistona drudei
liv-i-STO-na · DROO-dee-eye
PLATE 591

Livistona drudei is an endangered species native to northeastern Queensland, Australia, where it grows in low coastal plains, especially along streams and rivers, and just inland from mangrove swamps. The epithet honors C. G. Oscar Drude (1852–1933), a German botanist and ecologist, who, with H. Wendland, published an important early work on the taxonomy of Australian palms.

The trunk attains a height of 80 or more feet (24 m) in habitat and is 1 foot (30 cm) in diameter. It is mostly free of leaf bases and fibers except for its uppermost part and is light gray to almost white. The petiole bases are purple. The beautiful open leaf crown is semicircular, as leaves are held well above the horizontal. The deep green leaves are 4–5 feet (1.2–1.5 m) long on petioles 5–7 feet (1.5–2.1 m) long, and their segments are free for two-thirds of their length. The individual segments are cleft for 60 percent of their length and are conspicuously pendent. The globular ½-inch (13-mm) fruits are a semiglossy purplish black.

This palm is exquisite when mature and is certainly one of the most beautiful, if not the most beautiful, in the genus. Its silhouette is thrilling and yet diaphanous because of the open, circular crown with its long leaf petioles. The species is adaptable to zones 10 and 11 and marginally so in 9b. It needs full sun, regular, adequate moisture, and a rich soil. It is fast growing if given the proper conditions.

Livistona endauensis
liv-i-STO-na · en-dow-EN-sis
PLATE 592

Livistona endauensis is endemic to peninsular Malaysia, where it grows in low, mountainous rain forest, usually on top of limestone hills and ridges. The epithet is Latin for "of Endau," a region of Malaysia.

The trunk of this relatively small species grows to 30 feet (9 m) tall with a maximum diameter of 6 inches (15 cm). It is deep olive green in all but its oldest parts and graced with widely spaced light rings of leaf base scars. The leaf crown is open and spherical or nearly so. The glossy green leaves are 3 feet (90 cm) across, nearly circular, and weakly costapalmate on petioles 5 feet (1.5 m) long that are armed throughout with teeth. The lanceolate, pointed segments are free for 60 percent of their length and are stiff and straight. The fruits are slightly oblong, ½ inch (13 mm) in diameter, and bluish green.

Although formally described in 1987, the species is still rare in cultivation. Fairchild Tropical Botanic Garden collected a large number of seeds in 2002, and these have grown into lovely young palms. This palm is probably not hardy to cold and certainly needs copious, regular moisture. It appreciates light shade as a juvenile but will take full sun when mature.

Livistona fulva
liv-i-STO-na · FOOL-va
PLATE 593

Livistona fulva is endemic to Blackdown Tableland in central eastern Queensland, Australia, where it grows in open woodland in sandstone areas near the bases of cliffs, in gullies, and along streams at 1200–2000 feet (370–610 m). The epithet is Latin for "tawny colored," an obvious allusion to the feltlike tomentum on the new leaves. Until Tony Rodd described it as a species in 1998, this palm was known as *L.* 'Blackdown Tableland'.

One of the smaller species, *L. fulva* has a trunk that grows to 35 feet (11 m) in habitat and around 1 foot (30 cm) in diameter. Older stems are usually free of leaf bases, dark gray, and heavily and closely ringed, whereas younger trunks are mostly covered with cinnamon-brown leaf bases and fibers. The leaf crown is usually semicircular and is open because of the long leaf petioles, which are each 6–8 feet (1.8–2.4 m) long. Individual leaves are almost flat, almost circular, and 3 feet (90 cm) wide, with segments that are free for half their length. The segments are very shallowly cleft and are rigid, not pendulous. Leaf color is green above with a grayish hue beneath. The undersides of new leaves are covered with a beautiful coppery or light golden brown tomentum, which usually disappears as the short hairs are worn off, especially with older trees.

This palm is slow growing but relatively hardy to cold, being adaptable to zones 9 through 11 and possibly marginal in 8b, especially of drier climes. It does well in tropical and Mediterranean climates but not hot desert areas. It seems to thrive in partial shade or full sun and is not fussy about soil type as long as it is freely draining. It is drought tolerant but also languishes and grows more slowly if not provided with adequate, regular moisture.

Livistona humilis
liv-i-STO-na · HYOO-mi-lis
PLATE 594

Livistona humilis is endemic to the Northern Territory, Australia, where it grows in open, flat monsoonal *Eucalyptus* woods of low elevations, generally on deep, sandy soil. The epithet is Latin for "small." The common name is sand palm.

The solitary trunk grows to 20 feet (6 m) tall in habitat but is often shorter. It has a diameter of 2–3 inches (5–8 cm) and is gray and covered with petiole stubs. The leaf crowns are circular in older trees and open. The small, flat, circular leaves are 12–18 inches (30–45 cm) in diameter on petioles 2 feet (60 cm) long. The stiff, nonpendulous segments are free for about three-quarters of their length, linear-lanceolate, and glossy medium green on the upper surface, lighter below. The trees are dioecious. Female inflorescences are long and project well beyond the leaf crown with thin yellow branches, while male inflorescences are shorter and arching. The oblong fruits are ¾ inch (19 mm) long and dark purplish black.

Although it is the smallest Australian species of *Livistona*, this palm is elegantly picturesque with its neat and tidy round crown and knobby trunks and should be much more widely planted. Its only faults are that it is slow growing and difficult to transplant. It needs sun and, while drought tolerant, it grows and looks better with adequate, regular moisture. It is probably not cold hardy.

Livistona inermis
liv-i-STO-na · i-NUR-mis
PLATE 595

Livistona inermis occurs naturally in the Northern Territory, Australia, with a few collections known from adjacent Queensland. It grows in spectacular settings on or at the base of red sandstone cliffs at low elevations. It is sometimes called the small fan palm in Australia, and this moniker is unfortunate as it implies something not overly desirable. The epithet is Latin for "unarmed" and alludes to the petiole margins.

The solitary stem attains a maximum height of 30 feet (9 m) in habitat with a diameter no greater than 4 inches (10 cm). It is gray and its lower parts usually have persistent petiole stubs. The leaf crown is airy and open because of the small, fine-textured leaves and relatively long petioles. The nearly circular leaves are 1–2 feet (30–60 cm) wide, with narrow, light green to grayish green segments that are free almost to the petiole. The petioles are 2–3 feet (60–90 cm) long, have a reddish hue, and are unarmed but may have a few tiny prickles on the margins near the stem. The leaf segments are mostly stiff but not rigid and are often pendent near their apices; they are split to three-quarters of their length. The oblong fruits are black when ripe and ½ inch (13 mm) long.

This species is reportedly slow growing and resents transplanting. It needs sun and a fast-draining soil. Alas, it is not hardy to cold and is adaptable only to zones 10 and 11. The palm is elegance personified.

Livistona jenkinsiana
liv-i-STO-na · jen'-kinz-ee-AH-na
PLATE 596

Livistona jenkinsiana occurs naturally in northeastern India (Sikkim and Assam), Nepal, Myanmar, northern Thailand, and southern China, including the island of Hainan, where it grows in mountainous rain forest at 400–3800 feet (120–1160 m). The epithet honors Major Francis Jenkins (1793–1866), British commissioner of Assam and collector of the type specimen.

The solitary trunk attains a height of 30 feet (9 m) in habitat and is 6–10 inches (15–25 cm) in diameter, with flattened petiole stubs persisting on its lower portions. Petioles are 4–6 feet (1.2–1.8 m) long and are armed throughout with stout teeth. The circular leaf crown consists of many 8-foot (2.4-m) nearly circular leaves with stiff segments that are free for half of their length; the segment tips are cleft for about half of their length. Juvenile plants have larger and much less segmented leaves. Leaf color is shiny green above and a glaucous grayish green below. The fruits are distinctively kidney shaped, wider than long, 1 inch (2.5 cm) in diameter, and deep, dull blue.

This noble species is attractive at all stages. Young plants, especially those grown in partial shade, have leaves that much resemble those of *Saribus rotundifolius* but are smaller. The palm is surprisingly hardy to cold considering its tropical origins and is adaptable to zones 10 and 11 and marginally so in warmer microclimates of 9b. It relishes partial shade when young and needs regular, adequate moisture at all ages. It appreciates a well-drained, humus-laden soil but is known to grow in impoverished soils.

Livistona lanuginosa
liv-i-STO-na · la-noo'-ji-NO-sa
PLATE 597

Livistona lanuginosa is a rare and endangered species endemic to stream margins and washes in eastern central Queensland, Australia. Until it was described in 1998 by Tony Rodd, it was known as *L.* 'Cape River'. The epithet is Latin for "woolly," which alludes to the dense and lengthy white tomentum on the petioles and the inflorescence bracts.

The trunk attains a height of 60 feet (18 m) in habitat with a diameter of 1 foot (30 cm). The base of the stem is usually quite swollen and is covered with persistent petiole stubs. The leaf crown is spherical but beautifully open and airy because of the long leaf petioles, 6 feet (1.8 m) long. The grayish green leaves are 4–6 feet (1.2–1.8 m) in diameter and have segments that are free for one-third their length. The segments are further divided for one-quarter of their length and are slightly pendent only at their tips. The young petioles as well as the rachis are covered in a dense white wool on their upper surfaces, and the bottom surfaces of the blades are covered in a thick coating of white wax which gives them a light bluish gray cast. The inflorescences are half hidden in the leaf crown, and the round purplish black fruits are about 1 inch (2.5 cm) in diameter.

This species is slow growing but does best in full sun from an early age and in a free-draining soil. While it is drought tolerant when mature, it cannot be expected to flourish unless given regular, adequate moisture. It is adaptable to zones 10 and 11 and is marginal in warm microclimates of 9b, especially in drier climes.

Livistona mariae
liv-i-STO-na · MAH-ree-eye
PLATE 598

Livistona mariae is endemic to sandstone canyon bottoms in mostly desert regions of central Australia. Although the habitat appears arid, the palms have permanent access to underground water. The epithet honors Grand Duchess Marie Alexandrovna (1853–1920), daughter of Tsar Alexander II of Russia and daughter-in-law of Queen Victoria. It is sometimes called the Central Australian cabbage palm.

The solitary trunk grows to 90 feet (27 m) in habitat, 40 feet (12 m) in cultivation, and is nearly 1 foot (30 cm) in diameter. Older trunks are deep to light gray and are closely set with rings of leaf base scars. Younger trunks are covered in a dense, tight fiber mass and in old leaf bases that impart a dark brown coloration to the stems. The leaf crown is 15 feet (4.5 m) wide and tall. The reddish brown petioles are 6–8 feet (1.8–2.4 m) long and have sharp 1-inch (2.5-cm) teeth on their lower margins. The circular leaves are 6 or more feet (1.8 m) in diameter, with linear-lanceolate segments that are free for half their length. The segments are essentially split halfway and pendent and, like the rest of the blade, are grayish green on both sides, with a waxy coating on the undersides. The leaves of young palms are usually tinted reddish purple or deep rose, and their leaf segments are stiff and not pendent. The inflorescences are 6 feet (1.8 m) long, mostly erect, and much branched. They bear creamy yellow flowers. The round ½-inch (13-mm) fruits are semiglossy black when mature. Tony Rodd (1998) considered this species to consist of 3 subspecies, but they were treated as 3 separate species in the monograph by John Dowe (2009).

This palm adapts to various well-drained soils. It is drought tolerant when established but grows much faster with regular, adequate moisture. It is adaptable to zones 10 and 11 but is probably safe in 9b, especially in drier climates. It needs full sun from youth to old age. If given good care in a tropical or nearly tropical climate, this species is fast growing once past the juvenile stage: 25 feet (7.6 m) of trunk in 15 years. It has the same landscape uses as *L. australis* and *L. chinensis* but is nobler appearing than either of them. It does not look bad even when used as a specimen tree surrounded by space but is much more pleasing in groups of 3 or more individuals of varying heights. It needs room and extremely good light to be grown indoors.

Livistona muelleri
liv-i-STO-na · MYOOL-ler-eye
PLATE 599

Livistona muelleri is indigenous to far northern and northeastern Queensland, Australia, as well as southern New Guinea, where it grows in monsoonal savannas, on the edges of rain forest, and in open and drier woodlands, always at low elevations. The epithet honors German botanist, Sir Ferdinand J. H. von Mueller (1825–1896), government botanist for the colony of Victoria, Australia. It is called Cairns fan palm in Australia.

Mature trees in habitat are known to form trunks as tall as 30 feet (9 m) but most are shorter. They are 6–10 inches (15–25 cm) in diameter, dark gray, and heavily and closely ringed, the lowest parts covered in rounded stubs of the old petioles, if not blackened and burned by grass fires. The leaf crown is hemispherical and dense. The circular, flat leaves are 2 feet (60 cm) wide on petioles 2–3 feet (60–90 cm) long. The segments are free for over half their length and are stiff, not pendulous; they are a dark grayish green above and a lighter bluish gray-green beneath. The inflorescences, which do not exceed the leaves, are notably colorful: red branches bearing flowers with maroon sepals and yellow petals. The fruits are ellipsoid or pear shaped, ½ inch (13 mm) long, and deep brown to black.

This palm is exceptionally slow growing but is adaptable to zones 10 and 11 and may be marginal in 9b, especially in drier climates. It needs sun and copious, regular moisture but is not fussy about soil type, growing in alkaline and acidic soils, and even those that are on occasion waterlogged. This neat and attractive species is especially nice in groups of 3 or more individuals.

Livistona nitida
liv-i-STO-na · NI-ti-da
PLATE 600

Livistona nitida is endemic to southeastern Queensland, Australia, where it grows near permanent springs and along banks of rivers and streams at low elevations. Until it was described in 1998 by Tony Rodd, it was known as *L.* 'Carnarvon'. The epithet *nitida* is Latin for "shiny" and alludes to the glossy fruits.

The trunk attains a height approaching 100 feet (30 m) with a diameter of 1 or more feet (30 cm), is light gray to light brown, and is distinctly ringed with leaf base scars. Some petiole bases persist on the lowest part of the trunk. The leaf crown is circular and open. The leaves are 5–6 feet (1.5–1.8 m) wide on petioles 5 feet (1.5 m) long. The yellowish green leaf segments are free from one another for two-thirds their length, and each one is cleft at the tip for two-thirds its length. The segment tips are narrow, limp, and pendent, creating a weeping or curtain effect. The inflorescences are shorter than the leaves. The round ½-inch (13-mm) fruits are glossy black.

This species is unusually hardy to cold considering its near tropical origins and is adaptable to zones 9 through 11 and possibly is marginal in 8b, especially in drier climes. It is a beautiful palm, similar in most respects to *L. australis* but more robust.

Livistona rigida
liv-i-STO-na · RI-ji-da
PLATE 601

Livistona rigida is endemic to northwestern Queensland and the Northern Territory, Australia, where it grows on river banks at low elevations. The epithet is Latin for "rigid," an obvious allusion to the stiff inflorescence branches.

The species is closely related to *L. mariae*, and Tony Rodd (1998) made it a subspecies of that taxon. Dowe (2009) returned it to full species status, a view shared by most taxonomists. Compared to *L. mariae*, the leaves of *L. rigida* are much more rigid, especially in juvenile plants; are much more grayish green or even bluish gray-green; are covered in a thin coating of wax on their undersurfaces; and are smaller. The leaf segments, when mature, are pendent; they are free for half their length, and the segment tip is further split for one- to two-thirds the length of the segment. The inflorescence does not exceed the leaves. The globose ½-inch (13-mm) fruits are semiglossy black.

This palm is amazingly cold tolerant, considering its tropical origins. It survives (with protection when young) in warm microclimates of zone 8b and is dependable for the most part in 9 through 11. It is fast growing and not particular about soil type but does better in a good one. It is drought tolerant but grows slowly and does not look as good under such conditions.

Livistona saribus
liv-i-STO-na · SAR-i-bus
PLATES 602–603

Livistona saribus is among the most widespread species in the genus, from southeastern China, Vietnam, Laos, Cambodia, Thailand, and peninsular Malaysia to Borneo and the island of Luzon in the Philippines, where it grows in swampy rain forest, clearings, and monsoonal savannas. It is usually called taraw palm. The epithet is a Latinized form of a Moluccan name for the species.

Mature trunks grow to 100 feet (30 m) in their native haunts but are usually 50 feet (15 m) tall under cultivation, with a diameter of 1 foot (30 cm). Old plants have trunks free of leaf bases and fibers. The trunks are then pale gray, with deep but incomplete circles of leaf base scars. Petiole stubs may be present on the lowest part of the trunk. The leaf crown is dense and hemispherical to spherical, depending on whether the old dead leaves are present, and 15 feet (4.5 m) wide and tall. Leaves are generally 4 feet (1.2 m) wide on petioles that are 5–6 feet (1.5–1.8 m) long, green or a beautiful deep orange to deep red, and armed with stout, black, curved teeth that may be as long as 3 inches (8 cm) and have bulbous bases. The segments are free for one- to two-thirds of their length, exhibit drooping tips, and are grouped together with deeper divisions between groups (an unusual feature for the genus). Leaf color is deep green on both surfaces. The leaves of juvenile trees, with less than 6 feet (1.8 m) of trunk height, are radically different and consist of fewer, wider segments, which are deeply split once or twice, and are not pendent. The yellow inflorescences are 5 feet (1.5 m) long and sparsely branched. They bear small, yellow flowers. The round 1-inch (2.5-cm) fruits are an outstandingly beautiful shiny blue or purple.

There seem to be 2 slightly different forms of this species: one with green petioles and one with red or deep orange petioles in younger plants. The red petiole forms are more attractive and significantly more tender to cold. No one seems to know where the more highly colored form originated or why it is more tender to cold, but it is safe only in zones 10 and 11.

This species is relatively fast growing when younger and if given regular, adequate moisture and a humus-laden soil. It luxuriates in partial shade when young but needs full sun as it ages if it is to grow as fast as it should and look as good as it can. It is adaptable to zones 9 through 11 and is marginal in 8b. The palm has visual similarities to *L. australis* when older. When juvenile, it is much more interesting because of the larger, less segmented, flat leaves and the beautiful petiole thorns. The leaf segments of older individuals are not as pendulous as those of *L. australis* and this characteristic gives it another attraction in the landscape. It should not be planted as a single specimen surrounded by space, but groups of 3 or more individuals of varying heights are beautiful indeed. This palm is a good candidate for indoor cultivation only when young.

Livistona victoriae
liv-i-STO-na · vik-TOR-ee-eye
PLATE 604

Livistona victoriae, described by Tony Rodd in 1998, was formerly known informally as *L.* 'Victoria River'. It occurs naturally in far northeastern Western Australia and northwestern Northern Territory, Australia, where it grows in sandstone ravines and intermittent river courses, always where groundwater is available. The epithet is Latin for "of Victoria," a reference to the Victoria River in the palm's habitat.

The solitary trunk grows to 50 feet (15 m) high with a diameter of 8 inches (20 cm). It is light gray, straight, and nearly smooth in its older parts. The leaf crown is spherical and open. The petiole is 3–6 feet (90–180 cm) long, with teeth at the base. The leaves are 3–5 feet (90–150 cm) long, strongly costapalmate with segments that are free for more than half their length. Each segment is cleft at the tip for about two-thirds its length; the tips are rigid, not pendulous. They are a beautiful bluish gray-green to silvery green and are thick and leathery. The round ½-inch (13-mm) fruits are reddish brown to black.

Because of its leaf color, this species is an especially attractive one. It is amazingly hardy to cold considering its tropical origins and is probably adaptable to zones 9 through 11 in the drier climates. It needs sun and a fast-draining soil.

Lodoicea
lo-do-EE-see-a

Lodoicea is a monotypic genus of massive, solitary-trunked, palmate-leaved, dioecious palm. The name honors King Louis XV of France. Seeds can take 3 months to one year to germinate. The germination stalk can go down 6 feet (1.8 m), so it is best to sow seed directly in the landscape rather than in a container. Depending on how warm the ground is, a year or more can elapse between the time the stalk first goes down to when the first leaf emerges from the ground. For further details of seed germination, see *Bismarckia*.

Lodoicea maldivica
lo-do-EE-see-a · mal-DIV-i-ka
PLATES 605–608

Lodoicea maldivica is endemic to Curieuse and Praslin, 2 small islands in the Seychelles, where it grows in large stands in forests and moist valleys at low elevations. It was probably much more widespread in the Seychelles in the past. The epithet is Latin for "of Maldives" and was bestowed when only the fruit of the palm was known and that from beaches of the Maldives, southwest of India in the Indian Ocean.

The solitary trunk grows to a height of 80 or more feet (24 m) with a diameter of 20 inches (50 cm) maximum. The stem is straight, columnar, and light tan in its younger parts but light gray to dark gray in its older parts. It is swollen at the base and, in habitat, the bulbous stem base forms a depression in the ground from which it grows. Male trees reportedly grow taller than female trees, which must invest more of their resources into reproduction rather than growth. The leaf crown consists of 15–20 leaves and is mostly hemispherical but sometimes exhibits pendent dead leaves at its base. The leaves are held on robust petioles 6–12 feet (1.8–3.6 m) long and are deeply costapalmate, with blades 18 feet (5.4 m) long and 12 feet (3.6 m) wide. The leaf segment divisions are relatively shallow in younger trees but much deeper in older ones. Segments are lanceolate, coriaceous, and mostly stiff, but slightly pendent at their tips. Leaf color is glossy deep green on both surfaces.

The inflorescences grow from among the leaf bases. The male inflorescences are fleshy, pendent spikes 6 feet (1.8 m) long and 3 inches (8 cm) thick, with sessile starlike whitish blossoms. The female inflorescences are shorter, with several thick, fleshy branches, also bearing sessile but extremely large, fleshy, white cup-shaped blossoms; they are the largest of all palm flowers. The infamous fruits are immense and top shaped, 20 inches (50 cm) long and 12–15 inches (30–38 cm) wide; they are nearly black and are similar to a coconut. Pollination is thought to be affected by small flies, but may be assisted by wind or even by the indigenous geckos. The fruits require 5–7 years to reach maturity, at which time they can weigh up to 45 pounds (20 kg). The seeds inside these fruits are almost as large as the fruit itself and have 2 hemispherical lobes connected by a deep groove; they are the largest seeds in the world and the subject of much lore.

Today the endosperm is marketed in Asia, and the nonviable nuts are sold to tourists. Unfortunately, the harvest of nuts is not sustainable. Not enough seeds are allowed to germinate in the forests, and scientists predict that not enough seedlings will be on hand to replace adult palms that die from cyclone strikes or other natural causes.

The trees are extremely slow growing, especially when younger: it may take 20 years or even longer for the trunk to start forming. The species does not tolerate frost and is adaptable only to zones 10b and 11. It needs regular, adequate moisture, partial shade when in seedling stage, and sun thereafter. It is not fussy about soil type as long as it is fast draining.

Loxococcus
lahx-o-KAHK-kus

Loxococcus is a highly endangered, monotypic genus of solitary-trunked, pinnate-leaved, monoecious palm. The genus name is from Greek words meaning "slanting" and "seed." Seeds germinate within 30 days. For further details of seed germination, see *Areca*.

Loxococcus rupicola
lahx-o-KAHK-kus · roo-PI-ko-la
PLATE 609

Loxococcus rupicola is endemic to Sri Lanka, south of the Indian subcontinent, where it grows in mountainous rain forest at 1000–5000 feet (300–1520 m). It is critically endangered because of destruction of the rain forest. The epithet is Latin for "rock dwelling" and refers to the habitat in which the species grows.

The slender gray trunk slowly attains a height of 20 feet (6 m) or slightly more. It is nearly smooth in its older parts and indistinctly ringed in its younger parts. The crownshaft is 2 feet (60 cm) tall, deep green, smooth, cylindrical, and barely thicker than the trunk. The leaf crown is hemispherical or slightly more rounded, open, and airy with 10 or fewer leaves. Each leaf is 10 feet (3 m) long and is ascending but beautifully arching when new and spreading when older. The stout petiole is 10 inches (25 cm) long and holds many widely spaced, narrow, deep green leaflets that are 2 feet (60 cm) long; in young leaves, the leaflets grow from the rachis at an angle which creates a V-shaped leaf but older leaves have spreading leaflets that are almost in a single flat plane. The inflorescences grow from beneath the crownshaft and are short, with stiff but spreading flowering branches that bear both male and female blossoms. The egg-shaped fruits are 1 inch (2.5 cm) long and red to reddish brown when mature.

This beautiful palm is almost absent from cultivation outside of a few botanical gardens. It is reportedly slow growing, needs a constantly moist but well-drained acidic soil, and does not tolerate cold.

Lytocaryum
ly-to-KAHR-ee-um

Lytocaryum includes 4 solitary-trunked, pinnate-leaved, monoecious palms in southern Brazil. They are very closely allied to *Syagrus*; in fact, DNA evidence now shows that *Lytocaryum* should be sunk into *Syagrus*. This is another case where the classification needs to catch up with the molecular evidence. The inflorescences grow from among the leaf bases and are branched once with numerous short flowering branches bearing small, unisexual blossoms of both sexes. The brown ellipsoid fruits are 1 inch (2.5 cm) long. Three species are unusual (but not unique) among palms in that the fruits split open at maturity to release the endocarps. The genus name is derived from Greek words meaning "loose" and "nut," a reference to the split fruits. Seeds germinate within 180 days but can take up to a year. Germination is sporadic. For further details of seed germination, see *Acrocomia*.

Lytocaryum hoehnei
ly-to-KAHR-ee-um · HUR-nee-eye
PLATE 610

Lytocaryum hoehnei is endemic to a small area of wet, mountainous forest in southern Brazil, where it grows at 2600–3200 feet (790–980 m) and is nearly extinct. The epithet honors Brazilian botanist and conservationist Frederico C. Hoehne (1882–1959). The species differs from *L. weddellianum* mainly in its longer leaflets that are even more silvery on their undersurfaces, and in its slightly larger fruits. It is also perhaps even more beautiful than *L. weddellianum* but is virtually unknown in cultivation. It would have the same cultural requirements, with possibly slightly more cold tolerance.

Lytocaryum insigne
ly-to-KAHR-ee-um · in-SIG-nee
PLATE 611

Lytocaryum insigne is found in the Brazilian states of Rio de Janeiro and Espírito Santo, where it grows in low forests at elevations above 3000 feet (900 m). It tolerates full sun and drier conditions than its kin. The epithet is Latin for "distinguished" or "conspicuous," a reference to its overall size, which is the largest in the genus. It also is the tallest species in the genus, with a stem that reaches 30 feet (9 m). It has a stiffer, shorter leaflet than *L. weddellianum*. The petiole and rachis are densely covered with blackish brown hairs. The fruits are greenish brown. This species is likely a bit less delicate and sensitive in its horticultural requirements, so let us hope it becomes more available in the future. At present, it is found only in specialist collections and botanical gardens.

Lytocaryum weddellianum
ly-to-KAHR-ee-um · wed-del'-ee-AH-num
PLATE 612

Lytocaryum weddellianum is endemic to southeastern coastal Brazil, where it occurs in the understory of humid, ever-wet rain forest to an elevation of 2500 feet (760 m). It is endangered because of development along Brazil's coast but is commonly cultivated, especially in containers in Europe, and this may be its only hope of survival. The epithet honors Hugh A. Weddell (1819–1877), a British botanist and collector in South America.

The solitary trunk slowly grows to a height of 10 feet (3 m) with a diameter of 4 inches (10 cm) in habitat, but in cultivation is seldom taller than 6 feet (1.8 m). The stem is usually covered in dark fibers from the leaf sheaths and with the pale stubs of the dead petioles. The open, spherical leaf crown has beautifully arching leaves 2–3 feet (60–90 cm) long on petioles that are 1 foot (30 cm) long and covered in chestnut brown hairs. The leaves resemble those of *Phoenix roebelenii*, but are smaller. The newer leaves often show a slight twist of the rachis at its midpoint. The numerous linear-lanceolate leaflets are 4–6 inches (10–15 cm) long and regularly spaced along the arching rachis, where they grow in a single flat plane. They are deep green above and grayish green to almost completely gray beneath. The fruits are green when ripe.

This palm cannot withstand full sun, especially in hot climates. It has a modicum of cold tolerance and is adaptable to zones 10 and 11 and, if occasionally protected, can be grown in 9b. It does not tolerate drought and needs regular, adequate moisture as well as a rich, friable, well-drained soil that is slightly acidic. It grows in South Florida if protected from the sun and if the soil is either amended or heavily mulched. This deliciously elegant miniature palm should be sited in an intimate, under-canopy site in the garden. It makes one of the finest house or container plants. Its only fault is that it is slow growing.

Manicaria
ma-ni-KAHR-ee-a

Manicaria consists of 2 pinnate-leaved, monoecious palms in tropical America. The genus name is Latin for "gloves" and refers to the large, tough, fibrous primary bracts of the inflorescences. Bernal and Galeano (2010) discussed the differences between the 2 species. Only the larger species is (occasionally) cultivated. Seeds of *Manicaria* germinate sporadically over several months. Seeds allowed to dry out before sowing do not germinate as well as those kept slightly moist. For further details of seed germination, see *Areca*.

Manicaria saccifera
ma-ni-KAHR-ee-a · sa-KIF-er-a
PLATES 613–614

Manicaria saccifera occurs naturally in 2 large and disjunct regions. The northern and western part of its range is centered in Central America (from southern Belize and eastern Guatemala, through the Caribbean coastal regions of northern Honduras, eastern Nicaragua, Costa Rica, and Panama) and the Pacific coastal lowlands of Colombia and extreme northwestern Ecuador, where the species is a denizen of coastal swamps. The other area is east of the Andes in northern South America and includes eastern and southern Venezuela, Guyana, Surinam, and French Guiana; here the species grows in seasonally inundated rain forest, always as a large undergrowth subject in wet areas or along the banks of rivers, often forming dense stands or colonies. The epithet is Latin for "sack-bearing" and alludes to the sacklike peduncular bract.

The palm forms a solitary or forking trunk to 30 feet (9 m) tall and 10 inches (25 cm) in diameter but is usually 6 feet (1.8 m) high. In nature, dead leaves and fibrous leaf bases persist and mostly hide the stem. The erect, elliptical leaves ascend from the trunk and may be 25 feet (7.6 m) long but usually are 20 feet (6 m) long; they are 4–6 feet (1.2–1.8 m) wide and mostly entire and undivided until the wind or age separates them into segments of greatly varying widths. The blade is a rich, deep emerald green on both surfaces and is heavily corrugated and serrated on its outer margins with the evidence of their apices. When they are older, the soft petioles, which are 4–6 feet (1.2–1.8 m) long, easily break under the weight of the immense leaf blades. The inflorescences grow from among the leaf bases, are branched once, and are 3 feet (90 cm) long. In nature they are usually hidden (because of their short length) by the leaf bases and the litter trapped therein. These flower clusters are unique because of the large sacklike primary bract that covers them like a hood. It is a conical sheath of "woven" fibers that is pushed away only as the peduncle of the flowering branches elongates and the fragrant flowers start to open, but it usually remains hanging in the leaf crown. The hard, woody fruits are round and are covered in pyramidal tubercular, blunt projections. The fruits are durable and, in coastal or lowland stands of the species, detach from the trees and float down the rivers to the sea to be often seen on distant shores, including Florida's Atlantic coast.

The species is, outside of botanical gardens, rare in cultivation, mainly because of its tropical requirements: it cannot withstand temperatures below 50°F (10°C). It needs year-round warmth, copious and constant moisture, a humus-laden acidic soil, and shade or partial shade when young. It grows in the open in full sun, but the rich color and integrity of its blades are compromised: the leaves are always tattered and unsightly when exposed to the elements. If the proper growing conditions can be provided, it makes one of the most startlingly beautiful and impressive plantings possible.

The palm has, in the past, had many human uses. Its great leaves are used in the habitat for thatch and, at one time, were used as the sails for canoes. The remarkable peduncular bract is still used to make hats that need no weaving, and nowadays for making hats and other souvenirs for tourists who visit the Amazon.

Marojejya
mahr-ro-JAY-zhee-a

Marojejya comprises 2 solitary-trunked, pinnate-leaved, monoecious (or possibly polygamous) palms endemic to eastern Madagascar, one of them extremely rare in habitat, both of unusual beauty and magnificence. The palms have funnel-shaped crowns that catch litter and impound water, at least as juveniles in the forest understory. The inflorescences are short with thick ropelike flowering branches and are almost hidden by the leaf bases. The oblong to nearly round fruits are formed in clusters, are less than 1 inch (2.5 cm) in diameter, and are red or black when mature. The genus name is taken from the Marojejy Mountains in northeastern Madagascar, where the palm was first recorded. Seeds of *Marojejya* germinate within 45 days. Seeds allowed to dry out before sowing do not germinate as well as those kept slightly moist. For further details of seed germination, see *Areca*.

Marojejya darianii
mahr-ro-JAY-zhee-a · dar-ee-AH-nee-eye
PLATES 615–616

Marojejya darianii is an extremely rare and endangered species endemic to one small site in Madagascar, a swampy valley bottom in dense rain forest at 1300 feet (400 m) elevation. The epithet honors Mardy Darian, a palm collector in Southern California and the first person to introduce this palm into cultivation.

In nature, the species forms a solitary trunk to 50 feet (15 m) tall with a diameter of 1 foot (30 cm). The leaf crown resembles a gigantic shaving brush as the leaves grow erect and ascending from the top of the trunk, without benefit of petioles. In individuals past the juvenile stage the leaves are 12–15 feet (3.6–4.5 m) long, 4 feet (1.2 m) wide at their midpoints, and are undivided when new, with a linear-obovate outline that is apically bifid. Both surfaces are deep green and deeply grooved with the outlines of the fused or cryptic pinnate segments. The rachis is stout but soft and is much lighter green. Time and the elements usually tatter the blades into pinnae of variable widths.

Although none are in cultivation, mature palms are startling in their beauty and splendor because of the dimensions of their immense, undivided new leaves. The cultural requirements are water and more water, in an acidic, fast-draining soil. This palm is reportedly immune to temperatures at or even slightly below freezing once past the seedling stage. It also seems adaptable to sun from an early age. It is slow growing in its seedling stage but reportedly much faster thereafter. It makes a stunningly beautiful container subject when young and is adaptable to a greenhouse if given enough light and moisture.

Marojejya insignis
mahr-ro-JAY-zhee-a · in-SIG-nis
PLATE 617

Marojejya insignis is endemic to mountainous rain forest on the east coast of Madagascar from sea level to 3700 feet (1130 m). The epithet is Latin for "distinguished" or "conspicuous," an apt descriptor for this remarkable palm. This species is smaller than *M. darianii*, its trunk attaining a maximum height of 20 feet (6 m) in habitat, but its leaves are slightly larger. The petioles are short, to 4 feet (1.2 m) long, and the leaves are segmented in their apical halves into many narrow, regularly spaced dark green leaflets, each 3 feet (90 cm) long, stiff but not rigid, and growing from the rachis at only a slight angle, thus resulting in a nearly flat leaf. This species is as beautiful and spectacular as *M. darianii*, but it is not nearly as sought after. It has the same cultural requirements but is probably slightly more hardy to cold.

Masoala
mah-so-AH-la

Masoala is composed of 2 massive solitary-trunked, pinnate-leaved, monoecious palms in Madagascar. One has a stout trunk and leaves bearing many linear leaflets; the other has ascending, erect leaves that are scarcely divided. The large inflorescence consists of a long, bract-covered peduncle with long, wormlike yellowish flowering branches at its end. The fruits are round and brownish yellow. Both species are rare and endangered in habitat, and both are still rare in cultivation and reportedly slow growing. They are intolerant of freezing temperatures. The genus is named for Masoala Peninsula in Madagascar, where one species (*M. madagascariensis*) has its habitat. Seeds of *Masoala* germinate within 60 days. Seeds allowed to dry out before sowing do not germinate as well as those kept slightly moist. For further details of seed germination, see *Areca*.

Masoala kona
mah-so-AH-la · KO-na

Masoala kona is endemic to central southeastern Madagascar, where it grows in low mountainous rain forest. The epithet is the aboriginal name for the palm. The trunk grows to a maximum height of 25 feet (7.6 m) in habitat but is usually much shorter, with a maximum diameter of 8 inches (20 cm). It is covered in old leaf bases in its younger parts. The leaf crown looks like a narrow feather duster and contains 15 ascending, stiff, unarching leaves 10–12 feet (3–3.6 m) long. The leaflets are fused into 6–15 upswept, pleated segments, the longest and widest being near the bottom of the blade and 6 feet (1.8 m) long and 2 feet (60 cm) wide. The leaflets grow from the rachis at angles that create a V shape to the blade, and their color is bright emerald green on both surfaces. This impressive species has strong visual affinities to its relative *Marojejya darianii*.

Masoala madagascariensis
mah-so-AH-la · mad-a-gas'-kar-ee-EN-sis
PLATE 618

Masoala madagascariensis is endemic to northeastern Madagascar, where it grows in lowland rain forest. The epithet is Latin for "of Madagascar." In habitat, the trunk grows to a maximum height of 30 feet (9 m) but is usually much shorter; its diameter is 8 inches (20 cm). The leaf crown is dense, less than hemispherical, and shaped like a feather duster, with 24 ascending, straight, and unarching leaves 10–12 feet (3–3.6 m) long on stout petioles 2 feet (60 cm) long. The rachis is often twisted near its base or midpoint so that the plane of the leaf is positioned vertically. The leaflets are 2 feet (60 cm) long, regularly spaced along the rachis, and grow from it in a single flat plane but are forward pointing towards the apex of the blade. They are narrowly elliptic, long pointed, and medium green. This impressive species is so slow growing that one plants it for future generations.

Mauritia
maw-RI-tee-a

Mauritia includes 2 massive solitary-trunked, palmate-leaved, dioecious palms in South America. The immense leaves have many deeply cut segments, which, in individuals past the juvenile stage, grow at slightly different angles from the petiole. The inflorescences are also massive and consist of distichous, horizontally spreading woody branches held at right angles to the stem and from which a curtain of smaller flowering branches descends with either male or female blossoms. The round or oblong fruits are covered with large scales and are red to reddish brown. The oily fruits are edible and rich in vitamin A. The genus name honors Johan Maurits (1604–1679), governor in Recife, Brazil, of the Dutch Chartered West India Company. Seeds of *Mauritia* germinate within 45 days. Seeds that are allowed to dry out before sowing do not germinate as well as those kept slightly moist. For further details of seed germination, see *Calamus*.

Mauritia flexuosa
maw-RI-tee-a · flex-yoo-O-sa
PLATES 619–620

Mauritia flexuosa occurs naturally in a great area of northern South America east of the Andes, including all of Amazonia except its eastern part, always in open sites. It grows along rivers and streams and in swamps, from elevations of a few hundred to 3000 feet (900 m) and, in swamps of the flatlands, forms immense colonies to the exclusion of most other trees. The epithet is Latin for "flexible" and alludes to the pendulous leaf segments.

The trunk attains a height of 80 or more feet (24 m) and a diameter of 1–2 feet (30–60 cm) in habitat. The younger portions are grayish green to tan with widely spaced, undulating, dark rings of leaf base scars, while the older parts become light gray or nearly white and are relatively smooth. The giant leaf crown is hemispherical as the living leaves rarely descend beneath the horizontal, but there are usually several (nearly) pendent, dying and dead leaves which create, especially from any distance, the visual effect of a spherical mass of foliage. The leaves are 15 feet (4.5 m) wide on large light green petioles to 30 feet (9 m) long, with bases 4 feet (1.2 m) wide and of an even lighter hue. The great leaves are basically circular and divided into 200 or more narrow, stiff but not rigid segments, each 6–7 feet (1.8–2.1 m) long and with a pendent tip. The segments grow at slightly different angles from the leafstalk to give a 3-dimensional, almost pinnate, visual effect, especially in leaves that are strongly costapalmate. The leaves of seedlings and younger individuals are not nearly so massive and are flat, with many fewer segments. The inflorescences are single, giant woody branches held horizontal from among the leaf bases; they are at least 6 feet (1.8 m) long. From these distichous branches grow the pendent, cream-colored, subsidiary branches, 1–2 feet (30–60 cm) long, that descend from the main branch like a curtain and bear the small, unisexual blossoms. In female trees, the heavy clusters of red fruits are pendent on the single remaining flowering branch that also becomes pendent. The individual fruits are 3 inches (8 cm) long, round or oblong, deep red, and covered in large overlapping scales.

This is one of the most grandiose, impressive, and extravagantly beautiful palms. Its majesty must be seen to be appreciated. Alas, the species is cold tender and is adaptable only to zones 10b and 11. It is also nearly aquatic and, while it survives in drier sites of tropical climates, it does not fulfill its potential unless provided with constant and copious moisture as well as nearly constant warmth. It prefers acidic soil.

This species is greatly significant to the local peoples. The fruit pulp is mixed with sugar and water to make highly prized drinks, sorbets, and other desserts. In Brazil, the oil expressed from the fruits is an important commercial industry. The fibers of the young leaves are used for making ropes, hammocks, and utensils.

Mauritiella
maw-ri′-tee-EL-la

Mauritiella consists of 3 clustering, palmate-leaved, dioecious palms in South America. The genus is closely related to *Mauritia*, and its name means "little *Mauritia*." *Mauritiella* differs from *Mauritia* in its smaller stature, its clustering habit, and its much smaller inflorescences with nondistichous branches. The juvenile leaves of both genera are radically different from those of mature individuals, with many fewer segments growing in a flat plane. The fruits of the 2 genera are similar, but those of *Mauritiella* are usually smaller. The 3 *Mauritiella* species are compared in a publication by Bernal and Galeano (2010).

These lovely species do not tolerate cold and are nearly aquatic, requiring constant and copious moisture. In habitat, they often grow either in streams and rivers or in land that is regularly flooded by the watercourses. They are not so exacting about acidic soil preferences as the *Mauritia* species and do well on several soils, as long as they are constantly moist. They are adapted to full sun even in youth but luxuriate in partial shade. In a much more diminutive way, they are as beautiful as the species of *Mauritia*. Seeds germinate within 45 days. For further details of seed germination, see *Calamus*.

Mauritiella aculeata
maw-ri′-tee-EL-la · a-kyoo′-lee-AH-ta
PLATE 621

Mauritiella aculeata is indigenous to northern South America, where it grows along the banks of the Orinoco and its tributaries at low elevations, in extreme southern Venezuela, western Colombia, and northwestern Brazil. The epithet is Latin for "prickly," an allusion to the leaflet margins with their tiny teeth as well as to stem base with its small root spines.

The numerous clustering stems grow to 25 feet (7.6 m), if that, with a maximum diameter of 4 inches (10 cm). They are light, almost white, and half reclining, sometimes twisting, never straight, and often lean out over the water when they grow along streams and rivers. The lower portions of the trunks bear small root spines and have a small mound of buttressing aerial roots at their bases. The species forms pseudo-crownshafts with its silvery glaucous, large leaf sheaths. The leaf crowns are open because of the long petioles and are hemispherical to spherical. The few leaves in the crown are each 30 inches (76 cm) wide and circular with many narrow segments. The segments are dark green above but a beautiful silvery green beneath. They are ½ inch (13 mm) wide and pendent in their apical halves, growing from the petiole at slightly different angles. The tips of the segments are pendulous. The oblong fruits are 1½ inches (4 cm) long, and olive colored, maturing to reddish brown.

This clumper is simply beautiful. Its enchantingly thin, usually sinuous stems and its open, weeping, silvery shimmering crowns are diaphanously exquisite. It is not in cultivation outside of botanical gardens.

Mauritiella armata
maw-ri'-tee-EL-la · ahr-MAH-ta
PLATES 622–623

Mauritiella armata is the most widespread species in the genus and covers almost all of Amazonia and adjacent uplands in northern South America. It grows in forest swamps, mostly at low elevations but, in the Guyanas, can be found on mountainous rain forest to 4500 feet (1370 m). The epithet is Latin for "armed" and refers to the stout root spines on the lower trunk.

The trunks attain a height of 60 feet (18 m) and a diameter of 6–12 inches (15–30 cm). Like the trunks of other species, they have small root spines on their lower portions and a small mound of stilt roots at their bases. The stems are usually straight and almost white but, in their older parts, are indistinctly ringed; their younger parts are distinctly ringed with widely spaced leaf base scars and are reminiscent of the beautiful stems of *Dypsis lutescens* or *D. cabadae*. The leaf crowns are hemispherical because of the mostly ascending leaves and are open because of the long petioles. The leaves are hemispherical to nearly circular and 3–4 feet (90–120 cm) wide. The many stiff, narrow segments are glossy deep green above and silvery beneath. The fruits are similar to those of *M. aculeata*. The clumps of *M. armata* are usually more sparse than those of *M. aculeata* and young clumps are endearing while mature palms are noble in appearance.

Maxburretia
max-bur-RET-ee-a

Maxburretia comprises 3 small, clustering, spiny, palmate-leaved, polygamous palms in Thailand and peninsular Malaysia. One species has subterranean stems. The leaf bases are fibrous, and the fibers usually persist as spinelike projections among the petioles and on the stem. Inflorescences are sparsely branched, erect, and usually project out of the leaf crown. The flowers are either bisexual or unisexual on different individuals. The small, rounded fruits are black when mature. The species are slow growing, and perhaps for that reason are rare in cultivation. The genus, which is related to *Rhapis* and *Guihaia*, was named for Karl E. Maximilian Burret (1883–1964), German botanist and palm specialist. Seeds of *Maxburretia* are not generally available, so little information is known about their germination. Since the genus belongs in the same subtribe as *Chamaerops*, *Rhapis*, and *Trachycarpus*, the seeds might germinate sporadically. For further details of seed germination, see *Corypha*.

Maxburretia furtadoana
max-bur-RET-ee-a · fur-tah'-do-AH-na
PLATES 624–625

Maxburretia furtadoana is a rare species endemic to open, wet but scrubby forest on limestone cliffs and hills of peninsular Thailand at low elevations. The epithet honors Caetano X. Furtado (1897–1980), a palm botanist and taxonomist working in Singapore.

The sparsely clustering trunks slowly attain a maximum height of 10 feet (3 m) in habitat. They are 8 inches (20 cm) in diameter with persistent, spiny leaf bases only 2 inches (5 cm) in diameter without the sheaths. The leaf crowns are open and hemispherical to almost spherical. The leaves are held on unarmed petioles 2 feet (60 cm) long, are hemispherical to nearly spherical, and are 30 inches (76 cm) wide. The many narrow, half limp, long-tipped segments are free nearly to the petiole. Leaf color is glossy deep green above and grayish green beneath.

These little palms, although extremely rare in cultivation outside of their habitat, are picturesque and, if the spiny leaf bases are kept, a real curiosity in the garden. They look especially good in a rock garden, although they are not drought tolerant. They thrive on limestone soils as well as neutral or even slightly acidic ones and grow equally well in full sun or partial shade. They are adaptable only to zones 10 and 11.

Medemia
me-DEE-mee-a

Medemia is a monotypic genus of rare and endangered, solitary-trunked, palmate-leaved, dioecious palm endemic to Sudan. The origin of the name is obscure. Seeds can take up to 18 months to germinate and do so sporadically. Germination is virtually identical to that of *Bismarckia*, which see for further details.

Medemia argun
me-DEE-mee-a · ahr-GOON
PLATES 626–627

Medemia argun grows in wadis of true desert, the lower portions of its trunk often covered in wind-blown sand. The species epithet is a Bedouin name for the palm.

The trunk attains 30 or more feet (9 m) in height and a diameter of 1 foot (30 cm) in habitat. The stems of younger individuals are covered in persistent leaf bases, but older trees have brown trunks with closely spaced rings of leaf base scars. The leaf crown is dense and spherical and usually has a short shag of dead leaves at its base. The strongly costapalmate, bluish or grayish green leaves lack a hastula (unlike a *Hyphaene* leaf, which has a prominent hastula), are 4–5 feet (1.2–1.5 m) long, and are held on stout, yellowish petioles that are 3 feet (90 cm) long and in younger individuals are armed on their margins with black teeth; older palms have unarmed leafstalks. The yellowish orange rib (costa) extends well into the blade and causes it to recurve. The divisions between segments extend deep into the blade, and the segments are stiff, rigid, and wide-lanceolate, with long tapering and almost threadlike tips. The inflorescences are similar to those of *Bismarckia*. The fruits are egg shaped, under 2 inches (5 cm) long, and glossy black when mature.

The palm is related to *Bismarckia* and *Hyphaene* and looks much like the latter. The species was known first from "mummified" fruits left as offerings to the dearly departed in ancient Egyptian tombs. It was presumed by many botanists to be extinct until scientists discovered living plants in 1837. It remained a poorly known and almost mythical palm until 1995, when Martin Gibbons of England and Tobias Spanner of Germany found the species in Wadi Delah in Sudan. They returned to the desert in 1996, found more locations for the species, and brought back viable seeds, which they have since distributed through their retail nurseries in England and Germany. Several young plants now grow around the world, including in South Florida. The palm has the same cultural requirements as *Bismarckia* as far as minimum temperatures go (Gibbons 1993). It also is slow to moderately fast growing, drought tolerant, sun loving, and accepting of alkalinity and salinity.

Metroxylon
me-TRAHX-i-lahn

Metroxylon is composed of 5 large to massive clustering and solitary-trunked, pinnate-leaved, polygamous palms in the Moluccas, New Guinea, the Solomon Islands, the Vanuatu islands, Fiji, the Carolines, and Samoa. All but one species are monocarpic (hapaxanthic) and have large inflorescences with branches that are either tiered and shaped like a Christmas tree or are upward sweeping and look like gigantic insect antennae but are horizontally spreading or even pendent when the fruits mature. The fruits are relatively large and covered in tough, overlapping scales; they make beautiful knick-knacks when dry. One universally distinguishing character is the massive leaf sheaths and petioles, all of which exhibit half-ringed markings consisting of rows or combs of long, soft spines. In addition, all the species have spines on the petioles and leaf rachises as well as incipient aerial roots growing from the younger rings of leaf base scars.

All species are extremely tender to cold and adapted only to zone 11, often being marginal in 10b. They are also all water lovers and, while they grow in poor soil, really appreciate and luxuriate in a rich, organically based medium. All are adapted to full sun when past the seedling stage, and all are invariably fast growing, given enough moisture and a soil that is not too poor.

These great palms are the essence of the tropical look and are exotic and beautiful in the extreme. Some people find them overpowering, primitive, and grotesque looking. The genus name is derived from Greek words meaning "pith" and "wood" and alludes to the trunk's starchy pith, the source of true sago. Seeds germinate within 60 days. For further details of seed germination, see *Calamus*.

Metroxylon amicarum
me-TRAHX-i-lahn · am-i-KAH-rum
PLATE 628

Metroxylon amicarum is endemic to the Caroline Islands in Micronesia, where it grows in rain forest. The epithet is Latin for "of friends" and was bestowed when the species was mistakenly thought to hail from the Friendly Islands (now the Kingdom of Tonga).

This tall, magnificent species has a solitary trunk to 80 feet (24 m) but seeming to mostly top out at 60 feet (18 m). It is usually covered in its younger parts with massive old leaf bases but, in its older parts, is ringed with widely spaced leaf base scars. The leaf crown is hemispherical at most and is usually shaped like a gigantic shuttlecock because of the stiff, ascending, erect leaves, which are 15 feet (4.5 m) long on petioles 3–4 feet (90–120 cm) long. The leaflets are numerous, 3 feet (90 cm) long, lanceolate, and dark green, growing from the rachis at several angles to create a plumose leaf. This is the only non-monocarpic species; the inflorescences are much narrower than those of the monocarpic species, and they grow from amid the leaf bases and are erect until the large fruits make them pendent. The ovoid fruits are 3–4 inches (8–10 cm) long, covered in large, glossy brown scales.

The seeds are extremely hard and are still used as "vegetable ivory" as are the seeds of *Phytelephas* species; the palm is called the Caroline ivory-nut palm in certain parts of the world. It is as massive and as striking as the large *Raphia* species and is spectacular in the landscape. It is also probably the hardier to cold than any other species in the genus.

Metroxylon sagu
me-TRAHX-i-lahn · SAH-goo
PLATE 629

Metroxylon sagu is probably endemic to New Guinea but is now naturalized or cultivated all over the Asian tropics. In its native habitat it covers vast areas of swamps. This is the sago palm from which sago is extracted. The epithet is the aboriginal name for the starch in the stems.

The clumping stems sometimes attain a height of 60 feet (18 m) but more often are half that stature before flowering. They are 2 feet (60 cm) in diameter and are clothed in their upper parts with old leaf bases, their lower parts rough but graced with widely spaced rings of leaf base scars. The leaf crowns are shaped like gigantic shuttlecocks and hold erect, arching leaves that are 20 feet (6 m) long or more. The longest of the many dark green leaflets are 5 feet (1.5 m). The leaflets grow from the rachis at an angle that creates a slight V shape to the leaf. The great terminal inflorescences are 25 feet (7.6 m) tall with upward-thrusting primary branches and secondary branches. The fruits are the size of baseballs, light brown or greenish brown, and covered in large, overlapping scales.

The form of this species varies according to how tall the stems grow, the length of the leaves, and the amount of spininess thereon. Some forms have spiny leaf sheaths, petioles, and rachises with partial rings of flexible, but vicious, golden spines, each of which can be 3 inches (8 cm) long, while others are nearly spineless. This variability is doubtless due to the palm's long history of cultivation. It has been grown for millennia for the sago starch in its trunks; the stems were felled before flowering, as the formation and maturation of the flowers and fruits tend to drain the stem of its starch reserves.

This palm is one of the world's most impressive. Because of its suckering habit, it can make a great veritable, tiered wall of sumptuously beautiful and tropical-looking foliage. Alas, all forms of the palm are marginal if not impossible even in zone 10b. In addition they need constant and copious moisture and a rich, humus-laden soil; their favored habitat is a mucky swamp. The species can be propagated by seed and by removal of the suckers.

Metroxylon salomonense
me-TRAHX-i-lahn · sal-o-mo-NEN-see
PLATES 630–631

Metroxylon salomonense occurs naturally in eastern Papua New Guinea, Vanuatu, and the Solomon Islands, where it grows mostly in low, swampy areas but also in lowland rain forest to 2600 feet (790 m). The epithet is Latin for "of Solomon Islands," a reference to one of its habitats.

The massive, light brown trunk attains a height of 50 feet (15 m) before flowering and may be 4 feet (1.2 m) in diameter. It is covered in its upper parts with massive leaf bases, but with widely spaced rings of leaf base scars otherwise. The great leaf crown is less than hemispherical, generally shaped like a wide shuttlecock. The leaves are 30 feet (9 m) long on massive petioles 4–5 feet (1.2–1.5 m) long. The sheaths, petioles, and even the leaf rachises bear large, yellowish spines. The dark green leaflets grow from the rachis at only a slight angle and create a leaf that is nearly flat. The terminal inflorescence is 12 feet (3.6 m) tall, with horizontally spreading branches from which descend curtains of whitish tertiary flowering branches. The yellowish green, scaly fruits resemble 3-inch (8-cm) apples. This species is perhaps the most impressive of the solitary-trunked palms.

Metroxylon vitiense
me-TRAHX-i-lahn · vee-tee-EN-see
PLATES 632–633

Metroxylon vitiense is endemic to Fiji, where it grows in rain forest and in low, swampy areas. The epithet is Latin for "of Fiji." The solitary trunk attains a height of 50 feet (15 m) before flowering and a diameter of 3 feet (90 cm). The stem is mostly free of leaf bases but is strongly ringed with widely spaced leaf scars. The leaf crown is relatively narrow and shaped like a gigantic shaving brush, with erect leaves 12–15 feet (3.6–4.5 m) long. The stiff, dark green leaflets are 4 feet (1.2 m) long and grow from the massive rachis in a nearly single flat plane in older leaves. The upper parts of the tree are spiny, as well as the sheaths, petioles, leaf rachises, and leaflets.

Metroxylon warburgii
me-TRAHX-i-lahn · wahr-BUR-gee-eye
PLATE 634

Metroxylon warburgii occurs in Figi, Samoa, and the Solomon Islands, mostly in low, swampy areas but also on hillsides, always at low elevations. The epithet honors Otto Warburg (1859–1938), German plant collector and botanist. This is the smallest of the solitary-trunked species with a stem only 25 feet (7.6 m) tall and 1 foot (30 cm) in diameter. The leaf crown is hemispherical and the leaves are 10 feet (3 m) long. Despite its stature, this species has the most beautiful leaves in the genus as they are more arching, and the deep green leaflets are not as stiff with pendent apices. The leaflets grow from the rachis at an angle that creates a slightly V-shaped leaf.

Myrialepis
mir'-ee-a-LEE-pis

Myrialepis is a monotypic genus of clustering, pinnate-leaved, dioecious climbing palm with individual stems that die after fruiting. It occurs in Myanmar, Thailand, Cambodia, Laos, Vietnam, peninsular Malaysia, and Sumatra. The genus name is from Greek words meaning "ten thousand" (and, by extension, "innumerable") and "scale" and alludes to the myriad tiny scales on the outer skins of the fruits. Seeds are not commonly available, so germination information is nonexistent.

Myrialepis paradoxa
mir'-ee-a-LEE-pis · par-a-DAK-sa
PLATE 635

Myrialepis paradoxa grows at the edges of rain forest and in clearings therein as well as along the banks of rivers and streams from sea level to 3500 feet (1070 m), often forming large thickets and colonies. The epithet is Latin for "paradox."

The clustering stems can climb to great heights, are 3 inches (8 cm) in diameter, and are ringed with whorls of vicious, golden spines. The leaves are widely spaced along the reddish brown stems, to 10 feet (3 m) long, with widely spaced linear and long pointed, dark green, pendent leaflets. The rachises extend beyond their leaves to form cirri 4 feet (1.2 m) long that have a few pairs of backward-pointing barbs. The inflorescences grow from the uppermost leaf axils, including the terminal one, and are 2 feet (60 cm) long and much branched. The round 1-inch (2.5-cm) fruits are greenish brown and covered in minute scales that impart a glistening luster to the skin of each one.

The species is similar in appearance to many *Calamus* species and other rattan genera but is most closely related to *Plectocomiopsis*. It is not in cultivation outside of botanical gardens but, because of its densely clustering habit, would make an effective, tropical-looking wall of foliage from a safe distance. It can, like most other rattans, be pruned severely to try to keep it in bounds, but in a tropical climate, which it requires, its growth is rampant.

Nannorrhops
NAN-nor-rahps

Nannorrhops is a monotypic genus of palmate-leaved, monoecious, branching palm in the Middle East. It has an unusual growth habit: the aerial stems fork, with one branch developing a terminal inflorescence and the other continuing in vegetative growth. The branch bearing the flower stalk will eventually die, leaving a flattened scar on the main trunk of the palm. The genus name is a combination of Greek words meaning "dwarf shrub." Seeds germinate within 30 days and should not be allowed to dry out before sowing. For further details of seed germination, see *Corypha*.

Nannorrhops ritchieana
NAN-nor-rahps · rich-ee-AH-na
PLATES 636–637

Nannorrhops ritchieana occurs naturally in dry mountainous areas of the southern Arabian Peninsula, Iran, Afghanistan, and Pakistan, where it grows usually in stream beds or on hillsides. The only common name seems to be mazari palm, which is the vernacular name in the palm's habitat. The epithet likely commemorates Joseph Ritchie (1788–1819), British naturalist and explorer of northern Africa.

The clustering trunks are mostly subterranean but sometimes emerge from the ground to reach heights of 15 feet (4.5 m). The Orto Botanico of Rome has a very old specimen whose main trunk creeps along the ground for 30 feet (9 m). Emergent trunks are covered in brown leaf bases and orange-brown fiber and are 8 inches (20 cm) in diameter. A clump can reach 20 feet (6 m) tall and wide after many years, but most mature clumps are 6 feet (1.8 m) tall and 8 feet (2.4 m) wide. The leaves are 4 feet (1.2 m) wide and are wedge shaped to semicircular. There are 30 stiff, narrowly lanceolate segments divided almost to the petiole, each one split to half its length. The leaves are held mostly erect on petioles that are 3 feet (90 cm) long and that usually but not always bear tiny teeth along their margins; they have, when new, patches of light brown, fine, woolly fibers. Leaf color is dull grayish green to chalky blue-gray. The inflorescence is much branched, 6 feet (1.8 m) long or tall, and is held erect, bearing small, white, bisexual flowers. The rounded ½-inch (13-mm) fruits are orange-brown and reportedly edible.

This species is slightly susceptible to lethal yellowing disease. In dry climates, it is hardy to cold, being adaptable to zones 8 through 11 and marginal in 7. It is drought tolerant and thrives on many soils, even calcareous ones. It needs full sun at all ages and is slow growing. Because of its form, color, and hardiness, this palm is valuable as a strong accent along sunny paths and in sunny borders of the cactus and succulent garden. It makes a nice potted specimen for outdoors, but the container must have unimpeded drainage. It is not known to be grown indoors and is probably not a good candidate.

Nenga
NEN-ga

Nenga includes 5 clustering or solitary-trunked, pinnate-leaved, monoecious palms in Myanmar, Thailand, Vietnam, peninsular Malaysia, Sumatra, Java, and Borneo. The genus name is based on the Javanese vernacular name. The species, often with stilt roots, have crownshafts of varying dimensions and tightness. The inflorescences grow from beneath the crownshafts and are small, pendent, and spicate or branched once; they bear spirally arranged, unisexual flowers of both sexes. The tips of the infructescence branches are persistent and bare. The fruits are mostly small and are invariably egg- or teat shaped and range from brilliant red to purple or black.

All the species are intolerant of frost and need partial shade, especially in hot climates, abundant and regular moisture, and a decent, fast-draining soil. Only one species (*N. pumila*) is in general cultivation outside of botanical gardens, but all the species are deliciously gorgeous palms of great elegance. Seeds germinate within 30 days. For further details of seed germination, see *Areca*.

Nenga pumila
NEN-ga · POO-mi-la
PLATE 638

Nenga pumila is widespread from southern Thailand through peninsular Malaysia to Java, Sumatra, and Borneo. It grows in the understory of low mountainous rain forest. The epithet is Latin for "small," although the species is not particularly diminutive when mature.

The clustering stems attain a height of 18 feet (5.4 m) with a diameter of 2 inches (5 cm). They are light brown in their older parts but deep green in their newer parts with beautiful, distinct rings of leaf scars. The crownshafts are 2 feet (60 cm) tall and are yellowish to deep green but are overlain with a brown tomentum. The leaf crown is hemispherical and open, with 6 leaves, each of which is 6 feet (1.8 m) long and held on a brown felty petiole that is 18 inches (45 cm) long. The many leaflets, also 18 inches long, are emerald green on both surfaces, linear-lanceolate, S shaped, limp, and half-pendent. The tan inflorescences grow from beneath the crownshafts and are 18 inches long and pendent. The flowers are scented. The brownish orange fruits are ½ inch (13 mm) in diameter.

Nenga pumila var. *pachystachya* [pa-kee-STAY-kee-a] is, if possible, even more attractive than the type as it grows taller and has brilliant orange-red to scarlet fruits that are slightly larger. The varietal epithet is from Greek words meaning "thick spike." The ripe fruits are said to emit the aroma of stale garlic.

There is scarcely a more beautiful palm. The tiers of leaf crowns and the elegant, reedlike stems are beyond enchanting. The only problem with the palm is that it is not fast growing.

Neonicholsonia
nee'-o-nik-ol-SO-nee-a

Neonicholsonia is a monotypic genus of small, solitary-trunked, pinnate-leaved, monoecious palm in Central America. The name honors George Nicholson (1847–1908), curator of the Royal Botanic Gardens, Kew, and author of *The Illustrated Dictionary of Gardening*. The prefix, meaning "new," prevents confusion with *Nicholsonia*, an earlier name used for a genus of legumes. Seeds germinate within 30 days.

Neonicholsonia watsonii
nee'-o-nik-ol-SO-nee-a · waht-SO-nee-eye
PLATE 639

Neonicholsonia watsonii is indigenous to eastern central Honduras, central Nicaragua, most of Costa Rica, and southwestern Panama, where it occurs in the undergrowth of lowland rain forest to an elevation of 800 feet (240 m). The epithet honors William

Watson (1858–1925), British horticulturist, author, and successor to Nicholson at Kew.

This small palm has a short, usually underground, trunk. The leaves, which appear to grow directly from the ground, are 6 feet (1.8 m) long, ascending, and slightly arching when new and later spreading. They are held on short, soft petioles and bear widely spaced narrowly elliptical, long-tipped, emerald green leaflets that are 1 foot (30 cm) long and grow opposite each other on either side of the rachis in a single flat plane. The inflorescences seem to arise directly from the ground in individuals with no aerial stem but grow from among the leaf bases. They are tall, erect spikes that usually rise above the leaves and bear both male and female blossoms. The fruits are ovoid, ½ inch (13 mm) long, black when mature, and attached directly to the spike.

This little palm is beautiful as a large groundcover or as an under-canopy specimen, looking something like a giant fern or a tropical cycad. It is tender to cold and is slow growing, most individuals never forming an aerial stem; those that do take many years to do so. It needs partial shade at all stages, especially in hot climates, constant moisture, and a humus-laden soil.

Neoveitchia
nee-o-VEETCH-ee-a

Neoveitchia consists of 2 rare, pinnate-leaved, monoecious palms in Fiji and Vanuatu. Despite its name, which means "new *Veitchia*," this genus is unrelated to *Veitchia*, but the genera have some overall similarities in form. *Neoveitchia* species have significantly stouter and straighter trunks. They also do not form true crownshafts, but this is not so visually apparent as the large leaf sheaths clasp the top of the stem and are almost closed. The inflorescences grow from beneath the leaf bases and consist of a thick, succulent-looking peduncle from which a spray of cream-colored flowering branches spread, each one 2 feet (60 cm) long.

Seeds of *Neoveitchia* are slow to germinate, but some growers report that carefully cracking and removing the endocarp before sowing speeds up germination and increases the percentage of seeds that germinate. Because the seeds are more prone to disease and insect attacks without their endocarps, they should be planted in a sterilized potting medium such as perlite. Seeds allowed to dry out before sowing do not germinate as well as those kept slightly moist. For further details of seed germination, see *Areca*.

Neoveitchia brunnea
nee-o-VEETCH-ee-a · BROON-nee-a

Neoveitchia brunnea was described in 1996 and is a rare and endangered species endemic to Vanuatu, where it grows in mountainous rain forest at 1000 feet (300 m). The epithet is Latin for "brown" and alludes to the color of the female flowers and the anthers of the male blossoms. This species differs from *N. storckii* in being slightly shorter, in details and color of the flowers, in lacking hairy scales on the undersides of the leaflets, and in leaf bases that are green rather than black. Otherwise, it is as impressive and as beautiful as *N. storckii*. It is not known to be in cultivation.

Neoveitchia storckii
nee-o-VEETCH-ee-a · STOR-kee-eye
PLATES 640–641

Neoveitchia storckii is endemic to the island of Viti Levu in Fiji, where it grows in lowland rain forest. It is critically endangered due to the trunks being felled for use in house construction but is now protected by the government. The epithet honors the original collector of the species, Jacob Storck (1838–1893), a German planter in Fiji and assistant to botanist Berthold Seemann during his study of the Fijian flora.

The trunk attains 50 feet (15 m) of height in habitat and 1 foot (30 cm) in diameter except at its base where it is much expanded. It is light brown in its upper parts, where it is also distinctly ringed with undulating deeper brown leaf base scars; its older parts are gray and indistinctly ringed. The partial crownshaft is amazingly distinct despite not being a true closed shaft: it is slightly thicker than the stem and nearly black except at the summit where it shades into a deep green, the color of the petioles. The leaf crown is hemispherical or slightly more rounded and contains 10 stiff leaves that arch only near their apices and are 15–18 feet (4.5–5.4 m) long. The many deep green leaflets are 30 inches (76 cm) long, linear-lanceolate with a long tip, limp, and mostly pendent. There is a distinct twist to the leaf from its midpoint on which orients the apical half to the vertical plane. The ellipsoid fruits are not quite 2 inches (5 cm) long and are orange when mature.

Because of its great leaves, which actually look bigger than they really are, with their beautiful vertical twists, the palm is among the most impressive, beautiful, and tropical looking in the family. It has visual affinities with *Veitchia*, *Dictyosperma album*, and the coconut. It is not tolerant of cold, being adaptable only to zones 10b and 11. While it survives in frostless Mediterranean climes like coastal Southern California, it languishes there and does not grow well. It is a water lover and must not be subjected to drought at any stage. It appreciates partial shade when young but needs full sun when older to fulfill its landscape potential. It is surprisingly tolerant of various soils and even seems to flourish in the limey ones of South Florida.

Nephrosperma
nef-ro-SPUR-ma

Nephrosperma is a monotypic genus of solitary-trunked, pinnate-leaved, monoecious palm in the Seychelles. The genus name is from Greek words meaning "kidney" and "seed," an allusion to the shape of the seeds. Germination of seeds occurs within 30 days. For further details of seed germination, see *Areca*.

Nephrosperma vanhoutteanum
nef-ro-SPUR-ma · van-hoot′-tee-AH-num
PLATE 642

Nephrosperma vanhoutteanum grows mainly on wet mountain slopes but sometimes in the undergrowth of rain forest at elevations from sea level to 2000 feet (610 m). The epithet honors Louis van Houtte (1810–1876), a Belgian nurseryman, plant explorer, and publisher of horticultural works.

The trunk attains a height of 30 feet (9 m) in habitat and is 6 inches (15 cm) in diameter except at its base where it is swollen. It is a light gray and, in its youngest parts, is indistinctly ringed. The leaf crown is hemispherical to nearly spherical and open. It forms no crownshaft, but the grayish leaf sheaths are dramatically large and prominent. The gracefully arching leaves are 6 feet (1.8 m) long in mature individuals but to 10 feet (3 m) long in younger ones; the petioles are 3–4 feet (90–120 cm) long. The individual leaflets are 4 feet (1.2 m) long, generally linear-lanceolate but of varying widths, strongly ribbed, and closely spaced along the rachis. They are a light grayish green to bright grassy green and are leathery but soft, limp, and pendent. All parts of the leaves are spiny on young plants, but the leaves may be spineless on older plants. The extraordinary inflorescences are spikes 10–12 feet (3–3.6 m) long that arch out of the leaf crown and, at their apices, bear spreading flowering branches 2 feet (60 cm) long with both male and female blossoms. The spherical cherry red fruits are ½ inch (13 mm) in diameter.

This species is tropical looking and appealing, especially before it is mature, when the leaves are longer and can be viewed up close. Its drawbacks are that it is extremely sensitive to cold and it is slow growing unless planted in a rich but fast-draining soil and watered constantly. Even then, it is not fast. It is adaptable only to zone 11, although it may be possible in the warmest microclimates of 10b when it is small and can be protected.

Normanbya
nor-MAN-bee-a

Normanbya is a monotypic genus of solitary-trunked, pinnate-leaved, monoecious palm indigenous to northeastern Queensland, Australia. The name honors George A. C. Phipps (1819–1890), the Marquess of Normanby, who was governor of Queensland when the palm was named. Seeds of *Normanbya* generally germinate within 45 days if they are not allowed to dry out before sowing, but they can take up to 180 days. Germination is sporadic. For further details of germination, see *Areca*.

Normanbya normanbyi
nor-MAN-bee-a · nor-MAN-bee-eye
PLATE 643

Normanbya normanbyi grows in wet rain forest, usually along streams or rivers but also in low, swampy areas at low elevations. Like the genus name, the epithet honors the Marquess of Normanby.

In habitat, the trunk attains a height of 60 feet (18 m), a diameter of 6 inches (15 cm), and is light tan in its younger parts and graced with widely spaced, nearly white rings of leaf base scars. The crownshaft is 2–3 feet (60–90 cm) tall, bulging at its base, cylindrical, and silvery green to light gray. The leaf crown is hemispherical to nearly spherical and has about 10 leaves on short petioles. The leaves are 8 feet (2.4 m) long and gracefully arching. Leaf color is medium to deep green above but silvery or even bluish silvery green beneath. Older plants have many wedge-shaped leaflets with oblique, jagged apices; the leaflets are split to their bases and radiate at different angles from the rachis to give a plumose aspect to the leaf. Younger palms also have wedge-shaped leaflets with jagged apices, but the leaflets grow in almost a flat plane off the rachis. The inflorescences grow from beneath the crownshaft and are sprays of flowering branches 3 feet (90 cm) wide bearing both male and female flowers. The fruits grow in pendent clusters and are 2 inches (5 cm) long, ovoid, and pinkish to red or brownish purple.

This distinctive and extremely attractive species is close to *Wodyetia* in appearance but more elegant and refined looking because of its thinner trunk and more diaphanous leaf crown. It is slower growing than *Wodyetia* and more finicky in its cultural requirements, needing constant moisture and a deep, slightly acidic soil that is rich in organic matter. It is possible in the limestone soils of South Florida with a thick mulch, copious, constant moisture, and an amended soil, but it grows more slowly there than it does in its native Queensland. It must have partial shade when young, especially in hot climates, but is almost as hardy to cold as *Wodyetia*, surviving short spells of temperatures slightly below freezing. Its great beauty seems worth whatever heroic measures need to be taken if the gardener can provide such; it is the rich man's *Wodyetia*.

Nypa
NEE-pa

Nypa is a monotypic genus of solitary-trunked, pinnate-leaved, monoecious palm indigenous to a wide area of the Asian tropics and eastwards to Micronesia and Melanesia, as well as the Solomon Islands. It has naturalized in parts of Africa and the Americas. The genus name is a transliterated form of the Malaysian name for the palm.

Nypa is the only member of the subfamily Nypoideae. The large seeds germinate easily and quickly, within days after harvesting, and with the husk intact. In fact, germination may commence while the fruits are still on the club-shaped infructescence, and seeds shipped in a bag for several days often begin germinating before reaching their destination. Seeds should be kept wet but well drained with half the seed exposed to the air. Seeds should not be allowed to dry out before sowing as this adversely affects germination rate.

Nypa fruticans
NEE-pa · FROO-ti-kanz
PLATES 644–645

Nypa fruticans is widespread in the Asian and Pacific tropics, from Sri Lanka, northeastern India and Bangladesh, the coasts of Myanmar, Thailand, peninsular Malaysia, Sumatra, Vietnam, Cambodia, southeastern China, the Philippines, the Ryukyu Islands of Japan, Indonesia, New Guinea, northeastern Australia, the Bismarck Archipelago, and the Solomon Islands. It has naturalized in Nigeria, Panama, and Trinidad. It grows always along the coasts, usually in mangrove estuaries or lagoons, always at low elevations, and often forms vast colonies along the banks of coastal rivers, inland if the rivers are in a flat floodplain. It is called nipa or mangrove palm. The epithet is Latin for "shrubby" or "bushy."

This solitary-trunked species looks clustering because of its prostrate or subterranean stem that branches dichotomously. It forms adventitious roots when and where it is prostrate on top of the soil and exhibits widely spaced rings of knobby, lighter leaf scars if its thick, succulent trunk is uncovered. The leaf bases are usually covered in mud but, when exposed, are seen to be stout, succulent looking, and a deep chocolate or purplish brown, which characteristics apply to the petioles and leaf rachises as well. The newer leaves are mostly ascending and erect, but older ones are more spreading. They are 20 feet (6 m) long, including the robust but soft petiole, which may be a third of the length. The many lanceolate leaflets are 3 feet (90 cm) long, regularly spaced, and growing from the rachis at only a slight angle, which gives the leaf only a hint of a V shape. The leaflets are stiff but not rigid, usually pendent at their tips, and may be light green or even yellowish green to deep emerald green, depending on the soil and amount of moisture the plants get. There are conspicuous, chaffy, brown hairs along the midvein of the underside of the leaflet.

The species has remarkable inflorescences and fruits. The former grow from among the leaf bases and are erect and usually 3–5 feet (90–150 cm) high. They consist of a spikelike portion at the top of which is a globular mass of dull purplish female flowers subtended by short catkinlike yellow branches of male blossoms, the whole affair enclosed at first in leathery, ochre bracts. The mahogany-colored infructescence forms atop the spike and is round and clublike with many angular fruits packed into the rounded mass. In habitat, these individual fruits separate when ripe, fall into a stream or lagoon, and float off to start new colonies. As in many mangrove species, the seeds often germinate while the fruits are floating, before they become established in the mud.

The most important cultural consideration is constant water, which, contrary to popular belief, need not be saline. In fact, this palm dies in undiluted salt water, although it luxuriates in brackish to mainly fresh water. A lake or pond is not necessary for growing the palm if its soil can be kept constantly moist. It is adaptable to many soils including calcareous ones and needs full sun when past the seedling stage. The second most important cultural factor is temperature: this palm needs a tropical or nearly tropical climate and is adaptable only to zones 10b and 11. It is fast growing.

This fantastic and fantastic-looking species has a tropical but wild and primitive demeanor. It creates a true "jungle" aspect like no other species can when planted around ponds or small lakes. Fortunate is the gardener who has room to grow it. It would only succeed in a large conservatory or atrium that has full or nearly full sun exposure, an area wide enough, and a soil deep enough to accommodate the thick trunks.

The species has many uses in its native haunts. The primary one is for thatch and weaving baskets and other utensils, but the leaflets are also used for rolling cigarettes. The dried petioles, because they are filled with tiny air sacs, are used for construction of huts, for fuel, and for fishing net floats. The immature fruits and seeds are considered a delicacy in parts of the palm's range, and the peduncles of the inflorescences yield a sugary sap that is used in sugar, alcoholic beverages, and vinegar. The young, unfolding leaves are also sometimes eaten as a salad. The most dramatic use is, however, as a stabilizing plant to thwart erosion of banks along streams, rivers, and ponds or lagoons. Healthy stands of *N. fruticans* are credited with protecting some coastlines and interior areas during the Indian Ocean tsunami of December 26, 2004.

Oenocarpus
ee-no-KAHR-pus

Oenocarpus comprises 9 solitary or clustering, pinnate-leaved, monoecious palms in Central America and northern and northeastern South America. Most species are indigenous to lowland rain forest, but a few grow on the slopes of the Andes to 3500 feet (1070 m). Some species form loose, open crownshafts with large, prominent leaf sheaths. The pinnate leaves are large, often plumose, the leaflets with silvery undersides, and new growth is usually reddish. The inflorescences grow from beneath the leaf sheaths and consist of pendent sprays of relatively short flowering branches that resemble a whisk broom or a horse's tail; they bear unisexual blossoms of both sexes.

The large species have strong visual affinities with the large species of *Attalea* and could be similarly used in the landscape. None of the *Oenocarpus* species are tolerant of cold, although reportedly several have survived temperatures slightly below freezing. All are water lovers and appreciate a humus-laden soil that is not too alkaline. They are sun lovers also once past the seedling stage.

The genus name is from Greek words meaning "wine" and "fruit," the allusion being to the edible fruits from which wine, among other comestibles, is made. Seeds germinate within 90 days.

Oenocarpus bacaba
ee-no-KAHR-pus · bah-KA-bah
PLATE 646

Oenocarpus bacaba is indigenous to northern Amazonia in the Guyanas, the southern half of Venezuela, eastern Colombia, and northeastern Brazil, where it grows in lowland rain forest to 3500 feet (1070 m) but is not found in the seasonally inundated forests. The epithet is the aboriginal name for the palm in Brazil.

The solitary stem attains a height of 60 feet (18 m) and a diameter of 1 foot (30 cm) or less. The grayish green pseudo-crownshaft is distinct, like an inverted pyramid, and 4 feet (1.2 m) tall. The leaf crown is shaped like a gigantic shuttlecock, with 12 erect and ascending, stiff, slightly arching leaves 15–20 feet (4.5–6 m) long on stout, grayish green petioles 1–2 feet (30–60 cm) long. The many leaflets are each 3–4 feet (90–120 cm) long, linear, deep olive green to pure emerald green, limp, and pendent; they grow from the rachis at different angles to create a loosely plumose leaf. The gorgeous inflorescence branches, 4 feet (1.2 m) long, turn brilliant scarlet as the fruits, ¾ inch (19 mm) long, mature to deep purplish black.

This palm is stunningly attractive and grows fast with lots of water and a decent soil. It is adaptable to the full equatorial sun past the seedling stage.

Oenocarpus bataua
ee-no-KAHR-pus · bah-ta-OO-a
PLATE 647

Oenocarpus bataua has the largest range in the genus: from extreme eastern Panama into the western slopes of the Andes in Colombia and Ecuador, into northern Bolivia, the lowland rain forest of northern Colombia, eastern Colombia, northeastern Ecuador, Peru east of the Andes, northeastern Brazil, Venezuela, and the Guyanas. The epithet is a Latinized form of the aboriginal name in Brazil. This species grows on the eastern Andean slopes to 3500 feet (1070 m), through the seasonally wet *llanos* (savannas) of Venezuela, the seasonally inundated forests of Amazonia, and along the edges of lowland rain forest as well as the banks of streams and rivers. The epithet is an aboriginal name.

The solitary stem attains a height of 80–90 feet (24–27 m) in habitat but has a diameter of 18 inches (45 cm), is light tan to light gray, relatively slender and, in its younger parts, is graced with widely spaced darker rings of leaf base scars. The pseudo-crownshaft is 6 feet (1.8 m) tall, like an inverted pyramid, and deep olive green to brown, with many long, stiff, almost needlelike black fibers on its margins. The leaf crown resembles a gigantic shaving brush and usually contains 12 long, stiff, erect leaves, each 30 feet (9 m) long on a short petiole 1 foot (30 cm) long. The leaves arch only near their tips, if at all, and bear many, regularly spaced, long-tipped olive green to emerald green, silvery backed leaflets that are 4–5 feet (1.2–1.5 m) long and pendent but grow in a single plane from the rachis, creating a hanging curtain effect. The new growth is usually dark red or maroon. The pendent inflorescences are 6 feet (1.8 m) long and bear purple-black fruits 1 inch (2.5 cm) long. The flowering branches turn rusty red as the fruits develop.

There are 2 recognized varieties of this species, but their differences are in details of the inflorescences and not important in the landscape. This species is one of the world's most impressive palms. It grows robustly and fast in a tropical or nearly tropical climate if given ample, regular moisture and a decent soil. It is adapted to full sun once past its seedling stage and has extremely colorful new growth when young.

Oenocarpus distichus
ee-no-KAHR-pus · DIS-ti-kus
PLATE 648

Oenocarpus distichus is indigenous to northern central Brazil, from the northeastern tip of Bolivia eastward to the mouth of the Amazon River, where it grows in lowland rain forest. The epithet is Latin for "distichous" and alludes to the 2-ranked disposition of the leaves in the crown.

The solitary trunk attains a height of 60 feet (18 m) in habitat and is 1 foot (30 cm) in diameter. It is light gray to nearly white and is graced with widely spaced dark rings of leaf base scars. The grayish green to grayish brown pseudo-crownshaft is 3 feet (90 cm) tall and shaped like a flat, inverted pyramid. The leaf crown is hemispherical to nearly spherical but is also nearly in one plane because the leaves grow only from opposite sides of the trunk. Individual leaves are mostly erect and slightly arching, 10–15 feet (3–4.5 m) long, on short petioles usually less than 1 foot (30 cm) long. The deep olive green leaflets are 3 feet (90 cm) long, grow from the rachis in different planes, and are pendent, creating a heavy curtain effect. The pendent reddish brown inflorescences have flowering branches 3 feet (90 cm) long. The round ½-inch (13-mm) fruits are purple-black.

Because of the flat leaf crown, this is the most distinctive solitary-trunked species. It is also possibly the most attractive.

Oenocarpus mapora
ee-no-KAHR-pus · ma-POR-a
PLATE 649

Oenocarpus mapora has a wide distribution from western Costa Rica, through Panama, the western slopes of the Andes in Colombia and Ecuador, northern Colombia and western Venezuela, central Colombia, eastern Ecuador, Peru east of the Andes, and northern Bolivia, where it grows in lowland rain forest. The epithet is the aboriginal name of the palm in Venezuela.

The clustering stems attain a height of 50 feet (15 m) and a diameter of 6 inches. The pseudo-crownshaft is shaped like a narrow, inverted pyramid, 3 feet (90 cm) tall in older plants, and is deep olive green to greenish brown. The leaf crown is hemispherical or nearly spherical and consists of 6 spreading leaves. Individual leaves are 4–15 feet (1.2–4.5 m) long on petioles 1–3 feet (30–90 cm) long and usually show a slight twist near their midpoint that orients the apical half of the leaf into a vertical plane. The leaflets are 2–3 feet (60–90 cm) long, broad, and regularly spaced, and grow from the rachis mostly in a flat plane; they are limp and pendent. Leaf color is medium to deep olive green to almost emerald green. The inflorescences consist of pendent, reddish flowering branches 2 feet (60 cm) long. The round to ovoid, blackish purple fruits are 1 inch (2.5 cm) long.

This species is extremely attractive because of the tiers of large leaves and slender, graceful whitish trunks. It looks especially good planted near water.

Oncocalamus
ahn'-ko-KAL-a-mus

Oncocalamus is composed of 5 clustering, pinnate-leaved, spiny, monocarpic, monoecious, climbing palms in western tropical Africa. They grow in lowland rain forest, clearings therein, swamps, and along banks of rivers and streams.

Seedlings have undivided, apically bifid leaves; adult plants have numerous regularly spaced leaflets mostly growing in a single flat plane. The upper parts of the stems are covered in tightly clasping leaf sheaths that bear spines, and there are spines on the short petioles as well as the rachis and even the leaflet margins. The plants climb mainly by the elongated cirri with their paired, backwards-pointing spines at the apices of the adult leaves. The inflorescences grow directly from the tops of the stems and consist of male and female flowers on pendent flowering branches, each blossom enclosed by a papery bract, these bracts overlapping on the branches. The fruits are small, globular, yellowish brown to reddish brown, and covered in overlapping scales. The individual climbing stems gradually die after flowering and fruiting.

This genus is unknown in cultivation outside of a few botanical gardens. The name is derived from the Greek word meaning "tumor" or "mass," probably in reference to the intrusion of the seed coat into the endosperm, and *Calamus*, a related genus of rattan palms. Seeds are not generally available, so germination information is nonexistent.

Oncosperma
ahn-ko-SPUR-ma

Oncosperma includes 5 large clustering, spiny, pinnate-leaved, monoecious palms in the Philippines, Borneo, Sulawesi, Java, peninsular Malaysia, Sumatra, and Sri Lanka. The younger parts of the trunks are covered in rings of downward-pointing black spines, and there are spines on the short crownshafts as well as the petioles and leaf rachises. The inflorescences grow from beneath the crownshaft and consist of many pendent flowering branches bearing both male and female blossoms, the whole enclosed in woody, spiny bracts when developing. The round fruits are dark purple or black.

The growing points of all species are edible and considered a delicacy in their habitats. The trunks of most species are durable and are used locally for construction. Only 2 species are commonly in cultivation. All require tropical or nearly tropical climates and copious, regular moisture but are not finicky about soil type as long as it is well drained. They are adaptable to full sun from an early age.

The genus name comes from Greek words meaning "tumor" or "mass" and "seed" and alludes to the seeds, in which the endosperm is penetrated by ingrowths of the seed coat. Seeds of *Oncosperma* germinate within 45 days. For further details of seed germination, see *Areca*.

Oncosperma fasciculatum
ahn-ko-SPUR-ma · fa-sik'-yoo-LAH-tum

Oncosperma fasciculatum is endemic to Sri Lanka in low mountainous rain forest. The epithet refers to the grouped (fascicled) leaflets along the leaf rachis. This densely clustering species is similar to *O. tigillarium* but has fewer stems per clump, does not grow as tall, and has shorter leaves with less pendent leaflets.

Oncosperma gracilipes
ahn-ko-SPUR-ma · gra-SI-li-pees
PLATE 650

Oncosperma gracilipes is endemic to the Philippines, where it grows in low mountainous rain forest on several islands. The epithet is Latin for "graceful foot" and alludes to the long petioles. This species is smaller than most others in the genus, its stems only growing to 30 feet (9 m) high, with a diameter of 4 inches (10 cm). The crownshafts are deep reddish yellow to orange with striations of small black spines. The leaf crowns are hemispherical with 6 spreading leaves 6–8 feet (1.8–2.4 m) long on petioles 2 feet (60 cm) long. The leaflets are pendent in older leaves but slightly so in younger ones, which characteristic but lends to the overall beauty of the different tiers of foliage. The species is wonderfully colorful and elegantly graceful, the most ethereal looking in the genus.

Oncosperma horridum
ahn-ko-SPUR-ma · HOR-ri-dum

Oncosperma horridum is a sparsely clustering species often found with a single trunk. It has a wide natural distribution in peninsular Malaysia, Sumatra, Borneo, and some islands of the Philippines, where it grows in mountainous rain forest from sea level to 1800 feet (550 m). The epithet is Latin for both "bristling" and "frightening," and alludes to the spiny younger parts of the trunks.

The stems grow to a height of 60 feet in their native haunts with a diameter of 1 foot (30 cm). The crownshafts are 3 feet (90 cm) tall and prominent, being conspicuously swollen near their bases, and tan or greenish brown. The leaf crowns are hemispherical to nearly spherical, and the leaves are 12–18 feet (3.6–5.4 m) long, spreading, stiff, and barely arching. The younger leaves usually exhibit a twist near the middle of the rachis, which results in the apical half of the leaf being oriented vertically. The leaflets are numerous, linear-lanceolate, long tipped, and deep emerald green on both surfaces, growing in a nearly flat plane from the rachis with only their apices pendent.

This species is not as beautiful as *O. tigillarium* because of the fewer, stockier trunks that usually attain similar statures, leaving the leaf crowns looking congested. Selecting and pruning out one or 2 stems could probably remedy this.

Oncosperma tigillarium
ahn-ko-SPUR-ma · ti-ji-LAH-ree-um
PLATES 651–652

Oncosperma tigillarium is indigenous to peninsular Thailand, peninsular Malaysia, Sumatra, Java, and Borneo, where it grows in low, swampy areas near the coasts, often landward of mangrove forests. The epithet is Latin for "of little stems" and must refer to the relative slenderness of the stems, certainly not their height.

The stems grow to a height of 80 or more feet (24 m) with a diameter of 6 inches (15 cm). They are light tan and, in their younger parts, exhibit widely spaced darker brown rings of leaf scars accompanied by rings of downward-pointing, long black spines. The crownshafts are 5 feet (1.5 m) tall in older individuals, cylindrical and slightly thicker than the stems beneath them. They are pale green or sometimes grayish green. The hemispherical leaf crown contains as many as 36 leaves, each of which is 10 feet (3 m) long, gently and gracefully arching, held on a petiole 1–2 feet (30–60 cm) long. The numerous linear-lanceolate leaflets are 2 feet (60 cm) long and are pendent from the rachis, appearing like a curtain; the lowermost ones are usually accompanied by long, whitish reins when newly unfurled. Leaf color is medium green.

The inflorescences grow from nodes beneath the crownshaft and consist of pendent, bright yellow flowering branches bearing both male and female blossoms. The fruits are small, rounded, and dark purple with a waxy, chalky bloom.

The most important cultural requirements are a tropical or nearly tropical climate and copious, regular moisture. The species seems almost indifferent to soil type, even intermittently soggy ones, and it is at home on calcareous soils, especially if the trees are mulched. In regions with distinct dry seasons, such as South Florida, the clumps suffer, and the wonderful leaves become unsightly unless irrigated. This palm will resprout when frozen to the ground if the soil itself does not freeze, but this practice is futile as the great beauty of the large clumps is lost and the plant dies out if subjected to such stress in consecutive winters. The clumps are adapted to full sun from an early age and, with enough water, grow quite fast.

No palm species is more beautiful than this one. The great height and density of the clumping stems, as well as the array of different heights of leaf crowns with their beautifully arching leaves and curtains of pendent leaflets, make this palm a veritable symphony, almost a world unto itself, of glorious forms. Some old individuals have one or a few much more dominant stems, which only add to the incredible beauty of the clump. The palm can stand alone as a specimen surrounded by space or be a splendid, large part of other vegetation. A large avenue with this species planted along either side would be one of the most magnificent sights possible, especially if the drive were curving! This palm is not for small gardens, but plots that are only somewhat larger than average could well accommodate a single clump in an outlying region. This palm is not known to be grown indoors and, because of its great size, would seem to be a poor choice for trying. In addition, it would need lots of light.

The trunks are reportedly immune to marine conditions and the wood is strong and durable and used for pier pilings as well as house construction. The seeds are also reportedly used as a betel-nut substitute in Malaysia.

Orania
o-RAN-ee-a

Orania consists of 28 solitary-trunked, pinnate-leaved, monoecious palms in Madagascar, peninsular Thailand, peninsular Malaysia, Sumatra, Java, Borneo, the Moluccas, Sulawesi, and the Philippines, with most species in New Guinea, including 11 species described in 2012 (Keim and Dransfield 2012). The genus name honors William, Prince of Orange (1840–1879).

The mostly tall, robust palms have distinctly stiff leaf rachises and ascending large leaves that exhibit little arching. Several species have distichous leaves, an arrangement that is very uncommon elsewhere in the palm family. The leaflets are often pendulous and are usually long, with their apices obliquely truncated and slightly jagged. The leaflets of the Madagascar species have chaffy scales on the undersides of their midveins; the leaflets of other species do not. The palms do not form crownshafts. The inflorescences grow from among the leaf bases, are branched 1–3 times, and produce large, unusual round fruits with 1–3 seeds. The number of seeds creates the form of the fruit: 2-seeded fruits with 2 adjoining globes and 3-seeded fruits with 3 adjoining spheres. The fruits of *O. sylvicola* are reputedly poisonous, although the veracity of the report and the chemical nature of the poison have yet to be ascertained.

The substantial variation in form among the species has yet to be horticulturally exploited: tall canopy or short subcanopy palms; spiral, distichous or subdistichous crowns; stiff, spreading or recurved leaves; leaflets in one plane or in more than one plane, held horizontally or pendent. All the species are extraordinarily impressive and tropical looking. The genus is unaccountably rare in cultivation in the Western Hemisphere. All the species are intolerant of cold and need copious, regular moisture as well as protection from the midday sun when young. They are, however, seemingly adapted to various soils, even calcareous ones.

Orania seeds are remote germinating. The germination stalk is up to 6 inches (15 cm) long, so seeds should be planted in a deep container. Seeds generally germinate within 90 days but can take a bit longer. Seeds allowed to dry out completely before sowing do not germinate as well as those kept slightly moist. For further details of seed germination, see *Areca*.

Orania disticha
o-RAN-ee-a · DIS-ti-ka
PLATE 653

Orania disticha is widespread across the island of New Guinea, where it grows in lowland rain forest. The epithet is Latin for "distichous," referring to the disposition of the leaves in 2 ranks at the top of the trunks.

The trunk of this large, magnificent, and elegant-looking species grows to a height of 60 feet (18 m) and a diameter of 9 inches (23 cm). The leaf crown is in the form of a gigantic flat green fan with the great leaves growing on opposite sides of the slender stem. They are erect, ascending, scarcely arching, and 8–10 feet (2.4–3 m) long on petioles 2–3 feet (60–90 cm) long; they bear many closely and regularly spaced leaflets, which are up to 5 feet (1.5 m) long and 3 inches (8 cm) wide. The pendent leaflets create the effect of a large green curtain hanging from the unbending leaf rachises. This is an incredibly exotic and appealing species, and is a complete mystery why it is not much more obvious in cultivation.

Orania lauterbachiana
o-RAN-ee-a · law-ter-bah-kee-AH-na
PLATE 654

Orania lauterbachiana is widespread throughout the island of New Guinea, where it grows in rain forest up to 2700 feet (820 m). The epithet honors Carl A. G. Lauterbach (1864–1937), German naturalist and explorer in New Guinea. The trunk attains a height of 60 feet (18 m) in habitat and a diameter of only 7 inches (18 cm). The crown is shaped like a gigantic shaving brush, and the leaves are 10–15 feet (3–4.5 m) long, erect, ascending, and nicely arching on long petioles. The many regularly spaced leaflets are 2–3 feet

(60–90 cm) long and are pendent; they are also broad in relation to their length, a characteristic which make this beautiful species unusual in the genus.

Orania longisquama
o-RAN-ee-a · lon-jee-SKWA-ma

Orania longisquama is a large palm endemic to the lowland rain forest of eastern Madagascar, up to 1800 feet (550 m) in elevation. The epithet is Latin for "long scale" and refers to the rusty brown, scaly pubescence of the leaf bases, petioles, rachises, and inflorescence branches.

The trunk grows to 65 feet (20 m) tall and 10 inches (25 cm) in diameter and is smooth and brown. The base of the trunk is expanded and produces surface roots. The leaves are arranged in a spherical crown of stiff leaves 6–7 feet (1.8–2.1 m) long on petioles 2–4 feet (60–120 cm) long. The stiff leaflets grow from the rachis at an angle that creates a V-shaped leaf and are medium green above but slightly glaucous white below. The tips of the leaflets are irregular, as if torn at an oblique angle. The inflorescences emerge from among the petioles enclosed in large bracts, which are covered in a rusty brown, scaly pubescence. The pale yellow flowers are followed by globose to elongate, green fruits up to 2 inches (5 cm) long and slightly narrower in diameter.

A striking palm for the garden, this species deserves to be more widely grown. It has an informal appearance that would lend itself well to a grouping of 3 or more palms. It is not known to be cold hardy. As with *Pseudophoenix sargentii*, seedlings of this species sometimes have a distichous leaf arrangement, but the arrangement becomes spiral as the palms mature.

Orania palindan
o-RAN-ee-a · PAL-in-dahn
PLATE 655

Orania palindan is widespread from the Philippines to the Moluccas, Sulawesi, and northern New Guinea, where it grows in low mountainous rain forest. The epithet is the aboriginal (Tagalog) name for the species. This palm is the tallest, most robust species in the genus, the trunk attaining a height of 90 feet (27 m) and a diameter of just over 1 foot (30 cm). The leaf crown is hemispherical with typical, unarching leaves 10 feet (3 m) long that bear long, narrow, pendent leaflets. The fruits are about 3 inches (8 cm) in diameter and greenish to golden yellow when ripe. This species seems to be the most cold-tolerant in the genus, surviving brief periods of temperatures slightly below freezing.

Orania ravaka
o-RAN-ee-a · rah-VAH-ka
PLATE 656

Orania ravaka is another distichous species, but this one is native to northeastern Madagascar. The epithet is the Malagasy word for "jewel" or "ornament" and is a comment on the profound beauty of this species.

The trunk grows to 65 feet (20 m) tall and 10 inches (25 cm) in diameter and is smooth and brown. The leaf crown is a distichous fan of stiff leaves, held on long petioles up to 3 feet (90 cm) long with the rachises another 6 feet (1.8 m) long. The stiff leaflets grow from the rachis at an angle that creates a V-shaped leaf. Leaflet color is medium green above but strongly glaucous white below. The tips of the leaflets are irregular, and the midveins have chaffy, reddish brown scales along their undersides. The pale yellow flowers are followed by flattened-globose, yellow or brown fruits 1½–2½ inches (4–6 cm) in diameter.

As its name signifies, this species of *Orania* is one of the most beautiful. Because of their unusual growth habit, the distichous species are not always easy to work into a landscape design. They would be an excellent choice for a narrow space next to a tall building. They need abundant shade and moisture as seedlings and juveniles.

Orania regalis
o-RAN-ee-a · ree-GAL-is

Orania regalis is a rare species endemic to western New Guinea and the Aru Islands, southwest of New Guinea, where it grows in lowland rain forest and swamps. The epithet is Latin for "regal."

The trunk of this massive species grows to at least 30 feet (9 m) high and a diameter of 8–10 inches (20–25 cm) at its base. It is light gray to tan and is heavily ringed with widely spaced darker leaf base scars. The leaf crown is hemispherical with several stiff and unarching leaves 16 feet (4.9 m) long on stout petioles 6 feet (1.8 m) long. The many pendent leaflets are 6 feet (1.8 m) long and, at their bases, 6 inches (15 cm) wide. The inflorescence has a long peduncle and many rachillae that sweep forward, are scarcely spreading, and look like brooms. The flowers are creamy white, and the fruits are red, 2–3 inches (5–8 cm) in diameter.

Until the year 2000, this had not been seen in the wild since the 1870s. It has been in cultivation in the Bogor Botanic Garden since the early 19th century.

Orania sylvicola
o-RAN-ee-a · sil-vi-KO-la
PLATE 657

Orania sylvicola is indigenous to peninsular Thailand, peninsular Malaysia, Singapore, Java, Borneo, and Sumatra, where it grows in low mountainous rain forest. The epithet is Latin for "inhabitant of the forest." The trunk attains a height of 50 feet (15 m) and a diameter of 6 inches (15 cm). It is gray and distinctly ringed with slightly darker leaf base scars in its upper parts. The leaf crown is spherical or nearly so and usually contains 18 stiff, unarching leaves, which are 10 feet (3 m) long on petioles 2 feet (60 cm) long. The pendent leaflets are 2–3 feet (60–90 cm) long, deep green above and brownish gray-green beneath. The flowers are creamy white, and the fruits are yellowish green when ripe and 2 inches (5 cm) in diameter.

Orania trispatha
o-RAN-ee-a · try-SPAY-tha
PLATE 658

Orania trispatha is one of 3 species endemic to Madagascar. This rare and endangered palm grows in lowland rain forest and swamps from sea level to 1500 feet (460 m). The epithet is Latin and Greek for "three spathes."

The trunk grows to a height of 65 feet (20 m) in habitat with a diameter of 1 foot (30 cm). There is no crownshaft, but the large leaf bases are distinct, prominent, and almost white. The leaf crown is a giant fan of distichous, stiff and unarching leaves, each 6–7 feet (1.8–2.1 m) long on an incredibly long 6-foot (1.8-m) petiole. The stiff leaflets grow from the rachis at an angle that creates a V-shaped leaf and are a deep green above but have a glaucous bluish gray-green hue on their undersides. The flowers are creamy white, and the fruits are yellowish green when ripe and 2 inches (5 cm) in diameter. Because of its distichously arranged leaves in the form of a gigantic flat fan, the species is one of the most exotic-looking palms in the world.

Oraniopsis
o-ran'-ee-AHP-sis

Oraniopsis is a monotypic genus of solitary-trunked, pinnate-leaved, dioecious palm in northeastern Queensland, Australia. The name translates as "like *Orania*," the genus in which it was once mistakenly placed based on its similar appearance. *Oraniopsis* has no relation to *Orania* and is, in fact, closely related to *Ravenea*. Seeds can take several months to a year or more to germinate and do so sporadically. For further details of seed germination, see *Ceroxylon*.

Oraniopsis appendiculata
o-ran'-ee-AHP-sis · ap-pen-dik'-yoo-LAH-ta
PLATES 659–660

Oraniopsis appendiculata is found naturally in mountainous rain forest at 1000–5000 feet (300–1520 m), but occasionally much lower. It grows in high-rainfall forests or along streams. The epithet is Latin for "appendaged" and refers to appendages on the petals, but the petals have no such appendages.

The trunk grows to 60 feet (18 m) in habitat but is usually shorter. It is typically 12–18 inches (30–45 cm) thick, straight, and light to dark gray. The leaf crown is shaped like a shuttlecock to almost hemispherical, usually with several dead leaves hanging beneath the crown. The stiff, ascending to spreading leaves are 12–15 feet (3.6–4.5 m) long on stout petioles 2 feet (60 cm) long. The many regularly spaced, narrowly linear leaflets are stiff and straight but grow in a single flat plane. Leaf color is deep emerald green above but silvery to coppery green beneath. The pendent inflorescences grow from the leaf bases and are 30 inches (76 cm) long with the flowering branches covered in a long brown tomentum and bearing small, white flowers. Female trees produce large clusters of round 1-inch (2.5-cm) yellowish orange fruits.

This palm looks like a cross between a coconut and an *Arenga* species, but it grows like neither palm. It is very slow growing and not likely to form a trunk for 20 or even 30 years. In addition, it needs constant, high humidity with copious, regular moisture and a humus-laden, quickly draining soil. The only good news relative to its cultural requirements is that it has survived unscathed temperatures below freezing. It is very uncommon in cultivation.

Parajubaea
pa'-ra-joo-BAY-a

Parajubaea comprises 3 solitary-trunked, pinnate-leaved, monoecious palms in the mountains of South America. The leaves are large and their sheaths are fibrous. The leaf rachis is usually twisted near its midpoint so that the apical half of the leaf is oriented vertically. The inflorescences grow from among the leaf bases and are generally similar to those of *Jubaea*, *Syagrus*, and *Cocos* (coconut palm). They are enclosed at first in large, woody boat-shaped bracts and, at anthesis, are branched once into stiff, short, erect flowering branches bearing both male and female blossoms. The large fruits are 3 inches (8 cm) long, have edible seeds, and are borne in compact, elongated masses. The leaf sheath fibers are used for weaving material, principally in making rope.

The 3 species are similar in appearance and are massive looking with fibrous stems when young, but, with one exception, quite tall and elegant, with smooth trunks when older. All are adapted only to Mediterranean or temperate climates with cool night temperatures and moderate to warm winters. They are impossible to grow in hot, humid tropical regions; indeed, they are among the world's highest-growing palm species. The irony is that, considering the great elevations at which they grow in habitat, some species are not as hardy to cold as would be expected.

The genus name translates as "close to *Jubaea*" and refers to the similarities in the fruits of the 2 genera. Seeds can take more than 2 years to germinate and do so sporadically. For further details of seed germination, see *Acrocomia*.

Parajubaea cocoides
pa'-ra-joo-BAY-a · ko-KO-i-deez
PLATE 661

Parajubaea cocoides was known for many decades only in cultivation. It is a relatively common ornamental in Quito, Ecuador, and is cultivated throughout Ecuador and Colombia, at elevations above 9000 feet (2700 m). Wild stands have recently been discovered in Peru. The epithet translates as "similar to *Cocos*."

The trunk grows to 50 feet (15 m) tall with a diameter of 18 inches (45 cm), and is heavily covered in dark cinnamon-colored leaf bases and fibers when young but is light to dark gray and nearly smooth in older trees. The leaf crown is hemispherical to almost spherical and usually consists of 24 leaves, each 10–12 feet (3–3.6 m) long on a petiole 3 feet (90 cm) long. The leaves are beautifully arching, divided into many linear, glossy leaflets 2 feet (60 cm) long, growing in a single flat plane from the rachis.

When young, the species has visual affinity with *Jubaea* but when older is more delicate appearing with a much more slender trunk. Its leaf crown is stunningly attractive and tropical looking, much like that of a coconut palm, and it is almost impossible to believe its nativity is so elevated and nontropical.

Parajubaea sunkha
pa'-ra-joo-BAY-a · SOON-ka
PLATE 662

Parajubaea sunkha is endemic to Bolivia in the foothills of the eastern slope of the Andes at 5600–6500 feet (1710–1980 m). It grows in deciduous forest of elevated, seasonally dry valleys. The epithet is the aboriginal name. This species is the smallest in the genus with a trunk no more than 25 feet (7.6 m) tall and 18 inches (45 cm) in diameter. The crown of leaves is nearly spherical. The upper part of the trunk is covered with abundant dark brown fibers, the products of the disintegrating leaf sheaths. The leaflets are not regularly spaced along the rachis, as they are in the other species. This species is reportedly also slower growing than the others.

Parajubaea torallyi
pa'-ra-joo-BAY-a · to-RAL-lee-eye
PLATE 663

Parajubaea torallyi is a rare and endangered species endemic to deciduous forest in the dry valleys of the Andes foothills of western Bolivia. The epithet honors a Dr. Torally, who was a physician in the Bolivian department of Chuquisaca in the early 19th century. The species is similar in stature and appearance to *P. cocoides* but has a slightly more open leaf crown because of the slightly longer petioles; it also has a slightly thinner trunk. Two varieties are recognized based largely on the size of the fruits, but in the landscape, they are indistinguishable.

Pelagodoxa
pel'-a-go-DAHK-sa

Pelagodoxa is a monotypic genus of rare and endangered, solitary-trunked, pinnate-leaved, monoecious palm. The name is derived from Greek words meaning "sea" and "glory," an allusion to the oceanic island origin and great beauty of this species. Seeds can take 30–180 days or longer to germinate and do so sporadically. Seeds allowed to dry out before sowing do not germinate as well as those kept slightly moist. For further details of seed germination, see *Areca*.

Pelagodoxa henryana
pel'-a-go-DAHK-sa · hen-ree-AH-na
PLATES 664–666

Pelagodoxa henryana is found in low mountainous rain forest in the Marquesas Islands, where it occurs in the undergrowth in a single valley. It is now almost extinct, with fewer than 12 adults trees. It is possible that the palms were established long ago at this site by islanders and that the true origin of the plants is unknown. The closest living relative is *Sommieria* of New Guinea. The epithet honors Charles Henry, an early 20th-century French botanist and collector of the first specimen from the Marquesas in 1916.

The trunk attains a maximum height of 25 feet (7.6 m) and a diameter of 6 inches (15 cm) in habitat. It is light brown, usually columnar and straight, and has widely spaced, raised rings of leaf base scars. The leaf crown is hemispherical to spherical. The leaves are linear-elliptical to linear-obovate, 6–7 feet (1.8–2.1 m) long, 3–4 feet (90–120 cm) wide, entire and undivided, and shortly bifid apically, with shallow indentations on the margins corresponding to the pinnately disposed veins in the blade. The leaves are borne on stout petioles 2 feet (60 cm) long, are heavily ribbed on both surfaces, and are deep green above but grayish green beneath. The inflorescences grow out of the leaf bases and are short, erect when new, 1–2 feet (30–60 cm) long, and branched twice; they bear both male and female blossoms. The fruits are round, covered with corky, warty projections, light greenish brown, and to 6 inches (15 cm) in diameter; they are borne in short, pendent clusters in the leaf crown. Fruit size varies considerably. Small-fruited palms are found in the Solomon Islands and the Vanuatu islands, but these are the result of 19th-century seed imports and are not a part of the original distribution of the species.

This palm is tropical in its requirements and is marginal in zone 10b. It must have copious, regular moisture, constantly high humidity, and protection from the full sun of hot climates. It does not tolerate hot or cold dry winds and needs protection from any strong breeze if its magnificent, undivided leaves are to remain pristine. It prefers a humus-laden soil but does well even in calcareous soil if organic matter or a mulch is added.

Few palms are as beautiful as this species. Its large undivided, deeply grooved leaves and its nearly rounded crown are more than exciting, and its curious fruits will always elicit comment. It should receive a protected and featured spot under a high canopy.

Phoenicophorium
feen'-i-ko-FOR-ee-um

Phoenicophorium is a monotypic genus of solitary-trunked, spiny, pinnate-leaved, monoecious palm in the Seychelles. The name combines *Phoenix*, in this case, a generalized prefix for any pinnate palm, and the Greek word for "stolen," a reference to the fact that one of the early (and few) specimens in Europe was purloined from the Royal Botanic Gardens, Kew. Seeds germinate within 30 days. For further details of seed germination, see *Areca*.

Phoenicophorium borsigianum
feen'-i-ko-FOR-ee-um · bor-sig'-ee-AH-num
PLATES 667–668

Phoenicophorium borsigianum grows from sea level to 1000 feet (300 m) in valleys and on hillsides and ridge tops in rain forest. It forms extensive stands where forest once stood and is able to colonize degraded land. The epithet honors Berlin industrialist August Borsig (1804–1854), who amassed an important collection of exotic plants.

In habitat, the gray trunk attains a height of 50 feet (15 m) and a diameter of 4 inches (10 cm). The hemispherical leaf crown contains 12 gracefully arching leaves 6 feet (1.8 m) long on stout petioles 8–10 inches (20–25 cm) long. The petioles extend into the blade as stout, thick, yellowish green midribs. In seedlings and juveniles, the petioles and leaf rachises are orange. The blade is entire but usually segmented by wind or other mechanical means and is deep emerald green on both surfaces. It is obovate and shortly bifid apical, and bears large indentations along its margins as well as distinct grooves in the blade, each corresponding to a fused pinna. Young leaves are distinctly V shaped, but older leaves are nearly flat. Young trees have many black spines on the leaf sheaths, petioles, and leaf rachises, but older plants usually have few, if any, spines. The erect inflorescences are 3 feet (90 cm) long, grow from among the leaf bases, are twice branched, and bear both male and female yellow flowers. The ovoid fruits are less than ½ inch (13 mm) long and orange-red when ripe.

This species does not tolerate cold and is adaptable only to zone 11. It also does not tolerate sustained temperatures below 45°F (7°C). It is adaptable to full sun once past the seedling stage and, although a true water lover, tolerates occasional short lapses in its moisture supply. It seems not too fussy about soil type as long as it is well drained.

This palm is one of the world's most beautiful. Its gorgeous leaves are nearly incomparable in beauty, especially in young trees that can be protected from wind. It is one of the few palms that can look good even as a single specimen surrounded by space but is much more effective in groups of 3 or more individuals of varying heights.

Phoenix
FEE-nix

Phoenix is composed of 14 pinnate-leaved, dioecious palms from the Canary and Cape Verde Islands, tropical and subtropical Africa, the Mediterranean region, the Middle East, the Arabian Peninsula, India, Indochina, and eastwards to Hong Kong, Taiwan, and the southern Philippine islands. They grow in habitats as diverse as deserts, mangrove sea coasts, and swampy areas. Most species hail from semiarid regions but grow near springs, watercourses, or high groundwater levels.

Five species always have clustering trunks, 5 have solitary trunks, and 4 can be either clustering or solitary. While most are medium sized to massive or robust in stature, a few are dwarf and one has a prostrate, creeping trunk. Several trunked species do not form an aboveground stem for several years and sometimes refuse to do so under drought or other adverse conditions. The petioles are short or nonexistent, and in all species the lowermost leaflets are modified into often long, vicious spines. These spines are one of the defining characteristics of the genus; they are extremely dangerous and pose a hazard to people pruning, moving, or disposing of leaves. Puncture wounds are prone to infection, and spines that penetrate bone tissue can lead to "pseudo-tumors" in the bone.

All the species have leaflets that, unlike almost all other pinnate-leaved palms, have a V shaped lengthwise fold. The inflorescences of male and female plants are similar in appearance but are always found on separate trees. They are white or cream colored as are the unisexual flowers and emerge from a leathery, usually boat-shaped bract which falls away shortly thereafter. The fruits are usually in large pendent clusters and mature orange, brown, or sometimes black.

All species freely hybridize with one another, and seed harvested from trees grown in proximity with other *Phoenix* species (especially in Florida) are likely to be hybrids. The genus name is a Latinized form of the Greek word for "date palm," which itself was possibly derived from the Phoenicians, who may have carried the palm from one location to another.

Phoenix is the only member of the tribe Phoeniceae. All the species are remote germinating. The germination stalk ranges from 1 to 3 inches (2.5–8 cm), so seeds should be planted in a container at least 6 inches (15 cm) deep. Seeds allowed to dry out before sowing do not germinate as well as those kept slightly moist. Fresh seeds germinate easily within 30 days and should be planted soon after harvesting.

Phoenix acaulis
FEE-nix · ay-KAW-lis
PLATE 669

Phoenix acaulis occurs naturally in northern India and Nepal, where it grows in scrublands and savannas and in the undergrowth of pine forests from 1200 to 5000 feet (370–1520 m) elevation. The epithet is Latin for "trunkless."

The stem remains mostly underground and, even when emergent, is only a few inches in height. The erect leaves are 2–5 feet (60–150 cm) long and grayish green with widely spaced sharply pointed leaflets 1 foot (30 cm) long. The leaflets grow in clusters and from different angles off the rachis; the outermost leaflets are not rigid and in older plants are often semipendulous. The fruits are borne on very short infructescences nestled at ground level among the leaves. This species can often be confused with acaulescent individuals of *P. loureiroi*, but the latter has fruits borne above the leaves.

This palm is more of a curiosity than ornamental and doubtless has some hardiness to cold. It would fit into a cactus or succulent garden or other type of xeriscape as it is drought tolerant, its roots growing quite deep in search of moisture.

Phoenix canariensis
FEE-nix · ka-nar'-ee-EN-sis
PLATE 670

Phoenix canariensis is endemic to the Canary Islands in moist and semiarid areas from sea level to 2000 feet (610 m), always in open areas with abundant groundwater. The common name is Canary Island date palm. The epithet is Latin for "of the Canaries."

Mature trunks reach 70 feet (21 m) but are usually shorter in cultivation. They are massive, usually straight as an arrow (unless planted on slopes where, if they do not fall over, become leaning and curved) and columnar, with a diameter of 2–3 feet (60–90 cm); they often sit atop a mound of densely packed aerial roots in older palms. Young trunks are decidedly bulbous, almost like a pineapple, and are covered with persistent, large triangular leaf bases. Old trunks are almost free of the leaf bases except below the leaf crown, where they form a "bulb" or "pineapple," and the stems show a beautiful pattern of horizontally elongated diamond-shaped leaf base scars in closely set spiraling rings. The trunks are light or dark brown. The immense leaf crown is densely packed with often well over 100 leaves and is 25–40 feet (7.6–12 m) wide and as tall. The leaves are 10–20 feet (3–6 m) long, narrowly elliptical, and dull olive green to deep green; the pure and unhybridized species never shows a bluish hue. The slightly arching leaves have stiff, sharp-pointed leaflets, which grow from the rachis in a slight V shape. The leaves are held on short pseudo-petioles that bear long, vicious spines. The inflorescences are 6 feet (1.8 m) long or longer with many deep yellow to light orange single branches that bear small, white, unisexual flowers. The oblong 1-inch (2.5-cm) fruits are brilliant orange and densely packed into broad, heavy, pendent clusters. They are edible but mostly unpalatable.

The species is susceptible to lethal yellowing disease. It is also highly susceptible to wilt disease caused by *Fusarium oxysporum*. Both diseases have taken a lethal toll on the species; wilt seems to spread more rapidly, generally on unsterilized pruning implements. The palm is suited to most fast-draining soils. It is adaptable to zones 9 through 11 in areas subject to wet freezes in winter but is usually good in zones 8 through 11 in drier ones, especially when older. It needs full sun from youth to old age and, while drought tolerant when established, grows so slowly under this condition as to be frustrating.

Among the most massive palms, this species has a stately presence that no other palm can surpass and, at the same time, seems warm and almost motherly and protective. There is no finer plant for lining avenues, and a grove of trees is a wonder to behold. The palm is beautiful and welcome sight in all but small, cramped areas. It tolerates salt-laden winds but not saline soil. Although it is one of the finest trees for planting near the seashore, it is definitely not a strand plant. With some practice, one can learn to recognize male plants by their more flat-topped crowns and shorter leaves, in contrast to the females, which have more rounded crowns and longer leaves.

The practice of removing older hanging fronds is esthetically abominable, but some people prefer it, or are taught to think they should by landscapers who need work or by homeowner associations suffering from a pathological obsession with "neatness." The usual excuse of both these groups is to expose the "bulb" or "pineapple" beneath the subsequently marred crown of leaves; the bulb is unattractive, especially when compared to the magnificence of the full crown! The only valid excuse for the severe bob is in transplanting or moving the tree. This species is generally too large for conservatories, although it has been grown in European glasshouses since the early 19th century (Zona 2008).

In the palm's habitat the leaves are used for weaving baskets and other utensils and for Christian religious services, and the stem provides sap for syrup. Sasha Barrow (1998) pointed out that the greatest threat to the species in habitat is the importation and cultivation of exotic species of *Phoenix* which hybridize with the endemic. These exotics may also expose the endemics to lethal yellowing disease.

Phoenix dactylifera
FEE-nix · dak-ti-LI-fe-ra
PLATES 671–672

Phoenix dactylifera would seem indigenous to all dry regions of North Africa, the Middle East, Turkey, Pakistan, and northwestern India, but was probably originally native only to North Africa or the Middle East, having been long ago transported by people to regions eastwards and westwards. It grows at oases, along rivers and streams, or anywhere underground water is available in desert areas. The species is universally known as the date palm or true date palm. The epithet is from both Greek and Latin and translates as "finger-bearing," an allusion to the clusters of thumb-size fruits.

This sparsely clustering species only suckers with age. Mature trunks grow to 90 feet (27 m) in habitat, but, under cultivation, do not usually exceed 70 feet (21 m), and that would be for old palms. Production palms (always female) are sometimes sold to landscapers when they exceed the height at which the fruits can be efficiently and safely harvested. The trunk diameter is never more than 2 feet (60 cm) and is usually 18 inches (45 cm). Younger trunks are usually clothed with the narrowly triangular leaf bases, but older trunks are often free of them and show a closely set pattern of spiraling, wide, flat "knobs." The leaf crown is 20 feet (6 m) tall and wide and, with 20–40 leaves, is much sparser than *P. canariensis*, usually less than full and rounded. The leaves are 10–20 feet (3–6 m) long, stiff, and slightly arching, held on pseudo-petioles 3–4 feet (90–120 cm) long. The pseudo-petioles are armed with long, sharp acanthophylls. The stiff, nonpendent leaflets are 1–2 feet (30–60 cm) long, gray-green to almost bluish gray-green, and grow from the rachis at several angles to give the blade a slight V shape. The orange inflorescences grow from among the leaves and are 4 feet (1.2 m) long. They bear many small, whitish, unisexual blossoms. The fruits are 1–3 inches (2.5–8 cm) long, oblong or cylindrical, orange when mature, and are borne in large, spreading handlike pendent masses.

The sweet, edible fruits have been a staple of Middle Eastern peoples for millennia and a source of much commerce for thousands of years. There are now hundreds, if not thousands, of local cultivars. The date palm does not produce quality fruits in humid climates, no matter how much heat is apparent during the summer; however, 2 cultivars can produce fruits in such regions as the Gulf coast of the United States: *P. dactylifera* 'Medjool' and *P. dactylifera* 'Zahedii' (Grant Stephenson, pers. comm.).

The species is slightly susceptible to lethal yellowing disease, and in warm, humid climes the older leaves are usually fungus ridden and removed because of their unsightliness. The date palm is hardy to cold, especially in dry climates where it is adaptable to zones 7 through 11, although marginal in zone 7. In areas subject to wet freezes in winter, it is only good in zones 8b through 11 and is marginal in 8b. Because of its long association with deserts, the date palm is popularly thought to have water requirements similar to that of a cactus. Nothing could be further from the truth. It is always found near groundwater and, while drought tolerant when established, it languishes and even dies under true drought conditions. It seems indifferent to soil type as long as its drainage is unimpeded. The date palm also tolerates saline soil and air and thus is a good choice for seaside plantings except on the seaward side of dunes. It needs full sun from youth to old age.

Phoenix dactylifera cultivars became, in the last quarter of the 20th century, one of the most frequently planted landscape palms in the tropical and subtropical areas of the United States, particularly in the southwestern desert areas. This phenomenon is partly due to the rapid development of date plantations in California and Arizona into ever-expanding tracts of suburbia, those great old palm groves giving way to acres of new houses; many of the former occupants of the groves were transported to adorn further developments around the country and the world. We are assured, however, by Henry Donselman (pers. comm.), that there are now as many commercial date palms as ever, the groves younger and in different locations.

Most date palms are unfortunately being planted in straight rows of trees of equal height and then trimmed "for neatness" to look like colossal, stiff shaving brushes. This esthetically abominable habit makes these naturally stiff-looking palms even more stiff- and sententious-looking, and the visual effect is harsh, hot, and overly energetic, almost frenetic. Considering that these assaults on the sensibilities are often perpetrated in hot, arid climates, the effect is all the more disturbing; instead of lushness and quietude, coolness and gracefulness, these new developments and plantings have an austere and irascible, regimented and hostile, tiring and overbearing demeanor.

Unfortunately, the female palms available in the landscape trade will bear fruits in the summer and drop these same fruits on the sidewalks and pavements below their crowns. The smell of trodden and rotting dates is not a pleasant one.

These noble trees are capable of providing wonderfully graceful landscape accents with their full crowns giving form and texture to the skyline and, in groups of full-crowned trees of varying heights, can add much beauty to warm-climate gardening, whether arid or not. It is not a particularly good candidate for growing indoors, although given a bright or sunny situation and good air circulation, it sometimes grows well in large conservatories and atriums.

The date palm's role in the history of the Middle East is analogous to those of the coconut palm (*Cocos nucifera*) and the African oil palm (*Elaeis guineensis*) in the tropics. It has been in cultivation for at least 5000 years. Due to the mystery of its place of origin and widespread cultivation, the date palm cannot be considered threatened in the traditional sense; however, as Sasha Barrow (1998) pointed out, the number of traditional cultivars is decreasing and, once lost, these varieties are irretrievable.

Phoenix loureiroi
FEE-nix · loo-ray-RO-ee
PLATE 673

Phoenix loureiroi is indigenous to India, southern Bhutan, and eastwards to Hong Kong, Taiwan, and the southern islands of the Philippines, where it grows from sea level to 5300 feet (1620 m). It is found in low mountainous deciduous or evergreen and open forests but also in cleared areas. The epithet honors João de Loureiro (1717–1791), a Portuguese Jesuit missionary to Southeast Asia, naturalist, and author of an early account of the flora of Indochina.

This small species occurs as both a solitary-trunked and a clustering palm. The stems are 3–12 feet (90 cm to 3.6 m) tall and to 1 foot (30 cm) in diameter, usually covered in all but their oldest parts with a dark brown mat of fiber and the short, stubby remains of the leaf bases. The leaves are variable in morphology but are mostly 6 feet (1.8 m) long and arching, with closely set narrow dark green leaflets that have sharply pointed apices. The leaflets, which grow from the rachis at different angles to create a stiffly plumose leaf, may also be much wider at their bases, even more crowded and bluish gray-green. The erect straw-colored inflorescences are borne on long stalks among or above the leaves (the similar *P. acaulis* holds its fruits at ground level). The mature fruits are bluish black.

The variability of *P. loureiroi* is pronounced and seems to be related to types of exposure: in full sun the species tends to be short, slow growing, and more readily clustering, with shorter, bluer leaves and more crowded and fatter leaflets, while as an undergrowth subject in moist, partially shady forests, it grows taller with markedly fewer clustering individuals and with the leaflets less stiff, more widely spaced, and pure green. The two varieties, *P. loureiroi* var. *loureiroi* and *P. loureiroi* var. *pedunculata* [pe-dun'-kyu-LAH-tah], are distinguished by anatomical characteristics of the leaflet margins and midribs.

Because of its wide range and many differing habitats, the little palm has unusual hardiness to cold for a species confined mainly to the tropics. It has withstood temperatures in the mid-20s Fahrenheit (around −4°C) unscathed and is probably adaptable to zones 8 through 11. It grows in partial shade or full sun with the exposure variations listed above to be expected. As for soil type, it seems to almost run the gamut, thriving in calcareous as well as acidic media, although always becoming more robust on good soils. It is not fast growing.

Phoenix paludosa
FEE-nix · pa-loo-DO-sa
PLATE 674

Phoenix paludosa occurs naturally along the coasts of northeastern India, Bangladesh, the Andaman and Nicobar Islands, Myanmar, Thailand, northern Sumatra, and peninsular Malaysia, where it grows landward of the mangrove forests in swampy areas that are periodically invaded by salt water. In peninsular Malaysia it is threatened by habitat destruction. The epithet is Latin for "swampy" and refers to the habitat. The common name is mangrove date palm.

The stems of this clustering species are mostly erect, growing to 15 feet (4.5 m) high. The clumps are dense and usually a wall of foliage. The trunks are covered in a mass of narrow leaf bases, tightly adhering dark brown fibers, and long needles from the leaf bases; the upper portions of the trunks are not visible unless the thatch of dead leaves is trimmed. Each leaf crown is sparse when it can be seen among the competing ones. The leaves are 6–8 feet (1.8–2.4 m) long and arching, giving the crown a full and rounded aspect. The soft but erect and nonpendent leaflets grow from the rachis in clusters and are arrayed at different angles, giving the leaf a full plumose effect. The leaflets are grayish on the underside and have dark spots where cells have accumulated tannins. The embryo is at the base of the seed, rather than at the middle, as in all other species of *Phoenix*.

This palm is not hardy to cold and needs warmth year-round, making it difficult to maintain in most Mediterranean climates like those of Southern California. It does not tolerate drought and must have full sun after the seedling stage. As for soil, as long as it remains moist it works. The species is similar in appearance to *P. reclinata*, but differs in the softer leaflets with glaucous undersides, a smaller stature, and much more supple and arching leaves. It can be pruned to look like a miniature *P. reclinata*, although with difficulty and sometimes pain. It is probably best treated in the landscape as a large hedge or barrier. It is ideally suited to wet, swampy sites.

Phoenix pusilla
FEE-nix · poo-SIL-la
PLATE 675

Phoenix pusilla is indigenous to southern India and Sri Lanka, where it grows in monsoonal lowlands that are seasonally inundated, at the edges of swamps, and on drier ridges and hills. The common name is Ceylon date palm. The epithet is Latin for "tiny" in reference to the small seeds.

The species has a solitary trunk for the most part, but clustering individuals are not rare in habitat. The trunk usually grows no taller than 15 feet (4.5 m) but may be 1 foot (30 cm) in diameter and is usually covered with numerous, closely set, gray and narrow leaf bases, a distinguishing if not diagnostic characteristic. The leaves are 8–10 feet (2.4–3 m) long, erect and ascending, arching only near their apices. The shiny, light green leaflets are thin, acutely pointed, and 1 foot (30 cm) long, growing from several angles off the rachis to give the leaf a plumose but stiff appearance. The base of each leaflet is marked by a yellow-orange swelling.

Attractive mainly because of the wicker-patterned trunks, this species is eminently worthy of cultivation. Alas, it is not cold hardy, being adaptable to zones 10 and 11, with an occasional nice specimen found in the more favorable microclimates of 9b. Like most other species in the genus, it is drought tolerant but grows more slowly and less robustly in dry soil. It grows slowly even in favorable conditions.

Phoenix reclinata
FEE-nix · rek-li-NAH-ta
PLATE 676

Phoenix reclinata grows along rivers and streams in tropical western, eastern, and central Africa, in Zambia, Malawi, and Mozambique in southern Africa, and in Madagascar and the Comoro Islands, from sea level to 10,000 feet (300 m) elevation, in clearings in the rain forest, seasonally flooded savannas, monsoonal forests, and even rocky mountainsides and grasslands. Common names are Senegal date palm and African wild date palm. The epithet is Latin for "reclining," an allusion to the angle at which the outer stems of a clump grow.

The trunks of this large clustering species reach 40 feet (12 m) high but more often, especially in cultivation, 30 feet (9 m). Each trunk is 4–7 inches (10–18 cm) in diameter and is covered until old with short leaf bases and matted brown fibers; older trunks show closely set rings of leaf base scars in their lower portions. The palm freely suckers, and a clump may contain 25 separate stems and be as much as 50 feet (15 m) tall and 45 feet (14 m) wide. The trunks are never straight but rather gracefully leaning outwards from the center of the cluster. The leaves are 12–15 feet (3.6–4.5 m) long, narrowly elliptical, and gracefully arching. The leaflets are 1 foot (30 cm) long and grow from the rachis at several different short angles, but the leaf appears from any distance flat. Leaf color is bright to deep, glossy green on both sides, and the blade is held on short pseudo-petioles, 1 foot (30 cm) long and armed with long, vicious acanthophylls. The full, rounded crown has 20–40 leaves. The fruits are ½–1 inch (13–25 mm) long, ovoid, reddish brown, and edible but mostly unpalatable.

This species is slightly susceptible to lethal yellowing disease. It is not hardy to cold, and the "pure" species is adaptable only to zones 10 and 11. Most plants in cultivation in the United States, especially in Florida, are hybrids. This is both good and bad: good because the hybrids are usually more tolerant of cold; bad because some of the hybrids do not have the gracefulness of the unadulterated species. It is impossible to give accurate data for the hardiness of individual hybridized specimens as the degree of cold tolerance depends on how much of the type is there and also what other species might be involved; in general, the hybrids that look enough like the type to be tagged with the binomial are damaged below 25°F (−4°C) and usually die from temperatures much below 20°F (−7°C), while the pure species is killed at 25°F (−4°C) and below. The palm is indifferent to soil type as long as it is not highly acidic or extremely alkaline. It needs full sun from youth to old age and, with regular, adequate moisture, grows moderately fast; prolonged drought conditions set it back.

Although sometimes disdained in tropical or subtropical regions because of its ubiquity and weediness—it is sometimes even regarded as invasive—this palm makes a beautiful, large clump. The clumps are made even more graceful and dramatic if a few trunks are thinned out as the mass develops so that the remaining trunks are of differing heights and so that their individual beauty can be seen and appreciated to its fullest. One of us (RLR) has seen this palm as a hedge with the many young trunks of evenly spaced plants obscured by a tangled mass of foliage, the visual effect being anything but attractive. If a hedge or barrier is wanted, it is infinitely more esthetically pleasing to let a few trunks of varying heights form first and then allow the many subsequent suckers to form the desired barrier at the base of the larger ones—this can be a captivating tableau. One of the most beautiful landscape objects is a "wall" of these palms with tall trunks of varying heights above the tiers of lower leaves resulting from the offshoots. There is probably no more perfect candidate for planting near ponds, lakes, or swimming pools, the gracefully reclining, thin trunks lending a tropical and exotic essence only matched by something like the coconut. While the clumps are beautiful enough standing isolated in a wide expanse of lawn, they look much better if incorporated into masses of lower vegetation so that their crowns serve as canopy-scapes. This date palm is only good indoors if great space and high light levels are available.

The trunks are used for construction, the leaves for thatch and for weaving baskets and other utensils. The fruits are often eaten, and the seeds ground into a flour by the native peoples in the palm's habitats.

Phoenix roebelenii
FEE-nix · ro-be-LEE-nee-eye
PLATES 677–678

Phoenix roebelenii occurs naturally in northern Laos, Vietnam, and southern China, where it grows along rivers in areas that are periodically flooded. Common names are pygmy date palm and miniature date palm. The epithet honors the German-born horticulturist Carl Roebelen (1855–1927), orchid collector employed by Sanders of St. Albans, England, and first collector of the species in Laos.

The form and habit of the species in cultivation are different from those of wild plants. Cultivated specimens have solitary, erect stems, with large, fully rounded leaf crowns, while wild plants are clustering, with curving, twisting, thinner trunks, and sparser crowns. Sasha Barrow (1994) hypothesized that the form in cultivation is "the product of a series of hybridization events with other cultivated *Phoenix* species, and somewhere along this series the clustering ability was lost." She also surmised, and rightly so we believe, that the clustering habit and sparser crowns of the wild populations allow the individual palms to withstand the periodic flood stages of the large rivers along which they grow. This palm is, in fact, a rheophyte.

In cultivation, mature trunks usually grow 6–8 feet (1.8–2.4 m) tall but may reach 12 feet (3.6 m) in very old individuals. The stem is 6–8 inches (15–20 cm) in diameter immediately below the leaf crown, where it is usually covered in triangular leaf bases and a mat of brown fiber. Lower down, the stem is 3 inches (8 cm) in diameter and has closely set spirals of knobby leaf base scars. Older trunks may have a small mass of aerial roots at their bases. The full, rounded leaf crown is usually 6 feet (1.8 m) wide and tall and has 50 gracefully arching leaves, each 3–5 feet (90–150 cm) long. The leaflets are 10–12 inches (25–30 cm) long, grow from the rachis almost in one flat plane, and are supple, thin, and pendent. The pseudo-petiole is 5 inches (13 cm) long and spiny. Leaf color is glossy deep green in mature leaves, but new leaves are covered with a fine chalky bloom that gives them a beautiful grayish green cast. The much-branched inflorescences are never more than 2 feet (60 cm) long, and the paddle-shaped bracts from which they emerge are almost as long. The small, chaffy-looking flowers are light yellow to almost white and produce fruits that are ½ inch (13 mm) long, oblong, and black.

This palm is surprisingly cold hardy for a tropical species, withstanding temperatures slightly below freezing without damage. It makes a good candidate in zones 10 and 11 and is marginal in 9b in areas subject to wet freezes in winter. Geoff Stein (pers. comm.) reported that, in Southern California, the species usually does well even in zone 8b. The cultural requirement most often neglected with this species is its moisture needs: it is thirsty and requires constant, adequate water, even withstanding occasional flooding. In addition, it relishes a good, friable, and slightly acidic soil. It grows significantly slower in partial shade than in full sun.

The cultivated form of this species is unexcelled for using as a silhouette, a small patio tree alongside a garden water feature, or along a walkway where its elegance of form and detail can be appreciated up close—but ever mindful of the sharp spines. *Phoenix roebelenii* is often sold in containers as a "double palm," meaning that there are 2 or more individuals to a pot; this practice is heartily encouraged as few things are lovelier than a planting of 2 or 3 palms of varying heights. The pygmy date does not look good in groups of 3 or more individuals of varying heights if the palm(s) are isolated specimens surrounded by space; its leaves are too delicate and gossamer to be so used unless near a wall or other contrasting form where the silhouette may be appreciated. This is an excellent choice for growing indoors if enough light can be provided.

Phoenix rupicola
FEE-nix · roo-PI-ko-la
PLATE 679

Phoenix rupicola grows in wet or dry mountainous forests and clearings therein in India and Bhutan from elevations of 1000 to 4000 feet (300–1200 m), usually on hillsides, cliffs, or otherwise sloping ground. The only common name seems to be cliff date palm. The epithet is Latin for "inhabitant of rocks" and refers to the palm's general habitat.

The mature height of this solitary-trunked species is 25 feet (7.6 m) with a trunk diameter of 10 inches (25 cm). The trunk is usually free of leaf bases except near the crown and has closely set spirals of knobby leaf base scars. The leaf crown is full and round to elongated, 15 feet (4.5 m) wide and 20 feet (6 m) tall. Leaves are 8–10 feet (2.4–3 m) long with leaflets 18 inches (45 cm) long and growing from the rachis in a single flat plane. The pseudo-petiole is 2–3 feet (60–90 cm) long and armed with spines, which are shorter, fewer, more pliable, and less vicious than those of any other *Phoenix* species. The evenly spaced leaflets are deep olive to glossy emerald green on both surfaces, thin, and limp. The newer leaves usually have a twist to the rachis that has the leaflets oriented vertically for half the length of the blade. The inflorescences are remarkably similar to those of *P. dactylifera* in appearance and in location on the tree, and the oblong 1-inch (2.5-cm) fruits are yellow to orange, ripening to a purplish brown.

This species is one of the most cold tender in the genus and is adaptable only to zones 10 and 11 in areas subject to wet freezes in winter and is often marginal in 10a. Geoff Stein (pers. comm.) indicated that it is adaptable to zone 9b in the drier Mediterranean climate of Southern California. It is adapted to partial shade when young but needs full sun when older to look its best and grow its fastest. It is drought tolerant when established but hardly grows under such conditions. In poor soil, it is subject to yellowing and disease. Rather, it requires a humus-laden, fast-draining, and mostly moist soil. It is slightly susceptible to lethal yellowing disease.

This is arguably the most beautiful solitary-trunked date palm. It combines the grace and elegance of the pygmy date palm (*P. roebelenii*) with the nobility of the Canary Island date (*P. canariensis*), and a group of these palms with trunks of varying heights is one of the most beautiful sights the plant world has to offer. It is attractive enough to stand alone as an isolated specimen tree, and it is small enough to serve as a focal point in a large patio or courtyard, where its lushness and grace remind one of a small coconut. It is excellent as a canopy-scape. It needs space and high light levels if grown indoors.

Phoenix sylvestris
FEE-nix · sil-VES-tris
PLATE 680

Phoenix sylvestris is indigenous to the lower elevations in southern Pakistan and most of India, where it grows in monsoonal plains and scrublands. Common names are toddy palm, wild date palm, silver date palm, and sugar date palm. The epithet is Latin for "of the forest."

Mature trunks of this solitary-trunked species may reach 50 feet (15 m) high and have a diameter of 18 inches (45 cm). It is covered until old in broadly triangular leaf bases and thence in the closely set spirals of stretched out diamond-shaped leaf base scars. Old trunks may produce, like those of *P. canariensis*, a dense mass of aerial roots near their bases. The leaf crown is 30 feet (9 m) wide and 20–30 feet (6–9 m) tall, full and founded; next to the crown of *P. canariensis*, it has more leaves than any other *Phoenix* species except the Canary Island date palm (to 100). The gently arching leaves are 8–10 feet (2.4–3 m) long on spiny petioles 3 feet (90 cm) long. The stiff leaflets are 18 inches (45 cm) long, irregularly spaced, and grow from the rachis at several different angles, giving the leaf a somewhat plumose appearance. Leaf color is waxy grayish to bluish green. The inflorescence is 3 feet (90 cm) long, with several yellow branches bearing many small, whitish blossoms. The oblong 1-inch (2.5-cm) fruits are borne in large, wide, pendent clusters and are orange but mature to a lovely purplish red.

The wild date palm is slightly susceptible to lethal yellowing disease. It is hardy to cold in zones 9b through 11 and marginal in most areas of 9a. It is drought tolerant when established but grows faster and looks better with adequate, regular moisture. It thrives on most fast-draining soils but needs sun, especially when past the juvenile stage.

This species is similar to *P. canariensis* but not as massive, and has a different leaf color and a stiffer, more compact crown. Though not as cold hardy as *P. canariensis*, it serves the same landscape purposes and usually grows significantly faster. It is not known to be grown indoors and is not recommended.

Phoenix theophrasti
FEE-nix · thee-o-FRAS-tee
PLATE 681

Phoenix theophrasti is indigenous to the Mediterranean coasts of Crete and Turkey. The epithet honors the ancient Greek philosopher and botanist Theophrastus. The species has been known for centuries but was not technically described until 1967. It is similar to *P. dactylifera* and difficult to distinguish from it. It is, however, more clustering, of smaller stature, and has shorter, more bristly leaves. Its cultural requirements and attributes are identical to those of *P. dactylifera*, and its landscaping uses are, with the exception related to its stature, the same. This species is highly salt tolerant.

Pholidocarpus
fo-li′-do-KAHR-pus

Pholidocarpus includes 6 large solitary-trunked, spiny, palmate-leaved, hermaphroditic palms from peninsular Thailand, peninsular Malaysia, and Sumatra eastwards through the Moluccas. All occur in lowland rain forest along watercourses or in swamps.

These are magnificently tall palms. Some are reported to form trunks more than 100 feet (30 m) high. They are usually straight, columnar, and almost white in their older parts, while light tan with closely spaced rings of ridged leaf base scars in their younger parts. The leaf crowns are large, globular, and open because of the long, spiny petioles and consist of numerous large, circular leaves with many wide segments extending nearly to the petiole, each of these again divided or bifurcated, much as in *Saribus*. In overall appearance, these species look like glorified crosses between *Borassodendron* and *Saribus*. The inflorescences grow from the leaf bases and are long and much branched, and the flowering branches bear bisexual blossoms. The large fruits are round or ovoid, usually light brown, with a corky surface shallowly divided into a mosaic of contiguous squares or short pyramidal projections.

No fan-leaved palms are more beautiful, and it is perplexing why these are not much more widely cultivated in tropical climates. All are similar in general appearance. They require tropical conditions and are possible only in zones 10b and 11. They are also true water lovers and need copious, regular moisture, as their swamp-dwelling habits suggest. They appreciate a humus-laden soil and need full sun when past the juvenile stage, at which time they prove to be robust and relatively fast growers in sufficiently warm climes.

The genus name is 2 Greek words meaning "scaly" and "fruit." Seeds germinate within 30 days and should not be allowed to dry out before sowing. For further details of seed germination, see *Corypha*.

Pholidocarpus macrocarpus
fo-li'-do-KAHR-pus · mak-ro-KAHR-pus
PLATES 682–683

Pholidocarpus macrocarpus is indigenous to peninsular Thailand and peninsular Malaysia, where it grows in lowland rainforest swamps and along streams and rivers. The epithet is from Greek words meaning "large fruit" and alludes to the globular 3-inch (8-cm) light brown warty fruits. The trunk attains a height of nearly 100 feet (30 m). The leaves are 6 feet (1.8 m) wide on stout petioles 6–7 feet (1.8–2.1 m) long. The lower margins of the petioles bear a distinct yellow stripe and the upper margins bear curved spines that are 3 inches (8 cm) long and have bulbous bases.

Pholidocarpus majadum
fo-li'-do-KAHR-pus · my-YAH-dum
PLATE 684

Pholidocarpus majadum is endemic to Borneo, where it grows in lowland swamps, along the edges of rivers, and in clearings in the rain forest. The epithet is a Latinized form of *maias*, the name used for orangutans in Sarawak, and given to this palm by Odoardo Beccari to commemorate the place of origin of this palm. The species is similar in appearance to *P. macrocarpus* except for its stature, which is 70 feet (21 m).

Pholidostachys
fo-li'-do-STAY-kees

Pholidostachys consists of 4 pinnate-leaved, monoecious palms in tropical America. They are solitary trunked except for occasional clustering in one species. Although they are delicate in appearance and all are undergrowth subjects in rain forest, they are not small palms. Their leaf crowns are open and generally hemispherical, the leaves on long petioles and spreading. Newly emerged leaves are usually colored orange. The inflorescences grow from among the leaves and are mostly spikelike or branched once. The flowering branches are distinctive in bearing the tiny flowers in pits, each of which is covered by a small bract. The many bracts along the branches overlap one another in a fish-scale pattern. The fruits are borne in congested clusters and are rounded and purplish, brown, or black.

These palms are related to and visually similar to those in *Geonoma*. All species are cold sensitive and adaptable only to zones 10b and 11. They require constant moisture, high humidity, and a humus-laden soil. All are adaptable to partial shade, which they absolutely require in hot climates and when young. They are relatively slow growing even under optimum conditions. The genus name is 2 Greek words meaning "scale" and "spike" and alludes to the fish-scale pattern on the branches of the inflorescences. Seeds germinate within 60 days. For further details of seed germination, see *Asterogyne*.

Pholidostachys dactyloides
fo-li'-do-STAY-kees · dak-ti-LO-i-deez
PLATE 685

Pholidostachys dactyloides is indigenous to eastern coastal Panama, western coastal Colombia, and western coastal Ecuador, where it grows in rain forest at low to moderate elevations. The epithet is Latin for "fingerlike," the meaning of which was unexplained, but perhaps refers to the 5 rachillae on the original specimen. The trunk attains a height of 30 feet (9 m) and a diameter of 2–3 inches (5–8 cm). The sparse leaf crown is in the form of a feather duster or, at most, is hemispherical. The mostly erect leaves are 5–6 feet (1.5–1.8 m) long and bear widely spaced, elliptical leaflets that are, after maturing, green on both surfaces.

Pholidostachys kalbreyeri
fo-li'-do-STAY-kees · kal-BRAY-er-eye
PLATE 686

Pholidostachys kalbreyeri is a rare species found in wet areas of Panama and Colombia in rain forest at low elevations. It is named for Guillermo Kalbreyer (1847–1912), a German plant collector for James Veitch and Sons of Chelsea, London. The trunk reaches only 10 feet (3 m) tall and is 2½–4½ inches (6–11 cm) in diameter. The erect leaves are 4–6 feet (1.2–1.8 m) long with widely spaced, elliptical leaflets. The bract of the inflorescence never falls away, but instead disintegrates during flowering and decays in place as the fruits develop.

Pholidostachys pulchra
fo-li'-do-STAY-kees · POOL-kra
PLATE 687

Pholidostachys pulchra is indigenous to extreme southeastern Nicaragua, the Caribbean slopes of Costa Rica and Panama, and the western slope of the Andes in Colombia, where it grows from sea level to 2000 feet (610 m) in mountainous rain forest. The epithet is Latin for "beautiful," which is an apt description of this delicate-looking species. The slender trunk attains a height of 25 feet (7.6 m) but a diameter of only 1–2 inches (2.5–5 cm) in habitat. Rarely is it found as a sparsely clustering specimen. The open leaf crown is hemispherical in older plants but usually in the form of a shuttlecock in younger ones. The leaves are 3–5 feet (90–150 cm) long on thin, wiry petioles 2–3 feet (60–90 cm) long and bear a few widely and irregularly spaced wide leaflets, which grow in a single flat plane from the rachis. The new growth is usually bronzy orange.

Pholidostachys synanthera
fo-li'-do-STAY-kees · sin-AN-the-ra
PLATE 688

Pholidostachys synanthera grows naturally in the intermontane valleys of the Andes of northern and western Colombia and in mountainous rain forest of western Amazonia in eastern Ecuador, northern Peru, and extreme western Brazil. It occurs from sea level to 5000 feet (1520 m), always in wet forest. The epithet is from Greek words meaning "joined" and "flower," the significance of which was not explained.

The trunk of this robust and beautiful species attains a maximum height of 20 feet (6 m), but is usually much shorter, and is 3 inches (8 cm) in diameter. It is covered in many fibrous leaf bases in its younger parts but is smooth with closely spaced, indistinct rings of leaf base scars in its older parts. The dense leaf crown is hemispherical to nearly spherical and has leaves 7 feet (2.1 m) long on thin petioles 4 feet (1.2 m) long. The relatively few leaflets are wide and irregularly spaced in a flat plane along the rachis; they measure 7 inches (18 cm) across, 30 inches (76 cm) long, and have pendent tips.

Physokentia
fy'-so-KEN-tee-a

Physokentia comprises 7 solitary-trunked, pinnate-leaved, monoecious palms in low mountainous rain forest of the Bismarck Archipelago, the Solomon Islands, Fiji, and the islands of Vanuatu in the western Pacific. The beautifully ringed trunks are green in their younger parts and sit atop prominent, prickly, and usually branched stilt roots. All species have distinct crownshafts above which the short-petioled leaves are mostly ascending and beneath which the well-branched inflorescences form, bearing both male and female flowers that produce clusters of red or black cherrylike fruits.

These are some of the loveliest feather-leaved palms, their large leaves with large, wide pinnae and their stilt roots the essence of the tropical look. Alas, they are rare in cultivation. All are intolerant of cold and adaptable only to zones 10b and 11. They require copious moisture and humidity at all times and a humus-laden soil. Dry air is lethal. As most are undergrowth palms, they relish partial shade and a sheltered site, the larger species tolerating full sun except in the hottest climates. They make some of the most beautiful greenhouse subjects.

The genus name combines the Greek word for "bubble," perhaps a reference to the spherical fruits, and *Kentia*, an out-of-date name for *Hydriastele*, but used in compounds to signify an understory pinnate palm of delicate beauty. Seeds of *Physokentia* germinate within 60 days. For further details of seed germination, see *Areca*.

Physokentia dennisii
fy'-so-KEN-tee-a · den-NI-see-eye

Physokentia dennisii is endemic to mountainous rain forest of the Solomon Islands, where it generally grows at elevations below 1000 feet (300 m). The epithet honors Geoffrey F. C. Dennis of the Forestry Department of the Solomon Islands, who assisted Harold E. Moore, Jr., in finding the species.

The trunk grows to 30 feet (9 m) tall and 6–8 inches (15–20 cm) in diameter. It is a beautiful dark green in its younger parts with widely spaced light brown rings of leaf base scars, while in its older parts is light gray. It is supported by a dense cone of stilt roots 4 feet (1.2 m) tall and is topped by a yellowish to gray-green crownshaft that is 3 feet (90 cm) tall and distinctly bulging at its base. The hemispherical leaf crown contains 10 spreading leaves 6 feet (1.8 m) long on short petioles. The many leaflets are regularly spaced and grow from the rachis in a single, nearly flat plane. The central pinnae are more than 2 feet (60 cm) long, 3 inches (8 cm) wide, linear-elliptic, and dark green above and below, with long tapering and slightly pendent tips. The pendent, greenish white, single inflorescence is 3 feet (90 cm) long and produces rounded ½-inch (13-mm) fruits that are deep orange or red when ripe.

Physokentia insolita
fy'-so-KEN-tee-a · in-so-LEE-ta
PLATE 689

Physokentia insolita is endemic to the Solomon Islands, where it grows in mountainous rain forest from elevations of 1700 to 3000 feet (520–900 m). The epithet is Latin for "unusual" or "strange" and was bestowed by Harold E. Moore, Jr., because the species is atypical for the genus in details of its flowers and fruits.

The slender trunk attains a height of 50 feet (15 m) and a diameter of 6–8 inches (15–20 cm) in habitat. It is deep green in its younger parts, with widely spaced gray rings of leaf base scars but, in its older parts, is light brown to deep gray. It is supported on a narrow, sparse cone of stilt roots that is 5 feet (1.5 m) tall. The crownshaft is 18–24 inches (45–60 cm) tall, slightly bulged at its base, and deep olive green but covered in tan scales which, unless rubbed off, impart their hue to the shaft. The leaf crown is spherical, with 8 spreading leaves, each 6 feet (1.8 m) long on a petiole 1–12 inches (2.5–30 cm) long. The few broad, heavily ribbed leaflets are irregularly spaced, grow from the rachis in a single flat plane, and have obliquely truncated, jagged apices. The rounded ½-inch (13-mm) fruits become deep red when ripe.

Physokentia petiolata
fy'-so-KEN-tee-a · pee'–tee-o-LAH-ta
PLATE 690

Physokentia petiolata is endemic to Fiji, where it grows in mountainous rain forest from 2500 to 4000 feet (760–1200 m) elevation on Viti Levu, as well as Gau Island. The epithet is Latin for "having a petiole," which seems to be a statement of the obvious. *Physokentia rosea* is a synonym.

The dark green trunk attains a maximum height of 25 feet (7.6 m) with a diameter of 4 inches (10 cm) and shows irregularly spaced, undulating, tan rings of leaf base scars. A narrow cone of stilt roots 3 feet (90 cm) tall supports it. The gray-green crownshaft is 18 inches (45 cm) tall, slightly bulged at its base and cylindrical above that point. The hemispherical leaf crown has 8 spreading leaves 6 feet (1.8 m) long on petioles 8 inches (20 cm) long. The linear-elliptic, deep green leaflets are regularly spaced and grow in a single flat plane from the rachis, the longest ones 2 feet (60 cm) long. The flowers are deep pink to red in bud. The round ½-inch (13-mm) fruits turn glossy black when mature.

Phytelephas
fy-TEL-e-fas

Phytelephas is composed of 6 solitary-trunked and clustering, pinnate-leaved, dioecious palms in South America. The stems are mostly short and stout, covered in leaf bases and fibers, and are often subterranean or recumbent and creeping on the ground. The leaves are large and impressive but undistinguished. In contrast, the inflorescences and fruits are distinctive and unusual. The male inflorescences are pendent, ropelike, unbranched spikes densely covered with many small flowers, each bearing dozens or even hundreds of stamens. The female inflorescences are large congested heads of large flowers with extremely long, fleshy, curling, ribbonlike sepals and petals. The dark brown fruits are formed in immense, congested heads. The rounded heads are covered in woody facets with spiky projections protruding from the center of each facet. Individual fruits contain several seeds which, when mature, have a hard, pure white or cream-colored endosperm that has been and is used as ivory for carving *objets d'art*, buttons, and other utilitarian commodities; it is generally called *tagua* in habitat and vegetable ivory in English-speaking regions.

These exotic palms are unusual yet appealing and tropical looking with their large leaves and many leaflets, somewhat resembling gigantic tropical cycads. All are intolerant of frost and need copious, regular moisture; they seem indifferent to soil type although they grow faster and look better in a humus-laden soil. The species are commonly referred to as ivory-nut palms. The genus name is from Greek words meaning "plant" and "elephant" or, by extension, "ivory."

Phytelephas is a member of the tribe Phytelepheae. All the genera in this subfamily are considered vegetable ivory palms and germinate in the same way. Seeds can take 30 days to 3 years or even longer to germinate and do so sporadically. They can be stored for some time but are best planted soon after harvest.

Phytelephas aequatorialis
fy-TEL-e-fas · ek′-wa-tor-ee-AL-is
PLATE 691

Phytelephas aequatorialis is endemic to western Ecuador, where it grows in lowland rain forest and on the wet western slopes of the Andes to 5000 feet (1520 m). The epithet is Latin for "equatorial" and alludes to the palm's habitat in a country whose boundaries straddle the equator.

This solitary-trunked species is the tallest in the genus, the stem attaining a height of 50 feet (1.5 m) and a diameter of 1 foot (30 cm). The immense leaf crown is spherical, mainly because of the persistent dead leaves that hang beneath the live ones. Individual leaves are linear-elliptical and 25–30 feet (7.6–9 m) long on short, stout petioles. The leaflets are extremely numerous and grow from the rachis in small groups and often at slightly different angles. The central leaflets are 2–3 feet (60–90 cm) long, but the basal and apical leaflets are much shorter, the ones in between showing a continuum of diminishing lengths. They are all narrowly linear, a deep green on both surfaces, and slightly pendent. This stupendous and magnificent species deserves to be much more widely planted in tropical regions.

Phytelephas macrocarpa
fy-TEL-e-fas · mak-ro-KAHR-pa
PLATES 692–693

Phytelephas macrocarpa occurs naturally in the lowland rain forest of western Amazonia in eastern Peru, extreme western Brazil, and northwestern Bolivia. The epithet is Greek for "large fruit." The solitary trunk is subterranean or, if aerial, short and usually recumbent. The leaf crown resembles a gigantic shuttlecock or is nearly hemispherical, and has 18 leaves, each 20 feet (6 m) long, erect, but gently arching. The leaflets are regularly arranged in one plane. The male inflorescences often reach the ground because of the short or subterranean trunks.

Phytelephas seemannii
fy-TEL-e-fas · see-MAHN-nee-eye
PLATE 694

Phytelephas seemannii is indigenous to eastern Panama and northwestern Colombia, where it occurs in the undergrowth of lowland rain forest in areas with seasonal flooding. The epithet honors German botanist and plant explorer Berthold C. Seemann (1825–1871).

This unusual species forms a solitary aerial trunk 1 foot (30 cm) in diameter and to 40 feet (12 m) long; it is prostrate for most of its length with the apical part and leaf crown erect. The lower portions of the stem usually die off and all parts form aerial roots that grow into the soil on the recumbent parts of the trunk. As the trunk slowly grows long, the weight of the leaf crown, which can measure more than 50 feet (15 m) in diameter, pulls the upper part of the trunk to the ground where it roots and eventually grows erect again. After the base of the stem dies off, the process starts again and, in time, results in the peregrination of the palm over short distances, the path usually circular. Because the trunk constantly grows longer but never exceeds 15 feet (4.5 m) high before it returns to a recumbent position and always produced new roots, the species has the potential of being immortal, eschewing the risks related to great height that cause the eventual death of erect-growing palms. The leaflets are regularly arranged in one plane.

Phytelephas tenuicaulis
fy-TEL-e-fas · te-nu-ee-KAW-lis
PLATE 695

Phytelephas tenuicaulis is indigenous to extreme south central Colombia, northeastern Ecuador, and northwestern Peru, where it grows along rivers and streams and in swamps at low elevations. The epithet is Latin for "thin stemmed." This clustering species has beautiful spreading but scarcely arching leaves 6–8 feet (1.8–2.4 m) long with numerous, regularly arranged leaflets growing in a single flat plane.

Pigafetta
pi-ga-FET-ta

Pigafetta consists of 2 tall solitary-trunked, spiny, pinnate-leaved, dioecious palms in Sulawesi, the Moluccas, and Papua Province of Indonesia, where they grow as pioneer species in clearings, at the edges of the mountainous rain forest, former lava flows, and along wide riverbanks, at elevations from sea level to 3000 feet (900 m). The 2 species are similar to each other from a distance and when studied from the preserved, fragmentary specimens found in museums. In 1998, John Dransfield clarified the taxonomic confusion surrounding the 2 species, distinguished, among other characters, by their geographical distribution.

There are no more beautiful or magnificent palms. The trunks reportedly attain heights of 120 feet (37 m) and more but are 18 inches (45 cm) in diameter, which gives old, tall individuals a grand elegance but imparts to younger ones a remarkably beautiful robustness. The unusually smooth stems are green in their upper half or even two-thirds, with widely spaced, distinct light gray rings of leaf base scars, and are light gray in their oldest parts. The leaf crowns are open in spite of the large number of leaves therein and are hemispherical or nearly spherical. The leaves are 20 feet (6 m) long, beautifully arching to nearly recurved, on stout petioles 6 feet (1.8 m) long with fat sheathing bases. These leaf sheaths vary from deep olive green to pale green or almost white and are covered with closely set rows or "combs" of dark gray or golden flexible spines 3 inches (8 cm) long. The leaflets are 3–4 feet (90–120 cm) long, evenly and closely spaced along the rachis, and grow from it at a slight angle that creates a shallow V-shaped blade. They are deep green on both surfaces, linear-lanceolate, long tipped and pendent at their apices, and armed with bristles on their midribs.

The male and female inflorescences are similar in general appearance and grow from among the leaf bases. They are 5–6 feet (1.5–1.8 m) long, and their main axis grows outwards at right angles to the trunk, with the flowering branches descending from it like a curtain; they resemble the great palmate-leaved palm of South America, *Mauritia*. The female trees produce tiny fruits generally less than ½ inch (13 mm) in diameter, creamy white when mature, and covered in overlapping, diamond-shaped scales.

Both species are famous for their rapid growth, which may be the fastest of all palms. After the seedling stage, they add 3 feet (90 cm) of trunk height per year. This growth rate may be an adaptation to colonizing open ground. They require full sun even as seedlings and copious, regular moisture. They survive in poor soil if supplied with enough water but come into their own only in a rich medium with humus; an organic mulch and a palm fertilizer with micronutrients help much, the latter especially important in the warmer parts of the year. *Pigafetta elata* (mistakenly named *P. filaris* at the time) reportedly withstood temperatures in the upper 20s Fahrenheit (around −2°C) with little damage in a protected site in Miami, Florida. It cannot be expected, however, to withstand such temperatures regularly as it is doubtless similar to the coconut in that it stops photosynthesizing when temperatures fall below 60°F (16°C). These palms are reportedly shallow rooted, and young plants are definitely top-heavy and need protection from high winds.

These palms are of incredible majesty and are among the few species that look wonderful planted alone as specimens, even surrounded by space. They look even better with other vegetation near or at their bases and, when planted in groups of 3 or more individuals of varying heights, the tableau is beyond thrilling. These palms are not for small gardens or intimate sites. The genus name, which is sometimes misspelled as "*Pigafettia*," honors Antonio Pigafetta (ca. 1491–ca. 1534), a comrade and biographer of the explorer Ferdinand Magellan. Seeds germinate within 30 days. For further details of seed germination, see *Calamus*.

Pigafetta elata
pi-ga-FET-ta · ee-LAH-ta
PLATES 696–697

Pigafetta elata is indigenous to Sulawesi from elevations of 1500 to 3000 feet (460–900 m). The epithet is Latin for "tall." The leaves of juvenile plants tend to be spreading. The leaf sheaths and leaf rachises are deep dark olive green and densely covered with closely set rows of spines, each spine 3 inches (8 cm) long and gold in young plants but nearly black in older individuals. The flowering branches of the inflorescences are deep brownish orange to cinnamon in color, and the fruits are globose or ovoid. This species is reportedly more tolerant of cold than *P. filaris*.

Pigafetta filaris
pi-ga-FET-ta · fi-LAR-is
PLATE 698

Pigafetta filaris occurs naturally in the Moluccas and Papua Province, Indonesia, from sea level to 1000 feet (300 m). The epithet is Latin for "threadlike." The leaves of young individuals tend to be ascending and erect. The leaf sheaths and leaf rachises have a chalky coating of wax and few spines compared to *P. elata*; the spines are golden in both juvenile and mature trees. The inflorescences are similar to those of *P. elata* but light yellow. The fruits are ovoid. The species is slightly less cold hardy than *P. elata*.

Pinanga
pi-NANG-ga

Pinanga includes 131 solitary-trunked and clustering, pinnate-leaved, monoecious palms in southwestern and northeastern India, Sri Lanka, Nepal, Bhutan, Myanmar, southern China, Vietnam, Laos, Thailand, Malaysia, Indonesia, the Philippines, and New Guinea. Although a few species are tall, most are small, undergrowth inhabitants of rain forest, and some are acaulescent. Most grow at relatively low elevations, but a few extend into the cloud forests of tropical mountains, always in wet environments.

Some species show mottled and often beautiful leaf variegation, especially when young. Some have entire leaves with the pinnate segments fused together, while others have highly segmented leaves. The elegantly slender, beautifully ringed trunks look a lot like bamboo culms. The smaller species usually form aerial roots on their stems and, like some *Chamaedorea* species, can be propagated by allowing these stems to root in soil and then severing them from the mother plant. Several species form stilt roots at the bases of their stems. Many of the larger species have prominent crownshafts that are often colorful and beneath which are formed the short inflorescences, although some of the smaller species have inflorescences that grow from among the leaf sheaths. A few species have spikelike inflorescences but most are branched once. All have separate male and female flowers on the same plant, and these blossoms are often ivory colored or shades of pink, red, and purple. The small fruits are usually red or black.

There are some truly rare species in the genus with natural ranges limited to a certain soil type. All can grow in semishady spots, and indeed, most prefer such sites; a few large species are adaptable to full sun when past the juvenile stage. The species with mottled or variegated leaves are generally more colorful in shade and poor soil. None of the species are drought tolerant, and some are true water lovers that, nevertheless, usually require a fast-draining soil. None are tolerant of cold. These palms are unaccountably neglected in American tropical and subtropical gardens. The genus name is a Latinized form of the Malay word for palms. Seeds of *Pinanga* germinate within 30 days. For further details of seed germination, see *Areca*.

Pinanga adangensis
pi-NANG-ga · ah'-dahn-GEN-sis
PLATE 699

Pinanga adangensis is a rare and endangered species indigenous to peninsular Thailand and peninsular Malaysia, where it occurs in the undergrowth of rain forest and along streams and rivers at low elevations. The epithet is Latin for "of Adang," a group of islands off the coast of Thailand, in the Straits of Malacca.

The stems of this clustering species grow to 18 feet (5.4 m) high with a diameter of slightly more than 1 inch (2.5 cm), and are dark green with prominent, widely spaced light brown rings of leaf base scars. The loose crownshafts are 2 feet (60 cm) tall, slightly thicker, if at all, than the stems, and light tan with purplish brown hues due to a short tomentum. The gracefully arching leaves are 6 feet (1.8 m) long on petioles 2 feet (60 cm) long. The regularly spaced, dark green linear-lanceolate leaflets are 18 inches (45 cm) long, the apical pair wider than the others. The teat-shaped fruits are purplish black when mature and ½ inch (13 mm) in diameter.

This species is more beautiful than most *Chamaedorea* species and makes an elegant house or greenhouse plant. As an undercanopy subject in the garden, it is superb lining paths or as an accent in other vegetation.

Pinanga aristata
pi-NANG-ga · ar-i-STAH-ta
PLATE 700

Pinanga aristata is endemic to Kalimantan (Indonesian Borneo), where it occurs in the undergrowth of mountainous rain forest at 500–3000 feet (150–900 m). The epithet is Latin for "having an awn or bristle," an allusion to the leaflets that taper into narrow, awnlike tips.

The stems of this sparsely clustering species grow slowly to 6 feet (1.8 m) high. A clump seems to always have a single dominant and taller stem, with subsidiary and much shorter ones at its base. The short crownshafts are scarcely thicker than the slender trunks and are grayish green. The leaves are apparently of 2 types—entire with a bifid apex, or segmented with broad leaflets—but always light yellowish green with heavy, deep green or purplish green mottling and always heavily ribbed. New growth is bluish green or purple greenish with dark purplish or green mottling. The little fruits are scarlet.

It is almost impossible to overwater this little beauty and it even tolerates waterlogged soil for short periods. It is intolerant, however, of full sunlight, especially in hot climates.

Pinanga batanensis
pi-NANG-ga · ba-tah-NEN-sis
PLATE 701

Pinanga batanensis is endemic to the island of Batan, Batanes Islands (of the Philippines) north of Luzon, where it occurs in low mountainous rain forest. The epithet is Latin for "of Batan." The solitary trunk grows to 20 feet (6 m) tall with a diameter of 4–5 inches (10–13 cm) and is light brown with widely spaced light gray rings of leaf base scars. The crownshaft is 2 feet (60 cm) tall, light silvery or grayish green, and slightly bulging at its base. The leaf crown is hemispherical to nearly spherical, with spreading, gracefully arching leaves 6–8 feet (1.8–2.4 m) long. The numerous and regularly spaced linear-lanceolate, soft, dark green leaflets are 2 feet (60 cm) long and grow from the rachis in a single flat plane. The little teat-shaped fruits are ½ inch (13 mm) in diameter and deep orange. This species is not colorful, but it is beautiful, looking like a cross between a large *Chamaedorea* species and *Howea forsteriana*.

Pinanga bicolana
pi-NANG-ga · bee-ko-LAH-na
PLATE 702

Pinanga bicolana is a solitary-trunked species endemic to the island of Luzon in the Philippines, where it occurs in lowland rain forest. The epithet is Latin for "of Bicol" and honors Bicol National Park, Luzon, Philippines.

The green trunk grows to 10 feet (3 m) tall with a diameter of 1½ inches (4 cm), and is deep green with widely spaced rings of leaf base scars. The cylindrical crownshaft is 2 feet (60 cm) tall, slightly thicker than the stem, and colored light green with a brownish red suffusion. The sparse leaf crown is hemispherical to nearly spherical, with 6 spreading leaves, each 3 feet (90 cm) long on a petiole 10 inches (25 cm) long. The leaflets of adult trees are unequally spaced along the rachis, are unequal in size, and are mostly obliquely ovate with S-shaped ribs and jagged apices; they are bluish to deep olive green above with mottling of lighter green and are grayish green beneath. The teat-shaped fruits are ½ inch (13 mm) in diameter and red, maturing to purplish black.

Pinanga caesia
pi-NANG-ga · SY-see-a
PLATES 703–704

Pinanga caesia is a solitary-trunked species endemic to Sulawesi, where it grows in low mountainous rain forest. The epithet is Latin for "lavender blue" and alludes to the striking color of the leaf rachis.

The trunk attains a height of 12 feet (3.6 m), a diameter of 3 inches (8 cm), and is colored deep olive green with light brown, undulating rings of leaf base scars. The plump crownshaft is 18 inches (45 cm) tall, bulging at its base, and light to dark orange-brown, shading to purple at its summit. The leaf crown is hemispherical or nearly so. The slightly arching leaves are 5–6 feet (1.5–1.8 m) long and bear numerous regularly spaced, yellowish or bluish green, linear, wedge-shaped leaflets with jagged tips; the leaflets grow from the rachis in a single, nearly flat plane and, as they age, become pendent at their tips. The new growth is a beautiful rusty- or orange-brown. The flowering branches of the inflorescences are light, bright pink at anthesis and turn to coral or scarlet as the little fruits also mature to deep red. This species is very colorful.

Pinanga chaiana
pi-NANG-ga · chy-AH-na
PLATE 705

Pinanga chaiana is endemic to low mountainous rain forest in Sabah. The epithet honors Paul Chai, a 20th-century forest botanist in Borneo. This distinctive little palm is solitary trunked, with a stem to 10 feet (3 m) high and slightly less than 1 inch (2.5 cm) in diameter. It is silvery green with widely spaced dark brown or reddish brown rings of leaf base scars. The narrow crownshaft is 2–3 feet (60–90 cm) tall, tapering from its purplish brown summit to its light pinkish tan base. The leaf crown is almost hemispherical. The ascending and spreading, stiff leaves are 3–4 feet (90–120 cm) long, obovate, entire and undivided, apically bifid, and emerald green, with deep pleats in the blade and a light yellowish green, stout midrib.

Pinanga copelandii
pi-NANG-ga · kop-LAN-dee-eye
PLATE 706

Pinanga copelandii is endemic to several islands in the Philippines, where it grows in lowland rain forest. The epithet honors Edwin B. Copeland (1873–1964), American botanist, agriculturist, and founder of the Philippines College of Agriculture (now the University of the Philippines) at Los Baños, Laguna. The species is similar in overall appearance to *P. maculata* but is much larger, its solitary trunk attaining a height of 25 feet (7.6 m) with a diameter of 4 inches (10 cm). Its leaves are not as colorful as *P. maculata* and what little mottling they show is green. It is nevertheless a beautiful species.

Pinanga coronata
pi-NANG-ga · kor-o-NAH-ta
PLATES 707–708

Pinanga coronata occurs naturally in the undergrowth of Indonesia's rain forest. The only common name seems to be ivory cane palm. The epithet is Latin for "crowned," an allusion to the shape of the leaf crown. The species was previously known as *P. kuhlii*.

The stems of this clustering species may grow to 20 feet (6 m) or sometimes more but are usually half this tall, with a maximum diameter of 2 inches (5 cm). They are light green (occasionally gray in the oldest parts) and are beautifully ringed with brown leaf base scars. The leaf crown is sparse and consists of usually no more than 6 leaves. A mature clump is 12 feet (3.6 m) tall and 8 feet (2.4 m) wide, although in habitat the palms may be approximately twice as large. The small crownshaft is less than 1 foot (30 cm) high, light green to nearly yellow, smooth, and of a slightly greater diameter than the trunk. The leaf bases often turn brown before the leaf falls, which makes the crownshaft a papery brown. The leaves are 4–5 feet (1.2–1.5 m) long but 3 feet (90 cm) wide and are held on petioles 1 foot (30 cm) long that are covered in small, cinnamon-colored scales. The leaflets grow from the rachis at a slight angle and are heavily veined, the veins depressed above, resulting in a grooved look. The leaves grow from the trunk almost erect but later are spreading, and the stout petioles are slightly arching. The shape of the leaflets varies from elliptic-lanceolate and pointed to wide-oblong and truncate with toothed apices, some individuals having only one or the other shape, others having mixed types on the same plant. Leaf color is light green to emerald green on both surfaces, new leaves a beautiful light pink in some clones. The inflorescences are 1 foot (30 cm) long, grow from beneath the crownshaft with small, pendent branches, and bear small, white flowers of both sexes; the branches turn a deep red as the fruits mature. The fruits are ovoid, ½ inch (13 mm) long, and jet black when ripe.

The species does not tolerate full sun, especially in hot climates. It needs constant, abundant moisture and a rich but fast-draining soil. It is also intolerant of cold and is adaptable only to zones 10b and 11, although there are nice specimens in warm microclimates or protected sites in 10a. It is comfortable only in partially shaded sites except in areas with mild summers. This palm is easily overwhelmed by other plant forms and needs to be planted where its graceful aspect can be seen—near companions with rounded or contrasting, nonpinnate foliage. It makes a beautiful small hedge or wall in semishady sites and is a near perfect accent in borders. It also is unsurpassed in intimate patios and courtyard plantings. This is a fine choice for growing indoors if given enough light (but not full sun) and water.

Pinanga crassipes
pi-NANG-ga · KRAS-si-peez
PLATES 709–710

Pinanga crassipes is a small palm from forests near Kuching, Sarawak, on the island of Borneo. The epithet is Latin for "thick foot" and refers to thick peduncle at the base of the inflorescence.

The trunk of this solitary dwarf palm is 2½ inches (6 cm) thick and grows to about 1 foot (30 cm) tall; it is closely ringed with leaf scars. The leaves of juveniles are bifid, with prominent raised veins, and are strikingly mottled light and dark green, making this palm one of the most beautiful in the genus—until the palm outgrows it spotty youth and produces evenly green or very faintly mottled leaves as an adult. Mature leaves bear 4 or 5 pairs of slightly S-shaped segments, with the terminal pair wider than the others. The inflorescence is borne below the crown, and is comprised of 5–8 rachillae, each 8–10 inches (20–25 cm) long, cascading from a short, stocky peduncle. The white flowers are followed by ellipsoid fruits, about ½ inch (13 mm) long, that turn from red to shiny black. As a seedling or juvenile, this palm is unparalleled. It demands shade and moisture.

Pinanga curranii
pi-NANG-ga · kur-RAN-ee-eye
PLATE 711

Pinanga curranii is endemic to the Philippines, where it grows in lowland rain forest. The epithet commemorates Hugh M. Curran (1875–1960), American forestry officer and collector in the Philippines.

The stems of this mostly clustering, occasionally solitary-trunked species grow to a height of 20–25 feet (6–7.6 m) with a diameter of 2 inches (5 cm). The crownshafts are 3 feet (90 cm) tall, slightly thicker than the trunk, cylindrical, and olive green. The open leaf crown is hemispherical with 6 spreading leaves 2½ feet (75 cm) long on petioles 8 inches (20 cm) long. The numerous leaflets grow from the rachis in a single flat plane but are of unequal widths and are spaced unequally along the rachis. They are generally linear-elliptic but are curved forwards into a falcate form. They are deep emerald green above but yellowish or grayish green beneath with jagged apices. The teat-shaped fruits are ½ inch (13 mm) in diameter and turn deep red when ripe. This palm has a beautiful silhouette with its large, open crown of deep green, stiff-leafleted leaves.

Pinanga densiflora
pi-NANG-ga · den-si-FLOR-a
PLATE 712

Pinanga densiflora is a densely clustering species endemic to low mountainous rain forest in Sumatra. The epithet is Latin for "densely flowered."

The stems grow to 12 feet (3.6 m) high but have a diameter of less than 1 inch (2.5 cm). The crownshafts are hardly thicker than the stems and are usually tan. The leaf crowns are not obvious unless the clumps are thinned. The leaves of mature plants are 4 feet (1.2 m) long on petioles 2 feet (60 cm) long, with a few broad leaflets of unequal sizes and shapes, the inner ones generally broadly elliptic with pointed tips, the apical ones much broader with truncated, jagged tips; all are hazily ribbed. Younger leaves are light green with heavy, dark mottling, and new growth is pinkish hued. Older leaves generally are deep green with little mottling. The young leaves are attractive and unusual, and the mature clumps are dense with a mix of young and old leaves. The small fruits are jet black when ripe.

Pinanga dicksonii
pi-NANG-ga · dik-SO-nee-eye
PLATE 713

Pinanga dicksonii is endemic to southern India in wet forests of the Western Ghat Mountains, growing at an elevation of 1000 feet (300 m). The epithet honors a Dr. Dickson, a surgeon in southern India in the 1830s.

The stolons of this clustering species may form stems some distance from each other as well as at the base of the mother plant. The stems grow to 25 feet (7.6 m) high with a diameter of 2 inches (5 cm) and are deep green with widely spaced light gray rings of leaf base scars. The crownshafts are 2 feet (60 cm) tall, nearly the same diameter as the trunks, and a rich gold. The hemispherical leaf crowns contain 6 stiffly spreading leaves 4 feet (1.2 m) long on short petioles. The numerous leaflets are 2 feet (60 cm) long, regularly spaced along the rachis, linear-lanceolate, and curved or falcate, a pure light green on both surfaces. In younger plants the leaflets are stiff and grow from the rachis at an angle to create a slightly V-shaped leaf, but in older plants they are more lax, even pendent, and grow in a single flat plane. The flowering branches of the inflorescences are golden when new but later turn coral colored with pink flowers that produce small, red fruits. This beautiful species is reportedly more tolerant to cold than most other *Pinanga* species.

Pinanga disticha
pi-NANG-ga · DIS-ti-ka
PLATES 714–715

Pinanga disticha is a small, clustering species indigenous to Sumatra and peninsular Malaysia, where it occurs in mountainous rain forest from sea level to 4000 feet (1200 m), often forming dense colonies. The epithet is Latin for "distichous" and alludes to the disposition of the fruits in 2 rows on either side of the infructescence. The stems grow to a maximum of 4 feet (1.2 m) high with a diameter of ¼ inch (6 mm). The leaf crowns are sparse, with deep green leaves that are 1 foot (30 cm) long and mostly obovate with deep apical notches, but in some plants have a few segments. The leaves are mottled to varying degrees with light green to almost white splotches in the best clones. The inflorescence is an unbranched spike, and the little fruits are red when ripe. This species, especially the more highly colored individuals, makes a stunning high groundcover in the shade of tropical or nearly tropical climates.

Pinanga geonomiformis
pi-NANG-ga · jee-o-no'-mi-FOR-mis
PLATE 716

Pinanga geonomiformis is native to the forests of Luzon, Philippines. The epithet means "in the form of *Geonoma*," a genus of tropical American palms, and refers to the leaf shape. This small, elegant species forms clusters of stems up to 4 feet (1.2 m) tall and

¼ inch (6 mm) in diameter. The leaves are about 16 inches (41 cm) long and have 2–4 pairs of S-shaped segments that are uneven and irregular in their width and placement along the rachis. Leaf color is dull green on both sides. The small, unbranched inflorescence bears white flowers, followed by slender, slightly curved fruits ½ inch (13 mm) long.

Pinanga gracilis
pi-NANG-ga · GRA-si-lis
PLATE 717

Pinanga gracilis is native to montane forests of Assam, Tibet, Bangladesh, and Myanmar, where it occurs at elevations up to 3500 feet (1070 m). The epithet is Latin for "slender." This species is sometimes called Himalayan pinanga. It has slender, clustering stems up to 15 feet (4.5 m) tall but only ½ inch (13 mm) in diameter. The stems have prominent leaf scars and resemble culms of bamboo. The leaves are up to 4 feet (1.2 m) long and 2 feet (60 cm) wide. The leaflets are few, the terminal pair being wider than the others. The inflorescence is unbranched. Fruits are in 3 rows and are bright scarlet when ripe. As this species occurs at higher elevations than most species, it is believed to tolerate cooler temperatures. It is, however, intolerant of frost. Like most *Pinanga* species, it needs shade and moisture.

Pinanga heterophylla
pi-NANG-ga · he′-ter-o-FIL-la
PLATES 718–719

Pinanga heterophylla is a colony-forming species from the Philippines, where it is found in forests on the island of Negros. The Latin epithet means "different leaf" and refers to the different sizes and irregular spacing of the leaflets within a leaf. The clustering stems grow to 8 feet (2.4 m) tall but are a slender ½ inch (13 mm) in diameter. The leaf crown is sparse, the leaves about 3 feet (90 cm) long, and evenly dark green. The flowers and fruits are spirally arranged on 4–6 rachillae. This species would make an elegant and effective screen or hedge in a shady, moist situation.

Pinanga insignis
pi-NANG-ga · in-SIG-nis
PLATE 720

Pinanga insignis is indigenous to several islands of the Philippines and the Palau islands to the east, where it occurs in low mountainous rain forest. The epithet is Latin for "remarkable."

The solitary trunk grows to 40 feet (12 m) high with a diameter of 8 inches (20 cm), making this species the tallest in the genus. The stem is deep olive green with light brown rings of leaf base scars. The crownshaft is 4 feet (1.2 m) tall, bulging at its base, and an even deeper olive green to brownish green. The leaf crown is hemispherical, with as many as a dozen stiff, spreading leaves, 10 feet (3 m) long, on short, almost nonexistent petioles. The deep green leaflets are 3 feet (90 cm) long and linear-lanceolate, growing in a single flat plane from the rachis. They have shortly bifid apices and are stiff and straight when new but gradually soften and become pendent. The flowering branches of the inflorescences are 2 feet (60 cm) long and produce pendent clusters of deep red, ovoid fruits.

This species is tender to cold and needs partial shade when young, especially in hot climates. It is magnificent in the landscape and has visual affinities with *Dictyosperma* and even *Archontophoenix*.

Pinanga isabelensis
pi-NANG-ga · ee′-sa-bel-EN-sis

Pinanga isabelensis is yet another delicate, understory *Pinanga* from the Philippines. This one, however, occurs in coastal forest, very close to the sea. The epithet means "of Isabela," a province of Luzon, where it was first collected. This small palm has slender stems about ¾ inch (19 mm) in diameter. It grows slowly and the distances between the leaf scars on the stem are only 1 inch (2.5 cm) or so. Leaves are about 3 feet (90 cm) long and carry 6 or 7 pairs of widely spaced leaflets. The leaflets are narrow, slightly S shaped, and have 3 or 4 main veins, which are pubescent on the lower surface. The basal and terminal segments are narrower than the segments from the middle of the leaf. The inflorescence has 5 or 6 branches, and bears flowers in 2 rows. The fruits are about ½ inch (13 mm) long.

Pinanga javana
pi-NANG-ga · ja-VAH-na
PLATES 721–722

Pinanga javana is endemic to mountainous rain forest of western Java, where it is in danger of extinction because of land clearing for agricultural expansion. The epithet is Latin for "of Java."

The mature stem of this solitary-trunked species grows to 30 or more feet (9 m) high and 4–5 inches (10–13 cm) in diameter. It is free of leaf bases and fiber, is light gray to almost white except for the youngest parts which may be green, and is wonderfully ringed with dark green leaf base scars. The leaf crown is 12 feet (3.6 m) wide and 6 feet (1.8 m) tall. The crownshaft is 3–4 feet (90–120 cm) tall, emerald to olive green, bulging at its base, and smooth and shiny. The leaf crown is open and hemispherical with no more than 10 leaves, each 4–6 feet (1.2–1.8 m) long and 4 feet (1.2 m) wide, slightly but gracefully arching on a short green petiole that is usually no more than 6 inches (15 cm) long. The leaflets are 2–3 feet (60–90 cm) long, uniformly shaped, and regularly spaced along the rachis, growing in a single flat plane. They are slightly limp and pendent, narrowly lanceolate and tapering, and a vibrant, glossy green on both sides. The 2 terminal leaflets are wider than the others. The inflorescences grow from beneath the large crownshaft and are 2 feet (60 cm) long with pendent yellowish green branches bearing small, yellowish white flowers of both sexes. They usually form a "skirt" around the trunk. The fruits are 1 inch (2.5 cm) wide, rounded, and deep orange.

This palm is a true water lover and must never suffer drought. It also needs a rich, humus-laden, well-drained soil but luxuriates in full sun when mature. It does not tolerate freezing temperatures but is well adapted to frostless Mediterranean climes if its moisture requirements can be met. It is one of the choicest, most beautiful palms in the genus but is not a candidate for planting as a single specimen isolated with space around itself. It looks especially nice with a background of darker green foliage and has an almost overwhelmingly attractive silhouette. It is well suited for a large patio or courtyard, especially if planted in groups, and it is among the finest canopy-scapes.

Pinanga limosa
pi-NANG-ga · lee-MO-sa
PLATE 723

Pinanga limosa is endemic to peninsular Malaysia, where it occurs in wet areas of lowland rain forest and in swamps. The epithet is Latin for "swampy." It is a small species that grows deep in the shady understory. This solitary-trunked palm rarely exceeds 6 feet (1.8 m) in height and is only about ½ inch (13 mm) thick. The leaves are deep green, about 18 inches (45 cm) long, and are either undivided, deeply cleft at the apex, and somewhat undulating, or divided into about 4 pairs of S-shaped segments that are unequal in size. The terminal pair is usually the largest. The inflorescence is a short spike or pair of rachillae that hangs down from below the crownshaft. The round ¼-inch (6-mm) fruits are shiny and black, borne on bright scarlet branchlets. No other small palm is more charming than this one. It needs deep shade and moist soil at all times, and it cannot tolerate cold.

Pinanga maculata
pi-NANG-ga · mak-yoo-LAH-ta
PLATES 724–725

Pinanga maculata is endemic to several islands in the Philippines, where it occurs in the undergrowth of dense rain forest and along riverbanks, from sea level to 5000 feet (1520 m). The epithet is Latin for "spotted" and alludes to the mottling of the leaves.

The solitary trunk attains a height of 25 feet (7.6 m) in habitat. It is dark green with widely spaced light gray rings of leaf base scars. The crownshaft is 18 inches (45 cm) tall, scarcely thicker than the trunk, and brownish purple to reddish brown. The leaf crown is hemispherical, and the individual, spreading leaves are 4 feet (1.2 m) long on petioles 1 foot (30 cm) long. The stiff and strongly ribbed leaflets are borne in 8–10 pairs and are broad, unequally sized, and unequally spaced on the rachis. The upper leaflets are deeply incised apically into 4–6 pointed lobes. Leaf color is deep green above, with a silvery hue beneath. The new growth in young plants is light yellowish green mottled with orange, pink, and red.

This species is one of the world's most beautiful small palms. Each individual is a little symphony of color and form and needs to be placed where its intimate beauty can be appreciated up close.

Pinanga malaiana
pi-NANG-ga · ma-lay-AH-na
PLATE 726

Pinanga malaiana is indigenous to mountainous rain forest in peninsular Thailand, Sumatra, peninsular Malaysia, and western Borneo, from sea level to 3000 feet (900 m). The epithet is Latin for "Malayan."

The solitary-trunked, rarely clustering species has a stem to 20 feet (6 m) tall and 1½ inches (4 cm) in diameter. The stems sits atop a short cone of stilt roots and is deep olive green with widely spaced reddish brown rings of leaf base scars. The crownshaft is 18–20 inches (45–50 cm) tall, light green overlain with golden and light to dark purplish hues. The open, nearly spherical leaf crown consists of 6 spreading, slightly arching leaves 6 feet (1.8 m) long on yellowish petioles 4 feet (1.2 m) long. The leaflets are deep emerald green, 3 feet (90 cm) long, regularly spaced along the rachis, and narrowly lanceolate and long tipped. The leaves resemble those of *Howea forsteriana*. The inflorescences are 1 foot (30 cm) long and turn coral red as the beaked fruits, 1 inch (2.5 cm) long, start out pinkish and ripen to a shiny blackish purple. This is the quintessentially tropical palm. It is outstanding in every way.

Pinanga modesta
pi-NANG-ga · mo-DES-ta
PLATE 727

Pinanga modesta is a small palm from the forests of Mindanao, Philippines. The epithet is Latin for "mild" or "restrained" and refers to the palm's delicate habit. The clustering stems can reach 6 feet (1.8 m) tall but are only about ½ inch (13 mm) in diameter, with leaf scars 1–2 inches (2.5–5 cm) apart. The crownshaft is light green overlain with pink. Each leaf is about 3 feet (90 cm) long with numerous slender, curved leaflets, light green and regularly arranged. The leaflets are finely pubescent on the underside. The fruits are ovoid, borne in 2 rows along the branches of the infructescence.

Pinanga negrosensis
pi-NANG-ga · neg-ro-SEN-sis
PLATE 728

Pinanga negrosensis is endemic to the Philippine island of Negros, where it grows in low mountainous rain forest. The epithet is Latin for "of Negros." The sparsely clustering trunks of this large species attain a height of 30 feet (9 m) with a diameter of 3 inches (8 cm). The stems are light green in their younger parts, nearly black in their older parts, and beautifully ringed with brown leaf base scars. The leaf crown looks like a feather duster or is, at most, hemispherical. The leaves are 6–7 feet (1.8–2.1 m) long and have elliptical leaflets 1 foot (30 cm) long that grow from the rachis at an angle to create a slightly V-shaped leaf. The fruits are ½ inch (13 mm) long and yellow, ripening to a deep red or shiny black. This species is definitely among the most beautiful in the genus.

Pinanga paradoxa
pi-NANG-ga · par-a-DAHK-sa
PLATE 729

Pinanga paradoxa is native to the forests of peninsular Malaysia. The Latin epithet means "paradox" and refers to the seeds which are unlike the seeds of other *Pinanga* species known at the time. This palm was once the basis for a new genus, but nowadays it is classified with other *Pinanga* species. It closely resembles some divided-leaf forms of *P. subintegra*, which are native to Thailand and the northern Malay Peninsula. The clustering stems grow to a maximum of 7 feet (2.1 m) high with a diameter of ¼ inch (6 mm). The leaf crowns are sparse with usually just 4 or 5 leaves, each 18 inches (45 cm) long. The leaves are deep green on the upper surface, gray-green on the underside, mostly obovate with deep apical notches. Some plants have leaves with up to 5 pairs of segments. The new leaf is pink. The inflorescence is an unbranched spike. The little fruits are elongate, first red, then black when ripe.

Pinanga patula
pi-NANG-ga · PAT-yoo-la
PLATES 730–731

Pinanga patula occurs naturally in Sumatra, where it occurs in mountainous rain forest from sea level to 4000 feet (1200 m). The epithet is Latin for "spreading," an allusion to the leaf crown. This variable species has both solitary-trunked and clustering forms. The trunk attains a height of 12 feet (3.6 m) with a diameter of slightly more than 1 inch (2.5 cm). The crownshaft is 2 feet (60 cm) tall, hardly thicker than the stem, and light orange to nearly white. The leaf crown is nearly spherical, with 10 spreading and gently arching leaves, each 3 feet (90 cm) long. The S-shaped, light to deep green leaflets are 18 inches (45 cm) long and may be broad or narrow but are always deeply ribbed and regularly and widely spaced along the rachis. The little fruits are brown, red, or black when ripe. This attractive species brightens up shady areas with its form and color.

Pinanga philippinensis
pi-NANG-ga · fil-li-pee-NEN-sis
PLATE 732

Pinanga philippinensis is endemic to a few islands in the Philippines, where it occurs in dense rain forest at low to moderate elevations. The epithet is Latin for "of Philippines." A synonym is *P. elmeri*.

The clustering stems grow to 8 feet (2.4 m) tall with a diameter of 1 inch (2.5 cm). They are olive green with widely spaced, dark green rings of leaf base scars. The crownshafts are 18–24 inches 45–60 cm) tall, bulging near their summits, and bluish gray to yellow. The leaf crowns are hemispherical with large, spreading, slightly arching leaves 6 feet (1.8 m) long on golden petioles 8 inches (20 cm) long. The numerous leaflets are deep green, narrowly linear, long tipped, and regularly spaced along the yellowish rachis. They grow in a single flat plane, and their long tips are slightly pendent. This species is one of the more beautiful in the genus because of its colors and form. It is delightful in any partially shaded site.

Pinanga rumphiana
pi-NANG-ga · rum-fee-AH-na
PLATE 733

Pinanga rumphiana is endemic to Papua New Guinea, where it occurs in low mountainous rain forest. The epithet honors G. E. Rumphius (ca. 1628–1702), a German-born botanist working in Ambon, Indonesia, whose work, *Herbarium Amboinense*, is one of the earliest, first-hand accounts of the Indonesian flora. This species was formerly known as *P. punicea*.

The solitary trunk grows to nearly 30 feet (9 m) high and 4–5 inches (10–13 cm) in diameter. It is light gray to light brown except for the youngest portion, which is green, and it is strongly ringed with darker leaf base scars. The crownshaft is 3 feet (90 cm) tall, slightly thicker than the trunk, cylindrical, and light reddish brown to almost red. The hemispherical leaf crown has 6 stiff, spreading leaves 8 feet (2.4 m) long with linear-lanceolate leaflets 2 feet (60 cm) long. The deep olive green leaflets grow from the rachis in a single flat plane, and the older ones are pendent. The fruits are brilliant scarlet. This species is among the finest and most beautiful in the genus, almost perfect in form and color. It is, alas, also one of the most tender to cold.

Pinanga salicifolia
pi-NANG-ga · sah-li'-si-FO-lee-a
PLATE 734

Pinanga salicifolia is known from Borneo. The epithet means "willow-leaf" and refers to the narrow, curved leaflets. The stem is solitary or clustering. The leaves have up to 12 pairs of regularly arranged, narrow, S-shaped leaflets borne on a gray- and brown-pubescent rachis. The leaflets have a single midvein, except for one of the lowermost, which bears 2 midveins, and the terminal leaflet pair, which has 2 or 3 midveins on each side. Leaf color is dark green, but lighter on the underside. The inflorescence is comprised of about 3 pendulous branches and bears pink flowers on red rachillae. The fruits are slender and curved. This palm is remarkably variable in its leaf form from seedling to adult, which means that it will bring interest and delight wherever it is grown.

Pinanga sclerophylla
pi-NANG-ga · skle-ro-FIL-la
PLATE 735

Pinanga sclerophylla is endemic to the Philippine island of Mindoro, where it grows in mountainous rain forest from elevations of 1000 to 5000 feet (300–1520 m). The epithet is from Greek words meaning "hard" and "leaf" and alludes to the stiffness of the leaves.

The robust, solitary trunk attains a height of 20 feet (6 m) and a diameter of 2–3 inches (5–8 cm). It is deep green and graced with prominent light gray rings of leaf base scars. The crownshaft is 3 feet (90 cm) tall, deep olive green to bluish green, smooth and mostly cylindrical, slightly thicker than the trunk. The leaf crown is shaped like a shuttlecock. The steeply ascending, dark green leaves are 6 feet (1.8 m) long with regularly spaced, linear, stiff leaflets 2 feet (60 cm) long growing in a flat plane from the rachis. The oblong fruits in large pendent masses are deep orange when ripe.

This species is one of the larger species in the genus. It is tolerant of full sun once it reaches adulthood, but adequate irrigation will always be necessary.

Pinanga scortechinii
pi-NANG-ga · skor-te-KEEN-ee-eye
PLATE 736

Pinanga scortechinii is indigenous to peninsular Thailand and peninsular Malaysia, where it grows in mountainous rain forest from elevations of 1000 to 3000 feet (300–900 m). The name honors Benedetto Scortechini (1845–1886), Italian priest and botanist who collected plants in Australia and Malaysia. *Pinanga fruticans* is a synonym.

The clustering stems attain a height of 15 feet (4.5 m) with a 1-inch (2.5-cm) diameter. They are deep olive green with widely spaced reddish brown to gray rings of leaf base scars. In some specimens, a single stem is dominant, and the other stems are suppressed. The crownshafts are 2 feet (60 cm) tall, plump, elongatedly elliptical, and a beautiful golden yellow to brownish orange. The leaf crowns are sparse and open, with mostly ascending leaves 6–7 feet (1.8–2.1 m) long on yellowish green petioles 2 feet (60 cm) long. The dull green leaflets are widely spaced along the rachis, 18 inches (45 cm) long, narrow, long tipped, and slightly S shaped, and grow from the rachis in a single flat plane. The leaflets are softer than those of *P. malaiana*, and on some individuals are lightly mottled. The inflorescences are short, erect, and golden. They produce congested, erect masses of fruits that are white when new, turning pink and later black when ripe.

Pinanga simplicifrons
pi-NANG-ga · sim-PLI-si-frahnz
PLATE 737

Pinanga simplicifrons is indigenous to peninsular Thailand, peninsular Malaysia, Singapore, Sumatra, and Borneo, where it grows in lowland rain forest near water and swamps, often forming colonies. The epithet is Latin for "simple frond," an allusion to the mostly undivided leaves.

The stems of this dwarf, clustering species sometimes attain a height of 4 feet (1.2 m) and are usually less than ½ inch (13 mm) in diameter. The crownshafts are rudimentary, loose, and ill defined but extended along the upper part of the stems. They are mostly light brown because the sheaths remain on the little trunks until they rot away. The leaves are 1 foot (30 cm) long, mostly unsegmented, linear-obovate, deeply bifid at the apex, the outer margins of the division toothed, glossy deep green, and strongly ribbed; a few plants have leaves with several variably shaped segments. The thin petioles are as long as the leaves, making the little leaf crown airy and open. The inflorescences and infructescences are unusual in that they are generally mature inside the persistent bracts and papery sheaths of the ill-defined crownshafts through which they eventually break through. The ellipsoid or globose fruits are ½ inch (13 mm) long and are red when ripe.

This strikingly beautiful palm is still rare in cultivation. Its fondness for wet habitats indicates it is a water-lover. The Malaysian populations with pinnately divided leaves have been separated out as *P. simplicifrons* var. *pinnata* [pin-NAH-ta].

Pinanga sobolifera
pi-NANG-ga · so-bo-LI-fe-ra
PLATE 738

Pinanga sobolifera is another delightful species from the Philippines, this one from the hill forests of Luzon. The epithet means "bearing an underground creeping stem."

This clustering species sends out long stolons or runners, allowing the palm to colonize a large area. The stems grow to about 7 feet (2.1 m) tall and up to 1 inch (2.5 cm) in diameter. The crownshaft is slightly swollen and 14 inches (36 cm) long. The leaves, borne on petioles 3 inches (8 cm) long, have 10–13 pairs of leaflets, each of which tapers to a narrow point. The terminal pair of segments is wider than the others. The inflorescence has 8–11 rachillae with spirally arranged flowers. The ellipsoid fruits, also spirally arranged, are ½ inch (13 mm) long and ripen from green through pink to shiny black.

This species is similar to *P. heterophylla* but differs in producing long runners. It would make a lovely, informal, semitransparent screen in a shady area of the garden. Because of its tendency to run, this palm is not suitable for gardens with limited space.

Pinanga speciosa
pi-NANG-ga · spee-see-O-sa
PLATE 739

Pinanga speciosa is endemic to the island of Mindanao in the Philippines, where it grows in mountainous rain forest at 1200–4000 feet (370–1200 m). The epithet is Latin for "beautiful."

This large, solitary-trunked species spends its youth as an undergrowth subject but usually emerges above the canopy as an adult. The trunk attains a height of 30 feet (9 m) and a diameter of 4–5 inches (10–13 cm), is deep bluish green, and has widely spaced, whitish rings of leaf base scars. The crownshaft is 4–5 feet (1.2–1.5 m) tall, slightly thicker than the trunk at the base of the shaft, tapers up to the leaf crown, and is an even deeper bluish green, almost purple. The leaf crown is open, sparse, and hemispherical to nearly spherical. The leaves are 8 feet (2.4 m) long on light-colored petioles 2–3 feet (60–90 cm) long. The stiff leaflets grow from the rachis in a single flat plane and are 3 or more feet (90 cm) long, linear-lanceolate, long tipped, and deep green on both surfaces. The small fruits are blackish purple and formed in 2 rows along pendent branches.

This species is certainly among the more magnificent species in the genus and is thrilling in groups of 3 or more individuals of varying heights. It is capable of adapting to full sun once past the juvenile stage, except in hot, dry climates.

Pinanga sylvestris
pi-NANG-ga · sil-VES-tris

Pinanga sylvestris hails from Vietnam, Cambodia, Laos, and Thailand. It occurs in mountain forests above 3000 feet (900 m). The Latin epithet means "of the forest." It is sometimes listed under its synonym *P. cochinchinensis*.

The stems grow to a height of 15 feet (4.5 m) and a diameter of 2 inches (5 cm). They are deep green with a rusty brown, scaly covering on the youngest parts. The leaf scars are widely spaced light gray rings. The elongated, narrow, pale yellow to almost white crownshafts are 3 feet (90 cm) long and scarcely thicker than the trunks. The spreading, slightly arching leaves are 6 feet (1.8 m) long and bear numerous linear-lanceolate, limp and slightly pendent leaflets that are 2 feet (60 cm) long, emerald green above and pale silvery green beneath. The leaves in some clones are bronze-colored as they emerge, later turning green.

This species is stunning in its beauty, and its tiers of foliage are attractive in almost any site. It withstands more sun than most undergrowth species.

Pinanga urosperma
pi-NANG-ga · yoo-ro-SPUR-ma
PLATE 740

Pinanga urosperma is from the Babuyan Islands archipelago, north of Luzon island, in the Philippines. It occurs in forests on steep slopes. The epithet combines the Greek words for "appendage" and "seed," an allusion to the narrow, tail-like base of the endocarp and seed.

This species is one of the larger ones in the genus. The trunk is up to 15 feet (4.5 m) tall, 6 inches (15 cm) in diameter, and green with prominent leaf scars about every 4 inches (10 cm). The leaves are 5 feet (1.5 m) long, regularly and evenly divided into narrow pinnae 20–22 inches (50–56 cm) long. The pinnae are slightly silvery on the underside. The black ellipsoid fruits are arrayed in 2 rows on either side of the rachillae. The seed is teardrop shaped, with the narrow part at the base.

Still rare in cultivation, this magnificent palm deserves wider use in gardens. It needs shade in its youth but can take full sun as an adult as long as adequate moisture is available.

Pinanga veitchii
pi-NANG-ga · VEECH-ee-eye
PLATE 741

Pinanga veitchii is endemic to the lowland rain forest in Sarawak. The epithet honors James Veitch, Jr. (1815–1869), British nurseryman whose nursery at Chelsea was one of the epicenters of palm introduction and horticulture in 19th-century England. Mature plants are similar to *P. simplicifrons*, but young individuals have distinctive, almost startling leaf color: dark green blades mottled with varying amounts and shades of brown, pink, yellow, dark green, and even purple, the blades sometimes almost entirely of one color. The species is reportedly slow growing and difficult to make happy, even in warm, moist, tropical climates. It requires shade or partial shade at all stages of life and a wet, humus-rich soil.

Pinanga watanaiana
pi-NANG-ga · wa'-tan-ny-AH-na
PLATE 742

Pinanga watanaiana is found only on the island of Phuket, Thailand, where it is a rare palm of the hill forest above 450 feet (140 m). The epithet honors Watana Sumawong, palm enthusiast and collector of Bangkok, Thailand.

This extremely beautiful palm has greenish brown clustering stems that are prominently ringed like bamboo culms. The stems are slender, less than 1 inch (2.5 cm) in diameter, but reach heights of 8 feet (2.4 m). The leaves are few per stem, but what splendid leaves they are: regularly divided into narrow segments, except for the broad terminal pair. All of the segments are dark green, lighter on the underside, and prominently mottled with light and dark spots on the upper side. The crownshaft is yellow. The rachillae are pinkish red, and the fruits ripen from green through red to black. Like most *Pinanga* species, this extraordinary palm requires shade or partial shade at all stages of life and a wet, humus-rich soil.

Plectocomia
plek'-to-KO-mee-a

Plectocomia comprises about 16 clustering and solitary-trunked, pinnate-leaved, climbing, dioecious palms in northern India, Bhutan, Myanmar, southern China including the island of Hainan, Laos, Vietnam, Thailand, Sumatra, peninsular Malaysia, Java, Borneo, and the southern islands of the Philippines. These are, for the most part, large, thick-stemmed, spiny palms, which ascend the heights by cirri that may be half again as long as the long leaves. The stems die after flowering.

The leaves have mostly widely spaced, linear-elliptic, pendent leaflets on thorny rachises. The leaf sheaths covering the upper portion of the stem are always spiny, sometimes incredibly so. The inflorescences are formed from the topmost nodes or leaf axils of the far-ranging stems and are long, stiff primary branches growing at right angles from the vegetative stems. They bear long curtains of pendent flowering branches with remarkable, distichous, boat-shaped bracts enclosing either male or female flowers. The fruits are usually oblong, brown or red, always covered in overlapping scales that have small, spinelike apices. The great, pendent infructescences are often spectacular, especially in the red-fruited species.

Unlike many other rattan species, these have no economic importance as the thick stems are pithy and shrink upon drying, but like most other rattans, their leafy portions are often so far up in the trees that viewing them is difficult. Since they are most attractive when they can be seen, the clustering ones can be pruned to stay within reasonable bounds. Most species are tropical in their temperature requirements, but a few from montane regions, like *P. himalayana*, have some hardiness to cold; this species seems to be the only one in cultivation and it is sought after because of its reputed hardiness.

The genus name is from 2 Greek words meaning "plaited hair," an allusion to the form of the inflorescences with their overlapping bracts. Seeds are not generally available, so germination information is nonexistent.

Plectocomia elongata
plek'-to-KO-mee-a · ee-lahn-GAH-ta
PLATE 743

Plectocomia elongata is the most widespread species in the genus, occurring from Vietnam, Thailand, peninsular Malaysia, Sumatra, Java, Borneo, and the Philippines, where it grows in monsoonal, mountainous rain forest at elevations to 6500 feet (1980 m). It typically inhabits disturbed edges of forest, on poor soil. The epithet is Latin for "elongate," in reference to the long stem. *Plectocomia griffithii* is a synonym. The stems are 150 feet (46 m) long and 8 inches (20 cm) in diameter. The gently arching leaves are 18 feet (5.4 m) long. The rachis extends 8 feet (2.4 m) into a cirrus. The curtainlike pendent leaflets are 2 feet (60 cm) long, dark green above and grayish or even greenish white beneath. The red fruits are highly ornamental. Two varieties are recognized: *P. elongata* var. *elongata*, which is solitary, and *P. elongata* var. *philippinensis* [fi-li-pee-NEN-sis], which is clustering.

Plectocomia himalayana
plek'-to-KO-mee-a · him-a-lay-YAH-na
PLATES 744–745

Plectocomia himalayana is endemic to northeastern India (Bengal), where it grows in the Himalayan foothills at an elevation of 7500–8500 feet (2290–2590 m). The epithet is Latin for "Himalayan." The stems are 90 feet (27 m) long. The stiff leaves are 6 feet (1.8 m) long and bear widely spaced whorls of light green, elliptic, and pendent leaflets. Although the species epithet evokes images of snow-covered Mt. Everest, growers report that this palm is hardy only to the upper or, at most, mid-20s Fahrenheit (around −2 to −4°C). It is doubtless a wise choice for a rattan in cool but nearly frostless Mediterranean climates—it thrives in the San Francisco Botanical Garden—and may do well in warm, tropical or nearly tropical climes.

Plectocomiopsis
plek-to-ko'-mee-AHP-sis

Plectocomiopsis is composed of 5 pinnate-leaved, spiny, climbing, dioecious palms in peninsular Thailand, peninsular Malaysia, Myanmar, Cambodia, Laos, Sumatra, and Borneo. The genus name translates as "similar to *Plectocomia*," the two genera being alike except for their inflorescences. *Plectocomiopsis* stems are as long as those of *Plectocomia* but not usually as thick, and the leaflets are generally not pendent as are those of *Plectocomia* species. The plants climb by cirri. The inflorescences grow from the topmost nodes of a stem with one thick, upright main axis and 2 pendent secondary branches bearing unisexual flowers. Female plants produce fig-shaped, scaly fruits on stems that die off after flowering. None of the species seems to be in cultivation. Seeds are not generally available, so germination information is nonexistent.

Plectocomiopsis corneri
plek-to-ko'-mee-AHP-sis · KOR-ner-eye
PLATE 746

Plectocomiopsis corneri occurs in Sumatra and peninsular Malaysia, where it grows in lowland rain forests and swamps. The epithet honors British botanist E. J. H. Corner (1906–1996), former assistant director of the Singapore Botanic Gardens and author of *The Natural History of Palms*. The stems of this high-climbing rattan are almost an inch (2 cm) in diameter. The leaf sheaths have few or no spines, and the part above the petiole (the ocrea) is golden yellow. The leaflets are narrow, bright shining green, with acuminate and "hooded" or "cupped" tips. This distinctive palm is not known to be in cultivation, but it is especially handsome and would be worth the effort to accommodate it wherever conditions are tropical.

Podococcus
po-do-KAH-kus

Podococcus includes 2 species of clustering, pinnate-leaved, monoecious palms in tropical western Africa. The genus name is derived from Greek words meaning "foot" and "berry" and refers to the fruits on short stalks. These diminutive palms are attractive because of the color and leaflet form but are not known in cultivation outside of a few botanical gardens. Both species are reportedly slow growing and need copious, regular moisture. They are doubtless tender to cold. Seeds are not generally available, so germination information is nonexistent.

Pogonotium
po-go-NO-tee-um

Pogonotium consists of 3 very spiny, pinnate-leaved, dioecious palms in low mountainous rain forest of peninsular Malaysia and Sarawak on the island of Borneo. They are closely related to the rattans (*Calamus*, *Ceratolobus*, and *Daemonorops*) but do not climb, and they lack both cirri and flagella. The short stems are almost completely covered in the spiny leaf sheaths, and there are spines on the petioles and leaf rachises as well as short bristles on the leaflets. The sheath spines are long, thin, and hairlike. The most unusual characteristic of the genus is the unisexual inflorescence, which is spikelike, erect, and, at first, enclosed within 2 long, narrow, spiny, earlike extensions (auricles) of the bases of the petioles. The small fruits are half hidden in the persistent auricles, are red, purplish, or brown, and are covered in small scales. The genus is not known to be in cultivation. Its name is from a Greek word meaning "bearded little ear" and alludes to the hairlike spines on the auricles. Seeds are not generally available, so germination information is nonexistent.

Ponapea
po-na-PAY-a

Ponapea comprises 3 species of medium-sized, solitary-trunked, pinnate-leaved, monoecious palms closely related to *Ptychosperma*. The genus was described in the early 20th century but was for many years synonymized under *Ptychosperma*. Recent molecular and morphological evidence allows the genus to be recognized once again. The species are native to Palau, Pohnpei, and Kosrae, which are parts of the Caroline Islands in the western Pacific Ocean, north of New Guinea. The genus gets its name from the island of Pohnpei, formerly known as Ponape. Seeds germinate within 60 days. For further details of seed germination, see *Areca*.

Ponapea hosinoi
po-na-PAY-a · ho-si-NO-ee
PLATE 747

Ponapea hosinoi is endemic to the island of Pohnpei, where it grows in low mountainous rain forest. The epithet honors an early 20th-century Japanese agronomist, Hoshino Shutaro, who introduced hundreds of useful crops to Japan.

The stem attains a height of 50 or more feet (15 m) and a diameter of 8 inches (20 cm), and is light green in its younger parts, where it is also faintly but beautifully ringed, and light gray to tan in its older parts. The crownshaft is 2–3 feet (60–90 cm) tall, cylindrical, slightly thicker than the trunk, and a silvery olive green. The leaf crown is hemispherical or slightly more so, with 10–12 beautifully arching leaves 10 feet (3 m) long. The stiff leaflets are regularly spaced along the rachis and grow from it at an angle that creates a V-shaped blade. The leaflets are linearly to widely wedge shaped, 40 inches (102 cm) long and 10 inches (25 cm) wide, and have mostly obliquely truncated apices that are always jagged. The fruits are 1 inch (2.5 cm) long and deep red when mature. This species is one of the world's most beautiful palms, but it is tender to cold and needs copious, regular moisture as well as high humidity; its enemies are cold, drying winds.

Ponapea ledermanniana
po-na-PAY-a · lee'-der-man-nee-AH-na
PLATE 748

Ponapea ledermanniana is endemic to the island of Pohnpei, where it grows in low mountainous rain forest. The epithet honors Carl L. Ledermann (1875–1958), Swiss horticulturist and botanist who collected plants in West Africa and New Guinea.

The trunk grows to 60 feet (18 m) but is usually half that height, with a diameter of 4 inches (10 cm). The crownshaft is 18 inches (45 cm) tall, bulging at its base, and light olive green to silvery green. The sparse, open leaf crown is hemispherical or slightly less so. The leaves are 10 feet (3 m) long on petioles 1 foot (30 cm) long, and are beautifully arching but do not usually descend beneath the horizontal. The stiff leaflets are at least 1 foot (30 cm) long and regularly and closely spaced along the rachis, growing from it at an angle that creates a V-shaped blade; their shape is generally a narrow wedge, with an obliquely truncated, jagged apex. The spindle-shaped fruits are more than 1 inch (2.5 cm) long and deep red when ripe.

This attractive species is stout and robust appearing but also has an elegance and refinement that is nearly formal. A group of 3 or more individuals of varying heights is one of the handsomest pictures of the tropical world.

Prestoea
pres-TO-ee-a

Prestoea is composed of 10 pinnate-leaved, monoecious palms in tropical America. The genus is closely related to *Euterpe* but is distinguished by not forming crownshafts or forming only partial (false) crownshafts and in having broader, less pendent leaflets. A few *Prestoea* species are small understory palms with undivided leaves. Most are clustering to some extent and are of only moderate height except for one salient exception. The inflorescences grow from among the leaf base sheaths and are similar to those of *Euterpe* but are not hirsute. They are white when new but change to pink or red as the fruits mature; those of *Euterpe* never change color. The fruits are blackish purple, and the endosperm is ruminate.

Few of these palms seem to be in cultivation, and this is a shame considering their great beauty. None of them are tolerant of frost, but those from montane habitats are adaptable to frost-free Mediterranean climates. All of them require copious moisture and a humus-laden soil. They luxuriate in partial shade, and some are adaptable to full sun except in hot, dry climates. None are tolerant of calcareous soils and usually fail, even with an organic mulch.

The genus name honors Henry Prestoe (1842–1923), a British botanist and horticulturist who was superintendent of the botanic gardens in Trinidad. Seeds germinate within 60 days.

Prestoea acuminata
pres-TO-ee-a · a-kyoo'-mi-NAH-ta
PLATES 749–750

Prestoea acuminata is a variable species occurring naturally over a vast area of the American tropics, from eastern Cuba, southwards through the entire Antilles except for Jamaica and Trinidad, into central Guatemala, northern Nicaragua, through all of Costa Rica and Panama, and into northern Venezuela, and the Andean slopes of Colombia, Ecuador, and Peru, growing from elevations of 1200 to 8000 feet (370–2440 m). It is most abundant in the mountains of Puerto Rico, where it forms vast colonies on the slopes. The epithet is Latin for "acuminate" and refers to the tips of the leaflets.

This palm is found both as a clustering and a solitary-trunked species. The stems grow to maximum height of 50 feet (15 m) with a maximum diameter of 8 inches (20 cm). They are light brown, with distinct and widely spaced olive green rings of leaf base scars in the newer parts of the trunks. The crownshafts are 2–3 feet (60–90 cm) high, cylindrical to a narrow inverted pyramid, and deep green to purplish green or even violet. The leaf crowns are hemispherical or nearly hemispherical and hold 4–10 spreading, gently arching leaves 4–8 feet (1.2–2.4 m) long on petioles 1 foot (30 cm) long. The leaf often exhibits a twist of the rachis at its midpoint, the apical part of the blade being thus oriented vertically. The numerous leaflets are 2–3 feet (60–90 cm) long, regularly spaced along the rachis in a single flat plane, light to deep green, and broadly lanceolate to linear-lanceolate. They are mostly spreading but are pendent in the older leaves.

Three naturally occurring varieties are distinguished by inflorescence, fruit details, and geographical origin that do not seem important to horticultural or landscaping considerations; ironically *P. acuminata* var. *montana* [mon-TA-na], the mountain cabbage palm of Puerto Rico, actually occurs naturally at lower elevations than the other 2 varieties and is usually single stemmed. It is usually found colonizing former landslides and other disturbed areas.

This beautiful species is adapted only to frost-free regions that have nights cooler than those of the lowland tropics, such as are found in South Florida. It languishes in calcareous soils, even when mulched. It needs partial shade and protection from the midday sun when young, wherever it is grown.

Prestoea carderi
pres-TO-ee-a · KAR-der-eye

Prestoea carderi is a highly variable species from montane rain forests of Colombia, Venezuela, Ecuador, and Peru, at 3000–6000 feet (900–1800 m). It was named for John Carder (?–1908), an experienced collector of orchids in Latin America for the nursery of William Bull of Chelsea. Several names are now considered synonyms of this species, including *P. simplicifrons*, *P. latisecta*, *P. brachyclada*, and *P. humilis*.

This palm usually has clustering stems with one or two of them dominant, but occasionally it appears solitary, in which case the slender stem is about 1½ inches (4 cm) in diameter and may grow to 12 feet (3.6 m). The sparse crown of leaves is made up of regularly and evenly pinnate leaves up to 6 feet (1.8 m) long. Occasionally the leaves are undivided, and if so, then much shorter. The leaf sheath is open and does not form a distinct crownshaft. The evenly green leaflets are held horizontally from the rachis and are up to 20 inches (50 cm) long. The inflorescence is erect, held just above the crown, but arches over when laden with fruits. The rachillae are covered with rich, reddish brown felt. The flowers are minute. The round, shiny, purplish black fruits are less than ½ inch (13 mm) in diameter.

This species has been in cultivation in England since the 1870s, and the past 140 years have not diminished this palm's abundant charms. The forms from Colombia and Peru with undivided leaves are especially choice. This palm makes a fine conservatory plant, given adequate humidity, warmth, and shade. Despite its montane habitat, this species is not likely to be hardy outside zones 10 and 11.

Prestoea decurrens
pres-TO-ee-a · dee-CUR-renz
PLATE 751

Prestoea decurrens is indigenous to southern Nicaragua, Costa Rica, Panama, and the western slopes of the Andes in Colombia and Ecuador, where it grows in rain forest, especially along streams and rivers from sea level to 5000 feet (1520 m) but mostly below 3000 feet (900 m). The epithet is Latin for "running down," in this case referring to the persistent leaf bases that create an elongated but loose false crownshaft.

The clustering stems attain a height of 25 feet (7.6 m) but are usually shorter, and their maximum diameter is 4 inches (10 cm). They are green in all but their oldest parts and are graced with grayish rings of leaf base scars. The open leaf crown is more-or-less hemispherical and usually holds 6 leaves, each 6 feet (1.8 m) long on a petiole 1–2 feet (30–60 cm) long. The numerous, light green leaflets are 2 feet (60 cm) long, grow in a single flat plane from the rachis, and are neither stiff nor pendent.

Although rare in cultivation, this especially beautiful species has the overall look of one of the giant tree ferns with which it grows in Costa Rica. It has a handsome clumping habit, which usually includes one, possibly two, dominant stems, the others subsidiary and much shorter.

Pritchardia
prit-CHAHR-dee-a

Pritchardia includes about 27 solitary-trunked, palmate-leaved, hermaphroditic palms in the islands of Hawaii and the South Pacific. It is the only genus of palm native to Hawaii, where all but 3 species are endemic. Many species are now critically endangered. Most species grow in wet tropical rain forest above 1000 feet (300 m), but a few are naturally found in drier areas, always at or near natural springs or water seepages.

The leaves are large and stiff, and the blade usually much pleated and costapalmate. Many species have wide, nearly flat leaves, but others have strongly undulate leaves. The segments are united for much of their length, giving the leaves a grand appearance akin to leaves of the larger *Licuala* species with undivided blades. The petioles are invariably long, stout, and unarmed, and are often covered in a dense, almost feltlike, tan or silvery tomentum, as are the sheaths. The inflorescences, which sometimes extend beyond the leaf crown, consist of a tubular series of bracts from which the stiff and short flowering branches project, bearing white, yellow, or orange flowers. The fruits are large, mostly spherical, and greenish brown or black when ripe; a few are elliptical or ovoid.

Most species can stand neither cold nor hot drying winds. All are intolerant of freezing temperatures and adapted to zones 10b and 11. Some of them grow near the seashore and can withstand salt-laden winds and slightly saline soil. All are susceptible to the lethal yellowing disease.

These are some of the world's most beautiful palm species and should find a place in any tropical or subtropical garden that is large enough to accommodate them and in a region not affected by lethal yellowing. The genus name honors William T. Pritchard (1829–1907), British consul to Fiji. Donald Hodel (2007, in press) has provided an excellent framework for appreciating this genus.

Seeds of *Pritchardia* germinate within 30 days and should not be allowed to dry out before sowing. For further details of seed germination, see *Corypha*.

Pritchardia arecina
prit-CHAHR-dee-a · ar-e-SEE-na
PLATE 752

Pritchardia arecina is endemic to the Hawaiian island of Maui, where it grows in wet forest at 1400–4000 feet (430–1200 m). The epithet is Latin for "little *Areca*" because the fruits are about the same general shape as those of the betel nut, *A. catechu*. The trunk attains a height of 40 feet (12 m) and a diameter of 10–20 inches (25–50 cm). The leaf crown is hemispherical, with large, stiff, wedge-shaped leaves 3 feet (90 cm) long on petioles that are 4 feet (1.2 m) long and have fine fibers at their bases. The inflorescences are as long as the petioles, but they elongate as the fruit develop so that they nearly equal the leaf blades. The rachillae are covered in reddish brown, feltlike or woolly hairs. The black or brown ovoid fruits are 1½ inches (4 cm) wide and almost 2 inches (5 cm) long.

Pritchardia beccariana
prit-CHAHR-dee-a · bek-kahr´-ee-AH-na
PLATE 753

Pritchardia beccariana is endemic to mountainous rain forest on the island of Hawaii, where it grows to 4200 feet (1280 m). The epithet honors the great Italian naturalist and palm specialist Odoardo Beccari (1843–1920).

Mature trunks can grow to 65 feet (20 m) but are usually no more than 40 feet (12 m) tall, with a diameter of 8–10 inches (20–25 cm). The large leaf crown is full and rounded because of the adhering dead leaves and is 15 feet (4.5 m) wide and tall. The semicircular to circular leaves are about 6 feet (1.8 m) in diameter on petioles 6 feet (1.8 m) long; the blade is nearly flat. The stiff segments are united for three-fourths of their length, and their tips do not droop. Leaf color is bright, clear green on both sides of the blade, and a light dusting of tan scales covers the underside. The many-branched inflorescences are 5 feet (1.5 m) long and bear small, yellowish flowers. The glossy black ellipsoid or spherical fruits are 1½ inches (4 cm) long and about as wide.

This palm thrives in partial shade to full sun except in the hottest climates where it needs protection from the midday sun. It is a water lover and must not suffer drought conditions; it also needs a humus-rich, well-drained soil. It is one of the tallest species in the genus, and old palms are noble in appearance and make wonderful canopy-scapes. As is true for most other *Pritchardia* species, the younger plants are incredibly attractive up close because of the near perfection of the heavy leaves.

Pritchardia glabrata
prit-CHAHR-dee-a · glah-BRAH-ta
PLATE 754

Pritchardia glabrata is endemic to mountains on the Hawaiian islands of Lanai and Maui, where it grows on very steep slopes and ridges at 2000–3000 feet (610–900 m). It is endangered. The epithet is Latin for "becoming hairless." This species was formerly known as *P. lanaiensis* and as a variant of *P. remota*. The stout trunk slowly grows to a maximum height of 15 feet (4.5 m) and is 8–12 inches (30 cm) in diameter. The leaf crown is nearly spherical. The undulate leaves are 3 feet (90 cm) wide on stout petioles less than 3 feet (90 cm) long and are colored medium green on both surfaces. The segments are united for half their length, and their tips are either stiff or drooping. The flower stalks are about as long as the petioles. The shiny black fruits are either spherical or slightly elliptical and just over 1 inch (2.5 cm) long.

Pritchardia hardyi
prit-CHAHR-dee-a · HAHR-dee-eye
PLATE 755

Pritchardia hardyi is endemic to the Hawaiian island of Kauai, where it grows in wet rain forest at 1600–2300 feet (490–700 m). The epithet commemorates the original discoverer, W. V. Hardy, field assistant for the Division of Hydrography in Hawaii in the early 20th century. The trunk attains a height of 30 or more feet (9 m) and is about 1 foot (30 cm) in diameter. The leaf crown is spherical or nearly so, and the stiff leaves are large, flat semicircles 3 feet (90 cm) long, glossy green on the upper surface and silvery gray to nearly white below. The inflorescences are as long or longer than the crown. The rachillae are densely covered with reddish brown felt. The black fruits are about 1½ inches (4 cm) long and about ½ inch (13 mm) wide. This is among the finest-looking palms in the genus with its large, flat deep green leaves. It is fast growing.

Pritchardia hillebrandii
prit-CHAHR-dee-a · hil-le-BRAN-dee-eye
PLATES 756–757

Pritchardia hillebrandii has long been known from cultivated plants in Hawaii. It is found on spectacular rocky islets off the north coast of Molokai and was probably once native to the mainland of Molokai as well. The epithet honors William Hillebrand (1821–1886), a German physician to the Hawaiian royal family and author of an early account of the Hawaiian flora; his property and plantings later became the Foster Botanical Garden in Honolulu.

The trunk attains a height of about 20 feet (6 m) and a diameter of 10 inches (25 cm). The spherical leaf crown has undulate leaves 4 feet (1.2 m) long on equally long petioles. The long-tipped segments have stiff or pendent tips. The leaves are grayish green on both surfaces, but in some individuals, they are almost white with wax. The inflorescences are about as long as the petioles. The spherical 1-inch (2.5-cm) fruits are shiny brown or black.

Judging from its natural habitat, we can presume that this palm is tolerant of salt spray. The species is slow growing but well worth growing for its silvery leaves. A well-known specimen in the Ho'omaluhia Botanical Garden is a gorgeous silvery blue.

Pritchardia kaalae
prit-CHAHR-dee-a · kah-AH-lee
PLATE 758

Pritchardia kaalae is a rare and endangered species endemic to Waianae Mountains of western Oahu in the Hawaiian chain, where it grows in moist to dry forests at elevations up to 2500 feet (760 m). The epithet is Latin for "of Kaala," a mountain in the palm's habitat. The trunk of this slow-growing smaller species attains a height of about 25 feet (7.6 m) and a diameter of 6–10 inches (15–25 cm). The leaf crown is hemispherical or slightly more so, and the leaves are slightly undulate, deep green on both surfaces, with long-tipped segments whose apices are pendulous. The inflorescences extend well beyond the leaf crown. The spherical 1-inch (2.5-cm) fruits are black.

Pritchardia lanigera
prit-CHAHR-dee-a · la-NI-je-ra
PLATE 759

Pritchardia lanigera is endemic to the island of Hawaii, where it grows in wet forest at elevations from 1500 to 4300 feet (460–1310 m). The epithet is Latin for "wool-bearing." Two species are included in the synonymy of this species, *P. eriostachys* and *P. montis-kea*. This small palm grows 20–45 feet (6–14 m) tall with a trunk diameter of 6–20 inches (15–50 cm). Its beauty comes from its large, emerald green leaves, 4 or more feet (1.2 m) wide, which are perfect semicircles and borne on stout petioles 3 feet (90 cm) long. The segments are united for half their length and have stiff tips. The inflorescences are as long as or a bit longer than the petioles. The short rachillae are completely covered in pinkish brown, woolly hairs. The shiny black fruits are ellipsoid to nearly spherical, over 2 inches (5 cm) long and just over 1½ inches (4 cm) wide.

Pritchardia maideniana
prit-CHAHR-dee-a · may-den'-ee-AH-na
PLATE 760

Pritchardia maideniana was described from a plant cultivated in Sydney, Australia, but is in fact native to the island of Hawaii, where it is critically endangered. It is found in coastal forests at sea level to mesic forests at an elevation of 2000 feet (610 m). The epithet honors Joseph H. Maiden (1859–1925), author, botanist, and director of the Royal Botanic Gardens, Sydney, Australia. This species now includes *P. affinis*.

The trunk usually grows to 35 feet (11 m) high, but some individuals attain 70 feet (21 m). The stem has a diameter of 6–12 inches (15–30 cm) and is light to dark tan. The leaf crown in older trees is open and hemispherical unless the dead leaves are retained, which is often the case, and then the crown is spherical. The leaves are 3–4 feet (90–120 cm) long on petioles 4 feet (1.2 m) long, diamond shaped, undulating, and deeply costapalmate with stiff segments that are deep green above and slightly paler green beneath. The inflorescences are shorter than the leaf crown and produce round 1-inch (2.5-cm) black fruits. This salt-tolerant palm is well suited for coastal gardens.

Pritchardia martii
prit-CHAHR-dee-a · MAHR-tee-eye
PLATE 761

Pritchardia martii is endemic to the Hawaiian island of Oahu, where it grows in mountainous wet forest at 1000–2700 feet (300–820 m). The epithet honors the great German botanist Carl F. P. von Martius (1794–1868). Included in this species are palms formerly known as *P. gaudichaudii*, *P. kahanae*, and *P. rockiana*, among others. The many names applied to this species attest to its high degree of variability.

This slow-growing, short, and stocky species attains a maximum height of 30 feet (9 m) and a diameter of 1 foot (30 cm), but it may flower and fruit when much, much smaller. The leaf crown is open and nearly spherical because of the skirt of dead leaves. The petioles are 3–4 feet (90–120 cm) long, stout, and covered in felty pubescence. The leaves are 3–4 feet wide, semicircular, broad, and flat or slightly undulate, with shallow and broad, stiff segments. Leaf color is deep olive to nearly bluish green above but silvery beneath. The inflorescence is shorter than, or sometimes as long as, the petioles. The shiny black fruits are 2 inches (5 cm) long and nearly as broad. This palm is among the world's most beautiful small palms; it seems the essence of what a palm should look like.

Pritchardia minor
prit-CHAHR-dee-a · MY-nor
PLATE 762

Pritchardia minor is endemic to the Hawaiian island of Kauai, where it grows in wet forest along fog-enshrouded ridges and steep slopes, at 1000–4200 feet (300–1280 m). The epithet is Latin for "smaller," a reference to the slender stem and diminutive habit. *Pritchardia eriophora* is a synonym.

The trunk grows to about 30 feet (9 m) tall but is only 3–5 inches (8–13 cm) in diameter. The gorgeous crown is small, spherical, and dense. The petioles are 3 feet (90 cm) long and carry semicircular leaves 2–3 feet (60–90 cm) wide with stiff segments. The blades are glossy green above and silvery, almost white below. The inflorescence is about as long as the petioles and covered with pinkish brown, woolly hairs. The rachillae are 3 inches (8 cm) long and densely woolly as well. The shiny black fruits are round or ellipsoid, about 1 inch (2.5 cm) long and ½ inch (13 mm) wide. This palm is another of the world's most beautiful small palms and seems perfect and jewel-like.

Pritchardia mitiaroana
prit-CHAHR-dee-a · mee'-tee-ah-ro-AH-na

Pritchardia mitiaroana was described by in 1995 from Mitiaro Island in the remote Cook Islands of the South Pacific, where it grows near sea level in scrubby forest on limestone. The epithet is Latin for "of Mitiaro." The light brown trunk attains a maximum height of 25 feet (7.6 m) and is spindle shaped. The leaf crown is nearly spherical and dense, with semicircular, slightly undulate, light green leaves 3–4 feet (90–120 cm) wide on greenish white, stout petioles 3 feet (90 cm) long. The leaflets tips are stiff or slightly drooping. The inflorescence is about as long as the petiole or a bit shorter. The shiny black fruits are spherical and less than ½ inch (13 mm) in diameter. Although probably not yet in cultivation, this beautiful, robust species should be nearly perfect for areas with calcareous soils.

Pritchardia munroi
prit-CHAHR-dee-a · mun-RO-ee
PLATE 763

Pritchardia munroi is a critically endangered species from the Hawaiian islands, where a handful of individuals are left in dry, evergreen scrub forest at about 2000 feet (610 m) on Molokai, and in wetter forest at 3000 feet (900 m) on Maui. The epithet honors George Campbell Munro (1866–1963), a New Zealand naturalist and ornithologist who arrived in Hawaii in 1890. The trunk of this small, slender species grows to 15 feet (4.5 m) high but is only 4–6 inches (10–15 cm) in diameter. It is light gray and covered in dense, brown leaf sheath fibers in its younger parts. The nearly spherical leaf crown is open because of the stout petioles 4 feet (1.2 m) long. Individual leaves are deep green, 3 feet (90 cm) long, semicircular to wedge shaped, and strongly undulate. The segments have pendent apices. The inflorescence is about as long as the petiole. The rachillae are 3 inches (8 cm) long and have a dense covering of grayish brown hairs. The round 1-inch (2.5-cm) fruits are shiny black or brown.

Pritchardia pacifica
prit-CHAHR-dee-a · pa-SI-fi-ka
PLATES 764–765

Pritchardia pacifica is known only from cultivated specimens or escaped and naturalized populations from islands in the South and West Pacific. It has been erroneously reported as native to the Solomon Islands and Tonga (Hodel 2007). Common names are Fiji fan palm and Pacific fan palm. The epithet is Latin for "of the Pacific." The name "*P. woodfordiana*," a binomial with no botanical standing, has been applied to individuals of this species growing in the Solomon Islands.

Mature trunks can grow to 45 feet (14 m) high with a diameter up to 1 foot (30 cm). The leaf crown is 12 feet (3.6 m) wide and tall, full and rounded, with as many as 40 leaves. Each leaf is 3–6 feet (90-180 cm) wide on a stout petiole, which is 3–4 feet (90–120 cm) long and brownish green, covered with a beautiful scurfy white wax coating that imparts a lovely tan coloration. The leaf is semicircular to diamond shaped, and the segments are divided a quarter of the way into the blade and are stiff. The leaf blade is undulate. Leaf color is bright, light green to emerald green. The inflorescences are 2–3 feet (60–90 cm) long and much branched, and bear small, yellow flowers. The round ½-inch (13-mm) fruits are glossy black.

This palm is adaptable to most free-draining soils, including calcareous ones. It needs average but regular moisture and thrives in full sun or partial shade. It is especially suited to plantings in groups and loses some of its "warmth" and charm as an isolated specimen tree surrounded by space. It is small enough to be used as a courtyard or patio subject, and young plants are overwhelmingly beautiful. Because it tolerates salt-laden winds and slightly saline soil, it is one of the finest landscape subjects for planting near the seashore. It needs space, high light levels, and good air circulation to be grown indoors.

Pritchardia remota
prit-CHAHR-dee-a · ree-MO-ta
PLATE 766

Pritchardia remota occurs in Hawaii on the islands of Nihoa and Niihau, where it grows at 250–800 feet (75–240 m) elevation in dry, barren locales, often at the base of cliffs where there is water seepage. It is endangered. The epithet is Latin for "remote," an allusion to the small island of Nihoa, which is some distance from the more famous and larger islands of the Hawaiian chain.

Included as a synonym of this species is *P. aylmer-robinsonii*. *Pritchardia napaliensis* was previously believed to be a variant but is recognized as a distinct species.

The trunk grows to 30 feet (9 m) tall and 6–12 inches (15–30 cm) in diameter and bears a larger, spherical crown of leaves. The petioles are 4 feet (1.2 m) long and hold blades that are 4 feet (1.2 m) long and wide, undulate, and deeply divided into segments with drooping tips. Leaf color is grayish green. The nonpubescent inflorescences are as long as or almost as long as the petioles. The round 1-inch (2.5-cm) fruits are shiny black or brown. This species responds well to cultivation, relishing the fertility and moisture that are absent in its natural habit. In gardens, expect rapid growth and a large, shaggy crown.

Pritchardia schattaueri
prit-CHAHR-dee-a · shat-TOW-er-eye
PLATE 767

Pritchardia schattaueri is a rare and endangered species endemic to the southwestern coast of the island of Hawaii, where it grows at 2000–2600 feet (610–790 m). The epithet honors George A. H. Schattauer (1920–2005) of Kona, Hawaii, the discoverer of the species in 1960. Only a dozen individuals are known in the wild.

This palm is one of the tallest in the genus, the trunk ascending to the astounding height of 80 or more feet (24 m), with a diameter of 1 foot (30 cm). The leaf crown is spherical and dense. Each of the 30 leaves is 5–6 feet (1.5–1.8 m) wide, semicircular, and glossy deep green on both surfaces. The petioles are 6–7 feet (1.8–2.1 m) long, covered in a light brown chalky tomentum on their lower surfaces, and have abundant fibers at their bases. The divisions between the leaf segments extend 2 feet (60 cm) into the blade, and the tips of older leaves are pendent. The inflorescences are as long as, or nearly as long as, the petioles. The shiny black fruits are 2 inches (5 cm) long and about 1½ inches (4 cm) wide. Although this incredibly elegant and beautiful species is now in cultivation, it is still rare.

Pritchardia thurstonii
prit-CHAHR-dee-a · thur-STO-nee-eye
PLATES 768–769

Pritchardia thurstonii is endemic to coastal regions of the islands of Fiji and Tonga. It is sometimes called Thurston palm. The epithet honors Sir John B. Thurston (1836–1897), a British colonial officer and governor of Fiji.

Mature trunks grow to 25 feet (7.6 m) tall, are 8 inches (20 cm) in diameter, and are free of leaf bases and fiber except near the leaf crown. Trunk color is light brown to deep gray, with closely set rings of leaf base scars and narrow vertical fissures. The leaf crown contains 20 flat leaves that are 4–6 feet (1.2–1.8 m) wide on petioles 3 feet (90 cm) wide and light green but covered in a gray to white scurfy and waxy bloom. Leaf shape is round to diamond shaped, and the divisions between segments extend almost half way into the center of the blade. Leaf color is light to deep green on both sides, but the undersides have distinctive parallel lines of scales. The tips of the segments are stiff. The inflorescences are 10 feet (3 m) long, arching up and out of the leaf crown, and branched near the tip. They bear small, yellow flowers. The round ½-inch (13-mm) fruits are purplish black.

This species is adaptable to various free-draining soils, including calcareous ones. It needs full sun and regular, adequate moisture but is tolerant of salt spray as well as slightly saline soil. The palm is visually similar to *P. pacifica* and has the same landscape uses. It has an even more beautiful silhouette and is more tolerant of seaside conditions. It is difficult indoors as it needs high light levels and good air circulation.

Pritchardia viscosa
prit-CHAHR-dee-a · vis-KOS-a
PLATE 770

Pritchardia viscosa is an endangered species endemic to the Hawaiian island of Kauai, where it grows in very wet forest at 1800–2500 feet (550–760 m). Only 3 individuals are known to survive in the wild. The epithet is Latin for "viscous," an allusion to the sticky secretions on the inflorescences.

The trunk grows to 30 feet (9 m) high with a diameter of 8–18 inches (20–45 cm). The leaf crown is open and nearly spherical with semicircular leaves 4 feet (1.2 m) wide. The tips of the leaf segments are stiff. Leaf color is glossy and light, grassy green above and a beautiful silvery green beneath. The inflorescence is shorter than the petioles, which are 3 feet (90 cm) long, and is densely covered with pinkish brown hairs and silvery wax. The rachillae are covered in a sticky, vanishlike secretion. The shiny black fruits are about 1½ inches (4 cm) long and ½ inch (13 mm) wide. Cultivation in botanical gardens may be the last hope for this species.

Pritchardia waialealeana
prit-CHAHR-dee-a · wy-ah'-lai-ah-lai-AH-na
PLATE 771

Pritchardia waialealeana is endemic to wet mountainous rain forest on the Hawaiian island of Kauai at 1600–2400 feet (490–730 m). The epithet is Latin for "of Waialeale," a mountain in the palm's habitat. The trunk of this tall, robust species attains a minimum height of 60 feet (18 m) and a diameter of 10–18 inches (25–45 cm). The leaf crown is massive, dense, and spherical, with semicircular leaves 4 feet (1.2 m) wide on petioles 2 feet (60 cm) long. The blade is slightly undulate, and the segment tips are either stiff or lax. The inflorescence is about as long as the petiole. The shiny black, round or ellipsoid fruits are less than 1 inch (2.5 cm) in diameter.

Pseudophoenix
soo-do-FEE-nix

Pseudophoenix consists of 4 solitary-trunked, pinnate-leaved, hermaphroditic palms in the Florida Keys, the Bahamas, Cuba, Hispaniola, the Yucatán Peninsula, coastal Belize, and the island of Dominica. All the species have large leaves with numerous, narrow, pointed pinnae that diminish in size near the base of the leaf. The very narrow leaflets at the base of the leaf are reminiscent of the modified leaflet-spines (acanthophylls) of *Phoenix*, although the 2 genera are unrelated. The basal leaflets were the inspiration for the genus name, which means "false *Phoenix*."

All the species form a short crownshaft, although it is not as well defined as that of *Veitchia* or *Roystonea*. The flowers are greenish yellow and attractive to many kinds of insects. Fruits are red when ripe. They are spherical if only one seed matures but will be 2-lobed if 2 seeds form or even 3-lobed with 3 seeds.

All the species are intolerant of cold and adapted only to zones 10 and 11, although specimens are to be found in favorable microclimates of 9b. These palms are denizens of coastal environments where the soil is calcareous and poor, the sun is relentless, and the erratic rainfall is confined mainly to the summer; thus, they are well adapted to seaside planting and hot regions with limey soils. They are all slow growing, some extremely so.

This genus is the only member of the tribe Cyclospatheae. A hard shell (endocarp) covers the seed, making germination slow, sporadic, and seldom successful. A germination rate of 75 percent or better is possible when the seeds are allowed to dry out for 2–4 weeks, after which it is easy to crack and remove the outer shell. The seed should then be soaked in room temperature water for 2 days. Because the seeds are more prone to disease and insect attacks without their shells, they should be planted in a sterilized potting medium such as coarse perlite. The seeds should be barely covered, then watered thoroughly, before placing the container in a semishady area. Bottom heat is important, as is regular water. The seeds must not be allowed to dry out. Within 60–120 days after sowing, a large number of the seeds should germinate, with some stragglers continuing to sprout for up to a year.

Pseudophoenix ekmanii
soo-do-FEE-nix · ek-MAH-nee-eye
PLATE 772

Pseudophoenix ekmanii is a rare species endemic to the Barahona Peninsula of the Dominican Republic, where it grows in open scrub and cactus woodland on karst limestone where soils are thin and limited to pockets and fissures in the limestone. The epithet honors the great Swedish botanist Erik Ekman (1883–1931), who worked extensively in the Caribbean.

The white, waxy trunk attains a height of 20 feet (6 m) and bulges above its midpoint. It is distinctly and beautifully ringed with dark brown or gray leaf scars. Older individuals lose the waxy coating and have dull gray trunks. The leaf sheaths are dark green and form a crownshaft only 1–2 feet (30–60 cm) high. The leaf crown is spherical but sparse and open, and the gray-green leaves are 6–8 feet (1.8–2.4 m) long, ascending, and mostly erect—especially when young—and spreading when older. The inflorescence is held among the leaves and is much branched. The round ½-inch (13-mm) fruits are red and have a watery flesh.

Young palms with a few stiff, gray-green leaves look more like cycads than palms. The species is slow growing as a seedling, but the widely space leaf scars on the trunks of young adult plants suggest that the growth rate increases in mid-life. It is certainly one of the most beautiful palms in the Caribbean and would be an asset to any garden.

Pseudophoenix lediniana
soo-do-FEE-nix · leh-din'-ee-AH-na
PLATE 773

Pseudophoenix lediniana is endemic to a single locality in southwestern Haiti, where it is endangered. The epithet honors R. Bruce Ledin (1914–1959), botanist, palm enthusiast, and former vice-president of the Palm Society, the predecessor to the International Palm Society.

Mature trunks may attain a height of 65 feet (20 m), but this is rare, especially under cultivation, and only half this stature is the norm. The mature stem has a bottle shape: narrow at the base, broadening and swelling in the middle, and then tapering to a narrow neck on top. Trunk color is an outstandingly beautiful greenish gray, usually with a thin covering of wax and beautiful, widely spaced deep green rings of leaf base scars. There are 15–17 leaves in the crown, which is open and graceful, with a pleasing architectural look. The bluish green, olive green, or grayish green crownshaft is up to 2 feet (60 cm) tall, smooth, waxy, and tapering. The lax leaves are to 12 feet (3.6 m) long on stout petioles 6 inches (15 cm) long. The thin, linear, and deep green leaflets grow from the rachis at slightly different angles, but the angles are short and the overall appearance of the leaf is almost flat. The leaflets are limp and, especially on older leaves, beautifully pendent. The inflorescences grow from among the leaves but are pendulous and 4 feet (1.2 m) long, hanging below the crown. The round fruits are about an inch (2.5 cm) in diameter with firm flesh.

This species is drought tolerant but grows faster and looks better with regular, adequate moisture. Under such conditions, it is probably the fastest-growing species in the genus. It requires a well-drained soil and full sun at all stages. Because of its relatively fast growth, it is also more capable of recovering from freeze damage than the other species.

The visual aspects of the taller species, like *P. ledinii* and *P. vinifera*, are often compared with those of the royal palms (*Roystonea*) because of the sturdy, mostly light colored trunks and large dark green pinnate leaves. The practice really does neither genus credit but, if one must indulge in it, this species would seem to invite the comparison as the trunks are tall and the leaves gracefully arching with thin, lissome leaflets. This species is the fastest growing in its genus, and its most salient feature is the wonderfully beautiful trunk and crownshaft, which almost demand that the palm be planted where its stem can be viewed up close. It is especially beautiful against a large background of deep green foliage that shows off the light-colored trunks with their wonderful dark rings. The palm is not known to be grown indoors but there would seem no reason it could not be, given high light levels and good air circulation.

Pseudophoenix sargentii
soo-do-FEE-nix · sar-JEN-tee-eye
PLATES 774–775

Pseudophoenix sargentii, the most widespread species in the genus, is indigenous to the Florida Keys, the Bahamas, Cuba, islands off Hispaniola, Belize, the Mexican state of Quintana Roo, Navassa Island (where only one palm remains), and Dominica. Throughout its broad range, it grows near the sea on sandy, limestone soils. Common names are buccaneer palm and cherry palm. The epithet honors Charles S. Sargent (1841–1927), former director of the Arnold Arboretum and one of the original collectors of the species.

Mature trunks attain a height of 25 feet (7.6 m) and a diameter of 4–10 inches (10–25 cm). The stems are grayish green with broad, brown, closely set rings of leaf base scars when young, but eventually turn light to dark gray on the older parts. They have a spindle-shaped bulge at the halfway point. The leaf crown is 10 feet (3 m) wide and 6 feet (1.8 m) tall. There are 7–16 leaves in a crown, and the crown is seldom spherical. The tapering crownshaft is about 1 foot (30 cm) tall, smooth, waxy, and bluish to grayish green. The leaves are 7 feet (2.1 m) long and gracefully arching or stiff and upright on stout green petioles 1–3 feet (30–90 cm) long. The bluish green leaflets grow from the rachis at a slight angle to give a shallow V shape to the leaf. The inflorescences emerge from among the leaves and are 3–4 feet (90–120 cm) long and erect or pendulous with many short branches bearing small, greenish yellow blossoms. The round ½-inch (13-mm) fruits are deep red and have a watery pulp.

This palm is usually slow growing, even under ideal cultural conditions, but some plants grow much faster than others (for example, the Navassa Island genotype), an observation indicating some genetic variability in the species. It is drought tolerant but looks better with average and regular moisture. Full sun is a necessity but the species is adaptable to most free-draining soils. This species has the same landscape uses as *P. lediniana* but is smaller, much slower growing, and more generally suitable for intimate sites like courtyards and patios. It is one of the most salt-tolerant palms and does well near the sea, even occasionally tolerating inundation by salt water. This palm is not known to be grown indoors, but there would seem no reason it cannot be if given high light levels and good air circulation.

Pseudophoenix vinifera
soo-do-FEE-nix · vi-NI-fe-ra
PLATES 776–777

Pseudophoenix vinifera is endemic to dry hillsides over a wide area of Hispaniola. Common names are cherry palm and wine palm. Common names are cherry palm and wine palm. The epithet translates from the Latin as "wine-bearing," a reference to the sap that is used for making wine.

Mature trunks may attain a height of 70 feet (21 m) or even more in habitat but usually only half this height under cultivation. Their diameter is 1 foot (30 cm) at the base, gradually expanding as the stem bulges, which is usually near its midpoint; above the bulge, the stem tapers to a narrow neck, giving the whole trunk a bottle shape. Trunk color is brown to deep gray and the stems bear closely set wide, dark rings of leaf base scars until they are old, at which time the rings are indistinct. The leaf crown is 15 feet (4.5 m) wide and 8 feet (2.4 m) tall. There are usually about 24 leaves to a crown, which is often full and rounded because of the arch of the leaves. The crownshaft is short, up to 2 feet (60 cm) tall, grayish green to almost silver, and tapers from its base to its tip. The leaves are 7–10 feet (2.1–3 m) long and gracefully arching on silvery green petioles 4–12 inches (10–30 cm) long. The leaflets grow from the rachis at several different angles, but the angles are small and the dark green leaflets are limp and pendulous, giving a slightly plumose aspect to the leaf. The inflorescence is 4–6 feet (1.2–1.8 m) long, pendent, with lax branches, and bears small, greenish yellow flowers. The round 1-inch (2.5-cm) brilliant scarlet fruits make a spectacular display, hanging in long pendent branches from the leaf crown.

This palm is adaptable to full sun and most well-drained soils. It is always slow growing. The species has the same landscape use as *P. lediniana* with the proviso that its trunks are a more strongly bottle shaped. It is not known to be grown indoors but there would seem no reason it cannot be if given high light levels and good air circulation.

Ptychococcus
ty-ko-KAHK-kus

Ptychococcus comprises 2 moderate to tall solitary-trunked, pinnate-leaved, monoecious palms in New Guinea. The crownshafts sit atop relatively slender trunks. The inflorescences grow from beneath the crownshafts, are massive and much branched, and bear flowers of both sexes. The large, ovoid fruits are orange to red when ripe and have watery flesh. The ridged seeds are enclosed in hard, ridged endocarps.

This genus is closely related to *Ptychosperma*, as evidenced by its jagged ended leaflets. Neither species is common in cultivation, which is a shame as they are wonderfully attractive. In the landscape, they are reminiscent of the larger *Veitchia* species. They are tender to cold and need constant, copious moisture. The good news is that they do not seem fussy about soil type as long as it is fast draining.

The genus name comes from Greek words meaning "folded berry," alluding to the ridged endocarps and seeds. The seeds generally germinate within 45 days and occasionally longer. For further details of seed germination, see *Areca*.

Ptychococcus lepidotus
ty-ko-KAHK-kus · le-pi-DO-tus
PLATE 778

Ptychococcus lepidotus is endemic to montane rain forest of central New Guinea, from 2500 to 5300 feet (760–1620 m), perhaps as high as 10,000 feet (3000 m). The epithet is Latin for "scaly" and alludes to the covering of brown scales on the flower petals as well as the petiole and leaf rachis.

The trunk grows to 25 feet (7.6 m) tall or more and is 4–12 inches (10–30 cm) in diameter. The crownshaft is 2–4 feet (60–120 cm) tall, cylindrical, slightly bulging at its base, and light green with dark scales at the apex. The leaf crown is spherical or nearly so and contains a dozen beautifully arching leaves, each 10 feet (3 m) long on a petiole 6 inches (15 cm) long. The leaflets are regularly spaced along the rachis and grow in a single flat plane from it. They are mostly linear-oblong, with obliquely truncated, jagged apices and are dark green on both surfaces. New leaves emerge in shades of red or rose. The fruits are about 1 inch (2.5 cm) wide and up to 2 inches (5 cm) long. The endocarp is brown and thick-walled.

This beautiful palm does not like hot, humid tropical conditions, which is to be expected considering its elevated habitat. It survives in tropical conditions but is better in frostless Mediterranean climes, especially those in which there is a significant drop in temperature between night and day. It is probably hardy in zones 10a.

Ptychococcus paradoxus
ty-ko-KAHK-kus · par-a-DAHK-sus
PLATE 779

Ptychococcus paradoxus is endemic to lowland rain forest in New Guinea, from sea level to 2800 feet (850 m). The epithet is Latin for "paradox," so named because botanists were perplexed by this species, which did not seem to belong in any known genus, before classifying it in a genus of its own. This species has been known as *P. elatus*, *P. archboldianus* and *P. arecinus*, but these names are now regarded as synonyms.

The trunk is straight and columnar and grows to 80 feet (24 m) tall with a diameter of 4–10 inches (10–25 cm). It is nearly white and graced, in its younger parts, with widely spaced, tan rings of leaf base scars. The crownshaft is 2 feet (60 cm) tall, slightly bulging at its summit, and light silvery green. The leaf crown is mostly spherical, with stiff, spreading leaves 8–10 feet (2.4–3 m) long on stout petioles 1 foot (30 cm) long. The leaflets are 2–3 feet (60–90 cm) long, grow from the rachis in a single flat plane, and vary from linear-oblong with undulate margins and truncated, jagged tips to linear-lanceolate. Leaf color is uniformly dark, dull green on both surfaces, and new growth has a pinkish tinge. The fruits are 1½–2½ inches (4–6 cm) long and up to 1½ inches (4 cm) wide. The endocarp is black and thin-walled.

This species is incredibly beautiful because of its large, oversized crown in relation to the slender stem. It also is strictly tropical and impossible outside of zones 10b and 11.

Ptychosperma
ty-ko-SPUR-ma

Ptychosperma is composed of about 29 solitary-trunked or clustering, pinnate-leaved, monoecious palms in the Solomon Islands, Bismarck Archipelago, Louisiade Archipelago, Kai and Aru Islands, and northern Australia, with the majority of species in New Guinea, especially Papua New Guinea. All are indigenous to wet areas, whether rain forest, lowland swamps, or low mountain valleys.

The slender, ringed trunks, prominent crownshafts, and leaflets with jagged or toothed apices distinguish these species. Many of them have pinkish, maroon, or bronzy colored new foliage. The inflorescences grow from beneath the crownshafts and are comprised of flowering branches that bear both male and female flowers. The round or ellipsoid fruits are orange-red, red, or black. Some of the important characters for identifying the species include the disposition of the leaflets in the middle of the leaf (regularly versus irregularly spaced), the color of the fruit at maturity (red or orange versus black), and the condition of the endosperm (homogeneous versus ruminate).

The various species apparently hybridize with one another if grown in close vicinity. Although putative hybrids can be extremely handsome, they are not a good thing if one is concerned with maintaining what is (or was) originally in nature. Under such circumstances, care should be taken when distributing seeds produced in gardens where different species are grown together. The genus name is a combination of Greek words meaning "folded seed."

Seeds of most species germinate within 30 days, but seeds of other species, such as *P. waitianum*, can take 6 months or longer. Seeds allowed to dry out before sowing do not germinate as well as those kept slightly moist. For further details of seed germination, see *Areca*.

Ptychosperma burretianum
ty-ko-SPUR-ma · bur-ret´-ee-AH-num
PLATE 780

Ptychosperma burretianum is endemic to eastern Papua New Guinea, where it grows in low mountainous rain forest. The epithet honors Karl E. Maximilian Burret (1883–1964), German botanist and palm specialist.

This clumping species (rarely solitary) has stems 25 feet (7.6 m) tall and 1 inch (2.5 cm) in diameter. The slender little crownshafts are 1 foot (30 cm) tall, scarcely thicker than the stems, and silvery green near the apex. The leaf crowns are open and hemispherical with a few gently arching leaves 5–6 feet (1.5–1.8 m) long on petioles 1 foot (30 cm) long. The leaflets are a lustrous medium green on both surfaces and are regularly arranged in the middle of leaf but irregularly arranged at the base or apex. They are wedge shaped, the longest 10 inches (25 cm), and have an uneven and jagged apex. The new leaves are pinkish bronze, the fruits are black, and the endosperm is homogeneous. This species is doubtfully distinct from *P. waitianum* (which see).

Ptychosperma caryotoides
ty-ko-SPUR-ma · kar´-ee-o-TO-i-deez
PLATE 781

Ptychosperma caryotoides is endemic to central and southeastern Papua New Guinea in mountainous rain forest from 600 to 4000 feet (180–1200 m). The epithet translates as "like *Caryota*," in reference to the broad leaflets, but the resemblance to *Caryota* is only superficial.

This solitary-trunked species is highly variable, but in general the trunk attains a height of 30 feet (9 m) with a diameter of 1–4 inches (2.5–10 cm). The crownshaft is 1 foot (30 cm) tall, usually bulging at its summit rather than at its base, and light green. The leaf crown is sparse, open, and less than semicircular, with a few, erect and ascending leaves, each one 2–5 feet (60–150 cm) long on a petiole 4–24 inches (10–60 cm) long. The few leaflets are regularly spaced and grow in a single flat plane from the rachis. They are wedge shaped with an obliquely truncated, jagged apex; the leaflets of the terminal pair are usually much broader than the others. The fruits are bright red, and the endosperm is ruminate.

This species is one of the most variable in the genus, and consequently, it can be hard to identify and characterize. It is, in all its forms, graceful and an asset to any garden.

Ptychosperma cuneatum
ty-ko-SPUR-ma · kyu-nee-AH-tum
PLATE 782

Ptychosperma cuneatum is found in eastern Papua Province, Indonesia, and in adjacent West Sepik Province, Papua New Guinea, where it grows in lowland rain forest. The epithet is Latin for "cuneate," referring to the wedge-shaped leaflets. The clustering trunks grow to 15 feet (4.5 m) high with a maximum diameter of 1 inch (2.5 cm). The crownshaft is 18 inches (45 cm) tall, barely thicker than the stem, and cylindrical. The leaf crown is hemispherical or nearly so, with a few spreading, and slightly arching leaves 3–5 feet (90–150 cm) long. The leaflets are arranged irregularly in one plane, the largest leaflets 1 foot (30 cm) long, wedge shaped, with deeply incised apices resulting in 2 "tails" to each apex. The fruits are ½ inch (13 mm) long and become black when ripe. The endosperm is homogeneous. This lovely species is best grown in shade and is intolerant of cold.

Ptychosperma elegans
ty-ko-SPUR-ma · EL-e-ganz
PLATE 783

Ptychosperma elegans is endemic to Australia, where it grows in the coastal lowland rain forest of eastern Queensland. Common names are solitaire palm and Alexander palm. The epithet is Latin for "elegant."

The mature solitary trunk may grow to nearly 40 feet (12 m) in its native haunts but is usually no more than half this height under cultivation. The stem is 2–6 inches (5–15 cm) in diameter, light gray, and distinctly ringed with widely spaced, darker leaf base scars on its younger parts. The crownshaft is 2 feet (60 cm) tall, silvery, olive green to nearly gray, smooth, waxy and slightly bulging at its base. The leaf crown is hemispherical or slightly more so and up to 10 feet (3 m) wide and 6 feet (1.8 m) tall. The leaves are 3–8 feet (90–240 cm) long on petioles 1 foot (30 cm) long and have regularly arranged leaflets growing from the rachis at an angle to create a slightly V-shaped leaf. Each leaflet is 2 feet (60 cm) long, narrowly wedge shaped to oblong, with an abruptly squared off end that is slightly jagged or toothed. Leaf color is a dark to olive green above and a grayish green beneath. There are usually no more than 8–10 leaves in the crown, and they are gently arching but do not stray much below the horizontal plane and fall cleanly after aging. The inflorescence grows from beneath the crownshaft and is 2–3 feet (60–90 cm) long, with many greenish branches bearing small, white blossoms. The ovoid 1-inch (2.5-cm) fruits are bright red, and the endosperm is ruminate.

This palm is adaptable to most well-drained soils but only looks its best in a deep, rich, humus-laden medium. It needs average and regular moisture and does not tolerate freezing temperatures, being adaptable only to zones 10b and 11. It thrives in either partial shade or full sun. Although it is solitary, it is often sold as a "triple"—with 3 palms planted in one pot.

The tree is of unsurpassed grace, symmetry, and beauty of form and is even more exquisite when planted 2 or 3 to a group of varying heights of trunk. Its silhouette is lovely. Among the most splendid sights one of us (RLR) saw was a large Spanish-style courtyard with a small group of solitaire palms of varying heights in its center surrounded at their bases with small elephant ears (*Alocasia* species) and ginger relatives; pygmy date palms (*Phoenix roebelenii*), self-heading *Philodendron bipinnatifidum*, and small ferns were planted around the perimeter of the courtyard. The solitaire palm is an excellent choice for growing in large indoor spaces if given enough light and water.

Two forms or *P. elegans* are in cultivation: a slender, more refined form with a small crown, and a larger, more vigorous form with large leaves and a thick trunk up to 6 inches (15 cm) in diameter. The latter may, in fact, be a stabilized hybrid of unknown parentage that arose in horticulture. The slender form, with impossibly tall stems not more than 3 inches (8 cm) in diameter, is by far the more graceful and desirable of the two.

Ptychosperma furcatum
ty-ko-SPUR-ma · fur-KAH-tum
PLATE 784

Ptychosperma furcatum is endemic to southeastern Papua New Guinea, where it grows in lowland rain forest. The epithet is Latin for "forked" and alludes to the apices of the leaflets. In nature, this palm can be clustering or solitary trunked, the stems reaching a maximum height of 15 feet (4.5 m) and a diameter of 1–2 inches (2.5–5 cm). The crownshafts are 1 foot (30 cm) tall. The leaf crowns are hemispherical, with arching leaves 4–5 feet (1.2–1.5 m) long on petioles 1 foot (30 cm) long. The regularly spaced leaflets are narrowly wedge- or strap-shaped and unequally bifid at their apices. Leaf color is glossy deep green, and the longest leaflets are 18 inches (45 cm) long. The ovoid 1-inch (2.5-cm) fruits turn orange to red when ripe. The endosperm is homogeneous. This species is attractive, especially the clumping forms, but is tender to cold. It is very uncommon in cultivation.

Ptychosperma keiense
ty-ko-SPUR-ma · kee-EN-see
PLATE 785

Ptychosperma keiense is widely distributed from West Papua Province, Indonesia, through the Kai and Aru Islands, in lowland rain forest, often along river banks. The epithet means "of Kei [Kai Islands]."

Mature clumps may be 20 feet (6 m) tall and 10 feet (3 m) wide with trunks up to 2½ inches (6 cm) in diameter. They are light gray with prominent, widely spaced darker leaf scars. The crownshaft is almost 2 feet (60 cm) tall, smooth, and light green to olive green. The leaf crown is hemispherical and can be 10 feet (3 m) wide but is usually no more than 5 feet (1.5 m) tall. The leaves are 3–6 feet (90–180 cm) long on petioles usually less than 1 foot (30 cm) long. The leaflets are up to 18 inches (45 cm) long, arranged regularly in one plane, and narrowly wedge shaped to linear-lanceolate with obliquely squared, toothed ends. Leaf color is medium to deep green on both surfaces. The inflorescences grow from beneath the crownshaft and are 2 feet (60 cm) long and much branched. The ovoid ½-inch (13-mm) fruits are red, and the endosperm is homogeneous.

This species is similar to and may be confused with *P. macarthurii* and *P. propinquum*. In the landscape setting, the 3 are interchangeable, although *P. keiense* is more robust. *Ptychosperma keiense* is readily recognized by its endocarp, which has 3 deep grooves and 2 shallow grooves.

Ptychosperma lauterbachii
ty-ko-SPUR-ma · law-ter-BAH-kee-eye
PLATE 786

Ptychosperma lauterbachii is endemic to eastern Papua New Guinea, where it grows in lowland rain forest and swamps. The epithet honors Carl A. G. Lauterbach (1864–1937), German naturalist and explorer in New Guinea.

This solitary-trunked and clustering species has stems as tall as 40 feet (12 m) with a diameter of 2–4 inches (5–10 cm). They are light gray to light tan and graced with widely spaced, darker rings of leaf base scars in all but their oldest parts. The crownshafts are 18–24 inches (45–60 cm) tall and light, silvery green. The leaf crowns are hemispherical or slightly less so, sparse, and open. The leaves are up to 9 feet (2.7 m) long on petioles 1 foot (30 cm) long. The medium and matte green leaflets are arranged regularly in one plane. They are linear to somewhat wedge shaped, the longest ones 70 inches (178 cm) long, all with obliquely truncated and jagged, often bifid apices. The fruits are orange to red when mature, and the endosperm is slightly ruminate.

This species, like *P. caryotoides*, is polymorphic and difficult to characterize. The beautiful trunks are similar to giant bamboo culms, and the clumping forms are exquisite with their levels of foliage, creating beautiful silhouettes. Alas, the palm is tender to cold and is demanding of copious, regular moisture.

Ptychosperma lineare
ty-ko-SPUR-ma · li-nee-AH-ree
PLATE 787

Ptychosperma lineare is endemic to eastern Papua New Guinea, where it grows in lowland rain forest and coastal swamps. The epithet is Latin for "linear" and alludes to the narrow leaflets.

The clustering stems grow to 50 feet (15 m) high with a diameter of 1–2 inches (2.5–5 cm). They are light olive green in their younger parts, graced with widely spaced rings of whitish leaf base scars. The slender crownshafts are 2 feet (60 cm) tall and silvery green. The open, nearly spherical leaf crowns have arching leaves, each one 6 feet (1.8 m) long on a petiole 4–20 inches (10–50 cm) long. The leaflets are regularly and closely spaced along the rachis from which they grow in a single, nearly flat plane, although the lowermost leaflets are often clustered together. Leaflet shape is generally narrowly lanceolate, with an obliquely truncated and unevenly jagged tip. The fruits are jet black and contrast beautifully with the orange or red branches of the infructescence. The endosperm is homogeneous.

This species is one of the most elegant, the tiers of leaf crowns absolutely arresting in appeal. In overall appearance, it can be confused with *P. furcatum*, which has red fruits.

Ptychosperma macarthurii
ty-ko-SPUR-ma · mak-AHR-thur-ree-eye
PLATE 788

Ptychosperma macarthurii occurs naturally in rain forest and low swamps of southern Papua New Guinea and northern Australia. It is called Macarthur palm. The epithet honors Sir William Macarthur (1800–1882), Australian plant breeder, viticulturist, and amateur botanist. *Ptychosperma bleeseri*, a name applied to Australian populations, is a synonym.

Mature clumps may be 25 feet (7.6 m) tall and 20 feet (6 m) wide and produce a few dominant stems up to 2 inches (5 cm) in diameter. They are light tan to light gray with widely spaced, prominent darker rings of leaf base scars. The crownshaft is 1–2 feet tall (30–60 cm), bulging at its base, smooth, and light green to olive green. The leaf crown is hemispherical and can be 10 feet (3 m) wide but is usually no more than 5 feet (1.5 m) tall. The leaves are 3–6 feet (90–180 cm) long on petioles that are usually less than 1 foot (30 cm) long and are held erect and arch slightly. The leaflets are up to 18 inches (45 cm) long, arranged irregularly in one plane, and narrowly wedge shaped to linear-lanceolate with obliquely squared, toothed ends. Leaf color is medium to deep green on both surfaces. The inflorescences grow from beneath the crownshaft and are much branched but 2 feet (60 cm) long. The branches are yellowish green and bear small, white blossoms. The ovoid ½-inch (13-mm) fruits are red, and the endosperm is homogeneous.

This palm is a water lover and should not be subjected to drought conditions. It does not tolerate frost and is adaptable only to zones 10b and 11. The Macarthur palm survives in poor dry soil but only looks worthwhile in a rich, humus-laden, moist but well-drained medium. It seems to thrive in partial shade or full sun.

There are 2 basic ways of using the Macarthur palm effectively in the landscape: one is to grow the plant only with all tall stems, planting smaller vegetation (such as *Chamaedorea* palms) at and around the base of the trunks; the other way is to have a mass, a wall, of tiered leaves by letting the suckers grow. Both methods are lovely and create wonderful tableaux of tropical luxuriance. The former method exhibits the beautifully ringed trunks more so than the second method, but the tiered layers of tropical-looking pinnate leaves are sumptuous. A clump of the palm is beautiful and elegant enough to stand alone, even isolated in space, but it is as a courtyard or patio feature that it is at its best, a site in which the wonderful trunks (like those of timber bamboo) can be seen close up. The plant's silhouette is as beautiful as that of the *P. elegans*. This is a good choice for growing indoors if enough light and space can be provided.

This species is doubtfully distinct from *P. propinquum*. Some plants in cultivation under the name *P. macarthurii* may in fact be *P. kiense* (which see).

Ptychosperma microcarpum
ty-ko-SPUR-ma · my'-kro-KAHR-pum
PLATE 789

Ptychosperma microcarpum is endemic to eastern Papua New Guinea, where it grows in lowland rain forest, usually along river banks, from sea level to 2000 feet (610 m). The epithet is from Greek words meaning "small fruit," although the fruit of this species are no smaller than most other species in the genus.

The stems of this usually clustering species attain a height of 30 feet (9 m) and a diameter of 1–2 inches (2.5–5 cm). The crownshafts are 18–24 inches (45–60 cm) tall and silvery olive green. The leaf crowns are hemispherical or slightly less so, with 6 leaves. Each slightly arching leaf is 4–6 feet (1.2–1.8 m) long on a petiole 1 foot (30 cm) long. The leaflets are 1–2 feet (30–60 cm) long, grow in clusters, and are generally narrowly wedge shaped with obliquely truncated, unevenly jagged apices; the leaflets of the terminal pair are usually no wider than the others. All the leaflets grow from the rachis at slightly different angles, which gives the leaf a plumose effect. Leaf color is light green on both surfaces. The ovoid fruits are ½ inch (13 mm) long and turn orange to red when ripe. The endosperm is homogeneous.

The most distinctive feature of this palm is the form of the leaves. The leaflets are irregularly arranged, clustered, and diverge from the rachis in more than one plane, giving the leaf a fox-tail look. Also, the female flowers have a delightful fragrance reminiscent of jasmine. This plant should be used in intimate, closed settings, such as courtyards or atriums, where the scented flowers can be appreciated.

Ptychosperma propinquum
ty-ko-SPUR-ma · pro-PIN-kwum

Ptychosperma propinquum occurs naturally on the islands of the Aru Archipelago, south of the western end of the island of New Guinea, where it grows in lowland rain forest. The epithet, Latin for "closely related," was not explained but may refer to the similarity, noted in the original description, of this species to *P. ambiguum*. The 2 species differ in fruit color.

In habitat, the clustering stems grow to a height of 25 feet (7.6 m) with a diameter of almost 1 inch (2.5 cm). The plump crownshafts are 1 foot (30 cm) tall and olive green. The leaf crowns are hemispherical or slightly less so, and the leaves are 2–3 feet (60–90 cm) long on petioles 1 foot (30 cm) long. The dark green leaflets grow in closely set clusters of 2 or 3 but arise off the rachis in a nearly flat plane. They are generally 1 foot (30 cm) long with obliquely truncated, jagged apices. The fruits are ½ inch (13 mm) long and turn bright red when ripe. The endosperm is homogeneous.

Further study may reveal that this species is not distinct from *P. macarthurii* (which see). See also the similar-appearing *P. keiense*.

Ptychosperma salomonense
ty-ko-SPUR-ma · sal-o-mo-NEN-see
PLATE 790

Ptychosperma salomonense is endemic to the Solomon Islands from sea level to 1500 feet (460 m). The epithet is Latin for "of the Solomons."

The solitary trunk grows to 40 feet (12 m) high with a maximum diameter of 2 inches (5 cm). The crownshaft is 18 inches (45 cm) tall, cylindrical, slightly thicker than the stem, and deep olive green. The hemispherical leaf crown has 5–10 spreading, slightly arching leaves, each one 5–8 feet (1.5–2.4 m) long on a petiole 5–12 inches (13–30 cm) long. The leaflets are regularly spaced and grow from the rachis at an angle that gives young leaves a slight V shape. They vary in shape and length, most being strap- to wedge shaped and from a few inches long near the base to 20 inches (50 cm) long near the midpoint of the leaf, always with obliquely truncated, jagged apices. Leaflet color is medium to deep green on both surfaces. The ½-inch (13-mm) fruits become orange or red when ripe, and the endosperm is ruminate.

This species is visually and genetically similar to *P. elegans* but is, if possible, even more beautiful. Its elegantly thin stems and luscious, large crown of leaves are breathtakingly handsome. The 2 species can be distinguished by the shape of the endocarp, which is nearly spherical with shallow grooves in *P. salomonense*.

Ptychosperma sanderianum
ty-ko-SPUR-ma · san-der-ee-AH-num
PLATES 791–792

Ptychosperma sanderianum is unknown in the wild but likely came from southeastern Papua New Guinea in lowland rain forest. The epithet honors Henry F. C. Sander (1847–1920), the famed orchid nurseryman of St. Albans, England, who first exhibited a young plant of this species in 1898 at a meeting of the Royal Horticultural Society where the plant won a First Class Certificate. The species has been in cultivation ever since.

The stems of this small, clustering species grow to 10 feet (3 m) high with a diameters of 1–2 inches (2.5–5 cm). The crownshafts are deep olive green, cylindrical, and 19 inches (48 cm) tall. The leaf crown is hemispherical to slightly more, with no more than 12 arching leaves, each 4–5 feet (1.2–1.5 m) long. The dark green leaflets are 18 inches (45 cm) long and are regularly and closely spaced along the rachis from which they grow in a single flat plane. They are narrowly wedge shaped with obliquely truncated, jagged apices. The ovoid ½-inch (13-mm) fruits become deep red when ripe. The endosperm is homogeneous.

This species is extremely graceful, with visual affinities to *P. macarthurii* but more delicate appearing. It is readily grown in a conservatory with bright light and adequate humidity. The cultivar with undivided leaves is extremely handsome.

Ptychosperma schefferi
ty-ko-SPUR-ma · SHEF-fer-eye
PLATE 793

Ptychosperma schefferi occurs naturally in north-central New Guinea, from the vicinity of Jayapura, Indonesia, to West Sepik Province, where it grows in lowland rain forest. The epithet honors Rudolph H. C. C. Scheffer (1844–1880), Dutch botanist and former director of the Buitenzorg (now Kebun Raya) Botanic Garden in Bogor.

In nature, this species has clustering stems to 20 feet (6 m) high and almost 3 inches (8 cm) in diameter. The plump crownshafts are 18 inches (45 cm) tall, deep olive green, and bulging near their midpoints. The leaf crowns are hemispherical or slightly less, with 10 barely arching, spreading leaves, each 5–6 feet (1.5–1.8 m) long on a petiole 18 inches (45 cm) long. The leaflets, also 18 inches long, are regularly and closely spaced along the rachis from which they grow at an angle that gives younger leaves a slight V shape. The fruits are ½ inch (13 mm) long and turn black when ripe. The endosperm is weakly ruminate. This wonderfully attractive, robust species has the essence of the tropical look.

Ptychosperma waitianum
ty-ko-SPUR-ma · wait-ee-AH-num
PLATES 794–795

Ptychosperma waitianum is endemic to southern Papua New Guinea, where it grows in lowland rain forest. The epithet honors Lucita H. Wait (1903–1995), who, from 1957 to 1972, was secretary of the Palm Society (now the International Palm Society) and manager of its seed bank. This species may be conspecific with *P. burretianum*; they seem to be 2 extremes along a continuum of overall size.

The stems of this clumping species reach 15 feet (4.5 m) tall and less than 1 inch (2.5 cm) in diameter. The plump, silvery green crownshafts are 1 foot (30 cm) tall. The leaf crown is hemispherical or slightly more so, with 8 arching, spreading leaves, each 2–3 feet (60–90 cm) long on a petiole 1 foot (30 cm) long. The leaflets are 6–8 inches (15–20 cm) long, deep green on both surfaces, broadly wedge shaped, and grow in a single flat plane from the rachis. They are irregularly clustered near the base of the leaf, but regularly spaced in the middle of the leaf. Their apices are deeply cleft into 2 irregularly notched main lobes. The new growth is brilliant deep orange to deep salmon. The fruits are black, and the endosperm is homogeneous. This species is prized for the colorful new leaf that adds a luminous splash of color to a shady garden.

Raphia
RAF-ee-a

Raphia includes about 20 large pinnate-leaved, monoecious, monocarpic palms in Africa and Madagascar and one in tropical America. Most are denizens of low, swampy areas, but one, *R. regalis*, grows on upland sites in rain forest and has the largest leaves in the plant kingdom.

There are both clustering and solitary-trunked species, and a few have subterranean stems. Many species form upward-pointing aerial roots from nodes on the stem that are assumed to aid in oxygenating the palm when growing in water; these roots are formed even on dry land. The great leaves are erect, ascending, and usually slightly arching near their tips; they are borne on massive, orange petioles arising from broad leaf bases that are soon woody and are often accompanied by masses of curling black fibers. The leaflet margins as well as the midribs of the leaflets have short bristles. In several species, the leaflets grow from the rachis at angles to create a plumose leaf. The inflorescences of these palms are commensurably large and are much branched. The thick flowering branches bear both male and female flowers and all branches are subtended by bracts of varying sizes. The fruits are also large and are covered in beautifully shiny orange-brown overlapping scales.

In their respective habitats, all the species have various human uses: the epidermis of the leaflets are used for making baskets and twine, which was formerly important to the Western world as raffia; the pith in the stems is used to make flour; the long, heavy leaf base fibers of *R. hookeri* are used for making brooms and other utilitarian products; the petioles are used for construction; the leaves are used for thatch; and the sap is tapped from emerging inflorescences to make fermented alcoholic drinks.

All the species are intolerant of cold and need copious water. The species listed below are not fussy about soil type and flourish even in calcareous soils. The genus name is a Latinized form of *rofia*, the Malagasy name for *Raphia farinifera*. Seeds germinate within 60 days. For further details of seed germination, see *Calamus*.

Raphia australis
RAF-ee-a · aw-STRAH-lis
PLATE 796

Raphia australis is indigenous to southeastern Mozambique and extreme northeastern South Africa, where it grows in lowland swamps, often in permanent shallow fresh water. The epithet is Latin for "southern." The solitary stem sometimes attains a height of 40 feet (12 m) with a diameter of 18 inches (45 cm). The leaves are immense, to 60 feet (18 m) long, on huge petioles to 3 feet (90 cm) long. The leaflets are 2–3 feet (60–90 cm) long and deep green to bluish green, and grow from the rachis in pairs, each leaflet in the pair at a different angle from the rachis but always stiff and erect, thus giving the leaf a slightly plumose, V-shaped effect. The single inflorescence grows terminally erect from the top of the trunk, is 10–12 feet (3–3.6 m) tall, and looks like a much smaller version of a *Corypha* inflorescence.

Raphia farinifera
RAF-ee-a · fa-ri-NIF-er-a
PLATES 797–798

Raphia farinifera is indigenous to Uganda, Kenya, and Tanzania, where it grows in rain forest or swamps from sea level to 2600 feet (790 m); it is probably naturalized, rather than native, in Madagascar. The epithet is Latin for "flour-bearing," a reference to the meal obtained from the starchy pith of the stems. Included within this species is the synonym *R. ruffia*.

This species is similar to *R. australis* but may be clustering or solitary trunked, has longer, slightly more plumose leaves, and has pendent inflorescences. The stems are 5–30 feet (1.5–9 m) high with diameters of 2 or more feet (60 cm). The leaves can attain a length of 70 feet (21 m), including an orange petiole 20 feet (6 m) long. The leaflets are 8 feet (2.4 m) long, deep green above and a glaucous bluish green beneath. The inflorescences are 10 feet (3 m) long and ropelike, with many congested branches subtended by large triangular bracts.

Mature specimens are nearly overwhelming in their grandeur. They should be kept moist to achieve their incredible potential.

Raphia hookeri
RAF-ee-a · HOOK-er-eye
PLATES 799–800

Raphia hookeri is found naturally in tropical western Africa, where it grows in lowland rain forest and swamps. The epithet honors British botanist Sir Joseph D. Hooker (1817–1911), director of the Royal Botanic Gardens, Kew. This solitary-trunked and clustering species has stems 30 feet (9 m) high and 1 foot (30 cm) in diameter. The leaves are 40 feet (12 m) long, including petioles 6–8 feet (1.8–2.4 m) long, and are much more spreading and arching than leaves of the 2 previously described species. The leaf bases are obscured by large, dense masses of long, curling, dark gray or black fibers. The soft, pendent leaflets are 5 feet (1.5 m) long and grow from different angles off the rachis to create a fully plumose leaf. They are glossy deep green above but glaucous silvery green beneath. The inflorescences are 7 feet (2.1 m) long and similar to those of *R. farinifera*.

Raphia sudanica
RAF-ee-a · soo-DAN-i-kah
PLATE 801

Raphia sudanica is widespread in tropical western Africa, where it grows mainly along rivers and streams at low elevations. The epithet is Latin for "of the Sudan," which refers to the Sudan Region of tropical west Africa, not the country of Sudan. The species was formerly known as *R. humilis*. This relatively small, clustering palm has either subterranean or short trunks. The leaf crown is shaped like a shuttlecock, with ascending, slightly arching leaves 15–20 feet (4.5–6 m) long. The dark green leaflets are stiff and mostly erect, growing from the rachis at slightly different angles but not creating a truly plumose effect. The large leaf looks like that of a large *Phoenix* species. This is one of the smaller species in the genus, making it more suitable to the average-size garden. It is a thirsty palm whose needs are almost aquatic. It is simply beautiful in or around pools, lakes, or streams.

Raphia taedigera
RAF-ee-a · te-DID-je-ra

Raphia taedigera has a peculiar distribution in Africa and tropical America: Nigeria, Cameroon, the Caribbean slopes of Nicaragua, Costa Rica, Panama and into northwestern Colombia, as well as the Amazon River estuary of northeastern Brazil. Throughout its range it grows in lowland rain forest and swamps. The epithet is

Latin for "torch-bearing," an allusion to the use of the wood for torches in the Americas.

The stems of this clustering species attain a maximum height of 20 feet (6 m) and a maximum diameter of 1 foot (30 cm). The leaves are typical of most other *Raphia* species: erect and ascending, with only the tips arching. They are 30 feet (9 m) long, with plumosely arranged leaflets, each of which is 4 feet (1.2 m) long, deep green on both surfaces, and soft and slightly pendent. The pendent inflorescences are 8 feet (2.4 m) long and similar in general appearance to the other pendulously flowered species.

The unusual distribution—on both sides of the Atlantic—has led to speculation that the palm has been introduced by humans into one continent or the other, but either scenario seems unlikely.

Ravenea
ra-VEN-ee-a

Ravenea consists of 20 pinnate-leaved, dioecious palms endemic to Madagascar and the Comoro Islands off the southeastern coast of Africa. Most species of this diverse genus are denizens of the rain forest, but a few are found in open drier woods and one in semiarid conditions. One species is clustering, the others solitary-trunked. They form no crownshafts, and the inflorescences grow from the leaf bases and are sparsely branched, the male panicles usually longer, more slender, and the branches more congested than those of the females. The genus name honors Jacques Louis Fréderic Ravené (1823–1879), a wealthy Berlin industrialist and patron of the arts. All species have adjacent germination, except *R. julietiae* and *R. louvelii*, which have remote germination. Seeds generally germinate within 30–60 days depending on the species. For further details of seed germination, see *Ceroxylon*.

Ravenea albicans
ra-VEN-ee-a · AL-bi-cans
PLATE 802

Ravenea albicans is a rare species from northeastern Madagascar, where it grows in rain forest on the slopes of mountains in acidic soil rich in heavy metals. The epithet is Latin for "whitish." The solitary trunk grows to 15 feet (4.5 m) tall and 5 inches (13 cm) in diameter, and is covered with persistent leaf bases making for a shabby appearance. It has a sparse crown of erect leaves up to 12 feet (3.6 m) long. The leaf rachis has lengthwise rows of dark brown scales that give it a striped appearance. The leaflets are evenly spaced, stiffly spreading, and lanceolate. They have a dense whitish coating on their undersides. Even the leaves of seedlings have the characteristic white undersides. Still rare in cultivation, this species requires shade and moisture.

Ravenea dransfieldii
ra-VEN-ee-a · dranz-FEEL-dee-eye
PLATE 803

Ravenea dransfieldii is endemic to eastern Madagascar, where it grows in rain forest on mountain slopes and crests. It is named for the eminent John Dransfield (1945–), English botanist and co-author of *Palms of Madagascar*, *Genera Palmarum*, and other seminal works on palms. The solitary, pale brown trunk grows to 20 feet (6 m) tall and 4–7 inches (10–18 cm) in diameter. The leaf bases are green but covered in tan tomentum. The leaves are mostly erect, with regularly arranged, lanceolate leaflets, which are a bright green on both surfaces. The lowermost leaflets hang down limply. The leaves have a 90-degree twist in their rachis. The bright orange fruits are about ½ inch (13 mm) in diameter. This species requires partial shade and moisture, especially when young.

Ravenea glauca
ra-VEN-ee-a · GLAW-ka
PLATE 804

Ravenea glauca is an endangered species endemic to southern central Madagascar, where it grows in dry, evergreen forests and ravines at 2200–4000 feet (670–1200 m). The epithet is Latin for "glaucous" and alludes to the chalky color of the undersides of the leaflets.

The slender, light brown, indistinctly ringed, solitary trunk attains a maximum height in habitat of 25 feet (7.6 m) with a diameter of 3–4 inches (8–10 cm). The leaf crown is hemispherical, with 18 spreading leaves covered with white scales on the underside. The leaves are 4–6 feet (1.2–1.8 m) long on petioles 3–6 inches (8–15 cm) long. The leaf sheath is white with fibrous margins. The rachis is white on the underside. The regularly spaced leaflets are 2 feet (60 cm) long, narrowly linear-lanceolate, and glossy deep green on both surfaces or lightly glaucous on the underside; new growth is stiff while the older leaflets are more lax and slightly pendent. The inflorescence is pendulous, borne on a long stalk. The rounded ½-inch (13-mm) fruits become yellow when mature.

Some growers have named the species mini-majesty palm because of its supposed visual affinities to *R. rivularis*, but this species is much more delicate, not to mention smaller in all its parts, and its leaves do not exhibit the characteristic midpoint twist that those of *R. rivularis* do. In addition, it is much slower growing than the majesty palm and does not require nearly as much water. It is as hardy to cold as *R. rivularis* and appreciates a decent soil but does not seem as finicky about the quality of the medium. It flourishes in partial shade, especially when young, but is adaptable to full sun when older except in hot climates.

Ravenea hildebrandtii
ra-VEN-ee-a · hil-de-BRAND-tee-eye
PLATES 805–806

Ravenea hildebrandtii is a critically endangered species endemic to low mountainous rain forest of the Comoro Islands, where it is an undergrowth subject. The epithet honors Johannes M. Hildebrandt (1847–1881), a German explorer, botanist, and the original collector of the species.

The solitary trunk grows to a maximum height of 25 feet (7.6 m) in habitat with a diameter of 2 inches (5 cm); under cultivation, the stature seems to be one-third the habitat height. The leaf crown is hemispherical with about 20 small, mostly spreading leaves that are 2–3 feet (60–90 cm) long on petioles 18 inches (45 cm) long. Younger leaves sometimes exhibit a twist of the rachis at its midpoint, which gives to the apical half of the blade a vertical orientation. The light green leaflets, which are 12–18 inches (30–45 cm) long, are regularly spaced along the rachis and grow in a single flat plane from it. The ellipsoid fruits are ½ inch (13 mm) long and turn a deep yellow when ripe.

This beautiful little miniature palm is slightly susceptible to lethal yellowing disease but flourishes on calcareous soils. It is reportedly slightly hardier to cold than *R. rivularis*, but not by much, being safe only in zones 10 and 11. It is drought tolerant but looks better and grows faster with regular, adequate moisture, although it is never fast growing. Indeed, specimens 25 feet (7.6 m) tall must be exceedingly old. It needs partial shade and protection from the midday sun at all ages. This species was a popular conservatory subject in Europe in Victorian times.

Ravenea julietiae
ra-VEN-ee-a · joo′-lee-ET-ee-eye
PLATE 807

Ravenea julietiae is another rare and endangered species endemic to lowland rain forest in eastern Madagascar. The epithet honors Juliet Beentje, who was the first to point a wild individual out to her husband, Henk, co-author of *Palms of Madagascar*.

In habitat, the solitary trunk attains a maximum height of 30 feet (9 m) with a diameter of 6 inches (15 cm). The leaf crown is hemispherical or slightly less so and is elongated in younger plants. The 20 leaves per crown are ascending and mostly erect when young, 4–8 feet (1.2–2.4 m) long, and gracefully arch near their apices; the older leaves are more spreading. The dark green narrowly linear-lanceolate leaflets are 2–3 feet (60–90 cm) long, regularly but widely spaced along the rachis, and grow from it at angles to create a V-shaped leaf. The extraordinarily long female inflorescences extend well beyond the crown.

The silhouette of this palm is extraordinarily beautiful; it resembles a small *Pigafetta*. Reportedly the species is slow growing and difficult.

Ravenea lakatra
ra-VEN-ee-a · la-KAH-tra
PLATE 808

Ravenea lakatra is yet another rare and critically endangered species endemic to the lowland rain forest of eastern Madagascar. The epithet is the aboriginal name for the palm.

In habitat, the solitary trunk can attain a maximum height of 45 feet (14 m) with a diameter of 6 inches (15 cm). The stems are extremely hard and durable and have erect, woody, tonguelike leaf base remnants that are 3 inches (8 cm) long along the entire length of the trunk. The leaf crown is less than hemispherical because of the ascending leaves. Individual leaves, 12 feet (3.6 m) long and slightly arching, exhibit the characteristic twist to the vertical at their midpoints and are borne on petioles 3–5 feet (90–150 cm) long. The narrowly lanceolate deep green stiff leaflets are 2 feet (60 cm) long and regularly spaced along the rachis, growing from it in a single flat plane. The female inflorescences are compact, held among the leaves, and bear round ½-inch (13-mm) fruits that are black when ripe. The seed is unique in the genus for having a pointed, rather than rounded, apex.

Ravenea madagascariensis
ra-VEN-ee-a · mad-a-gas′-kar-ee-EN-sis
PLATE 809

Ravenea madagascariensis is endemic to the central and southern half of eastern Madagascar, where it grows in rain forest and dry, evergreen forests from sea level to 5300 feet (1620 m). The epithet is Latin for "of Madagascar." *Ravenea madagascariensis* var. *monticola* is a synonym, as it does not differ from typical *R. madagascariensis*.

The solitary trunk attains a height of 60 feet (18 m) in habitat and a diameter of 6–8 inches (15–20 cm). The leaf crown is usually less than hemispherical unless the dead leaves are retained. The leaves are ascending, straight or spreading, and 6–10 feet (1.8–3 m) long, and often exhibit a twist to the vertical at their midpoints. The stout, light-colored petioles are 1–2 feet (30–60 cm) long. The leaflets are 2–3 feet (60–90 cm) long, medium to dark green, and narrowly lanceolate; they either grow off the rachis at an angle to create a V-shaped leaf, or grow in a single flat plane and are often pendent, especially when older. The margins of the leaflets often curl upwards, giving the appearance of induplicate plication. The small, round ¼-inch (6-mm) fruits are orange when ripe. This imposing but rather disheveled palm reportedly has a slight tolerance to cold, having withstood unscathed temperatures below freezing.

Ravenea moorei
ra-VEN-ee-a · MOR-ee-eye
PLATE 810

Ravenea moorei is another rare and endangered species endemic to low-elevation secondary forest on the Comoro Islands. The epithet honors palm taxonomist Harold E. Moore, Jr. (1917–1980), of Cornell University.

The solitary trunk attains a height of 60 feet (18 m) in habitat with a diameter of 5–12 inches (13–30 cm) and is much swollen at the base. The leaf crown is hemispherical or slightly less so, with many stiff, straight, leaves, each 4–7 feet (1.2–2.1 m) long on a short petiole 4–10 inches (10–25 cm) long. Near the tip of the leaf, the rachis twists 90 degrees. The many leaflets are regularly disposed along the rachis and grow from it in a single flat plane. They are 2 feet (60 cm) long, deep green on both surfaces, and are narrowly lanceolate and pendent on older leaves. The round ½-inch (13-mm) fruits turn orange when ripe.

This majestic species has visual affinities to the Asian *Orania* species. It is still rare in cultivation and little is known of its cultural requirements, but it is definitely neither cold hardy nor drought tolerant but probably is adaptable, except in hot climates, to full sun, even when young.

Ravenea musicalis
ra-VEN-ee-a · moo-zi-KAL-iss
PLATES 811–812

Ravenea musicalis is a rare species endemic to a single river in southeastern Madagascar at low elevations. The epithet is Latin for "musical," a whimsical allusion to the sound of the fruits falling into the river as Henk Beentje collected the first specimen. This is a truly aquatic palm species; the seeds sprout and grow beneath the water, living as submerged aquatic plants until they are large enough to break the surface of the water.

The bottle-shaped, solitary trunk attains a height of 25 feet (7.6 m) in habitat and is much swollen in its lower half and narrow in the upper half. The leaf crowns of young plants are similar to those of *R. rivularis*, including the twists of the leaf rachises at their midpoints, and the leaf dimensions of the 2 species are nearly identical. The spongy, orange fruits are ½ inch (13 mm) in diameter and split open easily. This species has proven difficult to cultivate. Seeds sprout easily, but seedlings seem to require flowing water to flourish.

Ravenea rivularis
ra-VEN-ee-a · riv-yoo-LAR-iss
PLATES 813–814

Ravenea rivularis is a vulnerable species endemic to southern and central Madagascar, where it grows along the banks of streams and rivers or in swamps on the edge of the rain forest from elevations of 1200 to 2500 feet (370–760 m). It often forms large colonies along riverbanks. The epithet is Latin for "of the rivers." Majesty, or majestic, palm is its common name.

Mature trunks can grow to 80 feet (24 m) in their native haunts but are seldom more than half that stature under cultivation. The solitary stems are swollen at their bases and exhibit widely spaced leaf scar rings. There are usually 15–25 leaves in a crown, and palms with trunks usually have a full, rounded crown as the older leaves become pendent. The leaves are 6–8 feet (1.8–2.4 m) long on stout petioles usually no more than 1 foot (30 cm) long. The leaflets grow from the rachis in a flat plane and are 2 feet (60 cm) long, narrowly elliptic and tapering, and regularly spaced along the rachis. They are thin and supple but not usually pendent. The leaves are erect in young plants but, as the tree develops a trunk, they begin to arch and are usually twisted at the middle of the rachis so that the leaflets along the apical half of the blade are oriented vertically or nearly so. Leaflet color is a smooth deep green on both sides. The inflorescences are 3 feet (90 cm) long and much branched, and bear small, white flowers. The round ½-inch (13-mm) fruits are bright red when ripe.

The palm is nearly aquatic in its native habitat, so it is difficult, if not impossible, to give it too much moisture in cultivation. It also wants a rich, even mucky soil and, when older, full sun. It is fast growing when younger if given enough water and a soil that is decent, but it is not hardy to cold and its success is limited to zones 10 and 11.

Majesty palm is attractive enough to stand alone as a specimen tree but looks better in groups of 3 or more individuals of varying heights. It flourishes in partial shade when young and is a superb subject for planting in such sites where many other large pinnate-leaved palms languish, but the site should be moist. Too often one sees this palm languishing on the tops of mounds of earth or surrounded by concrete and asphalt. This species has gained popularity as an interiorscape subject as it grows fast indoors and looks good if given lots of light and constant moisture. These conditions are difficult to maintain in the average home. It should not dry out while in a container and should be watered when the top inch (2.5 cm) or so of the container's soil is dry; it also prefers high relative humidity and becomes ragged looking when grown indoors with artificial dry heat.

Ravenea robustior
ra-VEN-ee-a · ro-BUS-tee-or
PLATE 815

Ravenea robustior is relatively widespread in Madagascar, where it grows in rain forest and clearings therein as well as in moist evergreen forests from near sea level to almost 6000 feet (1800 m). The epithet is Latin for "more robust," the comparison presumably with other *Ravenea* species.

The solitary trunk grows to 100 feet (30 m) in habitat with a diameter to 2 feet (60 cm) and a swollen base, but these dimensions are rare and the trees are usually half this stature. The leaf crown is dense and shaped like a shuttlecock in young palms because of the ascending, erect leaves 5–12 feet (1.5–3.6 m) long on petioles 2–4 feet (60–120 cm) long. The leaves are mostly stiff and unarching, but older leaves may be spreading and some show the characteristic twist to the vertical at their midpoints. The dark green leaflets are 2–4 feet (60–120 cm) long, regularly and closely spaced along the rachis, and grow from it in a single flat plane. They are somewhat pendulous in older specimens. The fruits are olive shaped and are orange when ripe.

The palm is majestic and beautiful but reportedly slow growing so that one plants it for the benefit of future generations; the largest individuals in habitat must be very old indeed. Amazingly, this palm is also reported to take some cold, surviving unscathed temperatures as low as 20°F (−7°C) in Mediterranean climes.

Ravenea sambiranensis
ra-VEN-ee-a · sahm-bir'-a-NEN-sis
PLATE 816

Ravenea sambiranensis is widespread in Madagascar, from the west central part of the island to its northwestern part, with scattered locations up and down the eastern coast. It grows from sea level to 6000 feet (1800 m), in lowland rain forest as well as on drier slopes of the interior. The epithet is Latin for "of Sambirano," a river in northwestern Madagascar.

In habitat, the solitary trunk attains a maximum height of

almost 100 feet (30 m) but is usually much shorter and is never more than 1 foot (30 cm) in diameter. The leaf crown is dense and shaped like a shuttlecock because of the erect, ascending leaves. The leaves of young palms are straight and unarching, with leaflets growing from the rachis in a single flat plane; the leaves of older, trunked individuals are beautifully recurved, with dark green, rigid, narrowly lanceolate leaflets 2–3 feet (60–90 cm) long growing off the rachis at a steep angle to create a deeply V-shaped leaf. In all cases, however, the crown is never more than hemispherical. The fruits are reddish orange when ripe.

This is another species whose adult silhouette is extraordinarily beautiful. It is reportedly cold resistant and not overly slow growing but needs fast-draining soil.

Ravenea xerophila
ra-VEN-ee-a · ze-ro-FI-la
PLATE 817

Ravenea xerophila is a rare and endangered species endemic to southern Madagascar, where it grows at 600–2600 feet (180–790 m), mostly on the tops of hills, in dry thorn forests characterized by exotic-looking, giant, cactuslike *Alluaudia* and *Didierea* species, unique to the island. The epithet is derived from 2 Greek words meaning "dry loving," alluding to the habitat.

The solitary trunk attains a maximum height of 25 feet (7.6 m) but is usually shorter, with a diameter to 1 foot (30 cm). The younger parts of the stem are covered in dense, persistent, dark gray or black sheets of leaf sheath remnants. The small leaf crown is hemispherical, with 20 ascending, arching, deep to bluish green leaves 5–6 feet (1.5–1.8 m) long on petioles 2 feet (60 cm) long. The narrow, stiff leaflets are numerous, closely and regularly spaced along the rachis and growing from it at angles that create a deeply V-shaped blade. Fruits are about 1 inch (2.5 cm) in diameter and yellowish.

This palm is attractive but agonizingly slow growing. It is, however, unusually hardy to cold, having withstood unscathed temperatures in the upper 20s Fahrenheit (around −2°C). It is drought tolerant but grows and looks better with adequate and regular moisture; in habitat its aggressive root system goes deep to find underground moisture. It would not be out of place in a cactus or succulent garden.

Reinhardtia
ryn-HARD-tee-a

Reinhardtia comprises 6 solitary-trunked and clustering, pinnate-leaved, monoecious palms in southern Mexico, Central America, northern Colombia, and the Dominican Republic. All are undergrowth plants in the rain forest and are mostly small palms. Only one of them ever becomes a canopy tree, and, interestingly enough, it is the only species whose natural range is not in Central and South America, where the other 5 are found.

The leaves of these palms are unusual in that they may be entire and undivided or divided into several pinnae, and some have openings or "windows" at the base of the clustered pinnae, which phenomenon has led to their being called window palms or windowpane palms. The species do not form crownshafts, and the leaf sheaths wear down into a network of dark fibers that adhere to the stems, similar to *Caryota* species, although the 2 genera are not closely related. The inflorescences are spikelike or a cluster of sparsely branched, relatively short flowering branches at the ends of long primary branches. The flowering branches bear both male and female flowers. The fruits are mostly small and blackish purple.

All the species are in danger of extinction because the forests in which they dwell are being cut and burned, and they cannot exist in nature outside these forests. For such interesting, attractive palms, these species are inexplicably uncommon in cultivation, especially considering their reported ease of cultivation. Their main requirements are a tropical or nearly tropical climate, abundant and regular moisture, a humus-rich soil that is perfectly drained, and partial to near total shade.

The genus name is believed to commemorate either Johannes C. H. Reinhardt (1778–1845) or his son, Johannes T. Reinhardt (1816–1882), both of whom were important Danish biologists. Seeds germinate within 45 days.

Reinhardtia gracilis
ryn-HARD-tee-a · gra-SI-lis
PLATE 818

Reinhardtia gracilis is a threatened undergrowth species in rain forest of southern Mexico, Central America, and northern Colombia, from elevations near sea level to 4300 feet (1310 m). The common name is window palm or windowpane palm. The epithet is Latin for "graceful." Four varieties with slight vegetative differences have been described. All are beautiful.

The species is found in nature mostly with clustering stems but also rarely as a solitary-trunked individual. Mature stems sometimes grow to 10 feet (3 m) high but usually are no more than 4–5 feet (1.2–1.5 m), and their diameter is never more than ½ inch (13 mm). The upper parts of the stems are usually covered in a wrapping of the old leaf bases which are green, turning brown, with the lower parts of the stems green to gray and ringed. If clustered, the clumps are never more than 4 feet (1.2 m) wide. The leaves are evenly pinnate and 6–12 inches (15–30 cm) long, with 4–8 pairs of segments. The petiole is 3–24 inches (8–60 cm) long, thin, and green. Each segment is 3–6 inches (8–15 cm) long, wedge shaped, with its apex squared or more usually obliquely truncated, and with small indentations at the tip, each point corresponding to a fused pinna. The apical segments are usually much broader than the lower ones, and each segment has small, elliptical "windows" at its base, near the leaf rachis. The inflorescences are composed of peduncles 2–3 feet (60–90 cm) long, from whose ends the short flowering branches are green but usually turn red when mature. They bear small, whitish flowers of both sexes. The oblong fruits are slightly more than ½ inch (13 mm) long and turn black when mature. The endosperm is homogeneous.

This palm needs constant moisture, relishes high relative

humidity, and wants a rich, humus-laden, fast-draining soil. It does not tolerate full sun or freezing temperatures. It is suitable only for an intimate site in which it can be seen up close in any other situation it is lost. The plant is perfect for containers on a shaded patio or courtyard. It needs good light without direct sun, and languishes and dies in heavy shade.

Reinhardtia latisecta
ryn-HARD-tee-a · la-ti-SEK-ta
PLATE 819

Reinhardtia latisecta occurs naturally in 5 disjunct populations in Central America: central Belize, Guatemala, northeastern Honduras, southern Nicaragua, and 2 locales along the Pacific slope of Costa Rica, always in lowland rain forest. The epithet describes the leaflets and is derived from Latin words meaning "wide" and "segment." In the United States, this large, clustering species is called giant windowpane palm.

The stems attain a height of 25 feet (7.6 m) with a diameter of slightly more than 2 inches (5 cm). The leaves are 3–5 feet (90–150 cm) long, obovate but divided into 4–6 segments of varying widths; the leaflets of the terminal pair are always much larger than the others. The margins of all segments are toothed and exhibit little elliptical "windows" at their bases adjacent to the leaf's rachis. The inflorescences consist of peduncles 3 feet (90 cm) long from the ends of which emerge the short flowering branches only 8 inches (20 cm) long. Creamy white at first, the flowering branches turn reddish before the ovoid, ½-inch (13-mm) black fruits mature. The endosperm is ruminate.

This species is the most beautiful in the genus. Its size assures that it will not be lost among other vegetation and its large, matte green, shapely leaves are a delight up close and from afar. Its only faults are its intolerance of cold and its demand for high humidity at all times.

Reinhardtia paiewonskiana
ryn-HARD-tee-a · py-ay-wons'-kee-AH-na
PLATE 820

Reinhardtia paiewonskiana is endemic to steep ridges in low mountainous rain forest in the southern Dominican Republic. It co-occurs with *Prestoea acuminata* var. *montana*, to which it bears a striking resemblance. The species was named in 1987 for Benjamin Paiewonsky, president of the Fundación Pro-Flora Dominicana, which operates the botanical garden in Santo Domingo, Dominican Republic.

This solitary-trunked species is the tallest in the genus, the stem attaining a maximum height of 40 feet (12 m) with a diameter of 5–6 inches (13–15 cm). The leaf crown is hemispherical or slightly more so and contains a dozen leaves, each 6 feet (1.8 m) long on a petiole 12–18 inches (30–45 cm) long. The emerald green leaflets are as long as 2 feet (60 cm) and are narrowly elliptical and long tipped. They are regularly and closely spaced along the rachis from which they grow in a single flat plane, becoming pendent with age. The inflorescence peduncles are 2–3 feet (60–90 cm) long with terminal flowering branches 18 inches (45 cm) long. The round ¾-inch (19-mm) fruits are black, and the endosperm is ruminate.

The species has a beautiful silhouette, reminiscent of *Howea forsteriana*. It dislikes dry air and warm nighttime temperatures and so has proven difficult to cultivate in South Florida. It will likely be a real winner in the wetter parts of Hawaii and other places that can mimic its natural habitat.

Reinhardtia simplex
ryn-HARD-tee-a · SIM-plex

Reinhardtia simplex occurs naturally in northeastern Honduras, along the Caribbean slopes of Nicaragua, Costa Rica, and Panama to extreme northwestern Colombia, where it grows in low mountainous rain forest. The epithet is Latin for "simple."

The stems of this small, clustering species attain a maximum height of 4 feet (1.2 m) with a diameter less than ½ inch (13 mm). The leaves are glossy deep green above, duller, lighter green beneath, and 8 inches (20 cm) long on wiry petioles 6 inches (15 cm) long. The leaves are either simple and undivided or segmented into 3 leaflets, the terminal one much larger and linearly oval or diamond shaped, the other 2 at right angles to the rachis and linear wedge shaped, all with toothed margins. The inflorescences consist of peduncles 2 feet (60 cm) long from the ends of which the short flowering branches produce obovoid blackish purple fruits ½ inch (13 mm) in diameter. The endosperm is homogeneous. Another charmer for an intimate space in a garden or conservatory, this species is, like others in the genus, unaccountably rare in cultivation.

Retispatha
re-ti-SPAY-tha

Retispatha is a monotypic genus of rare pinnate-leaved, dioecious, nonclimbing or shortly climbing rattan palm endemic to Borneo, where it grows in low mountainous rain forest and is widespread but not common throughout the island, often forming thickets. The genus name is derived from Greek words meaning "network" and "spathe," an allusion to the netlike bracts subtending the flowering branches. Seeds are not generally available, so germination information is lacking.

Retispatha dumetosa
re-ti-SPAY-tha · doo-me-TO-sa
PLATE 821

Retispatha dumetosa is densely spiny but lacks climbing organs (cirrus and flagella). It is not known in cultivation outside of a few tropical botanic gardens but would seem to be eminently worthy of being planted. Its stature would allow it to contrast beautifully with other vegetation. It needs a tropical or nearly tropical climate and copious, regular moisture in partial shade to full sun. The epithet is from the Latin word for "thicket."

Rhapidophyllum
ra-pi'-do-FIL-lum

Rhapidophyllum is a monotypic clustering, palmate-leaved, polygamous or dioecious palm in southeastern United States. The genus name combines *Rhapis*, the lady palm genus, with the Greek word for "leaf," as both genera have a similar manner of splitting of the leaflets. Fresh seeds can take 90 days to a year or longer to germinate and do so sporadically. For further details of seed germination, see *Corypha*.

Rhapidophyllum hystrix
ra-pi'-do-FIL-lum · HIS-trix
PLATE 822

Rhapidophyllum hystrix is endemic to southwestern Mississippi, southern Alabama, southern Georgia, northern peninsular Florida, and extreme southern coastal South Carolina, where it occurs in the undergrowth on the coastal plain in deciduous forests or in swampy lowlands, often on limestone soils but always in moist sites. The common name is needle palm. The epithet is from the Greek word for "porcupine" and alludes to the needlelike spines on the leaf sheaths. Except for a few sites in Florida, the needle palm is very rare in habitat. This lamentable situation has arisen lately because of collection of specimens for the nursery trade as the seeds are slow to germinate and slow growing once they sprout. In addition, the species reproduces poorly in the wild. It would seem to be a vanishing relict species, even in Florida.

The short stems are mostly less than 3 feet (90 cm) high and are 4 inches (10 cm) in diameter but usually appear much thicker because of the mass of petiole bases and fibers that cover them. The stems may also be densely clustered or sparsely suckering, the latter condition seemingly associated with shadier sites. In spite of the diminutive trunk stature, the total height of a palm can reach 12 feet (3.6 m). Old trunks often start to disintegrate near their bases and begin to fall over but have the ability to root from almost any point on the stem, thus slowly creeping along the ground. The summits of the stubby trunks are covered in soft leaf base fibers and in dense clusters of upright black or dark brown needlelike spines, each of which is usually 6–8 inches (15–20 cm) long but may be 16 inches (41 cm). The leaves are 4 feet (1.2 m) wide and range from semicircular to slightly more so, with 15–24 segments that are nearly 1 inch (2.5 cm) wide and free for three-quarters of their length. The petiole may be 3–5 feet (90–150 cm) long, depending on how much sun the palm receives. The segments are deep green above and deep green with a silvery sheen beneath. The inflorescences are short and stubby like the trunk itself and are formed near the apex of the stem among the mass of dense fibers. The flowers are most often either male or female but some may also be hermaphroditic. In any case the flowers are yellowish white or pale lavender and have a slight musky odor, especially at night. Individual populations never seem to bloom regularly, but rather they are prolific in some years and, in other years, may produce few inflorescences. The round ½-inch (13-mm) fruits are reddish purple to brown and covered in yellowish tomentum. They are densely packed in a tight cluster atop the trunk and sometimes germinate among the leaf fibers of the crown before they break apart and fall. They have a strong, sickeningly sweet odor when ripe.

The species may be hardier to cold than any other palm. It is adaptable to zones 6 through 10 and reportedly has survived in zone 5 without protection. It luxuriates in partial shade; although it can adapt to full southern sun, it does not look as good—the leaves are smaller, the petioles shorter, and the clustering or suckering phenomenon much more dense. It is slow growing under all conditions, especially shady ones. The palm is adapted to several soils from heavy to sandy as long as they are rich and moist. The most important cultural requirement is probably regular, adequate moisture.

This palm has extraordinarily beautiful leaves, especially in shady sites, and is a wonderful close-up specimen. It looks as good mixed in with contrasting vegetation against which its half-pinwheel leaves offer much needed visual relief. Some people disagree, but this palm looks its best if grown in shade with most of its suckers removed to reveal the beautiful form of the leaf crown. Its infamous needles are usually no problem as they are almost hidden in the center of the leaf crown. Given adequate moisture, this species might make a wonderful conservatory or atrium specimen.

Rhapis
RAY-pis

Rhapis is composed of about 11 mostly densely clustering, palmate-leaved, dioecious palms in southern China including Hainan Island, Southeast Asia, Thailand, and northern Sumatra. The mostly small undergrowth species grow in dry or monsoonal evergreen forests at low to moderate elevations, usually on limestone soils. The leaves are borne in usually elongated crowns, are very deeply divided (usually to the petiole), and often are circular or nearly so; the segments are pleated and their apices bear small indentations. The petioles are long and slender, and the upper parts of the stems are clothed in a tightly adhering, dense network of dark leaf sheath fibers. The inflorescences are short and congested, and the flowering branches bear either male or female blossoms. Female plants produce 1- to 3-lobed, small, usually rounded fruits that are white or yellowish when mature.

Two species are extremely common in cultivation, but the others are rare. Most species are surprisingly hardy to cold, considering their tropical or nearly tropical origins. They do not look good in the full sun of hot climates but, at the same time, are generally tolerant of neglect and drought. The most widely grown species, *R. excelsa*, makes one of the finest indoor subjects. Deep containers work best; *Rhapis* roots go deep and will lift a plant out of a too-shallow container.

The taxonomy of *Rhapis* is still somewhat contentious and unsettled, despite Hastings's 2003 revision. Two new species have been described since then, and the problem surrounding the identity of many of the forms in cultivation is intractable. For example, the cultivated entity called *Rhapis* 'Super Dwarf', which grows like a tuft of grass only 6 inches (15 cm) tall, does not match any of the described species. The possibility that some cultivated forms may be of hybrid origin cannot be ruled out, and some cultivated entities may represent undescribed species. The genus name is Greek for "rod," an allusion perhaps to the slender stems.

Seeds germinate within 120 days and should not be allowed to dry out before sowing. For further details of seed germination, see *Corypha*.

Rhapis excelsa
RAY-pis · ek-SEL-sa
PLATES 823–824

Rhapis excelsa is native to a wide swath of territory from Vietnam through southern China and Hainan, on limestone hills and cliffs and in dry, evergreen forests at moderate elevations. Common names are lady palm and bamboo palm, the latter sobriquet unfortunate as several other palm species share it; a better alternative vernacular is the one used by Lynn McKamey, a Texas grower who specializes in the genus: large lady palm. The epithet is Latin for "tall" and is ironic since the species is not, by far, the tallest in the genus.

The species forms dense and often large clumps (of 100 or more stems) with elongated leaf crowns. The new stems mostly grow along the perimeter of a clump, which phenomenon results in leaves from ground level up to the tops of the clumps. The stems grow to a height of 10 feet (3 m) with a maximum diameter of 1 inch (2.5 cm). They are covered in all but their oldest parts with a loosely adhering network of dark gray to black fibers. The leaves are 18 inches (45 cm) wide on thin petioles also 18 inches (45 cm) long, and are circular or nearly circular. The 4–14 segments of slightly varying widths are free almost to the petiole and are generally linear-elliptic and a deep, glossy green above and below. They are also leathery, bluntly truncated at their apices with small indentations, mostly stiff, and almost never pendent in any part. The short inflorescences grow from the upper leaf axils and are cream colored, often with a pinkish tinge. The rachillae are hairless and bear small, round fruits that are greenish yellow when ripe.

A plethora of cultivars exist, most of which have originated in Japan and are dwarf with variegated leaves. Many of these are extravagantly beautiful up close; their use in the landscape seems limited to intimate sites. They are, however, unparalleled container plants.

The species is amazingly hardy to cold considering its purported tropical and subtropical origins. It is unscathed by temperatures in the low 20s Fahrenheit (around −7°C), even in regions where winter freezes are accompanied by moisture. It is also known to sprout from its rhizomes after the stems have been killed to the ground by cold. It seems to thrive on almost any soil type as long as it is well drained and neither too acidic nor too alkaline, and it luxuriates in partial shade, which is a requirement in hot climates if the deep green leaf color is to be preserved. Variegated cultivars should be kept from direct sun. About the only fault of this species is that it is slow growing, especially if not given regular, adequate moisture.

That the leaves grow from the trunks for most of the trunk length and that the new canes or trunks grow mostly from the perimeter of a clump make this little palm species near perfection for informal tall hedges in partially shaded sites. It is also almost unexcelled as a close-up specimen where its grace can be fully appreciated; like many other clumping palms, it is more graceful and artistic if some of the crowded trunks are removed to better show off its wonderful silhouette. Few plants are as beautiful near water as *R. excelsa*. The dwarf forms, both variegated and of single hue, make exceptionally beautiful plantings along a walkway or among stones in partially shaded rock gardens. There is hardly a finer candidate for growing indoors. The lady palm endures neglect and abuse similar to *Howea forsteriana*, and asks only for a site with bright light (but not sun) and regular watering.

Rhapis humilis
RAY-pis · HYOO-mi-lis
PLATE 825

Rhapis humilis is apparently unknown in the wild, all known plants having come from cultivated individuals in China. It is assumed to have originated in southern China. The common name is slender lady palm. The epithet is Latin for "humble" or "small," an even greater irony than that of the above species since the stems of this clumper attain a height of 20 feet (6 m).

The trunks have generally less fiber than those of *R. excelsa*, are also thinner, and are beautifully and closely ringed in the parts lacking fiber. The clumps are not as massive and the stems are significantly fewer in number. In addition, the leaf crowns are not nearly so elongated as in *R. excelsa*, and the leaves are wider but not usually as circular. The segments are generally more numerous and are longer, softer, pendent, and definitely pointed at their apices. The inflorescences are similar to those of *R. excelsa*, but the rachillae are pubescent.

The slender lady palm is more cold tolerant than the large lady palm and is adaptable to zones 8b through 11. It is not at home in hot climates, like that of South Florida. It flourishes in Mediterranean climates, where the nights, even in summer, have lower temperatures. Like *R. excelsa*, it needs protection from the midday sun in hot climates. It appreciates the same soils as does *R. excelsa*, and its moisture requirements are the same and possibly less strict. The only method of propagation is by removing the suckers, as no individual palm produces fruit. It is as slow growing as the large lady palm. There are a few cultivars with smaller stature or with variegated leaves, and these are more useful indoors than in the landscape.

This species is, if possible, more beautiful than the large lady palm, taller and more elegant looking with softer, larger leaves that are even more appealing. As a large hedge or barrier, it is unexcelled, and its silhouette is as lovely as that of any palm species. The only landscape situation it is not perfectly suited to is that of a specimen planting surrounded by space, and even there it is good if its silhouette is against a wall or contrasting vegetation.

The species is almost as amenable to indoor culture as is *R. excelsa*, the mitigating factor being its size.

Rhapis laosensis
RAY-pis · lah-o-SEN-sis
PLATE 826

Rhapis laosensis is a rare and endangered species occurring naturally in Laos and adjacent northeastern Thailand and Vietnam, where it grows in lowland rain forest. The epithet is Latin for "of Laos." The stems of this small, sparse clumper are thin and reach 9 feet (2.7 m) high. The leaves are composed of 2–5, but usually 3, diverging, pleated segments, each of which is 8 inches (20 cm) long, strap shaped or oblong, with a small, abruptly pointed, "hooded" tip. They are soft, slightly recurved, and glossy medium green above and paler, duller green beneath. This delicate, attractive palm makes a beautiful tall groundcover. It is less tolerant of cold than *R. excelsa* or *R. humilis*, and needs more moisture than them. It seems adaptable only to zones 10 and 11 and possibly in warm microclimates of zone 9b. It also seems to flourish in humid regions. Propagation is by removal of the suckers.

Rhapis multifida
RAY-pis · mul-TIF-i-da
PLATE 827

Rhapis multifida is endemic to southern China (and probably adjacent Vietnam), where it grows in monsoonal evergreen forests at low to moderate elevations, usually on limestone soils. The epithet is derived from Latin words meaning "many" and "division."

The slender stems of this sparse clumper grow to 8 feet (2.4 m) tall. The mature leaves are 2 feet (60 cm) wide, semicircular to slightly more so, and contain 12–24 narrow segments that are free almost their entire length. The segments of young leaves are stiff, but those of older leaves are lax, slightly arching, and pendent. They are glossy medium green on both surfaces. The inflorescences are several branched and erect, with yellowish flowering branches covered in short, felty hairs. The round fruits are less than ½ inch (13 mm) in diameter.

This palm has the same cultural requirements as *R. excelsa* except it is not as hardy to cold, being adaptable to zones 9b through 11. It has the most beautiful leaves of the known species and is extraordinarily attractive as a silhouette against contrasting vegetation or other objects.

Rhapis siamensis
RAY-pis · sy-a-MEN-sis

Rhapis siamensis is endemic to limestone outcrops in southern Thailand. The epithet means "native to Siam," Siam being the classical name for Thailand. The stems grow in dense clusters, maturing to 15 feet (4.5 m) tall and 1 inch (2.5 cm) in diameter. The semicircular leaves are 8–18 inches (20–45 cm) across and have 5–12 leaflets, which are oblong with abruptly pointed tips. The leaflets are joined to one another up to one-fourth their length and are glossy dark green above and paler green below. The rachillae are not hairy. The whitish fruits are less than ½ inch (13 mm) in diameter. This species is not as widely known in U.S. gardens, but it is highly desirable with many fine attributes. Its graceful stature would be an asset to any garden, especially in a shady, moist situation. Its cold hardiness is still untested, but this tropical species probably will not thrive outside zones 10 and 11.

Rhapis subtilis
RAY-pis · SUB-ti-lis
PLATE 828

Rhapis subtilis occurs naturally in Laos, Thailand, Vietnam, and Sumatra, where it grows in monsoonal rain forest at low elevations on soils derived from sandstone. The epithet is Latin for "thin," an allusion to the slender canes and petioles. It is usually called Thailand lady palm.

This small but dense clumper has stems 6–8 feet (1.8–2.4 m) tall. The leaf segments vary in form, from many narrow segments in a nearly circular blade, to single and unsegmented, wide strap-shaped or ovate leaves with "hooded," pointed tips. All are glossy dark green above and below, and the segments are pleated and are borne on thin petioles 1 foot (30 cm) long. The forms with many, narrow segments look much like dwarfed versions of *R. humilis* and, when first introduced into cultivation by Watana Sumawong, were thought to be forms of it.

The species is not as cold hardy as most of the others, being adaptable only to zones 10 and 11. It seems to greatly resent cold, drying winds and needs nearly constantly high humidity. The clumps are so dense and the rhizomes so brittle that it is nearly impossible to successfully remove suckers for propagation; but seeds are readily produced from male and female plants.

Rhopaloblaste
ro-pa-lo-BLAS-tee

Rhopaloblaste includes 6 pinnate-leaved, monoecious palms in low mountainous rain forest of the Nicobar Islands, peninsular Malaysia, Singapore, the Moluccas, New Guinea, and the Solomon Islands. All but one species is solitary trunked. All have slender, distinctly ringed trunks, handsome crownshafts, and large leaves, most with beautifully pendent leaflets. All species have dark brown scales covering the leaf rachis. The much-branched inflorescences grow from beneath the crownshafts. Two distinctive features of most species are that the 2 lowermost branches of the inflorescence bend backwards toward the trunk and that the rachillae are twisted and sinuous. Clusters of orange or red fruits are produced, and the endosperm is deeply ruminate.

These palms are among the world's most beautiful, but they are unaccountably rare in cultivation, even in tropical regions, possibly because they are slow growing. All are tender to cold and possible only in zones 10b and 11. They are true water lovers and much resent dry conditions, including desiccating winds.

The genus name comes from Greek words meaning "club" and "bud," a reference to the club-shaped embryo. Seeds of *Rhopaloblaste* germinate within 45 days. For further details of seed germination, see *Areca*.

Rhopaloblaste augusta
ro-pa-lo-BLAS-tee · aw-GUS-ta
PLATE 829

Rhopaloblaste augusta is a rare and endangered solitary-trunked species endemic to the Nicobar Islands, where it grows in low mountainous rain forest. The epithet is Latin for "noble" or "majestic." This species and *R. ceramica* are the tallest in the genus.

The solitary trunk attains a height of 75 feet (23 m) in habitat but seldom half that height under cultivation. It is never more than 1 foot (30 cm) in diameter above its swollen base and is light brown to gray with widely spaced darker rings of leaf base scars. The crownshaft is 2 feet (60 cm) tall, not much thicker than the trunk, slightly bulging at its midpoint, and grayish light brown and velvety to the touch. The leaf crown is hemispherical or more so, composed of 11 stiff, unarching (except at their tips), spreading leaves 12 feet (3.6 m) long on petioles 3–4 inches (8–10 cm) long. The regularly and closely spaced narrow leaflets are 2 feet (60 cm) long and pendent, forming curtains on each side of the rachis. The ellipsoid fruits are deep orange to red and less than 1 inch (2.5 cm) long. This palm is enthrallingly beautiful with a silhouette that is beyond compare, and there is arguably no more beautiful sight in the plant world than 3 or more individuals of greatly varying heights.

Rhopaloblaste ceramica
ro-pa-lo-BLAS-tee · ser-RAHM-i-ka
PLATE 830

Rhopaloblaste ceramica occurs naturally on the islands of the Moluccas, as well as the island of New Guinea, where it grows in low mountainous rain forest. The epithet is Latin for "of Ceram," one of the Moluccas.

The solitary trunk attains a height of 100 feet (30 m) in habitat with a diameter of less than 1 foot (30 cm) above the swollen base. It is light gray or tan and graced with widely spaced darker rings of leaf base scars. The crownshaft is 3 feet (90 cm) tall, light tan to almost silver, and distinctly bulged at its base. The leaf crown is hemispherical or slightly more so and contains 15–17 straight, slightly arching leaves, each 10 feet (3 m) long. The narrow leaflets are 3 feet (90 cm) long, regularly and closely spaced along the rachis, and are pendent in older leaves, creating twin curtains of green along both sides of the rachis. The beautiful scarlet fruits are ovoid and more than 1 inch (2.5 cm) long. This attractive species is akin visually to *R. augusta* but taller.

Rhopaloblaste elegans
ro-pa-lo-BLAS-tee · EL-e-ganz
PLATE 831

Rhopaloblaste elegans is endemic to the Solomon Islands, where it grows in lowland rain forest on limestone soils. The epithet is Latin for "elegant."

The solitary stem attains a height of 40 or more feet (12 m) and a diameter of 6–8 inches (15–20 cm) above its swollen base. The crownshaft is 3 feet (90 cm) tall, scarcely thicker than the trunk, cylindrical, and brownish green. The leaf crown is spherical or nearly so with leaves 10 feet (3 m) long or slightly more, pendent and slightly arching. The numerous, narrow leaflets are 2 feet (60 cm) long, light green, regularly and closely spaced along the rachis, and partially pendent but not forming the curtain effect of other species; rather, they gracefully arch downwards. The globose 1-inch (2.5-cm) fruits are waxy and a beautiful cinnamon color when ripe.

Its name says it all. The species is certainly among the most elegant of any genus, with its slender, straight trunk and large, shimmering, light green arching leaves.

Rhopaloblaste ledermanniana
ro-pa-lo-BLAS-tee · lee'-der-man-nee-AH-na

Rhopaloblaste ledermanniana is endemic to New Guinea, where it grows in mountainous rain forest from sea level to 3500 feet (1070 m). The epithet honors Carl L. Ledermann (1875–1958), Swiss horticulturist and botanist who collected extensively in New Guinea. The species was formerly known as *R. brassii*.

The stem of this small, solitary-trunked species grows to 45 feet (14 m) high with a diameter of 2 inches (5 cm). The crownshaft is 2 feet (60 cm) tall, scarcely thicker than the trunk but bulging near its midpoint, cylindrical, a strange purplish green, and covered in a scurfy tomentum. The leaf crown is hemispherical or slightly less so and contains 11 straight, ascending, slightly arching leaves. Each leaf is 8–10 feet (2.4–3 m) long on a gray to tan rough-textured petiole of 8 inches (20 cm) long. The dark green leaflets are 2 feet (60 cm) long, stiff and nonpendent, growing at a slight angle off the rachis. In young leaves, the arrangement creates a shallow-shape leaf, but in older leaves, which are much more lax, spreading, and half-pendent, the arrangement of the leaflets creates a flat leaf. The inflorescences seem relatively gigantic and produce large clusters of ellipsoid, ½-inch (13-mm) bright red fruits. Although rare in cultivation, the species is deserving of much greater distribution as it is one of the most beautiful in the genus.

Rhopaloblaste singaporensis
ro-pa-lo-BLAS-tee · sing-a-por-EN-sis
PLATE 832

Rhopaloblaste singaporensis occurs naturally in the undergrowth of lowland rain forest of peninsular Malaysia and Singapore. The epithet is Latin for "of Singapore." The clustering stems grow to 10–15 feet (3–4.5 m) tall with a diameter of about 1 inch (2.5 cm). The brown, dead leaf bases persist and obscure the crownshaft. The straight, spreading leaves are 3 feet (90 cm) long. The linear-lanceolate, long-tipped leaflets are 18 inches (45 cm) long and regularly and closely spaced along the rachis from which they grow in a single flat plane. They are spreading, nonpendent, and glossy medium green above but paler green beneath. The globose ½-inch (13-mm) fruits are deep orange to red when ripe. This small palm makes a startlingly beautiful accent among other vegetation. It is the most distinctive species in the genus and looks nothing like the other species, yet it has a charm and grace all its own.

Rhopalostylis
ro-pa'-lo-STY-lis

Rhopalostylis consists of 2 tall solitary-trunked, pinnate-leaved, monoecious palms in the South Pacific islands. Some authors recognize 3 species, others, 2 and a variety. We see 2 species. The distinctly smooth, ringed trunks topped by narrow, erect crowns shaped like a shaving brush define the term "shaving brush palm." The crownshafts are prominent, distinct, and often bulbous or greatly swollen. The inflorescences grow from beneath the crownshafts and consist of short flowering branches on short peduncles. The fruits are rounded and red when ripe.

Both species are inhabitants of moist, cool, subtropical coastal areas at low to moderate elevations where the temperature does not vary much from season to season or from night to day; as a result they are at home in cool, nearly frostless Mediterranean climates but are almost impossible to grow in hot or tropical climates. In their habitats, they are considered hardy to cold, which they are, but the cold conditions of Australia and New Zealand are different from those of the United States, where great masses of chilly air can descend from the arctic in the winter, suddenly dropping the temperature from balmy to frigid conditions. The palms have been killed, even in coastal Southern California, when these arctic outbreaks occur.

The genus name is from Greek words meaning "club" and "pillar," an allusion to the club-shaped sterile pistil (pistillode) in the male flowers. Seeds germinate within 60 days. For further details of seed germination, see *Areca*.

Rhopalostylis baueri
ro-pa'-lo-STY-lis · BOW-er-eye
PLATE 833

Rhopalostylis baueri occurs naturally on Norfolk and Raoul Islands in the South Pacific, where it grows in subtropical rain forest at 1000 feet (300 m). Australians call it Norfolk Island palm. The epithet honors Ferdinand Lucas Bauer (1760–1826), an Austrian botanical illustrator who spent several months on Norfolk Island in 1803. *Rhopalostylis baueri* var. *cheesemanii* (sometimes treated as a distinct species) is included as a synonym.

The trunk attains a maximum height in habitat of 50 feet (15 m) with a diameter of 8 inches (20 cm). It is light to dark green in all but its oldest parts and is distinctly ringed with light gray, closely set leaf base scars. The plump crownshaft is 3 feet (90 cm) tall, light to dark green, and bulged at its base. The leaves are 12 feet (3.6 m) long on short, almost nonexistent petioles. In trees exposed to the sun, the leaves are stiff and erect, with regularly and closely spaced, narrowly elliptical, light green, stiff leaflets 2–3 feet (60–90 cm) long; the leaflets grow from the rachis at a slight angle to create a slightly V-shaped leaf. Palms grown in shadier conditions have slightly more spreading leaves with more lax and pendent leaflets. The ovoid ½-inch (13-mm) fruits are reddish brown when ripe.

The species is hardy in zones 9b through 11 but only for the cooler Mediterranean climes; it refuses to grow in hot climates, although it is reportedly more tolerant to these conditions than *R. sapida*. Regular, adequate moisture is almost as important as climate, and specimens subjected to drought, drying winds, or too much sun become ratty looking and are not worth having. The species is fast growing in sun or partial shade. It prefers a slightly acidic, moist, and fast-draining soil but seems able to adapt to several types if they are not too alkaline or too acidic and if they have some humus. This palm reportedly makes a good container or indoor plant if given bright light and regular irrigation.

Rhopalostylis sapida
ro-pa'-lo-STY-lis · sa-PEE-da
PLATE 834

Rhopalostylis sapida is endemic to New Zealand and the Chatham Islands to the east, where it grows in dense, wet forests at elevations up to 2000 feet (610 m). The epithet is Latin for "good to eat," an allusion to the growing point, which was used by the Maori as a cabbage. This species is similar to *R. baueri* but shorter, the trunk attaining a maximum height of 30 feet (9 m), with a more bulbous crownshaft, and with brighter, purer red fruits. The leaves are 16 feet (4.9 m) long. This palm is more tolerant of cold than *R. baueri* and more intolerant of hot climates. In the Chatham Islands, it is the world's southernmost naturally occurring palm species. Individuals native to the Chathams reportedly have wider leaflets, more robust stems, and pinkish petioles. In Australia and New Zealand, the palms are called nikau palms, feather duster palms, shaving brush palms, and brush palms.

Roscheria
ro-SHER-ee-a

Roscheria is a monotypic genus of solitary-trunked, pinnate-leaved, monoecious palm endemic to the Seychelles. The name honors Albrecht Roscher (1836–1860), German explorer in East Africa. Seeds germinate within 30 days. For further details of seed germination, see *Areca*.

Roscheria melanochaetes
ro-SHER-ee-a · mel'-a-no-KEET-eez
PLATE 835

Roscheria melanochaetes is an endangered species endemic to Mahé and Silhouette Islands, where it occurs in the undergrowth of low mountainous rain forest. The epithet is from the Greek and translates as "black bristle," an obvious allusion to the spines.

The trunk in habitat grows slowly to 25 feet (7.6 m) with a diameter of 3 inches (8 cm). A small cone of stilt roots supports it. Young trees have rings of black spines at each node of the stem but, in older plants, these are sparse or even nonexistent; the crownshaft and lower portions of the petioles exhibit short black spines, even in mature plants. The crownshaft is 1 foot (30 cm) tall, scarcely thicker than the trunk, slightly bulging at its midpoint, and light green overlain with tiny brown scales, especially near its summit. The leaf crown is open, sparse, and rounded with 6 arching leaves 4–6 feet (1.2–1.8 m) long on petioles 6–8 inches (15–20 cm) long. The leaves are unusual in that the leaflets vary in shape and size; no 2 on a given tree seem to have the same dimensions.

Some are elliptic with a single rib or midvein and with long-pointed apices, while others are much broader, with several ribs and obliquely truncated apices bearing indentations between each rib or midvein. All have several folds or pleats, especially prominent at the juncture of rachis and blade. The widths of the segments are mostly random as to their placement along the rachis, but in all cases, the leaflets of the terminal pair are wide, often the widest. The thin, membranous leaflets are light olive to emerald green above, with a duller, brownish hue beneath. Juvenile leaves are undivided except for their bifid apices. Newly emerging leaves are usually pink to red. The inflorescences are borne from the leaf axils, unlike most other crownshaft palms in which they are borne beneath the crownshafts. They are 2–6 feet (60–180 cm) long and are much branched, the pendent flowering branches bearing both male and female flowers. The small, egg-shaped fruits are ¼ inch (6 mm) long and are red when ripe.

The species is as unusual in appearance as it is beautiful. It is slow growing, intolerant of drought, and needs partial shade when young and at all ages in hot climates. It is tender to cold, as are all Seychelles species, being adaptable to zone 11 and marginal in 10b. It must have a moist soil with humus. These conditions are difficult to provide in many areas, even tropical ones.

Roystonea

roy-STON-ee-a

Roystonea comprises 10 large solitary-trunked, pinnate-leaved, monoecious palms in tropical America, principally in and around the Caribbean Basin. The species are similar and often difficult to distinguish one from the other. Scientific differentiation is by inflorescence shape, flower color, and other floral details, as well as characteristics of the fruit.

The trunks are moderate in height to very tall, usually massive, often bulging at some point and, except for a few species, white or nearly so in all but their youngest parts, which are green and prominently ringed. The bases of the stems are usually much enlarged. All species form prominent, often large crownshafts. The leaf crowns are generally full, dense, and rounded, and the leaves large to very large, commensurate with the height of the trunk. Falling leaves can damage smaller plants growing beneath them. No species has adult individuals with leaflets growing in a single flat plane, but some species have leaves that are much more plumose than others; juvenile trees tend to have much flatter leaves than do adults of the same species. The inflorescences grow from beneath the crownshafts and are first enclosed in large, leathery, bat- or horn-shaped bracts. The flowering branches are many, spreading, and white when first released from the bracts. They bear unisexual blossoms of both sexes. The fruits are round or ellipsoid and curved. They ripen from green through red to dark purple or dull black.

Some of the world's most famous palms are in this genus. One of these, *R. regia*, is considered by many people to be the world's most beautiful; it is certainly among the most popular. All have grandeur with which few other genera can compete. All are unsuitable for arid regions unless well irrigated, and all are tender to cold, the hardiest being adaptable to zones 10 and 11. They are almost without exception adapted to calcareous soils, and yet grow faster and look better in a medium that incorporates humus. A thick mulch is second best. Under optimum conditions, all are fast growing once past the seedling stage, and some of the larger species are fast growing even when young. All are also adapted to full sun once past the seedling stage, and they grow more slowly if not given enough light.

The genus name honors General Roy Stone (1836–1905), a U.S. Army engineer in Puerto Rico. Fresh seeds germinate within 90–120 days, but dried seeds can take 6 months to a year before starting to germinate and then do so sporadically for another year. Long-term viability is fairly good, but germination time is much longer for older seed.

Roystonea altissima

roy-STON-ee-a · al-TIS-si-ma

PLATE 836

Roystonea altissima is endemic to Jamaica, where it grows on hillsides and mountain slopes in the interior of the island from elevations of 200 to 2600 feet (60–790 m). The epithet is Latin for "highest," but the species is not the tallest in the genus.

The trunk attains a height of 60 feet (18 m) in habitat with a diameter of 1 foot (30 cm). It gently tapers from bottom to top, is usually light tan and, in its younger parts, is graced with widely spaced darker rings of leaf base scars. The crownshaft is 5 feet (1.5 m) tall, deep olive green, not much thicker than the trunk, and generally cylindrical. The leaf crown is spherical and holds 15 slightly arching but spreading leaves 12 feet (3.6 m) long on short petioles. The medium green leaflets are 30 inches (76 cm) long, closely and regularly spaced along the rachis, and grow from it at faintly different angles to create a slightly plumose leaf. The inflorescences bear violet male flowers, and the fruits are obovoid and curved, ½ inch (13 mm) long, and a dark, blackish purple when ripe.

Roystonea borinquena

roy-STON-ee-a · bo'-reen-KAY-na

PLATE 837

Roystonea borinquena is indigenous to Hispaniola, Puerto Rico, and Vieques and St. Croix in the U.S. Virgin Islands, where it grows in savannas, cleared areas, and at the bases of limestone hills from sea level to 2600 feet (790 m). It is depicted on Haiti's coat of arms. The epithet is Latin for "Borinquen," a Spanish transliteration of the Taíno name for Puerto Rico.

The trunk grows to 50 feet (15 m) high with a diameter of 18 inches (45 cm). It is sometimes bulging near its midpoint, is light gray to tan, and is distinctly ringed in its younger parts with darker leaf base scars. The crownshaft is 5 feet (1.5 m) tall, not much thicker than the trunk, and deep emerald green. The spherical leaf crown holds 15 arching leaves, each one 15 feet (4.5 m) long on a petiole 18 inches long. The deep green leaflets are 3–4 feet (90–120 cm) long and grow off the rachis at angles which create a fully plumose leaf. The inflorescences are 4 feet (1.2 m) long and bear colorful male flowers that are yellow with purple anthers. The rounded ½-inch (13-mm) fruits are brownish black.

Roystonea maisiana
roy-STON-ee-a · my-see-AH-na
PLATE 838

Roystonea maisiana is a rare endemic in extreme eastern Cuba, where it grows in open savannas at 1200–1500 feet (370–460 m). The epithet is Latin for "of Maisí," the easternmost tip of Cuba.

The light gray to white trunk attains a height of 60 feet (18 m) in habitat with a diameter of only 1 foot (30 cm) and seldom bulges at any point above its enlarged base. The crownshaft is deep olive green, 4 feet (1.2 m) tall, mostly cylindrical, and nearly the same thickness as the trunk. The leaf crown is hemispherical to nearly spherical and contains 20 arching leaves 12–15 feet (3.6–4.5 m) long. The dark green leaflets are 3 feet (90 cm) long and grow from the rachis at several angles to create a fully plumose leaf. The inflorescences bear creamy white male flowers and produce elongate, curved, blackish purple fruits.

This species is distinguished by its curved fruits and slender trunks. It is still rare in cultivation, but because of its elegant proportions, it is more than worthy of being planted.

Roystonea oleracea
roy-STON-ee-a · o-le-RAY-see-a
PLATES 839–840

Roystonea oleracea occurs naturally in the Lesser Antilles, northern Venezuela, and extreme northeastern Colombia, where it grows along the edges of rain forest and seasonally flooded savannas from sea level to 5300 feet (1620 m). The epithet is a Latin word meaning "pertaining to a vegetable garden" and alludes to the palm's edible growing point.

This species is the only one from South America and the tallest in the genus. The trunk attains the remarkable height of 130 feet (40 m) or even more in habitat, with a diameter of 2 feet (60 cm). It is a light to medium gray, slightly bulging, if at all, at any point above its swollen base. The crownshaft is 6 feet (1.8 m) tall, deep green, and gently tapering to its summit. The leaf crown is hemispherical and has 20 gently arching leaves, each 15 feet (4.5 m) long. The deep green leaflets are 2–3 feet (60–90 cm) long and grow at 2 faintly different angles from each side of the rachis to create a slightly plumose leaf. The branches of the inflorescence are sinuous when they first emerge from the enclosing bract. The male flowers are white, and the fruits are typically a blackish purple.

There is no more majestic palm, but it is reportedly also the least hardy to cold. Photographs seldom give a true sense of its grandeur, and an avenue of this species, such as the one in the Rio de Janeiro Botanical Garden, is a grand sight.

Roystonea princeps
roy-STON-ee-a · PRIN-seps
PLATE 841

Roystonea princeps is endemic to southwestern Jamaica, where it grows in swampy areas near sea level. The epithet is Latin for "prince," an allusion to its stateliness. The trunk grows to 60 feet (18 m) high with a diameter of 1 foot (30 cm). It is light gray to white and seldom bulging at any point above its swollen base. The crownshaft is 5 feet (1.5 m) tall, silvery green, and cylindrical. The leaf crown is spherical or nearly so with 15 leaves, each 12 feet (3.6 m) long on a petiole 12–18 inches (30–45 cm) long. The emerald green leaflets are 30 inches (76 cm) long and grow from the rachis at several angles to create a plumose leaf. The male flowers are white, and the fruits are elongate, curved, and blackish purple. The species is especially beautiful because of the long, supple, and mostly pendent leaflets and airy inflorescence.

Roystonea regia
roy-STON-ee-a · REE-jee-a
PLATE 842

Roystonea regia is widely distributed from the shores of the Gulf of Campeche in Mexico (in the states of Campeche, Tabasco, Veracruz) and along the Caribbean coast of the Yucatán in Mexico (in the state of Quintana Roo), along the Caribbean coasts of Belize and northwestern Honduras, the Cayman Islands in the Gulf of Mexico, all of Cuba (where it is the national tree and appears on the coat of arms), the westernmost Bahamas, and the southern tip of Florida (in the counties of Collier, Monroe, and Miami-Dade). It grows naturally along the edges of rain forest, in wooded savannas, along rivers and streams, and in swamps from sea level to 3000 feet (900 m). Were it not for its great beauty, it would be considered weedy as it readily naturalizes in tropical regions. Common names are royal palm, Cuban royal palm, and Florida royal palm. The epithet is Latin for "regal."

The trunk attains a height of 100 feet (30 m) and a diameter of nearly 2 feet (60 cm). It is light gray to white and distinctly ringed in its younger parts with darker leaf base scars. It usually exhibits at least one slightly swollen point along its columnar structure, and its base is enlarged. The crownshaft is 6 feet (1.8 m) tall, emerald green. The leaf crown is spherical or nearly so and holds 15 leaves, each 12 feet (3.6 m) long with a petiole 8 inches (20 cm) long. The deep green leaflets are 2–4 feet (60–120 cm) long and grow off the rachis at different angles to create a fully plumose leaf. The male flowers are white, and the rounded fruits are the typical blackish purple when ripe.

The palm wants regular, adequate moisture (but not standing water), a tropical or nearly tropical climate, and a humus-laden soil but grows in nearly pure limestone, especially if heavily mulched. Freezes and droughts are recorded in survivors by constrictions of the trunk. This palm often survives a freeze and has been known to recover from temperatures in the mid-20s Fahrenheit (around −4°C) after being defoliated. It should not be planted in the shade. In full sun and with good irrigation it can add 1 or more feet (30 cm) of new trunk per year.

This species is the most widely planted one in the genus, and with good reason. It is magnificently beautiful and a veritable symbol of the wet tropics and the tropical look. It is hard to misplace it in the landscape as it looks good even as a single specimen plant surrounded by space. Plantings of 3 or more individuals of varying heights are a vision of paradise itself, and its canopy-scape floating over lower vegetation is not only enthralling but also is how it looked in much of its habitat, like the Everglades of South Florida. There is no more beautiful palm species. It is not known to be grown indoors, even in large atriums.

Roystonea violacea
roy-STON-ee-a · vee-o-LAY-see-a
PLATE 843

Roystonea violacea is endemic to the extreme eastern tip of Cuba, where it grows in open savannas and on hills at 1200–1500 feet (370–460 m). The epithet is Latin for "violet" and alludes to the color of the male flowers and the young trunks.

The trunk grows to 50 feet (15 m) in habitat and, especially when young, is an unusual dark tan to almost milk-chocolate color overlain with a slight purplish hue, with distinct light tan rings of leaf base scars; older trees usually have dark gray to dark tan stems. The crownshaft is 6 feet (1.8 m) long, light green, cylindrical, and slightly thicker than the trunk. The spherical leaf crown contains about 15 leaves, each 10–12 feet (3–3.6 m) long. The deep green leaflets are 3 feet (90 cm) long and grow from different angles off the rachis to create a plumose leaf. The male flowers are violet colored, and the fruits are round and the typical blackish purple when ripe.

This extraordinarily beautiful species is still rare in cultivation. It is, by far, the most colorful species in the genus.

Sabal
SAY-ball

Sabal is composed of 16 variable, solitary-trunked, palmate-leaved, hermaphroditic palms in the warm temperate, tropical, and subtropical Americas. Most inhabit open, dry areas, but a few are denizens of swamps or forests. None grow above 5000 feet (1520 m) in the wild. Some species are weedy, meaning that their seedlings spring up quickly after fire or other agents of destruction have cleared the land. The seeds are carried by birds, so seedlings occur in fencerows. Many species are difficult to distinguish by their leaves and nearly impossible to identify as juveniles. Because the trunks of some species vary so much, even when palms are viewed in their entirety their identification can be confusing, especially in regions where more than one species occurs or is planted.

Some species are dwarf with subterranean trunks but most have aboveground stems from 10 to 80 feet (3–24 m) tall. On some individuals the leaf bases persist for many years if not removed and often this look is considered ornamental, as with *S. palmetto* trunks and their crisscrossed, old "boots." In garden settings, the "boots" are ideal perches for epiphytic ferns, orchids, and bromeliads. *Sabal pumos* and *S. rosei* have trunks with "knobs" indicating the locations of former leaf bases. Rings of leaf base scars are usually indistinct, but in a few tropical species can often be discerned from a reasonable distance.

The leaves are borne on spineless petioles and are shallowly to very deeply costapalmate, sometimes appearing almost pinnate because of the costa which curves downwards and causes the bases of the segments to arch out from it at a steep angle, making the leaf distinctly V shaped. A hastula is always present at the top of the leaf juncture but may be long or short, narrow or wide. The apices of the leaf segments are sometimes deeply bifid, and there are often thin, curling, lighter-colored filaments between the segments. The inflorescences grow from the leaf crown, are much branched, and bear slender panicles of whitish, fragrant flowers. The round or pear-shaped, black or brown (when mature) fruits are usually much less than 1 inch (2.5 cm) long.

Some of the cold-hardiest palms are in this genus. Indeed, *S. minor*, if not the hardiest, is certainly near to being so. Almost without exception, these palms are adaptable to various soils, and a few can even thrive in waterlogged media. Most of them need a lot of heat in summer and are poorly adapted to regions with cool summers and colder winters. Most are slow growing, and a few are exceptionally slow. In seedling stage, all grow the saxophone type of underground stem and root system and, for many of these, form no aerial trunk for years.

Some "mystery" *Sabal* species are known from cultivation, such as *S.* 'Riverside', a robust, blue-green-leaved palmetto propagated from a single specimen found in Riverside, California, in the 1950s. These may be hybrids involving *S. bermudana* and *S. mexicana* (Hodel 2008), but other robust species, such as *S. causiarum*, may be involved, if in fact, the cultivar is a hybrid. Hybridization in *Sabal* has never been conclusively documented. Then there is the arborescent Brazoria palm, also called Brazoria hybrid palm, from Brazoria County, Texas. It may be a hybrid between *S. mexicana* and *S. minor*, *S. palmetto* and *S. minor*, or between other, as yet undetermined species. They remain a mystery.

The etymology of the genus name was never explained by its author. Liberty H. Bailey conjectured that it may be a form of an aboriginal name for a species from South America. Almost all species in the genus are called palmetto of one sort or another. Fresh seeds germinate within 90 days. Seeds that are allowed to dry out and are then rehydrated may take 9 months or longer to germinate. For further details of seed germination, see *Corypha*.

Sabal bermudana
SAY-ball · ber-myoo-DAH-na
PLATE 844

Sabal bermudana is endemic to Bermuda, where it occurs in lowland marshes or on drier soil above the low spots, sometimes near the ocean. It is scarce in habitat and, if development increases, will be in danger of extinction. Common names are Bermuda palm and Bermuda palmetto. The epithet is Latin for "of Bermuda."

This palm grows slowly to a height of 20 feet (6 m) in habitat, increasing less than ½ inch (13 mm) per year (Zona 1990). It has a tan or gray, indistinctly but closely ringed trunk, the diameter of which is 1 foot (30 cm). The leaf crown is full, rounded, and dense with the deeply costapalmate leaves which are deep green, sometimes with a bluish cast. The inflorescences are shorter than the leaf crown and mostly hidden; they produce small, black, pear-shaped fruits.

The species looks much like *S. palmetto* but is usually more stout and robust appearing because of its short height, its large crown, and its thick trunk. It grows in partial shade but prefers full sun and thrives on calcareous as well as slightly acidic soils. In spite of its subtropical origin, its cold tolerance is probably on a par with *S. palmetto*; it is among the few *Sabal* species that tolerate areas with cool summer temperatures. It is extremely slow growing.

Sabal causiarum
SAY-ball · kow-see-AHR-um
PLATE 845

Sabal causiarum is indigenous to Hispaniola, Puerto Rico, and the small island of Anegada in the British Virgin Islands, where it grows in open places on sandy soil at low elevations, often in large groves. The common name is Puerto Rican hat palm. The epithet is a Latinized form of the Greek word *kausia*, which referred to a hat worn by ancient Macedonians, and succeeds in referencing, albeit obliquely, the traditional use of this species: the leaves have been used for many centuries to weave hats, as well as baskets and other utensils.

When this palm is mature or nearly so, it is among the distinctive and more easily identified species in the genus. The trunks attain heights of 30 feet (9 m) and are the stoutest in the genus, often with diameters of 2 feet (60 cm) or even more. They are mostly smooth and light gray and look like stout columns atop which the perfectly round, dense crown of leaves resides. The crown is 10 feet (3 m) tall and 15 feet (4.5 m) in diameter. Young leaves have floppy, straw-colored, tonguelike auricles at the bases of their petioles. The large leaves on petioles 5 feet (1.5 m) long are deeply costapalmate and deep green or bluish green with some whitish, thin, curling threads between the segments. The inflorescences are as long as or slightly longer than the leaf crown, and the round, black fruits are ½ inch (13 mm) in diameter, causing the infructescence to hang beneath the crown.

This palm is drought tolerant once established but responds with faster growth and better form and color if given adequate irrigation. It needs as much sun as possible, even in youth, and a free-draining soil. When young, the palm is tender to cold and usually does not survive temperatures below 20°F (−7°C); however, older trees have been known to survive 10°F (−12°C), although damaged, sometimes severely. The species is adaptable to zones 8 through 11, except in regions where temperature regularly reaches 10°F (−12°C); under these conditions, even large specimens will surely die. Once past the seedling stage, the palm grows relatively fast.

This noble palmetto looks best in large groups or groves. To stand in such a grove is akin to being among the great columns of *Copernicia baileyana*. The other great landscaping use for this palm is to line a walk or promenade, even a thoroughfare, the resulting visual appeal being a living colonnade of smooth, stout trunks. Montgomery Botanical Center has planted a large circle of these palms, which in time will form a ring of Doric columns. This species is also great as a canopy-scape but is certainly one of the least likely candidates for indoor use.

Sabal domingensis
SAY-ball · do-meen-GEN-sis
PLATE 846

Sabal domingensis is native to western Cuba and interior areas of northern Hispaniola, where it grows on hills, in savannas, and in mountainous cut-over forests up to 3000 feet (900 m). The epithet is Latin for "of [Santo] Domingo," a reference to the colonial name for the Dominican Republic.

The palm attains the same stature as *S. causiarum*, and its trunk is almost as massive; in fact, it is often difficult to distinguish the 2 species. The principal differences between them are the presence of auricles and the shape of the fruits. The fruits of *S. domingensis* are slightly larger than those of *S. causiarum* and are pear shaped rather than round. *Sabal domingensis* lacks auricles at the bases of the petioles of young leaves; *S. causiarum* has auricles.

Saba domingensis and *S. causiarum* have the same cultural requirements and landscape use, except *S. domingensis* does well without too much summer heat. The palm in habitat has the same human uses as does *S. causiarum*, namely, the weaving of hats from the leaves. In fact, one of the vernacular names in Haiti for the palm (*latanier-chapeau*) translates as "hat palm."

Sabal etonia
SAY-ball · ee-TO-nee-a
PLATE 847

Sabal etonia is endemic to Florida, where it is an important component of the sand pine scrub community and is abundant on deep, sandy soils in the central and north central peninsula, along most of the eastern coast, and with one isolated population on the western coast near Bradenton. The common name is scrub palmetto. The epithet refers to a site known as Etonia Scrub, near Eustis, Florida, where the species was first collected.

This palm is, for the most part, a dwarf species with a subterranean trunk; it occasionally forms an aerial stem to 6 feet (1.8 m) high. The leaves are 2 feet (60 cm) wide, distinctly light green to yellow-green, and costapalmate but not arching. The leaf segments, except in old, dying leaves, are stiff and extend almost to the petiole; there are white curling threads among them. The inflorescence is shorter than the leaf crown but is thicker or bushier than that of most other *Sabal* species. The round ½-inch (13-mm) fruits are black when mature and usually bend the infructescence to the ground in trunkless individuals and beneath the leaf canopy in arborescent individuals.

The seedlings are difficult to distinguish from seedlings of *S. palmetto*, but are usually not as dark green and are less costapalmate. Adults have larger fruits. This little palm is drought tolerant and luxuriates in sandy, calcareous to slightly acidic soils. It flourishes in partial shade or full, hot sun and is adaptable to zones 8 through 11 and marginal in 7. It is quite refreshing as a border to taller and, especially, darker vegetation.

Sabal maritima
SAY-ball · ma-RI-ti-ma
PLATE 848

Sabal maritima occurs naturally in Cuba and Jamaica, where it grows near the coasts, in the low hills, savannas, and clearings, in forested areas from sea level to 2000 feet (610 m). The epithet is Latin for "maritime," an allusion to the palm's coastal habitat. The species looks like a much more robust version of *S. palmetto*, the distinguishing features being its thick, smooth, and nearly white trunk and its larger leaves on longer petioles, which create a larger, more open crown. The young petioles are covered with light tan scales, which gradually fall off as the petiole weathers. Also, the flowers are crowded on the rachillae, such that they touch one another when open. This species is doubtless not as hardy to cold as *S. palmetto*; it reportedly has survived 27°F (−3°C) unscathed. Otherwise the palm's cultural requirements are the same as those of *S. palmetto*.

Sabal mauritiiformis
SAY-ball · maw-rit′-tee-i-FOR-mis
PLATES 849–850

Sabal mauritiiformis is indigenous to southeastern Mexico, northern Guatemala, Belize, northeastern Costa Rica, the isthmus region of Panama, northwestern and northern Colombia, northern Venezuela, and Trinidad, where it grows in rain forest, savannas, and cleared forest areas, from sea level to 3000 feet (900 m). The epithet is Latin for "in the form of *Mauritia*," a similar-appearing but unrelated genus of palmate-leaved palms from South America.

This species is readily distinguishable from all other *Sabal* species. The trunk attains a height of 80 feet (24 m) in habitat and is straight as an arrow and less than 1 foot (30 cm) in diameter, making it an elegant sight when older. It is light tan or light gray in its older parts but a beautiful deep green in the younger parts. Adding to the great beauty of the stem are the closely set rings of darker leaf base scars. The leaf crown is also handsome, as the shape is a large obovate one, with the beautiful leaves silhouetted against the sky because of their unusually long petioles. Each leaf is 6–7 feet (1.8–2.1 m) long and shallowly costapalmate, which would make it almost flat were it not for its long, pendent segments joined in groups of 3. Leaf color is olive to deep green above and silvery, lighter green, sometimes with an almost bluish hue beneath. The segments in juvenile palms are unusually wide, deeply pleated, lanceolate, and joined in groups of 3. They are all much more deeply divided, much narrower, and pendent at their apices in individuals past the juvenile stage. The inflorescences are 6–8 feet (1.8–2.4 m) long, erect but arching in fruit, and extend well beyond the crown. The round to pear-shaped fruits are black when mature and less than ½ inch (13 mm) in diameter.

Reports of this species withstanding temperatures in the mid-20s Fahrenheit (around −4°C) unscathed are rare and probably unreliable. The palms are defoliated by temperatures in the low 20s (around −7°C), and most are killed by temperatures below that. The palm grows well in partial shade when young but is slower and, when older, needs full sun. It seems to thrive on slightly calcareous soils but accepts any well-drained soil; a good one results in faster, more robust growth and better color. The more water, the better, although the tree is somewhat drought tolerant.

Few palms are more beautiful. The epithet tells a lot about its beauty, as *Mauritia flexuosa*, the plant after which it was named, is arguably the most beautiful palmate-leaved species in the Americas. The silhouette of this sabal is galvanizing and makes an arresting canopy-scape. No palm is lovelier in groups of 3 or more individuals of varying heights. It thrives indoors, at least for a while, with good light.

Sabal mexicana
SAY-ball · mex-i-KAH-na
PLATE 851

Sabal mexicana occurs naturally from the extreme southeastern tip of Texas (Brownsville area), southwards through eastern Mexico, Guatemala, Honduras, and El Salvador, where it grows in drier lowlands. At one time, it was common along the Rio Grande almost as far north as Laredo but has since been destroyed except for a small area near the mouth of the Rio Grande for agricultural purposes. In Texas, it is called Texas palmetto, Rio Grande palmetto, or Texas sabal. The epithet is Latin for "of Mexico." *Sabal texana* and *S. guatemalensis* are synonyms.

Very old individuals in habitat can grow trunks 50 feet (15 m) tall and 1 foot (30 cm) in diameter. These are always straight and columnar. In cultivation the stems are sometimes covered in the typical crisscrossing leaf bases; in nature they seldom retain these boots below a height of 25 feet (7.6 m), the remainder of the stem being gray to dark gray to dark tan with closely set, nearly indistinct rings. If not pruned, the leaf crown is rounded and massive; the deeply costapalmate leaves are uniform dark green, sometimes very dark green. The black, rounded fruits are about ½ inch (13 mm) in diameter.

The species looks much like *S. palmetto* but is distinguished principally by its larger leaf crown, usually darker green leaves, always straight and columnar trunk, and its more robust appearance—because its trunk is shorter than that of *S. palmetto*, its seems thicker. The trunks are occasionally used for construction, and the leaves much used for thatch in the Yucatán. This species is as hardy as *S. palmetto* and is usually hardier in drier climates, in which it has withstood 10°F (−12°C) unscathed; in areas subject to wet freezes in winter it seldom is undamaged by that temperature. Otherwise, it has the same cultural requirements as *S. palmetto* except that it is more drought tolerant. Because it appears so robust, it readily lends itself to planting along avenues, and it makes wonderfully handsome specimens.

Sabal minor
SAY-ball · MY-nor
PLATE 852

Sabal minor is found throughout the southeastern United States in Texas, Oklahoma, Arkansas, Louisiana, Mississippi, Alabama, Georgia, South Carolina, North Carolina, and Florida, where it occurs in the undergrowth of evergreen and deciduous forest at low elevations in often swampy conditions. Curious populations occur in the Texas Hill Country in arid grasslands. The species also occurs in northeastern Mexico, in the state of Nuevo León, at an elevation of 1000 feet (300 m) in dry, subtropical pine-oak forest on rocky hillsides (Goldman 1991). It is the most northerly ranging palm in the United States and in its genus. Common names are dwarf palmetto, bush palmetto, little blue stem, and swamp palmetto. The epithet is Latin for "smaller."

This species forms subterranean and aerial trunks according to habitat and environmental conditions. In the western parts of its range, it is often arborescent, sometimes with stems to 18 or more feet (5.4 m), while in the northern and southeastern parts of its habitat it usually has no aboveground trunk or only a short one. In the Big Thicket region of eastern Texas, it sometimes forms arborescent stems that are 10 feet (3 m) long but basically prostrate and creeping along the ground under the bald cypresses (*Taxodium distichum*), American beeches (*Fagus grandifolia*), southern magnolias (*Magnolia grandiflora*), and other giant forest trees; little of this ecosystem still exists. In southern Louisiana the beautiful arborescent forms in Delta Country south of New Orleans, especially in the southern and natural segment of Jean Lafitte National Park, show erect stems that are 8–10 feet (2.4–3 m) tall and grow in swampland that is flooded for most of the year. In contrast, some populations from the Florida Panhandle seem to be almost dwarf, with smaller overall dimensions of their leaves. They never form aerial stems.

The leaves are shallowly costapalmate and deep, dark green, with deep, stiff, linear-lanceolate segments that are not cleft apically. The segments are free for four-fifths of their length. The segments are often in 2 groups separated by a gap in the middle of the blade that extends to the petiole, and there are usually no curling threads among them. The inflorescences are erect and slightly arching at their tips but are taller than the leaf crown. The round ¼-inch (6-mm) fruits are black or dark brown.

This palm is extremely hardy to cold, being adaptable to zones 6 through 11. Some forms (not the arborescent ones from Texas or Louisiana) are known to survive in zone 5, however, and this implies that the provenance of an individual palm has a lot to do with its cold hardiness. This species is a true water lover and one that relishes partial shade. Under such conditions, it reaches its full beauty, although it can survive (but not look good) being planted high and dry or in the full, blazing sun of southern latitudes. It prefers fertile, moist soil whether acidic or calcareous.

The species is as beautiful as any purely tropical palm when used as a groundcover under canopy. Otherwise, it seems totally out of place, especially as a specimen plant surrounded by space and in full sun. There is hardly a better or more beautiful palm for low, swampy areas under canopy, where it can give a tropical look to decidedly nontropical regions. It can grow perfectly well indoors with bright light and much water.

Sabal palmetto
SAY-ball · pahl-MET-to
PLATE 853

Sabal palmetto occurs as far north as Cape Fear in North Carolina and is found in coastal southern South Carolina, all of Florida except the western panhandle region, the Bahamas, the Florida Keys, and western Cuba. It grows mostly in low, swampy areas and along streams and rivers, but also in hammocks and open pine forests as well as along the coasts just inland of the dunes. Common names are palmetto, cabbage palm, or cabbage palmetto. The epithet is a corruption of the Spanish word for "little palm."

The mature trunks can attain heights of more than 80 feet (24 m). The tallest individuals in habitat seem to be in Florida. The trunks of these old patriarchs are never covered in leaf bases, and they are usually light gray to light tan. Younger individuals, especially in cultivation, are often covered with the distinctive and light-colored, crisscrossing "boots." The leaf crowns of large specimens are relatively small compared to the total height of the palm and are always (unless pruned) a complete globe. The leaves are deeply costapalmate, the costa recurved, and are uniformly deep green. The arching inflorescences are usually slightly longer than the radius of the leaf crown and project beyond it. The fruits are round, black, and nearly ½ inch (13 mm) in diameter. In the past, the leaves were cut for broom and brush fibers, and the growing point was harvested for food.

This palm is one of the world's hardiest to cold, thriving in zones 8 through 11, and marginal in 7b; but, as is the case with most *Sabal* species, it is not good in these zones where summer temperatures are cool, and it struggles to grow in areas such as the U.S. Pacific Northwest. The second most important factor for its successful culture is adequate, regular water. As for soil type, it tolerates but does thrive in saline conditions. It luxuriates for many years as an undergrowth subject and, in some ways, is more beautiful in partial shade than full sun, although it never reaches the great stature it is capable of without eventually having full sun. It is slow growing when young and moderately slow when older.

There is a landscaping phenomenon in which this native jewel is dug from its pastures and hammocks and planted in rows along Florida thoroughfares like telephone poles. In the dry season, they are drought stricken, and they show their suffering in the form of abnormally small crowns throughout the rest of the year, no matter how much rain falls in the rainy season. The final assault on their dignities is that their unnaturally small crowns are constantly pruned. They should, instead, be planted as a canopy-scape. Otherwise it is much better to leave them in their beautiful habitat. This palm is not known to be used indoors past its seedling stage.

Sabal palmetto 'Lisa' is an unusual cultivar that has been available in Florida for several years. It is distinguished by having its leaf segments joined in pairs. The threads between the segments are absent, and the leaf is much less costapalmate than typical *S. palmetto*. This cultivar is propagated by seeds, but a small percentage of offspring revert back to the typical leaf form.

Because the cabbage palmetto readily naturalizes, it is unthreatened in its natural range; however, there are but few mature specimens in Florida, where development, urbanization, and transplantation from the wild eat away relentlessly at their numbers. This decimation of the old beauties and their habitat is hard to grasp for those who are too young, but, if they could only see the former magnificence of groves of towering palmettos, say from old photographs, they might help effect the return of this palm and its habitat in the future.

Sabal pumos
SAY-ball · POO-mos
PLATE 854

Sabal pumos is endemic to central western Mexico, where it grows in tropical deciduous forest of mountainous areas and in cleared pastures at 2000–4000 feet (610–1200 m). The epithet is the aboriginal name for its edible fruits, which are nearly 1 inch (2.5 cm) in diameter and are the largest in the genus. The slender trunks are usually 6–8 inches (15–20 cm) in diameter but grow to 50 feet (15 m) high in old individuals and, in nature, are clean with a knobby pattern of leaf base scars. This palm resembles a slender *S. palmetto* and seems extremely rare in cultivation; in fact, it is a poorly known species in general. No data are available for its cold hardiness, but it is said to be drought tolerant and is probably as hardy as *S. mexicana*. Coming from high elevations, it may be more adaptable to Mediterranean climates. It would doubtless need full sun and a fast-draining soil.

Sabal rosei
SAY-ball · RO-zee-eye
PLATE 855

Sabal rosei is endemic to Mexico along its northwestern coast from Culiacán southwards into the state of Jalisco, where it grows in tropical deciduous forests from sea level to 2500 feet (760 m). The epithet honors Joseph N. Rose (1862–1928), American botanist specializing in the flora of western North America, author of *The Cactaceae*, and the palm's original collector. The species has similarities to *S. pumos* and is related to it but more attractive.

The slender trunk is 8–9 inches (20–23 cm) in diameter but, with age, attains a height of 40 or more feet (12 m). In habitat, the stem is clean, straight, and covered with picturesque knobby leaf base scars, while in cultivation it often has a covering of crisscrossed leaf bases. The leaf crown is beautifully globular, much like that of *S. palmetto*. The leaves are deeply costapalmate and a uniform deep green with stiff, erect segments.

This species is reported to have withstood 22°F (−6°C) unscathed in Florida, where the climate is wetter than that of the palm's habitat; its cold tolerance is probably comparable to that of *S. palmetto*. It is a sun lover and is probably not particular about soil as long as it is freely draining.

Sabal uresana
SAY-ball · oo-re-SAH-na
PLATE 856

Sabal uresana is endemic to northwestern Mexico in the states of Chihuahua and Sonora, where it grows in the foothills of the Sierra Madre Occidental in dry subtropical thorn forests from sea level to 4500 feet (1370 m). The epithet is Latin for "of Ures," a town in western Sonora and a part of the palm's habitat.

Old individuals in habitat reach a height of 60 feet (18 m) with robust trunks to 18 inches (45 cm) in diameter. The leaves are a distinctive gray to blue-green on both surfaces; in some populations, they are almost silvery white. The petioles are twice as long as the deeply costapalmate blades, which are 3–4 feet (90–120 m) long. The many deeply divided, thin segments have pendent ends. The leaf crown is usually dense and globular despite the length of the petioles.

This is one of the finest-looking trees in the genus, although it is slow growing. It is also evidently hardy to cold, having survived unscathed temperatures in the low 20s Fahrenheit (around −7°C) in a dry climate like that of Tucson, Arizona; it is probably not so resistant to cold in areas subject to wet freezes in winter where the low temperatures usually last longer and are accompanied by precipitation.

Sabal yapa
SAY-ball · YAH-pa
PLATES 857–858

Sabal yapa is indigenous to southwestern Cuba, the Isle of Youth, the Yucatán Peninsula of Mexico, and northern Belize, where it grows at low elevations in swampy forests, open savannas, and dry, exposed sites on limestone soil. The epithet is a corruption of one of the vernacular names for the palm in Cuba. Except for the leaf size and the color of the leaf underside, the palm is similar to *S. mauritiiformis*; the leaves of *S. yapa* are a uniform green on both sides and are slightly smaller. The 2 species differ in the shape of their flowers; *S. yapa* has a bell-shaped calyx that is unique in the genus. Like *S. mauritiiformis*, it is not hardy. It is more drought tolerant and is perfectly adapted to calcareous soils. Its silhouette and canopy-scape are as beautiful as *S. mauritiiformis*.

Salacca
sa-LAHK-ka

Salacca includes about 20 clustering, pinnate-leaved, dioecious palms in Myanmar, Thailand, Laos, Vietnam, peninsular Malaysia, Sumatra, Java, the Moluccas, Borneo, and the Philippines. All are undergrowth subjects in lowland tropical rain forest and all are quite spiny. Most are small. Many have subterranean trunks. Several have undivided, apically bifid leaves, and these are especially beautiful. The leaves can be large relative to the short or subterranean trunks, and they grow on long, stout, spiny petioles. The blades themselves or the leaflets of the segmented leaves bear bristles and are a lighter color on their undersides. The inflorescences grow from among the leaf bases, are usually branching, and are accompanied by papery, quickly disintegrating bracts.

Most of them are long enough to be pendent and lie partially on the ground where they are probably pollinated by beetles and where they sometimes form roots to create new rosettes. The fruits are usually round, red, and covered in overlapping scales, each with a sharp tip, showing the relationship of this genus to the rattans. Several species produce edible fruits.

Except for the species with undivided leaves, most species are rare in cultivation, especially in the Western Hemisphere. They are without exception intolerant of cold, being adaptable only to zones 10b and 11, but many species seem to tolerate well frostless Mediterranean climates if provided with enough water. Although several of them are known to resprout from their roots if frozen back, they are so slow growing that they usually die off if subjected to successive freezes. They are as intolerant of drought as they are of frost. They prefer a moist, fast-draining soil with organic matter, and they do not flourish on calcareous or poor soils. The larger species, especially those with segmented leaves, adapt to full sun, but the smaller species, especially those with undivided leaves, need partial shade at all stages.

The genus name is a Latinized form of the name in Malaysia, *salak*. Seeds germinate within 60 days. For further details of seed germination, see *Calamus*.

Salacca magnifica
sa-LAHK-ka · mag-NI-fi-ka
PLATE 859

Salacca magnifica is endemic to Borneo, where it grows in mountainous rain forest from sea level to 3000 feet (900 m). The epithet is Latin for "magnificent." This acaulescent clumper is truly and wonderfully magnificent in appearance. The great undivided leaves are linear-obovate, apically bifid, and 15 feet (4.5 m) long or longer. They are ascending, stiff, and little arched on spiny, stout petioles 2 feet (60 cm) long. The blade has the typical deeply grooved surfaces due to the margins of the fused pinnate segments and is deeply toothed along its margins, especially in its apical section. Leaf color is a glossy bright medium green above but a beautiful silvery green to nearly white beneath. The blade is usually V shaped. The female inflorescences are erect and spikelike but mostly pendent in fruit. Male inflorescences are short and much branched. The pear-shaped fruits are 2 inches (5 cm) long and deep rose or yellowish brown.

Salacca wallichiana
sa-LAHK-ka · wahl-lik´-ee-AH-na
PLATE 860

Salacca wallichiana occurs naturally from Myanmar southwards and eastwards through Indochina into peninsular Thailand and peninsular Malaysia, where it grows in swamps, clearings of monsoonal rain forest, along rivers and streams, and in the undergrowth of monsoonal evergreen forest. Its distribution has probably been greatly influenced by humans cultivating it. The epithet honors Nathaniel Wallich (1786–1854), superintendent of the botanical gardens in Calcutta, India, and enthusiastic collector of South Asian plants.

The stems become aerial but are generally creeping and 6 feet (1.8 m) long with whorls of vicious black spines 3 inches (8 cm) long. The ascending, erect leaves are unarching but become spreading, are 25 feet (7.6 m) long, and are borne on petioles 6–10 feet (1.8–3 m) long. The limp, arching leaflets are 3 feet (90 cm) long and grow in groups of 2–5 at differing angles from within each group to create a densely plumose leaf; the leaflets of the terminal pair are usually broader than the others. All the leaflets are linear-elliptic, with truncated, jagged apices, and are a deep, almost bluish green on both surfaces. The large egg-shaped fruits are 3 inches (8 cm) long, reddish brown, and covered in overlapping scales with upturned bristly pointed tips. They are edible and considered a delicacy in habitat. This species is different in aspect from *S. magnifica* but is also magnificent because of its immense size and heavily plumose leaves.

Salacca zalacca
sa-LAHK-ka · za-LAHK-ka
PLATES 861–862

Salacca zalacca occurs naturally in Sumatra and Java, where it grows in lowland rain forest and swamps, often forming impenetrable thickets. It is widely cultivated throughout Indonesia and Malaysia. The epithet is a Latinized form of the aboriginal name for the species. This acaulescent, densely clustering species has large, erect leaves springing from the ground in immense rosettes. The leaves are 15 feet (4.5 m) long, ascending but arching, and borne on spiny petioles 4 feet (1.2 m) long. The many leaflets grow from the rachis in a nearly flat plane but are arranged into closely set groups of 3–5. Individual leaflets are 2 feet (60 cm) long, narrowly elliptic, dark green above and silvery green or pure silver beneath. The fruits are reddish brown to nearly red and are edible. In habitat they are relished and an item of commerce. The species is beautiful because of the silvery-backed leaflets and makes a wonderful accent when planted among other vegetation, in either partial shade or full sun.

Saribus
SER-i-bus

Saribus is an old genus newly resurrected and reorganized to accommodate 8 species of palms formerly included in *Livistona* and one in *Pritchardiopsis*. They are large, hermaphroditic fan palms, distributed throughout the Philippines, Sulawesi, Moluccas, New Guinea, Solomon Islands, and New Caledonia.

The species of *Saribus* have, until recently, been included in *Livistona* and the monotypic genus *Pritchardiopsis*. Comparative studies of DNA have shown *Saribus* to be distinct, with *Pritchardiopsis* embedded within it. Thus, *Saribus* is reinstated and *Pritchardiopsis* is sunk as a synonym. *Saribus* has 3-pronged inflorescences and fruits that are orange, orange-brown, or red (except *S. jeanneneyi*, whose fruits pass through an orange stage but are said to ripen to a purplish color); whereas, *Livistona* has inflorescences with a single main axis and fruits that are black, brown, blue, blue-green, or purple. Confusingly, *Livistona saribus* is not part of the genus *Saribus* and remains in *Livistona*.

Saribus is the aboriginal name used for these palms in eastern Indonesia. Seeds generally germinate within 90 days unless they are allowed to dry out before sowing. Dried seeds take longer to germinate and do not germinate as well as those kept slightly moist. For further details of seed germination, see *Corypha*.

Saribus jeanneneyi
SER-i-bus · jah-NE-nee-eye
PLATE 863

Saribus jeanneneyi is the only palmate-leaf palm native to New Caledonia, where it grows in rain forest at 600 feet (180 m) elevation on soils rich in heavy metals. The species is highly endangered: only one mature individual and a handful of seedlings remain in the wild. This species was formerly placed in its own genus, *Pritchardiopsis*. The name honors a Mr. Jenneney of New Caledonia, who brought the palm to the attention of botanists. Nothing is known of Mr. Jenneney.

With great age, the solitary trunk reportedly grows to 35 feet (11 m) high with a diameter of 6 inches (15 cm). The leaf crown is hemispherical and open because of the relatively long petioles of 4–5 feet (1.2–1.5 m). The leaves are 5 feet (1.5 m) wide, deep green on both surfaces, semicircular to nearly circular, with many stiff segments. The petioles are unarmed on mature plants, but juveniles have some teeth. The inflorescences are 3 feet (90 cm) long and grow from the leaf bases. The round 2-inch (5-cm) fruits are purplish brown when ripe.

The species is beautiful, especially when older, but is slow growing. It needs partial shade when young, copious and regular moisture. Seedlings are difficult to grow, but once plants have established palmate leaves, they seem to do better and can be grown more easily. Young plants established in the ground can take poor soil conditions as well as full sun and wind.

Saribus merrillii
SER-i-bus · mer-RIL-lee-eye
PLATE 864

Saribus merrillii is highly endangered and endemic to the Philippines, where only a handful of wild individuals remain. It grows, or grew, along the edges of and in clearings in the rain forest at low elevations. The epithet honors Elmer D. Merrill (1876–1956), botanist, plant explorer of the Asian tropics, and former director of the New York Botanical Garden and the Arnold Arboretum.

The trunk attains a maximum height of 60 feet (1.8 m) and is 6–12 inches (15–30 cm) in diameter. The leaf crown is open and spherical. The circular leaves are 3–4 feet (90–120 cm) in diameter on petioles 3–6 feet (90–180 cm) long. The leaves are medium to deep green on both surfaces. The leaf segments are free for one-third their length, with an equally deep apical cleft. The segment tips can be either rigid or pendulous. The inflorescence is 3-pronged, and the fruits are shiny dark red.

This beautiful and fast-growing species is rare in cultivation. It is amazingly hardy to cold considering its tropical origins and is adaptable to zones 9b through 11. It needs regular, adequate moisture and flourishes in full sun even from youth.

Saribus rotundifolius
SER-i-bus · ro-tun'-di-FO-lee-us
PLATE 865–866

Saribus rotundifolius extends from southern peninsular Malaysia to Sarawak, Brunei, and Sabah; as well as Sulawesi, the Moluccas, and many Philippine islands, where it grows in low mountainous rain forest. It is often called footstool palm. The epithet is Latin for "round leaf" and alludes to the shape of the leaves, mainly on juvenile plants. This species includes those formerly recognized as *Livistona rotundifolia* var. *luzonensis* and *L. robinsoniana*.

The solitary trunk attains a height of 100 feet (30 m) in habitat but is no more than 1 foot (30 cm) in diameter. Older parts of the trunks are smooth and pale gray to almost white, and show regularly spaced reddish brown rings of leaf base scars except for the oldest parts. In several younger individuals the leaf crown is extended and even reaches to the ground in specimens that are 20 feet (6 m) tall overall, but it is always rounded, much shorter, and open in older palms. The petioles are 8 feet (2.4 m) long and bear curved spines along their lower margins. The leaves are 5–6 feet (1.5–1.8 m) wide and are circular on juvenile palms. In older palms, the leaves are significantly smaller, do not form complete circles, and are divided into segments for half the length of the blade. Each segment has a short split at its end, and the segment tips may be rigid or pendulous. Leaf color is glossy deep green on both surfaces. The 3-pronged inflorescences are 8 feet (2.4 m) long with many branches bearing small, yellow flowers. The round 1-inch (2.5-cm) fruits are bright orange-red, maturing to black.

The species is slightly susceptible to lethal yellowing disease. It is the least cold-tolerant species in the genus, adaptable only to zones 10b and 11, and marginal in 10a. It wants copious, regular moisture but is not fussy about soil type. It grows well in partial shade when young but needs full sun when older. Because of the long petioles, the leaf crown is open and exquisitely graceful, and is among the finest canopy-scapes. When planted in groups of 3 or more individuals of varying heights, it is even more attractive.

Young plants have larger, less deeply segmented leaves than do adults. The leaves are stunningly attractive up close, looking more like disks than fan leaves, and are choice for intimate areas. Alas, they, like other young palms in nature, grow up and reach considerable heights. The juvenile aspect of the leaves is maintained to some extent if the palm is in partial shade. The species is a good indoor candidate when young.

Saribus woodfordii
SER-i-bus · wood-FOR-dee-eye
PLATE 867

Saribus woodfordii occurs in the Solomon Islands and Papua New Guinea, where it grows over a wide range of habitats, from low mountainous rain forest to costal scrub. The epithet honors Charles M. Woodford (1852–1927), a British naturalist and first Resident Commissioner of the Solomon Islands.

The solitary trunk attains a height of 40 feet (12 m) and a diameter of 6–8 inches (15–20 cm). It is gray to tan and heavily ringed with leaf base scars in its younger parts but nearly smooth in its older parts. The leaf crown is spherical and dense with leaves 3 feet (90 cm) wide. The leaf segments are free for one-half to three-quarters their length, with slightly pendulous tips. They are deep green on both surfaces and are borne on slender petioles, 4–5 feet (1.2–1.5 m) long and armed near their bases. Juvenile trees have rounded leaves with wide segments, whereas the leaves of older trees are only hemispherical. The flowers are red. The orange to reddish brown fruits are globose and about ½ inch (13 mm) in diameter.

This species is one of the most elegant in the genus, its lollipop form making an exquisite silhouette and, planted in groups of 3 or more trees of varying heights, it is almost unparalleled in beauty. It is surprisingly hardy to cold, being adaptable to zones 10 and 11 and possibly marginal in warm microclimates of 9b. It needs sun when past the juvenile stage and is a true water lover, but it is adaptable to either limey or slightly acidic soils, as long as they are freely draining.

Satakentia
sah-ta-KEN-tee-a

Satakentia is a monotypic genus of solitary-trunked, pinnate-leaved, monoecious palm allied to *Carpoxylon* and *Neoveitchia*, sharing some visual similarities with the latter. The genus name is a combination of 2 words, one the surname of Toshihiko Satake (1910–1998), Japanese industrialist, amateur botanist, and lover of palm trees, and the second word, *Kentia*, an archaic genus name used in combination to mean any sort of solitary, elegant, pinnate-leaf palm. Seeds of *Satakentia* germinate within 6 months if they are going to at all. If they are allowed to dry out before sowing, they do not germinate well. Furthermore, seeds picked slightly green seem to germinate better than fully ripe seeds. For further details of seed germination, see *Areca*.

Satakentia liukiuensis
sah-ta-KEN-tee-a · lee-oo′-kiu-EN-sis
PLATE 868

Satakentia liukiuensis is endemic to the islands of Ishigaki and Iriomote in the Ryukyu archipelago, where it grows on hills and near sea level in moist forests. The epithet is Latin for "of Liukiu" (Ryukyu).

The trunk attains a height of 60 feet (18 m) and a diameter of 1 foot (30 cm) in habitat, and usually has a mass of adventitious roots at its base. It is a light brown to grayish brown and exhibits closely set rings of leaf base scars. The smooth, dark olive green, reddish brown, or purplish brown crownshaft is 2½ feet (75 cm) tall, sometimes bulging at its base, otherwise cylindrical. The leaf crown is spherical or nearly spherical and usually contains 12–14 spreading leaves, each 8–10 feet (2.4–3 m) long on a short petiole only 4 inches (10 cm) long in older palms, juveniles usually having longer leafstalks. The leaves are gracefully arching, sometimes twisting, and bear regularly spaced, lanceolate, dark green leaflets 2 feet (60 cm) long. The leaflets grow from the rachis in a single, nearly flat plane and are limp and slightly pendent. The inflorescences usually form a ring around the nodes below the crownshaft, each consisting of several stiff, horizontally spreading, whiskbroomlike branches on stout, ringed, short, and squat peduncles. The oblong fruits are ½ inch (13 mm) in diameter and turn black when mature.

The palm needs a nearly tropical climate and withstands only a touch of frost. It also wants regular, adequate moisture, a humus-laden soil, and full sun when past the juvenile stage. The crown and leaves look much like those of a coconut palm, giving this species the same landscaping uses. While still not common in cultivation, the species is rapidly gaining acceptance in South Florida where it seems to do exceptionally well.

Satranala
sah-tra-NAH-la

Satranala is a monotypic genus of solitary-trunked, palmate-leaved, dioecious palm endemic to northeastern Madagascar. The name is a derived from Malagasy words meaning "forest fan-palm." Seeds can take up to 18 months to germinate and do so sporadically. Because the germination stalk can go down 8 or more inches (20 cm), the container in which seeds are sown should be at least 14 inches (36 cm) deep. The endocarp splits into 2 halves like a walnut, a unique mode of germination within the palm family. For further details of seed germination, see *Bismarckia*.

Satranala decussilvae
sah-tra-NAH-la · dek-oo-SIL-vee
PLATES 869–870

Satranala decussilvae is rare in habitat, growing in mountainous rain forest below 1000 feet (300 m) on soils derived from quartzite or from rock rich in heavy metals and devoid of calcium. The epithet translates from the Latin as "splendor of the forest," an allusion to its beauty and habitat.

In the wild, the trunk attains a height of 50 feet (15 m) and a diameter of 7 inches (18 cm). The leaf crown is open and less than hemispherical. The leaves are clear, medium green on both surfaces, semicircular to nearly circular, and 8 feet (2.4 m) wide on minutely armed petioles 5 feet (1.5 m) long. The linear-lanceolate leaf segments are free for half their length and are stiff and non-pendent. The leaf is weakly costapalmate and virtually flat. The round 2-inch (5-cm) fruits are black, and the endocarp is variously ridged and ornamented.

The species is related to *Bismarckia*, which is evident by the form of the inflorescences. The palm is slow growing. It cannot be hardy to cold and is certainly a water lover. It seems to dislike any hint of calcium at its roots. In the years since its discovery, this palm has not proven itself amenable to cultivation, and its promise as an ornamental palm has not been fulfilled.

Schippia
SHIP-pee-a

Schippia is a monotypic genus of solitary-trunked, palmate-leaved, hermaphroditic palm. The name honors William A. Schipp (1891–1967), an Australian collector in Belize. Seeds germinate within 90 days and should not be allowed to dry out before sowing. For further details of seed germination, see *Corypha*.

Schippia concolor
SHIP-pee-a · KAHN-ku-lor
PLATES 871–872

Schippia concolor is endemic to Belize and adjacent Guatemala, where it grows in lowland rain forest and mountainous pineland from sea level to 1600 feet (490 m). The epithet is Latin for "single colored," which describes the leaves that lack silvery undersides.

The trunk slowly reaches to 30 feet (9 m) high with a maximum diameter of 4 inches (10 cm). The beautiful leaf crown is open and hemispherical or slightly more so with 12 semicircular to nearly circular 3-foot (90-cm) leaves on slender, unarmed petioles. The leaf segments are linear-lanceolate, free for two-thirds their length, and shortly bifid at their tips. They are glossy medium green above and below. The much-branched inflorescences are 1 foot (30 cm) long and white, and bear white, bisexual flowers. The round 1-inch (2.5-cm) succulent fruits are white when ripe and split open to reveal the dark brown endocarp.

This is a silhouette palm, its airy leaf crown of exquisitely elegant proportions. It does not look good as an isolated specimen surrounded by space but is more than charming as a small canopy-scape or against a highly contrasting background so that its form can be easily seen. It is adaptable to zones 10 and 11 and possibly in the warmest microclimates of 9b. It needs regular, adequate moisture to look good and grow other than slowly but thrives on calcareous as well as slightly acidic soils. It flourishes in partial shade or full sun. It can be grown in a container only if extra calcium is added to the potting medium.

Sclerosperma
skle-ro-SPUR-ma

Sclerosperma consists of 3 densely clustering, pinnate-leaved, monoecious palms in western tropical Africa. They are poorly known and are not in cultivation except possibly in a few tropical botanical gardens. The trunks are creeping or subterranean. The ascending, unarmed leaves on slender petioles are up to 12 feet (3.6 m) long and pinnately segmented into several wide leaflets or are undivided and apically bifid. The inflorescences are short and spikelike, borne among the leaf bases. The flowers are densely congested on the spike. The fruits are small, top shaped, and red when ripe. The few drawings and photographs available show wonderfully beautiful small palms with relatively large leaves, similar in general appearance to some *Salacca* species. The genus name is derived from Greek words meaning "hard seed." Seeds have not been readily available for this genus, so data are absent.

Sclerosperma mannii
skle-ro-SPUR-ma · MAN-nee-eye
PLATE 873

Sclerosperma mannii is native to rainforest swamps in Cameroon, Liberia, Gabon, Congo, Equatorial Guinea, and Nigeria. The epithet commemorates the German botanist Gustav Mann (1836–1916), who collected in tropical Africa for the Royal Botanic Gardens, Kew. Perhaps because it demands boggy conditions, it is virtually unknown in cultivation. The species is clustering and forms very short aboveground stems. The leaves are held stiffly upright and are irregularly divided into rhomboidal segments, each segment with multiple folds. The undersides of the leaves are silvery white. The flowers are borne on short spikes near the ground. The fruits are round and red. The striking contrast of green leaves with ghostly white undersides should make this palm much in demand. We are mystified that it is not more widely grown in gardens that can provide hot, wet conditions.

Serenoa
ser-e-NO-a

Serenoa is a monotypic genus of clustering, palmate-leaved, hermaphroditic palm. The name honors Sereno Watson (1826–1892), American botanist and curator of the Gray Herbarium at Harvard University. Seeds germinate within 120 days and should not be allowed to dry out before sowing. For further details of seed germination, see *Corypha*.

Serenoa repens
ser-e-NO-a · RE-penz
PLATES 874–876

Serenoa repens is endemic to the United States in extreme southeastern Louisiana, southern Mississippi, southern Alabama, southern and coastal Georgia to the southeastern South Carolina coast, and peninsular Florida, where it grows mostly on the coastal plain regions in sandy, open (mostly pine) woodland and near the coast on sand dunes. It is widely known as saw palmetto. The epithet is Latin for "creeping."

The trunks are subterranean until they are of some age, at which time they usually emerge above ground and can grow to 20 feet (6 m) long. The creeping stem branches sporadically, which can be clearly seen in palms that have recently experienced a forest fire. When above ground, the stems are covered in all but the oldest parts with a mass of fibers and leaf bases and are always creeping to some extent. The leaves are 3 feet (90 cm) wide on petioles 3–5 feet (90–150 cm) long and are edged with small, backward-pointing teeth. The segments are stiff and narrowly triangular or lanceolate, with a short notch apically. Leaf color varies from dull, light yellowish green to medium, pure green to almost pure silver or bluish silver. The really silvery individuals are found only in the southern two-thirds of peninsular Florida on a narrow strip along the Atlantic; no one knows why this is the case, but it may have to do with adaptation to a sunnier habitat on the east coast. These eastern forms usually have distinctly yellowish to almost orange petioles and leaf bases. The inflorescences appear in spring, and are slender, branched, and shorter than the leaf crown. The small, white, bisexual blossoms produce ellipsoid fruits that are ½–1 inch (13–25 mm) long and turn shiny black when mature. The flowers and fruits are mostly hidden by the densely packed leaves in younger plants, but the fruits are noticeable by their unpleasant odor when they are ripening.

The saw palmetto is hardy to cold and is adaptable to zones 8 through 11 and, with protection or a very favorable microclimate, to zone 7. It is drought tolerant and fire tolerant, at least to the swiftly running fires that occur in habitat, leaving leafless but live, scorched, and reclining trunks that were protected by the fibers and leaf bases. The stems sprout new leaves with the return of moisture. Fire-defoliated palms often bloom with the onset of spring rains.

Saw palmetto grows in partial shade as well as the full blazing sun of South Florida. It loves sandy soil and has great saline tolerance. In truth it is not particular about the type of soil, which in the palm's habitats may be acidic to very alkaline. As for moisture requirements, it grows in seasonally dry areas as well as swampland. Because of its great fibrous root system, the palm is almost impossible to transplant from the wild, but container-grow plants grow rapidly after planting out, even with only periodic care. In Southern California, the species is so slow growing that many growers eschew it as a landscape subject (Geoff Stein, pers. comm.).

The forms with silvery blue or blue-green leaves are usually considered more attractive than the plain green forms, but all forms make a dramatic, tropical-looking accent in the landscape. Great masses of the plants are tiresome to contemplate, however, even in nature. None of the colors seem attractive when isolated with surrounding space, and the great clumps look much better when integrated into a shrub border or mixed with tall but not overly umbrageous trees. Chemicals in the fruits of saw palmetto are advantageous in treating enlarged prostate, and the fruits are gathered for pharmaceutical use.

Socratea
sok-rah-TEE-a

Socratea comprises 5 tall, mostly solitary-trunked, pinnate-leaved, monoecious palms in tropical America. Most are denizens of lowland tropical rain forest, but a few occur naturally in mountainous rain forest to a maximum elevation of 6000 feet (1800 m). The most salient characteristics of these species are their large but open cones of spiny stilt roots at the bases of the trunks, and their wide, mostly plumose leaves with jagged apices. They form prominent crownshafts beneath which the relatively short once-branched inflorescences grow. The thick flowering branches bear white, unisexual flowers of both sexes. The genus is obviously related to *Dictyocaryum*, *Iriartea*, *Iriartella*, and *Wettinia*.

The species are completely tropical in their temperature requirements and are often killed by sustained temperatures below 55°F (13°C). They need constant, copious moisture as well as constantly high relative humidity. They are adapted to partial shade when young but, to reach their full potential, need sun when older.

The species have many human uses: the lower parts of the durable trunks are used for construction, the leaves are used for thatch, and the spiny stilt roots are used for grating and grinding manioc or tapioca (*Manihot*). Only one species is in cultivation, and it is not common outside of tropical botanical gardens. The genus name honors the ancient Greek philosopher Socrates (ca. 470–399 BC). Seeds germinate within 60 days and should not be allowed to dry out before sowing. For further details of seed germination, see *Iriartea*.

Socratea exorrhiza
sok-rah-TEE-a · ex'-or-RY-za
PLATE 877

Socratea exorrhiza is the most widespread species in the genus, occurring naturally from southern Nicaragua, southwards through Costa Rica, Panama, and through most of Colombia, Ecuador, Peru, northwestern Brazil, the southern half of Venezuela, and all of the Guianas, where it grows in lowland and mountainous rain forest from sea level to 3500 feet (1070 m). The epithet is from Greek words meaning "outside" and "root," an allusion to the stilt roots.

The solitary trunk grows to 70 feet (21 m) high and is 6–7 inches (15–18 cm) in diameter at its thickest point. It is light tan in its younger parts and is graced with widely spaced rings of darker leaf base scars. In its older parts it becomes gray to nearly white and is supported by an open cone, 6–10 feet (1.8–3 m) tall, of brown, thick stilt roots that bear short and fat but sharply pointed, light yellow spines. The crownshaft is 6 feet (1.8 m) tall, cylindrical except at its base where it is slightly swollen, and is deep bluish green to nearly blue. The leaf crown is hemispherical to nearly spherical and contains 6 leaves 6–8 feet (1.8–2.4 m) long on petioles 1 foot (30 cm) long. The stiff, straight, spreading leaves are slightly arching near their tips but are wide because of the leaflets, which are 2 feet (60 cm) long and linearly to broadly wedge shaped with a slanting, multitoothed or jagged apex. The leaves are split to their bases into almost equally wide, narrow segments that are arrayed in slightly different planes. Their color is medium to deep green to bluish green, and they grow from the rachis in bundles and at different angles to create a wide and plumose leaf. The flowering branches of the inflorescences are 1 foot (30 cm) long and produce ellipsoid deep yellow fruits that are 1 inch (2.5 cm) long.

This fascinating and beautiful species requires a wet tropical climate, is adaptable only to zone 11, and is marginal in 10b. It is impossible to maintain in frostless Mediterranean climates as a healthy individual. It has one of the most beautiful silhouettes of any palm and is enthralling as a canopy-scape palm.

Socratea rostrata
sok-rah-TEE-a · ro-STRAH-ta

Socratea rostrata is indigenous to the slopes and foothills of the western Andes in Ecuador and Peru, where it grows in wet mountainous rain forest from sea level to 6000 feet (1800 m). The epithet is Latin for "beaked," a reference to the projection at the end of the yellowish ellipsoid 1-inch (2.5-cm) fruits. The solitary trunk attains a height of 80 feet (24 m) and a diameter of 1 foot (30 cm), making it the largest in the genus. It is otherwise similar to *S. exorrhiza*.

Socratea salazarii
sok-rah-TEE-a · sal-a-ZAHR-ee-eye

Socratea salazarii occurs naturally in Peru along the eastern slopes and foothills of the Andes, in extreme western Brazil, and extreme northwestern Bolivia, where it grows in mountainous rain forest from 1000 to 2300 feet (300–700 m). The epithet honors Adolfo Salazar of the Peruvian Forest Service, who aided Harold E. Moore, Jr., in finding the species. It is similar in appearance and dimensions to *S. exorrhiza*, differing principally in its wide, unsplit leaflets and in its rarely clustering habit. The leaflets grow in only 2 slightly differing planes, creating a plumose leaf, and the terminal pair of leaflets are broad.

Solfia
SOL-fee-a

Solfia is a monotypic genus of solitary-trunked, pinnate-leaved, monoecious palm from Samoa. It had been included in the genus *Drymophloeus*, but its closest relative appears to be *Balaka*. The genus name honors Wilhelm H. Solf (1862–1936), one-time governor of German (Western) Samoa.

Solfia samoensis
SOL-fee-a · sah-mo-EN-sis

Solfia samoensis is found at elevations above 1500 feet (460 m) in wet rain forest. The epithet means "of Samoa." A synonym is *Drymophloeus samoensis*. This moderate-sized palm has a sparse, open crown with leaves up to 6 feet (1.8 m) long. The crownshaft is green, with dark scaly hairs near the apex. The evenly pinnate, evenly spaced leaflets are borne on one plane and are dark green on both surfaces. The leaflets have jagged, irregular tips. The inflorescence has a long peduncle and bears greenish white flowers. Male flowers have many stamens. The small, oblong fruits are red when ripe. The endocarps and seeds are round in cross-section, and the endosperm is homogeneous.

This species is unaccountably rare in cultivation. It is strictly tropical in all its requirements, and perhaps its susceptibility to cold or dry air limits its use in horticulture. It is a handsome palm, with visual similarities to the small Fijian species of *Veitchia*, one of the single-trunked species of *Ptychosperma*, or one the larger species of *Balaka*. It could be interchanged with any of these in a landscape setting. Fresh seeds germinate within 90 days. Seeds should not be allowed to dry out.

Sommieria
som'-mee-ER-ee-a

Sommieria is a monotypic genus of small, solitary-trunked, pinnate-leaved, monoecious palm from New Guinea. Other species have been named in the past, but at present only one species is given recognition. It is rare in cultivation even though it has been known since 1877. It is an elegant palm with leaves that look like gigantic seedling leaves of many other genera and have great visual affinities with mature *Asterogyne* species. The genus name honors Italian botanist Stephen Sommier (1848–1922). Seeds germinate within 60 days.

Sommieria leucophylla
som'-mee-ER-ee-a · loo-ko-FIL-la
PLATES 878–879

Sommieria leucophylla is endemic to Papua Province, Indonesia, where it is a small undergrowth palm in dense, wet, and humid rain forest. The stem is short but sturdy and distinctly ringed when aerial. The leaf crown is spherical because of the short-petioled but long, arching leaves that are undivided and deeply bifid apically or with 4–6 unequally sized leaflets. The leaflets are invariably linear-obovate, deeply pleated with the margins of the fused pinnae on both sides, have prominent, dark-colored midribs, and are glossy deep green above but chalky white or silver overlaid with a tan suffusion beneath. The inflorescences grow from among the leaf sheaths and consist of long, pendent peduncles at the ends of which flowering branches radiate in a starlike fashion. The tiny fruits are round, with blunt, warty projections, pinkish or red, and much resemble diminutive litchi fruits.

This palm is finicky, needing constant warmth, copious, regular moisture, a humus-laden soil, partial shade, and constantly high relative humidity. The epithet is derived from 2 Greek words meaning "white" and "leaf," an allusion to the silver-backed leaves.

Syagrus
sy-AG-rus

Syagrus is composed of about 53 solitary-trunked and clustering, pinnate-leaved, monoecious palms, mostly in South America, with a single species in the Lesser Antilles. Several species have subterranean stems and mimic grasses; their underground growing points are unharmed by the periodic fires that sweep through the grasslands in which they grow. None of the species form crownshafts. The inflorescences grow within the leaf crown and are spicate or branched once. The individual branches are often ropelike, pendent, and white or straw-colored, bearing unisexual flowers of both sexes. They are accompanied by a sometimes large, usually tough, woody bract that is shaped like a spoon or a boat. The endocarps have 3 germination pores (as in a coconut). The endosperm can be ruminate or homogeneous and in some species possesses a central cavity (also as in a coconut).

The genus name is derived from a Latin word that referred to some palm tree but definitely not one from this genus. Several genera are now subsumed under *Syagrus*, including *Arecastrum*, *Arikury*, *Arikuryroba*, *Chrysallidosperma*, and *Rhyticocos*.

Syagrus is closely related to *Cocos* (coconut) and has many species with edible seeds that are similar in taste to those of the coconut. In Brazil, several species grow together, and there are 5 naturally occurring hybrids. In addition, there are the beautiful man-made hybrids, such as the one between *S. romanzoffiana* and *Butia odorata*, known as ×*Butiagrus nabonnandii*. Seeds of *Syagrus* can take more than 2 years to germinate and do so sporadically. For further details of seed germination, see *Acrocomia*.

Syagrus amara
sy-AG-rus · a-MAH-ra
PLATE 880

Syagrus amara occurs naturally on the Windward Islands of the Lesser Antilles, where it grows in the hills and tropical forests along the coasts. It is one of the tallest species in the genus and is the only one not indigenous to South America. It is called overtop palm in its habitat, because mature individuals always grow significantly taller than the canopy of coastal vegetation among which they develop. The epithet is Latin for "bitter" and refers to the nearly liquid endosperm of the immature seeds.

The solitary trunks grow to 60 feet (18 m) or even more in habitat but are 8–9 inches (20–23 cm) in diameter. The leaves are 10 feet (3 m) long on petioles 1 foot (30 cm) long. The many dark green leaflets are 3 feet (90 cm) long, arranged in closely set clusters, and grow from the rachis in almost a single plane to slightly different angles from the rachis, creating a flat or a slightly plumose leaf; they are stiff in younger leaves but limp and pendent in older ones. The orange fruits are 2–3 inches (5–8 cm) long.

This species is not hardy to cold and is adaptable only to zones 10b and 11, although nice specimens are to be found in the warmer microclimates of 10a. It grows in partial shade, especially when young, but, when past the juvenile stage, needs sun to grow faster and achieve its coconut-like appearance. While not fussy about soil type, it does better and grows much faster with a rich soil and with regular, adequate moisture. Mature palms look like slender coconut palms and even the fruits of the 2 palms are similar. While nothing can match the beauty of a coconut, this one goes a long way trying and should be treated in the landscape as if it were *Cocos*. It is tolerant of salty soils.

Syagrus botryophora
sy-AG-rus · bo-tree-OFF-o-ra
PLATES 881–882

Syagrus botryophora is endemic to Brazil, where it grows in lowland rain forest along the central Atlantic coast. The epithet is from Greek words meaning "cluster-bearing," an allusion to the grapelike clusters of yellow-green fruits.

The solitary trunks reach heights of 50 or more feet (15 m) in habitat and diameters of less than 1 foot (30 cm). They are straight and columnar. Stem color is a beautiful green in the youngest parts, a light tan in the older parts, and a light gray in their oldest parts. The stems are also graced with distinct, widely spaced darker rings of leaf base scars in all but their oldest parts. The wonderful leaves are 10–12 feet (3–3.6 m) long but strongly arching, sometimes almost forming a semicircle in bending back toward the trunk. The stiff deep green leaflets are 2 feet (60 cm) long and grow from the rachis at an angle approaching 45 degrees, arching upwards to give the leaf a V shape.

This palm is one of the world's most beautiful, comparable in beauty to such archetypes of the palm world as *Actinorhytis calapparia*, *Pigafetta elata*, and some of the larger *Hydriastele* species. It is also among the world's fastest growing palms, and with a rich soil, ample moisture, and a tropical or nearly tropical climate can add 6 feet (1.8 m) of trunk per year until near its mature height. Under such optimum conditions, it grows so fast that its leaf crown tends to become elongated. It is one of the finest canopy-scapes, and specimen groups of 3 or more individuals of varying ages and heights are incomparable. Its only faults are that it requires a nearly tropical climate and it is easily uprooted in strong winds. Some of the literature indicates that it needs partial shade when young, but experience has shown that it flourishes in the sun from youth to old age. Although *Arecastrum romanzoffianum* var. *botryophorum* is now a synonym for this species, the palm with that name referred to by James C. McCurrach (1960) is simply a robust form of what we now know as *S. romanzoffiana*.

Syagrus cardenasii
sy-AG-rus · kahr-de-NAH-see-eye
PLATE 883

Syagrus cardenasii is endemic to central Bolivia, where it grows in the foothills of the eastern Andes in dry, open forest from 1200 to 4700 feet (370–1430 m) in elevation. In nature, it is highly variable, having solitary or clustering stems to 10 feet (3 m) tall and 2 inches (5 cm) in diameter or subterranean. The leaves are 8 feet long with irregularly spaced groups of 2–4, grayish green, stiff leaflets that grow at slightly different angles from the rachis. The short, erect, yellowish inflorescences produce greenish brown edible fruits. The species looks like a dwarf, untidy cross between a date palm and a clustering queen palm. It is probably not widely cultivated outside of a few botanical gardens.

Syagrus cearensis
sy-AG-rus · see-ah-REN-sis
PLATE 884

Syagrus cearensis has been cultivated in Florida since 1959, but always misidentified as other species of *Syagrus*. Larry Noblick (2004) sorted out the confusion and described this beautiful species as new. The epithet is Latin for "of Ceará," a state in the palm's habitat.

This mostly twin-stemmed species is endemic to northeastern Brazil in the states of Ceará, Pernambuco, Paraíba, and Alagoas, where it grows in mountainous deciduous forest at 300–2500 feet (90–760 m). The 2 or more stems reach 30 feet (9 m) high with a maximum diameter of 7 inches (18 cm) and are light tan to light gray with prominent darker rings of leaf base scars. The leaves are similar to those of *S. oleracea* but smaller. The nearly round fruits are almost 2 inches (5 cm) in diameter. The seeds have a central cavity.

The landscaping uses for this palm are as great as any in the genus, and it is beautiful up close or as a canopy-scape. It is adaptable to several soils and flourishes on calcareous ones. It grows well in partial shade, especially when young, but needs full sun when older for fastest growth and good form and color. It is not as fast growing as *S. romanzoffiana* or *S. botryophora* but is certainly not the slowest grower in the genus. It is drought tolerant but looks much better and grows faster with regular, adequate water. It is not cold hardy and is adaptable only to zones 10b and 11, but there are nice small specimens in warm microclimates of 9b.

Syagrus cocoides
sy-AG-rus · ko-KO-i-deez
PLATE 885

Syagrus cocoides is indigenous to southern Guyana and northern Brazil, where it grows mostly in lowland rain forest but also on the higher, open hills and grasslands. The epithet is Latin for "similar to *Cocos*," although the 2 species are easily distinguished. This solitary-trunked species attains a maximum height in habitat of 35 feet (11 m), and is slender, whitish, indistinctly ringed, and 4 inches (10 cm) in diameter. The leaves are 10–12 feet (3–3.6 m) long and have thin, dark green, narrow, pendent leaflets growing from the rachis at different angles to give the leaf a plumose appearance. The palm is not hardy to cold, grows in partial shade but prefers sun, needs a rich but well-drained soil, and is a water lover. It is elegant with a rounded, compact leaf crown that lends itself wonderfully to use as a canopy-scape and is exquisite as a specimen group of individuals of varying heights.

Syagrus comosa
sy-AG-rus · ko-MO-sa
PLATE 886

Syagrus comosa is indigenous to Brazil and the central eastern *cerrado* region, which is characterized by dry, monsoonal climate and low savanna and gallery forests. The epithet is derived from the Latin *coma*, meaning "tufted," and likely refers to the lovely tuft or crown of leaves. It is important to note that, while the cerrado has seasons of drought, it also has a high water table and the rainy season can deliver copious water in a short time that results in flooding in many areas. The species also grows along rivers, in gallery forests, and is apparently a water-lover.

The stem of this solitary-trunked (rarely clustered) species grows slowly to a maximum height of 30 feet (9 m) with a diameter of 5–6 inches (13–15 cm) and is covered just below the spherical crown with triangular, closely spaced leaf bases. The leaves are 4–5 feet (1.2–1.5 m) long, with closely set, wide, thick, and deep green to deep silvery green leaflets that grow from the rachis at more than one angle to give the leaf a semiplumose effect. The mature fruits are 2 inches (5 cm) long and deep yellow-green. They split open when ripe.

The species has unusual cold hardiness for such tropical origins; it seems to be adaptable to zones 9b through 11 and is possibly safe in 9a, especially with protection or in a favorable microclimate. While it is drought tolerant, it grows faster (albeit still slowly) and looks better with regular, adequate moisture. It should not be grown in the shade. This unusual looking but beautiful species is rare in habitat and in cultivation. It has an architectural and sturdy, almost heavy look, and the many, closely spaced narrow leaf bases are beautiful to behold up close.

Syagrus coronata
sy-AG-rus · kor-o-NAH-ta
PLATES 887–888

Syagrus coronata is endemic to eastern Brazil, where it grows in the low savannas and cerrados, often near streams, but also in the *caatinga*, a dry region of spiny shrubland with grasses and cactus species. It is sometimes called licuri palm. The epithet is Latin for "crowned" and refers to the large leaf crown.

The solitary trunk attains 40 feet (12 m) high with a diameter of 1 foot (30 cm). It is covered except in its oldest parts with large persistent old leaf bases and petiole parts that are arranged around the trunk in loosely spiraling rows; when these finally fall off, undulating rows of broad liplike projections remain that give the trunk an interesting appearance. The leaf crown is large and dense, and the leaves are 10 feet (3 m) long. The leaflets are 1 foot (30 cm) long each and grow from the rachis at different angles to give the leaf an almost fully plumose aspect; they are stiff, lanceolate, deep green above, and a waxy, silvery or whitish green beneath. The fruits are juicy, egg shaped, yellow or orange and covered in a brown felty hairs around their tips.

This species is adaptable to zones 10 and 11, although it often matures nicely in warmer regions of 9b. It needs sun when past the seedling stage. It is drought tolerant but grows faster and looks better with regular, adequate moisture. It is not particular about soil type as long as it is well drained. It is not fast growing even in full sun but, after forming a trunk, is moderate in its growth rate.

The palm is beautiful even as an isolated specimen surrounded by space but looks better in groups; a grove is glorious. Because of the almost unique pattern of leaf bases and scars on its trunk, it is also architectural in aspect and provides unending pleasure in the up-close contemplation of its stem. It makes a wonderful large tub plant. It works for a while indoors but needs much light and eventually space.

Syagrus ×costae
sy-AG-rus · KAHS-tee
PLATE 889

Syagrus ×costae is a hybrid of *S. coronata* and *S. cearensis*. It grows in abundance in the state of Pernambuco in northeastern Brazil. It is similar to *S. coronata* except that it does not usually have its leaves arranged in a spiral pattern. It has the same cultural requirements as *S. coronata* and seems to be the only naturally occurring *Syagrus* hybrid in general cultivation around the world. It is extraordinarily beautiful.

Syagrus flexuosa
sy-AG-rus · flek-soo-O-sa
PLATE 890

Syagrus flexuosa is endemic to southern and central Brazil, where it grows in the *cerrado* and open woodlands to 3800 feet (1160 m). The epithet is Latin for "flexible," an allusion to the pliable leaflets.

The trunks of this clustering species grow to a maximum height of 15 feet (4.5 m) and are usually covered in old leaf bases; a few individuals in habitat do not seem to cluster. The leaves are 4–5 feet (1.2–1.5 m) long, arching, and plumose, with many narrow, lanceolate, dark green, soft leaflets growing from the rachis at different angles. The leaflets are usually light or even whitish green beneath.

This nice little palm has the look of a soft, dwarf, spineless, and green *Phoenix* species. The solitary-trunked forms are especially beautiful. The species is planted for ornament in habitat but, alas, is little known outside its range. It needs full sun from youth to old age and a fast-draining soil. It seems hardy to cold, having survived temperatures in the upper 20s Fahrenheit (around −2°C) unscathed; were it to be cut back from a freeze, it might return (albeit slowly) from the root.

Syagrus glaucescens
sy-AG-rus · glaw-SES-senz
PLATE 891

Syagrus glaucescens is endemic to central eastern Brazil, where it grows in dry and rocky *cerrado* from elevations of 2200 to 4000 feet (670–1200 m). The epithet is Latin for "slightly glaucous" and alludes to the color of the leaves.

The solitary stems attain a maximum height of 12 feet (3.6 m) and are usually less than 6 inches (15 cm) in diameter without the leaf bases. The leaf bases are similar to and almost as characteristic and striking as those of *S. coronata*: they are in vertical rows and lend an angled, squared appearance to the stem. The leaves are 3–4 feet (90–120 cm) long and have stiff, lanceolate, bluish or silvery green, slightly waxy leaflets that grow from the rachis to create a V-shaped leaf.

The species is more picturesque than beautiful. It is slow growing and demands full sun and a fast-draining soil. It is moderately hardy to cold, being adaptable to zones 10 and 11 and marginal in 9b.

Syagrus harleyi
sy-AG-rus · HAR-lee-eye
PLATE 892

Syagrus harleyi is endemic to northeastern Brazil in the state of Bahia, where it grows in low mountains at 1200–4500 feet (370–1370 m), usually in rock crevices. The species was named after the botanist Raymond M. Harley (1936–) of the Royal Botanic Gardens, Kew, who collected the first specimen.

The stems of this clustering species rarely emerge above the ground. The erect, unarching, bright green, waxy leaves are 4–5 feet (1.2–1.5 m) long with widely spaced, narrow leaflets that grow from the rachis at a slightly upward angle. The leaflets are rigid or flaccid according to their provenance: those from lower elevations of the habitat are soft and usually pendent but those from higher elevations are rigid.

This palm is surprisingly attractive for such a diminutive species and, if it can be planted where its delicate and lacy leaves can be displayed against a background of contrasting color, it makes a beautiful specimen. It is drought tolerant, needs full sun from youth to old age, and is unusually hardy to cold, thriving in zones 9 through 11, and is probably possible, if marginal, in 8b. Its clustering and subterranean stems almost assure its return if the leaves are frozen back.

Syagrus inajai
sy-AG-rus · ee-na-HY

Syagrus inajai is indigenous to southern Guyana, Surinam, and French Guiana, and into northern Brazil, where it grows in lowland rain forest as well as clearings and other open sites under 1500 feet (460 m). The epithet is one of the Brazilian aboriginal names.

The solitary trunk reaches a maximum height of 50 feet (15 m) and is slender, no more than 6 inches (15 cm) in diameter. The leaves are 10 feet (3 m) long with many narrow, dark green, limp, and pendent leaflets that grow from different angles along the rachis to give the leaf a plumose effect. The crown is hemispherical. The orange fruits are 2 inches (5 cm) long, and the endocarp and seed are triangular in cross-section.

The visual aspect of this jewel of a palm is that of a slender-stemmed queen palm (*Syagrus romanzoffiana*). It is one of the most elegant-looking species in the genus but is not hardy to cold, being adaptable only to zones 10b and 11, although marginal in 10a. It requires a rich soil and constant moisture.

Syagrus macrocarpa
sy-AG-rus · mak-ro-KAHR-pa
PLATE 893

Syagrus macrocarpa is an uncommon species endemic to southeastern Brazil, where it grows in sandy soils in the transitional zone to Atlantic coastal forest. The epithet is Greek for "large fruit," a reference to the egg-shaped fruits that are 3 inches (8 cm) long.

The solitary trunk grows to a maximum height of 25 feet (7.6 m) and a diameter of 8 inches (20 cm). It is smooth and tan or light gray with widely spaced but indistinct darker rings of leaf base scars. The leaf crown is full and round, partly because of the persistent and pendent, old dead leaves. The beautifully arching leaves are 6–7 feet (1.8–2.1 m) long on smooth petioles 2 feet (60 cm) long. The leaflets are 2 feet (60 cm) long, medium to deep green, and soft and pendent at their tips; they grow from the felt-covered rachis at different angles but curl downward to create a plumose leaf. The edible fruits turn orange when mature.

This palm is similar to the queen palm (*Syagrus romanzoffiana*) but much shorter and thus has a more intimate and stockier, if not more robust appeal; its trunk also appears more solid and straighter and its crown larger. It is beautiful but is undeservedly rare in cultivation. It needs the same cultural practices as the queen palm but is not as hardy to cold, being safe only in zones 10 and 11.

Syagrus oleracea
sy-AG-rus · o-le-RAY-see-a
PLATE 894

Syagrus oleracea is indigenous to a large area of interior sub-Amazonian eastern Brazil, southwestwards to eastern Paraguay, where it grows in moist (monsoonal) semideciduous forests. The epithet is a Latin word meaning "pertaining to a vegetable garden," an allusion to the edible fruits, seeds, and growing point.

This large solitary-trunked species grows to a height of 60 feet (18 m) with a diameter of only 1 foot (30 cm). The trunk is usually straight, columnar, and almost white in its oldest parts. The leaf crown is full and round, and the individual arching leaves are 8–10 feet (2.4–3 m) long. The deep green leaflets are 1 foot (30 cm) long and grow either at angles along the rachis to create a V-shaped leaf, or at several different angles to create a semiplumose leaf.

This palm does not tolerate cold or drought, and it needs a rich, well-drained soil. It is adaptable only to zones 10b and 11, although it is marginal in 10a. It is certainly among the more beautiful species in the genus, with its tall, slender, and elegantly picturesque form. It is widely planted for ornament in its native haunts, especially lining the streets of larger cities, a landscaping use at which it excels. It is remarkably wonderful as a canopy-scape.

Syagrus orinocensis
sy-AG-rus · or′-i-no-SEN-sis
PLATES 895–896

Syagrus orinocensis is indigenous to northern and western Venezuela, eastern Colombia, and northwestern Brazil, where it grows in lowland rain forest but also in mountainous rain forest to slightly more than 1000 feet (300 m). The epithet is Latin for "of Orinoco," the great river between Colombia and Venezuela. *Syagrus stenopetala*, the name given to clustering individuals, is currently regarded as a synonym, but it may be resurrected in the future, at which time *S. orinocensis* will apply only to the single-stemmed palms.

The solitary or clustered stems grow to 40 feet (12 m) but are usually shorter. They are 6 inches (10–15 cm) in diameter, light gray, and mostly smooth and indistinctly ringed in their younger parts. The leaf crown is rounded but not dense, and the leaves are 8–10 feet (2.4–3 m) long, beautifully arching, and not very plumose. The medium to dark green leaflets are at least 1 foot (30 cm) long and are flaccid but not pendulous, growing at different angles from the rachis but mainly upwards to give the leaf a slight V shape. The orange, ellipsoid fruits are 1½ inches (4 cm) long.

This palm does not tolerate frost and is adaptable only to zones 10b and 11. It grows in partial shade and survives drought conditions, but it appreciates full sun, constant moisture, and a humus-laden, free-draining soil, under which conditions it is faster growing and more robust, although it is never really fast. The *stenopetala*-type grows enthusiastically in the thin, alkaline soils of South Florida, whereas the *orinocensis*-type grows rather grudgingly in those conditions.

It is impossible to understand why this species is not more common in cultivation as the single-stemmed form is one of the loveliest in the genus. It could serve as a paradigm for the pinnate-leaved palms. Its canopy-scape is among the most pleasing in the family, and it is wonderful even when young as a close-up specimen. Planting this palm in groups of 3 or more individuals of varying heights creates one of the world's most visually pleasing landscaping phenomena.

Syagrus picrophylla
sy-AG-rus · pik-ro-FIL-la
PLATE 897

Syagrus picrophylla is endemic to eastern Brazil in the central Atlantic coastal region, where it grows in mountainous rain forest at an elevation of 1000 feet (300 m). The epithet is Greek for "bitter leaf," an allusion to the taste of the palm's growing point.

It is one of the smaller solitary-trunked species, growing only to a maximum overall height of 30 feet (9 m). The gray to tan trunk tapers from base to tip and is 6 inches (15 cm) in diameter at its middle point. The stems is mostly smooth and shows indistinct darker rings of leaf base scars. The leaves are 5–6 feet (1.5–1.8 m) long and are mostly ascending with little arching evident. The leaflets are 1 foot (30 cm) long and widely spaced along the rachis but grow in clusters at different angles from it. Most are stiff but not rigid, but in some individuals, they are as limp as those of the queen palm (*S. romanzoffiana*). The leaflets are medium to deep green and grow at an angle which gives the leaf a slight V shape. The greenish yellow fruits split open when ripe.

This species needs a good soil fortified with humus and regular moisture. It survives in shade but will grow more slowly there. It does not tolerate cold and succeeds only in zones 10b and 11. This attractive species has the look of a stiffer, dwarfed queen palm. It is rare in cultivation outside its habitat but would make a stunning courtyard or patio subject as well as a miniature canopy-scape.

Syagrus pseudococos
sy-AG-rus · soo-do-KO-kos
PLATE 898

Syagrus pseudococos is endemic to southeastern Brazil, where it grows in low mountainous rain forest and clearings therein. The epithet is Greek for "false *Cocos*," an obvious allusion to its affinities with the coconut. The trunks grow to 50 feet (15 m) high and are usually less than 1 foot (30 cm) in diameter, gray, and smooth, with beautifully distinct, widely spaced darker rings of leaf base scars and a heavy, bulging base. The leaves are 6–7 feet (15 m) long on heavy petioles 1–2 feet (30–60 cm) long; they are beautifully arching and bear deep emerald green, lanceolate, and lax leaflets 1 feet (30 cm) long that grow from the rachis in clusters and at slightly different but always upright angles to give the leaf a distinct V shape.

The species does not tolerate frost, needs full sun when past the seedling stage, and requires a deep, humus-laden soil as well as regular, adequate moisture. In such conditions it is fast growing after its trunk is formed. It is almost as beautiful as the coconut, and it is nearly unfathomable why this species is not more widely cultivated. It is a veritable symphony of color with the intense green of its leaves against the pale trunks and, when in flower, its large, deep yellow inflorescences and great, brown, woody bracts. Its fully rounded crown is nearly perfect as a canopy-scape, and it is among the most dazzling specimen palms.

Syagrus romanzoffiana
sy-AG-rus · ro'-man-zof-ee-AH-na
PLATE 899

Syagrus romanzoffiana is indigenous to southern and southeastern Brazil, eastern Paraguay, northeastern Argentina, and northern Uruguay, where it grows in monsoonal forests and swampy areas, always at low elevations. It is almost universally called queen palm. At one time, the species was scientifically known as *Cocos plumosa*, and this now out-of-date binomial has almost reached the status of a common name; the palm is still known as *Arecastrum romanzoffianum* to most growers and nursery personnel. The palm was named in 1822 for Count Nicholas Romanzoff, Chancellor of the Russian Empire and patron of scientific explorations. A few varieties of this common and variable species have been described, but none is now accepted.

The trunks grow to 50 feet (15 m) in habitat but to 75 feet (23 m) in cultivation under optimal conditions. The trunk diameter is even more variable in habitat, with individuals from the wetter, more tropical areas usually exhibiting the thinnest stems—8 inches (20 cm) in diameter—while those from the drier, less tropical areas manifest thicker trunks—18–20 inches (45–50 cm) across. At one time taxonomists considered these regional forms to be separate varieties. The widely spaced rings of leaf base scars are always evident in the younger parts of a stem but are not overly distinct. The leaves are slightly less variable, 7–15 feet (2.1–4.5 m) long, and plumose. The soft leaflets are 1–2 feet (30–60 cm) long, and anywhere from light or yellowish green to verdant, deep green, the variation in intensity of the hue often due to nutritional factors. There is a wonderful arch to each leaf near its apex, and the leaf crown is thus full and round. In general, old dead leaves do not last long on the tree. The inflorescences are usually long and pendent but may also be erect and short. Most are 5 feet (1.5 m) long but some may be significantly longer; they all consist of many pendent yellowish branches bearing unisexual blossoms. The ovoid 1-inch (2.5-cm) fruits are orange when mature.

This species does not tolerate drought or calcareous soil. Unfortunately, it is commonly planted in South Florida (where both conditions are common) and in such large numbers that the average person tends to think the species is normally chlorotic and stunted. This is regrettable as the healthy palm is among the most beautiful that can be grown in nontropical regions, and even calcareous soils can be amended to the palm's liking with an organic mulch. It needs full sun once past the seedling stage and is adaptable in hot regions like southern Arizona if provided with regular moisture. Under optimum conditions, the species is one of the fastest growing palms and in its midlife can form at least 3 feet (90 cm) of trunk per year.

The queen palm is relatively hardy to cold and is usually unscathed by temperatures above 25°F (−4°C). It is adaptable to zones 9b through 11 in areas subject to wet freezes in winter and to 9a through 11 in the drier climes. It is doubtless true that, if they can be found, individuals from the southerly habitats have more cold tolerance than those from tropical Brazil, which fact has led to several names in the nursery trade for differing forms with accompanying claims of unusual hardiness. The problem is that the names are simply created on the spot and probably do not correspond with any given provenance.

One of the latest named forms has a slight silvery sheen to the undersides of the leaflets, robust trunks, and much smaller seeds. Its provenance probably is southerly but it cannot be documented. It is, however, significantly hardier to cold and survived the devastating freeze in central Florida of 1989 (David Witt, pers. comm.). These palms are the progeny of individuals grown by Dent Smith (founder of the International Palm Society), who obtained them from the Santa Catarina state of Brazil, whose southern boundaries extend to latitude 28° south.

Another popular cultivar has been given the name **'Abreojos'** after the place of its discovery, Punta Abreojos, Baja California Sur, Mexico. It is described as more robust, with longer, more densely plumose leaves, darker leaflets, more robust inflorescence, and larger fruits and endocarps. It has been distributed in Southern California.

According to Noblick (1996) the species is known to hybridize in habitat with *S. coronata* and *S. oleracea*, and some nursery forms are likely to be hybrids; unfortunately the other known potential parents in habitat are all more tender to cold than pure *S. romanzoffiana*. One man-made hybrid with *S. schizophylla* shows no diminution in cold hardiness and has gained widespread popularity in Southern California. The hybrid is extraordinarily graceful. In cultivation, the species has hybridized with *Butia odorata* (which see) to form ×*Butiagrus nabonnandii*, one of most beautiful palm hybrids.

The queen palm is among the most beautiful, tropical-looking palms that can be grown outside of tropical areas. It looks enough like the royal palms (*Roystonea*) that it can be used to line streets and avenues. Its canopy-scape is thrilling, and a specimen group of individuals of varying heights is among the most picturesque landscaping tableau. The species is widely planted worldwide in tropical and subtropical regions; in fact, it is a weedy species that tends to naturalize. But, like the coconut, its ubiquity is sorely missed by those whose climate prevents growing it. The queen palm is widely grown in large atriums where its size and light requirements can be met. Otherwise, it is not a good candidate for indoors.

Syagrus ruschiana
sy-AG-rus · rus-kee-AH-na
PLATE 900

Syagrus ruschiana is endemic to central eastern Brazil in the states of Espírito Santo and Minas Gerais, where it grows in open areas in the low granite hills on steep slopes at 300–1200 feet (90–370 m). The epithet honors Augusto Ruschi (1915–1986), Brazilian naturalist and ecologist.

The trunks of this clustering species can grow to 25 feet (7.6 m) tall with a diameter of 2–3 inches (5–8 cm) and are covered by persistent dead leaf bases. The leaves are about 10 feet (3 m) long, at least 3 feet (90 cm) wide, and beautifully arching, with closely and regularly spaced, narrow, lax, dark green leaflets 2 feet (60 cm) long growing in a single plane from the rachis. The long, slender, yellow inflorescences produce almost-rounded, 1-inch (2.5-cm), bright orange fruits.

This extraordinarily beautiful species has leaves reminiscent of those of *Howea forsteriana* (kentia palm) or even *Cocos* (coconut). The clustering trunks add to the overall appeal, and yet the species is undeservedly rare in cultivation, possibly because of its isolated and small habitat, in which it nevertheless is abundant.

Syagrus sancona
sy-AG-rus · san-KO-na
PLATE 901

Syagrus sancona is indigenous to 2 disjunct areas of northern South America: the first area is the Andean foothills of western Venezuela, central and western Colombia, and western Ecuador; the second region includes central and eastern Peru, west central Brazil, and northwestern Bolivia. The palm is found in lowland rain forest and clearings as well as in higher, drier areas to 3800 feet (1160 m). The epithet is a Latinized form of an aboriginal name for the species in Colombia.

This solitary-trunked species is the tallest in the genus, old trunks attaining heights of 100 feet (30 m) and a maximum diameter of 1 foot (30 cm) except at their bulging bases. They are always straight as an arrow and would be columnar were they of greater thickness. They are light gray to light tan and bear widely spaced but indistinctly darker rings of leaf base scars. The leaf crown is fully rounded and dense and, in old individuals, seems small. The leaves are 12 feet (3.6 m) long and strongly arching, almost recurved back to the trunks in some individuals. They are deep green with closely spaced, wide leaflets to 3 feet (90 cm) long. The leaflets are limp with usually pendent apices, and they grow from different angles off the rachis to give the leaf a plumose effect. The yellow, pendent inflorescences produce egg-shaped, yellow to orange fruits that are 1 inch (2.5 cm) long.

This beauty is not hardy to cold and is adaptable only to zones 10 and 11. It requires constant moisture, especially in dry climates, to look good and for fastest growth but is not fussy about soil as long as it is fast draining and not overly acidic or overly alkaline. It needs sun from youth to old age and, under optimum conditions, grows fast when past the seedling stage and up to near maturity.

The appearance of this palm is often compared to that of *S. romanzoffiana* (queen palm). While this may be true, it is hard to imagine anyone mistaking the 2 palms if they have both formed trunks. It is unsurpassed as a canopy-scape and is almost overwhelmingly beautiful as a specimen planting of 3 or more individuals of varying heights. It is not for small spaces. Only in large atriums or conservatories could this one be successful.

Syagrus schizophylla
sy-AG-rus · sky-zo-FIL-la
PLATE 902

Syagrus schizophylla is endemic to the northeastern Brazilian coastal forest known as the *restinga*. This is a vegetation type with low, shrubby, evergreen, woody, small trees and shrubs that grow on sandy soil—mostly prehistoric sand dune— near the coast as well as inland on the low hills. The epithet is Greek for "divided leaf." The only common name in English-speaking countries seems to be the arikury palm, derived from *Arikuryroba*, the genus into which the species was formerly placed, which itself was derived from a Brazilian common name.

The solitary trunk grows to 15 feet (4.5 m) but is usually 10 feet (3 m) high. It is slender but looks thicker because of the persistent leaf bases and fibers; even with these additions the stems are never more than 6 inches (15 cm) in diameter. The trunks are distinctive because these closely set, gray to tan, narrow leaf bases give them the look of fine wickerwork. The species is one of the few in the genus having spines, and these are limited to the long petioles and leaf bases. The leaves are 6 feet (1.8 m) long on petioles 2 feet (60 cm) long, spreading and slightly arching. Living leaves are never held below the horizontal. The leaflets are stiff but not rigid and in older leaves are pendent. They are linear-lanceolate, regularly spaced, dull to olive or even deep green, and usually grow from the rachis in a single flat plane but may also grow at a slight angle to give a V-shaped leaf. The inflorescences bear beautiful, egg-shaped, orange fruits that are 1 inch (2.5-cm) long. The palm's only human use seems to be its edible and sweet-tasting fruits. Unless one is ravenous for the taste, they are better left on the little trees because of their beauty.

The species is slightly susceptible to lethal yellowing disease. The palm is intolerant of frost and drought: its habitat is reputedly wetter than the Amazon Basin to the northwest. It is not fussy about soil as long as it drains well, and it luxuriates in full sun as well as partial shade, although it is slower growing, darker green, and usually has longer petioles in shadier sites; it is not fast growing in any situation.

This architectural-looking species looks its loveliest in up-close and intimate sites; it is the perfect patio palm. Its beauty is ruined if planted as a single specimen surrounded by space unless it is incorporated into lower vegetation above which its beautiful silhouette may be enjoyed. It does look much better as a specimen if planted in groups of 3 or more individuals of varying heights, but its small stature seems to mitigate against this situation also. It does well indoors for a while if given lots of light and good air circulation.

Syagrus smithii
sy-AG-rus · SMITH-ee-eye

Syagrus smithii is a rare and endangered species indigenous to southeastern Colombia, northeastern Peru, and western Brazil, where it grows in rain forest to 2000 feet (610 m). The epithet honors Earl E. Smith, a 20th-century American botanist.

This solitary-trunked species grows to a maximum overall

height of 50 feet (15 m) but is usually 30 feet (9 m), gray or tan, indistinctly ringed, and 2–4 inches (5–10 cm) in diameter. The leaf crown is full and rounded but not dense. The leaves are 8–10 feet (2.4–3 m) long with dark green, narrow, long-tipped leaflets 2 feet (60 cm) long and borne in clusters of 2–4 inserted at various angles. Amazingly, the leaves, which are 8 feet (2.4 m) long, can reportedly also be undivided and entire, even in adult trees, in which case they are an elongated wedge shape with deeply bifid apices. The yellow, egg-shaped fruits are 3 inches (8 cm) long and unusually large for the genus. The seed is nearly triangular in cross-section, and the endosperm is ruminate.

This elegant species is almost as rare in cultivation as it is in habitat. It is doubtless frost intolerant but would probably luxuriate in full sun or partial shade and would need regular, abundant moisture.

Syagrus vagans
sy-AG-rus · VAY-ganz
PLATE 903

Syagrus vagans is endemic to north central Brazil in the states of Bahia and Minas Gerais, where it grows in open, dry *caatinga* at 800–2900 feet (240–880 m). The epithet is Latin for "wandering," an allusion to the stem. Because the stem of this solitary species is underground, many plants growing together give the appearance of a clustering palm. The leaves grow directly from the rocky soil surface. The base of the petiole has a few teeth along its margin. They are 6 feet (1.8 m) long, stiffly erect, with narrow and rigid, dusky to deep green leaflets. The leaflets grow from the rachis at an angle, making the leaf V shaped. The fruits are green at maturity. This species looks like a mass of stiff leaflets; it is, nevertheless, sought after for cultivation. It would not be out of place in a xeric garden of succulents and cacti.

Syagrus vermicularis
sy-AG-rus · ver-mik-yoo-LAH-ris
PLATE 904

Syagrus vermicularis is a palm from central Brazil, where it occurs in semideciduous forests and in the transition zone to ever-wet forests, as well as persisting in secondary vegetation. Described in 2004, the species is easily recognized by its folded and twisted, slender rachillae. The epithet is Latin for "worm shaped," in reference to the rachillae.

The solitary trunk has a white, waxy coating when young, but the wax disappears with age. The fully rounded crown of leaves sits atop a stem that grows to 50 feet (15 m) tall and is 5–8 inches (13–20 cm) in diameter. The leaves are about 8 feet (2.4 m) long and bear stiff leaflets 2 feet (60 cm) long that are distributed in small groups and at different angles. The leaf has a slightly plumose shape but an overall V shape as well. The highly branched inflorescences seem to tumble out of the broad, woody bract. The rachillae, looking like masses of ramen noodles, are yellow at first, turning green as the fruits mature. The ovoid greenish yellow fruits are nearly 2 inches (5 cm) long.

This striking palm is still new to horticulture, but garden-worthy in many respects. The pale, slender trunk surmounted by a dark green crown of leaves makes for an iconic tableaux.

Synechanthus
sin-e-KAN-thus

Synechanthus includes 2 small, pinnate-leaved, monoecious palms in tropical America. One is solitary trunked, the other clustering. The species are related to and resemble those of *Chamaedorea*, the differences being in details of the flowers. Neither *Synechanthus* species is hardy to cold. Both need partial shade at all stages of growth, a friable, humus-laden soil that is constantly moist but also fast draining, and high humidity. The genus name is derived from Greek words meaning "joined flower" and refers to the arrangement of the flowers in rows on the branchlets. Seeds generally germinate within 90 days and do so in a fairly short time. For further details of seed germination, see *Hyophorbe*.

Synechanthus fibrosus
sin-e-KAN-thus · fy-BRO-sus
PLATE 905

Synechanthus fibrosus occurs naturally from the Mexican state of Veracruz, southwestwards through central Guatemala, northern Honduras, and thence southwards through eastern Nicaragua and into northeastern Costa Rica, where it occurs in the undergrowth of rain forest from sea level to 4000 feet (1200 m) in elevation. The epithet is Latin for "fibrous" and refers to the leaf sheaths.

Mature stems of this solitary-trunked species are 1 inch (2.5 cm) thick and can reach 15 feet (4.5 m) tall; they are dark green with beautiful white rings of leaf base scars. The leaves are ascending and seldom lie beneath the horizontal; they are 4 feet (1.2 m) long on petioles 1 foot (30 cm) long. The glossy medium green leaflets are arranged into groups along each side of the rachis and are offset to give the leaf a slightly plumose effect; they are slightly S shaped and the 2 apical leaflets are wider than the rest. The long inflorescences grow from among the leaves, are branched only apically, and are usually erect until fruit formation. They bear tiny yellow flowers that produce round ½-inch (13-mm) fleshy fruits that change from green to yellow to orange and finally red when mature. The endosperm is either homogeneous or minutely ruminate.

This delicate, ethereal-looking little palm needs to be sited where it can be appreciated up close. It's not always easy to accomplish, but a miniature landscape with this palm as a canopy-scape would be refreshing as an intimate site among larger areas of a garden or as a patio or courtyard feature. The species has for over a century been grown indoors in bright light or in greenhouses.

Synechanthus warscewiczianus
sin-e-KAN-thus · wahr'-say-wik-see-AH-nus
PLATE 906

Synechanthus warscewiczianus is a clustering species indigenous to southeastern Nicaragua, the east coast of Costa Rica, Panama, the west coast of Colombia, and northwestern Ecuador, where it grows in rain forest from sea level to 4000 feet (1200 m) in elevation. The epithet honors Józef Warszewicz (1812–1866), a Polish plant collector in Central and South America.

The trunks can grow to 15 feet (4.5 m) tall in sparse clumps that usually have one dominant stem and a few smaller, shorter trunks. The deep green, ascending leaves are 6 feet (1.8 m) long and are sometimes entire and apically bifid or, more often, irregularly divided, the segments of varying sizes and placement with the apical pair always much wider. The flowers and fruits are similar to those of *S. fibrosus*. The endosperm is deeply ruminate.

The clumping habit allows this palm to be used as a small specimen planting or as a component of a mixed shrub border.

Tahina
ta-HEE-nah

Tahina is a monotypic genus of very large solitary-trunked, palmate-leaved, hermaphroditic palm. Like *Corypha*, the plants die after producing enormous terminal inflorescences. The genus name is Malagasy for "blessed" or "protected" and is also the first name of the daughter of the palm's discoverer. At present (and for many years to come) all seeds come from the wild population. Because the population is small and the palms are hapaxanthic, seeds are available only sporadically, when an adult palm produces fruits before dying. Consequently, germination information is still sketchy. Fresh seeds appear to germinate quickly, within 30 days.

Tahina spectabilis
ta-HEE-nah · spek-TA-bi-lis
PLATES 907–908

Tahina spectabilis is a rare and massive palm endemic to a small, remote area of western Madagascar on seasonally flooded, deep soils often near limestone outcrops. Fewer than 100 juveniles and adults are known. The species epithet is Latin for "outstanding."

The solitary trunk grows to about 60 feet (18 m) tall and 20 inches (50 cm) in diameter. It is clean, gray and marked by conspicuous leaf scars 3–4 inches (8–10 cm) apart. The crown of enormous leaves is open and hemispherical, with some dead leaves retained at the bottom. The petiole, up to 15 feet (4.5 m) long, is unarmed, covered in white wax, and split at the base as in *Sabal*. The leaf is costapalmate, with a conspicuous hastula, and up to 15 feet (4.5 m) across. The individual segments are free for about half their length and their tips are stiff. The plications of the leaf are unusual in that there are major and minor folds, which may have something to do with packing such an enormous leaf into a spear-shaped bud. The inflorescence is terminal and covered with waxy white tubular bracts. The showy, greenish yellow flowers are followed by green ellipsoid fruits, about 1 inch (2.5 cm) long and just under 1 inch (2.5 cm) wide. The endosperm is strongly ruminate.

At the time of this palm's discovery, in 2006, fruits were ripening. By the following year, an arrangement was made so that the local villagers would protect the palms, harvest the seeds, and profit from their overseas sales. This resulted in seeds being sent to botanical gardens, palm collectors, and nurserymen all over the world. Profits from the seed sales allowed villagers to buy farming equipment and establish a conservation area around the remaining *Tahina* population. Flowering and fruiting occur sporadically in the population, and the next fruiting event cannot be predicted.

Tahina spectabilis is one of the most exciting palm discoveries in recent decades, although, it is still too new to judge what its contribution to horticulture might be. It is such a massive palm that only large gardens or parks are suitable sites for this species. From what we know of its natural habitat, we can surmise that it is a water-loving palm and that it is tolerant of alkaline soils. Its cold hardiness is untested. Juveniles from the 2007 seed distribution are handsome palms, with highly divided costapalmate leaves. They tolerate full sun at an early age if given ample water.

Tectiphiala
tek-ti-fee-AH-la

Tectiphiala is a rare and endangered monotypic genus of solitary-trunked or clustering, spiny, pinnate-leaved, monoecious palm endemic to central Mauritius. The genus name is derived from Greek words meaning "covering" and a type of vessel, and refers to the small bracts that tend to obscure the buds of the male flowers. Seeds are very difficult to germinate. The best results have been obtained by surface-disinfecting fresh seeds with dilute bleach, then sowing the seeds in sterilized medium, in a sealed plastic container, with bottom heat. Although germination is sporadic, with this technique, seeds will begin germinating within 30 days.

Tectiphiala ferox
tek-ti-fee-AH-la · FER-ahx

Tectiphiala ferox grows in moist, marshy scrub vegetation in heavy, acidic soil at an elevation of 2000 feet (610 m). The species was described as sparsely clustering, but no such individuals now exist, possibly because of the slow growth habit. The epithet is Latin for "fierce," an allusion to the spines.

The trunk grows to 20 feet (6 m) high with a diameter of 6–7 inches (15–18 cm). It is dark gray and exhibits closely set rings of leaf base scars. The rings in the youngest parts of the stems bear black spines 6 inches (15 cm) long. The crownshaft is 2 feet (60 cm) tall, greatly swollen at its base and tapering to its summit, and is a deep rust or cinnamon in color because of the dense, velvety covering of tomentum. It is also covered in soft but long, black, flexible, hairlike spines. The leaf crown is sparse and spherical because of the recurving leaves, each 6–8 feet (1.8–2.4 m) long on a short, stout, and furry petiole. The linear-lanceolate, long-tapering leaflets are 18 inches (45 cm) long and grow in closely set groups along the rachis. In each group, the leaflets grow at different angles to create a slightly plumose leaf. They are heavy and leathery but not stiff and are a medium to dark or even bluish green above but an astonishingly attractive silvery hue beneath. The inflorescences are initially covered in a spiny and hairy, elongated bract. They are branched once off a short, spiny peduncle, the flowering branches 18 inches (45 cm) long, few, and light yellow, bearing both male and female blossoms. The tiny ovoid fruits are blackish blue when ripe.

This unusually beautiful species is rare in habitat and in cultivation. It seems to need an acidic soil, constant moisture, and a tropical climate but is adaptable to full sun at an early age.

Thrinax
THRY-nax

Thrinax consists of 3 palmate-leaved, hermaphroditic palms in the Caribbean Basin. A few grow to 50 feet (15 m) tall but all have a delicate, almost ethereal quality with their very slender, solitary trunks and nearly circular leaves. The leaf crowns are relatively small, especially in older individuals, and are spherical. The trunks never have a shag of old dead leaves and are indistinctly ringed in their younger parts. The genus is closely related to *Coccothrinax*, *Hemithrinax*, and *Leucothrinax* but is distinguished by the unique combination of split leaf bases, stalked flowers, erect anthers in bud, and white fruits. The inflorescences grow from the leaf crown and bear white, bisexual flowers. The tiny round fruits are white when mature.

Most species are slow to exceedingly slow growing, but all are adaptable to various soils as long as they are freely draining, even the calcareous and salty types. These palms need a site in full sun when past the seedling stage. None are hardy to cold; most are adaptable in zones 10 and 11 and are marginal in 9b.

The palms are called thatch palms because their leaves are widely used in house construction, as the common name indicates. The genus name is Greek and translates as "trident," an allusion to the leaf segments.

Fresh seeds of *Thrinax* germinate within 90 days. Seed that is allowed to dry out completely and is then rehydrated may take 9 months or longer to germinate. For further details of seed germination, see *Corypha*.

Thrinax excelsa
THRY-nax · ek-SEL-sa
PLATE 909

Thrinax excelsa is endemic to Jamaica, where it grows in low mountains of the island's extreme eastern tip. It is called broad thatch palm in habitat and Jamaican thatch palm elsewhere. The epithet is Latin for "lofty, exalted, noble," allusions to both the palm's stature and its beauty.

Individual trunks grow to 35 feet (11 m) tall in habitat with a diameter of 8 inches (20 cm). The leaf crown is large and beautifully rounded. The large leaves are almost circular and 5–6 feet (1.5–1.8 m) in diameter with many linear-lanceolate segments that are free for half their length. They are a glossy medium to deep green above but a grayish and glaucous green beneath and are limp, with their apices usually pendent in the older leaves. The long inflorescences are difficult to see among the leaves, but the infructescences are pendent with round ½-inch (13-mm) fruits.

This palm is adaptable to most soils except for the quite acidic and waterlogged. It is not drought tolerant. It is tender to cold and adaptable in zones 10b and 11, although some nice individuals are found in warm microclimates of 10a. This species is faster growing than any other and makes a relatively quick and good show of tropical beauty. It is the most lush and most tropical-looking species in the genus because of its large leaves. As a younger plant it is beautiful and choice, and is captivating up close and in an intimate site. When older it is beautiful as a specimen, especially in groups of 3 or more individuals of varying heights. In truth, it is hard to misplace; it looks good in any site. None of the thatch palms are suited to growing indoors, but this one probably has the most potential if given room and bright light.

Thrinax parviflora
THRY-nax · pahr-vi-FLOR-a
PLATE 910

Thrinax parviflora is endemic to Jamaica, where it grows in open and rocky deciduous woods to 2800 feet (850 m). The epithet is Latin for "small flower." The palm attains a maximum overall height of 50 feet (15 m) in habitat, with a trunk diameter of no more than 6 inches (15 cm). The leaf crown is globular and the circular leaves are 3–4 feet (90–120 cm) in diameter with heavily veined, linear-lanceolate, often twisted and limp segments divided two-thirds of the way into the blade. The leaf apex is pendent. Leaf color is medium to deep green on both surfaces. The palm is extremely graceful, especially when older, because of its diaphanously slender trunk and rounded but sparse leaf crown. It is uncommon in cultivation and deserves much wider acceptance. A few taxonomists recognize 2 subspecies based on details of the leaf segment tips and the length of the inflorescences. Although it is sometimes called mountain thatch palm, it is not hardy to cold.

Thrinax radiata
THRY-nax · ray-dee-AH-ta
PLATE 911

Thrinax radiata is indigenous to the Florida Keys, Bahamas, western Cuba, Puerto Rico, Hispaniola, the northern and eastern coasts of the Yucatán Peninsula, and Belize, where it grows along the coasts. It is usually called Florida thatch palm in Florida. The epithet is Latin for "radiating" and alludes to the arrangement of the segments in the leaf.

The slender stem attains a maximum height of 40 feet (12 m) and a diameter of only 5 inches (13 cm). While the leaf petioles are 3 feet (90 cm) long, the leaf crown is rounded and dense in full sun, with circular 4-foot (1.2-m) leaves, which are medium green above and a lighter green below. The leaf segments are free two-thirds of their length and are pendent at their apices.

This relatively cold hardy species is adaptable to zones 10 and 11, with occasional handsome specimens seen in warm microclimates of 9b. It is slow growing in full sun and very slow growing in partial shade. When transplanted, it is slow to anchor itself in the ground and is vulnerable to wind damage. The palm demands a fast-draining soil. It is tolerant of saline soil and air, and its limp leaves are marred only by hurricane winds. It is a perfect candidate for the beach as well as sunny inland areas. Its crown is open and airy when young or when planted in partial shade but is tightly globular and dense when older and in full sun—so much so that the petioles are usually obscured, making it a distinctive and picturesque canopy-scape. It is a poor candidate for growing indoors past the juvenile stage.

Trachycarpus
tray-kee-KAHR-pus

Trachycarpus comprises 9 solitary-trunked, palmate-leaved, mostly dioecious palms in northern India, Nepal, Myanmar, Vietnam, southern China, and northern Thailand. The species hail from mountainous forests of considerable elevations and include some of the world's most cold hardy palms.

One species lacks an aboveground trunk; the rest form slender aerial stems that are covered in their younger parts with petiole and leaf base remains and the fibers associated with them. The leaf sheath and fibers provide important characters for species identification. The leaf crowns and leaves are relatively small and compact, and the slender petioles are mostly unarmed; a few species have petioles with only a few, fine teeth along their margins. The inflorescences grow from within the crown and are congested but much branched, the flowering branches mostly yellow, bearing either male or female blossoms on separate trees, occasionally with bisexual flowers. The fruits are small and generally black when mature. The shape of the fruits and seeds, whether round or reniform (kidney shaped), is useful in identifying species.

The taxonomy of the species awaits further detailed study. Stevens (2010) provided a useful overview of the species in cultivation. The genus name is derived from Greek words meaning "rough" and "fruit" and highlights the somewhat irregular shape of the fruits of some species. Fresh seeds germinate within 90 days. Seeds that are allowed to dry out and are then rehydrated may take 9 months or longer to germinate. For further details of seed germination, see *Corypha*.

Trachycarpus fortunei
tray-kee-KAHR-pus · for-TOON-ee-eye
PLATES 912–913

Trachycarpus fortunei is thought to be indigenous to central and eastern China, but its exact origin is unknown, so widely and for so long has it been in cultivation in that country. Common names are windmill palm, Chusan palm, and Chinese windmill palm. The epithet honors Robert Fortune (1812–1880), a Scottish horticulturist, explorer, and collector, who introduced hundreds of Chinese ornamental plants to Europe and established the tea industries in India and Sri Lanka. He introduced the Chusan palm to England in 1843.

The trunk grows to 45 feet (14 m) with time. The woody part is relatively slender, a maximum diameter of 8–10 inches (20–25 cm), but it is usually covered in its younger parts with old, woody leaf bases, petiole bases, and the dark brown fibers associated with them so that the diameter of these parts is often twice that of the lower portions. These leaf bases and fibers can be removed mechanically to reveal an elegantly slender, distinctly and closely ringed stem. The leaf crown is dense and spherical or often obovoid if the shag of dead leaves near its base is not removed. The leaves are deep green on both surfaces, 3 feet (90 cm) in diameter, on petioles 18–24 inches (45–60 cm) long. The depth of the divisions between the leaf segments is variable. The seeds are reniform.

This species is among the hardiest tall palms. It is adaptable to zones 7b through 11 and is even found in warm microclimates of 7a. It is extremely rare or absent from tropical or nearly tropical areas but is at its best in cool Mediterranean climates such as are found in the Pacific Northwest of the United States, the southern and low elevations of Switzerland, mild portions of the British Isles, and most of the Riviera region of Europe as well as southeastern Australia. In the right climate and with regular, adequate moisture and a decent soil, this species grows fast once past the juvenile stage and before it is mature. It is adapted to a range of soils, including slightly calcareous ones, and it flourishes in full sun except in hot climates. It should be protected from high winds, which spoil its elegance by tattering the leaves. The species is slightly susceptible to lethal yellowing disease.

The palm has a beautiful silhouette and makes a wonderful canopy-scape; it should not, however, be planted where its crown is regularly subjected to high winds. It looks nice when planted alone and surrounded by space but is even more satisfying in groups of 3 or more individuals of varying heights. While not as grand and majestic as the tropical royals (*Roystonea*), *Attalea*, *Jubaea*, *Livistona*, Canary Island date (*Phoenix canariensis*), or queen palms (*Syagrus romanzoffiana*), it is beautiful along streets or avenues. The species can be grown indoors if given lots of light and moisture.

Trachycarpus fortunei is the most commonly grown species of *Trachycarpus*. It is arguably the most beautiful and adaptable species in the genus but, alas, is ill adapted to hot climates in which the nighttime temperatures are consistently high.

Numerous cultivars have been selected, many informally named with their place of origin (*T. fortunei* 'Bulgaria', for example). Some are selected for their enhanced cold hardiness, while others are prized for variations in their crown shape, leaf texture, and the depth of divisions between the leaf segments. The most well-known cultivar is *T. fortunei* 'Wagnerianus' (formerly recognized as *T. wagnerianus*), which may have originated in cultivation in Japan. It is an easily identified cultivar with small, stiff leaves. The leaf crown is hemispherical and often elongated vertically, especially in younger individuals. The leaves of younger individuals are nearly circular, but those of older plants tend to be hemispherical; at all ages, they are only 18–26 inches (45–66 cm) wide on petioles 2–3 feet (60–90 cm) long. The approximately 40 segments are thick, leathery, stiff, and rigid, making the blade often "cupped."

Trachycarpus gemisectus
tray-kee-KAHR-pus · je-mi-SEK-tus

Trachycarpus gemisectus is from northern Vietnam (and perhaps adjacent China) and was described by Tobias Spanner, Martin Gibbons, and colleagues in 2003. It grows on steep slopes of limestone ridges in wet forest at 3600–5200 feet (1100–1580 m). The epithet is Latin for "twinned segments."

This species has a short, stout trunk about 6 feet (1.8 m) tall and 10 inches (25 cm) in diameter. The crown is sparse and open. The persistent leaf base fibers are coarse and wiry. The leaves, about 4 feet (1.2 m) wide, are green on the upper surface and waxy white on the underside. The approximately 40 leathery segments are joined in pairs, so that the leaves appear to have only 20 segments. The splits between pairs of segments go four-fifths of the way to the petiole. The fruits are reniform.

This species is still very new and rare in cultivation, in part because the remote, difficult terrain discourages seed collectors. Its cold hardiness remains untested, but it is likely to thrive in cool, frost-free, moist conditions.

Trachycarpus latisectus
tray-kee-KAHR-pus · la-ti-SEK-tus
PLATE 914

Trachycarpus latisectus is a rare and endangered species endemic to India in West Bengal and Sikkim, where it grows in mountainous forest at 4000–8000 feet (1200–2440 m). The epithet is derived from Latin words meaning "wide" and "segment," an allusion to the width of the leaf segments. The species was discovered and described by Tobias Spanner and Martin Gibbons, who proposed the vernacular Windemere palm because this species is cultivated on the grounds of a hotel of that name in Darjeeling, India. This species has been treated as a subspecies of *T. martianus*, but most botanists recognize it as a distinct species.

The trunk attains a height of 40 feet (12 m) and a diameter of 6 inches (15 cm). It is light gray and faintly ringed in its younger parts. The crown of living leaves is open and hemispherical or slightly more so without the small skirt of dead leaves but is spherical or obovoid with the skirt. The leaves are glossy light to deep olive green above and a lighter, glaucous hue beneath. They are circular or nearly so and 4½ feet (1.4 m) wide, with more than 65 segments on petioles 3–4 feet (90–120 cm) long and more than 1 inch (2.5 cm) wide. The thick, leathery segments are united for more than half their length, and some segments are united into groups of 2–4. Individual segments are 1½–2 inches (4–5 cm) wide and shortly bifid at their apices. The fruits and seeds are oblong.

Gibbons (1993) wrote that this species "is probably the only species in the genus which, owing to its wide elevation range from 3950 to 7900 feet (1200–2410 m), will adapt well to hotter regions." This palm is so new to horticulture that Gibbon's prediction has not yet been tested. The palms in cultivation in such areas are still juveniles. The species is not as cold hardy as *T. fortunei* but survives brief dips below freezing. In the wet tropical regions, such as Hawaii, it can take full sun, but it needs some shade in drier regions, such as Southern California. The leaf shape is the most beautiful in the genus.

Trachycarpus martianus
tray-kee-KAHR-pus · mahrt-ee-AH-nus
PLATE 915

Trachycarpus martianus is indigenous to northeastern India, Nepal, and northern Myanmar, where it grows in mountainous rain forest from 3000 to 8000 feet (900–2440 m). The epithet honors German botanist Carl F. P. von Martius (1794–1868). *Trachycarpus khasyanus* (also known as *T. martianus* subsp. *khasyanus*) is a synonym of this species.

The trunk grows to 45 feet (14 m) tall and is 7 inches (18 cm) in diameter. It is light to dark gray and, in its younger parts, is distinctly ringed with closely set ridged and darker rings of leaf base scars. The leaf crown is open and hemispherical or slightly more so. The leaves of older trees are semicircular, but those of younger ones generally are circular, with 65–75 segments. The leaves are 3–4 feet (90–120 cm) wide on slender petioles that are 4 feet (1.2 m) long, less than 1 inch (2.5 cm) thick, and margined with a thick white tomentum when new, as are the young leaves themselves. The rigid segments are joined for about half their length and are medium to dark green above and bluish gray-green beneath. The fruits and seeds are oblong.

Individuals from the northerly parts of the palm's range (Nepal) have slightly smaller leaves with fewer segments and are hardier to cold and more drought tolerant than are individuals from the southerly parts (Gibbons 1993). Because of its open crown and stiff leaf segments, the palm is beautiful, especially in silhouette. It is reportedly more tender to cold than *T. fortunei* and is probably adaptable only to zones 8 through 11. It seems to be more tolerant of warm to tropical climes and to need much more water than *T. fortunei*. The species prefers somewhat acidic soil.

Trachycarpus nanus
tray-kee-KAHR-pus · NA-nus
PLATE 916

Trachycarpus nanus is an endangered species endemic to southwestern China in Yunnan province, where it grows on dry, exposed, steep hills, at 6000–7500 feet (1800–2290 m). The epithet is Latin for "dwarf."

The short, stout trunk remains subterranean or, when aerial, only grows 1–2 feet (30–60 cm) high and is heavily covered in old leaf base fibers. The small, compact leaf crown consists of semicircular leaves 2 feet (60 cm) wide held on petioles 6–12 inches (15–30 cm) long and with tiny teeth along their margins. The 20–30 leaf segments are stiff and narrow and are united for one-fourth (or less) of their length. They are deep green to silvery bluish green, the latter hue being revealed only in full sun. Unlike all other species of *Trachycarpus*, this one has short, erect inflorescences and erect infructescences. The seeds are reniform.

This species is rare in cultivation. Seeds became available only after Martin Gibbons and Tobias Spanner rediscovered the species in the wild in 1993. It is reported to be at least as hardy to cold as *T. fortunei* and is probably more drought tolerant. Unlike *T. fortunei*, it prefers neutral to slightly acidic soil and requires excellent drainage.

Trachycarpus oreophilus
tray-kee-KAHR-pus · or-ee-o-FI-lus
PLATE 917

Trachycarpus oreophilus is endemic to the mountains of northwestern Thailand, growing in wet and cool monsoonal forests at 5000–7000 feet (1520–2100 m). The epithet is Latin for "mountain-loving." The species was described by Martin Gibbons and Tobias Spanner in 1997.

The trunk grows to 40 feet (12 m) high with a diameter of 6–8 inches (15–20 cm). The leaf crown is dense and hemispherical and contains 20 leaves, each 3 feet (90 cm) long on petioles 20 inches (50 cm) long, with tiny teeth along their margins. The leaves are semicircular to nearly circular, and the approximately 60 stiff, rigid segments are free for half their length. Leaf color is medium to deep green above and lighter grayish green beneath. The seeds are reniform.

This palm is rare in cultivation but is reportedly slow growing and needs a humus-laden, moist but fast-draining soil. It probably tolerates slightly calcareous soil conditions as it grows on limestone cliffs and slopes in habitat. It seems to be adaptable to zones 8 through 11, although it is still so new to cultivation that its adaptability has not been full assessed.

Trachycarpus princeps
tray-kee-KAHR-pus · PRIN-seps
PLATES 918–919

Trachycarpus princeps is endemic to southern central China, where it grows on limestone cliffs and ridge tops in monsoonal rain forest at 5000–6100 feet (1520–1860 m). The epithet is Latin for "prince" and alludes to "the stately bearing of this palm and the majestic way it looks down from its lofty position on the sheer cliff faces" (Gibbons 1993). The species was described in 1995 by Martin Gibbons, Tobias Spanner, and team.

The trunk grows to 30 feet (9 m) high with a diameter of 8 inches (20 cm). It is covered in all but its oldest parts by a thin mass of tight, dark brown fibers. The leaf crown is spherical or nearly so and contains 24 leaves 4 feet (1.2 m) wide on petioles 2–3 feet (60–90 cm) long and covered in a bluish white waxy substance when new and armed with tiny teeth along their margins. The leaves are semicircular with linear-lanceolate segments that are free for half their length and are bright medium green above and a beautiful glaucous, bluish white beneath. The seeds are reniform.

This species is still new to cultivation, but preliminary reports suggest that it will thrive in slightly alkaline to neutral soil, with moderate moisture and humidity, in sun or partial shade. It seems to be about as cold hardy as *T. fortunei*. It is slow-growing, but the adult leaf color is evident on very young plants, making them attractive garden features even at an early age.

Trachycarpus takil
tray-kee-KAHR-pus · TAH-keel

Trachycarpus takil is endemic to India in Uttar Pradesh province, where it grows in mountainous forest to 8000 feet (2440 m). It is nearly extinct. The epithet is a transliteration of the name of a mountain in the palm's habitat.

There is much confusion regarding this species, and many plants growing under this name in gardens are *T. fortunei*, which has been cultivated near the original locality of *T. takil* for over 100 years. Gibbons and Spanner (2009) provided a definitive description and illustrations of the true species.

This palm resembles *T. fortunei* in most respects, but has fewer leaves in the crown. *Trachycarpus takil* has shorter leaf sheath fibers, and the adult leaves are regularly divided into 45–62 segments. The very short leaf sheath that surrounds the trunk above the point of attachment of the petiole (the *ocrea*) is diagnostic for *T. takil*. The seeds are reniform. Because of the elevation of its habitat, this species is probably the most cold hardy in the genus; it is reportedly adaptable to zones 6b through 11 and probably marginal in 6a, although it is no good in hot climates. It is fast growing once past the seedling stage (Gibbons 1993).

Trithrinax
try-THRY-nax

Trithrinax is composed of 3 clustering, palmate-leaved, hermaphroditic palms in drier parts of southern South America. All 3 have persistent leaf sheaths that form downward-pointing spiny projections lasting sometimes for many years on the trunks. The white or cream inflorescences emerge from within the leaf crown, seldom exceed its length, and are much branched. The flowering branches are relatively short and are initially covered in several bracts. They bear bisexual flowers, which produce round white fruits that mature to yellow then to black or dark brown. All the species are drought tolerant, slow growing, and cold resistant. The genus name translates as "three *Thrinax*" and was not explained when it was coined. Fresh seeds germinate within 90 days. Seeds that are allowed to dry out and are then rehydrated may take 9 months or longer to germinate. For further details of seed germination, see *Corypha*.

Trithrinax brasiliensis
try-THRY-nax · bra-zil'-ee-EN-sis
PLATE 920

Trithrinax brasiliensis is endemic to far southeastern Brazil, where it grows in inland, open, dry savannas. The only common name seems to be spiny fiber palm. The specific epithet is Latin for "of Brazil."

Most individuals are solitary stemmed but occasionally a clustering specimen is seen with 2 or 3 trunks. Old palms form trunks to 50 feet (15 m) high but the norm, especially in cultivation, is half that height. They are covered in all but the oldest parts with a mat of dark gray fibers and with rows of blackish spines 2–4 inches (5–10 cm) long corresponding with the tops of the leaf bases. The leaves are 3–4 feet (90–120 cm) wide on petioles 2 feet

(60 cm) long and are slightly more than a half circle of deeply cut segments, each one being shallowly to deeply bifid apically. Leaf color is a deep green, light green, silver green, or even bluish green, and the segments are stiff but not overly rigid, especially in younger leaves; in older leaves the ends of the segments are often pendent.

This palm is cold hardy in zones 9 through 11 where wet freezes occur in winter but in zones 8 through 11 in drier climes; prolonged cold and precipitation are often lethal because the growing point becomes infected by fungus. The species is drought tolerant once established but grows faster and looks better with regular moisture, especially in drought conditions and periods of intense heat. It should not be planted in the shade where it grows extremely slowly and its growing point can become diseased in cold, wet weather. It requires a fast-draining soil.

The attraction of this species is its picturesque trunk, so pruning off the fiber mat and its accompanying spines results in a characterless stem. One might as well plant a windmill palm (*Trachycarpus fortunei*), which grows faster and looks basically like this species minus its trunk covering. Removing the old, dead bottom parts of the petioles can be effective, however.

Because of its slow growth rate, this palm serves well for years as a close-up specimen. It is especially nice combined with rocks, cacti, or succulents. Old clustering individuals make beautiful specimens, even when surrounded by space. This palm is not known to have been grown indoors and is probably a bad candidate for doing so.

Trithrinax campestris
try-THRY-nax · kam-PES-tris
PLATES 921–922

Trithrinax campestris is indigenous to northern interior Argentina and west central Uruguay, where it grows mostly along and in ravines, dry river beds, and canyons at low elevations, but also in open intermontane savannas. The epithet is Latin for "of the fields." While still abundant in habitat, the species is quickly losing ground to expanding agricultural interests in Argentina.

This palm often clumps, even when quite young. The trunks grow to 20 feet (6 m) high and are usually clothed, except for the oldest parts, in a shag of dead leaves, reminiscent visually of the genus *Washingtonia*. Young plants or those from which the shag has been removed show a characteristic mat of closely knit, curling leaf sheath fibers and downward-pointing spines. The distinctive leaves are less than hemispherical and diamond shaped or wedge shaped, and the entire blade is usually 2 feet (60 cm) wide. Leaf color is sometimes amazing, always beautiful: grayish green to almost silvery blue above and dull light or yellowish to grayish green beneath. The segments are united for half their length, and the petiole is 2 feet (60 cm) long. The leaf segments are possibly the most rigid of any palmate-leaved palm; they are shallowly bifid apically and the ends are pointed and sharp enough to be dangerous.

This species is hardy to cold and is adaptable to zones 8 through 11 in areas subject to wet freezes in winter but to zones 7 through 11 in drier areas. Its most important cultural requirements are sun and a soil with unimpeded drainage. There is hardly a more picturesque palm. Old clustering individuals are stunningly attractive even as specimens surrounded by space and, from a distance, look like small, silvery washingtonias. This denizen of the semiarid plains fits into any setting as long as it is sunny. It is hard to overpraise this palm. The species is not known to have been grown indoors and is probably a bad candidate for doing so.

Trithrinax schizophylla
try-THRY-nax · sky-zo-FIL-la
PLATE 923

Trithrinax schizophylla is indigenous to southeastern Bolivia, western Paraguay, southwestern Brazil, and northern Argentina, where it grows along rivers and streams in the dry, lowland subtropical thorn forests but always where it can find underground moisture. The epithet is from Greek words which translate as "divided" and "leaf."

This palm is found mostly as a clustering species but occasionally produces a single trunk. Individual stems grow to 20 feet (6 m) high in habitat and are usually free of leaf bases, fibers, and spines except near their summits. The leaves are beautiful because of their color and the width of their segments; they are silvery light green, and thin and divided nearly to the petiole. The petioles are 2 feet (60 cm) long and show off the leaves to good effect because of the thin, airy-looking segments.

The palm is adaptable to zones 9b through 11 and marginal in 9a. Its drought tolerance is on a par with the other 2 species, but it seems more adapted to partial shade than the others. The palm luxuriates if given regular, adequate moisture (David Witt, pers. comm.).

Veitchia
VEECH-ee-a

Veitchia consists of 8 solitary-trunked, pinnate-leaved, monoecious palms from wet forests of the western Pacific islands. These are without doubt among the most beautiful palms in the world and are an emblem of their homeland and all its exotic connotations.

The moderately tall to quite tall trunks are gray or tan in their older parts and beautifully ringed in all but the oldest parts. The exquisite deep green crownshafts are often silvery or almost gray-brown because of hairs or scales, especially near the apex. These attractive columns reside beneath sparse but elegantly handsome leaf crowns of large pinnate leaves. The greenish white inflorescences grow from beneath the crownshaft, the much-branched, large panicles consisting of thick, waxy flowering branches that bear both male and female blossoms. The fallen male flowers with their large tufts of stamens carpet the ground beneath the palm and add another interesting dimension to these palms. The large clusters of fruits are orange to red.

All the species are intolerant of cold and adaptable only to zones 10b and 11. Some individuals survive in favorable microclimates of 10a, but they grow more slowly there and are overall less robust.

The genus name honors James Veitch, Jr. (1815–1869), British nurseryman whose nursery, James Veitch and Sons, played a leading role in palm introduction and horticulture. *Veitchia* seeds germinate within 30 days. For further details of seed germination, see *Areca*.

Veitchia arecina
VEECH-ee-a · a-re-SEE-na
PLATE 924

Veitchia arecina is indigenous to the Vanuatu islands, where it grows in mountainous rain forest from sea level to 1000 feet (300 m). The epithet is Latin for "little *Areca*" and suggests a resemblance to the genus *Areca* (or its fruits). Synonyms include *V. montgomeryana* and *V. macdanielsii*. Common names are Montgomery palm, in honor of Robert H. Montgomery, the driving force behind Fairchild Tropical Botanic Garden, and sunshine palm.

The trunk can reach 80 feet (24 m) tall in habitat but has, above its expanded base, a maximum diameter of only 1 foot (30 cm) or less and is almost white in its older parts. In its youngest parts, just below the crownshaft, the trunk is often yellow-green. The beautiful, slightly bulging silvery light green crownshaft can be almost 5 feet (1.5 m) tall but seems almost irrelevant in older palms because of the great beauty of the leaves. They are slightly ascending and slightly arching, on short petioles, and are 6–12 feet (1.8–3.6 m) long with many deep green half-pendent leaflets 3 feet (90 cm) long growing from the rachis in one plane; they fall from the tree as they die and it is unusual to find an individual with live fronds that dip beneath the horizontal. The inflorescences are 3 feet (90 cm) long and usually form a white ring beneath the crownshaft that, from a distance, looks like a little starburst of white blossoms. The bright red fruits hang in clusters 4 feet (1.2 m) long.

This incredibly beautiful species is slightly susceptible to lethal yellowing disease. It is adaptable to a range of soils, from acidic to calcareous. It must have constant, abundant moisture. It thrives in partial shade when young but needs the full tropical sun when older. Because its trunk is slender and its leaves become detached in strong winds, this palm remains standing in a hurricane and usually survives. It is fast growing after it starts to form a trunk. This is the only *Veitchia* species that can be grown in southern coastal California (Geoff Stein, pers. comm.).

Not many palm species look good planted anywhere, but this one does. It is difficult to find words to describe its beauty as a canopy-scape, especially in groups of 3 or more individuals of varying heights; its silhouette is simply astonishing, especially in moonlight. Although it can be grown indoors, this palm grows large quickly and needs an extraordinary amount of light.

Veitchia filifera
VEECH-ee-a · fi-LI-fe-ra
PLATE 925

Veitchia filifera occurs naturally on 2 islands of Fiji, where it grows in mountainous rain forest from near sea level to 2300 feet (700 m). The epithet is Latin for "thread-bearing," a reference to the reins that often remain attached to the 2 lowermost leaflets of a leaf.

This palm is half the stature of *V. arecina*, and its crownshaft is an unusual olive green because of the covering of short brownish black hairs. The light brown trunk is rarely more than 6 inches (15 cm) in diameter even in mature individuals. The leaves are borne on petioles 8 inches (20 cm) long and are 6–7 feet (1.8–2.1 m) long. The many deep green linear-elliptic leaflets are 2 feet (60 cm) long, have jagged apices, and are slightly pendent with age. The reins are retained. The ellipsoid ½-inch (13-mm) fruits are red.

This species is not as fussy about moisture requirements as the other species in the genus and is adaptable to various soils. It thrives in partial shade as well as full sun, even when older, but it dislikes dry air. It is perfection itself and is suitable for almost any site in a smaller garden. It is surprisingly scarce in cultivation.

Veitchia joannis
VEECH-ee-a · jo-AN-nis
PLATE 926

Veitchia joannis occurs naturally in the islands of Fiji and Tonga, where it grows in mountainous rain forest from elevations near sea level to 2000 feet (610 m). The epithet honors John Gould Veitch (1839–1870), a plant explorer and son of nurseryman James Veitch, Jr.

This species is the tallest one in the genus. The nearly white trunks attain a height of 100 feet (30 m) in habitat and are usually 12 inches (30 cm) in diameter but never more than 16 inches (41 cm). The leaves are 10 feet (3 m) long on petioles 1 foot (30 cm) long. The deep green, limp, and pendent leaflets are 3 feet (90 cm) long. The beautiful red fruits are conical and 2 inches (5 cm) long and have a prominent apical "beak."

This species has the same cultural requirements as *V. arecina* and is interchangeable with that species in the landscape. It is probably the fastest-growing species if given adequate moisture and a decent soil. It may be more cold tolerant than the rest, but it is certainly not tolerant of frost.

It is among the few rivals of the coconut palm for absolute beauty and grace. There is no more beautiful silhouette or canopy-scape. It could not last too long indoors even if given optimum cultural conditions as it grows fast and tall.

Veitchia metiti
VEECH-ee-a · me-TEE-tee
PLATE 927

Veitchia metiti is a rare species from the Banks Islands of northern Vanuatu, where it grows in mountainous rain forest from near sea level to 3000 feet (900 m). The epithet is a corruption of the aboriginal name. The species was named and described by Odoardo Beccari in 1920 but lost to scientists for the next 76 years until rediscovered on the islands of Vanua Lava and Uréparapara.

It is one of the smaller species, growing to 40 feet (14 m) overall in habitat. The white trunks are 6 inches (15 cm) in diameter, and the slender, light silvery green crownshafts are 3 feet (90 cm) tall. The leaves are 8–10 feet (2.4–3 m) long on unusually long petioles at 2 feet (60 cm) long. The limp, narrow, dark green leaflets are 2–3 feet (60–90 cm) long. The inflorescence branches are stiff and upright and remain so even in fruit. The red ellipsoid fruits are 1½ inches (4 cm) long.

Because of its relatively long petioles, thin leaflets, and elegantly slender trunks, this species is the personification of grace and is almost diaphanous. It is in cultivation only in tropical botanical gardens such as Fairchild Tropical Botanic Garden, where it has grown well. Not much is known of its requirements other than its intolerance of frost; it probably is suited to various soils and full sun but demands constant moisture.

Veitchia spiralis
VEECH-ee-a · spy-RAL-iss
PLATE 928

Veitchia spiralis is endemic to the southern Vanuatu islands of Tanna and Anatom, where it grows in lowland rain forest. The epithet is Latin for "spiraling," a reference to the arrangement of the fibers in the fruits.

This species is almost identical to *V. arecina*, differing only in details of the fruit and seed, and thus is interchangeable with that species in the landscape. The gray trunks grow to 50 feet (15 m) high and are 8 inches (20 cm) in diameter. The crownshaft is 3 feet (90 cm) tall, tapering, and silvery green with dark scales at its apex. The leaves are 8 feet (2.4 m) long on petioles 6 inches (15 cm) long. The medium green leaflets are 2 feet (60 cm) long and measure 3 inches (8 cm) wide at their bases; they taper to a point or are squared off and jagged at their apices and are almost as pendent as those of *V. joannis*. The globose 1-inch (2.5-cm) fruits are red.

This outstandingly beautiful palm makes a beautiful canopy-scape. It requires a frost-free, warm, and moist climate but is adaptable to a range of soils from slightly acidic to calcareous. It needs full sun when past its juvenile stage.

Veitchia vitiensis
VEECH-ee-a · vit-ee-EN-sis
PLATES 929–930

Veitchia vitiensis is endemic to Fiji, where it grows in mountainous rain forest from near sea level to 3800 feet (1160 m). It is highly variable in the size and shape of leaflets and fruits, and over the years, various names have been applied to different populations. The epithet is Latin for "of Fiji."

The gray trunk of this medium species grows to 50 feet (15 m) tall in habitat and is less than 6 inches (15 cm) in diameter except at the base. The trunk is somewhat flexible and often leans or curves, depending on light conditions. The crownshaft is almost black with olive green mottling. It is 18 inches (45 cm) tall and sits beneath leaves 8–9 feet (2.4–2.7 m) long and usually beautifully and strongly arched (almost recurving) from their midpoint on. The leaflets are unusually widely spaced for the genus, medium to deep green, narrow and pointed at their bases but expanded to a width of 3 inches (8 cm) at their apices, which are squared off and shallowly and irregularly jagged; they are stiff and grow from the rachis at a slight angle which gives a shallow V shape to the leaf. The orange to red ellipsoid fruits are less than 1 inch (2.5 cm) long.

This palm has an elegant beauty when young and is as impressive as any when older. Its leaf crown, although comprising only a few leaves, is globular and is comparable to *Actinorhytis* and *Carpoxylon* species as a canopy-scape. It is among the most beautiful palms.

Veitchia winin
VEECH-ee-a · WIN-in
PLATE 931

Veitchia winin is indigenous to the central Vanuatu islands, where it grows in mountainous rain forest from sea level to 1700 feet (520 m). The epithet is one of the aboriginal names for the palm in Vanuatu. This species is similar to *V. arecina* but generally has a straighter and shorter trunk and a stiffer overall aspect. The gray or tan stem grows to 50 feet (15 m) tall and 1 foot (30 cm) or less in diameter. The most obvious difference between the 2 species is the size of the fruits, which in *V. winin* are small, ellipsoid but blunt at both ends, and no more than 1 inch (2.5 cm) long. This palm thrives on acidic as well as calcareous soils and is more drought tolerant than most other species. It is, however, no more frost tolerant and needs a site in full sun.

Verschaffeltia
ver-sha-FEL-tee-a

Verschaffeltia is a monotypic genus of solitary-trunked, spiny, pinnate-leaved, monoecious palm in the Seychelles. The name honors Belgian nurseryman, botanist, and author Ambroise Verschaffelt (1825–1886), who introduced many tropical plants into cultivation but whose main specialty was camellias. Seeds of *Verschaffeltia* germinate within 30 days. For further details of seed germination, see *Areca*.

Verschaffeltia splendida
ver-sha-FEL-tee-a · splen-DEE-da
PLATES 932–933

Verschaffeltia splendida grows on the steep slopes of mountainous rain forest at 1000–2000 feet (300–610 m). Some growers and nursery owners have coined monikers for this wonderful palm, such as splendid stilt palm, but the best one is probably Seychelles stilt palm. The epithet is Latin for "splendid."

The trunk can reach 80 feet (24 m) high when old but is never more than 12 inches (30 cm) in diameter. It is covered in its younger parts with rings of downward-pointing black spines that are 3 inches (8 cm) long and with the remains of the leaf bases; shorter black spines on the petiole and rachis tend to slough off with age. The bottom of the trunk is invariably a narrow cone of stilt roots that may be 6 feet (1.8 m) tall; the species is exceptional in that these aerial or prop roots are not spiny. The leaf crown usually has 12 or more light to deep green leaves, which, because of their gentle arching, give a rounded aspect, although they do not usually lie beneath the horizontal. Each leaf is 6–8 feet (1.8–2.4 m) long on a petiole 2 feet (60 cm) long and in younger palms is obovate with jaggedly toothed margins but in older specimens is distinctly oblong. The blade is undivided when new, except at its apex, and this is the main reason the species is splendid; in older palms mechanical factors invariably tear the leaves into segments of varying widths and numbers. The much-branched inflorescences grow from the midst of the leaf crown and are 6 feet (1.8 m) long with unisexual blossoms, becoming pendent when the clusters of round 1-inch (2.5-cm) brown fruits mature.

The species is intolerant of drought and hot, dry winds and is adaptable only in zone 11; it is marginal in 10b, although gorgeous specimens exist in warm, protected microclimates of that zone. The palm is not fast growing and needs a fertile, humus-laden, fast-draining soil and constant moisture. It relishes partial shade when young but readily adapts to full sun when older except in the hottest climates; it is almost impossible to maintain in tropical desert regions because of the winds rather than the temperatures or great amount of sunlight.

This is among the world's most beautiful palm species. Its large leaves, even when torn by wind, are indescribably lovely, and its overall form is perfect. It looks nice as a specimen plant, even with only one individual, but is simply stunning in groups. It seems to cry out for an intimate site where its beauty can be contemplated up close, although its ultimate size prevents such placement; when young, however, it is perfection seen up close. This enchanting species is easy to grow indoors when young. As it grows older, it wants such a large space and such good light that it usually becomes unmanageable.

Voanioala
vo-ah'-nee-o-AH-la

Voanioala is a rare and endangered monotypic genus of solitary-trunked, pinnate-leaved, monoecious palm from northeastern Madagascar. The genus name is the aboriginal name for the palm, which translates as "forest coconut." Seeds can take more than 3 years to germinate and do so sporadically. The shell is extremely thick and woody. For further details of seed germination, see *Acrocomia*.

Voanioala gerardii
vo-ah'-nee-o-AH-la · je-RAHR-dee-eye
PLATE 934

Voanioala gerardii is endemic to a small mountainous rain forest and in swampy valley bottoms on the Masoala Peninsula, where it grows at an elevation of 600–1200 feet (180–370 m). It is known from a very few, scattered populations, none of which have more than 10 individuals. The trees are still cut down for the edible growing point. In 1995, fewer than 10 individuals existed at the site of the original discovery of this palm; by 2003, only one adult specimen remained. The epithet honors Jean Gerard, one of the discoverers of the palm in the late 20th century.

In habitat, the trunk grows to a maximum height of 60 feet (18 m) with a diameter of 1 foot (30 cm) and is light to dark tan with prominent indented rings of leaf base scars. The leaf crown is less than semicircular as the great leaves, 15 feet (4.5 m) long, are stiff and ascending. The massive, heavy false petioles are 5 feet (1.5 m) long and are actually the extended leaf bases. The dark green leathery leaflets are as long as the petioles, stiff, and regularly spaced along the felt-covered rachis from which they grow in a single flat plane. The inflorescences, also 5 feet (1.5 m) long, grow erect from the leaf crown but are pendent with the ellipsoid reddish brown fruits that are 3 inches (8 cm) long. The fiber-covered endocarps resemble miniature coconuts.

This palm looks like a stiff, massive coconut tree. The seeds are sold worldwide at premium prices and, while this practice probably assures that the species does not disappear from the face of the earth, it is very sad that conservation measures could not have been implemented in Madagascar in time to assure its continuation there. The future of this palm on that island is very bleak. The palm is reportedly slow growing, so it will be a long time before the cultivated plants mature.

Wallichia
wahl-LI-kee-a

Wallichia includes 11 mostly small, clustering as well as solitary-trunked, pinnate-leaved, monoecious palms in India, Bangladesh, Nepal, Bhutan, Laos, Myanmar, Tibet, China, and Thailand. The genus is related to and similar to *Arenga* and *Caryota*. All 3 genera have monocarpic stems in which flowering commences from the leaf axils or nodes (rings) near the top of the stem, and moves downward on the trunk; the stem dies after all the inflorescences have produced fruit.

The branched, spidery inflorescences of *Wallichia* are composed of all male or all female flowers, but both types of inflorescence are found on the same plant. The small, round, purplish or reddish fruits contain irritating calcium oxalate crystals, so as with *Arenga* and *Caryota*, fresh fruits should be handled with extreme caution. All 3 genera have unusually shaped leaflets with induplicate plication. The leaves of *Wallichia* tend to be linear or lanceolate with asymmetrical lobes.

Most *Wallichia* species occur in the undergrowth of wet forests in the mountains from sea level to 6500 feet (1980 m) in elevation. None are hardy to cold, but many can withstand freezing or slightly lower temperatures, or recover nicely. Wallich palm, the widely used common name for the genus, is not very descriptive. The genus name honors Danish botanist and physician Nathaniel Wallich (1786–1854), superintendent of the botanical gardens in Calcutta, India, who supplied collections of tropical plants to botanists in Europe. Seeds of *Wallichia* germinate within 30–90 days depending on how fresh they are. For further details of seed germination, see *Arenga*.

Wallichia caryotoides
wahl-LI-kee-a · kar'-ee-o-TO-i-deez

Wallichia caryotoides is indigenous to Bangladesh, Myanmar, China and Thailand, where it occurs in the undergrowth of valley hillsides in wet forest at 2000–4000 feet (610–1200 m). The epithet is formed from the genus name *Caryota* and the Greek suffix for "similar to" and refers to the shape of the leaflets.

The stems of this clustering species are mostly short, less than 10 feet (3 m) tall, resulting in a 10-foot (3-m) square mass of petioles 4 feet (1.2 m) long springing from the ground and carrying leaves 8 feet (2.4 m) long. Widely spaced leaflets grow in a single plane from the beautifully arching rachis, which is 7 feet (2.1 m) long. The stiff leaflets are 1 foot (30 cm) long and have lobes that are variously and usually jaggedly toothed. Leaflet color is deep green above and lighter green beneath. The inflorescences are 3–4 feet (90–120 cm) long and produce pendent clusters of ovoid purple fruits ½ inch (13 mm) in diameter.

This palm wants a rich, well-drained soil, nearly constant moisture, and partial shade or at least protection from the midday sun in hot climates. It is hardy to cold, and temperatures of freezing or slightly below seem not to faze it. This attractive little thing is useful as a contrast in masses of other foliage, palms or otherwise, but does not look good as an isolated specimen surrounded by space.

Wallichia disticha
wahl-LI-kee-a · DIS-ti-ka
PLATE 935

Wallichia disticha occurs naturally in the Himalaya of northeastern India, Bhutan, Bangladesh, China, and Myanmar, and into Laos and northwestern Thailand, where it grows in wet valley forests at 2000–4000 feet (610–1200 m). The epithet is Latin for "distichous" and alludes to the arrangement of the leaves on opposite sides of the trunk in a single plane. This arrangement gives the palm, from one profile, the look of a ladder, and makes it the most unusual species in the genus and among the most unusual species in any genus. Flat and "two-dimensional," this palm is difficult to site in the landscape. It is also the largest species in the genus.

This usually solitary-trunked palm occasionally occurs as a sparsely clustering individual. The stem grows to a maximum height of 30 feet (9 m) overall and, when mature, is 1 foot (30 cm) in diameter and covered in all but its oldest parts in a tight, flat mat of dark brown or black fibers with a few lighter, looser striations arching between the persistent woody leaf bases. The leaf crown, if it can be called such, is extended, as in several *Caryota* species; in mature plants, leaves grow along half or more of the length of the trunk, while in younger plants leaves usually grow along the entire trunk. The leaf petiole is 2–3 feet (60–90 cm) long and holds a blade that is 8–10 feet (2.4–3 m) long on a stiffly ascending rachis, arching only near its apex. The stiff, linear leaflets grow in clusters from the rachis at several angles to give the leaf a plumose effect. Each leaflet is 8 inches (20 cm) long, with slightly wavy margins that bear one or more shallow lobes, and with an obliquely cut, jagged apex. Leaflet color is medium to dark green above and grayish green to silvery green beneath.

The species is not as hardy to cold as the previous 2 and is adaptable only to zones 10 and 11. It is not fussy about soil type, although it prefers one that is not too alkaline and it needs regular, adequate moisture. It grows in partial shade but prefers full sun, except in hot climates. It grows moderately fast, especially after its juvenile stage, but hardly lives more than 20–25 years.

This odd beauty is only good as a specimen plant, sited where its most unusual leaf arrangement can be appreciated, such as against the side of a building. Like the solitary-trunked species of *Caryota* and *Arenga*, this palm takes 4 years to complete flowering, after which the plant dies. This attribute is the main reason the palm is not more widely planted.

Wallichia oblongifolia
wahl-LI-kee-a · ob-lon-ji-FO-lee-a
PLATE 936

Wallichia oblongifolia is an undergrowth palm in wet mountain valleys of the Himalaya in northeastern India, Nepal, Bhutan, China, and Myanmar at elevations between 2000 and 4000 feet (610–1200 m). The epithet is from Latin words meaning "oblong leaf[let]." This species was formerly known as *W. densiflora*, but Henderson (2007) corrected the long-standing nomenclatural confusion.

This palm is similar in overall appearance and dimensions to *W. caryotoides* but has more uniform leaflets. The leaves are 10 feet (3 m) long with gracefully arching rachises. The equidistant leaflets are 18–24 inches (45–60 cm) long, 3 inches (8 cm) wide, and more numerous and more uniformly shaped than those of the above species. Each leaflet is narrowly oblong to broadly lanceolate, irregularly toothed, slightly undulate on its margins; the leaflets of the terminal pair are usually united into an obovoid with a deep notch apically. Leaflet color is bright silky green above and satiny silver with hints of green beneath.

This palm is more attractive than *W. caryotoides* but has the same culture requirements and the same landscaping uses. It has big satiny leaves, and forms clumps 12 feet (3.6 m) tall that are reminiscent of robust *Alpinia* clumps.

Wallichia siamensis
wahl-LI-kee-a · sy-a-MEN-sis

Wallichia siamensis is endemic to Thailand, where it grows in mountainous rain forest at 1500–3800 feet (460–1160 m). The epithet means "of Thailand." Henderson (2007, 2009) treated this name as a synonym of *W. caryotoides*. This palm is similar in general form and dimensions to *W. caryotoides* and *W. densiflora*. Its leaves are, however, even more attractive because of their color. The leaflets are a deep, satiny green above and a grayish or silvery and satiny, almost shimmering color beneath with no hint of green. This trait creates an incredibly lovely effect when a breeze touches the leaves. The palm is not hardy to cold and needs protection from the hot sun. It is also less forgiving of poor soil than the others and needs constant moisture.

Washingtonia
wah-shing-TO-nee-a

Washingtonia comprises 2 large solitary-trunked, palmate-leaved, hermaphroditic palms in southwestern United States and northwestern Mexico. The species are common in cultivation, especially in the United States, southern Europe, the Middle East, and other dry, mild regions of the world. Both species retain a skirt of dead leaves. Some people favor the natural look and welcome the skirt as roost to owls and mosquito-eating bats. Other people deplore the unkempt look and claim the skirt houses vermin.

The plants have supposedly hybridized in cultivation to the point that young palms for sale, especially in the United States, are as likely hybrid crosses as they are "pure" species. The putative hybrid is sometimes marketed as *W.* 'Filabusta'. The good thing about these hybrids, if indeed they exist, is that they may combine the rapid growth of *W. robusta* with the cold hardiness of *W. filifera*. The genus name honors the first U.S. president, George Washington (1732–1799).

Fresh seeds of *Washingtonia* germinate within 30 days. Seeds that are allowed to dry out and are then rehydrated may take 3 months or longer to germinate and do not germinate as well as fresh seeds. For further details of seed germination, see *Corypha*.

Washingtonia filifera
wah-shing-TO-nee-a · fi-LI-fe-ra
PLATE 937

Washingtonia filifera is indigenous to California, western Arizona, and northeastern Baja California, where it grows along streams and arroyos and near natural springs. Common names are California fan palm, desert fan palm, petticoat palm, and, in England, cotton palm. The epithet is Latin for "thread-bearing" and alludes to the whitish curling threads between the leaflets.

Mature trunks grow to 60 feet (18 m) tall and are 3–4 feet (90–120 cm) in diameter with swollen bases. They are gray and often narrowly fissured in their older parts but brown or even reddish brown elsewhere. Unless trimmed, the trunks have a skirt of dead leaves beneath the crown of living leaves and, in nature, this skirt sometimes covers the entire trunk, which characteristic has led to the vernacular name of petticoat palm. The silvery or grayish green leaves are 6–8 feet (1.8–2.4 m) wide and greater than hemispherical. The petioles are 6 feet (1.8 m) long and have sharp teeth along their margins. The segments are usually conjoined for less than half their length, are pendulous at their apices in older plants, and are accompanied by threads between each segment. The inflorescences grow from the leaf crown and extend well beyond it. The whitish, bisexual blossoms produce clusters of dark brown or black fruits that cause the infructescence to hang well below the canopy.

The species is hardy to cold and is, in dry climates, safe in zones 7 through 11. The trees are, like almost every other plant, much more tender to cold when young and, although they are usually unscathed by temperatures below 20°F (−7°C) when mature, may be severely damaged or even killed by the same temperatures when in the seedling stage, especially in wet climates. The species does much better in Mediterranean climates than humid tropical ones.

The palm is drought tolerant for short periods but looks better and grows faster with regular, adequate irrigation. Its "desert" home is misleading, as the palm grows only where its roots can access permanent water. It needs full sun from youth to old age and becomes etiolated without it. This species thrives especially in calcareous and alkaline soils as long as they are free draining.

The trunks of older palms are imposing and massive, and the trees look better as avenue borders or canopy-scapes where their wonderful crown silhouettes accent the horizon. Specimen groups of individuals of varying heights are stunningly attractive and tropical looking. This is not a good choice for indoor cultivation as it needs much light and grows so large so fast.

Washingtonia robusta
wah-shing-TO-nee-a · ro-BUS-ta
PLATE 938

Washingtonia robusta is indigenous to the southern half of the Baja California peninsula and to the adjacent mainland in the Mexican state of Sonora. It grows near streams, in arroyos, and at natural springs in the desert. Common names are Mexican fan palm and thread palm. The epithet is Latin for "robust" and refers to the palm's rapid growth rate.

This species differs from *W. filifera* in having taller (to 90 feet [27 m] in old individuals) and thinner trunks, fewer hairlike fibers on the leaves, and a tighter, more compact crown. While it is not an invariable diagnostic feature, a reddish brown area is often apparent on the bottom of the leaves of *W. robusta* near the juncture of blade and petiole (the colored area is more obvious in younger plants than in older ones); this patch is always lacking in *W. filifera*. The petioles of *W. robusta* are shorter and, in younger individuals, are reddish brown, especially near the base of the petiole, and always bear reddish brown spines, even when young. The older parts of the trunk are gray, and the newer parts are clothed in deep chestnut red or brown leaf bases, which in cultivation are often laboriously cut away to reveal the almost smooth and usually reddish younger trunk. The leaves are always bright green or yellowish green. The inflorescences, flowers, and fruits of the 2 species are similar.

Washingtonia robusta is decidedly more tender to cold than *W. filifera* and is safe only in zones 9b through 11. It is usually damaged by temperatures in the low 20s Fahrenheit (around −7°C), although it has been known to survive the low teens (around −11°C) in dry climates where freezing temperatures are short lived and not accompanied by precipitation. The species is tolerant of calcareous soils and seems to need only a well-drained medium, although it grows faster and looks better with a decent soil. It needs full sun at all ages. It also grows much more rapidly than *W. filifera*.

Mature palms are among the most elegant, graceful subtropical landscape subjects; their thin and extremely tall forms are ineluctably arresting in the landscape. This species is by far the most commonly planted palm in Southern California, far southern Texas, and Phoenix and Tucson in Arizona. It is not a good choice for indoor cultivation as it needs much light and grows so large so fast, although it is now often used in large atriums and other enclosures.

Welfia
WEL-fee-a

Welfia is a monotypic genus of solitary-trunked, pinnate-leaved, monoecious palm in Central and South America. The genus name commemorates the German house of Welf, an economically and politically powerful dynasty in medieval Europe. Seeds germinate within 60 days. For further details of seed germination, see *Asterogyne*.

Welfia regia
WEL-fee-a · REE-jee-a
PLATES 939–940

Welfia regia occurs naturally in northeastern Honduras, eastern Nicaragua, Costa Rica, Panama, northwestern and western Colombia, and into northern Ecuador, where it grows in mountainous rain forest from sea level to 5280 feet (1610 m). The epithet is Latin for "regal," an apt appellation. A synonym is *W. georgii*.

The gray trunk grows to 60 feet (18 m) in habitat and is 6 inches (15 cm) in diameter, making for an elegantly slender silhouette. The leaves are 12–18 feet (3.6–5.4 m) long on short, almost nonexistent petioles and are erect but arching near their apices, seldom lying beneath the horizontal on the tree. The evenly spaced, dark green, lanceolate leaflets are 2–3 feet (60–90 cm) long and grow in a single plane from the rachis, which is twisted from its midpoint to its apex. The new leaves are a beautiful cherry to deep red that lasts for a week. The inflorescences grow from beneath the leaf crown and are composed of massive single, thick, ropelike, orange branches 3 feet (90 cm) long that bear both male and female flowers.

This species is very rare in cultivation, mainly because it is one of the most cold sensitive palms; sustained temperatures below 55°F (13°C) kill it. It is difficult to overwater in a well-draining soil, which should also be humus laden and slightly acidic. It thrives in partial shade when young but appreciates sun when older. It does not tolerate frost but should do well in cool but frostless climates if provided with enough moisture in the air and in the soil. This is among the most beautiful canopy-scapes and, when young, is handsome in any situation other than a single specimen surrounded by space.

Wendlandiella
wend-lan'-dee-EL-la

Wendlandiella is a monotypic genus of small, clustering, pinnate-leaved, dioecious palm in South America. The name honors Hermann Wendland (1825–1903), German botanist and palm specialist; the diminutive suffix distinguishes this genus from another one dedicated to Wendland's grandfather. Seeds are not generally available as plants of only one sex are in cultivation. Propagation is easy through division of the clumps.

Wendlandiella gracilis
wend-lan'-dee-EL-la · GRA-si-lis
PLATE 941

Wendlandiella gracilis is indigenous to eastern Peru and extreme northwestern Brazil and Bolivia, where it occurs in the undergrowth of rain forest. The epithet is Latin for "graceful."

It is mostly a clustering species, but some individuals form solitary trunks. The little green trunks are dainty and grow to 3–4 feet (90–120 cm) tall with a girth of less than ½ inch (13 mm); they look like grass stems. The short pinnate leaves have 4–6 leaflets, and most individuals look, from any distance, as though they are palmate, similar to a *Rhapis* species; some individuals have undivided leaves that are deeply bifid at the apex and resemble certain *Chamaedorea* species.

This variation of leaf form corresponds to geographical distribution (Henderson et al. 1995) and has led to the erection of 3 varieties: *W. gracilis* var. *gracilis* has 8–12 separate but irregularly spaced leaflets; *W. gracilis* var. *polyclada* [pah-lee-KLAY-da] has 4 leaflets; and *W. gracilis* var. *simplicifrons* [sim-PLI-si-frahnz] has the undivided and deeply bifid leaf. Leaf color is medium to dark green and the blade is indistinctly ribbed. The inflorescences grow from beneath the tiny leaf crown, and the female plants produce tiny ovoid red fruits.

The palm is not hardy to cold and is safe without protection only in zones 10b and 11, but it is small enough to be easily protected. It requires constant moisture in a well-draining, rich, and humus-laden soil and does not tolerate full sun. It is rare in cultivation and is certainly not a spectacular commodity: it looks like a dwarf bamboo. This small palm is endearing, however, and is perfect as a groundcover for shady and intimate sites.

Wettinia
wet-TIN-ee-a

Wettinia is composed of 21 mostly solitary-trunked, stilt-rooted, pinnate-leaved, monoecious palms in South America. It is related to and has many similarities to *Socratea*, including stilt roots and fan-shaped leaflets that may be split longitudinally into several segments giving the leaf a plumose appearance. *Wettinia* species exhibit a curious flowering process: inflorescences emerge from brown- or orange-hooded, leathery spathes that resemble those of an aroid; the inflorescences are multiple at each node (trunk ring) beneath the crownshaft and bear either all male or all female blossoms that are thick, fleshy, and white. The middle and largest inflorescence is either male or female, but the lateral inflorescences, up to 15 in number, are usually male and are progressively smaller the farther they are from the center. The angular, tomentose fruits are usually densely packed into large sausagelike clusters. In only a few species are the fruits loose from one another.

None of the species is frost tolerant, although many of them, because of their elevated natural habitats, relish cool climates. All are true water lovers but also need a fast-draining, humus-laden, slightly acidic soil. They grow equally well in partial shade and full sun but need protection from midday sun in the hottest climates.

They require a consistently high relative humidity. Few of these species are in cultivation, which is a pity as they are sumptuously beautiful.

The trunks of all species are used in construction in South America because of the hard, durable wood. The genus is named after King Frederick Augustus I (1750–1827) of Saxony, of the house of Wettin, a dynasty of German royalty and nobility. Seeds germinate within 90 days and should not be allowed to dry out before sowing. For further details of seed germination, see *Iriartea*.

Wettinia aequalis
wet-TIN-ee-a · ek-WAHL-is

Wettinia aequalis is indigenous to Panama, Colombia, and western Ecuador. It occurs in lowland rain forest up to 1600 feet (490 m) in elevation. The epithet is Latin for "similar, equal, uniform." The allusion apparently refers to the 3 equal-size carpels in the female flower.

The singe stem, elevated on a cone of roots 1½ feet (45 cm) tall, grows to 25 feet (7.6 m) tall and up to 5 inches (13 cm) in diameter. The leaves are 11–12 feet (3.3–3.6 m) long; the leaflets are narrowly elliptical and in one plane. Marcescent leaves persist for some time and hide the crownshaft, which is 4 feet (1.2 m) long and warty. The ellipsoid fruits, about 2 inches (5 cm) long and 1 inch (2.5 cm) across, are loosely arranged on the infructescence, brownish yellow, and covered with short, gray hairs. The crown of leaves is almost like a tree fern and outstandingly beautiful but somewhat marred by the dead leaves that hang beneath it.

Wettinia augusta
wet-TIN-ee-a · aw-GUS-ta
PLATE 942

Wettinia augusta is indigenous to south central Colombia, Peru east of the Andes, western Brazil, and northern Bolivia, where it grows in lowland to mountainous rain forest to 2500 feet (760 m). The epithet is Latin for "noble" or "majestic." The trunks of this sparsely clustering species usually grow to 15 feet (4.5 m) tall but may reach 40 feet (12 m) and are never more than 8–9 inches (20–23 cm) in diameter; they grow atop a small cone of stilt roots. The leaves are 5 feet (1.5 m) long with evenly spaced deep green, linear, jagged-ended, limp, and pendent leaflets 1 foot (30 cm) long. The infructescences bear clusters of fruits that are densely covered with white hairs. The fruits are so closely packed that they take on an angular or prismatic shape. This lovely species has ethereally thin trunks and, in most individuals, tiers of leaf crowns. It makes an excellent specimen and canopy-scape.

Wettinia hirsuta
wet-TIN-ee-a · hir-SOO-ta
PLATE 943

Wettinia hirsuta is restricted to Colombia in rain forests between 1300 and 4300 feet (400–1310 m). The epithet is Latin for "hirsute" and alludes to the fruits. The solitary stem perches atop a cone of stilt roots. The cone is 12 inches (30 cm) tall and the stem is up to 40 feet (12 m) tall and 4 inches (10 cm) in diameter. The sparse crown of leaves sits above a gray-green crownshaft covered with purplish, stiff hairs. The leaves are up to 8 feet (2.4 m) long; the leaflets are narrowly fan shaped, undivided, and hang pendulously from the rachis. Fruits are densely packed, very hairy, and dark brown. The narrow trunk, sparse crown of leaves, and pendulous pinnae combine to form the very picture of elegance and refinement.

Wettinia maynensis
wet-TIN-ee-a · my-NEN-sis
PLATE 944

Wettinia maynensis is indigenous to the foothills of the eastern Andes in Peru through central Ecuador and into southernmost Colombia, where it grows in rain forest from elevations of 800 to 5200 feet (240–1580 m). The epithet is Latin for "of Maynas," a province in northern Peru.

A population of this species from southern Ecuador has clustering stems, but otherwise this palm has a single stem. The solitary stem attains a height of 40 feet (12 m) but is never more than 6 inches (15 cm) in diameter and is supported by a cone of spiny stilt roots 3 feet (90 cm) tall. The light to medium green or even bluish crownshaft is only as wide as the trunk, 2 feet (60 cm) tall, and gently tapering to its apex. The leaf crown is sparse with usually 6 or 7 leaves, but this characteristic emphasizes the beauty of the leaves. Borne on short petioles, the leaves are 6–8 feet (1.8–2.4 m) long and softly hairy, and have regularly, closely spaced, pendent leaflets, each 18 inches (45 cm) long, ovoid-linear with an oblique, jagged apex. Dead leaves hang beneath the crown for some time. The fruits are densely packed and covered with white hairs.

Wettinia praemorsa
wet-TIN-ee-a · pree-MOR-sa
PLATE 945

Wettinia praemorsa is indigenous to northeastern and central Colombia and northwestern Venezuela, where it grows in mountainous rain forest and cloud forest from 1300 to 8000 feet (400–2440 m). The epithet is Latin for "jagged" and alludes to the ends of the leaflets.

The stems of this usually clustering species grow to 50 feet (15 m) but are often half that height and are never more than 6 inches (15 cm) in diameter; they sit atop cones of stilt roots 3 feet (90 cm) tall and are crowned by silvery green cylindrical crownshafts, also 3 feet (90 cm) tall, that have a slight bulge at their bases. The leaf crown never has more than 6 large leaves, which are arranged to give an X shape to the crown. The leaves are 7–8 feet (2.1–2.4 m) long with many widely spaced, irregularly lanceolate, deep green, soft and pendent leaflets 2 feet (60 cm) long and ending with obliquely cut, slightly jagged tips. The leaflets usually grow in the same plane from the rachis, but many individuals have leaflets that are split lengthwise with each segment of the leaflet growing in a different plane, which then gives the leaf a plumose effect.

This is one of the handsomest species in the genus, its tiers of X-shaped leaf crowns and silhouettes creating an almost unbelievably beautiful effect. It is more than a pity that this palm does not like hot climates.

Wettinia quinaria
wet-TIN-ee-a · kee-NAHR-ee-a
PLATE 946

Wettinia quinaria occurs naturally in western Colombia and northwestern Ecuador, where it grows in rain forest from sea level to 3000 feet (900 m). The Latin epithet means "grouped in fives" and refers to the number of inflorescences per node.

The trunks of this mostly clustering species can grow as tall as 50 feet (15 m) but are more often half that height, above a mass of stilt roots 3–4 feet (90–120 cm) tall, and beneath an almost completely cylindrical and barely bulging, grayish green crownshaft 5–6 feet (1.5–1.8 m) tall. The leaf crown has 4–6 leaves, which, because of their disposition, create an appearance "like an X when seen from a distance" (Henderson et al. 1995); the arrangement also resembles the vanes of a windmill. Each leaf is 8 feet (2.4 m) long, steeply ascending, with a straight rachis, and a short petiole. The deep green, regularly spaced leaflets are 2 feet (60 cm) long, limp, and pendulous, giving the leaf blade an almost curtainlike appearance. The ellipsoid fruits are 1 inch (2.5 cm) long, loosely arranged, and dull, yellowish brown when ripe.

This is arguably the most beautiful species in the genus because of the great leaves and the tiers of leaf crowns that create a veritably breathtaking tableau; there is nothing quite like this look.

Wodyetia
wod-YET-ee-a

Wodyetia is a monotypic genus of solitary-trunked, pinnate-leaved, monoecious palm. The species was unknown to botanists until the early 1980s and is now in great demand because of its extraordinary beauty, fast growth, and adaptability. The genus name is a Latinized form of "Wodyeti," the name used by Johnny Flinders (1900?–1978), who was the last Aboriginal to have traditional knowledge of the flora and fauna of the region. Seeds germinate sporadically and generally within 90–180 days but can take as few as 30 days or as many as 270. In this way, the genus differs from most other members of the Areceae. Seeds allowed to dry out before sowing do not germinate as well as those kept slightly moist. For further details of seed germination, see *Areca*.

Wodyetia bifurcata
wod-YET-ee-a · by-fur-KAH-ta
PLATES 947–948

Wodyetia bifurcata is endemic to a remote area of the Cape York Peninsula in the Melville Range of Queensland, where it grows in monsoonal and rocky scrubland on sandy, granitic soils at elevations to 1200 feet (370 m). Although it is protected in Cape Melville National Park, it is vulnerable to extinction because of its very restricted native habitat. In the 1980s, the rampant collection of its seeds for ornamental horticulture was perceived by authorities as a threat to this palm. It was not, and nowadays, seeds are readily and abundantly available from cultivated specimens. The common name of foxtail palm alludes to the plumose leaves. The epithet is Latin for "twice forked" and refers to the branching of the fibers in the fruits.

The trunks attain 45 feet (14 m) of height in habitat. They are columnar, light gray to nearly white with widely spaced, prominent darker rings in their younger parts, and often slightly swollen near the middle of the stem. The smooth crownshaft is 3 feet (90 cm) tall in mature specimens, light green, the same diameter as the trunk at its base, and tapering from base to top. The leaves are borne on short petioles and are 8–10 feet (2.4–3 m) long with many medium to dark green leaflets growing from different angles around the rachis; the visual result is one of the most plumelike leaves in the family. Each leaflet is either divided longitudinally into 2 or more linear segments or lobed, the apical pair usually united. The much-branched inflorescences grow from beneath the crownshaft and hold small, yellowish green flowers of both sexes. The ovoid 2-inch (5-cm) fruits are deep orange-red when ripe. The black endocarps, with their flattened, forked fibers, are distinctive.

This palm can endure drought but grows slowly and looks stunted and usually chlorotic. It thrives when its deep root system taps underground moisture. It is slow growing in shade but wonderfully fast in full sun. While it survives in calcareous soils, it needs a slightly acidic, well-drained soil to avoid unsightly chlorosis of the leaves. The palm is not hardy to cold but usually survives short periods at 28°F (−2°C) unscathed.

This species is similar to *Roystonea regia*, also known as royal palm, but not as massive. It is among the best species for creating the "royal look" when planted in lines along an avenue or other promenades, and, as a canopy-scape, is nearly unrivaled. It resembles *Normanbya normanbyi*, but is heavier, more massive looking, easier to come by, and easier to grow. It is easily grown indoors if given enough light and space.

Zombia
ZAHM-bee-a

Zombia is a monotypic genus of clustering, spiny, palmate-leaved, hermaphroditic palm from the West Indies. The name is from the Haitian Creole name for the palm, *latanier zombi*, which literally means "ghost palm," but the palm's association with the undead is not obvious. Seeds of *Zombia* germinate within 90 days and should not be allowed to dry out before sowing. For further details of seed germination, see *Corypha*.

Zombia antillarum
ZAHM-bee-a · an-til-LAH-rum
PLATES 949–950

Zombia antillarum is endemic to Hispaniola, where it grows on dry hills at low elevations. It is endangered because of habitat destruction. The only common name seems to be zombie palm. The epithet is Latin for "of the Antilles."

The clumps are usually dense with spiny trunks to 10 feet (3 m) high and, with time, 10 feet (3 m) wide. The stems are covered except in their oldest parts with the beautifully intricate woven fibrous remains of the leaf sheaths and closely set rings of spiny, light or dark brown downward-pointing needlelike projections 2–4 inches (5–10 cm) long. The leaves are borne on petioles 2 feet (60 cm) long and the semicircular blade is 3 feet (90 cm) wide. The narrow, lanceolate segments are dull green above and silvery green beneath. The inflorescences grow from amid the leaves and are 18 inches (45 cm) long; they bear small, white flowers that produce globular 1-inch (2.5-cm) white fruits.

This palm is drought tolerant but looks better and grows faster, although it could never be called fast growing, with average and regular moisture. It is not cold tolerant and is adapted only to zones 10 and 11. It prefers full sun but will grow in partial shade though not as fast. It requires a free-draining soil and tolerates salt.

This species is exceptionally beautiful in a site where its almost unique stems can be seen and, since its spininess is confined to the stems themselves, it can be used in intimate or close-up situations. Its natural tendency is to form dense clumps that obscure the trunks, resulting in a large mound of foliage in which not only the trunks but also the shape of the leaves are obscured. Thus, some judicious pruning out of the stems reveals not only the wondrous woven fibers and their spiny projections but also the ghostly white fruits in season. It is possible indoors but would need much light and good air circulation.

LANDSCAPE LISTS

Drought-Tolerant Species

No palm is truly drought tolerant in the sense that most cactus species are, and even those palms growing in deserts do so at or near springs or other underground water sources. Furthermore, most palm species occur in tropical rain forests. Despite these disclaimers, a number of species thrive without copious moisture, can survive extended periods of deprivation, and can withstand drying winds. Tolerance is listed as either medium or high and applied to established palms. Newly transplanted palms almost always benefit from some irrigation until established.

High Drought Tolerance

- Allagoptera spp.
- Bismarckia nobilis
- Borassus spp.
- Brahea aculeata
- Brahea armata
- Brahea decumbens
- Butia archeri
- Butia campicola
- Butia paraguayensis
- Butia purpurascens
- Chamaerops humilis
- Coccothrinax spp.
- Cocos nucifera
- Copernicia ekmanii
- Copernicia rigida
- Dypsis decaryi
- Dypsis decipiens
- Gaussia spp.
- Hemithrinax ekmanii
- Hyphaene spp.
- Jubaea chilensis
- Leucothrinax morrisii
- Livistona mariae
- Livistona victoriae
- Medemia argun
- Nannorrhops ritchieana
- Phoenix spp. (except P. roebelenii)
- Pseudophoenix spp.
- Ravenea xerophila
- Rhapis laosensis
- Sabal etonia
- Sabal mexicana
- Sabal minor
- Sabal rosei
- Sabal uresana
- Serenoa repens
- Syagrus cardenasii
- Syagrus comosa
- Syagrus coronata
- Syagrus glaucescens
- Syagrus harleyi
- Syagrus macrocarpa
- Syagrus schizophylla
- Syagrus vagans
- Thrinax parviflora
- Thrinax radiata
- Trithrinax spp.
- Washingtonia filifera

Medium Drought Tolerance

- Acrocomia spp.
- Aiphanes lindeniana
- Arenga pinnata
- Brahea brandegeei
- Brahea dulcis
- Butia capitata
- Butia eriospatha
- Butia yatay
- Colpothrinax wrightii
- Copernicia spp.
- Corypha umbraculifera
- Guihaia spp.
- Latania spp.
- Livistona decora
- Maxburretia spp.
- Phoenix roebelenii
- Phoenix acaulis
- Phoenix reclinata
- Ravenea hildebrandtii
- Rhapis spp. (except R. laosensis)
- Sabal spp. (except S. minor)
- Syagrus ×costae
- Syagrus flexuosa
- Syagrus ruschiana
- Thrinax excelsa
- Trachycarpus nanus
- Trachycarpus princeps
- Washingtonia robusta
- Zombia antillarum

Water-Loving Species

Most palm species benefit from regular amounts of water, but several have extraordinary water requirements, even when established.

- Acoelorrhaphe wrightii
- Archontophoenix tuckeri
- Areca triandra
- Arenga microcarpa
- Arenga obtusifolia
- Astrocaryum murumuru
- Bactris militaris
- Calyptronoma spp.
- Chelyocarpus spp.
- Cyrtostachys renda
- Dypsis crinita
- Dypsis rivularis
- Elaeis guineensis
- Elaeis oleifera
- Eleiodoxa conferta
- Eremospatha wendlandiana
- Eugeissona tristis
- Euterpe oleracea
- Euterpe precatoria
- Geonoma interrupta
- Hydriastele costata
- Hydriastele ramsayi
- Hydriastele rheophytica
- Hydriastele wendlandiana
- Laccosperma spp.
- Licuala paludosa
- Licuala ramsayi
- Livistona australis
- Livistona benthamii
- Livistona decora
- Manicaria spp.
- Marojejya darianii
- Mauritia spp.
- Mauritiella spp.
- Metroxylon spp.
- Normanbya normanbyi
- Nypa fruticans
- Oncocalamus spp.
- Oncosperma tigillarium
- Orania trispatha
- Phoenix paludosa
- Phoenix roebelenii
- Pholidocarpus spp.
- Phytelephas tenuicaulis
- Ptychosperma lauterbachii
- Ptychosperma lineare
- Ptychosperma macarthurii
- Raphia spp.
- Ravenea musicalis
- Ravenea rivularis
- Rhapidophyllum hystrix
- Roystonea princeps
- Roystonea regia
- Sabal minor
- Salacca wallichiana
- Salacca zalacca
- Serenoa repens
- Tahina spectabilis
- Washingtonia robusta

LANDSCAPE LISTS

Fast-Growing Species

Very few palm species are fast growing compared to dicot trees. Most are exceedingly slow. The following list includes those that are faster than the average, assuming they are given optimum growing conditions.

Acanthophoenix spp.
Acrocomia spp.
Actinorhytis calapparia
Adonidia merrillii
Aiphanes aculeata
Archontophoenix alexandrae
Archontophoenix cunninghamiana
Areca catechu (very fast)
Areca triandra
Arenga microcarpa
Arenga obtusifolia
Arenga pinnata
Astrocaryum mexicanum
Bactris gasipaes
Burretiokentia vieillardii

Calamus spp.
Carpentaria acuminata (very)
Caryota spp. (very)
Chambeyronia macrocarpa
Cocos nucifera
Copernicia alba
Copernicia prunifera
Deckenia
Dypsis lutescens
Dypsis mananjarensis
Elaeis guineensis
Euterpe edulis
Euterpe oleracea
Heterospathe elata
Heterospathe phillipsii

Hydriastele costata
Hydriastele ramsayi
Hyophorbe indica
Licuala paludosa
Licuala peltata
Licuala spinosa
Livistona benthamii
Livistona chinensis
Livistona decipiens (very)
Livistona drudei
Livistona jenkinsiana
Livistona mariae
Livistona nitida
Livistona saribus
Metroxylon spp. (very)
Nypa fruticans

Oenocarpus spp.
Oncosperma spp.
Phoenix canariensis
Phoenix dactylifera
Phoenix reclinata
Phoenix rupicola
Pholidocarpus spp.
Pigafetta spp. (very)
Ptychosperma elegans
Ptychosperma macarthurii
Ptychosperma salomonense
Ravenea rivularis
Rhopalostylis spp.
Roystonea oleracea
Roystonea princeps
Roystonea regia (very)

Sabal mauritiiformis
Saribus merrillii
Saribus rotundifolius
Syagrus amara
Syagrus botryophora (very)
Syagrus cocoides
Syagrus inajai
Syagrus oleracea
Syagrus romanzoffiana (very)
Syagrus sancona (very)
Trachycarpus fortunei
Veitchia spp. (very)
Washingtonia filifera
Washingtonia robusta (very)
Wodyetia bifurcata

Unusually Slow Growing Species

Compared to dicot trees, palms are slow growing. Some are extremely slow growing and these are listed below.

Actinokentia divaricata
Areca multifida
Attalea spp.
Basselinia glabrata
Basselinia gracilis
Basselinia humboldtiana
Basselinia velutina
Brassiophoenix spp.
Ceroxylon spp.
Clinosperma spp.
Coccothrinax spp.
Colpothrinax spp.
Copernicia spp.

Cyphokentia cerifera
Cyphophoenix fulcita
Gaussia spirituana
Guihaia spp.
Hedyscepe canterburyana
Heterospathe elmeri
Howea belmoreana
Hyphaene spp.
Itaya amicorum
Johannesteijsmannia spp.
Jubaea chilensis
Jubaeopsis caffra

Lepidorrhachis mooreana
Licuala spp.
Livistona alfredii
Livistona humilis
Livistona inermis
Livistona muelleri
Lodoicea maldivica
Loxococcus rupicola
Lytocaryum weddellianum
Masoala spp.
Maxburretia spp.
Medemia argun

Nephrosperma vanhoutteanum
Oraniopsis appendiculata
Parajubaea spp.
Phoenix loureiroi
Pinanga veitchii
Pritchardia hillebrandii
Pritchardia kaalae
Pritchardia martii
Pseudophoenix spp.
Ravenea julietiae
Ravenea robustior

Ravenea xerophila
Rhapidophyllum hystrix
Roscheria melanochaetes
Sabal spp.
Saribus jeanneneyi
Syagrus comosa
Syagrus glaucescens
Tectiphiala ferox
Trachycarpus oreophilus
Trithrinax ekmanii
Voanioala gerardii
Zombia antillarum

LANDSCAPE LISTS

Hedge and Screen Palms

Whether practical or esthetic in their basic function, or sometimes both, hedges and screens require palms that hold their leaves from top to bottom.

Acoelorrhaphe wrightii
Allagoptera spp.
Areca macrocarpa
Areca triandra
Arenga australasica
Arenga caudata
Arenga engleri
Arenga microcarpa
Arenga porphyrocarpa
Arenga tremula
Bactris brongniartii
Bactris concinna
Bactris gasipaes
Bactris major
Bactris plumeriana
Brahea decumbens
Calyptrocalyx forbesii
Calyptrocalyx hollrungii
Calyptrocalyx polyphyllus
Caryota mitis
Caryota monostachya
Chamaedorea brachypoda
Chamaedorea cataractarum
Chamaedorea costaricana
Chamaedorea graminifolia
Chamaedorea hooperiana
Chamaedorea microspadix
Chamaedorea pochutlensis
Chamaedorea seifrizii
Chamaerops humilis
Chuniophoenix humilis
Chuniophoenix nana
Cyrtostachys elegans
Cyrtostachys glauca
Cyrtostachys renda
Daemonorops curranii
Dypsis cabadae
Dypsis lutescens
Dypsis pembana
Eleiodoxa conferta
Eugeissona tristis
Euterpe oleracea
Geonoma interrupta
Guihaia argyrata
Hydriastele kasesa
Hydriastele microspadix
Hydriastele rostrata
Licuala sallehana
Licuala spinosa
Mauritiella spp.
Metroxylon sagu
Nannorrhops ritchieana
Nypa fruticans
Oncosperma fasciculatum
Oncosperma horridum
Oncosperma tigillarium
Phoenix paludosa
Phoenix reclinata
Phytelephas tenuicaulis
Pinanga adangensis
Pinanga coronata
Pinanga densiflora
Pinanga dicksonii
Pinanga malaiana
Pinanga scortechinii
Pinanga sylvestris
Ptychosperma lauterbachii
Ptychosperma macarthurii
Ptychosperma sanderianum
Raphia sudanica
Raphia taedigera
Reinhardtia latisecta
Reinhardtia simplex
Rhapidophyllum hystrix
Rhapis excelsa
Rhapis humilis
Rhapis multifida
Rhapis subtilis
Salacca spp.
Serenoa repens
Syagrus cearensis
Syagrus flexuosa
Syagrus harleyi
Syagrus ruschiana
Syagrus vagans
Trithrinax campestris
Wallichia caryotoides
Wallichia densiflora
Wendlandiella gracilis
Zombia antillarum

Groundcovering Palms

Short palms that creep by aboveground or underground stems can be used in the landscape as groundcovers. Others are amenable for mass plantings because of their size and form.

Areca minuta
Arenga caudata
Arenga hookeriana
Asterogyne martiana
Brahea decumbens
Calyptrogyne ghiesbreghtiana
Chamaedorea adscendens
Chamaedorea brachypoda
Chamaedorea cataractarum
Chamaedorea elegans
Chamaedorea metallica
Chamaedorea pygmaea
Chamaedorea radicalis
Chamaedorea stolonifera
Chuniophoenix humilis
Chuniophoenix nana
Daemonorops curranii
Guihaia spp.
Lepidocaryum spp.
Licuala longipes
Licuala mattanensis
Licuala sarawakensis
Licuala triphylla
Neonicholsonia watsonii
Pinanga disticha
Pinanga geonomiformis
Pinanga polymorpha
Pinanga simplicifrons
Pinanga veitchii
Reinhardtia gracilis
Reinhardtia simplex
Rhapis laosensis
Sabal etonia
Sabal minor
Trachycarpus nanus
Wendlandiella gracilis

Climbing Palms

Vinelike palms exist in the tropics, and most of them have developed specialized organs for climbing. In almost all cases, these unique organs are formed only when the palm has attained some age and, more importantly, an aboveground stem. The following list is not exhaustive of all the climbing palms found in the world; moreover, in some of the genera listed below, not every species is a climber.

Calamus spp.
Ceratolobus spp.
Chamaedorea elatior
Daemonorops spp.
Desmoncus spp.
Eremospatha spp.
Korthalsia spp.
Laccosperma spp.
Myrialepis paradoxa
Oncocalamus spp.
Plectocomia spp.
Plectocomiopsis spp.
Retispatha dumetosa

LANDSCAPE LISTS

Large Species

These are very tall palms that normally attain a total height of 60 or more feet (18 m) or are otherwise massive in their proportions.

Astrocaryum chambira
Attalea spp. (most)
Beccariophoenix spp.
Bismarckia nobilis
Borassus spp.
Carpoxylon macrospermum
Caryota kiriwongensis
Caryota maxima
Caryota no
Caryota obtusa
Caryota rumphiana
Caryota urens
Ceroxylon spp. (most)
Cocos nucifera
Copernicia alba
Copernicia baileyana
Copernicia fallaensis
Copernicia gigas
Corypha spp.
Deckenia nobilis
Dypsis bejofo
Dypsis carlsmithii
Dypsis hovomantsina
Dypsis malcomberi
Dypsis mananjarensis
Dypsis pilulifera
Dypsis tokoravina
Dypsis tsaravoasira
Elaeis guineensis
Eugeissona utilis
Hydriastele costata
Hydriastele cylindrocarpa
Hydriastele longispatha
Hydriastele microcarpa
Hyphaene spp.
Iriartea deltoidea
Jubaea chilensis
Kentiopsis magnifica
Kentiopsis oliviformis
Lemurophoenix halleuxii
Livistona australis
Livistona carinensis
Livistona concinna
Livistona drudei
Livistona jenkinsiana
Livistona mariae
Livistona nitida
Livistona saribus
Lodoicea maldivica
Manicaria saccifera
Marojejya spp.
Mauritia flexuosa
Metroxylon spp.
Oenocarpus bataua
Oncosperma horridum
Oncosperma tigillarium
Orania sylvicola
Orania trispatha
Phoenix canariensis
Phoenix dactylifera
Phoenix reclinata
Phoenix sylvestris
Pholidocarpus spp.
Phytelephas aequatorialis
Pigafetta spp.
Pritchardia schattaueri
Pritchardia waialealeana
Ravenea rivularis
Ravenea robustior
Ravenea sambiranensis
Rhopaloblaste augusta
Roystonea spp.
Sabal causiarum
Sabal domingensis
Sabal mauritiiformis
Saribus rotundifolius
Syagrus sancona
Tahina spectabilis
Veitchia arecina
Veitchia joannis
Washingtonia robusta

Small Species

These palms are generally under 10 feet (3 m) total height and are not of great widths even if clustering.

Aiphanes ulei (some forms)
Allagoptera spp.
Areca guppyana
Areca ipot
Areca minuta
Arenga brevipes
Arenga caudata
Arenga hastata
Arenga hookeriana
Arenga porphyrocarpa
Asterogyne martiana
Bactris hondurensis
Balaka spp.
Barcella odora
Basselinia gracilis (some forms)
Basselinia vestita
Brahea decumbens
Brahea moorei
Butia archeri
Butia campicola
Butia microspadix
Calyptrocalyx arfakiensis
Calyptrocalyx awa
Calyptrocalyx doxanthus
Calyptrocalyx elegans
Calyptrocalyx hollrungii
Calyptrocalyx leptostachys
Calyptrocalyx micholitzii
Calyptrocalyx pachystachys
Calyptrocalyx pauciflorus
Calyptrocalyx polyphyllus
Calyptrogyne ghiesbreghtiana
Chamaedorea spp. (most)
Chuniophoenix humilis
Chuniophoenix nana
Copernicia cowellii
Dypsis beentjei
Dypsis bosseri
Dypsis catatiana
Dypsis coriacea
Dypsis hildebrandtii
Dypsis louvelii
Dypsis pachyramea
Dypsis remotiflora
Dypsis sanctaemariae
Dypsis simianensis
Geonoma cuneata
Geonoma deversa
Geonoma epetiolata
Guihaia spp.
Hemithrinax ekmanii
Heterospathe delicatula
Heterospathe longipes
Heterospathe philippinensis
Heterospathe scitula
Hyospathe elegans
Iguanura spp.
Iriartella spp.
Lepidocaryum tenue
Lepidorrhachis mooreana
Licuala spp. (most)
Linospadix spp.
Lytocaryum spp.
Maxburretia furtadoana
Neonicholsonia watsonii
Phoenix acaulis
Phoenix loureiroi
Phoenix roebelenii
Pholidostachys spp.
Pinanga spp. (many)
Ravenea hildebrandtii
Reinhardtia gracilis
Reinhardtia simplex
Rhapidophyllum hystrix
Rhapis excelsa
Rhapis laosensis
Rhapis multifida
Rhapis subtilis
Sabal etonia
Sabal minor
Sommieria leucophylla
Syagrus glaucescens
Syagrus harleyi
Syagrus schizophylla
Syagrus vagans
Trachycarpus nanus
Wallichia caryotoides
Wallichia densiflora
Wallichia siamensis
Wendlandiella gracilis

Species Tolerant of Alkaline Soil

This category is a tricky one. Almost any palm can be made to survive in almost any soil type with heroic efforts of amendment, irrigation, and fertilization. For example, *Caryota* species generally need a fairly rich, humus-laden soil that is slightly acidic, and yet peninsular Florida, where the soil is slightly to quite alkaline, is full of most of these species. The species listed here thrive with little to moderate, but not heroic, adjustments.

Acrocomia spp.
Adonidia merrillii
Allagoptera spp.
Aiphanes minima
Archontophoenix myolensis
Arenga pinnata
Arenga undulatifolia
Astrocaryum mexicanum
Astrocaryum standleyanum
Bismarckia nobilis
Borassodendron machadonis
Borassus spp.
Brahea spp.
Burretiokentia vieillardii
Butia spp.
Calyptronoma rivalis
Carpentaria acuminata
Caryota spp.
Chamaedorea glaucifolia
Chamaedorea graminifolia
Chamaedorea microspadix
Chamaedorea nationsiana
Chamaedorea neurochlamys
Chamaedorea oreophila
Chamaedorea sartorii
Chamaerops humilis
Chuniophoenix hainanensis
Coccothrinax spp.
Cocos nucifera
Copernicia spp.
Corypha spp.
Cryosophila spp.
Dictyosperma album
Dypsis cabadae
Dypsis decaryi
Dypsis lanceolata
Dypsis leptocheilos
Dypsis lutescens
Dypsis madagascariensis
Gaussia spp.
Guihaia spp.
Heterospathe elata
Hyophorbe spp.
Hyphaene spp.
Jubaea chilensis
Kentiopsis oliviformis
Latania spp.
Leucothrinax morrisii
Licuala spinosa
Livistona australis
Livistona benthamii
Livistona carinensis
Livistona chinensis
Livistona decipiens
Livistona endauensis
Livistona fulva
Livistona jenkinsiana
Livistona mariae
Livistona muelleri
Livistona nitida
Livistona rigida
Livistona saribus
Maxburretia furtadoana
Medemia argun
Nannorrhops ritchieana
Oncosperma tigillarium
Orania palindan
Phoenix spp.
Pritchardia mitiaroana
Pritchardia pacifica
Pritchardia thurstonii
Pseudophoenix spp.
Ptychosperma spp.
Raphia spp.
Ravenea hildebrandtii
Rhapidophyllum hystrix
Rhapis spp.
Roystonea regia
Sabal spp.
Saribus merrillii
Saribus rotundifolius
Saribus woodfordii
Schippia concolor
Serenoa repens
Syagrus botryophora
Syagrus cearensis
Syagrus coronate
Syagrus ×*costae*
Syagrus schizophylla
Tahina spectabilis
Thrinax spp.
Trachycarpus fortunei
Trachycarpus oreophilus
Trithrinax spp.
Veitchia spp.
Washingtonia spp.
Zombia antillarum

Salt-Tolerant Species

Palms tolerate various degrees of salinity in the soil and air. Those in the following list tolerate salt to varying degrees, from medium to extreme. Those species that occur naturally on coastal cliffs and beaches are most reliably salt-tolerant.

Acoelorrhaphe wrightii
Acrocomia spp.
Adonidia merrillii
Allagoptera spp.
Brahea aculeata
Chamaerops humilis
Coccothrinax spp.
Cocos nucifera
Copernicia brittonorum
Copernicia gigas
Copernicia macroglossa
Copernicia rigida
Hyophorbe lagenicaulis
Hyophorbe verschaffeltii
Hyphaene spp.
Leucothrinax morrisii
Licuala paludosa
Licuala spinosa
Livistona benthamii
Livistona carinensis
Livistona saribus
Medemia argun
Nannorrhops ritchieana
Nypa fruticans
Oncosperma tigillarium
Phoenix canariensis
Phoenix dactylifera
Phoenix loureiroi
Phoenix paludosa
Phoenix pusilla
Phoenix sylvestris
Phoenix theophrasti
Pritchardia hillebrandii
Pritchardia pacifica
Pritchardia thurstonii
Pseudophoenix ekmanii
Pseudophoenix sargentii
Sabal spp.
Saribus woodfordii
Serenoa repens
Syagrus amara
Syagrus schizophylla
Thrinax radiata
Veitchia spp.
Washingtonia spp.
Zombia antillarum

Species with Colored New Leaves

The new leaves of some species can be beautifully colored, but the coloration lasts only a few days, until the leaf turns green. Variation exists within each species in the list below. Individuals may vary in their color shade and intensity. Species that show colored new leaves only as seedlings or young plants but not as adults are omitted from the list.

Actinokentia divaricata
Archontophoenix tuckeri
Areca vestiaria (some forms)
Asterogyne martiana
Asterogyne spicata
Burretiokentia koghiensis
Calyptrocalyx spp.
Calyptrogyne spp.
Calyptronoma plumeriana
Ceratolobus spp.
Chambeyronia macrocarpa
Dypsis baronii
Dypsis coriacea
Dypsis crinita
Dypsis fibrosa
Dypsis hildebrandtii
Dypsis lantzeana
Dypsis louvelii
Dypsis marojejyi
Dypsis pinnatifrons
Dypsis remotiflora
Dypsis simianensis
Geonoma densa
Geonoma deversa
Geonoma epetiolata
Heterospathe delicatula
Heterospathe elata
Heterospathe intermedia
Heterospathe longipes
Heterospathe minor
Heterospathe scitula
Heterospathe woodfordiana
Hydriastele affinis
Hydriastele montana
Hydriastele pinangoides
Iguanura bicornis
Iguanura elegans
Iguanura wallichiana
Kentiopsis magnifica
Laccospadix australasicus
Lemurophoenix halleuxii
Livistona fulva
Manicaria saccifera
Metroxylon sagu (some forms)
Metroxylon warburgii
Nephrosperma vanhouttenaum
Oenocarpus spp.
Pholidostachys pulchra
Pinanga aristata
Pinanga caesia
Pinanga copelandii
Pinanga coronata
Pinanga curranii
Pinanga densiflora
Pinanga disticha
Pinanga maculata
Ptychosperma burretianum
Ptychosperma waitianum
Roscheria melanochaetes
Salacca zalacca (some forms)
Verschaffeltia splendida
Welfia regia

Species with Permanently Colored Leaves or Crownshafts

Acanthophoenix rubra
Adonidia merrillii golden cultivar
Allagoptera caudescens
Archontophoenix purpurea
Areca catechu some cultivars
Areca vestiaria
Arenga spp.
Astrocaryum spp.
Basselinia spp.
Bismarckia nobilis
Brahea armata
Brahea decumbens
Brahea moorei
Butia capitata (some forms)
Butia eriospatha
Butia yatay
Calyptrocalyx doxanthus
Caryota ophiopellis
Caryota zebrina
Ceroxylon spp.
Chamaedorea deneversiana
Chamaedorea glaucifolia
Chamaedorea metallica
Chamaerops humilis var. *argentata*
Chambeyronia macrocarpa (some forms)
Chelyocarpus ulei
Chuniophoenix hainanensis
Coccothrinax spp.
Colpothrinax spp.
Copernicia alba
Copernicia cowellii
Copernicia fallaensis
Copernicia hospita
Copernicia prunifera
Cryosophila spp.
Cyphokentia cerifera
Cyphophoenix alba
Cyrtostachys renda
Dypsis albofarinosa
Dypsis baronii
Dypsis hovomantsina
Dypsis lastelliana
Dypsis lutescens
Dypsis leptocheilos
Dypsis mananjarensis
Dypsis onilahensis
Dypsis paludosa
Dypsis pilulifera
Dypsis rivularis
Dypsis species 'Pink Crownshaft'
Euterpe catinga
Euterpe edulis
Euterpe oleracea (some forms)
Euterpe precatoria
Geonoma epetiolata
Guihaia argyrata
Hyphaene spp.
Itaya amicorum
Johannesteijsmannia magnifica
Kerriodoxa elegans
Latania loddigesii
Latania lontaroides
Leucothrinax morrisii
Lemurophoenix halleuxii
Licuala mattanensis 'Mapu'
Licuala radula
Mauritiella armata
Nannorrhops ritchieana
Nypa fruticans
Oncosperma gracilipes
Orania ravaka
Orania sylvicola
Orania trispatha
Oraniopsis appendiculata
Phoenix sylvestris
Pigafetta filaris
Pinanga adangensis
Pinanga aristata
Pinanga caesia
Pinanga copelandii
Pinanga coronate
Pinanga crassipes
Pinanga densiflora
Pinanga disticha
Pinanga maculata
Pinanga negrosensis
Pinanga scortechinii
Pinanga speciosa
Pinanga watanaiana
Pritchardia hillebrandii (some forms)
Pseudophoenix ekmanii
Ravenea albicans
Ravenea glauca
Ravenea xerophila
Sabal uresana
Salacca magnifica
Salacca zalacca
Satakentia liukiuensis
Serenoa repens (silver forms)
Sommieria leucophylla
Trachycarpus gemisectus
Trachycarpus princeps
Trithrinax campestris
Trithrinax schizophylla
Veitchia vitiensis
Wallichia densiflora
Wallichia siamensis

INTERNET SOURCES FOR INFORMATION ON PALMS

European Network of Palm Specialists
EUNOPS is an association of top palm researchers from around the world. Their website lists meetings and research projects taking place. The Glossary of Palm Terms is an extremely useful feature. http://eunops.org/content/glossary-palm-terms

Fairchild Guide to Palms
Fairchild Tropical Botanic Garden in Miami, Florida, has a searchable database of its many palm collections (living, photographs, herbarium specimens, and even DNA). The palm photograph gallery of palms growing at Fairchild is one of the best. www.palmguide.org

The International Palm Society
The IPS is a non-profit organization with members from over 80 countries. Affiliate societies in various parts of the US and other parts of the world hold meetings and functions where members can learn more about palms. Among the features to be found is PalmTalk, an interactive forum where members share photos and information and answer questions, as well as links to other affiliate societies with more information about palms. www.palms.org

The Palm Nut Pages
A new website by Paul Craft, the site offers photos, accepted names and synonyms, and horticultural information. www.palmnutpages.com

Royal Botanic Gardens, Kew
RBG Kew in England boasts the world's largest herbarium and library for palms, and is a Mecca for palm research. The World Checklist of Selected Plant Families is an excellent resource for finding out the accepted palm names, synonyms, and correct spelling. http://apps.kew.org/wcsp/home.do

An associated site is Palms of the World On-Line, which has a checklist of accepted names, synonyms, photographs, maps, and links to original publications in which palm genera and species were described in the botanical literature. This site is being constructed as a "Species Palmarum" in which every palm species known to science will be described and illustrated. www.palmweb.org

Tropicos, Missouri Botanical Garden
Tropicos is an excellent search feature for finding distribution and nomenclature data for any species. Images of herbarium specimens are also available. www.tropicos.org

University of Florida IFAS, Research and Education Center
The University of Florida has done a great deal of study on palm nutrition, diseases, pests, production and maintenance. Their website is a wealth of information on all these subjects: http://flrec.ifas.ufl.edu/palm_prod/palm_production.shtml

BOTANICAL GARDENS AND PUBLIC COLLECTIONS WITH SIGNIFICANT PALM COLLECTIONS

Australia

Flecker Botanic Gardens
Cairns, Queensland 4870
www.anbg.gov.au/chabg/bg-dir/044.html

George Brown Darwin Botanic Gardens
Palmerston, Northern Territory 0831
www.nt.gov.au/nreta/parks/botanic/index.html

The Palmetum
Townsville, Queensland 4810
www.anbg.gov.au/chabg/bg-dir/107.html

Royal Botanic Gardens
Mrs. Macquaries Road
Sydney, New South Wales 2000
www.rbgsyd.nsw.gov.au

Belize

Belize Botanic Gardens
P.O. Box 180
San Ignacio, Cayo
http://belizebotanic.org

Brazil

Jardim Botânico do Rio de Janeiro
Rio de Janeiro
www.jbrj.gov.br

China

South China Botanical Garden
Chinese Academy of Sciences
723 Xingke Road, Tianhe District
Guangzhou
Guangdong 510650
www.scib.ac.cn

Xishuangbanna Tropical Botanical Garden
Menglun, Mengla
Xishuangbanna
Yunnan 666303
www.xtbg.cas.cn

Colombia

Jardíns Botánico José Celestino Mutis
Avenida Calle 63 No. 68-95
Bogotá D.C.
www.jbb.gov.co/jardinbotanico

Costa Rica

The Robert and Catherine Wilson Botanical Garden
PO Box 73-8257
Sanvito, Coto Brus
Las Cruces
www.ots.ac.cr

Cuba

Jardín Botánico de Cienfuegos
Cienfuegos
Website with more information:
www.arboretum.harvard.edu/library/image-collection/cienfuegos-botanical-garden-cuba

Jardín Botánico de Las Tunas
Carretera del Cornito Km. 1
Las Tunas

Jardín Botánico Nacional
Havana
www.uh.cu/centros/jbn

Dominica

Roseau Botanic Gardens
Ministry of Agriculture
Forestry Division
Roseau
http://da-academy.org/dagardens.html

Dominican Republic

Jardín Botánico Nacional Dr. Rafael M. Moscoso
Apdo Postal 21-9
Santo Domingo
www.jbn.gob.do

France

Jardin d'Oiseaux Tropicaux
Route de Valcros - RD 559
83250 La Londe-les-Maure
www.jotropico.org

Jardins Botanique E. M. Heckel
48 Avenue Clot Bey
13272 Marseille, Cedex 8

Germany

Palmengarten der Stadt Frankfurt am Main
Siesmayerstrasse 61
60323 Frankfurt am Main
www.palmengarten-frankfurt.de

Indonesia

Bogor Botanic Garden
13 Jalan Otto Iskandardinata—Paledang
Bogor
www.bogor.indo.net.id/kri

Italy

Orto Botanico "Giardino dei Semplici"
Museo di Storia Naturale
Universita degli Studi di Firenze
Via P.A. Micheli 3
50121 Florence

Malaysia

Forest Research Institute Malaysia
52110 Kepong
Selangor
www.frim.gov.my

Rimba Ilmu Botanic Garden
University of Malaya
50603 Kuala Lumpur
http://rimba.um.edu.my

Philippines

Makiling Botanic Gardens
College of Forestry
University of the Philippines Los Banos
Laguna 4031
http://cal-nu.laguna.net/mcme

Portugal

Jardim Botânico da Universidade de Lisboa
Rua Escola Politécnica 58
Lisbon
Lisbon 1269-102
www.jb.ul.pt

Jardim Botânico Tropical
Largo de Belém
Lisbon 1400-209
www2.iict.pt/jbt

Parques de Sintra—Monte da Lua S.A.
Parque de Monserrate
Sintra 2710-405
www.parquesdesintra.pt

Singapore
Singapore Botanic Gardens
National Parks Board
1 Cluny Road
Singapore City 259569
www.sbg.org.sg

Spain
Jardín de Aclimatación de la Orotava
La Paz
38400 Puerto de la Cruz
Tenerife, Canary Islands

Palmetum
Parque Marítimo César Manrique
Santa Cruz de Tenerife
Tenerife, Canary Islands

Sri Lanka
Royal Botanic Gardens
Peradeniya

Thailand
Nong Nooch Tropical Botanical Garden
Pattaya, Chonburi Province
www.nongnoochgarden.com/home.html

United Kingdom
Royal Botanic Gardens, Kew
Richmond, Surrey TW9 3AB
www.kew.org

United States—California
Balboa Park
1549 El Prado
Balboa Park
San Diego, CA 92101
www.balboapark.org

Gana Walska Lotusland
Note: Lotusland is open by appointment only.
695 Ashley Road
Santa Barbara, CA 93108
www.lotusland.org

Huntington Botanical Gardens
1151 Oxford Road
San Marino, CA 91108
www.huntington.org

Lakeside Palmetum of Oakland
Lakeside Park Garden Center
666 Bellevue Avenue
Oakland, CA 94610
www.palmsnc.org/pages/palmetum.php

San Diego Botanic Garden
(formerly Quail Botanical Gardens)
230 Quail Gardens Drive
Encinitas, CA 92024
www.sdbgarden.org

San Diego Zoo Safari Park
(formerly Wild Animal Park)
15500 San Pasqual Valley Road
Escondido, CA 92027
www.sandiegozoo.org

San Francisco Botanical Garden at Strybing Arboretum
9th Avenue at Lincoln Way
San Francisco, CA 94122
www.sfbotanicalgarden.org

University of California Botanical Garden at Berkeley
200 Centennial Drive
Berkeley, CA 94720
www.botanicalgarden.berkeley.edu

United States—Florida
Ann Norton Sculpture Gardens
253 Barcelona Road
West Palm Beach, FL 33401
www.ansg.org

Fairchild Tropical Botanic Garden
10901 Old Cutler Road
Coral Gables, FL 33156
www.fairchildgarden.org

Flamingo Gardens
3750 South Flamingo Road
Davie, FL 33330
www.flamingogardens.org

Gizella Kopsick Palm Arboretum
901 North Shore Drive
St. Petersburg, FL 33713
www.stpete.org/parks/palm.asp

Harry P. Leu Gardens
1920 North Forest Avenue
Orlando, FL 32803
www.leugardens.org

Montgomery Botanical Center
Note: MBC is open by appointment only.
11901 Old Cutler Road
Coral Gables, FL 33156
www.montgomerybotanical.org

United States—Hawaii
Foster Botanical Garden
50 North Vineyard Boulevard
Honolulu, HI 96817
www.honolulu.gov/parks/hbg/fbg.htm

Harold L. Lyon Arboretum
University of Hawai'i-Manoa
3860 Manoa Road
Honolulu, HI 96822
www.hawaii.edu/lyonarboretum

Ho'omaluhia Botanical Garden
45-680 Luluku Road
Kaneohe, HI 96744
www.honolulu.gov/parks/hbg/hmbg.htm

Koko Crater Botanical Garden
at the end of Kokonani Street
Honolulu, HI 96825
www.co.honolulu.hi.us/parks/hbg/kcbg.htm

Pana'ewa Rainforest Zoo and Gardens
800 Stainback Highway
Hilo, HI 96720
www.hilozoo.com

Waimea Valley Audubon Center
59-864 Kamehameha Highway
Haleiwa, HI 96712

United States—Texas
Moody Gardens
One Hope Boulevard
Galveston, TX 77554
www.moodygardens.com

Venezuela
Jardín Botánico de Caracas
Caracas
www.fibv.org.ve/jardin

Palm Trunks

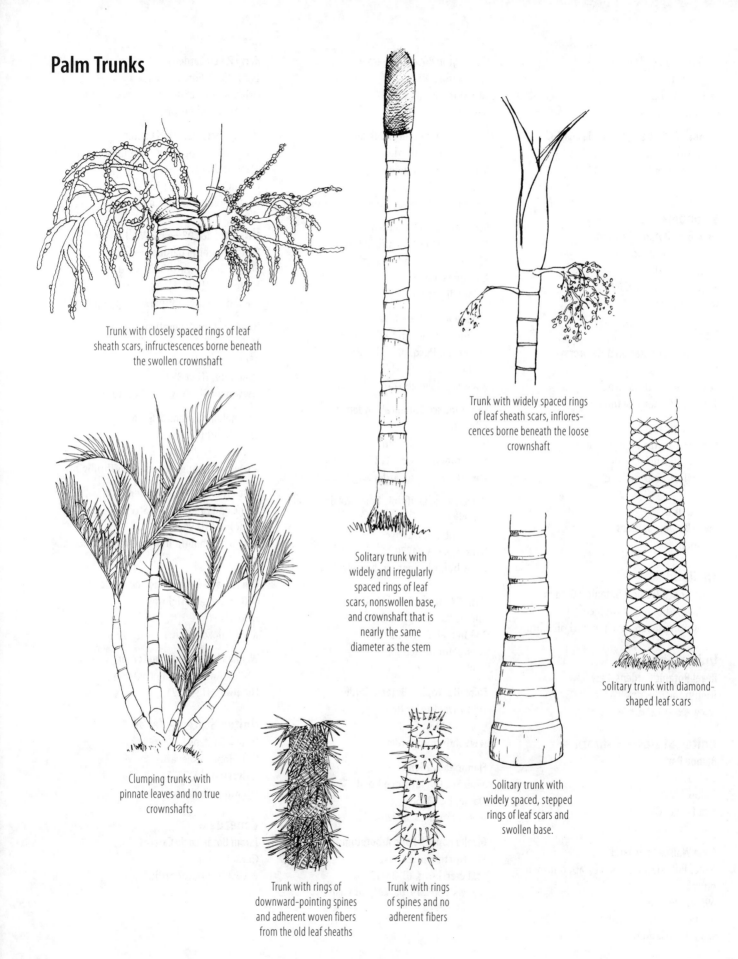

Palm Leaves

Pinnate leaf with closely spaced leaflets near the base of the leaf and widely spaced leaflets at the apex

Pinnate leaf with groups of partially fused irregularly shaped leaflets

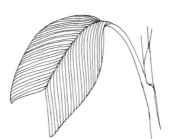

Unsegmented pinnate leaf with deeply bifid apex

Linear-acuminate leaflets

Plumosely pinnate leaf

Bipinnate leaf of *Caryota* species

Irregularly diamond-shaped leaflet with toothed apical margin

Linear-oblong leaflet with toothed apical margin

Nearly circular, unsegmented palmate leaf

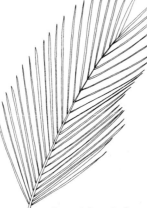

Narrowly linear leaflets

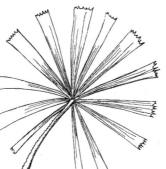

Circular, segmented palmate leaf with apically toothed segments

Ovate-acuminate leaflets

Diamond-shaped, alternating leaflets

Strongly costapalmate leaf

Linear-obdeltoid leaflet with toothed apical margin

Palmate leaf with apically bifid segments and small hastula

505

GLOSSARY

Acanthophyll A leaf or leaflet modified into a spine, such as is found at the base of the pinnate leaf of *Phoenix* species.

Acaulescent Without an aboveground stem or trunk.

Acuminate Tapering gradually to a narrow point or tip.

Adjacent germination A type of germination in which the radicle and shoot are formed adjacent to the palm seed. The contrasting condition is *remote germination*.

Anther The pollen-bearing structure of a male or bisexual flower.

Anthesis The period in the life of a flower during which the pollen is shed or the stigma is receptive to the pollen.

Auricle An earlike lobe, usually a part of the leaf sheath or petiole.

Axil The angle formed by the stem and the petiole.

Bifid Divided into 2 usually equal parts, most often used to describe the apex of a leaf.

Binomial A scientific name consisting of 2 parts: a genus name (also called generic name) and the specific epithet.

Bipinnate A pinnate leaf whose primary leaflets have been replaced by separate and smaller stalks that then bear leaflets.

Bisexual Said of flowers having both male and female parts; the contrasting condition is *unisexual*.

Boots The dead and usually split and woody leaf bases (or sheaths) adhering to a palm's stem or trunk, such as are common in the genus *Sabal*.

Bract A structure (botanically, it is a modified leaf) borne below a flower or an inflorescence. Palm inflorescences have one or more bracts, which may be woody, leathery or papery in texture.

Canopy-scape A term used in this volume to describe the use of a palm in the landscape wherein a palm's crown extends above adjacent or lower vegetation and, from any distance, is the primary if not the sole visual manifestation of the palm's whereabouts.

Carpel The ovule-bearing units that make up the pistil in a female or bisexual flower. Most palms have more than one carpel, but in many palms the carpels are united into a single pistil.

Caulescent Having an aboveground stem or trunk.

Cirrus, Cirri A modified and extended leaf rachis that allows a climbing palm to attach itself to a host.

Costa The projection of the petiole into the leaf blade (or lamina) of a costapalmate-leaved palm species; sometimes imprecisely called rib or midrib in the literature.

Costapalmate Said of palmate leaf with a discernible costa.

Cotyledon The first emerging leaf of a seedling; sometimes called the seed leaf.

Crownshaft A term used only with pinnate-leaved palm species to denote a tubular or cylindrical shaft above the woody part of the trunk; the tube is more or less columnar and consists of the expanded and tightly packed leaf bases (or sheaths) of the leaves presently on the palm.

Cultivar A cultivated variety of a species.

Deltoid Triangular shaped, with the broader part near the point of attachment.

Dichotomous Having a succession of two-forked divisions.

Dicot, Dicotyledon Said of a seed that produces seedlings with 2 seed leaves (cotyledons) as opposed to those of monocots (monocotyledon), which produce seedlings with only one seed leaf. Dicots have woody stems that increase girth over time; monocot stems do not increase in girth. In addition, the flowers of dicot (dicotyledonous) plants have their parts in 4s or 5s as opposed to those of monocots whose flower parts are in 3s or multiples of 3.

Dioecious Having male (staminate) and female (pistillate) flowers on separate plants. The term is from 2 Greek words meaning "two houses."

Distichous Arranged into 2 ranks on opposite sides of an axis or stem.

Endemic Confined to a particular region.

Endosperm The storage tissue in the seed. It can be oily, such as the "meat" of a coconut, or very hard, such as the "ivory" of a *Phytelephas* seed.

Epithet The word in a binomial that follows the genus name and denotes the species.

Falcate Sickle shaped.

Flagellum, Flagella A long, whiplike or tail-like modified and sterile inflorescence unique to the genus *Calamus* and used by the plant to climb a host.

Genus, Genera A collection of plants bearing similar characters. The taxonomic category that ranks below family and above species.

Glabrous Smooth, not hairy.

Glaucous Covered with a whitish or bluish bloom or a waxy or powdery substance that is easily rubbed away.

Hapaxanthic Of stems or trunks that flower and fruit and then die. Also called monocarpic when only one inflorescence is produced, as in *Corypha*.

Hastula A flap of tissue on some palmate leaves. This flap is variously shaped and protrudes from the point at which the petiole joins the leaf blade. The protruding organ may be tiny or large and may be found on either the upper or lower surface of the leaf, or on both surfaces.

Hermaphroditic Bearing bisexual flowers.

Homogeneous Said of endosperm when it is not intruded by the seed coat. Examples of homogeneous endosperm can be seen in the seeds of *Cocos nucifera*, *Dypsis lutescens*, *Phoenix canariensis*, and *Roystonea regia*. The contrasting condition is *ruminate*.

Induplicate The folding (plication) of the leaflet so that the margins are turned upwards and the leaflet has a V shape in cross-section. The contrasting condition is *reduplicate*.

Inflorescence The branch (and all accompanying bracts and subsidiary branches) that bears the flowers; the flower-bearing structure.

Infructescence The cluster of fruits formed on an inflorescence.

Intergeneric Between 2 genera; said of hybrids.

Lamina The blade of a leaf. In palms, the lamina may comprise several leaflets (as in pinnate leaves) or segments (as in palmate leaves) or a single expanse of tissue (as in undivided leaves).

Lanceolate Lance shaped. Having a shape that is longer than wide, with the widest part near the point of attachment.

Leaf base The lowest or bottom-most part of a leaf; a widened (often highly so) portion of the bottom of the leaf's petiole. Also called leaf sheath or sheath.

Leaflet The individual segment of a pinnate leaf.

Marcescent Withering but not falling. Used to describe leaves that persist after dying and hang (often in a skirt) beneath the crown of living leaves.

Mesic Of average rainfall, being neither wet nor arid.

Midrib The main vein(s) of a leaf. Also called midvein.

Midvein The main vein(s) of a leaf. Also called midrib.

Mogote A steep hill or small mountain of porous limestone in Cuba, full of solution holes and crevasses.

Monocarpic Of stems or trunks that flower and fruit, once (usually terminally) and then die. See also hapaxanthic.

Monocot, Monocotyledon Said of a seed that produces seedlings with only one seed leaf (cotyledon) as opposed to those of dicots, which produce seedlings with 2 seed leaves or cotyledons. Monocots do not usually produce woody tissues, as opposed to dicots, and, except for palms and a few other large monocots like the dragon tree (*Dracaena draco*), do not result in tree forms. In addition, the flowers of monocotyledonous plants have their parts in 3s or multiples of 3 as opposed to dicots whose flower parts are in 2s, 4s, 5s, or multiples thereof.

Monoecious Having separate male and female flowers on a single plant.

Monotypic Of one kind, such as a genus with a single species.

Node The point on a stem from which a leaf, a group of leaves, or an inflorescence grows. In palm species these points are usually also indicated by differently colored or differently textured rings (complete or incomplete) along the older parts of the stem or trunk.

Obdeltoid Triangular shaped, with the narrow end near the point of attachment.

Obovate Egg shaped with the narrow end at the point of attachment.

Obtuse Blunt. Not sharp or pointed.

Palmate Shaped like a splayed hand. Said of a compound leaf with leaflets arranged from a common (fanlike) point.

Panicle A branched and elongated inflorescence of flowers.

Peduncle The primary (basal) stem of an inflorescence.

Peltate Of a leaf or other structure attached to a petiole or stalk at its lower surface rather than to any part of its margin; the stalk is usually attached to the center of the leaf's underside and the leaf is usually rounded, like an umbrella.

Perfect Having male and female parts in the same flower.

Petiole The primary stalk of a simple or compound leaf. In palm species the stalk beneath the bottom-most leaves or leaflets, and above the leaf sheath or leaf base.

pH A measure of the alkalinity or acidity of a substance on a scale from 0 (most acid) to 14 (most alkaline), with 7 being neutral. Sulfuric acid has a pH of 0, while pure lye has a pH of 14. Most soils have a pH value of 5–9. Acid-loving plants usually need a value no less than 6, and alkaline-loving plants usually want one that is no higher than 8. Under all circumstances, organic or natural amendments to the soil are better than temporary chemical amendments; the organic amendments last longer and do not raise or lower the pH value suddenly and unnaturally.

Pinna, Pinnae A leaflet of a pinnate leaf.

Pinnate Featherlike. Of a leaf with more than 3 leaflets growing from the central axis (rachis). The leaflets may grow on opposite sides of a rachis in a single flat plane or they may grow at angles off the rachis to create a plumose effect.

Pistil The female reproductive organ of a flower consisting of the ovary, the style, and the stigma. The ovary is the portion that develops into the fruit. The style is usually a relatively short extension of the ovary atop which the stigma resides. The stigma receives pollen and stimulates the pollen grains to germinate and grow towards—by means of a pollen tube—the ovules in the ovary. The pollen cell unites with the egg cell in the ovule, which then becomes the embryo of the developing seed.

Pistillate Having a pistil. Said of a flower that is female or in the female phase.

Plicate Folded at least to some extent. Also called pleated.

Plumose Featherlike with the segments arising from the midrib in more than one plane and resulting in a bottlebrush appearance.

Pneumatophore A specialized root that grows from normal subterranean roots but which rises above the water or soil surface and serves to aerate the subterranean or subaquatic root system. Few palms produce such roots, but when they do, the roots are usually contiguous with the stem or trunk or nearly so.

Polygamous Having male and female flowers on the same plant, along with bisexual flowers (polygamo-monoecious) or male and female flowers on separate plants but also accompanied by bisexual flowers (polygamo-dioecious).

Prop roots Large, aerial roots at or near the base of a palm's trunk or stem, which usually provide support and stability to the tree and which are usually in the form of a cone. Also called stilt roots.

Pseudo-petiole In some palms, the margins of the leaf sheath can disintegrate in its upper portions to form a petiole-like stalk that is sometimes called a pseudo-petiole. It may intergrade and be nearly indistinguishable from a true petiole, so the distinction is of little value outside of botanical circles.

GLOSSARY

Rachilla, Rachillae The branches of an inflorescence that bear the flowers.
Rachis The primary and central stem of a compound leaf from which leaflets or subsidiary leafstalks arise, being above the petiole.
Radicle The first root formed by a sprouting seed.
Reduplicate The folding (plication) of the leaflet so that the margins are turned downwards and the leaflet has an inverted V shape in cross-section. The contrasting condition is *induplicate*.
Reins The threadlike pieces of leaf tissue that unite the tips of all the leaflets and are shed as the (usually pinnate) leaf unfolds. They medium times persist and dangle from the tips of the lowermost leaflets, one on each side of the leaf. In some palm species (for example, *Dypsis decaryi*), the reins last as long as the living leaf.
Remote germination A type of germination in which the embryo first forms a stalk that may extend for some distance from the palm seed and, at the end of which, the radicle and first shoot are formed. The contrasting condition is *adjacent germination*.
Reniform Kidney shaped.
Rheophyte A terrestrial plant adapted to grow on the banks of fast-moving rivers or streams; such plants can survive seasonal flooding. Literally from the Greek "stream flow" and "plant."
Rhizomatous Having rhizomes.
Rhizome An underground or on-ground stem growing horizontally and giving rise to roots, stems, and leaves at its nodes or growing tips.
Root pruning A technique used for transplanting palms that consists of digging a trench around the root perimeter of the palm some months in advance of the transplanting, so that the cut roots will, to some extent, regenerate and heal with the benefit of the surrounding soil before the plant is moved to its new location.
Ruminate Said of endosperm that is intruded or divided by ingrowths of the seed coat. Examples ruminate endosperm can be seen in the seeds of *Adonidia merrillii*, *Areca catechu*, *Chamaerops humilis* and *Ptychosperma elegans*. The contrasting condition is *homogeneous*.
Saxophone growth The seedling growth form of some palm species (for example, *Sabal*) in which the stem first grows downwards and then turns upwards to emerge aboveground. The resulting aerial stem often remains underground for many months or even years before emerging aboveground.
Serrate Sawlike; having a toothed margin.
Sheath The lowest or bottom-most part of a leaf; a widened (often highly so) portion of the bottom of the leaf's petiole where it clasps the stem. Also called leaf base or leaf sheath.
Spathe A large and usually woody bract which originally covers a palm inflorescence and from which the inflorescence later emerges. Also called peduncular bract.
Spatulate Spatula shaped; a modified oblong shape with the apex larger than the tapering base; an exaggerated obovate shape.
Spine A sharp protrusion on a leaf, branch, or stem. Used in this volume in its broad, nontechnical sense.
Stamen The male reproductive organ of a flower, consisting of the filament and the anther.
Staminate Having a stamen. Said of a flower that is male or in the male phase.
Stigma The part of the pistil that receives pollen grains and initiates their germination. In palms, it is always found of the apical (terminal) end of the pistil.
Stilt roots Large, aerial roots at or near the base of a palm's trunk or stem, which usually provide support and stability to the tree and which are usually in the form of a cone. Also called prop roots.
Stolon A rootlike stem that creeps along the surface of the soil and roots at specific nodes, creating new plants that are genetically identical to the parent plant.
Subtend To grow directly beneath.
Taxon, Taxa A taxonomic category of any rank, such as genus or species.
Tomentose Having tomentum.
Tomentum A covering of short, densely matted hairs.
Unisexual Said of single-sex flowers having either male or female parts; the contrasting condition is *bisexual*.
Xeric Of, from, or adapted to dry habitats.

BIBLIOGRAPHY

Bacon, C. D., and W. J. Baker. 2011. *Saribus* resurrected. *Palms* 55: 109–116.

Bailey, L. H. 1963. *The Standard Cyclopedia of Horticulture*. New ed. New York: Macmillan.

Bailey, L. H., and E. Z. Bailey. 1964. *Hortus Second*. New ed., rev. and reset. New York: Macmillan.

Bailey, L. H., and E. Z. Bailey. 2000. *Hortus Third*. New ed. New York: Barnes and Noble.

Baker, W. J., and I. Hutton. 2006. *Lepidorrhachis*. *Palms* 50: 33–38.

Banka, R., and W. J. Baker. 2004. A monograph of the genus *Rhopaloblaste*. *Kew Bulletin* 59: 47–60.

Barfod, A., and L. G. Saw. 2002. The genus *Licuala* (Arecaceae, Coryphoideae) in Thailand. *Kew Bulletin* 57: 827–852.

Barrow, S. 1994. In search of *Phoenix roebelenii*: the Xishuangbanna palm. *Principes* 38: 177–181.

Barrow, S. 1998. A monograph of *Phoenix* L. (Palmae: Coryphoideae). *Kew Bulletin* 53: 513–575.

Basu, S. K., and R. K. Chakraverty. 1994. *A Manual of Cultivated Palms in India*. Calcutta: Botanical Survey of India.

Bernal, R. 1998. The growth form of *Phytelephas seemannii*: a potentially immortal solitary palm. *Principes* 42: 15–23.

Bernal, R., and G. Galeano. 2010. Notes on *Mauritiella*, *Manicaria*, and *Leopoldinia*. *Palms* 54: 110–132.

Bernal, R., G. Ramírez, and R. I. Morales. 2001. Notes on the genus *Ammandra*. *Palms* 45: 123–126.

Blombery, A., and T. Rodd. 1989. *Palms of the World: Their Cultivation, Care, and Landscape Use*. London: Angus and Robertson.

Borchsenius, F., and R. Bernal. 1996. *Aiphanes*. *Flora Neotropica* 70: 1–95.

Borchsenius, F., H. Borgtoft Pedersen, and H. Balslev. 1998. *Manual to the Palms of Ecuador*. AAU Reports 37. Aarhus, Denmark: Aarhus University Press.

Boyer, K. 1992. *Palms and Cycads Beyond the Tropics*. Queensland, Australia: Palm and Cycad Societies of Australia.

Braun, A. n.d. *Las Palmas Cultivadas en Ciudades elevadas de la Parte Andina de América del Sur*. Caracas, Venezuela: August Braun.

Braun, A. 1995. *Las Palmas de las Sabanas de Venezuela*. Caracas, Venezuela: August Braun.

Broschat, T. K., and A. W. Meerow. 2000. *Ornamental Palm Horticulture*. Gainesville, Florida: University Press of Florida.

Corner, E. J. H. 1966. *The Natural History of Palms*. London: Weidenfeld and Nicholson.

Dowe, J. L. 1989a. *Palms of the South-West Pacific*. Queensland, Australia: Palm and Cycad Societies of Australia.

Dowe, J. L., ed. 1989b. *Palms of the Solomon Islands*. Queensland, Australia: Palm and Cycad Societies of Australia.

Dowe, J. L. 2009. A taxonomic account of *Livistona* R. Br. (Arecaceae). *Gardens' Bulletin Singapore* 60: 185–344.

Dowe, J. L. 2010. *Australian Palms: Biogeography, Ecology and Systematics*. Collingwood, Australia: CSIRO Publishing.

Dowe, J. L., and P. Cabalion. 1996. A taxonomic account of Arecaceae in Vanuatu, with descriptions of three new species. *Australian Systematic Botany* 9: 1–60.

Dowe, J. L., and M. D. Ferrero. 2001. A revision of *Calyptrocalyx* and the New Guinea species of *Linospadix* (Linospadicinae: Arecoideae: Arecaceae). *Blumea* 46: 207–251.

Dransfield, J. 1982. A day on the Klingklang Range. *Principes* 26: 19–33.

Dransfield, J. 1986. *Palmae (Flora of Tropical East Africa)*. Rotterdam/Boston: A. A. Balkema.

Dransfield, J. 1994. What's in a name? *Principes* 38: 145.

Dransfield, J. 1998. *Pigafetta*. *Principes* 42: 34–40.

Dransfield, J., and H. Beentje. 1995. *The Palms of Madagascar*. Kew: Royal Botanic Gardens and the International Palm Society.

Dransfield, J., B. Leroy, X. Metz, and M. Rakotoarinivo. 2008. *Tahina*: a new palm genus from Madagascar. *Palms* 52: 31–39.

Dransfield, J., N. W. Uhl, C. A. Asmussen, W. J. Baker, M. M. Harley, and C. E. Lewis. 2008. *Genera Palmarum: The Evolution and Classification of Palms*. Richmond, United Kingdom: Kew Publishing.

Dransfield, J., and S. Zona. 1997. *Guihaia* in cultivation: a case of mistaken identities. *Principes* 41: 70–73.

Ellison, D., and A. Ellison. 2001. *Betrock's Cultivated Palms of the World*. Sydney, Australia: UNSW Press.

Essig, F. B. 1978. A revision of the genus *Ptychosperma* Labill. (Arecaceae). *Allertonia* 1: 415–478.

Essig, F. B. 1982. A synopsis of the genus *Gulubia*. *Principes* 26: 159–173.

Evans, R. J. 1995. Systematics of *Cryosophila* (Palmae). *Systematic Botany Monographs* 46: 1–70.

Evans, R. J. 2001. A monograph of *Colpothrinax*. *Palms* 45 (4): 189.

Fairchild, D. 1943. *Garden Islands of the Great East*. New York: Charles Scribner's Sons.

Gibbons, M. 1993. *Palms*. Seacaucus, New Jersey: Chartwell Books.

Gibbons, M., and T. Spanner. 2009. *Trachycarpus takil*: lost and found, for now. *Palms* 53: 96–102.

Glassman, S. F. 1999. A taxonomic treatment of the palm subtribe Attaleinae (tribe Cocoeae). *Illinois Biological Monographs* 59: 1–414.

Goldman, D. H. 1999. Distribution update: *Sabal minor* in Mexico. *Palms* 43: 40–44.

Gunn, B. F. 2004. The phylogeny of the Cocoeae (Arecaceae) with emphasis on *Cocos nucifera*. *Annals of the Missouri Botanical Garden* 91: 505–522.

Guzman, E. D., and E. S. Fernando. 1986. *Philippine Palms*. Vol. 4, *Guide to Philippine Flora and Fauna*. Quezon City, Philippines: JMC Press.

BIBLIOGRAPHY

Hastings, L. 2003. A revision of *Rhapis*, the lady palms. *Palms* 47: 62–78.

Heatubun, C. D. 2002. A monograph of *Sommieria* (Arecaceae). *Kew Bulletin* 57: 599–611.

Heatubun, C. D., W. J. Baker, J. P. Mogea, M. M. Harley, S. S. Tjitrosoedirdjo, and J. Dransfield. 2009. A monograph of *Cyrtostachys* (Arecaceae). *Kew Bulletin* 64: 67–94.

Henderson, A. 1995. *The Palms of the Amazon*. New York: Oxford University Press.

Henderson, A. 2000. *Bactris* (Palmae). *Flora Neotropica* 79: 1–181.

Henderson, A. 2002. *Evolution and Ecology of Palms*. Bronx: New York Botanical Garden.

Henderson, A. 2007. A revision of *Wallichia* (Palmae). *Taiwania* 52: 1–11.

Henderson, A. 2009. *Palms of Southern Asia*. Princeton, New Jersey: Princeton University Press.

Henderson, A. 2011. A revision of *Geonoma* (Arecaceae). *Phytotaxa* 17: 1–271.

Henderson, A., and C. D. Bacon. 2011. *Lanonia* (Palmae), a new genus from Asia, with a revision of the species. *Systematic Botany* 36: 883–895.

Henderson, A., G. Galeano, and R. Bernal. 1995. *Field Guide to the Palms of the Americas*. Princeton, New Jersey: Princeton University Press.

Hodel, D. R. 1992. *Chamaedorea Palms*. Lawrence, Kansas: International Palm Society.

Hodel, D. R., ed. 1998. *The Palms and Cycads of Thailand*. Lawrence, Kansas: Allen Press.

Hodel, D. R. 2007. A review of the genus *Pritchardia*. *Palms* 51(Supplement): S1–S53.

Hodel, D. R. 2008. Suddenly *Sabal*, seriously. *The Palm Journal* (Southern California) 190: 4–31.

Hodel, D. R. 2011. A new nothospecies and two cultivars for the hybrids in cultivation between *Butia odorata* and *Jubaea chilensis*. *Palms* 55: 62–71.

Hodel, D. R. In press. *Loulu: The Hawaiian Palm*. Honolulu: University of Hawai'i Press.

Hodel, D. R., and J.-C. Pintaud. 1998. *The Palms of New Caledonia*. Lawrence, Kansas: Allen Press.

Johnson, D. V., ed. 1996. *Palms: Their Conservation and Sustained Utilization*. Gland, Switzerland: IUCN.

Johnson, D. V. 1998. *Tropical Palms*. Rome: Food and Agriculture Organization of the United Nations.

Jones, D. L. 1995. *Palms Throughout the World*. Washington, D.C.: Smithsonian Institution Press.

Kahn, F. 1997. *The Palms of Eldorado*. Marly-le-Roi, France: Orstom.

Keim, A. P., and J. Dransfield. 2012. A monograph of the genus *Orania* (Palmae: Oranieae). *Kew Bulletin*. In press.

Krempin, J. 1990. *Palms and Cycads Around the World*. Sydney, Australia: Horwitz Grahame Pry.

Langlois, A. C. 1976. *Supplement to Palms of the World*. Gainesville, Florida: University Press of Florida.

Lewis, C. E., and S. Zona. 2008. *Leucothrinax morrisii*, a new name for a familiar Caribbean palm. *Palms* 52: 84–88.

Lim, C. K. 2001. Unravelling *Pinanga* Blume (Palmae) in Peninsular Malaysia. *Folia Malaysiana* 2: 219–276.

Lockett, L. 1991. Native Texas palms north of the Lower Rio Grande Valley: recent discoveries. *Principes* 35: 64–71.

Lorenzi, H., L. R. Noblick, F. Kahn, and E. Ferreira. 2010. *Brazilian Flora Arecaceae (Palms)*. Nova Odessa, Brazil: Instituto Plantarum de Estudos da Flora.

Matatiken, D., and D. Dogley. 2006. *Guide to the Endemic Palms and Screw Pines of the Seychelles Granitic Islands*. Mahe, Seychelles: Plant Conservation Action Group.

McCurrach, J. C. 1980. *Palms of the World*. Stuart, Florida: Horticultural Books.

Meerow, A. W. 2002. *Betrock's Guide to Landscape Palms*. 4th ed. Cooper City, Florida: Betrock Information Systems.

Moore, H. E., Jr. 1972. *Chelyocarpus* and its allies *Cryosophila* and *Itaya*. *Principes* 16: 67–88.

Moraes Ramírez, M. 2004. *Flora de Palmeras de Bolivia*. La Paz, Bolivia: Universidad Mayor de San Andrés.

Muirhead, D. 1961. *Palms*. Globe, Arizona: Dale Stuart King.

Noblick, L. R. 1996. *Syagrus*. *The Palm Journal* (Southern California) 126: 12–45.

Noblick, L. R. 2004. *Syagrus cearensis*, a twin-stemmed new palm from Brazil. *Palms* 48: 70–76.

Noblick, L. R. 2006. The grassy *Butia*: two new species and a new combination. *Palms* 50: 167–178.

Noblick, L. R., and H. Lorenzi. 2010. New *Syagrus* species from Brazil. *Palms* 54: 18–42.

Pintaud, J.-C. 2000. An introduction to the palms of New Caledonia. *Principes* 44: 132–140.

Principes (later *Palms*). 1956–2011. Journal of the International Palm Society, Lawrence, Kansas. Vols. 1–55.

Rakotoarinivo, M., T. Ranarivelo, and J. Dransfield. 2007. A new species of *Beccariophoenix* from the high plateau of Madagascar. *Palms* 51: 63–75.

Rodd, A. N. 1998. Revision of *Livistona* in Australia. *Telopea* 8: 49–153.

Romney, D. H. 1997. *Growing Coconuts in South Florida*. Homestead, Florida: David H. Romney.

Sastrapradja, S. 1987. *Palms of Indonesia*. Queensland, Australia: Palm and Cycad Society of Australia.

Saw, L. G. 1997. A revision of *Licuala* (Palmae) in the Malay Peninsula. *Sandakania* 10: 1–95.

Schrock, D., ed. 2008. *Ortho All About Palms*. Des Moines, Iowa: Meredith Books.

Stearn, W. T. 2002. *Stearn's Dictionary of Plant Names for Gardeners*. 2nd ed. Portland, Oregon: Timber Press.

Stevens, C. 2010. Sorting out the many names of *Trachycarpus*. *The Palm Journal* (Southern California) 194: 4–37.

Stevenson, G. B. 1996. *Palms of South Florida*. Gainesville, Florida: University Press of Florida.

Stewart, L. 1994. *A Guide to Palms and Cycads of the World*. Sydney, Australia: Angus and Robertson.

Tomlinson, P. B. 1990. *The Structural Biology of Palms*. New York: Oxford University Press.

Tucker, R. 1987. *Pinanga*. *Palm Cycads* 16: 2–9.

Tucker, R. 1988. *The Palms of Subequatorial Queensland*. Queensland, Australia: Palm and Cycad Societies of Australia.

Uhl, N. W., and J. Dransfield. 1987. *Genera Palmarum*. Lawrence, Kansas: L. H. Bailey Hortorium and International Palm Society.

Vines, R. A. 2004. *Trees, Shrubs, and Woody Vines of the Southwest*. Caldwell, New Jersey: Blackburn Press.

Watling, D. 2005. *Palms of the Fiji Island*. Suva, Fiji: Environmental Consultants.

White, A. 1988. *Palms of the Northern Territory and Their Distribution*. Brisbane, Australia: Palm and Cycad Societies of Australia.

Whitmore, T. C. 1998. *Palms of Malaya*. 2nd ed. Bangkok, Thailand: White Lotus.

Zona, S. 1990. A monograph of *Sabal*. *Aliso* 12: 583–666.

Zona, S. 1996. *Roystonea*. *Flora Neotropica* 71: 1–36.

Zona, S. 1998. *Chuniophoenix* in cultivation. *Principes* 42: 198–200.

Zona, S. 2002. A revision of *Pseudophoenix*. *Palms* 46: 19–38.

Zona, S. 2005. A revision of *Ptychococcus*. *Systematic Botany* 30: 520–529.

Zona, S. 2008. The horticultural history of the Canary Island Date Palm (*Phoenix canariensis*). *Garden History* 36: 301–308.

Zona, S. 2011. The travels of *Jubaea*. *Pacific Horticulture* 72(1): 14–19

Zona, S., and F. B. Essig. 1999. How many species of *Brassiophoenix*? *Palms* 43: 45–47.

Zona, S., D. Evans, and K. Maidman. 2000. *La palma barrigona*. *Palms* 44: 85–87.

Zona, S., and D. Fuller. 1999. A revision of *Veitchia* (Arecaceae: Arecoideae). *Harvard Papers in Botany* 4: 543–560.

ACKNOWLEDGMENTS

A volume of this scope is not the work of a single person. It is rather analogous to film making in which the directors are but a part of the larger whole. For help with the original (2003) manuscript on which this book is based, the authors thank the following:

Neal Maillet of Timber Press, as executive editor suffered the many delays and missed deadlines in the production of this inclusive work.

John L. Dowe of Queensland, Australia, reviewed the manuscript for many genera on which he is expert and provided invaluable comments.

Martin Gibbons of Richmond, Surrey, United Kingdom, and Tobias Spanner of Munich, Germany, have traveled to most regions where palms grow.

Rolf Kyburz of K-Palm Nursery, Queensland, Australia, submitted several important photographs.

Larry Noblick of Montgomery Botanical Center, Miami, Florida, accompanied Bob on several tours of the grounds of that marvelous institution.

Chuck Hubbuch, formerly of Fairchild Tropical Garden and presently creating a garden for the zoo in Jacksonville, Florida.

Tim Hatch of Sarawak, Malaysia, contributed several photographs.

Dave Witt of Orlando, Florida, made many suggestions from his years of growing palm species in that city.

Geoff Stein of Thousand Oaks, California, kindly contributed some excellent drawings of palm parts.

Mike Dahme of Grant, Florida, and Puerto Rico reviewed parts of the manuscript and initiated the contact with John L. Dowe.

Bryan Laughland of Auckland, New Zealand, contributed a number of excellent photographs of palm species in New Caledonia and New Zealand.

Daryl O'Connor of Queensland, Australia, read much of the manuscript and offered insights on growing palms in his country.

Laura Tooley, horticulturist at Flamingo Gardens, Broward County, Florida, accompanied Bob on several tours through that wonderful old garden.

Katherine Maidman, former curator of palms at Fairchild Tropical Botanic Garden, Miami, Florida, came running out into the garden a number of times from her office to show Bob a particular plant.

Gaston Torres-Vera of Argentina, in spite of present economic and political turmoil in his country, managed to send several excellent photographs of *Trithrinax* species.

This book would never have been completed without the three women in our lives, and, although the book is dedicated to them, they deserve much more than a dedication.

For their valuable contributions present book, Paul Craft adds his thanks to following:

Bill Beattie, Cairns, Australia, contributed photographs and insights to the genus *Dypsis*.

Mike Harris of Caribbean Palms Nursery, Loxahatchee, Florida, submitted photographs and on travels has always been a keen observer of palms in habitat while sharing his insights.

Jeff and Suchin Marcus of Floribunda Palms, Hawaii, contributed photos, insights, and a fabulous garden in which to photograph many of the more uncommon species in this book.

Scott Zona extends his sincere acknowledgements to the following:

Jason Dewees, San Francisco, California, who helped make this project happen.

Two wonderful botanical gardens in Miami, Fairchild Tropical Botanic Garden and Montgomery Botanical Center, and their directors, Carl E. Lewis and Patrick Griffiths, for their leadership in the world of palms.

Nancy Korber, librarian at FTBG, who answered questions, checked the literature, and was always just an e-mail away.

The regulars at Palm Lunch at FTBG: Michael Davenport, Jack Fisher, Jay Horn, Christy Jones, Carl E. Lewis, Katherine Maidman, P. B. Tomlinson. I learned so much from you.

Friends and former colleagues from FTBG, especially Mary Collins, Don Evans, Marilyn Griffiths, Jason Lopez, Ken Neugent, and Lynka Woodbury, who helped me in innumerable ways.

Colleagues Beyte Barrios, Chad Husby, and Suzanne Koptur of Florida International University, for looking after the greenhouse while I was away chasing palms.

All my friends from the International Palm Society, who share my obsession for these amazing plants. I thank those who have opened their gardens to me and/or who have shared their expertise so freely, especially, Jeff Brusseau, Jim Cain, John DeMott, Dick Douglas, John Dransfield, Horace Hobbs, Tom Jackson and Kathy Grant, Bo-Göran Ludkvist, Jeff Marcus, Tim McKernan, Jill Menzel, Chris Migliaccio, and Randy Moore.

Tsyr Han Chow, to whom this book is dedicated, but who also deserves much more than a dedication. You are my North, my South, my East and West.

Finally, my parents, Patricia and Roger Zona, who supported and encouraged me every step of the way. At the very least, you deserve medals for enduring occasional blood-letting from my *Calamus* palm while I was away at college. Instead, I offer this simple acknowledgment of all you have done for me.

The authors gratefully acknowledge the many people who contributed photographs to this edition—Bill Beattie, John Dransfield, Grenville K. Godfrey, Mike Harris, Larry Noblick, Paul Latham, Carl E. Lewis, Jeff Marcus, Mijoro Rakotoarinivo, Lauren Raz, Tony Rodd, Peter Richardson, Leng Guan Saw, Jack Sayers, Morland Smith, and Kyle Wicomb.

And last, we thank Juree Sondker, acquisitions editor at Timber Press, and Linda Willms, project editor, who have been ever-helpful in bringing out this book.

PHOTOGRAPHY CREDITS

Bill Beattie: plates 399, 418
Paul Craft: plates 1–3, 6–9, 11, 13, 14, 16–21, 23–25, 28, 29, 31–33, 36–41, 43, 46, 48–54, 56, 60, 61, 63, 66, 68–71, 73–77, 79–81, 83, 84, 88, 90–94, 96, 98–100, 102, 104–110, 113–116, 121, 123–141, 143, 149, 150, 152, 154, 155, 158, 159, 162–165, 167–173, 175, 177, 181, 183–186, 188–194, 198, 199, 204–208, 210, 211, 213–232, 234, 236–245, 248–256, 258–262, 265, 267–277, 279–281, 283–293, 295–299, 301, 303–308, 310–315, 318, 319, 323–327, 330–332, 334, 335, 337, 341, 342, 345, 348, 349, 351–353, 356, 357, 359–366, 368, 369, 371, 373–380, 382–394, 397, 398, 403–417, 419–426, 428–436, 438–440, 442–447, 449–457, 461, 463, 465, 467–469, 471, 473–481, 483–487, 489, 491–493, 495, 496, 498–502, 505–510, 513, 515–518, 521–523, 525, 526, 528, 529, 532–536, 538, 541–544, 546–548, 550, 553–556, 558, 559, 562–565, 567–574, 578–580, 582, 583, 585, 588–591, 596–600, 603, 605, 608–619, 621–625, 627–632, 636, 638, 640–644, 648, 651, 652, 654–656, 659, 660, 664–671, 673, 674, 677, 679–683, 686, 687, 689–694, 696, 697, 699–701, 703–705, 707–713, 715–720, 724–727, 729–743, 750–754, 756–771, 773–776, 780–784, 787–791, 796–800, 802–804, 806, 808, 810, 812, 813, 818–820, 822–827, 829–835, 837, 839, 841, 842, 844–851, 858–861, 866, 869, 874, 876, 878–880, 882, 884, 885, 887, 888, 890, 891, 893–895, 897–899, 901–906, 908, 910–913, 920–926, 928–930, 932–937, 939, 940, 942, 943, 945–950. Also pages 2–3, 9, 12.
John Dransfield: plate 907
Martin Gibbons and Tobias Spanner: plates 22, 196, 197, 200, 202, 203, 339, 347, 549, 626, 647, 661–663, 685, 688, 744, 914–919
Grenville K. Godfrey: plate 247
Mike Harris: plates 264, 282
Tim Hatch: plates 58, 120, 512, 552, 576, 657, 684
Jerry Hooper: plate 749
Chuck Hubbuch: plates 78, 144, 320, 328, 329, 649, 695, 755, 868, 877, 944
Rolf Kyburz: plates 35, 64, 174, 358, 401, 497, 551, 606, 607, 658, 722, 779, 807, 815
Paul Latham: plate 873
Bryan Laughland: plates 101, 257, 459, 863
Carl E. Lewis: plates 195, 488, 747, 821
Jeff Marcus: plates 333, 367, 402, 721, 870
Celio Moya: plate 302
Larry Noblick: plates 142, 146–148, 151, 153, 883, 886, 892, 900
Daryl O'Connor: plates 340, 464, 778, 828
Mijoro Rakotoarinivo: plates 111, 112
Lauren Raz: plate 201
Peter Richardson: plates 26, 34, 156, 161, 336, 482, 540, 581, 584
Robert Lee Riffle: plates 4, 5, 57, 59, 62, 65, 85–87, 117, 119, 179, 180, 182, 233, 235, 263, 309, 316, 343, 344, 437, 441, 466, 470, 503, 560, 587, 593, 602, 633, 645, 672, 675, 676, 728, 801, 805, 853, 857, 864, 865, 871, 872, 881, 889, 909, 938
Tony Rodd: plates 586, 594, 595, 601
Leng Guan Saw: plates 67, 511, 514, 524, 527, 557, 561, 577
Jack Sayers: plates 520, 646
Morland Smith: plate 604
Kyle Wicomb: plates 10, 82, 209, 346, 566, 745, 840
Clayton York: plates 176, 653
Scott Zona: plates 12, 15, 27, 30, 42, 44, 45, 47, 55, 72, 89, 95, 97, 103, 118, 122, 145, 157, 160, 166, 178, 187, 212, 246, 266, 278, 294, 300, 317, 321, 322, 338, 350, 354, 355, 370, 372, 381, 395, 396, 400, 427, 448, 458, 460, 462, 472, 490, 494, 504, 519, 530, 531, 537, 539, 545, 575, 592, 620, 634, 635, 637, 639, 650, 678, 698, 702, 706, 714, 723, 746, 772, 777, 785, 786, 792–795, 809, 811, 814, 816, 817, 836, 838, 843, 852, 854–856, 862, 867, 875, 896, 927, 931, 941. Also pages 1, 5, 8, 15.

INDEX OF SYNONYMS AND COMMON NAMES

Common names and synonyms are cross-referenced to the scientific names by which the entries in the book are organized.

Acanthorrhiza. See *Cryosophila*
Acrocomia armentalis. See *A. crispa*
Actinorhytis poamau. See *A. calapparia*
African oil palm. See *Elaeis guineensis*
African wild date palm. See *Phoenix reclinata*
Aiphanes acanthophylla. See *A. minima*
Aiphanes aculeata. See *A. horrida*
Aiphanes caryotifolia. See *A. horrida*
Aiphanes corallina. See *A. minima*
Aiphanes erosa. See *A. minima*
Aiphanes luciana. See *A. minima*
Aiphanes vincentiana. See *A. minima*
Alex palm. See *Archontophoenix alexandrae*
Alexander palm. See *Archontophoenix alexandrae, Ptychosperma elegans*
Alexandra king palm. See *Archontophoenix alexandrae*
Alexandra palm. See *Archontophoenix alexandrae*
Alloschmidia glabrata. See *Basselinia glabrata*
Alsmithia longipes. See *Heterospathe longipes*
American oil palm. See *Elaeis oleifera*
Andes wax palm. See *Ceroxylon* spp.
Areca befaria. See *A. tunku*
Areca latiloba. See *A. montana*
Areca langloisiana. See *A. vestiaria*
Areca recurvata. See *A. montana*
areca nut palm. See *Areca catechu*
areca palm. See *Dypsis lutescens*
Arecastrum romanzoffianum. See *Syagrus romanzoffiana*
Arecastrum romanzoffianum var. *botryophorum.* See *Syagrus botryophora*
arikury palm. See *Syagrus schizophylla*
assaí palm. See *Euterpe oleracea*
Atherton palm. See *Laccospadix australasicus*
Australian cabbage palm. See *Livistona australis*
Australian fan palm. See *Livistona australis*

Bactris gasipaes var. *chichagui.* See *B. gasipaes*
Bactris neomilitaris. See *B. militaris*
Bailey fan palm. See *Copernicia baileyana*
Bailey palm. See *Copernicia baileyana*
bamboo palm. See *Chamaedorea* spp., *Rhapis excelsa*

bangalow palm. See *Archontophoenix cunninghamiana*
barbel palm. See *Acanthophoenix* spp.
bay-leaf palm. See *Sabal mauritiiformis*
Bermuda palm. See *Sabal bermudana*
Bermuda palmetto. See *Sabal bermudana*
betel-nut palm. See *Areca catechu*
betel palm. See *Areca catechu*
Bismarck palm. See *Bismarckia nobilis*
bitter coconut. See *Syagrus pseudococos*
black palm. See *Caryota maxima, Normanbya normanbyi*
blue hesper palm. See *Brahea armata*
blue latan palm. See *Latania loddigesii*
blue palmetto. See *Rhapidophyllum hystrix*
Borassus sambiranensis. See *B. aethiopum*
bottle palm. See *Hyophorbe lagenicaulis*
Brahea nitida. See *B. calcarea*
brittle thatch palm. See *Leucothrinax morrisii*
broad thatch palm. See *Thrinax excelsa*
Brongniartikentia. See *Clinosperma*
Brongniartikentia lanuginosa. See *Clinosperma lanuginosa*
broom palm. See *Leucothrinax morrisii*
brush palm. See *Rhopalostylis sapida*
buccaneer palm. See *Pseudophoenix sargentii*
buffalo-top. See *Leucothrinax morrisii*
bush palmetto. See *Sabal minor*
butterfly palm. See *Dypsis lutescens*

cabada palm. See *Dypsis cabadae*
cabbage palm. See *Sabal palmetto*
cabbage palmetto. See *Sabal palmetto*
cabbage tree. See *Livistona australis*
Cairns fan palm. See *Livistona muelleri*
calappa palm. See *Actinorhytis calapparia*
California fan palm. See *Washingtonia filifera*
Campecarpus fulcitus. See *Cyphophoenix fulcita*
Canary Island date palm. See *Phoenix canariensis*
caranday palm. See *Copernicia alba*
carnauba wax palm. See *Copernicia prunifera*
Caroline ivory-nut palm. See *Metroxylon amicarum*
Carpentaria palm. See *Carpentaria acuminata*
Caryota gigas. See *C. obtusa*
Caryota ochlandra. See *C. maxima*
cascade palm. See *Chamaedorea cataractarum*
cat palm. See *Chamaedorea cataractarum*

cataract palm. See *Chamaedorea cataractarum*
central Australian cabbage palm. See *Livistona mariae*
Ceylon date palm. See *Phoenix pusilla*
Chamaedorea chazdonii. See *C. dammeriana*
Chamaedorea erumpens. See *C. seifrizii*
Chamaedorea minima. See *C. pumila*
Chamaedorea sullivaniorum. See *C. pumila*
Chamaedorea tenella. See *C. geonomiformis*
Chamaerops humilis var. *cerifera.* See *C. humilis* var. *argentata*
cherry palm. See *Pseudophoenix sargentii*
Chilean palm. See *Jubaea chilensis*
Chilean wine palm. See *Jubaea chilensis*
Chinese fan palm. See *Livistona chinensis*
Chinese fishtail palm. See *Caryota ochlandra*
Chinese needle palm. See *Guihaia argyrata*
Chinese windmill palm. See *Trachycarpus fortunei*
Christmas palm. See *Adonidia merrillii*
Chrysalidocarpus. See *Dypsis*
Chrysalidocarpus lutescens. See *Dypsis lutescens*
Chusan palm. See *Trachycarpus fortunei*
cliff date palm. See *Phoenix rupicola*
clustering fishtail palm. See *Caryota mitis*
coconut palm. See *Cocos nucifera*
Cocos plumosa. See *Syagrus romanzoffiana*
cohune palm. See *Attalea cohune*
Cooktown fan palm. See *Livistona concinna*
Cooktown livistona. See *Livistona concinna*
Costa Rican bamboo palm. See *Chamaedorea costaricana*
cotton palm. See *Washingtonia filifera*
coyure palm. See *Aiphanes horrida*
creeping palmetto. See *Rhapidophyllum hystrix*
Cuban barrel palm. See *Colpothrinax wrightii*
Cuban belly palm. See *Colpothrinax wrightii, Acrocomia crispa*
Cuban bottle palm. See *Colpothrinax wrightii*
Cuban petticoat palm. See *Copernicia macroglossa*
Cuban royal palm. See *Roystonea regia*
Cyrtostachys brassii. See *C. loriae*
Cyrtostachys kisu. See *C. loriae*
Cyrtostachys lakka. See *C. renda*
Cyrtostachys peekeliana. See *C. loriae*

date palm. See *Phoenix dactylifera*
desert fan palm. See *Washingtonia filifera*
diamond joey. See *Johannesteijsmannia altifrons*
Diplothemium. See *Allagoptera*
Dominican silver thatch palm. See *Coccothrinax argentea*
doum palm. See *Hyphaene* spp.
Drymophloeus beguinii. See *D. litigiosus*
Drymophloeus samoensis. See *Solfia samoensis*
dwarf betel-nut palm. See *Areca macrocalyx*
dwarf palm. See *Chamaerops humilis*
dwarf palmetto. See *Sabal minor*
dwarf royal palm. See *Adonidia merrillii*
dwarf saw palmetto. See *Rhapidophyllum hystrix*
dwarf sugar palm. See *Arenga caudata*, *A. engleri*
Dypsis lastelliana 'Darianii'. See *D. leptocheilos*
Dypsis 'Mad Fox'. See *D. marojejyi*
Dypsis 'Mealy Bug'. See *D. mananjarensis*
Dypsis 'Orange Crush'. See *D. pilulifera*
Dypsis 'Stumpy'. See *D. carlsmithii*

European fan palm. See *Chamaerops humilis*
Everglades palm. See *Acoelorrhaphe wrightii*

fan palm. See *Livistona australis*
feather duster palm. See *Rhopalostylis sapida*
Fiji fan palm. See *Pritchardia pacifica*
fishtail lawyer cane. See *Calamus caryotoides*
fishtail palm. See *Caryota* spp.
flame-thrower palm. See *Chambeyronia macrocarpa*
Florence Falls palm. See *Hydriastele wendlandiana*
Florida royal palm. See *Roystonea regia*
Florida silver palm. See *Coccothrinax argentata*
Florida thatch palm. See *Thrinax radiata*
footstool palm. See *Saribus rotundifolius*
forest coconut. See *Voanioala gerardii*
Formosa palm. See *Arenga engleri*
fountain palm. See *Livistona decora*
foxtail palm. See *Wodyetia bifurcata*

Gastrococos crispa. See *Acrocomia crispa*
giant fishtail palm. See *Caryota kiriwongensis*, *C. maxima*, *C. no*
giant windowpane palm. See *Beccariophoenix* spp., *Reinhardtia latisecta*
gingerbread palm. See *Hyphaene thebaica*
give-and-take. See *Cryosophila stauracantha*
golden cane palm. See *Dypsis lutescens*
green coconut. See *Syagrus pseudococos*
Gronophyllum. See *Hydriastele*

Gronophyllum ledermanniana. See *Hydriastele ledermanniana*
Gronophyllum microcarpum. See *Hydriastele microcarpa*
Gronophyllum montanum. See *Hydriastele montana*
Gronophyllum pinangoides. See *Hydriastele pinangoides*
gru-gru palm. See *Acrocomia aculeata*
Guadalupe palm. See *Brahea edulis*
Gulubia. See *Hydriastele*
Gulubia costata. See *Hydriastele costata*
Gulubia cylindrocarpa. See *Hydriastele cylindrocarpa*
Gulubia hombronii. See *Hydriastele hombronii*
Gulubia macrospadix. See *Hydriastele macrospadix*
Gulubia microcarpa. See *Hydriastele vitiensis*
Gulubia palauensis. See *Hydriastele palauensis*
Gulubia ramsayi. See *Hydriastele ramsayi*
Gulubia valida. See *Hydriastele valida*

hardy bamboo palm. See *Chamaedorea microspadix*
hat palm. See *Sabal domingensis*
hedgehog palm. See *Rhapidophyllum hystrix*
hesper palm. See *Brahea aculeata*
Hexopetion alatum. See *Astrocaryum alatum*
Hexopetion mexicanum. See *Astrocaryum mexicanum*
Hispaniolan silver thatch palm. See *Coccothrinax argentea*
hurricane palm. See *Dictyosperma album*

Iguanura speciosa. See *Iguanura polymorpha*
Illawara king palm. See *Archontophoenix cunninghamiana* 'Illawara'
Iron Range king palm. See *Archontophoenix tuckeri*
ivory cane palm. See *Pinanga coronata*
ivory-nut palm. See *Phytelephas* spp.

jaggery palm. See *Caryota urens*
Jamaican thatch palm. See *Thrinax excelsa*
jelly palm. See *Butia capitata*
joannis palm. See *Veitchia joannis*
joey palm. See *Johannesteijsmannia* spp.

kaffir palm. See *Jubaeopsis caffra*
Kennedy River livistona. See *Livistona concinna*
kentia palm. See *Howea* spp.
king palm. See *Archontophoenix* spp.

lady palm. See *Rhapis excelsa*
large lady palm. See *Rhapis excelsa*

Latrum palm. See *Hydriastele wendlandiana*
Lavoixia. See *Clinosperma*
lawyer cane. See *Calamus australis*
Licuala dasyantha. See *Lanonia dasyantha*
Licuala delicata. See *L. scortechinii*
Licuala filiformis. See *L. triphylla*
Licuala "mapu." See *L. mattanensis* 'Mapu'
Licuala pygmaea. See *L. triphylla*
Licuala stenophylla. See *L. triphylla*
Linospadix. See *Calyptrocalyx*
lipstick palm. See *Cyrtostachys renda*
little bluestem. See *Sabal minor*
Livistona 'Blackdown Tableland'. See *L. fulva*
Livistona 'Cape River'. See *L. lanuginosa*
Livistona 'Carnarvon' or 'Carnarvon Gorge'. See *L. nitida*
Livistona decipiens. See *L. decora*
Livistona 'Eungella Range'. See *L. australis*
Livistona 'Paluma Range'. See *L. australis*
Livistona merrillii. See *Saribus merrillii*
Livistona robinsoniana. See *Saribus rotundifolius*
Livistona rotundifolia var. *luzonensis*. See *Saribus rotundifolius*
Livistona 'Victoria River'. See *L. victoriae*
Livistona woodfordii. See *Saribus woodfordii*
lontar palm. See *Borassus flabellifer*

Macarthur palm. See *Ptychosperma macarthurii*
macaw palm. See *Acrocomia aculeata*, *Aiphanes minima*
Mackeea magnifica. See *Kentiopsis magnifica*
Macrophloga. See *Dypsis*
Madagascar queen palm. See *Dypsis plumosa*
majestic palm. See *Ravenea rivularis*
majesty palm. See *Ravenea rivularis*
mangrove date palm. See *Phoenix paludosa*
mangrove fan palm. See *Licuala spinosa*
mangrove palm. See *Nypa fruticans*
Manila palm. See *Adonidia merrillii*
Markleya. See *Attalea*
Martinezia. See *Aiphanes*
mastodon palm. See *Aphandra natalia*
mat palm. See *Coccothrinax crinita*
Maximiliana. See *Attalea*
Maya palm. See *Gaussia maya*
mazari palm. See *Nannorrhops ritchieana*
mealy bug palm. See *Dypsis mananjarensis*
Mediterranean fan palm. See *Chamaerops humilis*
metallic palm. See *Chamaedorea metallica*
Mexican blue palm. See *Brahea armata*
Mexican fan palm. See *Washingtonia robusta*
miniature date palm. See *Phoenix roebelenii*
miniature fishtail palm. See *Chamaedorea metallica*

INDEX OF SYNONYMS AND COMMON NAMES

misty mountain palm. See *Laccospadix australasicus*
Montgomery palm. See *Veitchia arecina*
Moratia cerifera. See *Cyphokentia cerifera*
Mount Lewis king palm. See *Archontophoenix purpurea*
mountain cabbage palm. See *Euterpe precatoria, Prestoea acuminata* var. *montana*
mountain fishtail palm. See *Caryota gigas*
mountain thatch palm. See *Thrinax parviflora*
mule palm. See *Syagrus romanzoffiana*
Myola king palm. See *Archontophoenix myolensis*

neanthe bella. See *Chamaedorea elegans*
needle palm. See *Rhapidophyllum hystrix*
Nengella. See *Hydriastele*
Neodypsis. See *Dypsis*
Neodypsis decaryi. See *Dypsis decaryi*
Neophloga. See *Dypsis*
Neophloga 'Pink Crownshaft'. See *Dypsis* sp. 'Pink Crownshaft'
nikau palm. See *Rhopalostylis sapida*
nipa. See *Nypa fruticans*
Norfolk Island palm. See *Rhopalostylis baueri*
northern kentia palm. See *Hydriastele ramsayi*

old man palm. See *Coccothrinax crinita*
old man thatch palm. See *Coccothrinax crinita*
orange collar palm. See *Areca vestiaria*
Orbignya. See *Attalea*

Pacific fan palm. See *Pritchardia pacifica*
palmetto. See *Sabal palmetto*
palmyra palm. See *Borassus, B. aethiopum, B. flabellifer*
Panama hat palm. See *Sabal causiarum*
Paralinospadix. See *Calyptrocalyx*
Paralinospadix hollrungii. See *Calyptrocalyx hollrungii*
Parascheelea. See *Attalea*
parlor palm. See *Chamaedorea elegans*
paurotis palm. See *Acoelorrhaphe*
peaberry palm. See *Leucothrinax morrisii*
peach palm. See *Bactris gasipaes*
Peach River king palm. See *Archontophoenix tuckeri*
petticoat palm. See *Copernicia macroglossa, Washingtonia filifera*
Phloga. See *Dypsis*
Physokentia rosea. See *P. petiolata*
piccabeen palm. See *Archontophoenix cunninghamiana*

Pinanga cochinchinensis. See *P. sylvestris*
Pinanga elmeri. See *P. philippinensis*
Pinanga fruticans. See *P. scortechinii*
Pinanga kuhlii. See *P. coronata*
Pinanga punicea. See *P. rumphiana*
pindo palm. See *Butia capitata*
Plectocomia griffithii. See *P. elongata*
Polyandrococos. See *Allagoptera*
Polyandrococos caudescens. See *Allagoptera caudescens*
Polyandrococos pectinata. See *Allagoptera caudescens*
Pondoland palm. See *Jubaeopsis caffra*
poor man's Canary Island date palm. See *Elaeis guineensis*
potato chip palm. See *Chamaedorea tuerckheimii*
powder palm. See *Brahea moorei*
Prestoea brachyclada. See *P. carderi*
Prestoea humilis. See *P. carderi*
Prestoea latisecta. See *P. carderi*
Prestoea simplicifrons. See *P. carderi*
prickly palm. See *Brahea aculeata*
princess palm. See *Dictyosperma album*
Pritchardia affinis. See *P. maideniana*
Pritchardia aylmer-robinsonii. See *P. remota*
Pritchardia eriophora. See *P. minor*
Pritchardia eriostachys. See *P. lanigera*
Pritchardia gaudichaudii. See *P. martii*
Pritchardia kahanae. See *P. martii*
Pritchardia lanaiensis. See *P. glabrata*
Pritchardia montis-kea. See *P. lanigera*
Pritchardia rockiana. See *P. martii*
Pritchardia woodfordii. See *P. pacifica*
Pritchardiopsis jeanneneyi. See *Saribus jeanneneyi*
Ptychococcus archboldianus. See *P. paradoxus*
Ptychococcus arecinus. See *P. paradoxus*
Ptychococcus elatus. See *P. paradoxus*
Ptychosperma bleeseri. See *P. macarthurii*
Puerto Rican hat palm. See *Sabal causiarum*
purple crownshaft king palm. See *Archontophoenix purpurea*
purple king palm. See *Archontophoenix purpurea*
pygmy date palm. See *Phoenix roebelenii*

queen palm. See *Syagrus romanzoffiana*
Queensland black palm. See *Normanbya normanbyi*

Raphia humilis. See *R. sudanica*
Raphia ruffia. See *R. farinifera*
rattan. See *Calamus* spp.
rat's tail palm. See *Calyptrogyne ghiesbreghtiana*

red feather palm. See *Chambeyronia macrocarpa*
red latan palm. See *Latania lontaroides*
red-leaf palm. See *Chambeyronia macrocarpa*
redneck palm. See *Dypsis lastelliana, D. leptocheilos*
reed palm. See *Chamaedorea seifrizii*
Rhopaloblaste brassii. See *R. ledermanniana*
ribbon fan palm. See *Livistona decora*
ribbon palm. See *Livistona decora*
Rio Grande palmetto. See *Sabal mexicana*
rock palm. See *Brahea dulcis*
Rocky River king palm. See *Archontophoenix tuckeri*
rootspine palm. See *Cryosophila* spp.
round-leaf fan palm. See *Saribus rotundifolius*
royal palm. See *Roystonea regia*
ruffle palm. See *Aiphanes horrida*
ruffled fan palm. Scc *Licuala grandis*
ruffled palm. See *Chamaedorea tuerckheimii, Aiphanes horrida*

Sabal guatemalensis. See *S. mexicana*
Sabal louisiana. See *S. minor*
Sabal texana. See *S. mexicana*
sagisi palm. See *Heterospathe elata*
sago palm. See *Metroxylon sagu*
San José fan palm. See *Brahea brandegeei*
San José hesper palm. See *Brahea brandegeei*
San José palm. See *Brahea brandegeei*
sand palm. See *Livistona humilis*
saw palmetto. See *Serenoa repens*
Scheelea. See *Attalea*
scrub palmetto. See *Sabal etonia*
sealing-wax palm. See *Cyrtostachys renda*
seashore palm. See *Allagoptera arenaria*
Senegal date palm. See *Phoenix reclinata*
sentry palm. See *Howea* spp.
Seychelles stilt palm. See *Verschaffeltia splendida*
shaving brush palm. See *Rhopalostylis sapida*
silver date palm. See *Phoenix sylvestris*
silver palm. See *Coccothrinax argentata*
silver saw palmetto. See *Serenoa repens*
silver thatch palm. See *Coccothrinax argentata, Leucothrinax morrisii*
silvertop palm. See *Coccothrinax argentata*
Sinaloa hesper palm. See *Brahea aculeata*
Siphokentia. See *Hydriastele*
Siphokentia beguinii. See *Hydriastele beguinii*
Siphokentia dransfieldii. See *Hydriastele dransfieldii*
slender lady palm. See *Rhapis humilis*
small fan palm. See *Livistona inermis*
snakeskin palm. See *Caryota ophiopellis*
solitaire palm. See *Ptychosperma elegans*

INDEX OF SYNONYMS AND COMMON NAMES

solitary fishtail palm. See *Caryota urens*
spindle palm. See *Hyophorbe verschaffeltii*
spine palm. See *Rhapidophyllum hystrix*
spiny fiber palm. See *Trithrinax brasiliensis*
spiny licuala. See *Licuala spinosa*
splendid stilt palm. See *Verschaffeltia splendida*
sugar date palm. See *Phoenix sylvestris*
sugar palm. See *Arenga pinnata*
sunshine palm. See *Veitchia arecina*
swamp palmetto. See *Sabal minor*
Syagrus stenopetala. See *S. orinocensis*

talipot palm. See *Corypha* spp.
taraw palm. See *Livistona saribus*
teddy bear palm. See *Dypsis lastelliana, D. leptocheilos*
Texas palmetto. See *Sabal mexicana*
Texas sabal. See *Sabal mexicana*
Thailand lady palm. See *Rhapis subtilis*
thatch palm. See *Coccothrinax* spp., *Thrinax* spp.
thread palm. See *Washingtonia robusta*
Thrinax morrisii. See *Leucothrinax morrisii*
Thurston palm. See *Pritchardia thurstonii*
Tobago cane. See *Bactris guineensis*

toddy palm. See *Borassus flabellifer, Caryota urens, Phoenix sylvestris*
Trachycarpus khasyanus. See *T. martianus*
Trachycarpus wagnerianus. See *T. fortunei* 'Wagnerianus'
triangle palm. See *Dypsis decaryi*

vampire palm. See *Calyptrogyne ghiesbreghtiana*
vegetable porcupine. See *Rhapidophyllum hystrix*
Veillonia alba. See *Cyphophoenix alba*
Veitchia merrillii. See *Adonidia merrillii*
Veitchia macdanielsii. See *V. arecina*
Veitchia montgomeryana. See *V. arecina*
Vietnamese silver-backed fan palm. See *Guihaia argyrata*
Vonitra. See *Dypsis*
Vonitra crinita. See *Dypsis crinita*
Vonitra fibrosa. See *Dypsis fibrosa*
Vonitra utilis. See *Dypsis utilis*

wait-a-while. See *Calamus australis, C. caryotoides*
walking-stick palm. See *Linospadix monostachyos*
Wallich palm. See *Wallichia* spp.

Wallichia densiflora. See *W. oblongifolia*
Walsh river king palm. See *Archontophoenix maxima*
water rattan. See *Daemonorops angustifolia*
wax palm. See *Ceroxylon* spp.
weeping cabbage palm. See *Livistona decora*
Welfia georgii. See *W. regia*
white elephant palm. See *Kerriodoxa elegans*
wild date palm. See *Phoenix sylvestris*
Windemere palm. See *Trachycarpus latisectus*
windmill palm. See *Trachycarpus fortunei*
window palm. See *Beccariophoenix* spp., *Reinhardtia gracilis*
windowpane palm. See *Reinhardtia gracilis*
wine palm. See *Butia capitata, Caryota urens, Jubaea chilensis, Pseudophoenix vinifera*
wispy fan palm. See *Livistona inermis*
Wissmannia carinensis. See *Livistona carinensis*

yellow bamboo palm. See *Dypsis lutescens*
yellow latan palm. See *Latania verschaffeltii*
yellow wait-a-while. See *Calamus moti*

zebra fishtail palm. See *Caryota zebrina*
zombie palm. See *Zombia antillarum*